Lecture Notes in Computer Science

Lecture Notes in Computer Science

Vol. 352: J. Díaz, F. Orejas (Eds.), TAPSOFT '89. Volume 2. Proceedings, 1989. X, 389 pages. 1989.

Vol. 353: S. Hölldobler, Foundations of Equational Logic Programming. X, 250 pages. 1989. (Subseries LNAI).

Vol. 354: J.W. de Bakker, W.-P. de Roever, G. Rozenberg (Eds.), Linear Time, Branching Time and Partial Order in Logics and Models for Concurrency. VIII, 713 pages. 1989.

Vol. 355: N. Dershowitz (Ed.), Rewriting Techniques and Applications. Proceedings, 1989. VII, 579 pages. 1989.

Vol. 356: L. Huguet, A. Poli (Eds.), Applied Algebra, Algebraic Algorithms and Error-Correcting Codes. Proceedings, 1987. VI, 417 pages. 1989.

Vol. 357: T. Mora (Ed.), Applied Algebra, Algebraic Algorithms and Error-Correcting Codes. Proceedings, 1988. IX, 481 pages. 1989.

Vol. 358: P. Gianni (Ed.), Symbolic and Algebraic Computation. Proceedings, 1988. XI, 545 pages. 1989.

Vol. 359: D. Gawlick, M. Haynie, A. Reuter (Eds.), High Performance Transaction Systems. Proceedings, 1987. XII, 329 pages. 1989.

Vol. 360: H. Maurer (Ed.), Computer Assisted Learning – ICCAL '89. Proceedings, 1989. VII, 642 pages. 1989.

Vol. 361: S. Abiteboul, P.C. Fischer, H.-J. Schek (Eds.), Nested Relations and Complex Objects in Databases. VI, 323 pages. 1989.

Vol. 362: B. Lisper, Synthesizing Synchronous Systems by Static Scheduling in Space-Time. VI, 263 pages. 1989.

Vol. 363: A.R. Meyer, M.A. Taitslin (Eds.), Logic at Botik '89. Proceedings, 1989. X, 289 pages. 1989.

Vol. 364: J. Demetrovics, B. Thalheim (Eds.), MFDBS 89. Proceedings, 1989. VI, 428 pages. 1989.

Vol. 365: E. Odijk, M. Rem, J.-C. Syre (Eds.), PARLE '89. Parallel Architectures and Languages Europe. Volume I. Proceedings, 1989. XIII, 478 pages. 1989.

Vol. 366: E. Odijk, M. Rem, J.-C. Syre (Eds.), PARLE '89. Parallel Architectures and Languages Europe. Volume II. Proceedings, 1989. XIII, 442 pages. 1989.

Vol. 367: W. Litwin, H.-J. Schek (Eds.), Foundations of Data Organization and Algorithms. Proceedings, 1989. VIII, 531 pages. 1989.

Vol. 368: H. Boral, P. Faudemay (Eds.), IWDM '89, Database Machines. Proceedings, 1989. VI, 387 pages. 1989.

Vol. 369: D. Taubner, Finite Representations of CCS and TCSP Programs by Automata and Petri Nets. X. 168 pages. 1989.

Vol. 370: Ch. Meinel, Modified Branching Programs and Their Computational Power. VI, 132 pages. 1989.

Vol. 371: D. Hammer (Ed.), Compiler Compilers and High Speed Compilation. Proceedings, 1988. VI, 242 pages. 1989.

Vol. 372: G. Ausiello, M. Dezani-Ciancaglini, S. Ronchi Della Rocca (Eds.), Automata, Languages and Programming. Proceedings, 1989. XI, 788 pages. 1989.

Vol. 373: T. Theoharis, Algorithms for Parallel Polygon Rendering. VIII, 147 pages. 1989.

Vol. 374: K.A. Robbins, S. Robbins, The Cray X-MP/Model 24. VI, 165 pages. 1989.

Vol. 375: J.L.A. van de Snepscheut (Ed.), Mathematics of Program Construction. Proceedings, 1989. VI, 421 pages. 1989.

Vol. 376: N.E. Gibbs (Ed.), Software Engineering Education. Proceedings, 1989. VII, 312 pages. 1989.

Vol. 377: M. Gross, D. Perrin (Eds.), Electronic Dictionaries and Automata in Computational Linguistics. Proceedings, 1987. V, 110 pages. 1989.

Vol. 378: J.H. Davenport (Ed.), EUROCAL '87. Proceedings, 1987. VIII, 499 pages. 1989.

Vol. 379: A. Kreczmar, G. Mirkowska (Eds.), Mathematical Foundations of Computer Science 1989. Proceedings, 1989. VIII, 605 pages. 1989.

Vol. 380: J. Csirik, J. Demetrovics, F. Gécseg (Eds.), Fundamentals of Computation Theory. Proceedings, 1989. XI, 493 pages. 1989.

Vol. 381: J. Dassow, J. Kelemen (Eds.), Machines, Languages, and Complexity. Proceedings, 1988. VI, 244 pages. 1989.

Vol. 382: F. Dehne, J.-R. Sack, N. Santoro (Eds.), Algorithms and Data Structures. WADS '89. Proceedings, 1989. IX, 592 pages. 1989.

Vol. 383: K. Furukawa, H. Tanaka, T. Fujisaki (Eds.), Logic Programming '88. Proceedings, 1988. VII, 251 pages. 1989 (Subseries LNAI).

Vol. 384: G.A. van Zee, J.G.G. van de Vorst (Eds.), Parallel Computing 1988. Proceedings, 1988. V, 135 pages. 1989.

Vol. 385: E. Börger, H. Kleine Büning, M.M. Richter (Eds.), CSL '88. Proceedings, 1988. VI, 399 pages. 1989.

Vol. 386: J.E. Pin (Ed.), Formal Properties of Finite Automata and Applications. Proceedings, 1988. VIII, 260 pages. 1989.

Vol. 387: C. Ghezzi, J.A. McDermid (Eds.), ESEC '89. 2nd European Software Engineering Conference. Proceedings, 1989. VI, 496 pages. 1989.

Vol. 388: G. Cohen, J. Wolfmann (Eds.), Coding Theory and Applications. Proceedings, 1988. IX, 329 pages. 1989.

Vol. 389: D.H. Pitt, D.E. Rydeheard, P. Dybjer, A.M. Pitts, A. Poigné (Eds.), Category Theory and Computer Science. Proceedings, 1989. VI, 365 pages. 1989.

Vol. 390: J.P. Martins, E.M. Morgado (Eds.), EPIA 89. Proceedings, 1989. XII, 400 pages. 1989 (Subseries LNAI).

Vol. 391: J.-D. Boissonnat, J.-P. Laumond (Eds.), Geometry and Robotics. Proceedings, 1988. VI, 413 pages. 1989.

Vol. 392: J.-C. Bermond, M. Raynal (Eds.), Distributed Algorithms. Proceedings, 1989. VI, 315 pages. 1989.

Vol. 393: H. Ehrig, H. Herrlich, H.-J. Kreowski, G. Preuß (Eds.), Categorical Methods in Computer Science. VI, 350 pages. 1989.

Vol. 394: M. Wirsing, J.A. Bergstra (Eds.), Algebraic Methods: Theory, Tools and Applications. VI, 558 pages. 1989.

Vol. 395: M. Schmidt-Schauß, Computational Aspects of an Order-Sorted Logic with Term Declarations. VIII, 171 pages. 1989 (Subseries LNAI).

Vol. 396: T.A. Berson, T. Beth (Eds.), Local Area Network Security. Proceedings, 1989. IX, 152 pages. 1989.

Vol. 397: K.P. Jantke (Ed.), Analogical and Inductive Inference. Proceedings, 1989. IX, 338 pages. 1989 (Subseries LNAI).

Vol. 398: B. Banieqbal, H. Barringer, A. Pnueli (Eds.), Temporal Logic in Specification. Proceedings, 1987. VI, 448 pages. 1989.

Vol. 399: V. Cantoni, R. Creutzburg, S. Levialdi, G. Wolf (Eds.), Recent Issues in Pattern Analysis and Recognition. VII, 400 pages. 1989.

Vol. 400: R. Klein, Concrete and Abstract Voronoi Diagrams. IV, 167 pages. 1989.

Vol. 401: H. Djidjev (Ed.), Optimal Algorithms. Proceedings, 1989. VI, 308 pages. 1989.

Vol. 402: T.P. Bagchi, V.K. Chaudhri, Interactive Relational Database Design. XI, 186 pages. 1989.

Vol. 403: S. Goldwasser (Ed.), Advances in Cryptology – CRYPTO '88. Proceedings, 1988. XI, 591 pages. 1990.

Vol. 404: J. Beer, Concepts, Design, and Performance Analysis of a Parallel Prolog Machine. VI, 128 pages. 1989.

Vol. 405: C.E. Veni Madhavan (Ed.), Foundations of Software Technology and Theoretical Computer Science. Proceedings, 1989. VIII, 339 pages. 1989.

Vol. 406: C.J. Barter, M.J. Brooks (Eds.), AI '88. Proceedings, 1988. VIII, 463 pages. 1990 (Subseries LNAI).

Vol. 407: J. Sifakis (Ed.), Automatic Verification Methods for Finite State Systems. Proceedings, 1989. VII, 382 pages. 1990.

Lecture Notes in Computer Science

Edited by G. Goos and J. Hartmanis

458

J.C.M. Baeten J.W. Klop (Eds.)

CONCUR '90

Theories of Concurrency: Unification and Extension

Amsterdam, The Netherlands, August 27–30, 1990
Proceedings

Springer-Verlag
Berlin Heidelberg New York London
Paris Tokyo Hong Kong Barcelona

Editors

J. C. M. Baeten
Department of Software Technology, CWI
Kruislaan 413, 1098 SJ Amsterdam, The Netherlands
and
Programming Research Group, University of Amsterdam
Kruislaan 409, 1098 SJ Amsterdam, The Netherlands

J. W. Klop
Department of Software Technology, CWI
Kruislaan 413, 1098 SJ Amsterdam, The Netherlands
and
Department of Mathematics and Computer Science, Free University
De Boelelaan 1081, 1081 HV Amsterdam, The Netherlands

CONCUR '90

CR Subject Classification (1987): F.1.2, D.1.3, D.3.1, D.3.3, F.3.1

ISBN 3-540-53048-7 Springer-Verlag Berlin Heidelberg New York
ISBN 0-387-53048-7 Springer-Verlag New York Berlin Heidelberg

Printing and binding: Druckhaus Beltz, Hemsbach/Bergstr.
2145/3140-543210 – Printed on acid-free paper

Preface

The ESPRIT Basic Research Action 3006, CONCUR (Theories of Concurrency: Unification and Extension) started on September 1, 1989. The principal aims of the action are to explore the relationships among the different approaches to algebraic concurrency theory, and to develop a formalism applicable to a wide range of case studies. Verification of concurrent or distributed systems has up till now been undertaken only on a very small scale, in a haphazard way, and with a multitude of techniques and formal theories. For industrial applicability, it is essential that some unity emerge from the competing theories of concurrency, and that the verification process be supported by reliable software tools.

Coordinating partner of CONCUR is CWI (Centre for Mathematics and Computer Science) in Amsterdam, the other partners are the Universities of Edinburgh, Oxford, Sussex and Amsterdam, INRIA in Sophia Antipolis, and the Swedish Institute for Computer Science.

The conference CONCUR'90, hosted by CWI with the help of the University of Amsterdam, marks the end of the first year of CONCUR. In response to the call for papers, 54 papers were submitted. Of these, 31 were selected for presentation by the program committee. The selected papers appear in these proceedings, together with two articles by invited speakers. Unfortunately, due to lack of time, the other three invited speakers are only represented by abstracts.

The editors want to thank the members of the program committee with all their subreferees, and members of the organizing committee for all their efforts.

Amsterdam, July 1990

Program Committee

J.A. Bergstra (Amsterdam)
G. Berry (Sophia Antipolis)
E. Best (Hildesheim)
C.A.R. Hoare (Oxford)
J.W. Klop (chair, Amsterdam)
K.G. Larsen (Aalborg)
R. Milner (Edinburgh)
U. Montanari (Pisa)
E.-R. Olderog (Oldenburg)
J. Parrow (Stockholm)

Organizing Committee

J.C.M. Baeten (chair)
F. Snijders
L. Vasmel
W.P. Weijland

List of referees

S. Arnborg
J.C.M. Baeten
F. Baiardi
M. Bellia
J.A. Bergstra
G. Berry
E. Best
A. Bouali
G. Boudol
F. Boussinot
E. Brinksma
B. Buchberger
I. Castellani
Zhou Chao Chen
R. Cleaveland
J. Davies
R. De Nicola
R. De Simone
R. Devillers
A. Eliëns
J. Esparza
A. Fantechi
E. Fehr
G.L. Ferrari
A. Galligo
P. Gardiner
H.J. Genrich
R.J. van Glabbeek
S. Gnesi
J. Goguen
U. Goltz
R. Gorrieri
J.F. Groote
H. Hansson
C.A.R. Hoare
J. Hollman
P. Inverardi
D. Jackson
He Jifeng
B. Jonsson
M. Josephs
J.W. Klop
A.S. Klusener
E. Kranakis
K.G. Larsen
B. Lisper
H.H. Løvengreen
W.F. McColl
Q. Miller
R. Milner
B. Monien
U. Montanari
O. de Moor
P. Mussi
M. Nielsen
E.-R. Olderog
F. Orava
J. Parrow
A. Ponse
L. Priese
A. Ravn
J.J.M.M. Rutten
J.B. Saint
A. Maggiolo Schettini
S. Schneider
G. Sjödin
A. Skou
E. Smith
B. Steffen
E.P. de Vink
F.W. Vaandrager
L. Wallen
G. Winskel
J. Woodcock

Table of Contents

Partial Order Semantics of Concurrent Programs

Eike Best
Institut für Informatik, Universität Hildesheim, D-3200 Hildesheim

(Abstract of an Invited Lecture at CONCUR'90)

So-called arbitrary interleaving has traditionally been a popular method for describing the possible behaviours of a concurrent program (viz., its operational semantics). This method has been very successful, both because it is generally easy to formalise and because many (one might argue: all) interesting properties of a concurrent program can be expressed and analysed using interleaved sequences. Partially ordered sets have been suggested as an alternative description method. The concurrency that may be present in the behaviour of a concurrent program can be expressed in the associated partial order by means of the absence of any ordering. In addition and relating to these two approaches, a wide spectrum of intermediate notions, variations and extensions have also been investigated. There have occasionally been controversial discussions relating to the different possible ways of defining the operational semantics of concurrent programs. Sometimes, the interleaving approach has been rejected out of hand because it does not represent concurrency. At other times, the partial order approach has been rejected because it seemed technically too cumbersome to formalise.

I would like to argue in this talk that both interleaving semantics and partial order semantics (and any other appropriate semantics) can – and should, whenever appropriate – be defined peacefully side by side. In this view, the important problems become the following ones:

(i) What is the nicest way of formalising any of the desired semantics?

(ii) What is the precise relationship between the two (or more) semantics?

(iii) Which is the most appropriate semantics to analyse a given property, or class of properties?

A well-known instance of problem (iii) is the notion of fairness: while interleaving semantics suffices to express various interesting fairness notions, it turns out that some of them can be defined and analysed more lucidly in the partial order framework[1].

Recent investigations have produced ample knowledge about the questions (i)–(iii) claimed to be important above in the framework of fundamental models of concurrent systems, such as Bergstra/Klop's ACP, Hoare's CSP, Milner's CCS, Petri's Nets and others. These basic models are immediately useful to describe the standard flow of control of concurrent programs. As a rule, they need to be extended for the purpose of describing other important features of concurrent programs, such as the propagation of data values and non-standard control flow constraints, as might be induced by a priority operator or by the use of timers; in the sequel, these other features will be called 'non-basic'. In existing concurrent programming languages, non-basic features tend to play a prominent rôle; in particular, the correctness of programs written in languages with such features will in general depend crucially on their proper use.

This talk explores the partial order semantics of concurrent programs not only in the presence of standard flow of control, but also taking into account data structures, a priority construct and timing aspects. The following particular instances of the more general questions identified above will be discussed:

(i) What is a satisfactory way of defining a partial order semantics, paying particular attention to the non-basic features?

(ii) What is the relation to interleaving semantics, again with particular regard to the non-basic features?

(iii) What class of properties is likely to be amenable to analysis in terms of partial orders?

The discussion will be motivated and guided by an investigation underway within the Esprit Basic Research Action No.3148 DEMON (Design Methods Based on Nets). This study has two aims: (a) to give a partial order semantics to occam-2 using the Petri net model[2], and (b) to design a programming notation with a Petri net semantics.

---

[1] The transcription of this talk will contain pointers to the literature.

[2] occam-2 has all of the basic and non-basic features mentioned above.

SCONE : A SIMPLE CALCULUS OF NETS

Roberto Gorrieri Ugo Montanari
Dipartimento di Informatica — Università di Pisa
Corso Italia 40 — I - 56100, Pisa, Italy

ABSTRACT

A simple calculus of Place/Transition Petri Nets, called SCONE, is introduced. Relationships between SCONE and the subset of CCS without restriction and relabelling, called RCCS, are studied by showing that RCCS can be implemented onto the net calculus. The implementation is given by means of a suitable mapping from RCCS transitions to SCONE computations, resulting in a finite net representation for RCCS agents. By quotienting the transition system of RCCS with respect to the implementation mapping, we induce also a "true concurrent" semantics for RCCS. These results are developed in the framework of "graphs with algebraic structure" as explained in [MM88, DMM89, MY89, F90, FM90, Co90].

1. INTRODUCTION

Among the various approaches to the semantics of concurrency, we distinguish two: the so-called "interleaving" approach and the "true concurrent" one. The main merit of the former is its well-established theory. A concurrent system is described by a term of a language, which gives rise to a transition system. The states are themselves terms of the language and the transitions are defined by means of a deductive system in structural inductive form, as proposed by Plotkin [Plo81] with his *Structured Operational Semantics* (SOS for short). Equivalences among states / terms are defined according to a suitable notion of observation and the useful result is that observational congruences have a nice axiomatization. Unfortunately, there is a serious drawback: this approach relies on the well-known idea of describing system behaviours as sequences of transitions, a too simplistic view in many practical cases when information about distribution in space, about causal dependency or about fairness must be provided. On the other side, in the "true concurrent" approach, which started from the pioneering work of Petri [Pet62], this kind of information can be easily given, but net theory has not yet reached a completely satisfactory theoretical treatment if compared with the firm results coming from the interleaving side. Rephrasing and extending the ideas developed for the interleaving approach to the "true concurrent" case can be considered the main goal of a branch of concurrency theory. The present paper aims at giving a contribution in this direction.

The basic model of Place/Transition Petri Nets has recently received, by the second author in joint work with J. Meseguer [MM88], a simple algebraic description by showing that a P/T net can be *statically* described as an ordinary directed graph equipped with a commutative monoidal operation $\oplus$ on nodes, and *dynamically* as a graph with also two operations on transitions (the parallel composition operator $\otimes$ and the

Research supported in part by EEC Basic Research Action n.3011 CEDISYS.

sequential composition operator ;), together with suitable axioms for identifying those computations which are observationally identical. Unfortunately, Petri Nets, at least in their usual formulation, are not very suited for modular description of concurrent system: to get this capability we should have operators for building new nets from existing ones, but Petri Nets do not have any (finite) syntax generating them and so no general theory of composition and decomposition can be defined for them. Therefore, we should restrict our attention to a particular class of nets possessing such a syntax, which thus forms naturally a language for nets. To be more precise, we are interested not only in defining a language whose formulae specify distributed systems, but also in describing the nets representing their behaviour as a *calculus*. Hence, extending Plotkin's paradigm to distributed systems, formulae of the language would denote markings of the net, while net transitions would be defined by means of a syntax-driven deductive system. The title emphasizes this aspect of the paper.

By observing that transition systems, as well as P/T nets, are nothing but ordinary directed graphs, we discover that the notion of directed graph is a possible unifying mathematical tool for investigating the relationship between the two approaches to the semantics of concurrency. Moreover, it can be easily shown that an SOS specification, yielding the interleaving operational definition of a language as a transition system, can be described in algebraic form: the transition system is a two-sorted algebra with states and transitions as sorts [MY89, F90, FM90, Co90]. Therefore, the other common link between the two approaches is the algebraic structure for nodes and transitions. Indeed, SOS specifications and Petri Nets are both specializations of the graph concept obtained by adding (different) algebraic structure on nodes and transitions; in this way, graphs defined as two-sorted algebras represent the uniform framework we were looking for.

A calculus for nets can be introduced by defining an algebra for the nodes of the graph in such a way that it can be seen as a free algebra of markings, generated by the places. Therefore, the algebra must possess, among others, also the commutative monoidal operator $\oplus$. Here, as a case study, we have introduced a Simple Calculus Of NEts (SCONE). It has been admittedly chosen as simple as possible but with the necessary operators for considering it a real language for nets. The combinators generating places are prefixing, nondeterministic composition and the recursive definition construct. The combinators generating net transitions comprise the prefix, local choice, and the synchronization operators. The axioms of the calculus (act, sum-< and sum->, below) represent the set of the generators of the algebra and the inference rule (sync, below) is its sole operation, building a new transition from a pair of given transitions. In this way, the terms of the algebra denote the *proofs* of the transitions in the corresponding SOS specification. To help intuition, a transition is represented in the format $t : \nu{-}\mu{\rightarrow}\nu'$ where t is a proof term and $\nu{-}\mu{\rightarrow}\nu'$ is the corresponding SOS triple. SCONE is thus described by the following calculus in algebraic form, with axioms for associativity and commutativity of $\oplus$ on nodes and of $|$ on transitions:

act)	$[\mu,\nu\rangle$	:	$\mu\nu\ {-}\mu{\rightarrow}\nu$	
sum-<)	ν «+ ν'	:	$\nu{+}\nu'{-}\varepsilon{\rightarrow}\nu$	
sum->)	ν' +» ν	:	$\nu'{+}\nu{-}\varepsilon{\rightarrow}\nu$	
sync)	$t \mid t'$	:	$\nu_1{\oplus}\nu'_1\ {-}\tau{\rightarrow}\nu_2{\oplus}\nu'_2$	where $t{:}\nu_1{-}\lambda{\rightarrow}\nu_2$ and $t'{:}\nu'_1{-}\lambda^-{\rightarrow}\nu'_2$.

We want to stress the similarities of our construction with respect to Milner's [Mil89]. CCS is defined as a single whole transition system by means of an SOS specification. Similarly, our net calculus defines a single whole net. Moreover, if one is interested in the behaviour of a particular CCS agent, the relevant piece of transition system is the part reachable from the state corresponding to the agent; analogously, we can single out a sub-net corresponding to a SCONE marking. This is a fairly new result in the context of nets. As a matter of fact, several algebras have been proposed for Petri Nets [K78, GM84, W85, W87, Go88, Ch89] in such a way that a Petri net can be specified by a formula of the proposed algebra, but they lack the pleasant feature of having a single net comprising all the agent subnets. Recent ideas proposed by Degano, De Nicola and Montanari [DDM88a, DDM88b, DGM88, DDM89] go in the direction of transforming concurrent calculi, like CCS or CSP, into net calculi. In a sense, here we try to algebraically formalize some of the ideas developed there and in other related works [Old89, T89]. A first attempt in this direction is [MY89] where, however, the resulting algebraic structure does not correspond to a net.

According to Milner's paradigm, the next step in giving the semantics of a concurrent language consists of defining the equivalence classes of its computations according to an intended notion of observation. The possible observation out of a net computation is not unique: several notions have been developed, among which we mention *firing* or *step sequences* (sequences of steps each with *one* or *at least one* firing transition), *nonsequential processes* (unfoldings of the net N from an initial marking) [GR83], and *commutative processes* [BD87]. Defining the algebra of Petri net computations by means of the operation of parallel and sequential composition, Degano, Meseguer and Montanari have shown an elegant axiomatization of these notions in [DMM89], where actually a slight refinement of classical nonsequential processes, called *concatenable processes*, is considered. Out of these three different notions, we choose concatenable processes because they faithfully represent causal dependencies and, even more importantly, are equipped with a general operation of sequential composition of partial orders.

SCONE is an extremely simple language. Instead of defining a richer calculus of nets, one can design a truly concurrent calculus as an SOS specification, as usual, and then implement it with a suitable mapping from its transition system to the Petri Net of SCONE. This idea has been greatly influenced by the categorical formulation of Petri Nets, as proposed in [MM88], which provides a flexible tool for relating system descriptions at different levels of abstraction by means of suitable morphisms, called *implementation* morphisms, in the category of net computations. An implementation morphism, indeed, can map a net transition to an entire net computation. As a case study, we apply this paradigm to a sub-set of CCS not dealing with restriction and relabelling, we call RCCS. First, we introduce an algebraic presentation of the SOS specification for RCCS and then we show how to map the algebra of RCCS to the algebra of SCONE. The implementation morphism is another manner of looking at a denotational semantics for RCCS with SCONE as semantic domain. The combinators of SCONE are sometimes more elementary than those of the subset of CCS we model, so that, for instance, a RCCS transition must be mapped to a SCONE computation. Thus, the semantic morphism maps graph transitions to net computations by mapping basic operators of (the algebra of) RCCS to derived operators of (the algebra of) SCONE. The relevance of this result is that this mapping can be seen as an instance of a more general algebraic methodology for *implementing* concurrent languages (also in interleaving form) into others (possibly distributed).

As already observed, e.g., in [T89], the graph representation of an agent in interleaving semantics is usually larger than its net representation. Indeed, not all the RCCS agents have a finite transition system representation. As an example, the transition system reachable from the state "rec $x.\alpha x|\alpha x$" is infinite. Nonetheless, we prove that for any marking ν, the SCONE sub-net reachable from ν is always *finite*. Therefore, by means of the implementation mapping, we give a finite net representation to any RCCS agent. In the example above, the transition system for "rec $x.\alpha x|\alpha x$" has a natural net representation in SCONE as a self-loop transition labelled by α with two tokens in the unique place as initial marking. In the concluding section we will discuss the relationship with similar proposals [Go88, T89].

As a by-product of the implementation morphism, we get a "true concurrent" semantics for CCS as a quotient of states and computations of the RCCS transition system. Such a semantics is shown to be consistent with the classic interleaving one, and also with the "true concurrent" semantics given by *permutations* of transitions [BC89, F90, FM90].

The paper is organized as follows. An account of the algebraic formulation of Petri nets and of the axiomatization of processes is presented in Section 2. The SOS specification of RCCS in algebraic form is given in Section 3, while the proposed calculus of nets is introduced in Section 4. In Section 5 we describe the implementation mapping from RCCS to SCONE and then, in Section 6, we prove that the induced semantics is consistent with respect to Milner's and also faithfully represents causality. In Section 7, we give some hints about the extensions of the present approach needed for dealing with richer variants of CCS, i.e. to deal with restriction; then, we discuss the relations of our investigations with some related works [GMM88, DDM89, Go88, T89] and also with recent results on the connections between Petri Nets and Linear Logic [AFG90, GG89, MaM89]. Finally, appendices are added to help the reader not familiar with category theory and the algebraic construction of Petri processes. Appendix A is an introduction to the basic definitions of category theory used throughout the paper (see [ML71] for more details). In Appendix B we recall from [DMM89] the definitions of categories of symmetries, of processes and of concatenable processes.

2. Petri Nets and Processes

Here we recall the definition of Place/Transition Nets proposed in [MM88], and the $\mathcal{P}[N]$ construction of a slight refinement of Goltz-Reisig processes [GR83], called *concatenable processes*, introduced in [DMM89].

Definition 2.1. *(Graph)*

A *graph* G is a quadruple $N = (V, T, \partial_0, \partial_1)$, where V is the set of nodes (or states), T is the set of arcs (or transitions), and ∂_0, ∂_1 are two functions, called source and target respectively: $\partial_0, \partial_1: T \to V$. A *graph morphism* from G to G' is a pair of functions $\langle f,g \rangle$, $f: T \to T'$ and $g: V \to V'$ which preserve the source and the target functions: $g \cdot \partial_0 = \partial'_0 \cdot f$ and $g \cdot \partial_1 = \partial'_1 \cdot f$. This, with the obvious component-wise composition, defines the category Graph. ♦

Definition 2.2. *(Petri Nets)*

A *Place/Transition Petri Net* (net, in short) is a graph $N = (S^\oplus, T, \partial_0, \partial_1)$, where $S^\oplus$ is the free commutative monoid of nodes over a set of *places* S. The elements of $S^\oplus$, called also the *markings* of the net N, are represented as formal sums $n_1a_1 \oplus \ldots \oplus n_ka_k$ ($a_i \in S$, n_i is a natural number) with the order of the summands being immaterial, where addition is defined by $(\oplus_i\ n_ia_i) \oplus (\oplus_i\ m_ia_i) = (\oplus_i\ (n_i+m_i)a_i)$ and 0 is its neutral element.

A *Petri Net morphism h* from N to N' is a graph morphism – i.e. a pair of functions $\langle f,g \rangle$, $f: T \to T'$ and $g: S^\oplus \to S'^\oplus$, preserving source and target – where g is a monoid morphism (i.e. leaving 0 fixed and respecting the monoid operation $\oplus$). With this definition of morphism, nets form a category, called Petri, equipped with products and coproducts. ♦

In other words, a Petri Net is an ordinary graph where the nodes are defined as an algebra with the elements of the (possibly infinite) set S as generators and $\oplus$ as the only operation which is monoidal and commutative. Notice that finite multisets over a (possibly infinite) set S coincide with the elements of the free commutative monoid having S as set of generators.

To represent net computations observed as processes, we construct certain monoidal categories. Here we do not give a formal presentation of the construction, which can be found in Appendix B, but only an intuitive exposition of the relevant definitions. We first introduce a set of constant transitions, called *symmetries*, and axiomatize them. Given u in $S^\oplus$, a symmetry $p{:}u \to u$ is a transition expressing the fact that in a marking the tokens on the same place can be permuted. Since $n_1a_1 \oplus \ldots \oplus n_ka_k$ is the formal representation of any u in $S^\oplus$, a symmetry $p{:}u \to u$ can be represented as a vector of permutations $\langle \sigma_{a_1}, \ldots, \sigma_{a_k} \rangle$ where σ_{a_i} is a permutation of n_i elements. A suggestive graphical representation of a symmetry p on $3a \oplus 2b$ where $\sigma_a = \{1\to2, 2\to3, 3\to1\}$ and $\sigma_b = \{1\to2, 2\to1\}$ is depicted in the first operand of Figure

1.a. Three operations are defined on symmetries: parallel composition, sequential composition and the interchange operation. The intuition behind these operations can be easily grasped from Figure 1.

Supposing $p{:}u\rightarrow u$ and $q{:}v\rightarrow v$, the *parallel* composition $p\otimes q : u\oplus v \rightarrow u\oplus v$ of the two symmetries is obtained by putting side by side the permutations regarding the same place. To obtain the *sequential* composition $p\,;p'{:}\ u\rightarrow u$ of two symmetries on u, we have simply to follows the threads of the permutations. Both $\otimes$ and ; are associative but not commutative. If ι_{a_i} denotes the identity permutation, then for each $u\in S^{\oplus}$ the symmetry $\langle\iota_{a_1},\ldots,\iota_{a_k}\rangle{:}u\rightarrow u$, usually denoted by u (but also by id(u)), is an identity for sequential composition. Finally, exchanging the summands in a node gives rise to an *interchange* symmetry (Fig. 1.c); if u = 3a $\oplus$ 2b and v = 2a $\oplus$ b, then the interchange symmetry is formally denoted by $\pi(u, v) = \langle\sigma_a, \sigma_b\rangle$, where $\sigma_a = \{1\rightarrow 4, 2\rightarrow 5, 3\rightarrow 1, 4\rightarrow 2, 5\rightarrow 3\}$ and $\sigma_b = \{1\rightarrow 3, 2\rightarrow 1, 3\rightarrow 2\}$.

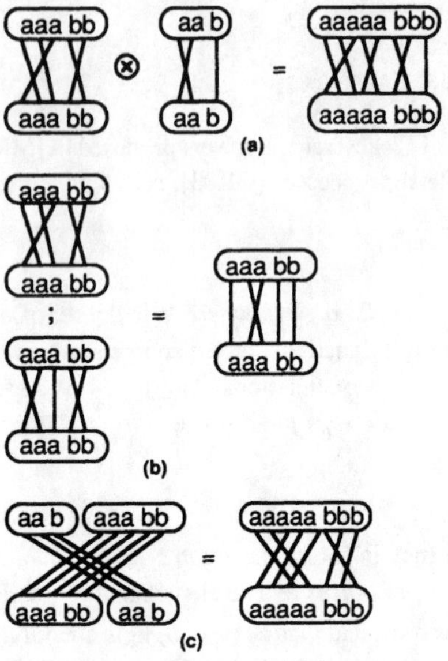

Figure 1. Three operations on symmetries

Definition 2.3. *(From a Net to the Category of its Processes)*

Given a net N = ($S^{\oplus}$, T, ∂_0, ∂_1), the category $\mathcal{P}$[N] of its processes is the quotient of the symmetric, strict monoidal category freely generated by the net N (where the monoidal operation is denoted by $\otimes$ on arrows and by $\oplus$ on objects, the operator of arrow composition is ;, and $\pi(u, v)$ is the natural isomorphism for commutativity), determined by the axioms below:

Objects:	$u\oplus v = v\oplus u$		
Isomorphisms:	$\pi(a, b) = id(a\oplus b)$	$\forall$ a, b $\in$ S, a $\neq$ b	(Ψ)
Transitions:	$t\,;p = t$	where t:$u\rightarrow v$ in N and p: $v\rightarrow v$ is a symmetry	
	$p\,;t = t$	where t:$u\rightarrow v$ in N and p: $u\rightarrow u$ is a symmetry.	♦

This definition states that we can freely generate the symmetric, strict monoidal category from the net N = ($S^{\oplus}$, T, ∂_0, ∂_1), considering a bifunctor ($\oplus$, $\otimes$) which is commutative up to isomorphism. The latter is specified by the natural isomorphism π. Then, we can identify the nodes $u\oplus v$ and $v\oplus u$ which differ only by such an isomorphism. However, the arrow $\pi(u, v){:}\ u\oplus v\rightarrow u\oplus v$ is not in general the identity: it represents a symmetry. The isomorphism axiom $\pi(a, b) = a\oplus b$ states that the symmetry $\pi(a, b)$ is in fact the identity of a$\oplus$b if a and b are different places. The coherence axiom $\pi(u, v\oplus w) = (\pi(u, v)\otimes w)\,;(v\otimes\pi(u, w))$ makes now arrows $\pi(u, v)$ to represent exactly all the symmetries (see Appendix B).

According to the above definition, category $\mathcal{P}[N]$ can be seen as a *two-sorted algebra*. The objects of $\mathcal{P}[N]$ are defined by the algebra of the nodes of N (i.e., $S^{\oplus}$) and the arrows by the algebra, having T and the symmetries as generators and ⊗ and ; as operations, satisfying axioms stating that ⊗ is a monoidal operation, that ; is arrow composition (associativity and existence of identities), that (⊕, ⊗) is a bifunctor

$$(\xi \otimes \xi') ; (\eta \otimes \eta') = (\xi ; \eta) \otimes (\xi' ; \eta') \qquad u \otimes v = u \oplus v \qquad \text{(functoriality axioms[1])}$$

that net transitions are not affected by symmetries (transition equations in (Ψ) – Def 2.3 –), and finally that the factors can be commuted in any parallel composition of computations provided that suitable symmetries generated by the interchange operation are sequentially composed before and after (naturality of the isomorphism). See Appendix B for more details.

These axioms define the arrows of $\mathcal{P}[N]$ as equivalence classes of net computations: an arrow ξ represents the *observation* out of any computation in the equivalence class of ξ. Such an observation turns out to be a slight refinement of Goltz-Reisig *nonsequential processes* [GR83], called *concatenable processes* [DMM89].

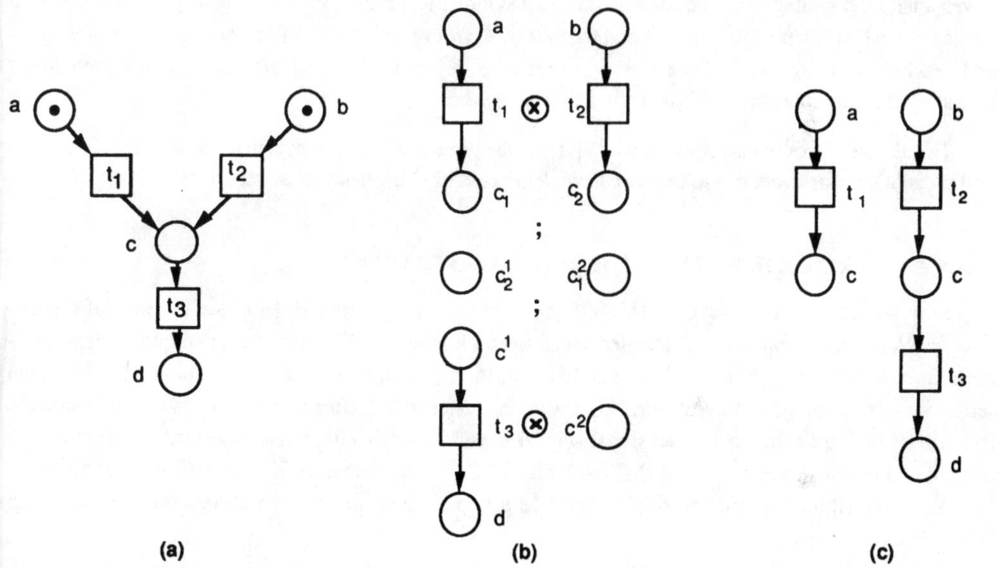

Figure 2. The evaluation of a term in $\mathcal{CP}$ [N] - in (b) - of the net N in (a), and the corresponding result in (c)

Definition 2.4. *(Concatenable Processes)*

A concatenable process for a net N is a triple $C = \langle \varphi, \theta, \kappa \rangle$, where:

- $\varphi = \langle f, g \rangle$ is a net morphism from a finite occurrence net[2] P to N.
- θ and κ are label-indexed ordering functions[3] on the minimal and maximal places of P (also called *origins* and *destinations*) respectively. ♦

---

[1] The second functoriality axiom states that the parallel composition of two identities is the identity of the composed state.

[2] An occurrence net is a net where the preset $^{\bullet}t$ and the postset $t^{\bullet}$ of any transition t are sets, F^* is a *partial ordering* and the preset $^{\bullet}a$ and the postset $a^{\bullet}$ of any place *a* contains at most one element, with xFy iff $x \in \partial_0(y)$ or $y \in \partial_1(x)$, $^{\bullet}x = \{y \mid yFx\}$ and $x^{\bullet} = \{y \mid xFy)$.

[3] Given a set S with a labeling function l: S → S', a label-indexed ordering function is a family $\delta = \{\delta_a\}$, $a \in S'$, of bijections, where $\delta_a : [a] \to \{1,\ldots,|[a]|\}$, with $[a] = \{b \in S \mid l(b) = a\}$. In the case of θ and κ, S is the set of minimal and maximal places in P, respectively, S' is the set of places in N and l is function *g*.

In other words, a concatenable process is a (finite) nonsequential process with, additionally, *ordered* labels on both origins and destinations. This means that, in the origins and destinations of a concatenable process, multiple tokens on the same place are distinguished by imposing an ordering on them.

Thanks to the ordering on origins and destinations, an operation of sequential composition can be defined for concatenable processes (which is not the case for nonsequential processes except when we have at most one token per place). Given two concatenable processes C and C', the sequential composition C ; C' is performed simply by matching the destinations of C with the origins of C', in the order specified by the label-indexed ordering functions. The parallel composition $C \otimes C'$ of two concatenable processes is (graphically) represented by putting side by side the two processes. The generators, i.e. the net transitions and the symmetries, are defined in a straightforward way. Thus, we have an algebra of concatenable processes. Figure 2 shows the evaluation of the term $(t_1 \otimes t_2) ; p ; (t_3 \otimes c)$ with $p = \langle \sigma_a, \sigma_b, \sigma_c \rangle$ and $\sigma_a = \sigma_b = \emptyset$ and $\sigma_c = \{1 \rightarrow 2, 2 \rightarrow 1\}$.

Theorem 2.5. *($\mathcal{P}[N] = \mathcal{CP}[N]$)*
Given a net N, concatenable processes on it form a symmetric, strict monoidal, category $\mathcal{CP}$[N] satisfying the equations (Ψ). There is a unique homomorphism H from $\mathcal{P}$[N] to $\mathcal{CP}$[N] preserving all the operations and leaving N fixed when viewed as a subnet of $\mathcal{P}$[N] and of $\mathcal{CP}$[N] via the obvious inclusions. Furthermore, homomorphism H is actually an *isomorphism.* ♦

This theorem proves that any arrow of $\mathcal{P}$[N] can be represented as a concatenable process and any term of the algebra can be *evaluated* to the concatenable process its equivalence class represents.

3. An Algebraic View of CCS Operational Semantics

Now we recall a subset of CCS [Mil89], together with its operational definition in terms of a graph $N_{CCS} = (V_{CCS}, T_{CCS}, \partial_0, \partial_1)$ with labelled transitions. Plotkin [Plo81] devised a technique, called *Structured Operational Semantics* (SOS), to specify a transition system N = (V, T) representing the operational semantics of a language in a purely structural way: the states in V are the terms of the language, while the transitions[1] in T are given in an intensional way by means of a deductive system defined by structural induction on the syntax. In the vein of [MY89, F90, FM90], we show that an SOS specification, and thus its associated transition system, can be described as a two-sorted algebra, where the sorts are states and transitions.

Let Λ be a countable set of action symbols, equipped with an involutive bijection $^- : \Lambda \rightarrow \Lambda$. The set Λ is called the set of observable actions, ranged over by $\alpha, \beta, \lambda, \ldots$. Let τ be a distinguished symbol not occurring in Λ, which stands for an internal, unobservable action. Symbol μ ranges over the set of all the actions. The states are the finite or *rational*[2] terms in the *continuous* algebra[3] CT_Σ [ADJ77] of the algebra Σ below:

- nil (a constant used to denote inaction)
- μ_ (a unary operator, one for each μ, used to prefix an action to a process)
- _+_ (the binary operator for nondeterminism)
- _|_ (the binary operator for parallelism)

being careful to discard the terms having at least one infinite branch, in the abstract syntax tree, with only a

[1] A transition in SOS format is a triple $\langle v, \mu, v' \rangle$, where v is its source, v' its target and μ its label. Indeed, a transition system is a particular kind of labelled directed graph.

[2] A rational (infinite) term is a term having a finite number of different sub-terms.

[3] Given a signature Σ, T_Σ is the *initial* (or word) *algebra.* With CT_Σ we denote the algebra obtained by completion of T_Σ, containing thus the infinite terms which are limits of chains of finite approximants.

finite number of prefixings. The set of such terms is the (not free[1]) algebra of nodes V_{CCS}. This algebra is expressive enough to represent the closed, guarded terms generated by the syntax below:

$$E ::= nil \mid \mu E \mid E+E \mid E|E \mid x \mid rec\ x.E$$

where only the restriction and relabelling operators of CCS are missing (see the conclusions for a discussion about this topic). Indeed, understanding the term rec x.E as a finite shorthand for the rational, possibly infinite term obtained by unwinding recursion, it can be shown that any closed, guarded term can be evaluated to an element of V_{CCS} and, viceversa, that any element of V_{CCS} can be represented as a closed, guarded term[2] of the syntax above.

This sub-set of CCS is called RCCS, and the closed, guarded terms above are called RCCS *agents*. From now on, for simplicity sake, we wil denote this fragment of Milner's calculus by CCS, whenever non ambiguous from the context.

The transitions of an SOS specification, having the format $v-\mu\rightarrow v'$, are defined by a set of axioms and inference rules, i.e. by a deductive system. Here, we characterize the set of transitions as an algebra: the axioms represent the set of the generators and the inference rules are the operations. In this way, the terms of the algebra T_{CCS} denote the *proofs* of the transitions in the corresponding SOS specification. Furthermore, every term of T_{CCS} is labelled with an action in $\Lambda \cup \{\tau\}$.
To help intuition, a transition is represented in the format:

$t : v-\mu\rightarrow v'$ where t is a proof term and $v-\mu\rightarrow v'$ is the corresponding SOS triple.

In CCS not all transitions can be synchronized, but only those labelled by complementary actions. On the other hand, we want to define an operation of synchronization which is total. Thus we introduce a special symbol $*$, labelling error transitions. The choice of a CCS-like synchronization algebra is part of our case study, but also different synchronization algebras (and different operators) might be considered as well.

Definition 3.1. *(Algebra of Transitions)*
Let $t : v_1-\mu\rightarrow v_2$ and $t' : v'_1-\mu'\rightarrow v'_2$ be two transitions and let $v \in V_{CCS}$. Then, T_{CCS} is the free algebra generated by the following constants and operations:

Act)	$[\mu,v\rangle$	:	$\mu v-\mu\rightarrow v$	for any $\mu \in \Lambda \cup \{\tau\}$
Sum<)	$t <+ v$	:	$v_1+v-\mu\rightarrow v_2$	
Sum>)	$v +> t$	:	$v+v_1-\mu\rightarrow v_2$	
Com-⌋)	$t \rfloor v$	:	$v_1\|v-\mu\rightarrow v_2\|v$	
Com-⌊)	$v \lfloor t$	:	$v\|v_1-\mu\rightarrow v\|v_2$	
Sync)	$t \| t'$	:	$v_1\|v'_1-\mu''\rightarrow v_2\|v'_2$	**with** $\mu'' := \text{if } \mu' = \mu^- \text{ then } \tau \text{ else } *$

where $*$ is a special symbol denoting error due to an impossible synchronization. ♦

As already mentioned, a term of the algebra above denotes a derivation of a transition in the SOS deductive system. Of course, in general an SOS transition has associated more than one proof term, since many derivations can give rise to the same SOS triple. As an example, the SOS transition $\alpha + \alpha -\alpha\rightarrow nil$ has two possible derivations denoted by $[\alpha,nil\rangle <+ \alpha$ and $\alpha +> [\alpha,nil\rangle$. As a matter of fact, graph $N_{CCS} = (V_{CCS}, T_{CCS}, \partial_0, \partial_1)$ can be seen as an object of the category of *CCS transition systems*, we call <u>CCS</u>. The objects of <u>CCS</u> are graphs with the (not necessarily free) CCS algebraic structure specified by the above definitions, and the <u>CCS</u> morphisms are graph morphisms which, furthermore, respect the algebraic structure on states and transitions.

---

[1] This set is an algebra because prefixing, nondeterministic and parallel operations are always defined on such a set; however, it is not free as shown by this simple counterexample: $\alpha.\alpha^\infty = \alpha^\infty$.

[2] The restriction to rational terms is driven by the will of having a finite recursive form for it. We have discarded terms having an infinite branch, in their abstract syntax tree, with only a finite number of prefixings because, by unwinding recursion, they would correspond to unguarded term. E.g., rec x.α|x cannot be evaluated to any term of the algebra.

In [F90, FM90] it is shown that CCS transition systems, factorized with respect to particular observational properties (e.g., strong congruence), can be characterized as final objects of suitable sub-categories of <u>CCS</u> (e.g., the sub-category whose morphisms are transition preserving morphisms).

4. SCONE : A SIMPLE CALCULUS OF NETS

In this section we introduce our Simple Calculus Of NEts (SCONE), following as much as possible the algebraic formulation of Plotkin's paradigm, exemplified in the previous section.

The terms of the algebra of markings (nodes) are the *rational* terms in the *continuous* algebra CT_Σ of the algebra Σ below:

- nil (a constant used to denote inaction)
- $\mu\_$ (a unary operator, one for each μ, used to prefix an action to a process)
- $\_+\_$ (the binary operator for nondeterminism)
- $\_\oplus\_$ (the operator for multiset union),

subject to the following axioms

$$u \oplus v = v \oplus u \qquad u\oplus(v\oplus w) = (u\oplus v)\oplus w \qquad u\oplus \mathrm{nil} = \mathrm{nil}\oplus u = u$$

being careful to discard the terms having at least one infinite branch, in their abstract syntax tree, with only a finite number of prefixings. The set of such finite and rational infinite terms is the (not free) algebra of nodes V_{SCONE}. Let S be the set of the terms in V_{SCONE} having a prefixing or a nondeterministic operation as top operation in the abstract syntax tree. Thus, $V_{SCONE} = S^{\oplus}$, i.e. V_{SCONE} is the free commutative monoid of nodes over a set of *places* S, having nil as neutral element.

With the same arguments of the previous section, this algebra is expressive enough to represent the closed, guarded terms, called SCONE *agents*, generated by the syntax below:

$$M ::= \mathrm{nil} \mid \mu M \mid M{+}M \mid M\oplus M \mid x \mid \mathrm{rec}\ x.M$$

with the identifications induced by the fact that $\oplus$ is a commutative monoid having nil as neutral element.

Intuitively, μv is the place from which a transition labelled by μ can be performed reaching marking v, $v{+}v'$ is a place from which two *pure choice* transitions reaching v and v', respectively, can be performed; $v\oplus v'$ is multiset union of the two markings v and v'. Term nil, being the neutral element of the monoidal operation, should not be considered a place; rather, it denotes absence of a place (see the examples below); indeed, no transition can exit from nil.

The general syntactical form of SCONE transitions is $t : v -\mu\rightarrow v'$, where v and v' are the source and the target of the transition, respectively, μ is its label, and t is a term of the algebra of transition proofs, whose operations are in a one-to-one correspondence with the inference rules of the calculus and whose generators are the axioms of the calculus.

Definition 4.1. *(Algebra of SCONE Transitions)*
Let $t : v_1 -\mu\rightarrow v_2$ and $t' : v'_1 -\mu'\rightarrow v'_2$ be two transitions and let $v \in V$. Then, the transitions in T_{SCONE} are generated by the following constants (*act*, *sum-<*, *sum->*) and operation (*sync*):

act)	$[\mu, v\rangle$	:	$\mu v -\mu\rightarrow v$	for any $\mu \in \Lambda \cup \{\tau\}$
sum-<)	v «+ v'	:	$v{+}v' -\varepsilon\rightarrow v$	
sum->)	v' +» v	:	$v'{+}v -\varepsilon\rightarrow v$	
sync)	$t \mid t'$	:	$v_1\oplus v'_1 -\mu''\rightarrow v_2\oplus v'_2$	**with** $\mu'' :=$ if $\mu' = \mu^-$ then τ else $*$

where ε is a special unobservable action and $*$ is the error symbol; furthermore, the operation of synchronization $\mid$ is subject to the following axioms of commutativity and associativity:

- t | t' = t' | t
- t | (t' | t") = (t | t') | t"

◆

SCONE is a Place/Transition Petri Net $N_{SCONE} = (V_{SCONE}, T_{SCONE}, \partial_0, \partial_1)$ because V_{SCONE} is the free commutative monoid over the set of places S, T_{SCONE} is the set of transitions and $\partial_0, \partial_1: T_{SCONE} \rightarrow V_{SCONE}$ are defined as follows: given a transition $t: \nu{-}\mu{\rightarrow}\nu'$, $\partial_0(t) = \nu$ and $\partial_1(t) = \nu'$. As an example, $\partial_0([\mu,\nu\rangle) = \mu\nu$ and $\partial_1([\mu,\nu\rangle) = \nu$. Notice that N_{SCONE} is not safe, because any place may contain a finite, unbounded number of tokens.

The generators of the algebra are of two kinds: action prefixing ($[\mu,\nu\rangle : \mu\nu\ {-}\mu{\rightarrow}\nu$) and local, internal choice transitions (ν «+ $\nu' : \nu{+}\nu'{-}\varepsilon{\rightarrow}\nu$ and ν' +» $\nu : \nu'{+}\nu{-}\varepsilon{\rightarrow}\nu$). The only operation for building new transitions from existing ones is "|", the synchronization operation. The intuition behind $t_1 | t_2$ is that it is a new transition, whose source and target are the multiset union of the two and whose label is the synchronization of the two, according to the fixed synchronization algebra. The commutativity and associativity axioms seem to be very natural in this perspective: indeed, $t_1 | t_2$ and $t_2 | t_1$ have the same preset, the same postset and the same label, and there is no apparent reason for considering them different. The choice of CCS synchronization algebra is motivated by our interest in showing the implementation mapping from CCS to SCONE in the next section.

It could be interesting to give a look at the shape of the transitions of the net. Those transitions which are not ∗-labelled may have several post-places (sometimes even none) and either one pre-place (action prefixing and internal choice transitions) or two pre-places (synchronization of two action prefixing transitions). Moreover, also loop transitions are allowed, due to recursion. Transitions which are ∗-labelled may have a more general shape, with more than two pre-places. Nonetheless, the study of an algebra of shape-constructors which represents all the possible net transitions (and in general the study of an algebra generating richer classes of nets) is outside the scope of the paper.

SCONE enjoys an interesting property: the sub-part of the global SCONE net reachable starting from a certain marking ν is finite, for any ν. For a finite net we mean a net with a finite set of places and a finite set of transitions.

Theorem 4.2. *(finite reachable sub-nets)*
Given a SCONE marking ν, the set S_ν of the places reachable from ν and the set T_ν of the reachable transitions are finite.

◆

Being SCONE a net, we can apply the algebraic construction of Section 2 to gain the symmetric strict monoidal category $\mathcal{P}[N_{SCONE}]$ of its computations observed as concatenable processes. Of course, the same construction can be also applied to the sub-nets of SCONE.

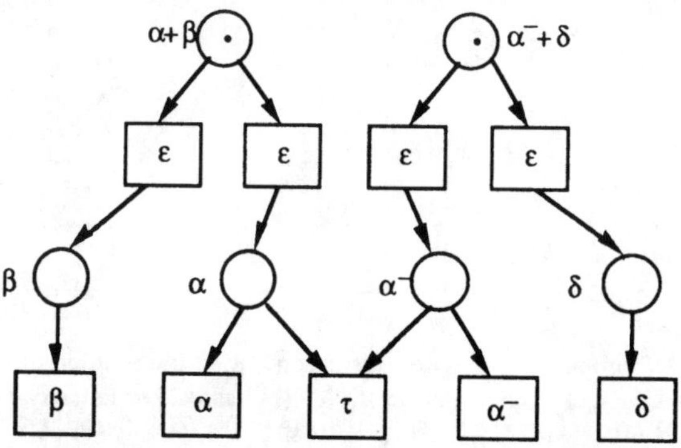

Figure 3. The relevant SCONE sub-net for the marking $(\alpha{+}\beta) \oplus (\alpha^{-}{+}\delta)$

Example 4.3.

The reachable sub-net for the marking (α+β) ⊕ (α⁻+δ) – more correctly, (αnil + βnil) ⊕ (α⁻nil + δnil) – is depicted in Figure 3. Of course, when describing nets graphically, we abandon the presentation of nets as graphs with algebraic structure for a more traditional representation as bipartite graphs. Note that transitions labelled by ε denote local choices. Notice also that, since nil is the neutral element of the monoidal operation, nil does not have a corresponding place.

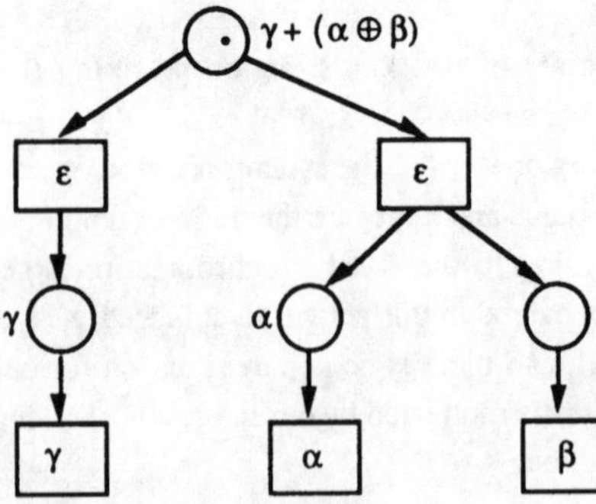

Figure 4. The SCONE sub-net for γ + (α⊕β)

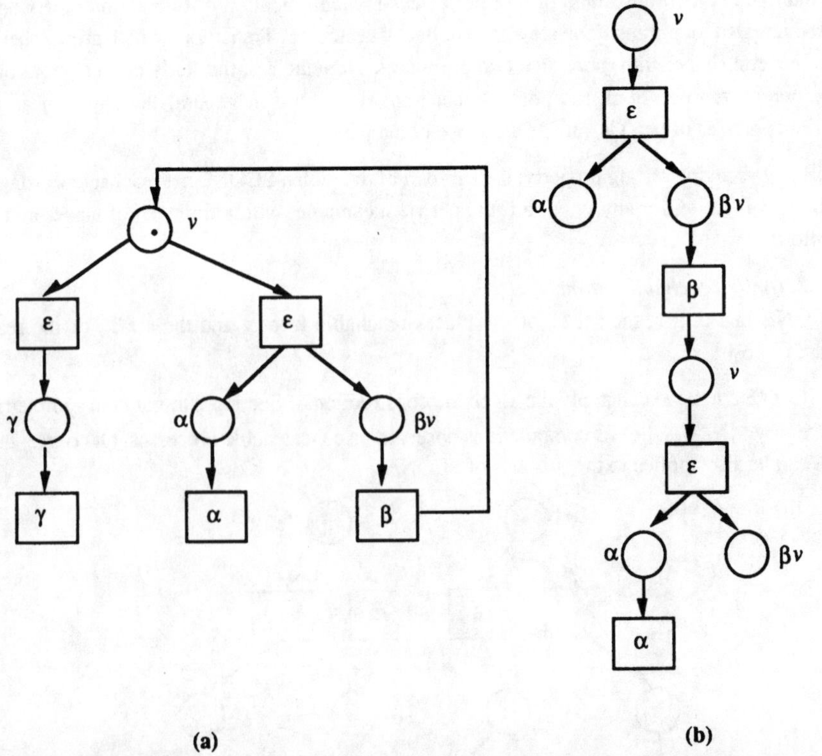

Figure 5. a) - The SCONE sub-net for the state ν corresponding to the solution of the recursive equation x = γ + (α ⊕ β.x). b) - The process in $\mathcal{P}[N_{SCONE}]$ associated to the computation (γ +» (α ⊕ βν)) ; (α ⊗ [β,ν>) ; (α ⊗ (γ +» (α ⊕ βν)) ; (α ⊗ ([α,nil> ⊗ βν))

Example 4.4.
The second example concerns the sub-net reachable from the place $\gamma + (\alpha \oplus \beta)$. This marking corresponds to the CCS agent $\gamma + (\alpha | \beta)$, which is considered a difficult agent to model in terms of net theory because of an interweaving of nondeterministic and parallel operators (see Example 6.4 for a discussion about this problem).

Example 4.5.
The third example shows a net (in Figure 5.a) reachable from the place corresponding to the evaluation of the recursive term $\nu = \text{rec } x.\gamma + (\alpha \oplus \beta x)$. In Figure 5.b it is shown a graphical representation of the process in $\mathcal{P}[N_{SCONE}]$ associated to the computation

$$(\gamma \text{ +» } (\alpha \oplus \beta\nu)) \,;\, (\alpha \otimes [\beta,\nu\rangle) \,;\, (\alpha \otimes (\gamma \text{ +» } (\alpha \oplus \beta\nu)) \,;\, (\alpha \otimes ([\alpha,nil\rangle \otimes \beta\nu)).$$

5. Mapping CCS Transitions to SCONE Computations

Relating different languages whose operational semantics has been defined in terms of graphs with algebraic structure is now an easy task: we have simply to give a denotational semantics of the first in terms of the second, i.e., we have to define a (two-sorted theory) morphism between the two graphs. In this way, not only the terms of the language (the nodes of the graph) are mapped, but also their operational behaviour is pointwise translated. Furthermore, if the target language is a Petri net language, then we get a distributed implementation for the language. This idea is related to the notion of implementation morphism introduced in [MM88]. The analogy is expressed by the fact that a denotational semantics maps basic operators to derived operators as well as an implementation morphism maps transitions to net computations. Here, we implement in SCONE our simplified version of CCS, thus providing it with a distributed implementation. This mapping can be seen as an evaluation function for CCS having SCONE as semantic domain.

Let us try to define the mapping of the CCS transition system N_{CCS} to the category $\mathcal{P}[N_{SCONE}]$ of SCONE computations. We start with an example, just to point out some technical problems. Let us consider the CCS agent $(\alpha+\beta) | (\alpha^-+\delta)$, which can be mapped to the SCONE marking $(\alpha+\beta) \oplus (\alpha^-+\delta)$ (see Example 4.3). The CCS transition

$$([\alpha,nil\rangle <+ \beta) \,|\, ([\alpha^-,nil\rangle <+ \delta)$$

which represents the synchronization of α and α^- (thus an elementary step in Milner's transition system), should be mapped to the net computation (thus to a *derived* operator)

$$(\alpha \text{ «+ } \beta \otimes \alpha^- \text{ «+ } \delta) \,;\, ([\alpha,nil\rangle \,|\, [\alpha^-,nil\rangle)$$

where first the local choices are executed in parallel and then the synchronization is performed. Formally, the mapping is a pair of functions $\langle f, g\rangle$, where function g maps CCS states to SCONE markings and function f maps CCS transitions to arrows of $\mathcal{P}[N_{SCONE}]$. The mapping has to be defined in a purely syntax-driven way. However, some CCS operations have no obvious counterpart in SCONE. To be more precise, we work out how the CCS transition $([\alpha,nil\rangle <+ \beta) | ([\alpha^-,nil\rangle <+ \delta)$ can be mapped to the net computation $(\alpha \text{«+ } \beta \otimes \alpha^- \text{ «+ } \delta) \,;\, ([\alpha,nil\rangle | [\alpha^-,nil\rangle)$. For CCS generators the mapping is trivial: $f([\alpha,nil\rangle) = [\alpha,nil\rangle$, $f([\alpha^-,nil\rangle) = [\alpha^-,nil\rangle$. Then, in the case of nondeterministic choice we have an interesting situation of non trivial mapping:

$$f([\alpha,nil\rangle <+ \beta) = \alpha \text{ «+ } \beta \,;\, f([\alpha,nil\rangle) = \alpha \text{ «+ } \beta \,;\, [\alpha,nil\rangle$$

$$f([\alpha^-,nil\rangle <+ \delta) = \alpha^- \text{ «+ } \delta \,;\, f([\alpha^-,nil\rangle) = \alpha^- \text{ «+ } \delta \,;\, [\alpha^-,nil\rangle.$$

The operator for nondeterministic choice is mapped to a derived operator, i.e. to a suitable combination of local choice and sequential composition, in such a way that a global choice CCS transition is implemented as a sequence of (at least) two transitions, the first of which is a local choice, resulting in a many step computation of SCONE. Indeed, any global choice can be seen as composed of, at least, two steps: the choice of the sub-components and the execution of an action from the selected components. However, in order to preserve the correct semantics of the language, these steps are to be executed *atomically* (see Section 7 and [GMM88, DDM89] for a detailed discussion about this topic), and the mapping f is the right mean to express this notion. Finally,

$$f(([\alpha,\text{nil}> <+ \beta) \mid ([\alpha^-,\text{nil}> <+ \delta)) = f([\alpha,\text{nil}> <+ \beta) \mid f([\alpha^-,\text{nil}> <+ \delta).$$

Now notice that $f([\alpha,\text{nil}> <+ \beta) \mid f([\alpha^-,\text{nil}> <+ \delta)$ is not defined in $P[N_{SCONE}]$, since the operator of synchronization is not defined *for computations*, but only for net transitions! Therefore, we should define an algebra G_{NSCONE}, obtained enriching the algebra of category $P[N_{SCONE}]$ with the operator of synchronization, and expressing which net computation the term $f(t_1) \mid f(t_2)$ should represent.

Defining an operator of synchronization for concatenable processes is a difficult task, since the operation seems to be intrinsically nondeterministic. As an example, it is not clear what should be the result of the synchronization of a transition t with the parallel composition $(t' \otimes t'')$. We might say that either the synchronization is not possible, or t can be synchronized with either t' or t'', or even with both, but apparently there is no sensible choice. Luckily, in the present case, only a restricted family of concatenable processes is interesting for synchronization: the processes in $P[N_{SCONE}]$ which are the targets of CCS transitions according to function f. For this family, it turns out that a *deterministic* operation of synchronization can be defined which exactly reflects our intuition about CCS synchronization. Therefore, first we characterize the class of the relevant processes, we call *transactions*, then we define an operation of synchronization on them. Hence, we add to the term algebra $P[N_{SCONE}]$ a corresponding operation together with an axiom expressing algebraically the effect of the synchronization. Finally, we show that $f(t)$ always evaluates to a particular kind of transaction, called *CCS transaction*.

Definition 5.1. (*Net Transactions and CCS Transactions*)

A *net transaction* for a net N is a concatenable process $C = \langle \varphi, \theta, \kappa \rangle$ where, additionally, the underlying occurrence net[1] has the property that there is a (unique) basic transition which is larger than all the others in the partial ordering. Such a transition is called the *commit* transition.

A CCS transaction C is a net transaction for the net of SCONE where, additionally, all the transitions are labelled by ε, with the exception of the commit transition which is labelled by a μ action. A CCS transaction with the commit transition labelled by μ is also called a μ-transaction. ◆

It is immediate to observe that any net computation ξ in $P[N_{SCONE}]$ which evaluates to a transaction can be algebraically represented in the format $\eta;(u \otimes t);p$ where η is a computation, t is the commit, u an identity and p a symmetry (but also net computations which are not transactions can have the same form). Indeed, it can be easily shown that ξ can be always reduced to the format $p_0;(u_1 \otimes t_1);p_1;\ldots;p_{n-1};$ $(u_n \otimes t_n);p_n$ by applying functoriality of $\otimes$. The commit transition, being caused by all the others, will always be the last. Now we want to prove that there is a standard representative for transactions. To this aim, we need some auxiliary definitions and results.

Definition 5.2. (*merge symmetry*)

A symmetry $p:u \oplus v \to u \oplus v$, $p = \langle \sigma_{a_1}, \ldots, \sigma_{a_k} \rangle$ is a (u,v)-*merge*, where $u = n_1 a_1 \oplus \ldots \oplus n_k a_k$ and $v = m_1 a_1 \oplus \ldots \oplus m_k a_k$, iff $\sigma_{a_h}(i) \le \sigma_{a_h}(j)$ whenever $1 \le i \le j \le n_h$ or $n_h+1 \le i \le j \le n_h+m_h$, $h = 1,\ldots,k$. ◆

A merge symmetry represents an arbitrary merge of the two identity symmetries, the former on u and the latter on v. Since the condition $\sigma_{a_h}(i) \le \sigma_{a_h}(j)$ is satisfied in the two intervals $1 \le i \le j \le n_h$ and $n_h+1 \le i \le j \le n_h+m_h$, we are sure that no exchange is possible within u or within v. Indeed, any symmetry on $u \oplus v$ can be seen as composed of two local exchanging symmetries followed by an (u,v)-merge.

---

[1] Remember that φ is a net morphism from a finite occurrence net P to N. An occurrence net is a net where the preset $\cdot t$ and the postset $t\cdot$ of any transition t are sets, F^* is a *partial ordering* and the preset $\cdot a$ and the postset $a\cdot$ of any place a contains at most one element, with xFy iff $x \in \partial_0(y)$ or $y \in \partial_1(x)$, $\cdot x = \{y \mid yFx\}$ and $x\cdot = \{y \mid xFy\}$.

Lemma 5.3. *(unique decomposition of symmetries)*
Given u and v, any symmetry $p{:}u\oplus v\to u\oplus v$ can be uniquely decomposed as $p = (p_1\otimes p_2);p'$ with $p_1{:}u\to u$, $p_2{:}v\to v$, and p' being a (u,v)-merge. ♦

Lemma 5.4. *(Unique decomposition of transactions)*
Any transaction ξ can be uniquely decomposed in (right-)standard form $\eta';(u\otimes \mathrm{t});p'$ where $\mathrm{t}{:}w\to v$ and p' is a (u,v)-merge. ♦

Of course, we are not forced to leave t on the right; indeed, for any transaction ξ there is also a *left-standard* form $\eta'';(\mathrm{t}\otimes u);p''$ where p'' is a (v,u)-merge.

With this notion in mind, a natural deterministic definition of synchronization between two transactions consists of putting in parallel the two processes but synchronizing the two commit transitions to become the commit for the resulting process. This operation on transactions is associative and, up to natural isomorphism, commutative. In the following definition we introduce a new algebra by enriching $\mathcal{P}[N_{SCONE}]$ with a derived operation $\square$ of synchronization defined on standard representatives of transactions, expressing the intuitive fact that the synchronization of two transactions is again a transaction.

Definition 5.5. *(from $\mathcal{P}[N_{SCONE}]$ to $\mathcal{A}_{NSCONE}$)*
$\mathcal{A}_{NSCONE}$ is the same graph as $\mathcal{P}[N_{SCONE}]$, with the extra operation $\square$ defined on transactions. Given two transactions $\xi = \eta;(u\otimes \mathrm{t});p$ and $\xi' = \eta';(\mathrm{t}'\otimes u');p'$ in standard form (right- and left- respectively), then

$$\xi\,\square\,\xi' = \eta\otimes\eta'\ ;\ u\otimes(\mathrm{t}\mid \mathrm{t}')\otimes u'\ ;\ p\otimes p'$$ ♦

Before entering into the details of the relevant results we want to prove, we will briefly discuss and clarify the definition of the synchronization operation. The simpler case is when the (CCS) transactions consist of exactly one step and there are no symmetries. In such a case, the operation states that identities do not participate to synchronizations.

$$(v\otimes \mathrm{t})\,\square\,(\mathrm{t}'\otimes v') = v\otimes(\mathrm{t}\mid \mathrm{t}')\otimes v' \qquad \mathrm{t}, \mathrm{t}' \in T.$$

This example also shows that for net transitions, which are transactions in standard form, $\mathrm{t}\,\square\,\mathrm{t}' = \mathrm{t}\mid\mathrm{t}'$.
The synchronization operation represents a kind of composition of transactions where the two commits are synchronized. This is one of the few deterministic ways of synchronizing two net computations, and certainly the only one meaningful for CCS transactions. The following theorems give some properties of this kind of cooperation.

Theorem 5.6. *(Synchronizations of Transactions are Transactions)*
i) Given two transactions ξ and ξ', $\xi\,\square\,\xi'$ is a transaction.
ii) Given a λ-transaction ξ and a λ^{-}-transaction ξ', $\xi\,\square\,\xi'$ is a τ-transaction. ♦

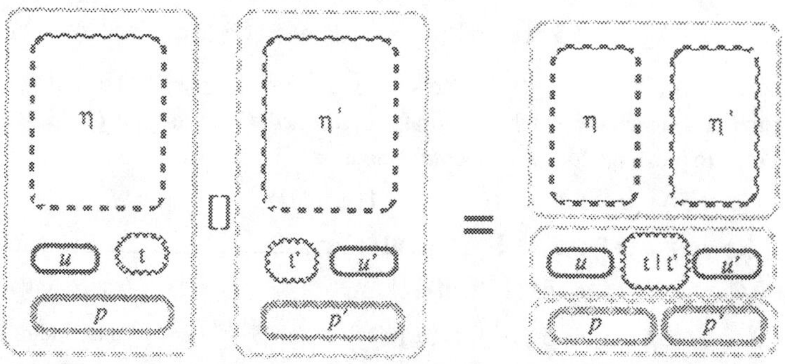

Figure 6. A graphical representation of the synchronization operation

Theorem 5.7. *(Commutativity up to symmetries of* $[]$*)*
Given two transactions $\xi_1:u_1\to v_1$, and $\xi_2:u_2\to v_2$, we have

$$\pi(u_1, u_2)\ ;\ \xi_2 [] \xi_1 = \xi_1 [] \xi_2\ ;\ \pi(v_1, v_2).$$ ♦

Now, we are ready to define the evaluation morphism from the two-sorted algebra of CCS to the two-sorted algebra of $\mathcal{A}_{NSCONE}$.

Definition 5.8. *(Implementing CCS in SCONE)*
Let $\mathcal{A}_{NSCONE} = (\mathfrak{V}, \mathfrak{T}, \partial_0, \partial_1)$. The pair $\langle f, g\rangle: N_{CCS} \to \mathcal{A}_{NSCONE}$, $f:T_{CCS} \to \mathfrak{T}$ and $g:V_{CCS} \to \mathfrak{V}$, is defined as follows, where $t:u\to w$:

- $g(nil) = nil$
- $g(\mu v) = \mu g(v)$
- $g(v+v') = g(v)+g(v')$
- $g(v \mid v') = g(v) \oplus g(v')$
- $f([\mu,v\rangle) = [\mu,g(v)\rangle$
- $f(t <+ v) = g(u) \ll+ g(v)\ ;f(t)$
- $f(v +> t) = g(v) +\gg g(u)\ ;f(t)$
- $f(v \lfloor t) = g(v) \otimes f(t)$
- $f(t \rfloor v) = f(t) \otimes g(v)$
- $f(t_1 \mid t_2) = f(t_1) [] f(t_2) = \eta_1 \otimes \eta_2\ ;\ u_1 \otimes (t_1 \mid t_2) \otimes u_2\ ;\ p_1 \otimes p_2$
  where $f(t_1) = \eta_1;(t_1 \otimes u_1);p_1$ and $f(t_2) = \eta_2;(t_2 \otimes u_2);p_2$ ♦

Proposition 5.9.
For each CCS transition t, $f(t)$ is always a CCS transaction. ♦

Proposition 5.10.
The pair $\langle f, g\rangle : N_{CCS} \to \mathcal{A}_{NSCONE}$ is a graph morphism, i.e., $g \cdot \partial_{iCCS} = \partial_{iSCONE} \cdot f$, $i = 0, 1$. ♦

Example 5.11.
Let us consider again the CCS agent $E = (\alpha+\beta) \mid (\alpha^-+\delta)$, its transitions which are not $*$-labelled and the SCONE sub-net in Figure 3. The initial marking of the sub-net we are interested in is exactly $g(E) = (\alpha+\beta)\oplus(\alpha^-+\delta)$. Transitions are mapped to computations as follows.

- $f([\sigma,nil\rangle) = [\sigma,nil\rangle$ for $\sigma\in \{\alpha, \beta, \alpha^-, \delta\}$
- $f([\alpha,nil\rangle <+ \beta) = \alpha \ll+ \beta\ ; f([\alpha,nil\rangle) = \alpha \ll+ \beta\ ;\ [\alpha,nil\rangle$ and similarly for the other choices,
- $f(([\alpha,nil\rangle <+ \beta) \rfloor(\alpha^-+\delta)) = f([\alpha,nil\rangle <+ \beta) \otimes g(\alpha^-+\delta) = (\alpha \ll+ \beta\ ;\ [\alpha,nil\rangle) \otimes (\alpha^-+\delta)$
  and similarly for other asynchronous moves,
- $f(([\alpha,nil\rangle <+ \beta) \mid ([\alpha^-,nil\rangle <+ \delta)) = f([\alpha,nil\rangle <+ \beta) [] f([\alpha^-,nil\rangle <+ \delta) =$
  $= (\alpha \ll+ \beta\ ;\ [\alpha,nil\rangle) [] (\alpha^- \ll+ \delta\ ;\ [\alpha^-,nil\rangle) = (\alpha \ll+ \beta \otimes \alpha^- \ll+ \delta)\ ;\ ([\alpha,nil\rangle \mid [\alpha^-,nil\rangle)$

Summing up, $f([\alpha,nil\rangle <+ \beta \mid [\alpha^-,nil\rangle <+ \delta) = (\alpha \ll+ \beta \otimes \alpha^- \ll+ \delta)\ ;\ ([\alpha,nil\rangle \mid [\alpha^-,nil\rangle)$, i.e. the choices are executed in parallel and then the synchronization is performed. Of course, the choices can also be done in any order; it is easy to prove the following identifications:

$$(\alpha \ll+ \beta\otimes (\alpha^-+\delta))\ ;\ (\alpha \otimes (\alpha^- \ll+ \delta))\ ;\ ([\alpha,nil\rangle \mid [\alpha^-,nil\rangle)$$
$$= ((\alpha \ll+ \beta\ ;\ \alpha)\otimes (\alpha^-+\delta\ ;\ \alpha^- \ll+ \delta))\ ;\ ([\alpha,nil\rangle \mid [\alpha^-,nil\rangle) \quad \text{(applying functoriality)}$$
$$= (\alpha \ll+ \beta\otimes \alpha^- \ll+ \delta)\ ;\ ([\alpha,nil\rangle \mid [\alpha^-,nil\rangle) \quad \text{(cancelling identities)}$$
$$= ((\alpha + \beta\ ;\ \alpha \ll+ \beta) \otimes (\alpha^- \ll+ \delta\ ;\ \alpha^-))\ ;\ ([\alpha,nil\rangle \mid [\alpha^-,nil\rangle) \quad \text{(introducing identities)}$$
$$= ((\alpha + \beta)\otimes \alpha^- \ll+ \delta)\ ;\ (\alpha \ll+ \beta\otimes \alpha^-)\ ;\ ([\alpha,nil\rangle \mid [\alpha^-,nil\rangle) \quad \text{(applying functoriality)}$$ ♦

Example 5.12.

A further example concerns the RCCS agent rec $x.(\gamma+(\alpha \mid \beta.x))$ and the net in Figure 5.a. In the literature there exist various proposals which deal with *finite* net semantics for CCS recursion, but all of them fail in correctly describing the possibility of initial parallelism of nondeterministic components inside recursion, like it appears in rec $x.(\gamma+(\alpha \mid \beta.x))$. E.g., [Go88] proposes a construction which associates the net depicted in Figure 7 to such a recursive term. In fact, the author herself shows this case as a counterexample for her construction. Although the proposed solution gives full account of the distributed nature of choice, it is incorrect, because action γ may be enabled by tokens which remain in place r. To be more precise, let ν = rec $x.\gamma+(\alpha \mid \beta.x)$, which corresponds in the net to the marking $r \oplus s$.
According to the operational semantics in Section 3, transition γ +> $(\alpha \lfloor [\beta,\nu>)$ is labelled by β and reaches state $\alpha \mid \nu$, corresponding to $2r \oplus s$. Then, transition $(\alpha \lfloor (\gamma$ +> $([\alpha,nil> \rfloor \beta.\nu))$ is labelled by α and reaches state $\alpha \mid nil \mid \beta.\nu$ from which actions α and β only can be performed. Simulating these two steps on the net leads erroneously to marking $r \oplus r' \oplus s$ where γ is enabled. Instead, our solution, depicted in Figure 5.b, correctly represents the intuitive causality and the possible conflicts among the three actions. ♦

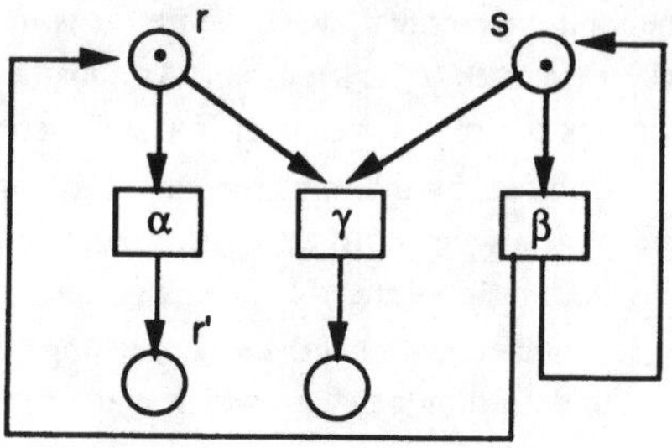

Figure 7. The net for rec $x.\gamma+(\alpha\oplus\beta.x)$ as proposed in [Go88]

In the previous section we have pointed out that, for any SCONE agent M, its reachable sub-net is finite. Here we have mapped the whole CCS transition system to the whole SCONE net. As an immediate corollary, we have that the reachable sub-net implementing any CCS agent is always finite. We want to stress the relevance of this result because, as already observed in [Go88, T89], in this way a finite net representation is given for a possibly infinite transition system representation. E.g., the CCS agent of Example 5.12 has an infinite transition system representation which becomes a finite net representation (Figure 5.a) when mapped to SCONE.

Corollary 5.13.

For any RCCS agent E, the sub-net implementing it *via* morphism $\langle f, g \rangle$ is finite. ♦

6. Distributed Semantics of RCCS

The semantics of our restricted CCS calculus is investigated here by exploiting the implementation mapping $\langle f,g \rangle : N_{CCS} \rightarrow \mathcal{G}_{NSCONE}$. Function g maps various different CCS agents to the same marking, due to the fact that $\oplus$ is a monoidal commutative operator. Thus, the quotient on the states is exactly induced by the axioms stating that $\mid$ forms a commutative monoid with nil as neutral element. Certain transitions are also identified, even if an axiomatic characterization within CCS is not so easy. Here we show that if we extend homomorphically f to CCS computations, the mapping will equate the computations obtained by permuting transitions generating independent events.

Proposition 6.1.

Let ν and ν' two states in V_{CCS}. If $g(\nu) = g(\nu')$ then ν and ν' are strongly equivalent [Mil89]. ♦

Proposition 6.2.
Let t and t' two transitions in T_{CCS}. If $f(t) = f(t')$ then t and t' have the same label. ♦

It is not easy to characterize the identifications on CCS transitions induced by f. Consider, for instance, two states v and v' which are mapped by g to the same marking; then, $f([\mu,v\rangle) = f([\mu,v'\rangle)$. E.g., $f([\mu,\alpha|\beta\rangle) = [\mu,\alpha\oplus\beta\rangle = [\mu,\beta\oplus\alpha\rangle = f([\mu,\beta|\alpha\rangle)$. Also associativity of $\otimes$ induces further identifications, e.g., $f((t \rfloor v) \rfloor w) = f(t \rfloor (v \mid w))$. Moreover, because of the synchronization axiom, we have that, e.g., $f((v \lfloor t) \mid t') = f(v \lfloor (t \mid t'))$. Indeed, it seems that we could axiomatize these identifications within the algebra of CCS transitions: a conditional axiom for action prefixing and nine axioms relating the three different CCS operators for parallel composition in all the possible ways for associativity. However, some form of commutativity is also possible. For instance, $f([\alpha,nil\rangle \mid [\alpha^-,nil\rangle) = f([\alpha^-,nil\rangle \mid [\alpha,nil\rangle)$ due to the commutativity of the SCONE synchronization operator. Nonetheless, in general $f(t_1 \mid t_2) \neq f(t_2 \mid t_1)$, e.g., $[\alpha,nil\rangle \mid (\alpha \rfloor [\alpha^-,nil\rangle)$ and $(\alpha \rfloor [\alpha^-,nil\rangle) \mid [\alpha,nil\rangle$ do not give rise to the same CCS transaction, because the synchronized α is different in the two transitions. Other identifications, which cannot be easily expressed within the algebra of CCS, are possible; e.g., $f(([\alpha,nil\rangle \mid [\alpha^-,nil\rangle) \rfloor \alpha) = f([\alpha,nil\rangle \mid (\alpha \rfloor [\alpha^-,nil\rangle))$. This example expresses the intuitive fact that, when the first α is to be synchronized with α^-, the relative position of the second α with respect to α^- is irrelevant. As a matter of fact, an axiomatization expressing the identifications on transitions is already available! It is sufficient to interpret the algebra of CCS inside the algebra of $\mathcal{G}_{N_{SCONE}}$ as specified by the implementation morphism $\langle f,g\rangle$. In this way, we exploit the finer grain of the operations in the algebra of $\mathcal{G}_{N_{SCONE}}$. Indeed, two CCS transitions t and t' are identified if and only if they give rise to the same concatenable process, or more precisely to the same CCS transaction, i.e. if we can prove $f(t) = f(t')$ in the equational theory consisting of the axioms of $\mathcal{P}[N_{SCONE}]$ together with the axioms arising from the definition of $\langle f, g\rangle$, which make explicit the description of the CCS operations as derived operations in the algebra of $\mathcal{P}[N_{SCONE}]$.

Having verified by Propositions 5.10, 6.1 and 6.2 that interleaving semantics is respected, we would like to check the "amount" of concurrency we are able to express; in other words, is the semantics induced by the implementation morphism correct with respect to the intuitive notion of causality?
To this aim, we can homomorphically extend the implementation morphism $\langle f, g\rangle$ also to CCS computations, and then observe what kind of identifications are made on them. The graph N_{CCS} can be made transitively closed by means of the partial operation _ ; _ on transitions. The basic idea is that the implementation morphism can be extended to CCS computations by $f(t_1 ; t_2) = f(t_1) ; f(t_2)$. What we would like to show is that whenever two CCS computations are different only for the ordering of two transitions which are causally independent, they are identified. First, we will show this on some examples; then, we formally prove our result.

Example 6.3.
Let us consider the CCS term $(\alpha+\beta)|(\alpha^-+\delta)$ and the net in Figure 3. The CCS computation

$$(([\alpha,nil\rangle <\!+ \beta) \rfloor (\alpha^-+\delta)) ; (nil \lfloor ([\alpha^-, nil\rangle <\!+ \delta)) \; : \; (\alpha+\beta)|(\alpha^-+\delta) \xrightarrow{\alpha\cdot\alpha^-} nil \mid nil$$

denotes the execution of an α followed by an α^-. It is mapped to the SCONE computation

$$((\alpha \ll\!+ \beta) ; [\alpha,nil\rangle) \otimes (\alpha^-+\delta) \; ; \; ((\alpha^- \ll\!+ \delta) ; [\alpha^-,nil\rangle)),$$

which, by functoriality and cancelling identities, is equivalent to the parallel execution of the two computations

$$((\alpha \ll\!+ \beta) ; [\alpha,nil\rangle) \otimes ((\alpha^- \ll\!+ \delta) ; [\alpha^-,nil\rangle)),$$

which, by introducing identities and applying functoriality, is equivalent to their execution in reverse order

$$(\alpha + \beta) \otimes ((\alpha^- \ll\!+ \delta) ; [\alpha^-,nil\rangle) \; ; \; ((\alpha \ll\!+ \beta) ; [\alpha,nil\rangle).$$

Notice that this net computation is the image of the CCS computation

$$((\alpha+\beta)\lfloor([\alpha^-,\mathrm{nil}\rangle <\!+\,\delta))\,;\,(([\alpha,\mathrm{nil}\rangle <\!+\,\beta)\rfloor\mathrm{nil}) : (\alpha+\beta)|(\alpha^-+\delta)\text{–}\alpha^-\cdot\alpha\rightarrow \mathrm{nil}\,|\,\mathrm{nil}$$

thus inducing an identification between the former and the latter CCS computations. ♦

Example 6.4.
An interesting test to measure the reliability of a true concurrent semantics is certainly represented by the CCS agent $E = \gamma + (\alpha|\beta)$, where an interweaving of nondeterministic and parallel operators may cause the possible loss of causal independency between the two concurrent actions α and β (see [DDM89] for a discussion about this problem).

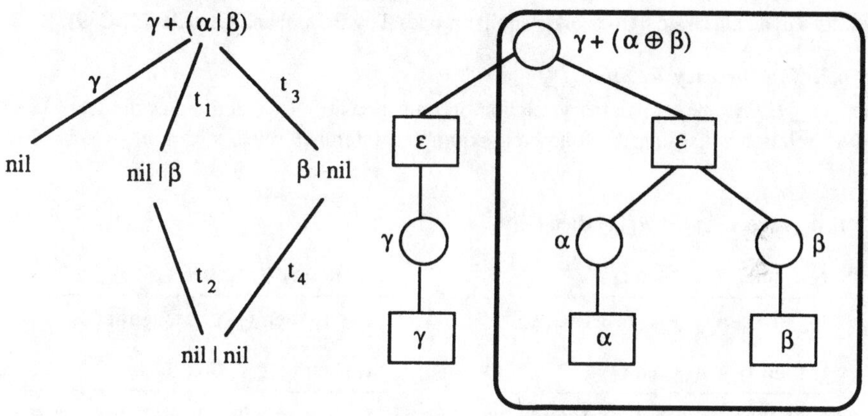

Figure 8. The transition system and the net for the agent $\gamma+\alpha\,|\,\beta$. Notice that both $t_1\,;t_2$ and $t_3\,;t_4$ are mapped to the same process, enclosed in the box.

The place $\gamma + (\alpha\oplus\beta)$ is the image of E, and the two CCS transitions

$$\gamma +\!> ([\alpha,\mathrm{nil}\rangle \rfloor\beta) \;:\; \gamma+(\alpha|\beta)\text{–}\alpha\rightarrow \mathrm{nil}|\beta$$

$$\gamma +\!> (\alpha\lfloor[\beta,\mathrm{nil}\rangle) \;:\; \gamma+(\alpha|\beta)\text{–}\beta\rightarrow \alpha|\mathrm{nil}$$

are mapped to the following two net computations, respectively:

$$(\gamma +\!\!\gg \alpha\oplus\beta)\,;\,([\alpha,\mathrm{nil}\rangle\otimes\beta)$$

$$(\gamma +\!\!\gg \alpha\oplus\beta)\,;\,(\alpha\otimes[\beta,\mathrm{nil}\rangle)$$

The image of the computation $(\gamma +\!> ([\alpha,\mathrm{nil}\rangle\rfloor\beta))\,;\,(\mathrm{nil}\lfloor[\beta,\mathrm{nil}\rangle) : \gamma+(\alpha|\beta)\text{–}\alpha\cdot\beta\rightarrow \mathrm{nil}|\mathrm{nil}$ is the net computation $(\gamma +\!\!\gg \alpha\oplus\beta)\,;\,([\alpha,\mathrm{nil}\rangle\otimes\beta)\,;\,[\beta,\mathrm{nil}\rangle$ where actions α and β are in fact causally independent. This net computation is equivalent to $(\gamma +\!\!\gg \alpha\oplus\beta)\,;\,(\alpha\otimes[\beta,\mathrm{nil}\rangle)\,;\,[\alpha,\mathrm{nil}\rangle$, which is the image of the CCS computation $(\gamma +\!> (\alpha\lfloor[\beta,\mathrm{nil}\rangle))\,;\,([\alpha,\mathrm{nil}\rangle\rfloor\mathrm{nil}) : \gamma + (\alpha|\beta)\text{–}\beta\cdot\alpha\rightarrow \mathrm{nil}|\mathrm{nil}$. Therefore, the two CCS computations are equivalent. ♦

Definition 6.5. *(Category of CCS Computations)*
Let **Cat**(N_{CCS}) denote the category obtained by adding an identity arc to each node of N_{CCS} and closing freely w.r.t. the (partial) operation _ ; _ of sequential composition of its transitions, adding the usual categorical axioms:

$$t\,;(t'\,;t'') = (t\,;t')\,;t'' \qquad\qquad u\,;t = t = t\,;v \qquad \text{where } t : u\rightarrow v.$$

♦

Note that the algebraic structure of CCS has not been extended to computations. The arrows of category **Cat**(N_{CCS}) are only computations composed of N_{CCS} transitions.

The mapping $\langle f,g\rangle : N_{CCS} \to \mathcal{A}_{NSCONE}$ can be extended homomorphically to become a mapping from **Cat**(N_{CCS}) to $\mathcal{A}_{NSCONE}$ by further adding the equation $f(t_1 ; t_2) = f(t_1) ; f(t_2)$. In this way we obtain a quotient of CCS computations as exemplified in the previous examples.

Definition 6.6.
Category $\mathbf{Con}_{CCS}$ is the category obtained from **Cat**(N_{CCS}) by the quotient map induced by $\langle f,g\rangle$. ♦

Now we prove that the identifications on CCS computations due to the implementation mapping are consistent with a set of axioms proposed in [F90, FM90], which defines in a self-evident way the truly concurrent semantics for CCS. A relation χ between computations of lenght two, called *concurrency relation*, relates computations differing just for "permuting" the order of independent transitions. This relation rephrases in an algebraic framework a similar proposal by Boudol and Castellani [BC89].

Definition 6.7. *(Concurrency Relation)*
Let $\_\ \textbf{then}\ \_\ \chi\ \_\ \textbf{then}\ \_$ be a quaternary relation on transitions of N_{CCS} defined as the least $1{,}2 \leftrightarrow 3{,}4$ commutative[1] relation satisfying the following axiom and inference rules, with $t_i : u_i \to v_i$, $i = 1, \ldots, 4$, and $t : u \to v$.

- $t_1 \rfloor u_2\ \textbf{then}\ v_1 \lfloor t_2\ \chi\ u_1 \lfloor t_2\ \textbf{then}\ t_1 \rfloor v_2$

- $$\frac{t_1\ \textbf{then}\ t_2\ \chi\ t_3\ \textbf{then}\ t_4}{t_1 {<}{+}w\ \textbf{then}\ t_2\ \chi\ t_3 {<}{+}w\ \textbf{then}\ t_4}$$

- $$\frac{t_1\ \textbf{then}\ t_2\ \chi\ t_3\ \textbf{then}\ t_4}{w{+}{>}t_1\ \textbf{then}\ t_2\ \chi\ w{+}{>}t_3\ \textbf{then}\ t_4}$$

- $$\frac{t_1\ \textbf{then}\ t_2\ \chi\ t_3\ \textbf{then}\ t_4}{t_1 \rfloor w\ \textbf{then}\ t_2 \rfloor w\ \chi\ t_3 \rfloor w\ \textbf{then}\ t_4 \rfloor w}$$

- $$\frac{t_1\ \textbf{then}\ t_2\ \chi\ t_3\ \textbf{then}\ t_4}{w \lfloor t_1\ \textbf{then}\ w \lfloor t_2\ \chi\ w \lfloor t_3\ \textbf{then}\ w \lfloor t_4}$$

- $$\frac{t_1\ \textbf{then}\ t_2\ \chi\ t_3\ \textbf{then}\ t_4}{t_1 \mid t\ \textbf{then}\ t_2 \rfloor v\ \ \chi\ t_3 \rfloor u\ \textbf{then}\ t_4 \mid t}$$

- $$\frac{t_1\ \textbf{then}\ t_2\ \chi\ t_3\ \textbf{then}\ t_4}{t \mid t_1\ \textbf{then}\ v \lfloor t_2\ \ \chi\ u \lfloor t_3\ \textbf{then}\ t \mid t_4}$$

- $$\frac{t_1\ \textbf{then}\ t_2\ \chi\ t_3\ \textbf{then}\ t_4\ \text{ and }\ t'_1\ \textbf{then}\ t'_2\ \chi\ t'_3\ \textbf{then}\ t'_4}{t_1 \mid t'_1\ \textbf{then}\ t_2 \mid t'_2\ \ \chi\ t_3 \mid t'_3\ \textbf{then}\ t_4 \mid t'_4}$$ ♦

Proposition 6.8.
Given four transitions t_1, t_2, t_3 and t_4 in N_{CCS} such that $t_1\ \textbf{then}\ t_2\ \chi\ t_3\ \textbf{then}\ t_4$, the following hold:
i) $t_1 ; t_2$ and $t_3 ; t_4$ are defined;
ii) $\partial_0(t_1) = \partial_0(t_3)$ and $\partial_1(t_2) = \partial_1(t_4)$;
iii) t_1 and t_4 (t_2 and t_3) have the same label. ♦

The concurrency relation singles out a "diamond" in the transition system N_{CCS} which is due to the different order of execution of independent transitions (for an example, see Figure 9). The axiom algebraically singles out the basic diamonds, and the other rules reproduces the diamonds in all the other possible contexts.

Theorem 6.9. *(consistency with the truly concurrent semantics for CCS)*
Given four basic transitions of $\mathbf{Con}_{CCS}$, i.e., t_1, t_2, t_3 and t_4 in N_{CCS}, then we have

$$t_1\ \textbf{then}\ t_2\ \chi\ t_3\ \textbf{then}\ t_4 \quad \text{implies} \quad t_1 ; t_2 = t_3 ; t_4$$ ♦

---

[1] Namely, $t_1\ \textbf{then}\ t_2\ \chi\ t_3\ \textbf{then}\ t_4$ *iff* $t_3\ \textbf{then}\ t_4\ \chi\ t_1\ \textbf{then}\ t_2$.

7. Concluding Remarks

In this section we will discuss three issues related to our work: first, the extensions needed to cope with the CCS operator of restriction (and relabelling); then, the relations with previous proposals in the area, namely about the finite net representation of CCS agents and about the issue of atomicity and transactions in modeling CCS; finally, some comments on an apparently different, actually strongly related, line of research based on the connections between Girard's Linear Logic and Petri Nets.

Extension to the case of full CCS

The operator $\_\backslash\alpha$ of restriction works as a filter for forbidding the execution of actions α and α^-. To be more concrete, the CCS agent $(\alpha \mid \alpha^-)\backslash\alpha$ cannot perform asynchronously the two actions α and α^-, but only the communication step, labelled by τ. We have to extend SCONE to have a restriction operation for nodes and transitions. The mapping for the restriction operator could be defined as follows:

$$g(v\backslash\alpha) = g(v)\backslash\alpha$$

in such a way that the SCONE marking associated to the CCS agent $(\alpha \mid \alpha^-)\backslash\alpha$ becomes

$$g((\alpha \mid \alpha^-)\backslash\alpha) = g(\alpha \mid \alpha^-)\backslash\alpha = (\alpha\oplus\alpha^-)\backslash\alpha.$$

On the other side, we would like to have in SCONE a distributive axiom for restriction,

$$(v\oplus v')\backslash\alpha = v\backslash\alpha \oplus v'\backslash\alpha$$

because otherwise $(v\oplus v')\backslash\alpha$ would represent a single place, in contrast with the intuition that actions performed by v and v' are neither causally dependent, nor in conflict; indeed, whenever a parallel composition "|" is present, we should get a multiset union of places from the two components. However, the distributive axiom induces equalities which are not true at all, e.g. $(\alpha \mid \alpha^-)\backslash\alpha = \alpha\backslash\alpha \mid \alpha^-\backslash\alpha$, where the latter is a completely deadlocked agent. Thus, our interpretation of | as multiset union is too simplistic in this case. Disjoint union is the answer to our problem. We can introduce two new unary operators for both nodes and transitions, $\mathsf{id}|\_$ and $\_|\mathsf{id}$, with the intuition that $v|\mathsf{id}$ makes v the left part of a larger system. The mapping becomes:

$$g(v \mid v') = (g(v)|\mathsf{id}) \oplus (\mathsf{id}|g(v'))$$

and now distributivity of restriction w.r.t. multiset union preserves the intended semantics. In our example, we get $g((\alpha \mid \alpha^-)\backslash\alpha) = (\alpha|\mathsf{id})\backslash\alpha \oplus (\alpha^-|\mathsf{id})\backslash\alpha$, which represents two places, each one independently stuck but able to cooperate for synchronization. The calculus for nets, obtained by enriching SCONE with the restriction operator and the two unary operators $\mathsf{id}|\_$ and $\_|\mathsf{id}$, generates 1-safe nets only[1].

Of course, also category $\mathcal{P}[N_{SCONE}]$ should be extended to cope with these new operators. To tell the truth, when multiset union is replaced by disjoint union, i.e., when considering 1-safe nets, the algebraic semantics can be more easily given by extending, instead of $\mathcal{P}[N]$, the strictly symmetric strict monoidal category obtained by the $\mathcal{T}[N]$ construction [MM88]. In fact, when considering 1-safe nets only, causality is correctly described also by $\mathcal{T}[N]$, which can therefore be used for giving a simpler semantics to the extended language.

Finite net representation of CCS agents

The construction we have presented in this paper for giving a finite (non safe) representation of a class of CCS agents has analogies with similar proposals in the literature. In particular, [Go88] was the first who generalized to the case of nets the construction for the recursive combinator given by Milner in [Mil84] for transition systems. Our proposal differs from [Go88] mainly with respect to the nondeterministic operation, which is centralized here (but still fully representing the intrinsic parallelism of agents) and distributed there. Moreover, the two solutions are different also because we do not replicate places with the same "name"; as

[1] A similar construction (however, without recursion and not yielding a Petri net, since the monoid of nodes is not free) is already available in [MY89]. In this paper we have preferred not to deal with these other aspects, mainly for focussing our attention on the implementation morphisms by means of a simple language and, more importantly, for allowing a finite representation of agents.

an example, let us consider the CCS agent $\alpha+(\alpha \mid \alpha)$. The nets resulting from our construction and from Goltz's are depicted in Figure 9.a and 9.b, respectively. Also for recursive agents there are some differences; the fact that SCONE markings are finite or rational infinite terms of a continuous algebra makes indistinguishable all the recursive terms evaluating to the same infinite term. For instance, rec x.αx and rec x.$\alpha\alpha$x are the same place α^∞ and the resulting net is composed of a loop transition only. Differently, the construction in [Go88], being more syntactical, gives two nets which are different but behaviourally equivalent (see Figure 10).
Other interesting proposals are presented in Taubner's Ph.D. thesis [T89], which however are less similar to our proposal than Goltz's one.

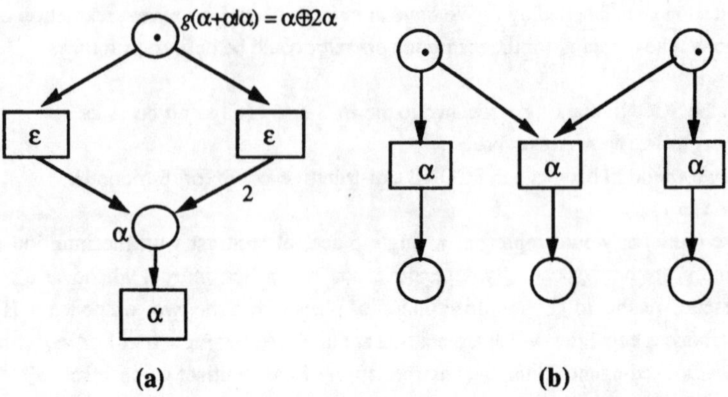

Figure 9. Two sub-nets for $\alpha+(\alpha\oplus\alpha)$: in (a) the SCONE solution and in (b) Goltz's solution. Note that only one transition α is needed in (a); notice also that the number 2 labelling the right arc ingoing to place α represents its weight. This example shows that our construction, in contrast with Goltz and Taubner's, gives non 1-safe nets also when recursion is not present.

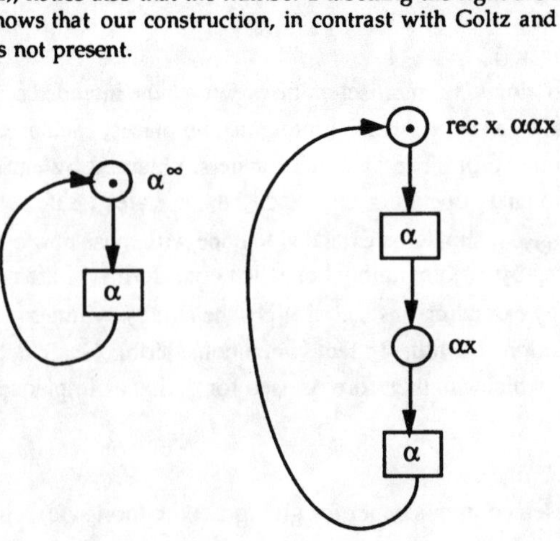

Figure 10. The SCONE sub-net for rec x.$\alpha\alpha$ x in (a) and Goltz' solution in (b)

Is it possible to find out a finite P/T net representation for any CCS agent ? It has been recognized [GM84, Go88, T89] that finite representations for all CCS do not exist since CCS with restriction is "Turing-powerful" whilst finite P/T nets are not. Nonetheless, since non 1-safe representations are usually smaller, it would be interesting to discover if it is possible to deal with restriction without introducing 1-safe nets.

Atomicity and Transactions

The idea of representing the sophisticated global nondeterministic operator of CCS in terms of local choice and sequentialization has been already proposed in the literature. The problem was first attacked in [GMM88], which described a new interleaving semantics for CCS where the inference rules for nondeterminism and recursion have been replaced by the corresponding axioms of local choice (like in SCONE) and recursion unwinding. Of course, in this way we got a calculus more generous than CCS where certain transitions, which are forbidden in Milner's transition system, are derivable. For instance, state $\gamma \mid nil$ is not reachable from state $(\gamma + \alpha) \mid \beta$ by performing an action β in CCS, while it is when replacing the inference rules with the axioms. The problem is that when a sub-agent makes a local choice, then it should have priority with respect to the other concurrent sub-agents willing to perform non choice actions. In this way, a sort of atomicity constraint is imposed on the calculus, which therefore becomes completely equivalent to CCS. In fact, it represents a lower level description of CCS where each local choice and recursion unwinding is seen as a separate move. The interesting point is that this atomicity constraint is ingrained in the syntax-driven deductive system.

In [DDM89], a truly concurrent semantics for CCS is proposed, which looks like the present proposal. Indeed, the modelled language has internal choice only. In order to recover the correct CCS semantics, a mechanism of atomicity has to be put on the net. The authors suggested a notion of μ-transaction, essentially coincident to the one proposed in Definition 5.1, to denote the *atomic* steps on the net, with the intuition that these are the sole feasible moves. In this way, a direct consistency with Milner's interleaving semantics is easily preserved and also the correct causal dependencies between concurrent actions are faithfully reproduced. Apart from the irrelevant fact that full CCS is taken into account there, the main difference between [DDM89] and the present proposal relies on the fact that restricting the behaviour of the net to μ-transactions is more naturally expressed by a morphism from the CCS transition system to the net. Of course, the definition of such a morphism is possible only in the completely algebraic framework we are working in. As a matter of fact, the implementation morphism can be seen as a way of defining a theory (the algebraic theory of RCCS) within the algebra of SCONE. From this point of view, the atomicity constraint is guaranteed by the definition of the CCS operations as terms of the theory. Of course, certain basic SCONE transitions, such as local choices, do not belong to this theory.

A first attempt in giving a morphism from the CCS transition system to a Petri net with local nondeterminism only is reported in [Bl88].

Linear Logic and Petri nets

A great deal of interest has been stirred up by the connections between Girard's linear logic [G87] and Petri Nets. In particular, Asperti [As87] and Gunther / Gehlot [GG89] showed that, for the tensor fragment, the correspondence is really tight: places are atomic propositions (or atoms), markings are tensor products of atoms, net transitions are extra-logical axioms (i.e. a net is a tensor theory) and net computations are proofs.

Martì-Oliet and Meseguer [MaM89] showed that the categorical interpretation of a tensor theory can be easily given by means of the strictly symmetric strict monoidal category obtained by the $\mathcal{T}$[N] construction [MM88]. Moreover, since linear categories are models for full linear logic, [MaM89] proposed to consider the linear category freely generated by the tensor theory N and then to try to interpret the complex arrows of the category as "idealized" net computations. As an example, they explain how to interpret the connectives of additive conjunction & and disjunction $\oplus$ as forms of nondeterminism (see also [GG89] about this example). Following this paradigm, [AFG90] gave a net interpretation of the connective of linear implication: given a tensor theory, a computation of the implicative fragment (i.e. of the tensorial fragment enriched with linear implication) can be seen as a lower level description, where the acquisition of *each* token for the firing of a transition is a *separate* move of an equivalent tensor computation (i.e. of a net computation).

The efforts in understanding the connections between linear logic and Petri nets has in part influenced our ideas, and our results can also be interpreted in this light. The operator of additive conjunction &, for instance, is analogous to SCONE operator of nondeterministic composition +. Therefore, our implementation of global nondeterminism by means of local nondeterminism and sequential composition can be considered a mean for giving a linear logic account of CCS nondeterministic operator. Furthermore, our generative algebraic method explained in Section 4 has the advantage of giving a finite intensional representation of an infinite net, that is a finite representation for an infinite tensor theory. This result has been achieved by introducing a synchronization operator which, given two net transitions (logical axioms) generates a new transition (axiom). Following the analogy, we have also that all finite tensor theories (with certain limitations) can be seen as sub-theories of this infinite theory.

Acknowledgements

We would like to thank Andrea Asperti, Carlos Blanco, Andrea Corradini, Pierpaolo Degano, Gian Luigi Ferrari, Ursula Goltz, Cosimo Laneve, Josè Meseguer and Daniel Yankelevich for stimulating discussions and helpful comments.

REFERENCES

[ADJ77] Gougen J.A., Tatcher J.W., Wagner E.G., Wright J.B., Initial Algebra Semantics and Continuous Algebras, *Journal of the ACM*, **24** (1), (1977), 68-95.

[As87] Asperti A., *A Logic for Concurency*, Tech. Report, Dipartimento di Informatica, November 1987.

[AFG90] Asperti A., Ferrari G., Gorrieri R., Implicative formulae in the "Proofs as Computations" Analogy, in Proc. *17th ACM Symp. on Principles of Programming Lang.* (POPL'90), San Francisco, January 1990, 59-71.

[BC89] Boudol G., Castellani I., Permutation of Transitions: An Event Structure Semantics for CCS, Proc. REX School: *Linear Time, Branching Time and Partial Order in Logics and Models for Concurrency*, LNCS 354, 1989, 411-427.

[BD87] Best E., Devillers R., Sequential and Concurrent Behaviour in Petri Net Theory, *Theoretical Computer Science*, **55** (1), (1987), 87-136.

[Bl88] Blanco C., Hacer Explícita la Elección Atómica de CCS Facilita la Construcción de Ordenes Parciales, Master thesis, ESLAI, Buenos Aires, December 1988.

[Ch89] Cherkasova L., Posets with Non-Actions: A Model for Concurrent Nondeterministic Processes, Arbeitspapiere der GMD n. 403, July 1989.

[Co90] Corradini A., *An Algebraic Semantics for Transition Systems and Logic Programming*, Ph.D. Thesis, TD 8/90, Dipartimento di Informatica, Pisa, March 1990.

[DDM88a] Degano P., De Nicola R., Montanari U., A Distributed Operational Semantics for CCS based on Condition/Event Systems, *Acta Informatica*, **26** (1988), 59-91.

[DDM88b] Degano P., De Nicola R., Montanari U., Partial Ordering Semantics for CCS, Internal Report 88-3, Dipartimento di Informatica, Univ. Pisa, 1988, to appear in *Theoretical Computer Science*.

[DDM89] Degano P., De Nicola R., Montanari U., Partial Ordering Description of Nondeterministic Concurrent Systems, in Proc. REX School: *Linear Time, Branching Time and Partial Order in Logics and Models for Concurrency*, LNCS 354, 1989, 438-466.

[DGM88] Degano P., Gorrieri R., Marchetti S., An Exercise in Concurrency: A CSP Process as a Condition/Event System, in *Advances in Petri Nets 1988*, LNCS 340, 1988, 85-105.

[DMM89] Degano P., Meseguer J., Montanari U., Axiomatizing Net Computations and Processes, in Proc. *Logic in Computer Science* (LICS '89), Asilomar, 1989, 175-185. Extended and revised version available as technical report of Dipartimento di Informatica, Pisa University.

[F90] Ferrari G., *Unifying Models for Concurrency*, Ph.D. Thesis, TD 4/90, Dip. di Informatica, Pisa, March 1990.

[FM90] Ferrari G., Montanari U., Towards the Unification of Models for Concurrency, in Proc. Coll. on Algebra and Trees in Prog. (CAAP'90), Copenhagen, LNCS 431, 162-176.

[G87] Girard J.Y., Linear Logic, *Theoretical Computer Science*, **50**, (1987), 1-102.

[Go88] Goltz U., On Representing CCS Programs by Finite Petri Nets, Proc. MFCS'88, LNCS 324, 339-350.

[GG89] Gunter C., Gehlot V., Nets as Tensor Theories, in *Advances in Petri Nets 1989*, LNCS 424.

[GM84] Goltz U., Mycroft A., On the Relationships of CCS and Petri Nets, in Proc. 11th ICALP, LNCS, 172, 1984, 196-208.

[GMM88] Gorrieri R., Marchetti S., Montanari U., A^2CCS: Atomic Actions for CCS, in Proc. Coll. on Trees and Algebras in Prog. (CAAP'88), LNCS 299, 1988, 258-270. Extended version to appear in *Theoretical Computer Science*, **72**, (1990).

[GR83] Goltz U., Reisig W., The Non-sequential Behaviour of Petri Nets, *Information and Co.*, **57**, (1983), 125-147.

[K78] Kotov V., An Algebra for Parallelism Based on Petri Nets, LNCS 64, 1978, 39-55.

[MaM89] Martì-Oliet N., Meseguer J., From Petri Nets to Linear Logic, in Proc. *3rd Conf. on Category Theory in Computer Science*, Manchester, LNCS 389, 1989, 313-340.

[Mil84] Milner R., A Complete Inference System for a Class of Regular Behaviours, *Journal of Computer and System Sciences*, **28**, (1984), 439-466.

[Mil89] Milner R., *Communication and Concurrency*, Prentice-Hall, 1989.

[ML71] Mac Lane S., Categories for the working Mathematicians, Springer-Verlag, 1971.

[MM88] Meseguer J., Montanari U., Petri Nets are Monoids: A New Algebraic Foundation for Net Theory, in Proc. *Logic in Computer Science*, Edinburgh, 1988, 155-164. Full version to appear in *Info. and Co*, also available as Tech. Rep. SRI-CSL-88-3, SRI International, January 1988.

[MY89] Montanari U., Yankelevich D., An Algebraic View of Interleaving and Distributed Operational Semantics for CCS, in Proc. *3rd Conf. on Category Theory in Comp. Scien.*, Manchester, LNCS 389, 1989, 5-20.

[Old89] Olderog E.-R., Strong Bisimilarity on Nets. in Proc. REX School: *Linear Time, Branching Time and Partial Order in Logics and Models for Concurrency*, LNCS 354, 1989, 549-573.

[Pet62] Petri C.A., *Kommunication mit Automaten*, Schriften des Institutes fur Instrumentelle Mathematik, Bonn, 1962.

[Plo81] Plotkin G., *A Structural Approach to Operational Semantics*, Technical Report DAIMI FN-19, Aarhus University, Department of Computer Science, Aarhus, 1981.

[T89] Taubner D., Finite Representation of CCS and TCSP Programs by Automata and Petri Nets, LNCS 369, 1989.

[W85] Winskel G., Categories of Model of Concurrency, in *Seminar on Concurrency*, LNCS 197, 1985, 246-267.

[W87] Winskel G., Petri Nets, Algebras, Morphisms and Compositionality, *Information and Computation* **72**, 1987, 197-238.

APPENDIX

A. Categorical Background

A *category* C is a graph (V, T, ∂_0, ∂_1), where the states in V and the transitions in T are usually called *objects* and *arrows*, respectively, with additionally:

- an operation $\mathrm{id} : V \to T$ called *identity*, such that $\partial_0(\mathrm{id}(v)) = v = \partial_1(\mathrm{id}(v))$,
- a (partial) operation $; : T \times T \to T$ called *composition*, assigning to each pair of arrows t and t', such that[1] $\partial_0(t')=\partial_1(t)$, an arrow t ; t' such that $\partial_0(t\,;t')=\partial_0(t)$, $\partial_1(t\,;t')=\partial_1(t')$,
- and the operations satisfy the two axioms below

  $\forall$ t, t', t'' such that $\partial_1(t) = \partial_0(t')$ and $\partial_1(t') = \partial_0(t'')$, $\quad t\,;(t'\,;t'') = (t\,;t')\,;t''$

  $\forall$ t such that $\partial_0(t) = u$ and $\partial_1(t) = v$, $\quad \mathrm{id}(u)\,;t = t = t\,;\mathrm{id}(v)$

  expressing associativity of composition and the fact that identities are units for it.

Let C, D be two categories. A (covariant) *functor* $F{:}C \to D$ is a pair of functions $F_V{:}V_C \to V_D$ and $F_T{:}T_C \to T_D$, such that for each t, t' $\in T_C$

- $\partial_{iD}(F_T(t)) = F_V(\partial_{iC}(t))\quad i = 0, 1$
- $F_T(\mathrm{id}(v)) = \mathrm{id}(F_V(v))$
- $F_T(t\,;t') = F_T(t)\,;F_T(t')$

Let F, G : $C \to D$ be two functors. A *natural transformation* $\tau : F \to G$ is a function from V_C to T_D assigning to a object u in C an arrow τ_u in D such that

- $\partial_0(\tau_u) = F_V(u)$ and $\partial_1(\tau_u) = G_V(u)$
- $\forall$ t in T_C with $\partial_0(t) = u$ and $\partial_1(t) = v$ $\quad \tau_u\,;G_T(t) = F_T(t)\,;\tau_v$

A transformation τ such that each component τ_u is invertible in D is called a natural *isomorphism*.

Let $F : D \to C$ and $G : C \to D$ be two functors. An *adjunction* from D to C is a triple ‹F, G, ψ› such that $\psi{:}T_C[F\_, \_] \to T_D[\_, G\_]$ is a natural isomorphism, i.e., for any pair of objects c in C and d in D, ψ is a natural isomorphism between the sets of arrows $T_C[F(d), c]$ and $T_D[d, G(c)]$. F is called *left adjoint* of G, and G is called *right adjoint* of F.

A *monoidal* category $(C, \otimes, e, \alpha, \lambda, \rho)$ consists of

- a category C;
- a bifunctor $\otimes : C \times C \to C$, thus satisfying the functoriality axioms

  $(t_1 \otimes t_2)\,;(t_1' \otimes t_2') = (t_1\,;t_2) \otimes (t_1'\,;t_2')$ $\quad \mathrm{id}(u)\otimes\mathrm{id}(v) = \mathrm{id}(u\otimes v)$
- a (left and right identity) object e;
- and three natural isomorphisms α, λ, ρ defined as follows:

  $\alpha_{u,v,w} : u\otimes(v\otimes w) \to (u\otimes v)\otimes w$

  $\lambda_v : e\otimes v \to v$

  $\rho_v : v\otimes e \to v$

  such that the following diagrams, representing the MacLane-Kelly coherence equations for associativity of $\otimes$ and identity of e, commute

[1] Notice that composition is defined following the interpretation of sequential composition of transitions (diagrammatic order), which is obvious in computer science, but is in contrast with the tradition in mathematics.

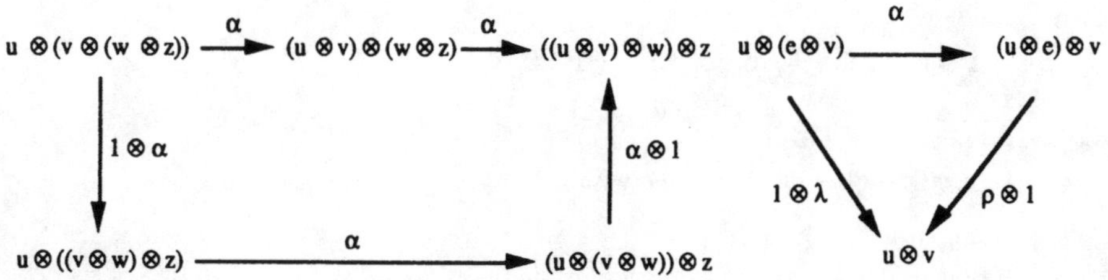

Let us consider two arrows $t:u \to v$, $t':u' \to v'$. An interesting consequence of stating that γ is a natural isomorphism is

$$\gamma_{u',u} ; (t' \otimes t) = (t \otimes t') ; \gamma_{v,v'}.$$

meaning that the *factors can be exchanged in any monoidal composition of transitions, provided that suitable exchanges are sequentially composed before and after.*

A *strict* monoidal category $(C, \otimes, e)$ is a monoidal category where α, λ, ρ are actually identities, i.e. $\otimes$ is associative with e as neutral element on objects and id(e) on arrows. No coherence axioms are needed.

A *symmetric* monoidal category $(C, \otimes, e, \alpha, \lambda, \rho, \gamma)$ is a monoidal category enriched with a natural isomorphism $\gamma_{u,v}:u\otimes v \to v\otimes u$ such that the following three equations hold (the last one in diagram form)

$$\gamma_{u,v} ; \gamma_{v,u} = \mathrm{id}(u\otimes v) \qquad \rho_v = \gamma_{v,e} ; \lambda_v$$

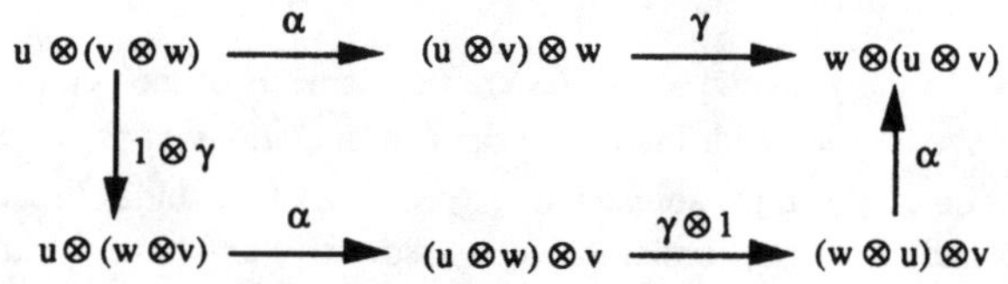

A symmetric strict monoidal category $(C, \otimes, e, \gamma)$ is a symmetric monoidal category where the bifunctor is a strictly monoidal operator. In this case, the coherence axioms are simply

$$\gamma_{u,v} ; \gamma_{v,u} = \mathrm{id}(u\otimes v) \qquad (\gamma_{u,v} \otimes \mathrm{id}(w)) ; (\mathrm{id}(v) \otimes \gamma_{u,w}) = \gamma_{u,v\otimes w}$$

A strictly symmetric strict monoidal category $(C, \otimes, e)$ is a symmetric strict monoidal category where $\gamma_{u,v}$ is actually an identity, i.e. $\otimes$ is a commutative monoidal operator. No coherence axioms are needed.

B. The Categories of Symmetries, of Processes and of Concatenable Processes

To define symmetries, let us consider a finite set I labelled by a function $l : I \to S$ which associates to every element x a label $l(x)$. When defined up to isomorphisms (i.e. up to label-preserving bijections), set I corresponds to an element $u = n_1a_1 \oplus ... \oplus n_ka_k$ in $S^{\oplus}$, where $n_i = |\{x \mid l(x)=a_i\}|$ $i = 1,\dots,k$. A *symmetry* p of the labelled set I is a bijective endofunction $p : I \to I$ which is label-preserving, i.e. such that $l(x) = l(p(x))$. We can associate it to v and write $p : u \to u$. It is clear that, by choosing a linear order for each of the sets $\{x \mid l(x)=a_i\}$, $i=1,\dots,k$, p can be expressed as a vector of permutations.

Given $u = n_1a_1\oplus...\oplus n_ka_k$ in $S^{\oplus}$, a symmetry $p:u\to u$ is a vector of permutations $\langle\sigma_{a_1}, \dots, \sigma_{a_k}\rangle$ with $\sigma_{a_i} \in \Pi(n_i)$, i.e., σ_{a_i} is a permutation of n_i elements ($|\sigma_{a_i}| = n_i$). We define two operations on symmetries: the parallel composition $p\otimes q$ and the sequential composition $p;q$ of two symmetries. Let $p:u \to u$ and

$q{:}v \to v$; then

$$p \otimes q{:}\ u \oplus v \to u \oplus v\ =\ \langle \sigma_{a_1} \otimes \sigma'_{a_1}, \ldots, \sigma_{a_k} \otimes \sigma'_{a_k} \rangle$$

where

$$\sigma \otimes \sigma'(x) = \begin{cases} \sigma(x) & \text{if } 0 < x \le |\sigma| \\ \sigma'(x - |\sigma|) + |\sigma| & \text{otherwise} \end{cases}$$

Let $p{:}u \to u$ and $q{:}u \to u$; then

$$p;q{:}u \to u\ = \langle \sigma_{a_1};\sigma'_{a_1},\ldots,\sigma_{a_k};\sigma'_{a_k} \rangle \qquad \text{where} \quad \sigma\,;\sigma'(x) = \sigma'(\sigma(x)).$$

Notice that both ⊗ and ; are associative but not commutative. If ι_{a_i} denotes the identity permutation, then for each u in $S^{\oplus}$ the symmetry $\langle \iota_{a_1}, \ldots, \iota_{a_k} \rangle : u \to u$, also denoted with abuse of notation by u, is an identity for sequential composition. Furthermore, ⊗ and ; satisfy the functoriality equations

$$(p \otimes p');(q \otimes q') = (p;q) \otimes (p';q') \qquad\qquad \mathrm{id}(u) \otimes \mathrm{id}(v) = \mathrm{id}(u \oplus v)$$

We need a third operation on symmetries, called *interchange*. Let π be in $\Pi(m)$, i.e. a permutation of m elements. Given m elements $u_i = n_{i_1}a_1 \oplus \ldots \oplus n_{i_k}a_k$, i = 1, ..., m, in $S^{\oplus}$, we define the *interchange symmetry* $\pi(u_1,\ldots,u_m){:}\ u_1 \oplus \ldots \oplus u_m \to u_1 \oplus \ldots \oplus u_m$, $\pi(u_1, \ldots, u_m) = \langle \sigma_{a_1}, \ldots, \sigma_{a_k} \rangle$ as follows:

$$\sigma_{a_j} = x - \sum_{i=1}^{h-1} n_{ij} + \sum_{\pi(i) \le \pi(h)} n_{ij} \qquad \text{if} \ \sum_{i=1}^{h-1} n_{ij} < x \le \sum_{i=1}^{h} n_{ij} \quad j = 1, \ldots, k.$$

As an example, let us consider Figure 1.c. Given two objects u = 2a ⊕ b and v = 3a ⊕ 2b, we have that $u \oplus v = v \oplus u$ being ⊕ commutative, but the permutation $\pi = \{1 \to 2, 2 \to 1\}$ obviously generates an *interchange* symmetry p on $u \oplus v$ = 5a ⊕ 3b, $p = \langle \sigma_a, \sigma_b \rangle$, where $\sigma_a = \{1 \to 4, 2 \to 5, 3 \to 1, 4 \to 2, 5 \to 3\}$ and $\sigma_b = \{1 \to 3, 2 \to 1, 3 \to 2\}$.

Given a set S, let Sym_S be the graph whose nodes are the elements of the commutative monoid $S^{\oplus}$ and whose transitions are symmetries with the above defined operations of ⊗ and ;. Then, Sym_S is a category, because identities do exist and ; is composition; Sym_S is strict monoidal, because the pair ‹⊕, ⊗› is a bifunctor (⊕ on nodes and ⊗ on transitions) which is associative and has neutral element; Sym_S is symmetric since the *interchange symmetry* associated to the permutation $\pi = \{1 \to 2, 2 \to 1\}$, $\pi(u, v) : u \otimes v \to v \otimes u$, is a natural isomorphism (equation $\pi(u, v)\ ;\ (p \otimes q) = (q \otimes p)\ ;\ \pi(v, u)$ holds) satisfying the coherence axioms of symmetric strict monoidal categories. Indeed, Sym_S is a symmetric strict monoidal category which is strictly symmetric on objects.

Given a net N = $(S^{\oplus}, T, \partial_0, \partial_1)$, the category $\mathcal{P}$[N] of its processes is defined as follows. The objects of $\mathcal{P}$[N] are the nodes of N, i.e. $S^{\oplus}$. $\mathcal{P}$[N] includes Sym_S as a subcategory, and has as additional arrows those defined by the following rules of inference:

$$\frac{t{:}u \to v \text{ in N}}{t{:}u \to v \text{ in } \mathcal{P}[N]}$$

$$\frac{\xi{:}u \to v \quad \xi'{:}u' \to v' \text{ in } \mathcal{P}[N]}{\xi \otimes \xi'{:}u \oplus u' \to v \oplus v' \text{ in } \mathcal{P}[N]} \qquad\qquad \frac{\xi{:}u \to v \quad \eta{:}v \to w \text{ in } \mathcal{P}[N]}{\xi;\eta{:}u \to w \text{ in } \mathcal{P}[N]}$$

and axioms expressing the fact that the arrows form a monoid:

$$(\xi \otimes \eta) \otimes \zeta = \xi \otimes (\eta \otimes \zeta) \qquad\qquad \xi \otimes 0 = \xi$$

the fact that $\mathcal{P}$[N] is a category[2]:

$$\xi\,;\partial_1(\xi) = \partial_0(\xi)\,;\xi = \xi \qquad\qquad (\xi\,;\eta)\,;\zeta = \xi\,;(\eta\,;\zeta)$$

and the fact that ⊗ is a bifunctor:

---

[2] Notice that there is no inference rule for generating identities, since they are already included in Sym_S. Recall that we represent with the same symbol both the nodes and their associated identities.

$$(\xi \otimes \xi') ; (\eta \otimes \eta') = (\xi ; \eta) \otimes (\xi' ; \eta') \qquad u \otimes v = u \oplus v$$

In addition, there is an axiom involving symmetries, stating that generators (i.e. the basic transitions of N) are symmetrical:

$$t ; p = t \quad \text{where } t : u \to v \text{ in N and } p : v \to v \text{ in } Sym_S$$
$$p ; t = t \quad \text{where } t : u \to v \text{ in N and } p : u \to u \text{ in } Sym_S.$$

Finally, the fact that *interchange symmetries* $\pi(u_1, \ldots, u_m)$, associated to any permutation π in $\Pi(m)$, define a natural isomorphism:

$$\pi(u_1, \ldots, u_m) ; (\xi_{\pi(1)} \otimes \ldots \otimes \xi_{\pi(m)}) = (\xi_1 \otimes \ldots \otimes \xi_m) ; \pi(v_1, \ldots, v_m).$$

The intuitive meaning of the monoidal operation ⊗ is parallel composition of arrows, and of course the meaning of the categorical operation ; is sequential composition of arrows. Since the generators of the arrows of $\mathcal{P}$[N] are the transitions of N and the symmetries, the inference rules above amount to closing them with respect to parallel and sequential composition, giving origin to a generalized notion of net computation. In other words, $\mathcal{P}$[N] is a *two-sorted algebra* whose objects are defined by the algebra of the nodes of N (i.e., $S^{\oplus}$) and the arrows by the algebra having T and the symmetries as generators and ⊗ and ; as operations satisfying the axioms above.

Category $\mathcal{P}$[N] of the processes of a net N can be obtained by first taking the left adjoint to the forgetful functor from symmetric strict monoidal categories to the category Petri of Petri nets, and then quotienting it with the axioms (Ψ) of Definition 2.3. The above free construction of $\mathcal{P}$[N] gives a more explicit description of these two steps.

Given a set S with a labeling function l: S → S', a *label-indexed ordering function* is a family $\beta = \{\beta_a\}$, $a \in S'$, of bijections, where $\beta_a : [a] \to \{1, \ldots, |[a]|\}$, with $[a] = \{b \in S \mid l(b) = a\}$.

A *plain process* for a net N is a morphism $\varphi = \langle f, g \rangle$ in Petri from a finite occurrence net P to N. The functions f and g map transitions and places of P to transitions and places of N, respectively. The places of P which are minimal in the partial ordering associated to it are called *origins*, the maximal places are called *destinations*.

We give plain processes a categorical structure, by taking as morphisms those Petri morphisms between the supporting occurrence nets that preserve mapping f and g.

A *concatenable process* for a net N is a triple $C = \langle \varphi, \theta, \kappa \rangle$, where:

- $\varphi = \langle f, g \rangle$ is a plain process for N;
- θ and κ are a label-indexed ordering function on the origins and destinations, respectively, where the labelling function is g restricted to the respective domains.

Isomorphic[3] concatenable processes are identified. Furthermore, we can associate to every concatenable process C of N two multisets of places, i.e. two elements of $S^{\oplus}$, as follows. The multisets O(C) and D(C) are defined as

$$O(C) = \sum_{i=1}^{k} n_i a_i \text{ and } D(C) = \sum_{i=1}^{k} m_i a_i$$

where a_i are places of N, and n_i and m_i are the numbers of origins b and destinations c of P, respectively, such that $g(b) = a_i = g(c)$.

We can picture a concatenable process C of N as an arrow C : O(C) → D(C). Also concatenable processes may easily be turned into a monoidal category, by defining operations on them.

[3] Two concatenable processes C and C' are isomorphic if there is a plain process isomorphism $\langle f'', g'' \rangle$ from P to P' preserving the label-indexed functions, namely with $\theta(b)=\theta'(g''(b))$ and $\kappa(b)=\kappa'(g''(b))$.

OPERATIONS ON CONCATENABLE PROCESSES

Net Transitions

Given a transition $t : n_1a_1\oplus...\oplus n_ka_k \rightarrow n'_1a_1\oplus...\oplus n'_ka_k$ in N, let P be the occurrence net with $I \cup I'$ as set of places, where $I = \{\langle i, j, 0\rangle \mid i = 1,...,k, j = 1,..., n_i\}$, $I' = \{\langle i, h, 1\rangle \mid i = 1,...,k, h = 1,..., n'_i\}$, and t' as unique transition with $\partial_0(t') = I$ and $\partial_1(t') = I'$. Then a concatenable process for t is $C = \langle \varphi, \theta, \kappa\rangle$ where $\varphi = \langle f, g\rangle$ is such that $f(t') = t$ and $g(\langle i, j, 0\rangle) = g(\langle i, h, 1\rangle) = a_i$, $i = 1,...,k$, $j = 1,..., n_i$, $h = 1,..., n'_i$, and the label-indexed ordering functions on the origins and destinations of P are given by $\theta_{a_i}(\langle i, j, 0\rangle) = j$ and $\kappa_{a_i}(\langle i, h, 1\rangle) = h$.

Symmetries

Given a symmetry $p = \langle \sigma_{a_1}, ..., \sigma_{a_k}\rangle : n_1a_1 \oplus... \oplus n_ka_k \rightarrow n_1a_1 \oplus...\oplus n_ka_k$, let P be the occurrence net having $I = \{\langle i, j\rangle \mid i = 1,...,k, j = 1,..., n_i\}$ as set of places, and no transitions. A concatenable process for p is $C = \langle \varphi, \theta, \kappa\rangle$ where $\varphi = \langle \emptyset, g\rangle$ is such that $g(\langle i, j\rangle) = a_i$, and the label-indexed ordering function on the origins and destinations of P are such that $\theta_{a_i}(\langle i, j\rangle) = j$ and $\kappa_{a_i}(\langle i, j\rangle) = \sigma_{a_i}(j)$, respectively.

Let in the following definitions of parallel and sequential composition of processes $C = \langle \varphi, \theta, \kappa\rangle : n_1a_1\oplus...\oplus n_ka_k\rightarrow m_1a_1\oplus...\oplus m_ka_k$ and $C' = \langle \varphi', \theta', \kappa'\rangle : n'_1a_1\oplus...\oplus n'_ka_k\rightarrow m'_1a_1\oplus...\oplus m'_ka_k$ be two concatenable processes for N, with P and P' as occurrence nets of φ and φ', respectively.

Parallel Composition

We define $C \otimes C' = \langle \varphi'', \beta'', \gamma''\rangle : (n_1+n'_1)a_1 \oplus ... \oplus (n_k+n'_k)a_k \rightarrow (m_1+m'_1)a_1 \oplus ... \oplus (m_k+m'_k)a_k$ where

φ'' is the coproduct of φ and φ' in the category of plain processes;

$\theta'' = \langle \theta_{a_1} \cup (n_1+\theta'_{a_1}), ...,\theta_{a_k} \cup (n_k+\theta'_{a_k})\rangle$ and

$\kappa'' = \langle \kappa_{a_1} \cup (n_1+\kappa'_{a_1}), ...,\kappa_{a_k} \cup (n_k+\kappa'_{a_k})\rangle$.

Sequential Composition

Finally, $C ; C'$ is defined only if $D(C) = O(C') = m_1a_1\oplus...\oplus m_ka_k$. Let us define a plain process $\varphi^- = \langle \emptyset, g^-\rangle$ from the occurrence net $P^- = (S^{-\oplus}, \emptyset, \emptyset, \emptyset)$ with $S^- = \{\langle i, j\rangle \mid i = 1,...,k, j = 1,..., m_i\}$ and $g^-(\langle i, j\rangle) = a_i$. Two plain process morphisms r, r' from φ^- to φ and φ', respectively, are induced by the two functions s and s' from S^- to the destinations of P and to the origins of P', respectively, that satisfy the following equations:

$$g(s(\langle i, j\rangle)) = a_i = g'(s'(\langle i, j\rangle)) \quad \text{and} \quad \kappa(s(\langle i, j\rangle)) = j = \theta'(s'(\langle i, j\rangle)).$$

It is not difficult to see that s and s', and therefore r and r', are defined uniquely. Also, it is easy to verify that in the category of processes the pushout construction involving r and r' exists, yielding a new process φ'' (which is obtained, roughly speaking, as the disjoint union of φ and φ', where for all x, the pairs s(x) and s'(x) have been identified). Then,

$$C ; C' = \langle \varphi'', \beta, \gamma'\rangle : n_1a_1 \oplus ... \oplus n_ka_k \rightarrow m'_1a_1 \oplus ... \oplus m'_ka_k.$$

VALUE-PASSING IN PROCESS ALGEBRAS

M. Hennessy

University of Sussex

ABSTRACT

Much of the theoretical development of process algebras in recent years has been focused on so-called "pure processes" whose only form of communication is through synchronisation signals. In this lecture I will review some recent work on adapting one such theory to processes which communicate values using channels.

In the standard theory, as applied to "pure processes", [He88], we have the reasonable identity

$$a.p + a.q = a.p \oplus a.q$$

where a is an uninterpreted synchronisation action and +, ⊕ represent external and internal choice respectively. In the extended language, with value-passing processes, this leads to the identities

$$c?x.p + c?x.q = c?x.p \oplus c?x.q$$

$$c!v.p + c!v.q = c!v.p \oplus c!v.q$$

where c?, c! represent input and output respectively along the channel c. In the basic language we do not have the identity

$$a.p + b.q = a.p \oplus b.q$$

but in the extended language we argue that such an identity is reasonable for output actions:

$$c!v.p + c!v'.q = c!v.p \oplus c!v'.q.$$

We present a behavioural theory for value-passing processes based on testing which justifies this identity and which, moreover, has a fully-abstract model in the form of an algebraic cpo based on the Acceptance trees of [He88]. We also discuss algebraic proof systems for value-passing processes which are complete with respect to this model.

[He88] M. Hennessy, <u>An Algebraic Theory of Processes</u>, MIT Press, 1988

[HI89] M. Hennessy, A.Ingolfsdottir "A Theory of Communicating Processes with Value-Passing", Technical Report No.3/89, University of Sussex, 1989. Also presented at ICALP 90.

[He89] M. Hennessy "A Proof System for Communicating Processes with Value-Passing". Technical Report No.5/89, University of Sussex, 1989. Also presented at FST-TCS'89.

Let's Make Models

C.A.R. Hoare

Summary

Science makes progress by constructing mathematical models, deducing their consequences, and testing them by experiment. Such models are the basis for engineering methods and codes of practice for design of reliable and useful products. Models can play a similar central role in the progress and practical application of Computing Science.

A model of a computational paradigm starts with choice of a carrier set of potential direct or indirect observations that can be made of a computational process. A particular process is modelled as the subset of observations to which it can give rise. Process composition is modelled by relating observations of a composite process to those of its components. Algebraic properties of such compositions are derived simply with the aid of set theory.

A model constructed as a family of sets is easily adapted as a calculus of design for total correctness. A specification is given by an arbitrary set containing all observations permitted in the required product. A product meets a specification if its potential observations form a subset of its permitted observations. Specifications of components can be composed mathematically in the same way as the components themselves. This permits proof of correctness of top-down designs even before their implementation begins. Algebraic properties and reasoning are helpful throughout development. Non-determinism is seen as no problem, but rather as a part of the solution.

Ideal Specification Formalism = Expressivity + Compositionality + Decidability + Testability + ⋯

Kim Guldstrand Larsen *
Swedish Institute of Computer Science †

Abstract

For the comparison of specification formalisms several criteria may be applied. In this paper we concentrate on the formalization of expressivity and compositionality. We apply these criteria to a number of specification formalisms ranging from behavioural formalisms (based on labelled transition systems) to logical formalisms (based on Hennessy–Milner logic). A main result of the paper is that a specification formalism must be at least as expressive as Hennessy–Milner Logic in order for specifications to be decomposable. Another main result is that *implicit* behavioural specifications are at least as expressive as logical specifications and do allow specifications to be decomposed. We also present specification formalisms for probabilistic processes, and evaluate these with respect to compositionality.

Introduction

The development of concurrent systems and processes constitutes a difficult and surprisingly subtle problem in computer science. Attempting to overcome this problem a number of specification formalisms and verification methods have been put forward in recent years. Two main categories of formalisms may be identified: the *logical* approach, in which a specification is a formula of some (temporal or modal) logic, and verification is a "model–checking" activity based on a denotational understanding of the specification; the *behavioural* approach, where specifications are objects of the same kind as implementations: in particular specifications have an operational interpretation. In this approach, verification is based on a comparison between the operational behaviours of the specification and the implementation.

In comparing specification formalisms we may apply several criteria some of which are the following:

Expressivity: A specification may be identified with the set of processes satisfying the specification. The relative expressivity of a specification formalism may then be measured in terms of the sets identified by (specifications of) the formalism: the more sets identified by the formalism the more expressive it is. High expressivity is important from a pragmatic point of view as it enables us to specify more precisely the set of acceptable implementations. In particular the specification formalism should ideally be sufficiently expressive that we are not forced to make

*On leave from Aalborg University, Denmark
†Address: SICS, Box 1263, S–164 28 Kista, SWEDEN. E-mail: kim@sics.se.

too early (design) decisions. Instead it should be possible to derive a final implementation from an initial (loose) specification through a series of refinement steps (i.e. a series of specifications decreasing in terms of processes identified). Related to the expressivity of a specification formalism is the induced equivalence on processes: two processes are equivalent if they satisfy the same specifications. It is clear that the more expressive a specification formalism is the stronger equivalence it will induce.

Compositionality: For the verification of large systems it is essential that the specification formalism support compositional verification. Firstly, the formalism must allow us to decompose the problem of correctness for a complex system into similar correctness problems for the components of the system. Moreover the correctness properties required of the components should preferably be as weak as possible in order not to restrict unnecessarily the choices of their implementation. Clearly these demands presume a certain expressivity of the specification formalism relative to the operators for building processes.

Secondly, it should be possible to infer properties of a combined system using only the specifications of its components, and without knowledge as to their implementation. In particular this would make "reusability of proofs" possible: if one replaces a component by another satisfying the same subspecification, then the properties inferred for the combined system remains valid. Moreover, we would obviously prefer the correctness properties inferred for the combined system to be as strong as possible. Again it is clear that a certain expressivity is required in order for the specification formalism to meet these demands.

Decidability: Through the study of examples it is rather clear that computer assistance is essential in order to make the verification of correctness feasible in practice. In order to develop automatic tools for carrying out this verification, the decidability and complexity of refinement (between specifications) and satisfaction (between processes and specifications) becomes key questions to answer. Existing verification tools are mostly based on a state–exploration of the implementing process. However, it is strongly believed that for compositional specification formalisms the state–exploration approach may be avoided, and better algorithms obtained.

Testability: At present formal verification of large systems is considered too costly and time consuming in most practical situations. Instead the validation of an implementation frequently takes the form of extensive *testing* against its specification. Although full correctness (in general) can never be achieved through testing, it ought to be the case that the more tests a system passes, the more *confidence* we may have in the correctness of the system. If fact we may hope that extensive testing can determine the correctness of a system with arbitrary high confidence. Also within the testing framework the notions of compositionality and complexity seem extremely important ones:

- Is it possible to conclude — with some confidence — anything about the correctness of a combined system through testing of its components?
- Is it possible to conclude — with some confidence — anything about the correctness of a component through the testing of the combined system in which it is used?
- The performance of a test will obviously have certain costs associated. Hence, given a desired level of confidence, we obviously want a test with lowest possible cost.

In this paper we concentrate on formalizing and applying the notions of expressivity and compositionality. In the next section, we introduce the notion of specification formalism and formalize various notions of expressivity and compositionality. Section 2 presents our framework of processes and contexts. In section 3, we present a number of behavioural specification formalisms and determine their relative expressivity. Section 4 introduces a logical specification formalism based on

Hennessy–Milner Logic and compare it in terms of expressivity with the behavioral specification formalisms of section 3. In section 5, we examine to what extent the various specification formalisms support compositionality. Finally, in section 6 we present specification formalisms for probabilistic processes, and discuss the question of compositionality.

1 Specification Formalism: Expressivity and Compositionality

To provide a basis for the remainder of this paper we introduce in this section the notion of specification formalism and formalize various notions of expressivity and compositionality.

Assume that Π is a set of processes or implementations. Then a *specification formalism* for Π is a structure $\mathcal{S} = \langle \Sigma, \mathbf{sat} \rangle$, where Σ is some set of specifications and **sat** is a subset of $\Pi \times \Sigma$. Whenever P **sat** S for P a process and S a specification, we say that P satisfies S or P is correct with respect to S; the relation **sat** is referred to as the satisfaction relation.

A specification formalism $\mathcal{S} = \langle \Sigma, \mathbf{sat} \rangle$ induces the following two functions $\mathsf{Mod}_{\mathcal{S}} : \Sigma \longrightarrow 2^{\Pi}$ and $\mathsf{Th}_{\mathcal{S}} : \Pi \longrightarrow 2^{\Sigma}$ relating specifications and processes:

$$\mathsf{Mod}_{\mathcal{S}}(S) =^{\Delta} \{P \in \Pi \mid P \ \mathbf{sat}\ S\} \qquad \mathsf{Th}_{\mathcal{S}}(P) =^{\Delta} \{S \in \Sigma \mid P \ \mathbf{sat}\ S\}$$

where $S \in \Sigma$ and $P \in \Pi$. Thus $\mathsf{Mod}_{\mathcal{S}}(S)$ is the set of all processes satisfying S, and $\mathsf{Th}_{\mathcal{S}}(P)$ is the set of all specification which is satisfied by P. Both $\mathsf{Mod}_{\mathcal{S}}$ and $\mathsf{Th}_{\mathcal{S}}$ can be extended to sets by intersecting the contributions of the elements; e.g. for a set of specifications A, $\mathsf{Mod}_{\mathcal{S}}(A) = \bigcap_{S \in A} \mathsf{Mod}_{\mathcal{S}}(S)$. The extended versions of $\mathsf{Mod}_{\mathcal{S}}$ and $\mathsf{Th}_{\mathcal{S}}$ constitutes a Galois connection between 2^{Σ} and 2^{Π} (with the appropriate orderings). A specification S is *consistent* or *satisfiable* if the set of models $\mathsf{Mod}_{\mathcal{S}}(S)$ is non–empty.

Refinement between specifications can now be defined simply as the inclusion between their models:

$$S_1 \Rightarrow S_2 \Leftrightarrow^{\Delta} \mathsf{Mod}_{\mathcal{S}}(S_1) \subseteq \mathsf{Mod}_{\mathcal{S}}(S_2)$$

i.e. S_1 refines S_2 provided any process satisfying S_1 also satisfies S_2. Dually, an equivalence is induced on processes based on their theories: two processes are equivalent in case they satisfy the same specifications [1]:

$$P \equiv_{\mathcal{S}} Q \Leftrightarrow^{\Delta} \mathsf{Th}_{\mathcal{S}}(P) = \mathsf{Th}_{\mathcal{S}}(Q)$$

Often $\equiv_{\mathcal{S}}$ can be shown to coincide with a behaviourally or semantically defined equivalence $\sim$ on processes, in which case the specification formalism $\mathcal{S}$ is said to be *adequate* with respect to $\sim$ (following the terminology introduced in [Pnu85]). Again following [Pnu85], a specification formalism is said to be *expressive* with respect to some process equivalence $\sim$, in case the formalism is powerful enough to express (as a single specification) the equivalence class of any given process; i.e.:

$$\forall P \in \Pi. \exists S \in \Sigma.\ \mathsf{Mod}_{\mathcal{S}}(S) = [P]_{\sim}$$

where $[P]_{\sim}$ denotes the set of processes $\sim$–equivalent with P.

The *relative expressivity* between specification formalisms can be formalized in the following way: For two specification formalisms $\mathcal{S} = \langle \Sigma, \mathbf{sat} \rangle$ and $\mathcal{S}' = \langle \Sigma', \mathbf{sat}' \rangle$ over a process set Π, we say that $\mathcal{S}$ is at least as expressive as $\mathcal{S}'$, if for any specification of $\mathcal{S}'$ one can find a specification of $\mathcal{S}$ allowing the same implementations; i.e.:

$$\forall S' \in \Sigma'. \exists S \in \Sigma.\ \mathsf{Mod}_{\mathcal{S}}(S) = \mathsf{Mod}_{\mathcal{S}'}(S')$$

[1]More generally a *preorder* between processes is induced in terms of the inclusion between their theories.

The notion of *compositionality* presumes the existence of certain operators $\mathcal{C}$ for combining processes (e.g. some operator for parallel composition of processes). In particular, if C is a (unary) operator (or *context*) and P is a process, we want to be able to interrelate properties of the combined process $C(P)$ with properties of the component process P. Clearly the interrelation has two directions, and we shall in the remainder of this section formalize notions of expressivity which will optimize the interrelation in each direction.

In one direction we are given an overall specification S and a combined process $C(P)$, and the problem is to *decompose* S into a subspecification T for P, such that it suffices to show P **sat** T in order to infer $C(P)$ **sat** S. To maximize the interrelation we want in addition the specification T obtained by the decomposition to be as weak as possible. Now, for C a context and S a specification define the set of processes $\mathsf{wip}(C, S)$ [2] as follows:

$$\mathsf{wip}(C,S) =^{\Delta} \{Q \in \Pi \mid C(Q) \textbf{ sat } S\}$$

then $P \in \mathsf{wip}(C, S)$ is obviously a both sufficient and necessary condition for $C(P)$ **sat** S to hold. This suggests the expressivity of all **wip**–sets as one criteria for compositionality: A specification formalism $\mathcal{S}$ *supports decomposition with respect to a set of contexts* $\mathcal{C}$ if the following holds:

$$\forall C \in \mathcal{C}. \forall S \in \Sigma. \exists T \in \Sigma.\ \mathsf{Mod}_{\mathcal{S}}(T) = \mathsf{wip}(C, S)$$

In the other direction we are given a subspecification T and the problem is to find a *composite* specification S such that $C(P)$ satisfies S for any component process P satisfying T. In order to maximize the interrelation we obviously want S to be as strong as possible. Now, for C a context and T a specification define the set of processes $\mathsf{sop}(C, T)$ [3] as follows:

$$\mathsf{sop}(C,T) =^{\Delta} \{Q \in \Pi \mid \exists P \textbf{ sat } T.\, Q \equiv_{\mathcal{S}} C(P)\}$$

that is, $Q \in \mathsf{sop}(C, T)$ if Q is equivalent to a process of the form $C(P)$, where P **sat** T. This suggests the expressivity of all **sop**–sets as another criteria for compositionality: A specification formalism $\mathcal{S}$ *allows composition with respect to a set of contexts* $\mathcal{C}$ if the following holds:

$$\forall C \in \mathcal{C}. \forall T \in \Sigma. \exists S \in \Sigma.\ \mathsf{Mod}_{\mathcal{S}}(S) = \mathsf{sop}(C, T)$$

Given a set of contexts $\mathcal{C}$, any specification formalism $\mathcal{S} = \langle \Sigma, \textbf{sat} \rangle$ induces a new specification formalism $\mathcal{I}_{\mathcal{S},\mathcal{C}}$ of *implicit specifications*. More precisely

$$\mathcal{I}_{\mathcal{S},\mathcal{C}} = \langle \mathcal{C} \times \Sigma, \textbf{sat}_{\mathcal{I}} \rangle, \quad \text{where } P \textbf{ sat}_{\mathcal{I}} (C, S) \Leftrightarrow^{\Delta} C(P) \textbf{ sat } S$$

i.e. P satisfies (C, S) in case the combined process $C(P)$ satisfies S. Now, it is easy to see that a specification formalism $\mathcal{S}$ allows decomposition with respect to $\mathcal{C}$ if and only if $\mathcal{S}$ is at least as expressive as the implicit specification formalism $\mathcal{I}_{\mathcal{S},\mathcal{C}}$.

If $\mathcal{C}$ contains an identity context I, then $\mathcal{I}_{\mathcal{S},\mathcal{C}}$ is at least as expressive as $\mathcal{S}$: for any specification S of $\mathcal{S}$, (I, S) provides an equivalent specification of $\mathcal{I}_{\mathcal{S},\mathcal{C}}$. Now, assume that $\mathcal{C}$ is closed under *composition* in the sense that for any contexts C and D there exists a combined context $C \circ D$ such that $C(D(P)) = (C \circ D)(P)$ for any process P. Then the implicit specification formalism $\mathcal{I}_{\mathcal{S},\mathcal{C}}$ always allows decomposition with respect to $\mathcal{C}$: simply note that the following equivalence holds for all contexts C, D, specifications S and processes P:

$$D(P) \textbf{ sat}_{\mathcal{I}_{\mathcal{S},\mathcal{C}}} (C, S) \quad \Leftrightarrow \quad P \textbf{ sat}_{\mathcal{I}_{\mathcal{S},\mathcal{C}}} (C \circ D, S)$$

[2] **wip** abbreviates *weakest inner property*.
[3] **sop** abbreviates *strongest outer property*.

2 Processes and Contexts

In this paper we follow the *reactive* view of concurrent systems advocated by Pnueli [Pnu85]; i.e. we assume that the semantics of concurrent systems (or *processes*) is given in terms of their interaction with their environment using the well–established model of *labelled transition systems* [Plo81].

Definition 2.1 *A* labelled transition system *is a structure* $\mathcal{P} = (\Pi, A, \longrightarrow)$ *where* Π *is a set of processes (states or configurations),* A *is a set of actions and* $\longrightarrow \subseteq \Pi \times A \times \Pi$ *is the transition relation.*

For $(P, a, Q) \in \longrightarrow$ we shall usually write $P \xrightarrow{a} Q$ which is to be interpreted: "P may perform the action a and become Q in doing so". We refer to Q as a (a–) derivative of P. We shall write $P \xrightarrow{a}$ as an abbreviation for $\exists Q.P \xrightarrow{a} Q$. We say that $\mathcal{P}$ is *image–finite* if the set $\{Q \mid P \xrightarrow{a} Q\}$ is finite for any process P and any action a. The set of *deterministic* processes is the largest set Π_d such that whenever $P \in \Pi_d$ then all derivatives of P belongs to Π_d and P has at most one a–derivative for any action a.

The notions of *simulation* [Lar87], *bisimulation* [Par81, Mil83] and 2/3–*bisimulation* [LS89] provide means of comparing processes at different descriptive levels of abstraction based on their operational behaviour.

Definition 2.2 $\mathcal{P} = (\Pi, A, \longrightarrow)$ *be a labelled transition system. Then a* simulation R *is a binary relation on* Π *such that whenever* $(P, Q) \in R$ *and* $a \in A$ *then the following holds:*

$$\textit{Whenever } P \xrightarrow{a} P', \textit{ then } Q \xrightarrow{a} Q' \textit{ for some } Q' \textit{ with } (P', Q') \in R.$$

A process Q *is said to simulate a process* P *in case* (P, Q) *is contained in some simulation* R*. We write* $P \leq Q$ *in this case.*

A bisimulation *is a binary relation* R *on* Π *such that both* R *and* $R^T = \{(Q, P) \mid (P, Q) \in R\}$ *are simulations. We write* $P \sim Q$ *in case* (P, Q) *is contained in some bisimulation.*

A 2/3–bisimulation *is a binary relation* R *on* Π *such that* R *is a simulation and whenever* $(P, Q) \in R$ *and* $Q \xrightarrow{a}$ *then also* $P \xrightarrow{a}$*. We write* $P \preceq Q$ *in case the pair* (P, Q) *is contained in some 2/3–bisimulation.*

The above defined notions enjoys a number of pleasant properties: $\leq$, $\sim$ and $\preceq$ is itself a simulation, bisimulation and 2/3–bisimulation respectively (in fact maximal ones), which admits a very elegant proof technique based on fixed point induction [Par81]. Moreover $\leq$ and $\preceq$ are preorders and $\sim$ is an equivalence relation on processes ordered by strength in the following way: $\sim \subseteq \preceq \subseteq \leq$. In fact all three relations are preserved by the usual CCS process constructs [Mil80] and indeed by any "natural" process construction [Lar86, dS85, GV89, Lar89].

Process algebra [Mil80, Mil89, Hoa78, BK85, Bou85] provides a framework for describing both the modular structure and the operational behaviour of reactive systems (or processes). In particular, a process algebra enables processes to be constructed (syntactically) through a number of operators (normally including some operator for parallel composition). Semantically, these operators are described through a number of inference–rules from which the operational behaviour of a composite process may be inferred from that of its components. In figure 1 is shown the well–known inference rules for the parallel composition operator $\mid$ and the restriction operator $\backslash L$ ($L \subseteq A$) of CCS [Mil80, Mil89].

$$\frac{P \xrightarrow{a} P'}{P \,|\, Q \xrightarrow{a} P' \,|\, Q} \qquad \frac{Q \xrightarrow{a} Q'}{P \,|\, Q \xrightarrow{a} P \,|\, Q'} \qquad \frac{P \xrightarrow{a} P' \quad Q \xrightarrow{\bar{a}} Q'}{P \,|\, Q \xrightarrow{\tau} P' \,|\, Q'} \qquad \frac{P \xrightarrow{a} P'}{P \backslash L \xrightarrow{a} P' \backslash L} a, \bar{a} \notin L$$

Figure 1: Inference rules for | and $\backslash L$ of CCS

In process algebra, derived operators (or *contexts*) are normally represented syntactically as terms with free variables possibly occurring. However, in order to facilitate a general investigation of the problem of equation solving, we developed in [Lar86, LX90a, LX90b] an operational theory of contexts in terms of action transducers. That is, a (unary) context is semantically viewed as an object which consumes actions provided by its internal process and in return produces actions for an external observer. We shall allow transduction in which the context produces actions on its own without involving the inner process, and also, we shall assume that the context may change during transductions.

Definition 2.3 *A context system $\mathcal{C}$ is a structure $\mathcal{C} = (K, A, \longrightarrow)$, where K is a set of* contexts, *A is a set of* actions, *$\longrightarrow \subseteq K \times (A_0 \times A) \times K$ is the* transduction *relation, $A_0 = A \cup \{0\}$ with 0 being a distinguished* no–action *symbol (i.e. $0 \notin A$).*

For $(C, (a, b), C') \in \longrightarrow$ we shall adopt the notation $C \xrightarrow[a]{b} C'$ and interpret this as: "by consuming the action a the context C can produce the action b and change into C'". For $a = 0$ the production of b does not involve consumption of any action. We shall use the following suggestive abbreviations $C \xrightarrow[a]{b}$, $C \xrightarrow[a]{}$ and $C \xrightarrow{b}$ (e.g. $C \xrightarrow[a]{}$ abbreviates $\exists C' \exists b.\, C \xrightarrow[a]{b} C'$). A context system $\mathcal{C}$ is *consumption–finite* if the set $\{(a, C') \mid C \xrightarrow[a]{b} C'\}$ is finite for any context C and any action a. The set of *deterministic* contexts is the largest set K_d such that whenever $C \in K_d$ then whenever $C \xrightarrow[a]{b} D$ and $C \xrightarrow[a]{b'} D'$ then $b = b'$ and $D = D' \in K_d$.

Example 2.4 The operational semantics of the CCS contexts $P \,|\, [\,]$ and $[\,]\backslash L$ [4] are given in figure 2 reflecting in a 1–to–1 fashion the inference rules of figure 1. The first rule of figure 2 states that

$$\frac{P \xrightarrow{a} P'}{P \,|\, [\,] \xrightarrow[0]{a} P' \,|\, [\,]} \qquad \frac{-}{P \,|\, [\,] \xrightarrow[a]{a} P \,|\, [\,]} \qquad \frac{P \xrightarrow{a} P'}{P \,|\, [\,] \xrightarrow[\bar{a}]{\tau} P' \,|\, [\,]} \qquad \frac{-}{[\,]\backslash L \xrightarrow{a} [\,]\backslash L} a, \bar{a} \notin L$$

Figure 2: Transduction rules for $P \,|\, [\,]$ and $[\,]\backslash L$

$P \,|\, [\,]$ may produce an action without consulting its argument process whenever P has transitions. The second rule allows the inner process to interact directly with the environment without involving the context (i.e. P). Finally, the third rule of figure 2 indicates that the context $P \,|\, [\,]$ may produce a τ–action as a result of an internal communication between the inner process (contributing $\bar{a}$) and the process P (contributing a). □

Now, given a (unary) context C and a process P we may syntactically form the *combined process* $C(P)$ by substituting P for the free variable (normally denoted $[\,]$) in C. The semantics of $C(P)$ is

[4] We use $[\,]$ as a free variable as this notation suggests the existence of a hole in which to place an argument process.

such that if $P \xrightarrow{a} P'$ and C has an a–consuming transduction $C \xrightarrow[a]{b} C'$ then $C(P) \xrightarrow{b} C'(P')$ should hold. Also, whenever $C \xrightarrow[0]{b} C'$, i.e. C has a transduction which does not involve any consumption, we require the transition $C(P) \xrightarrow{b} C'(P)$. Extending the transition relation for processes such that $P \xrightarrow{0} Q$ if and only if $P = Q$ [5], the above expectations are both met by the following (single) inference rule for combined processes:

$$\frac{C \xrightarrow[a]{b} C' \qquad P \xrightarrow{a} P'}{C(P) \xrightarrow{b} C'(P')} \tag{1}$$

In fact, together with transduction rules similar to those of figure 2, the above inference rule (1) yields a proof system which is sound and complete with respect to the standard operational semantics of CCS [Mil80].

3 Behavioural Specification Formalisms

3.1 Basic Formalisms

The notions of simulation, bisimulation and 2/3–bisimulation provide the basis of the three specification formalisms below. In all these formalisms specifications are themselves labelled transition systems (i.e. processes) and the satisfaction relation is simply defined as the corresponding process relation.

- $\mathcal{B} = \langle \Pi, \mathbf{sat}_{\mathcal{B}} \rangle$, where $P \ \mathbf{sat}_{\mathcal{B}} \ S \Leftrightarrow^{\Delta} P \sim S$
- $\mathcal{S} = \langle \Pi, \mathbf{sat}_{\mathcal{S}} \rangle$, where $P \ \mathbf{sat}_{\mathcal{S}} \ S \Leftrightarrow^{\Delta} S \leq P$
- $\mathcal{B}_{2/3} = \langle \Pi, \mathbf{sat}_{2/3} \rangle$, where $P \ \mathbf{sat}_{2/3} \ S \Leftrightarrow^{\Delta} S \preceq P$

For all three specification formalisms the equivalence induced on processes is the kernel of the underlying relation, i.e. $\equiv_{\mathcal{S}} = (\leq \cap \geq)$, $\equiv_{\mathcal{B}} = \sim$ and $\equiv_{\mathcal{B}_{2/3}} = (\preceq \cap \succeq)$. For the specification formalisms $\mathcal{S}$ and $\mathcal{B}_{2/3}$ the refinement ordering is obviously $\geq$ and $\succeq$ respectively.

For the specification formalism $\mathcal{B}$, note that for any process (= specification) P, $\mathsf{Mod}_{\mathcal{B}}(P) = [P]_{\sim}$. This shows that $\mathcal{B}$ is (trivially) expressive with respect to the equivalence $\sim$. More importantly, it implies that the refinement ordering between specifications of $\mathcal{B}$ degenerates to their $\sim$–equivalence! That is, specifications of $\mathcal{B}$ are either incomparable or equal in terms of allowed implementations. In particular $\mathcal{B}$ does not allow *loose* specifications to be expressed, and as a consequence compositional verification of complex systems becomes unduly complicated as it is not possible to abstract away the behavioural aspects of the components which is irrelevant in the particular context. With the specific purpose of allowing such abstractions while maintaining compatibility with $\mathcal{B}$ the notion of *relativized bisimulation* was introduced in [Lar86, LM87]. Here the equivalence between processes is relative to an *environment process* E describing the behavioural constraints which a context may impose upon its component. A transitions $E \xrightarrow{a} E'$ of the environment process may be read as: "E allows the component to perform a and afterwards change into E'". We can now formally present the notion of relative bisimulation:

[5]Note, that this extension does *not* change which processes are bisimular.

Definition 3.1 *Let $\mathcal{P} = (\Pi, A, \longrightarrow)$ be a labelled transition system. Then a* relative bisimulation *R consists of a family of $\langle R_E \rangle_{E \in \Pi}$ of binary relations on Π such that whenever $(P, Q) \in R_E$ and $E \xrightarrow{a} E'$ then:*

1. *Whenever $P \xrightarrow{a} P'$, then $Q \xrightarrow{a} Q'$ for some Q' with $(P', Q') \in R_{E'}$,*
2. *Whenever $Q \xrightarrow{a} Q'$, then $P \xrightarrow{a} P'$ for some P' with $(P', Q') \in R_{E'}$.*

We say that P and Q are bisimular relative to E, and write $P \sim_E Q$ if $(P, Q) \in R_E$ for some relative bisimulation R.

Thus, relative bisimulation is just like bisimulation except that only transitions permitted by the environment process are considered: we do not care how the processes may perform for transitions not permitted.

The notion of relative bisimulaton enjoys a number of pleasant properties: for any given environment process E, bisimularity relative to E (i.e. the relation $\sim_E$) is an equivalence weaker than that of bisimularity ($\sim$). In fact, the ordinary notion of bisimularity is just relative bisimularity with respect to a universal environment process U, which allows any action at any time (i.e. $U \xrightarrow{a} U$ for all $a \in A$), The weakest relative bisimulation is that with respect to a completely inactive environment process $\mathcal{O}$ (i.e. $\mathcal{O} \not\xrightarrow{a}$ for all $a \in A$), in which case relative bisimularity identifies all processes. A complete characterization of the relative strength of relative bisimularity in terms of the involved environments is given by the notion of simulation: for any two environment processes E and F, $\sim_E$ is weaker than $\sim_F$ if and only if $E \leq F$. We refer the reader to [Lar87] for the non–trivial proof of this simple and useful characterization. Now, the notion of relative bisimulation induces the specification formalism $\mathcal{R}$:

- $\mathcal{R} = \langle \Pi \times \Pi, \mathsf{sat}_{\mathcal{R}} \rangle$, where $P \; \mathsf{sat}_{\mathcal{R}} \; (S, E) \Leftrightarrow^{\Delta} P \sim_E S$

Here specifications are pairs (S, E), with S being an abstract process describing the desired behaviour of the implementation and E an environment process describing the constraints under which the implementation will be used. The satisfaction relation is based on relative bisimularity. From the discussion above, it is clear that $\mathcal{R}$ is at least as expressive as $\mathcal{B}$ (for any $S \in \Pi$, $\mathsf{Mod}_{\mathcal{B}}(S) = \mathsf{Mod}_{\mathcal{R}}(S, U)$). The equivalence induced by $\mathcal{R}$ is clearly that of bisimularity $\sim$, and it is obvious that $\mathcal{R}$ is expressive with respect to this equivalence. To see that $\mathcal{R}$ endeed does allow specifications with varying degree of looseness, we note that the ordering between specifications of $\mathcal{R}$ is (partially) determined by the simulation ordering on the environment components; i.e.:

$$(S_1, E_1) \Rightarrow (S_2, E_2) \quad \text{if} \quad E_2 \leq E_1 \wedge S_1 \; \mathsf{sat}_{\mathcal{R}} \; (S_2, E_2)$$

In the following we shall in particular make references to a restriction of $\mathcal{R}$, which contains only those specifications (S, E) of $\mathcal{R}$, where the environment process E is deterministic. We denote the restricted specification formalism $\mathcal{R}_d$.

3.2 Modal Specifications

The theory of *Modal Specifications* provides an alternative formalism for expressing *loose* specifications while maintaining compatibility with $\mathcal{B}$, and is studied at length in [LT88, HL89, Lar90, BL90, LX90b]. Modal Specifications is given an *operational interpretation* imposing restrictions on the transitions of possible implementations by describing which transitions are *necessary* and which are *admissable*. This is reflected by the notion of *Modal Transition System*, which is a labelled transition system with *two* transition relations: $\longrightarrow_{\Box}$ for describing the *required* transitions and $\longrightarrow_{\Diamond}$ for describing the *allowed* transitions:

Definition 3.2 *A modal transition system is a structure* $\mathcal{T} = (Q, A, \longrightarrow_\Box, \longrightarrow_\Diamond)$, *where* Q *is a set of (modal) specifications,* A *is a set of actions and* $\longrightarrow_\Box, \longrightarrow_\Diamond \subseteq Q \times A \times Q$, *satisfying the condition* $\longrightarrow_\Box \subseteq \longrightarrow_\Diamond$.

The condition $\longrightarrow_\Box \subseteq \longrightarrow_\Diamond$ says that anything required is also allowed, ensuring that any modal specification is consistent. Recall that the behaviour of processes is given in terms of a standard labelled transition system $\mathcal{P} = (\Pi, A, \longrightarrow)$. We will view processes as modal specifications, where all requirements are necessary ones, by considering processes as elements of the derived modal transition system $\mathcal{S}_\mathcal{P} = (\Pi, A, \longrightarrow_\Box, \longrightarrow_\Diamond)$, with $\longrightarrow_\Box = \longrightarrow_\Diamond = \longrightarrow$.

Now, the more a modal specification requires and the less it allows, the stronger we expect the specification to be. Using the derivation relations $\longrightarrow_\Box$ and $\longrightarrow_\Diamond$ this may be formalized by the following notion of *refinement*.

Definition 3.3 *A refinement* R *is a binary relation on* S *such that whenever* $(S, T) \in R$ *and* $a \in A$ *then the following holds:*

1. *Whenever* $S \xrightarrow{a}_\Diamond S'$, *then* $T \xrightarrow{a}_\Diamond T'$ *for some* T' *with* $(S', T') \in R$,
2. *Whenever* $T \xrightarrow{a}_\Box T'$, *then* $S \xrightarrow{a}_\Box S'$ *for some* S' *with* $(S', T') \in R$.

S *is said to be a refinement of* T *in case* (S, T) *is contained in some refinement* R. *We write* $S \lhd T$ *in this case.*

The defined relation $\lhd$ enjoys a number of interesting properties: $\lhd$ is itself a refinement; in fact the maximal one. Also, $\lhd$ is a *preorder* allowing looseness in specifications. As an example the weakest modal specification $\mathcal{U}$ is one which constantly allows any action, but never requires that any action must be performed. Operationally, $\mathcal{U}$ is completely defined by $U \xrightarrow{a}_\Diamond U$ for all actions a.

A straightforward generalization allows us to compare specifications from different modal transition systems (essentially by applying the above definition to disjoint sums of modal transition systems). Based on the introduced notions of modal transition system and refinement, we may now obtain the specification formalism

$$\mathcal{M} = \langle \Sigma_\mathcal{M}, \mathsf{sat}_\mathcal{M} \rangle$$

where $\Sigma_\mathcal{M}$ is the collection of all modal specifications and $P \ \mathsf{sat}_\mathcal{M} \ S$ is defined as $P \lhd S$, viewing P as a modal specification through the derived modal transition system $\mathcal{S}_\mathcal{P}$.

When using processes as specifications of $\mathcal{M}$ it is easy to see that $\mathsf{sat}_\mathcal{M}$ degenerates to $\mathsf{sat}_\mathcal{B}$ (simply note that for processes $\lhd$ and $\sim$ coincide). Moreover, processes are maximal modal specifications in the sense that whenever $S \lhd P$ then also $P \lhd S$. It follows that the process equivalence induced by $\mathcal{M}$ ($\equiv_\mathcal{M}$) is simply bisimularity ($\sim$), and that $\mathcal{M}$ is (trivially) expressive with respect to $\sim$.

In the following sections we will need to refer to certain restrictions of $\mathcal{M}$:

$\mathcal{M}_f$, where we only consider modal specifications from *finite* modal transition systems, and

$\mathcal{M}_t$, where we only consider modal specifications which are finite, and acyclic with the exception that the loosest (cyclic) modal specification U may be used at leaves.

$\mathcal{B}_t$, where the modal specifications are required to be finite and acyclic *processes*.

Modal specifications are best visualized by means of drawings, but for specifications of $\mathcal{M}_t$ it is sometimes convenient to use syntactically definitions based on the following very elementary language of terms:

$$S \ ::= \ \mathcal{O} \mid \mathcal{U} \mid a_{\Diamond}.S \mid a_{\Box}.S \mid S_1 + S_2$$

The terms of the above syntax defines the modal specifications of a modal transition systems, where $\longrightarrow_{\Diamond}$ and $\longrightarrow_{\Box}$ are the smallest relations satisfying the following rules:

$$\mathcal{U} \xrightarrow{a}_{\Diamond} \mathcal{U}\,(a \in A) \qquad a_{\Diamond}.S \xrightarrow{a}_{\Diamond} S \qquad a_{\Box}.S \xrightarrow{a}_{\Box} S \qquad \frac{S_i \xrightarrow{a}_m S_i'}{S_1 + S_2 \xrightarrow{a}_m S_i'}\ (i = 1,2;\ m = \Diamond, \Box)$$

The sublanguage of terms obtained from the constructs $\mathcal{O}$, $a_{\Box}.$ and $+$ will be used to define specifications of $\mathcal{B}_t$.

3.3 Results of Expressivity

In figure 3, we have visualized the complete ordering of relative expressivity between the specification formalisms introduced so far. The result is basically that $\mathcal{M}$ is as least as expressive as any of the basic specification formalisms; for the specification formalism $\mathcal{R}$ only the restricted version $\mathcal{R}_d$ is covered by the result.

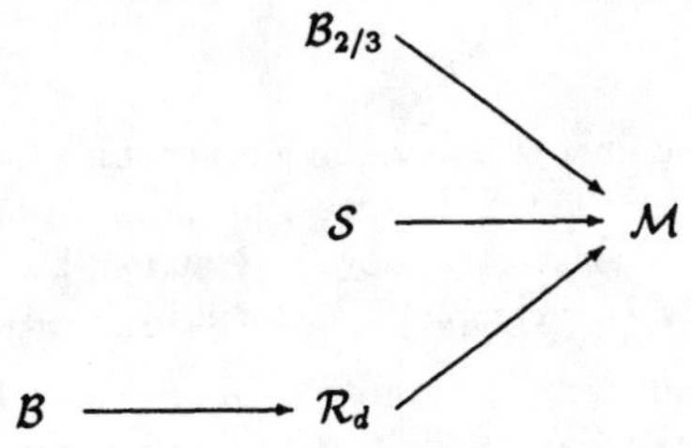

Figure 3: Relative Expressivity

Theorem 3.4 *$\mathcal{M}$ is at least as expressive as $\mathcal{S}$ and $\mathcal{B}_{2/3}$.*

Proof: For S a specification of $\mathcal{S}$ and $\mathcal{B}_{2/3}$ (i.e. $S \in \Pi$), let $S_{\geq}$ and $S_{\succeq}$ be modal specifications with behaviours according to the following inference rules:

$$\frac{S \xrightarrow{a} S'}{S_{\geq} \xrightarrow{a}_{\Box} S'_{\geq}} \qquad \frac{-}{S_{\geq} \xrightarrow{a}_{\Diamond} \mathcal{U}} \qquad \frac{S \xrightarrow{a} S'}{S_{\succeq} \xrightarrow{a}_{\Box} S'_{\succeq}} \qquad \frac{S \xrightarrow{a}}{S_{\succeq} \xrightarrow{a}_{\Diamond} \mathcal{U}}$$

then it is easy to see that for any process P, $P \geq S$ if and only if $P \lhd S_{\geq}$ and $P \succeq S$ if and only if $P \lhd S_{\succeq}$; or alternatively that P $\mathsf{sat}_{\mathcal{S}}$ S if and only if P $\mathsf{sat}_{\mathcal{M}}$ $S_{\geq}$ and P $\mathsf{sat}_{2/3}$ S if and only if P $\mathsf{sat}_{\mathcal{M}}$ $S_{\succeq}$. □

Theorem 3.5 *$\mathcal{M}$ is at least as expressive as $\mathcal{R}_d$.*

Proof: For (S, E) a specification of $\mathcal{R}_d$ (i.e. $S \in \Pi$ and $E \in \Pi_d$), let S^E be the modal specification with behaviour according to the following inference rules:

$$\frac{S \stackrel{a}{\longrightarrow} S' \quad E \stackrel{a}{\longrightarrow} E'}{S^E \stackrel{a}{\longrightarrow}_\Box S'^{E'}} \qquad \frac{E \stackrel{a}{\not\longrightarrow}}{S^E \stackrel{a}{\longrightarrow}_\Diamond \mathcal{U}}$$

Then the following equivalence holds for all processes P:

$$P \sim_E S \;\Leftrightarrow\; P \lhd S^E$$

Here we only carry out (part of) the proof of the *if*-direction in order to make explicit where the determinism of E is needed. Let R be the family defined by:

$$R_E = \{(P, S) \mid P \lhd S^E\}$$

We claim that R is a relative bisimulation. Thus let $(P, S) \in R_E$, $E \stackrel{a}{\longrightarrow} E'$ and $P \stackrel{a}{\longrightarrow} P'$. We must find a transition $S \stackrel{a}{\longrightarrow} S'$ such that $(P', S') \in R_{E'}$. However, as $P \lhd S^E$ we know that $S^E \stackrel{a}{\longrightarrow}_\Diamond S'^{E''}$ (only the first rule is applicable) with $P' \lhd S'^{E''}$. Now, using that E is deterministic, it follows that $E' = E''$ yielding a matching transition for S. □

4 Logical Specification Formalisms

In this section we introduce a logical specification formalism based on the classical Hennessy–Milner Logic, and we compare it with the graphical specification formalisms of the previous section in terms of expressive power.

4.1 Hennessy–Milner Logic

The logic we use is Hennessy–Milner Logic without negation [HM85]. The formulae of this logic is given by the following abstract syntax:

$$F \;::=\; \mathsf{tt} \mid \mathsf{ff} \mid F \wedge G \mid F \vee G \mid \langle a \rangle F \mid [a]F$$

where $a \in A$. Our interpretation of Hennessy–Milner Logic is an intuitionistic one relative to modal specifications extending the classical interpretation with respect to processes. More precisely, the semantics of a formula F of Hennessy–Milner Logic is a set of modal specifications $[\![F]\!]$ defined inductively on the structure of F as follows:

i) $[\![\mathsf{tt}]\!] = \Sigma_{\mathcal{M}}$

ii) $[\![\mathsf{ff}]\!] = \emptyset$

iii) $[\![F \wedge G]\!] = [\![F]\!] \cap [\![G]\!]$

iv) $[\![F \vee G]\!] = [\![F]\!] \cup [\![G]\!]$

v) $[\![\langle a \rangle F]\!] = \{S \mid \exists S' \in [\![F]\!].\, S \stackrel{a}{\longrightarrow}_\Box S'\}$

vi) $[\![[a]F]\!] = \{S \mid \forall S'.\, S \stackrel{a}{\longrightarrow}_\Diamond S' \Rightarrow S' \in [\![F]\!]\}$

Note, that for processes (for which the transition relations $\longrightarrow_\Box$ and $\longrightarrow_\Diamond$ are equal) our interpretation of Hennessy–Milner Logic coincides with the standard interpretation in [HM85]. In our extension, the $[a]$–modality expresses a universal quantification over $\stackrel{a}{\longrightarrow}_\Diamond$–transition, whereas the $\langle a \rangle$–modality expresses existential quantification over $\stackrel{a}{\longrightarrow}_\Box$–transitions.

When interpreted with respect to *processes*, Hennessy–Milner Logic yields the following specification formalism:

$$\mathcal{H} = \langle \Sigma_{\mathcal{H}},\ \mathsf{sat}_{\mathcal{H}} \rangle$$

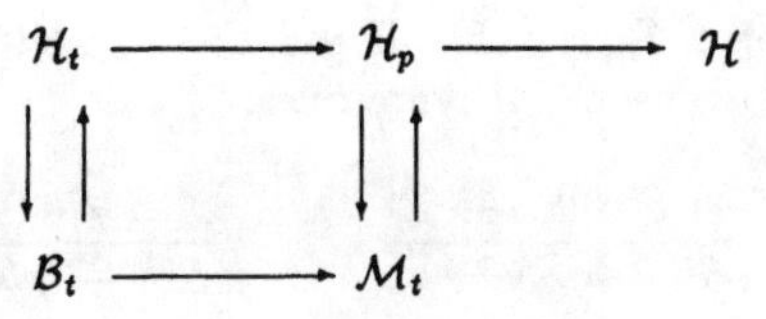

Figure 4: Relative Expressivity II

where $\Sigma_{\mathcal{H}}$ is the set of Hennessy–Milner formulae, and $P\ \mathbf{sat}_{\mathcal{H}}\ F$ if and only if $P \in [\![F]\!]$. The fundamental characterization theorem of [HM85] now extends in a natural way (see [BL90]); more precisely for any modal specifications S_1 and S_2 the following holds:

$$S_1 \lhd S_2 \iff \forall F.\ S_2 \in [\![F]\!] \Rightarrow S_1 \in [\![F]\!]$$

i.e. the more refined a modal specification is (with respect to $\lhd$) the more logical formulae it will satisfy [6]. From this characterization theorem it follows directly that $\mathcal{H}$ is adequate with respect to bisimularity ($\sim$): i.e. two processes are bisimular if and only if they satisfy the same specifications of $\mathcal{H}$

In the formulation of certain expressivity and testability results of the following sections, we will need restricted versions of $\mathcal{H}$ in terms of the formulae allowed:

$\mathcal{H}_{2/3}$: The use of the $[a]$–modality is very restrictive: in constructing legal formulae of $\mathcal{H}_{2/3}$ only $[a]$–subformulae of the form $[a]\mathrm{ff}$ are allowed. This sublogic of Hennessy–Milner Logic was introduced in [Blo88] under the name Limited Modal Logic, and in [LS89] the logic was shown adequate with respect to 2/3-bisimularity. Also, for specifications of $\mathcal{H}_{2/3}$ very strong testability results have been obtained [LS89].

$\mathcal{H}_p$: Only formulae which are *consistent* and *prime* are allowed. A formula F is consistent if $P \in [\![F]\!]$ for some process P. A formula F is prime if disjunctions can be resolved in the sense that whenever $[\![F]\!] \subseteq [\![H \vee G]\!]$ then either $[\![F]\!] \subseteq [\![H]\!]$ or $[\![F]\!] \subseteq [\![G]\!]$.

$\mathcal{H}_t$: Only formulae which are *consistent* and *maximal* are allowed. A formula F is maximal if $[\![G]\!] \subseteq [\![F]\!]$ implies $[\![F]\!] \subseteq [\![G]\!]$ for any consistent formula G.

4.2 Expressivity Results

In figure 4, we have visualized the complete ordering of relative expressivity [7] between the graphical specification formalisms $\mathcal{B}_t$ and $\mathcal{M}_t$ and the logical specification formalisms $\mathcal{H}_t$, $\mathcal{H}_p$ and $\mathcal{H}$. A main result (proved in details in [BL90]) is that the expressive power of (finite and acyclic) modal transition systems is precisely that of consistent and prime Hennessy–Milner formulae.

Theorem 4.1 *$\mathcal{H}_p$ and $\mathcal{M}_t$ are equally expressive.*

Proof: Here we just sketch the main ideas behind the proof and refer to [BL90] for a complete proof:

[6] This result is subject to the assumption that the transition relations of the modal specifications are image–finite.
[7] Except for transitive closures.

$\mathcal{M}_t \longrightarrow \mathcal{H}_p$: [8] For each specification S of $\mathcal{M}_t$ a (characteristic) Hennessy–Milner formula $\chi(S)$ is defined (inductively on the syntactical structure of S) such that for any modal specification T the following holds:

$$T \lhd S \;\Leftrightarrow\; T \in [\![\chi(S)]\!]$$

As a consequence it follows that for all processes P:

$$P \ \mathbf{sat}_{\mathcal{M}}\ S \;\Leftrightarrow\; P\ \mathbf{sat}_{\mathcal{H}}\ \chi(S)$$

demonstrating $\mathcal{M}_t \longrightarrow \mathcal{H}$. The desired expressivity result $\mathcal{M}_t \longrightarrow \mathcal{H}_p$ is obtained by observing that the characteristic formula $\chi(S)$ is both consistent and prime.

$\mathcal{H}_p \longrightarrow \mathcal{M}_t$: For any Hennessy–Milner formula F a finite set of $\mathcal{M}_t$ specifications $\Theta(F)$ is shown to exists (using a *strong* disjunctive normalform for formulae) such that for any modal specification S the following holds:

$$S \in [\![F]\!] \;\Leftrightarrow\; \exists T \in \Theta(F).\, S \lhd T$$

As a consequence it follows that for all processes P:

$$P\ \mathbf{sat}_{\mathcal{H}}\ F \;\Leftrightarrow\; \exists T \in \Theta(F).\, P\ \mathbf{sat}_{\mathcal{M}}\ T$$

Now, the desired result is obtained by proving that $\Theta(F)$ may be reduced to a singleton when F is consistent and prime. □

Theorem 4.2 *$\mathcal{B}_t$ and $\mathcal{H}_t$ are equally expressive.*

Proof:

$\mathcal{B}_t \longrightarrow \mathcal{H}_t$: It suffices to show that for any specification S of $\mathcal{B}_t$ the characteristic formula $\chi(S)$ is consistent and maximal. This follows easily from the fact that specifications of $\mathcal{B}_t$ (which are processes) are maximal with respect to $\lhd$ together with the (extended) characterization theorem for $\lhd$.

$\mathcal{H}_t \longrightarrow \mathcal{B}_t$: The proof is based on the following observations:

- For any process P and every $n \in N$, there exists a $\mathcal{B}_t$ specification P^n such that $P \sim_n P^n$ (we refer to [Mil80] for the definition of $\sim_n$). [9]
- For any formula F and process P, whenever $P \in [\![F]\!]$ then $P^n \in [\![F]\!]$ for any $n \geq d_F$, where d_F is the modal depth of F.

Now, let F be a consistent and maximal formula, and let P be some process satisfying F. We claim that the $\mathcal{B}_t$ specification P^{d_F} is equivalent to F; i.e. for any process Q:

$$Q \in [\![F]\!] \;\Leftrightarrow\; Q \sim P^{d_F}$$

The *if*-direction is obvious as $\mathcal{H}$ is adequate with respect to $\sim$. For the *only-if*-direction assume $Q \in [\![F]\!]$ but $Q \not\sim P^{d_F}$, Then [10] $Q \not\sim_m P^{d_F}$ for some $m \geq d_F$ and hence $Q^m \not\sim_m P^{d_F}$. Now, $[\![\chi(Q^m)]\!] \subseteq [\![F]\!]$ but $[\![F]\!] \not\subseteq [\![\chi(Q^m)]\!]$ as $[\![\chi(P^{d_F})]\!] \subseteq [\![F]\!]$ but $[\![\chi(P^{d_F})]\!] \not\subseteq [\![\chi(Q^m)]\!]$. Clearly this contradicts maximality of F. □

Obviously $\mathcal{B}_t$ is (strictly) less expressive than $\mathcal{M}_t$ (inducing a similar relationship between $\mathcal{H}_t$ and $\mathcal{H}_p$). Also, $\mathcal{H}_p$ is *strictly* less expressive than $\mathcal{H}$. To see this, simply observe that the formulae $\langle a\rangle\mathbf{tt} \vee \langle b\rangle\mathbf{tt}$ is *not* prime, as:

$$[\![\langle a\rangle\mathbf{tt} \vee \langle b\rangle\mathbf{tt}]\!] \subseteq [\![\langle a\rangle\mathbf{tt} \vee \langle b\rangle\mathbf{tt}]\!]$$

but neither of the following inclusions holds:

$$[\![\langle a\rangle\mathbf{tt} \vee \langle b\rangle\mathbf{tt}]\!] \subseteq [\![\langle a\rangle\mathbf{tt}]\!] \qquad [\![\langle a\rangle\mathbf{tt} \vee \langle b\rangle\mathbf{tt}]\!] \subseteq [\![\langle b\rangle\mathbf{tt}]\!]$$

[8] For specification formalisms $\mathcal{S}_1$ and $\mathcal{S}_2$, we will denote by $\mathcal{S}_1 \longrightarrow \mathcal{S}_2$ that $\mathcal{S}_2$ is at least as expressive as $\mathcal{S}_1$.

[9] This fact requires that the set of actions A is finite.

[10] Assuming image–finiteness of processes.

5 Compositionality

In this section we shall address the question of compositionality. In particular we shall examine to what extent, the various specification formalisms introduced in the previous sections support composition and decomposition of specifications under contexts as formalized in section 1. The main results are that (basically) all the specification formalisms allow specifications to be composed under contexts, whereas only the logical specification formalism $\mathcal{H}$ allows specifications to be decomposed. In particular we show that $\mathcal{H}$ is the smallest extension of $\mathcal{M}_t$ with this property.

5.1 Composing Specifications

To conclude that a specification formalism $\mathcal{S} = \langle \Sigma_S, \ \mathbf{sat}_S \ \rangle$ supports composition with respect to a set of contexts $\mathcal{C}$, it suffices to exhibit a *composition function* $\mathsf{C}_S : \mathcal{C} \times \Sigma_S \longrightarrow \Sigma_S$ such that the following holds for all processes Q:

$$Q \ \mathbf{sat}_S \ \mathsf{C}_S(C, S) \ \Leftrightarrow \ \exists P \ \mathbf{sat}_S \ S. \ Q \equiv_S C(P) \tag{2}$$

Theorem 5.1 *$\mathcal{B}$ supports composition with respect to any context system $\mathcal{C}$.*

Proof: Define the composition function $\mathsf{C}_\mathcal{B} : \mathcal{C} \times \Pi \longrightarrow \Pi$ in the following obvious way: $\mathsf{C}_\mathcal{B}(C, S) = C(S)$, then (2) amounts to show that for any process Q the following holds:

$$Q \sim C(S) \ \Leftrightarrow \ \exists P \sim S. Q \sim C(P)$$

The $\Rightarrow$–direction is trivially fulfilled by taking $P = S$, and the $\Leftarrow$–direction follows from fact that $\sim$ is preserved by all contexts [Lar89]. □

Theorem 5.2 *$\mathcal{M}$ supports composition with respect to any deterministic context system $\mathcal{C}$.*

Proof: For C a deterministic context and S a modal specification, let $\mathsf{C}_\mathcal{M}(C, S)$ be a modal specification with the following behaviour:

$$\frac{C \xrightarrow[a]{b} C' \qquad S \xrightarrow{a}_m S'}{\mathsf{C}_\mathcal{M}(C,S) \xrightarrow{b}_m \mathsf{C}_\mathcal{M}(C',S')} \ (m = \Diamond, \Box)$$

Note, that $\mathsf{C}_\mathcal{M}$ extends in a natural way the composition function $\mathsf{C}_\mathcal{B}$: for S a process $\mathsf{C}_\mathcal{B}(C,S) \sim \mathsf{C}_\mathcal{M}(C,S)$.

We now claim that $\mathsf{C}_\mathcal{M}$ is a composition function in the sense that the following instantiation of (2) holds for all processes Q:

$$Q \lhd \mathsf{C}_\mathcal{M}(C,S) \ \Leftrightarrow \ \exists P \lhd S. Q \sim C(P)$$

The $\Leftarrow$–direction follows from the fact that $\mathsf{C}_\mathcal{M}(C,S)$ is monotonic in S with respect to $\lhd$. Hence, whenever $Q \sim C(P)$ for some process P with $P \lhd S$, then $Q \sim \mathsf{C}_\mathcal{B}(C,P) \sim \mathsf{C}_\mathcal{M}(C,P) \lhd \mathsf{C}_\mathcal{M}(C,S)$.

For the $\Rightarrow$–direction, whenever $Q \lhd \mathsf{C}_\mathcal{M}(C,S)$ define a process $[Q,C,S]$ with the following behaviour:

$$\frac{Q \xrightarrow{b} Q' \quad C \xrightarrow[a]{b} C' \quad S \xrightarrow{a}_\Diamond S'}{[Q,C,S] \xrightarrow{a} [Q',C',S']} \qquad \frac{S \xrightarrow{a}_\Box S' \quad C \not\xrightarrow[a]{}}{[Q,C,S] \xrightarrow{a} \overline{S'}}$$

where $\overline{S}$ denotes an (arbitrary) process implementing S (which is guaranteed to exist due to the consistency of modal specifications). We now claim that whenever $Q \lhd \mathsf{C}_{\mathcal{M}}(C,S)$ then the following holds:

$$1)\ \ [Q,C,S] \lhd S \qquad\qquad 2)\ \ Q \sim C([Q,C,S])$$

1) follows directly from the inference rules for $[Q,C,S]$. For 2) we show that the following relation:

$$B = \{(Q, C([Q,C,S])) \mid Q \lhd \mathsf{C}_{\mathcal{M}}(C,S)\}$$

is a bisimulation. Thus assume $(Q, C([Q,C,S])) \in B$ and $Q \xrightarrow{b} Q'$. Then $C \xrightarrow[a]{b} C'$ and $S \xrightarrow{a}_\diamond S'$ with $Q' \lhd \mathsf{C}_{\mathcal{M}}(C',S')$ for some C', S'. But, then $[Q,C,S] \xrightarrow{a} [Q',C',S']$ and using the inference rule for combined processes $C([Q,C,S]) \xrightarrow{b} C([Q',C',S'])$ yielding a matching transition. Now consider a transition of $C([Q,C,S])$, i.e. $C([Q,C,S]) \xrightarrow{b} C'(X)$, where $C \xrightarrow[a]{b} C'$ and $[Q,C,S] \xrightarrow{a} X$ for some C', b and X. Then for the transition of $[Q,C,S]$ only the first inference rule could have been applied, and because C is deterministic the transition must be of the form $[Q,C,S] \xrightarrow{a} [Q',C',S']$ where $Q \xrightarrow{b} Q'$ and $S \xrightarrow{a}_\diamond S'$ and $Q' \lhd \mathsf{C}_{\mathcal{M}}(C'S')$. This demonstrates the existence of a matching transition for Q. □

Theorem 5.3 *$\mathcal{H}$ supports composition with respect to any deterministic context system $\mathcal{C}$.*

Proof: For C a deterministic context and F a Hennessy–Milner formula, let $\mathsf{C}_{\mathcal{H}}(C,F)$ be the formula defined as follows:

$$\mathsf{C}_{\mathcal{H}}(C,F) \ =^{\Delta} \bigvee_{S \in \Theta(F)} \chi(\mathsf{C}_{\mathcal{M}}(C,S))$$

We refer to the proof of theorem 4.1 for the "definition" and properties of the functions Θ and χ. The well–definedness of $\mathsf{C}_{\mathcal{H}}(C,F)$ follows from observing that for deterministic contexts C, $\lambda S.\mathsf{C}_{\mathcal{H}}(C,S)$ maps $\mathcal{M}_t$ specifications into $\mathcal{M}_t$ specifications.

We now claim that $\mathsf{C}_{\mathcal{H}}$ is a composition function in the sense that the following instantiation of (2) holds for all processes Q:

$$Q \in [\![\mathsf{C}_{\mathcal{H}}(C,F)]\!] \ \Leftrightarrow\ \exists P \in [\![F]\!].\, Q \sim C(P)$$

Using theorem 5.2 together with properties of χ and Θ the following sequence of equivalences proves our claim:

$$\begin{aligned} & Q \in [\![\mathsf{C}_{\mathcal{H}}(C,F)]\!] \\ & \quad \Leftrightarrow\ \exists S \in \Theta(F).\, Q \in [\![\chi(\mathsf{C}_{\mathcal{M}}(C,S))]\!] \ \Leftrightarrow\ \exists S \in \Theta(F).\, Q \lhd \mathsf{C}_{\mathcal{M}}(C,S) \\ & \quad \Leftrightarrow\ \exists S \in \Theta(F)\, \exists P \lhd S.\, Q \sim C(P) \ \Leftrightarrow\ \exists P \in [\![F]\!].\, Q \sim C(P) \end{aligned}$$

□

5.2 Decomposing Specifications

In figure 5 we have visualized the complete ordering of relative expressivity between the specification formalisms $\mathcal{M}_t$ and $\mathcal{H}$ together with the induced implicit specification formalisms $\mathcal{I}_{\mathcal{M}_t,\mathcal{C}}$ and $\mathcal{I}_{\mathcal{H},\mathcal{C}}$. From section 4.2 we know that $\mathcal{H}$ is at least as expressive as $\mathcal{M}_t$ inducing a similar relationship between $\mathcal{I}_{\mathcal{M}_t,\mathcal{C}}$ and $\mathcal{I}_{\mathcal{H},\mathcal{C}}$. Also [11] both $\mathcal{M}_t$ and $\mathcal{H}$ are no more expressive than their induced implicit

[11]Under the assumption that $\mathcal{C}$ contains an identity context.

specification formalisms. In addition two main expressivity results are proven in this section: Firstly it is shown that $\mathcal{H}$ is at least as expressive as $\mathcal{I}_{\mathcal{H},\mathcal{C}}$, which means that $\mathcal{H}$ supports decomposition (of specifications) with respect to contexts. Secondly we prove that $\mathcal{H}$ is no more expressive than $\mathcal{I}_{\mathcal{M}_t,\mathcal{C}}$ from which it follows that all the specification formalisms $\mathcal{H}$, $\mathcal{I}_{\mathcal{H},\mathcal{C}}$ and $\mathcal{I}_{\mathcal{M}_t,\mathcal{C}}$ are equal in terms of expressivity. Moreover, it follows as a particular consequence, that $\mathcal{H}$ is the smallest extension (in terms of expressivity) of $\mathcal{M}_t$ supporting decomposition.

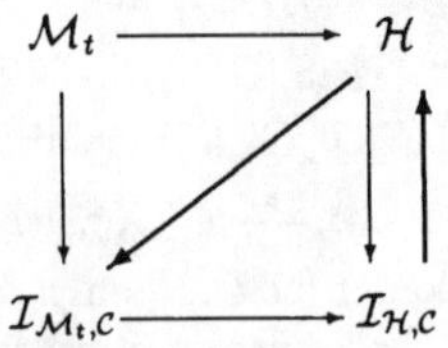

Figure 5: Relative Expressivity III

Using the results stated in section 4.2 it can be readily demonstrated that $\mathcal{M}_t$ does *not* support decomposition. Under the assumption that $A = \{a, \overline{a}, \tau\}$ consider the following implicit specification for a process P:

$$(\overline{a}.\mathrm{Nil} \,|\, P)\backslash a \lhd \tau_{\Box}.\mathcal{U}$$

It is clear that the set of processes described is precisely the models of the Hennessy–Milner formula $\langle a\rangle\mathrm{tt} \vee \langle\tau\rangle\mathrm{tt}$. However, we know from section 4.2 that this formula can not be expressed in $\mathcal{M}_t$; hence $\mathcal{M}_t$ does not support decomposition.

Theorem 5.4 *$\mathcal{H}$ supports decomposition with respect to any context system $\mathcal{C}$.* [12]

Proof: For C a context and F a Hennessy–Milner formula, we define inductively on the structure of F the formula $\mathsf{D}_{\mathcal{H}}(C, F)$:

$$\begin{array}{ll}
\mathsf{D}_{\mathcal{H}}(C, \mathrm{tt}) = \mathrm{tt} & \mathsf{D}_{\mathcal{H}}(C, \langle b\rangle F) = \bigvee_{C \xrightarrow[a]{b} C'} \langle a\rangle \mathsf{D}_{\mathcal{H}}(C', F) \\
\mathsf{D}_{\mathcal{H}}(C, \mathrm{ff}) = \mathrm{ff} & \\
\mathsf{D}_{\mathcal{H}}(C, F \wedge G) = \mathsf{D}_{\mathcal{H}}(C, F) \wedge \mathsf{D}_{\mathcal{H}}(C, G) & \mathsf{D}_{\mathcal{H}}(C, [b]F) = \bigwedge_{C \xrightarrow[a]{b} C'} [a] \mathsf{D}_{\mathcal{H}}(C', F) \\
\mathsf{D}_{\mathcal{H}}(C, F \vee G) = \mathsf{D}_{\mathcal{H}}(C, F) \vee \mathsf{D}_{\mathcal{H}}(C, G) &
\end{array}$$

It may now be proved that $\mathsf{D}_{\mathcal{H}}$ provides a *decomposition* function for $\mathcal{H}$ in the sense that the following holds for any process P:

$$P \ \mathsf{sat}_{\mathcal{H}}\ \mathsf{D}_{\mathcal{H}}(C, F) \ \Leftrightarrow\ C(P)\ \mathsf{sat}_{\mathcal{H}}\ F$$

We refer to [Lar86, LX90a] for the proof of this claim. □

Theorem 5.5 *$\mathcal{I}_{\mathcal{M}_t,\mathcal{C}}$ is at least as expressive as $\mathcal{H}$.* [13]

Proof: For F a Hennessy–Milner formula, figure 6 defines inductively on the structure of F a context C^F and a modal specification S^F. It is easily seen (using an inductive argument) that S^F is an $\mathcal{M}_t$ specification.

---

[12] Assuming that $\mathcal{C}$ is consumption–finite.

[13] The result obviously requires that $\mathcal{C}$ is sufficiently rich. In particular $\mathcal{C}$ should contain all the contexts of figure 6.

F	$\mathbf{C^F}$	$\mathbf{S^F}$
tt	$\mathcal{C}^{\mathbf{tt}}$ no transductions	$\mathcal{S}^{\mathbf{tt}}$ no transitions
ff	$\mathcal{C}^{\mathbf{ff}}$ no transductions	$\mathcal{S}^{\mathbf{ff}} \xrightarrow{a}_{\Box} \mathcal{O}$
$H \wedge G$	$C^{\mathcal{H}\wedge G} \xrightarrow[0]{x} C^F$ $C^{\mathcal{H}\wedge G} \xrightarrow[0]{y} C^G$	$S^{H\wedge G} \xrightarrow{x}_{\Diamond} S^H$ $S^{H\wedge G} \xrightarrow{y}_{\Diamond} S^G$
$\mathcal{H} \vee G$	$C^{H\vee G} \xrightarrow[0]{x} L^{H\vee G}$ $L^{H\vee G} \xrightarrow[0]{x} C^H$ $C^{H\vee G} \xrightarrow[0]{x} R^{H\vee G}$ $R^{H\vee G} \xrightarrow[0]{y} C^G$	$S^{H\vee G} \xrightarrow{x}_{\Box} T^{H\vee G}$ $T^{H\vee G} \xrightarrow{x}_{\Diamond} S^H$ $T^{H\vee G} \xrightarrow{y}_{\Diamond} S^G$ $S^{H\vee G} \xrightarrow{x}_{\Diamond} \mathcal{U}$
$\langle a \rangle H$	$C^{\langle a\rangle H} \xrightarrow[a]{x} C^H$	$S^{\langle a\rangle H} \xrightarrow{x}_{\Box} S^H$ $S^{\langle a\rangle H} \xrightarrow{x}_{\Diamond} \mathcal{U}$
$[a]H$	$C^{[a]H} \xrightarrow[a]{x} C^H$	$S^{[a]H} \xrightarrow{x}_{\Diamond} S^H$

Figure 6: The translation $\mathcal{H} \longrightarrow \mathcal{I}_{\mathcal{M}_t,\mathcal{C}}$

Jow we claim that (C^F, S^F) provides an $\mathcal{I}_{\mathcal{M}_t,\mathcal{C}}$ specification equivalent to F in the sense that for all rocesses Q:

$$Q \in [\![F]\!] \;\Leftrightarrow\; C^F(Q) \lhd S^F \tag{3}$$

3) is proved by induction on the structure of F. Here we only show the inductive argument for a isjunctive formula $F = H \vee G$; i.e. we show:

$$C^{H\vee G}(Q) \lhd S^{H\vee G} \quad\Leftrightarrow\quad C^H(Q) \lhd S^H \text{ or } C^G(Q) \lhd S^G$$

$\Rightarrow$: For the transition $S^{\mathcal{H}\vee G} \xrightarrow{x}_{\Box} T^{H\vee G}$ there are two possible matching transitions of $C^{H\vee G}(Q)$:

$$C^{H\vee G}(Q) \xrightarrow{x} L^{H\vee G}(Q) \quad\text{or}\quad C^{H\vee G}(Q) \xrightarrow{x} R^{H\vee G}(Q)$$

ut it is easy to see that $L^{H\vee G}(Q) \lhd T^{H\vee G}$ and $R^{H\vee G}(Q) \lhd T^{H\vee G}$ if and only if $C^H(Q) \lhd S^H$ and $^G(Q) \lhd S^G$ respectively.

$\Leftarrow$: Assume one of the disjuncts holds; $C^H(Q) \lhd S^H$ say. Clearly any (x–) transition of $C^{H\vee G}(Q)$ an be covered by $S^{H\vee G} \xrightarrow{x}_{\Diamond} \mathcal{U}$. Dually, the requirement $S^{H\vee G} \xrightarrow{x}_{\Box} T^{H\vee G}$ can be matched by $^{H\vee G}(Q) \xrightarrow{x} L^{H\vee G}(Q)$ as $L^{H\vee G}(Q) \lhd T^{H\vee G}$ if and only if $C^H(Q) \lhd S^H$. $\Box$

Example 5.6 Using the construction of theorem 5.5 we find that the Hennessy–Milner formula $\langle a\rangle \mathbf{tt} \vee \langle b\rangle \mathbf{tt}$ may be represented as the implicit specification (C, S), where the behaviours of C and S are given by the diagrams below (thick lines in the diagram for S indicate $\longrightarrow_{\Box}$-transitions, and thin lines indicate $\longrightarrow_{\Diamond}$-transitions):

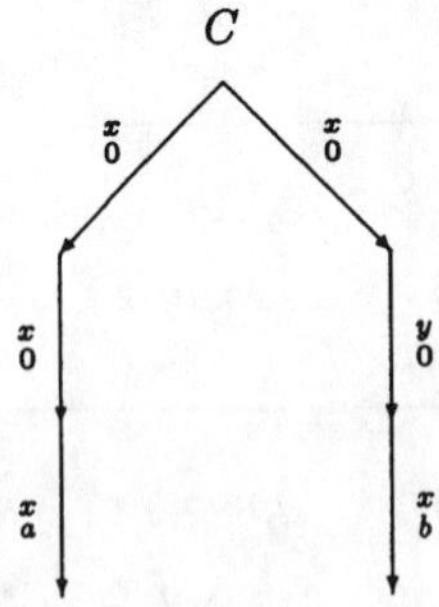

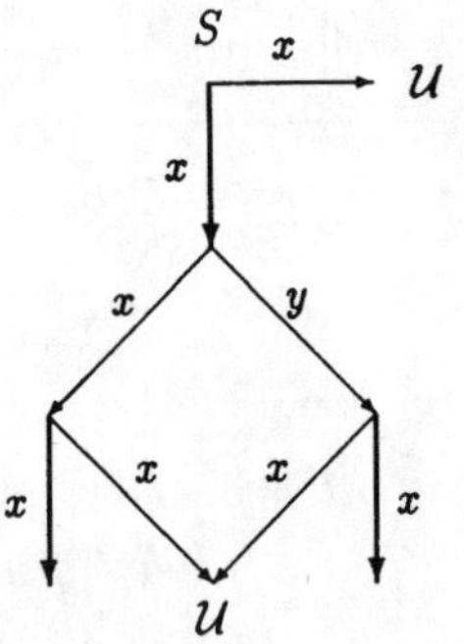

□

In terms of solving the question *behavioural versus logical specification formalisms*, theorem 5.5 may be interpreted in two ways:

- If you want specifications to be decomposable, you must use a logical specification formalism (or some specification formalism at least as expressive as $\mathcal{H}$),
- You can safely stay within the behavioural specification formalism $\mathcal{M}_t$ without loosing neither expressivity (relative to the logical specification formalism $\mathcal{H}$) nor the ability to decompose specifications. However, you may need to use *implicit* specifications [14].

The results of relative expressivity visualized in figure 5 may be extended in two ways as shown in figure 7: Adding a maximal fixed point construction to $\mathcal{H}$ yields a logical specification formalism $\nu\mathcal{H}$ supporting decomposition of specifications (see [LX90a]). However, $\nu\mathcal{H}$ is no more expressive than the implicit (behavioural) specification formalism induced by $\mathcal{M}_f$.

The result of theorem 5.5 remains valid with the specification formalism $\mathcal{B}_t$ (finite labelled transition systems) replacing $\mathcal{M}_t$. That is, $\mathcal{H}$ is the least extension of $\mathcal{B}_t$ supporting decomposition of specifications or equivalently, any specification of $\mathcal{H}$ may be expressed as a (implicit) $\mathcal{B}_t$ specification. To demonstrate the last formulation, we leave it to the reader to convince herself that the following equivalences hold [15]:

$$P \in [\![\langle a\rangle\langle b\rangle \mathbf{tt}]\!]$$
$$\Leftrightarrow \quad a.\mathrm{Nil} \,\|\, P \,+\, a.b.\mathrm{Nil} \,\|\, P \sim a.\mathrm{Nil} + a.b.\mathrm{Nil}$$

$$P \in [\![[a][b]\mathbf{ff}]\!]$$
$$\Leftrightarrow \quad a.\mathrm{Nil} \,+\, a.b.\mathrm{Nil} \,\|\, P \sim a.\mathrm{Nil}$$

[14] The specification of the scheduler in [Mil89] section 5.5 is in fact an implicit specification.

[15] $\|$ is the parallel operator of CSP [Hoa85]; i.e. $P \,\|\, Q \xrightarrow{a} R$ iff $R = P' \,\|\, Q'$ with $P \xrightarrow{a} P'$ and $Q \xrightarrow{a} Q'$.

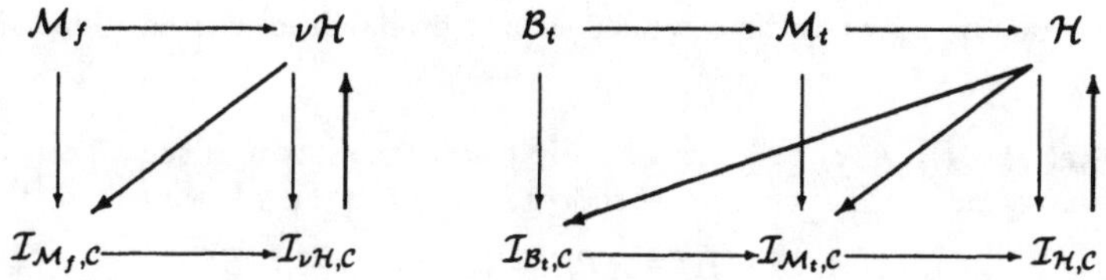

Figure 7: Relative Expressivity IV

6 Specification Formalisms for Probabilistic Processes

Recently there have been a growing interest in *probabilistic* models of processes; i.e. models where non–determinism in the behaviour of a process is assumed resolved on the basis of a probability distribution on the possible choices. We refer the interested reader to [RvGT90] for a selection of such probabilistic models.

The work carried out so far has mainly been focused on the development of calculi for defining probabilistic processes [GJS90, RvGT90], the identification of interesting behavioural equivalences between probabilistic processes [LS89, JS90, Chr90] and the axiomantization of these equivalences. In [LS89, Sko89], it is shown that the use of probabilistic models allows a *probabilistic testing framework* to be formed based on the classical statistical discipline of Hypothesis Testing.

Here we want to present some basic specification formalisms for probabilistic processes, and discuss the question of compositionality.

6.1 Probabilistic Processes. Probabilistic Bisimulation

Definition 6.1 *A* probabilistic transition system *is a structure* $\mathcal{P} = (\Pi, A, \pi)$, *where* Π *is a set of processes,* A *is a set of actions and* $\pi : \Pi \times A \times \Pi \longrightarrow [0,1]$ *is a* transition probability function *satisfying:*

$$\sum_{P' \in \Pi} \pi(P, a, P') = 1 \ \text{ or } \ \sum_{P' \in \Pi} \pi(P, a, P') = 0$$

for all processes P *and actins* a.

Note, that probabilities are used only to resolve *internal* non–determinism, as π defines a probability distribution for each process and *each action* [16]. In the following we shall apply the following notations:

$$\begin{array}{ll} P \xrightarrow{a}_{\mu} P' & \text{whenever } \pi(P, A, P') = \mu \\ P \xrightarrow{a} P' & \text{whenever } \pi(P, a, P') > 0 \\ P \xrightarrow{a}_{\mu} S & \text{whenever } \sum_{P' \in S} \pi(P, a, P') = \mu \\ P \xrightarrow{a} & \text{whenever } P \xrightarrow{a}_{1} \Pi \\ P \not\xrightarrow{a} & \text{whenever } P \xrightarrow{a}_{0} \Pi \end{array}$$

[16] Unless $\sum_{P'} \pi(P, a, P') = 0$.

The notion of bisimulation may now be extended so that probabilities of transitions are catered for [LS89]:

Definition 6.2 *Let $\mathcal{P} = (\Pi, A, \pi)$ be a probabilistic transition system. Then a probabilistic bisimulation $\equiv$ is an equivalence on Π such that whenever $P \equiv Q$ the the following holds* [17]*:*

$$\forall a \in A. \forall S \in \Pi/\equiv .\ P \xrightarrow{a}_{\mu} S \Leftrightarrow Q \xrightarrow{a}_{\mu} S$$

Two processes P and Q are said to be probabilistic bisimular in case (P, Q) is contained in some probabilistic bisimulation. We write $P \sim_p Q$ in this case.

The notion of probabilistic bisimularity has been shown to identify the *limit* of the probabilistic testing framework of [LS89]: two processes are probabilistic bisimular just in case they induce the same probability distribution on the observation set of *any* test. Moreover, probabilistic bisimularity has been axiomatized for a probabilistic calculus [GJS90] and studied in other probabilistic models [RvGT90].

Obviously, probabilistic bisimulation provides the basis for a behavioural specification formalism $\mathcal{B}_p$ being a a probabilistic version of $\mathcal{B}$. However, $\mathcal{B}_p$ suffers the same problems as $\mathcal{B}$; in particular $\mathcal{B}_p$ does *not*allow *loose* specifications to be expressed.

6.2 Probabilistic Modal Logic

Also Hennessy–Milner Logic has a probabilistic extension. In the version we propose below the $\langle a \rangle$ and $[a]$ modalities of Hennessy–Milner Logic have been replace by a continuum of modalities of the form $\langle a \rangle_{\mu}$, where a is an action and μ is a probability:

$$F \ ::= \ \mathbf{tt} \ | \ F \wedge G \ | \ \neg F \ | \ \langle a \rangle_{\mu} F$$

Our interpretation of this probabilistic modal logic is relative to a probabilistic transition system $\mathcal{P} = (\Pi, A, \pi)$. More precisely, the semantics of a formula F is a set of probabilistic processes $[\![F]\!]$ defined inductively as follows:

i) $[\![\mathbf{tt}]\!] = \Pi$ *iii)* $[\![\neg F]\!] = \Pi \backslash [\![F]\!]$

ii) $[\![F \wedge G]\!] = [\![F]\!] \cap [\![G]\!]$ *iv)* $[\![\langle a \rangle_{\mu} F]\!] = \{P \mid P \xrightarrow{a}_{\nu} [\![F]\!] \wedge \nu \geq \mu\}$

Note, that the modalities of standard Hennessy–Milner Logic are derivable as follows: $\langle a \rangle\mathbf{tt} \equiv \langle a \rangle_{\mu}\mathbf{tt}$ for any $\mu > 0$; $[a]F \equiv \langle a \rangle_1 F \vee \neg\langle a \rangle\mathbf{tt}$, and $\langle a \rangle F \equiv \neg[a]\neg F$.

The above probabilistic modal logic defines in an obvious way a logical specification formalism $\mathcal{PH} = \langle \Sigma_p, \mathbf{sat}_p \rangle$ for probabilistic processes. In [LS89] it is shown that the equivalence induced by $\mathcal{PH}$ is precisely that of probabilistic bisimularity [18].

We now focus on the question of compositionality. Obviously, this question only becomes meaningful given a set of operations for combining probabilistic processes. The work on calculi for probabilistic processes is only in its very beginning [GJS90, RvGT90] and suggests a number of difficulties: in particular it seems quite difficult to define a satisfactory probabilistic, asynchronous parallel operator. Therefore, we are not able to provide any general results, but will restrain the discussion of compositionality to the following probabilistic version of the CSP parallel operator:

$$\frac{P \xrightarrow{a}_{\mu} P' \qquad Q \xrightarrow{a}_{\nu} Q'}{P \,\|\, Q \xrightarrow{a}_{\mu\nu} P' \,\|\, Q'}$$

In particular we consider the problem of decomposing specifications relative to unary contexts of the form $Q \,\|\, [\,]$, where Q is a probabilistic process.

[17] For $\equiv$ and equivalence relation onΠ, we denote by $\Pi/\equiv$ the set of equivalence classes under $\equiv$.

[18] In fact this result is subject to an "image–finiteness"–like condition.

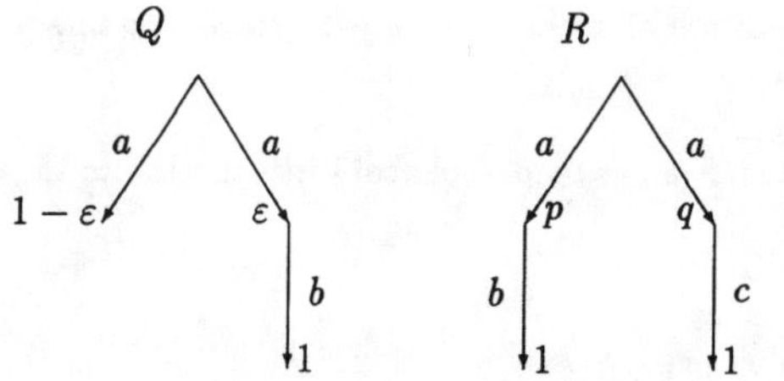

$(p+q=1)$

Figure 8: Probabilistic Processes

Example 6.3 Let Q be the probabilistic process shown in figure 8. Then for any probabilistic process P the following equivalence may be seen to hold:

$$Q \,\|\, P \;\mathsf{sat}_p\; \langle a\rangle_\mu \langle b\rangle_1 \mathbf{tt} \;\Leftrightarrow\; P \;\mathsf{sat}_p \begin{cases} \mathbf{ff} & \varepsilon < \mu \\ \langle a\rangle_{\mu/\varepsilon}\langle b\rangle_1 \mathbf{tt} & \text{otherwise} \end{cases}$$

□

Example 6.4 Let R be the probabilistic process shown in figure 8. Now consider the following implicit specification for a probabilistic process P [19]:

$$R \,\|\, P \;\mathsf{sat}_p\; \langle a\rangle_\mu (\langle b\rangle_1 \mathbf{tt} \vee \langle c\rangle_1 \mathbf{tt}) \tag{4}$$

Now, assume $P \xrightarrow{a}_x [\![\langle b\rangle_1 \mathbf{tt}]\!]$ and $P \xrightarrow{a}_y [\![\langle c\rangle_1 \mathbf{tt}]\!]$. Then (4) becomes equivalent to $px + qy \geq \mu$. That is, for P to satisfy the implicit specification of (4) the probabilities of *two types* of transitions of P should satisfy a certain relationship. Clearly, this type of property is *not* expressible in $\mathcal{PH}$, and hence $\mathcal{PH}$ *does not* support decomposition of specifications with respect to $\|$. □

We now increase the expressive power of our logic by replacing the modality $\langle a\rangle_\mu F$ with the following more general construct:

$$F \;::=\; \cdots \;\; [\langle a\rangle_{x_1} F_1, \ldots, \langle a\rangle_{x_n} F_n \;\mathsf{where}\; \phi(x_1,\ldots,x_n)]$$

Here $x_1,\ldots,x_n$ are formal variables ranging over probabilities and ϕ denotes an n–ary predicate over $[0,1]$. The semantics of the generalized modality is the following:

$$\begin{aligned} & P \in [\![[\langle a\rangle_{x_1} F_1, \ldots, \langle a\rangle_{x_n} F_n \;\mathsf{where}\; \phi(x_1,\ldots,x_n)]]\!] \\ \Leftrightarrow\;\; & \phi(\nu_1,\ldots,\nu_n) \text{ where } \forall i. P \xrightarrow{a}_{\nu_i} [\![F_i]\!] \end{aligned}$$

To see that the new modality endeed provides a generalization note:

$$\langle a\rangle_\mu F \equiv [\langle a\rangle_x F \;\mathsf{where}\; x \geq \mu]$$

Now, we claim that the extended probabilistic modal logic leads to a specification formalism $\mathcal{EPH}$ that *does support* decomposition:

[19]Here $F \vee G$ abbreviates as usual $\neg(\neg F \wedge \neg G)$.

Theorem 6.5 *$\mathcal{EPH}$ supports decomposition of specifications with respect to unary contexts of the form $Q \parallel [\,]$* [20].

Proof: For Q a probabilistic process and F an extended probabilistic modal formula, we define the formula $\mathsf{D}(Q,F)$ inductively as follows:

$$
\begin{aligned}
i)\quad & \mathsf{D}(Q,\mathrm{tt}) = \mathrm{tt}\\
ii)\quad & \mathsf{D}(Q,F\wedge G) = \mathsf{D}(Q,F)\wedge \mathsf{D}(Q,G)\\
iii)\quad & \mathsf{D}(Q,\neg F) = \neg \mathsf{D}(Q,F)\\
iv)\quad & \mathsf{D}(Q,[\langle a\rangle_{x_1}F_1,\ldots,\langle a\rangle_{x_n}F_n \text{ where } \phi(x_1,\ldots,x_n)])\\
& = [\ldots,\langle a\rangle_{y_{ij}}\mathsf{D}(Q_i,F_j),\ldots \text{ where } \phi(\textstyle\sum_i \mu_i y_{i1},\ldots,\sum_i \mu_i y_{in})]
\end{aligned}
$$

where in *iv)* $\{Q_1,\ldots,Q_m\}$ is the set of a–dervatives of Q and $Q \xrightarrow{a}_{\mu_i} Q_i$. We claim that D provides a decomposition function for $\mathcal{EPH}$ in the sense that the following holds for any probabilistic process P:

$$Q \parallel P \ \mathsf{sat}_p\ F \quad\Leftrightarrow\quad P\ \mathsf{sat}_p\ \mathsf{D}(Q,F)$$

□

Example 6.6 Let Q be the probabilistic process of figure 8. Then for any probabilistic process P the following holds:

$$
\begin{aligned}
& Q \parallel P\ \mathsf{sat}_p\ \langle a\rangle_\mu(\langle b\rangle_1\mathrm{tt} \vee \langle c\rangle_1\mathrm{tt})\\
\Leftrightarrow\quad & P\ \mathsf{sat}_p\ [\langle a\rangle_x\langle b\rangle_1\mathrm{tt}, \langle a\rangle_y\langle c\rangle_1 \text{ where } px+qy \geq \mu]
\end{aligned}
$$

□

We conjecture that $\mathcal{EPH}$ does support decomposition with respect to a general class of operations for combining probabilistic processes. Another open problem left for future work is to determine whether $\mathcal{EPH}$ is the smallest extension (in terms of expressive power) of $\mathcal{PH}$ supporting decomposition.

References

[BK85] J.A. Bergstra and J.W. Klop. Algebra of communicating processes with abstraction. *Theoretical Computer Science*, 37:77–121, 1985.

[BL90] Gérard Boudol and Kim G. Larsen. Graphical versus logical specifications. *Lecture Notes in Computer Science*, 431, 1990. In Proceedings of CAAP90.

[Blo88] Meyer Bloom, Istrail. bisimulation can't be traced. *Proceedings of Principles of Programming Languages*, 1988.

[Bou85] G. Boudol. Calcul de processus et verification. Technical Report 424, INRIA, 1985.

[Chr90] Ivan Christoff. Testing equivalences and fully abstract models for probabilistic processes. *Lecture Notes in Computer Science*, 1990. In Proceedings of CONCUR'90.

[dS85] R. de Simone. Higher–level synchronising devices in MEIJE–CCS. *Theoretical Computer Science*, 37, 1985.

[20]The result is subject to the condition that Q is image–finite.

[GJS90] A. Giacalone, C.C. Jou, and S.A. Smolka. Algebraic reasoning for probabilistic concurrent systems. 1990. In Proceedings of Working Conference on Programming Concepts and Methods, Sea of Gallilee, Israel, April 1990, IFIP TC 2.

[GV89] J.F. Groote and F. Vaandrager. Structured operational semantics and bisimulation as a congruence. *Lecture Notes in Computer Science*, 372, 1989. In Proceedings of ICALP'89.

[HL89] Hans Hüttel and Kim G. Larsen. The use of static constructs in a modal process logic. *Lecture Notes in Computer Science*, 363, 1989.

[HM85] M. Hennessy and R. Milner. Algebraic laws for nondeterminism and concurrency. *Journal of the Association for Computing Machinery*, pages 137–161, 1985.

[Hoa78] C.A.R. Hoare. Communicating sequential processes. *Communications of the ACM*, 21(8), 1978.

[Hoa85] C.A.R. Hoare. *Communicating Sequential Processes*. Prentice–Hall, 1985.

[JS90] C.C. Jou and S.A. Smolka. Equivalences, congruences, and complete axiomatizations for probabilistic processes. *Lecture Notes in Computer Science*, 1990. In Proceedings of CONCUR'90.

[Lar86] K.G. Larsen. *Context-Dependent Bisimulation Between Processes*. PhD thesis, University of Edinburgh, Mayfield Road, Edinburgh, Scotland, 1986.

[Lar87] K.G. Larsen. A context dependent bisimulation between processes. *Theoretical Computer Science*, 49, 1987.

[Lar89] Kim G. Larsen. Compositional theories based on an operational semantics of contexts. *Lecture Notes in Computer Science*, 430, 1989. In proceedings of REX workshop on Stepwise Refinement of Distributed Systems.

[Lar90] K.G. Larsen. Modal specifications. *Lecture Notes in Computer Science*, 407, 1990.

[LM87] K.G. Larsen and R. Milner. Verifying a protocol using relativized bisimulation. *Lecture Notes in Computer Science*, 267, 1987. In Proceedings of ICALP'87.

[LS89] K.G. Larsen and A. Skou. Bisimulation through probabilistic testing. *Proceedings of Principles of Programming Languages*, 1989.

[LT88] Kim G. Larsen and Bent Thomsen. A modal process logic. In *Proceedings on Logic in Computer Science*, 1988.

[LX90a] K.G. Larsen and L. Xinxin. Compositionality through an operational semantics of contexts. *Lecture Notes in Computer Science*, 1990. To appear in Proceedings of ICALP90.

[LX90b] K.G. Larsen and L. Xinxin. Equation solving using modal transition systems. In *Proceedings on Logic in Computer Science*, 1990.

[Mil80] R. Milner. *Calculus of Communicating Systems*, volume 92 of *Lecture Notes in Computer Science*. Springer Verlag, 1980.

[Mil83] R. Milner. Calculi for synchrony and asynchrony. *Theoretical Computer Science*, 25, 1983.

[Mil89] R. Milner. *Communication and Concurrency*. Prentice–Hall, 1989.

[Par81] D. Park. Concurrency and automata on infinite sequences. *Lecture Notes in Computer Science*, 104, 1981. in Proc. of 5th GI Conf.

[Plo81] G. Plotkin. A structural approach to operational semantics. FN 19, DAIMI, Aarhus University, Denmark, 1981.

[Pnu85] A. Pnueli. Linear and branching structures in the semantics and logics of reactive systems. *Lecture Notes in Computer Science*, 194, 1985. in Proc. of ICALP'87.

[RvGT90] S.A. Smolka R. van Glabbeek, B. Steffen and C.M.N. Tofts. Reactive, generative, and stratified models of probabilistic processes. *Logic in Computer Science*, 1990.

[Sko89] Arne Skou. *Validation of Concurrent Processes, with emphasis on testing.* PhD thesis, Aalbog University Center, Denmark, 1989.

An Axiomatization of Lamport's Temporal Logic of Actions

Martín Abadi*

1 Introduction

Lamport recently invented a temporal logic of actions suitable for expressing concurrent programs and for reasoning about their computations [Lam90]. In this logic, actions have syntactic representations, which can be combined and analyzed. Lamport views an action as a state transition, and a computation as a sequence of states. The basic construct for relating actions and computations is $[\ ]$; a computation satisfies the formula $[A]$ if either the computation has halted or the first action in the computation is an A action. The dual notation is $\langle A \rangle$, which means that the computation has not halted and that the first action is an A action. (Notice that $[A]$ and $\langle A \rangle$ are formulas, and not modalities as in dynamic logic [Pra76] and in Hennessy-Milner logic [HM85].) In addition, the language includes the temporal operators $\Box$ ("always") and $\Diamond$ ("eventually"), and thus it is easy to write both safety and liveness properties [Pnu77].

However, the temporal logic of actions is not very expressive in some respects (just expressive enough). One cannot define the "next" and the "until" operators of many previous temporal logics [Pnu81]. This is actually deliberate; formulas with nested occurrences of "until" are too often incomprehensible, and "next" violates the principle of invariance under stuttering, which is important for hierarchical and compositional reasoning [Lam89].

A proof system for the logic of actions might be obtained by translating into previous, richer formalisms. In this translation we forfeit the logic and two of its main advantages, understandable formulas and the possibility of reducing many arguments to simple calculations on actions. A new suit of rules for temporal reasoning with actions is therefore wanted. A complete axiomatization can provide some guidance in choosing and understanding the rules used in practice, and in particular the laws for reasoning about programs. (A decision procedure is less often helpful in this respect.)

At least two kinds of complete proof systems are possible: a propositional system and a first-order system. (In the first-order case, one can hope only for relative or nonstandard completeness results, of course.) In this paper, we study a proof system for a propositional temporal logic of actions, PTLA.

In the next section, we introduce the logic of actions through an example and discuss the underlying model very informally. We give precise definitions of the syntax and semantics of PTLA in Section 3. In Section 4, we present our proof system and prove its completeness.

*Digital Equipment Corporation, Systems Research Center, 130 Lytton Avenue, Palo Alto, CA 94301, USA.

2 An Example

In this example, we use the logic of actions for describing a trivial complete program with two processes. Process S repeatedly sends a boolean to process R and then waits for a signal; process R repeatedly receives a value from process S and then signals.

In a CSP-like notation, this program is:

$$[S :: *[[R!\,\mathit{true} \,\Box\, R!\,\mathit{false}]\ ;\ R?\,\mathit{ANY}]] \quad || \quad [R :: *[S?x\ ;S\,!\,\mathit{ANY}]]$$

(Here *ANY* is used as in Occam, for synchronization without message-passing [INM84][1].)

We take a program, such as this one, to denote the set of behaviors that it generates. In the logic of actions, a behavior is a sequence of states. It may help to view a state as a snapshot of a device that executes the program. All that matters is that each program variable has a value at each state.

Formally, behaviors are described in terms of actions. An action is a binary relation on program states. Intuitively, an action is the set of all pairs of states s and t such that the action can change the state from s to t; an action is enabled in s if it can change the state from s to t for some t. A predicate on primed and unprimed program variables expresses an action. For example, the action that negates the value of a variable y may be written $y' \equiv \neg y$ (or, equivalently, $y \equiv \neg y'$, or $(y \wedge \neg y') \vee (\neg y \wedge y')$); y' represents the value of y after the action.

In the semantics, then, states are primitive, and actions are not; this presents advantages and disadvantages in comparison with models with primitive actions. At any rate, the two approaches are valid, and they can provide the same sort of information (as one can translate between sequences of states and sequences of events). The properties of programs discussed in the logic of actions are interesting in either model.

Coming back to our example, we start the formal description of the program by listing its variables. In addition to the variable x, the program has two implicit control variables l_S and l_R, one for each process; a device executing the program would keep track of the values of these three variables. The boolean l_S is true when control is at the send command in process S, and false when control is at the receive command in process S. The boolean l_R is true when control is at the receive command in process R, and false when control is at the send command in process R. Thus, a state should assign values to x, l_S, and l_R.

Next we define an action for each communication event. The sending of *true* can change a state where l_S and l_R hold to a state where $\neg l_S$, $\neg l_R$, and x hold. Hence, for the sending of *true*, we write:[2]

$$A_t \triangleq l_S \wedge \neg l'_S \wedge l_R \wedge \neg l'_R \wedge x'$$

A similar formula expresses the sending of *false*:

$$A_f \triangleq l_S \wedge \neg l'_S \wedge l_R \wedge \neg l'_R \wedge \neg x'$$

The nondeterministic composition of these two actions is represented as a disjunction:

$$A \triangleq A_t \vee A_f$$

[1]Occam is a trade mark of the INMOS Group of Companies.

[2]The symbol $\triangleq$ means ***equals by definition***.

The other basic action of the program is the acknowledgement:

$$Ack \triangleq \neg l_S \wedge l'_S \wedge \neg l_R \wedge l'_R \wedge (x' \equiv x)$$

Thanks to the use of control variables, disjunction represents sequential composition, in addition to nondeterministic composition.[3] Thus, the sequential composition of A and Ack is represented as a disjunction:

$$N \triangleq A \vee Ack$$

The action N is the next-state relation of the complete program. A computation of the program, started from an arbitrary state, satisfies $\Box[N]$.

The program is enabled (that is, it can make progress) only when l_S and l_R have the same value. We define:

$$Enabled(N) \triangleq (l_S \equiv l_R)$$

The formula $\Box[N]$ allows some computations that are immediately deadlocked, when started in a state where $Enabled(N)$ does not hold. To restrict attention to the computations that start from the expected initial states, we define the predicate *Init*:

$$Init \triangleq l_S \wedge l_R$$

A computation satisfies $Init \wedge \Box[N]$ if it is a computation of the program that starts in a state where *Init* holds. One can prove formally that none of these computations deadlocks:

$$Init \wedge \Box[N] \Rightarrow \Box Enabled(N)$$

It is still possible for a computation that satisfies *Init* and $\Box[N]$ to halt, because we have not yet made liveness assumptions (both *Init* and $\Box[N]$ are safety formulas). The assumption of weak fairness for N suffices to guarantee continued progress. Weak fairness for N says that if N is always enabled after a certain point then eventually N takes place:

$$\mathrm{WF}(N) \triangleq \Box(\Box Enabled(N) \Rightarrow \Diamond\langle N\rangle)$$

The desired progress property follows:

$$(Init \wedge \Box[N] \wedge \mathrm{WF}(N)) \Rightarrow (\Box\Diamond l_R \wedge \Box\Diamond\neg l_R \wedge \Box\Diamond l_S \wedge \Box\Diamond\neg l_S)$$

A further requirement is that *true* and *false* are chosen fairly. Strong fairness for A_t says that if the transmission of *true* is enabled infinitely often (that is, $l_S \wedge l_R$ holds infinitely often), then the transmission of *true* happens eventually. Hence we set:

$$\begin{aligned} Enabled(A_t) &\triangleq l_S \wedge l_R \\ \mathrm{SF}(A_t) &\triangleq \Box(\Box\Diamond Enabled(A_t) \Rightarrow \Diamond\langle A_t\rangle) \end{aligned}$$

and strong fairness for A_f is written similarly:

$$\begin{aligned} Enabled(A_f) &\triangleq l_S \wedge l_R \\ \mathrm{SF}(A_f) &\triangleq \Box(\Box\Diamond Enabled(A_f) \Rightarrow \Diamond\langle A_f\rangle) \end{aligned}$$

[3]In Lamport's interleaving model, the action that corresponds to the parallel composition of two processes is the union of the actions that correspond to the processes, so disjunction represents parallel composition as well.

Under these strong-fairness assumptions, the program guarantees that the value of x is infinitely often *true* and infinitely often *false*:

$$(Init \wedge \Box[N] \wedge \mathrm{WF}(N) \wedge \mathrm{SF}(A_t) \wedge \mathrm{SF}(A_f)) \Rightarrow (\Box\Diamond x \wedge \Box\Diamond\neg x)$$

In introducing the logic through this example we have only exercised the notation, and not reasoned within it formally. The traditional approaches to safety and liveness verification are adequate for proving the properties that we have claimed in the example. Lamport has formalized these traditional approaches within the logic, and has exploited them in the study of moderately substantial algorithms (in particular with the general logic, mentioned in Section 5).

3 The Syntax and Semantics of PTLA

In this section, we give a precise definition of the syntax and semantics of a propositional temporal logic of actions, PTLA. This logic, although not introduced in [Lam90], is a formalization of Lamport's approach in a propositional setting.

3.1 Syntax

We have a countably infinite collection of ***proposition symbols*** $P_0, P_1, P_2, \ldots$ and a countably infinite collection of *action symbols* $A_0, A_1, A_2, \ldots$.

A *state predicate* is a boolean combination of proposition symbols. (We use the boolean connectives *false*, $\neg$, and $\wedge$, and view the connectives $\vee$, $\Rightarrow$, and $\equiv$ as abbreviations.) If P is a state predicate, then P' is a *primed state predicate.* An *action* is a boolean combination of state predicates, primed state predicates, and action symbols; thus, in particular, a state predicate is an action. This repertoire of actions is richer than that allowed in Hennessy-Milner logic (where only action symbols are considered); on the other hand, the regular expressions and the context-free grammars of dynamic logic do not seem necessary here.

A *formula* of the logic is:

- a state predicate;
- $[A]$, where A is an action;
- a boolean combination of formulas; or
- $\Box F$, where F is a formula.

We also write $\langle A \rangle$ for $\neg[\neg A]$, and $\Diamond F$ for $\neg\Box\neg F$.

Throughout, we use the letters O, P, Q, and R for state predicates, A and B for actions, and F and G for arbitrary formulas.

Lamport's logic also includes action formulas, for example of the form $A = B$. For simplicity, we do not allow these formulas, as it is possible to use paraphrases, such as $\Box([A] = [B])$ for $A = B$.

The primitive action symbols A_0, A_1, A_2, ... were not needed in the example of Section 2, and hence some motivation for them is in order. Often an action cannot be expressed as a boolean combination of state predicates and primed state predicates, because we have not been given a full specification of the action, or because the action is essentially first-order, as $x' = x + 1$. In these cases, having action symbols enables us to name the action and exploit any known propositional facts about it. For example, if A_0 stands for $x' = x + 1$, P_0 for $x = 0$, and P_1 for $x = 1$, then P'_1 is $x' = 1$, and we can write and use $P_0 \land A_0 \Rightarrow P'_1$; alternatively, $\Box[P_0 \land A_0 \Rightarrow P'_1]$ achieves the same effect.

3.2 Semantics

The semantics of the temporal logic of actions resembles those for other linear-time temporal logics. The novelties concern the meaning of the formulas of the form $[A]$.

An interpretation is a pair $(\mathbf{S}, I)$ where

- $\mathbf{S}$ is a non-empty set; an element of $\mathbf{S}$ is called a *state*, and $\mathbf{S}$ is called a *state space*;
- I is a pair of mappings I_ρ and I_α, which assign to each proposition symbol a subset of $\mathbf{S}$ and to each action symbol a subset of $\mathbf{S} \times \mathbf{S}$, respectively; intuitively, $I_\rho(P_i)$ is the set of states where P_i is true, and $I_\alpha(A_i)$ is the set of pairs of states related by A_i.

Sometimes we omit mention of $\mathbf{S}$, and simply refer to I as the interpretation.

We extend the mapping I_ρ to all state predicates, by setting:

$$\begin{aligned} I_\rho(\mathit{false}) &\triangleq \emptyset \\ I_\rho(\neg P) &\triangleq \mathbf{S} - I_\rho(P) \\ I_\rho(P \land Q) &\triangleq I_\rho(P) \cap I_\rho(Q) \end{aligned}$$

Then we extend the mapping I_α to all actions, by setting:

$$\begin{aligned} I_\alpha(P) &\triangleq I_\rho(P) \times \mathbf{S} \\ I_\alpha(P') &\triangleq \mathbf{S} \times I_\rho(P) \\ I_\alpha(\neg A) &\triangleq \mathbf{S} \times \mathbf{S} - I_\alpha(A) \\ I_\alpha(A \land B) &\triangleq I_\alpha(A) \cap I_\alpha(B) \end{aligned}$$

A *behavior* over $\mathbf{S}$ is an infinite sequence of elements of $\mathbf{S}$. If σ is the behavior $s_0, s_1, s_2, \ldots$, we denote s_i by σ_i and $s_i, s_{i+1}, s_{i+2}, \ldots$ by σ^{+i}. We say that σ *is halted* if $\sigma_0 = \sigma_i$ for all i. If σ is not halted, we write $\mu(\sigma)$ for the least i such that $\sigma_0 \neq \sigma_i$.

A *model* is a triple $(\mathbf{S}, I, \sigma)$, where $(\mathbf{S}, I)$ is an interpretation and σ is a behavior over $\mathbf{S}$. We define the *satisfaction relation* between models and formulas inductively, as follows:

$$\begin{aligned} (\mathbf{S}, I, \sigma) \models P_j &\triangleq \sigma_0 \in I_\rho(P_j) \\ (\mathbf{S}, I, \sigma) \models [A] &\triangleq \text{either } \sigma \text{ is halted or } (\sigma_0, \sigma_{\mu(\sigma)}) \in I_\alpha(A) \\ (\mathbf{S}, I, \sigma) \models \mathit{false} &\triangleq \text{false} \\ (\mathbf{S}, I, \sigma) \models \neg F &\triangleq (\mathbf{S}, I, \sigma) \not\models F \\ (\mathbf{S}, I, \sigma) \models F \land G &\triangleq (\mathbf{S}, I, \sigma) \models F \text{ and } (\mathbf{S}, I, \sigma) \models G \\ (\mathbf{S}, I, \sigma) \models \Box F &\triangleq \text{for all } i,\ (\mathbf{S}, I, \sigma^{+i}) \models F \end{aligned}$$

For example, $(\mathbf{S}, I, \sigma) \models [\mathit{false}]$ if and only if σ is halted.

The formula F is *satisfiable* if there exist $(\mathbf{S}, I, \sigma)$ such that $(\mathbf{S}, I, \sigma) \models F$. The formula F is *valid* if $(\mathbf{S}, I, \sigma) \models F$ for all $(\mathbf{S}, I, \sigma)$; we write this $\models F$.

4 A Complete Proof System

In the first subsection we give our axioms, and then we list some of their consequences. Finally, we prove the completeness of the axioms.

4.1 The System

The temporal logic of actions is an extension of the common temporal logic with the single modality $\Box$. Accordingly, we are going to base our axiomatization on a usual one, a system known as D (in [HC68]) or S4.3*Dum* (in [Gol87]). The axioms and rules for D are:

1. $\vdash \Box(F \Rightarrow G) \Rightarrow (\Box F \Rightarrow \Box G)$

2. $\vdash \Box F \Rightarrow F$

3. $\vdash \Box F \Rightarrow \Box\Box F$

4. $\vdash \Box(\Box F \Rightarrow G) \vee \Box(\Box G \Rightarrow F)$

5. $\vdash \Box(\Box(F \Rightarrow \Box F) \Rightarrow F) \Rightarrow (\Diamond\Box F \Rightarrow F)$

6. If $\vdash F$ then $\vdash \Box F$.

7. If F is an instance of a propositional tautology then $\vdash F$.

8. If $\vdash F$ and $\vdash F \Rightarrow G$ then $\vdash G$.

Axiom 4 is a classical way to express that time is linear—that any two instants in the future are ordered. Axiom 5, indirectly attributed to Geach in [HC68], is a simplification of the original $\Box(\Box(F \Rightarrow \Box F) \Rightarrow F) \Rightarrow (\Diamond\Box F \Rightarrow \Box F)$, due to Dummett and Lemmon; Axiom 5 expresses the discreteness of time.

We introduce some axioms about actions:

9. $\vdash [\mathit{false}] \Rightarrow [A]$

10. $\vdash \neg[\mathit{false}] \Rightarrow [P] \equiv P$

11. $\vdash \neg[\mathit{false}] \Rightarrow [\neg A] \equiv \neg[A]$

12. $\vdash [A \wedge B] \equiv [A] \wedge [B]$

13. $\vdash [(\neg P)'] \equiv [\neg P']$

14. $\vdash [(P \wedge Q)'] \equiv [P' \wedge Q']$

15. $\vdash \Box P \Rightarrow [P']$

16. $\vdash \Box(P \Rightarrow (([P'] \wedge G) \vee \Box G)) \Rightarrow (P \Rightarrow \Box G)$

Axiom 16 can be loosely paraphrased as follows: suppose that whenever P holds either G holds and P survives the next state change, or G is true forever; thus, G holds for as long as P holds, and becomes true forever if P stops holding; hence, if P is true initially then G is always true. All the other axioms are rather straightforward.

Our axiomatization could perhaps be simplified. It is worth recalling, however, that a less expressive logic does not always have a simpler proof system. For instance, the system for temporal logic with "next" is simpler than D [GPSS80], yet the "next" modality increases the expressiveness of the logic and its complexity (from coNP-complete to PSPACE-complete [SC85]).

4.2 Some Consequences of the Axioms

Some interesting consequences of the axioms are important in our completeness proof. We list and explain a few here.

- $\vdash [\mathit{false}] \Rightarrow (F \equiv \Box F) \wedge (F \equiv \Diamond F)$
  The formula expresses that once the computation has halted, all facts are permanent, meaning that F, $\Box F$, and $\Diamond F$ are equivalent for all F.

- $\vdash \langle P' \rangle \Rightarrow \Diamond P$
  In words, if the computation has not halted and the next action is P', then P holds eventually. This formula embodies the simplest method for proving liveness properties.

- $\vdash P \wedge \Box(P \Rightarrow [P']) \Rightarrow \Box P$
  This predictable induction principle follows directly from Axiom 16, when we instantiate G to P.

- $\vdash [P'] \wedge \Diamond G \Rightarrow G \vee \Diamond(P \wedge \Diamond G)$
  The theorem is another consequence of Axiom 16. It says if the action P' is about to take place (unless the computation halts) and G must hold eventually, then either G is true now, or P holds eventually and G holds later.

- $\vdash (P \wedge \Diamond G \wedge \Box(P \wedge \Diamond G \Rightarrow [P'])) \Rightarrow \Diamond(P \wedge G)$
  This is the dual to Axiom 16.

4.3 Soundness and Completeness

Theorem 1 (Soundness and Completeness) $\models F \Leftrightarrow \vdash F$

A simple induction on proofs shows that if $\vdash F$ then $\models \Box F$. It follows that if $\vdash F$ then $\models F$; thus, $\vdash$ is sound. The other direction of the claim (completeness) is more delicate, and the rest of this section is devoted to it.

Before embarking on the proof, we should recall a classical completeness theorem for D. The theorem says that if a formula G is not provable then $\neg G$ has a model. In fact, a model can be obtained from a structure of a very special form, known as a *balloon*. A balloon consists of a sequence of states $s_0, s_1, s_2, \ldots, s_m$, the *string*, and a set of states $\{t_0, t_1, t_2, \ldots, t_n\}$, the *bag*. (Without loss of generality, the states can be taken to be all distinct.) An interpretation gives values over these states to all the proposition symbols in G. With this interpretation, any sequence

$$s_0, \ldots, s_m, t_{i_0}, \ldots, t_{i_k}, \ldots$$

provides a model for $\neg G$, if all of $0, \ldots, n$ occur in $i_0, \ldots, i_k, \ldots$ infinitely often. Thus, a model is obtained from the balloon by linearizing the bag in any way whatsoever. This and similar constructions appear in [Gol87].

A formula *holds at a state* s in a behavior if it holds in the suffixes of the behavior that start with s. A formula *holds at a state* s in a balloon if it holds at s in all linearizations of the balloon. In both cases, we may also say that s *satisfies* the formula.

The basic strategy of our completeness proof is as follows. For every G, let G^* be the formula obtained from G by replacing each subformula of the form $[A]$ with a fresh proposition symbol; thus, $(\ )^*$ is a translation into a classical temporal formalism, with no actions. Assume that $\not\vdash F$; we want to show that $\not\models F$. That is, we want to find a model $(\mathbf{S}, I, \sigma)$ that satisfies $\neg F$. If $\not\vdash F$ then $\not\vdash (X \Rightarrow F)$, where X is any formula provable in PTLA (we will specify the choice of X below). A fortiori, $(X \Rightarrow F)$ cannot be derived using only the D axioms, and hence $(X \Rightarrow F)^*$ cannot be derived in D. Thus, by the completeness of D, there must be a balloon $\mathcal{B}$ and an interpretation that satisfy $\neg(X \Rightarrow F)^*$. The balloon and the interpretation will be useful in constructing a model for $\neg F$.

Notice that the proposition symbols that occur in F also occur in $\neg(X \Rightarrow F)^*$, and hence the interpretation that satisfies $\neg(X \Rightarrow F)^*$ over $\mathcal{B}$ must assign truth values to these proposition symbols. Naturally, if the proposition symbol $[A]^*$ is mentioned in our completeness argument then it will occur in X^* (otherwise we could not say much about $[A]^*$); therefore, the interpretation must also assign a truth value to $[A]^*$. These properties of the interpretation will serve in defining the desired I.

In the course of the proof, we rely on the fact that each state in $\mathcal{B}$ satisfies certain theorems of PTLA, or rather their translation under $(\ )^*$. The number of theorems needed is finite, and their choice depends only on the choice of F. (It suffices to consider instances of Axioms 9 to 16 for subexpressions of $\neg F$, and some simple boolean combinations of these, sometimes with primes.) We take for X the conjunction of all formulas $\Box T$, where T is one of these necessary theorems. For all practical purposes, from now on, we may pretend that we have all the theorems of PTLA at our disposal.

After these preliminaries, we are ready to start the necessary model construction.

Let $P_0, \ldots, P_k$ be a list of all the proposition symbols that occur in F. An *assignment* is a conjunction $Q_0 \wedge \ldots \wedge Q_k$ such that each Q_i is either P_i or $\neg P_i$. Given a state s in $\mathcal{B}$, there exists a unique assignment O_s that holds in s.

A state of $\mathcal{B}$ that satisfies $[\mathit{false}]^*$ is called a *halting* state. We have:

Proposition 1 *For every state s there exists an assignment R_s such that $[R_s']^*$ holds in s. Moreover, R_s is unique if and only if s is not a halting state.*

Proof To prove the existence of R_s, we first notice that $\vdash [P_j'] \vee [(\neg P_j)']$ for each proposition symbol P_j, and hence s must satisfy at least one of $[P_j']^*$ and $[(\neg P_j)']^*$. Therefore, let Q_j be one of

P_j and $\neg P_j$ such that $[Q_j']^*$ holds at s. If R_s is the conjunction of all these Q_j's, then $[R_s']^*$ holds at s, because the axioms yield $\vdash [Q_0'] \wedge \ldots \wedge [Q_k'] \equiv [(Q_0 \wedge \ldots \wedge Q_k)']$. If s is not halting then R_s is unique because $\vdash \neg[\mathit{false}] \Rightarrow [\neg P_i'] \equiv \neg[P_i']$ and $\vdash [\neg P_i'] \equiv [(\neg P_i)']$. If s is a halting state then R_s could be any assignment, since both $\vdash [\mathit{false}] \Rightarrow [P_i']$ and $\vdash [\mathit{false}] \Rightarrow [(\neg P_i)']$. ▮

Given a state s which is not halting, we say that t *follows* s if:

- t is the first state in the string strictly after s such that $O_t = R_s$, if such exists;
- else, t is a state in the bag and $O_t = R_s$, if such exists;
- else, $t = s$ if $O_s = R_s$.

Proposition 2 *If s is not halting, then some t follows s.*

Proof The axioms yield $\vdash [R_s'] \wedge \neg[\mathit{false}] \Rightarrow \Diamond R_s$, so s must satisfy this formula. Thus, R_s must hold at a state in the string strictly after s, or at a state in the bag, or at s itself. ▮

We proceed to construct the desired behavior σ inductively.

- Let σ_0 be the first state in the string of the balloon, or an arbitrary state in the bag if the string is empty.
- If σ_i is a halting state, let $\sigma_{i+1} = \sigma_i$.
- If σ_i is not a halting state, let σ_{i+1} be a state that follows σ_i. If possible, always pick a state that has not been visited previously. Otherwise, pick the state first visited the longest time ago, and start cycling.

The behavior σ ends with a cycle. It may be that this cycle is of length one, but the state s in it does not satisfy $[\mathit{false}]^*$. In this case, we modify σ trivially: we make a copy $\hat{s}$ of s and have σ cycle between these two different states. (Of course, s and $\hat{s}$ are different only formally, as they satisfy the same formulas.) This modification is convenient in giving a proper meaning to $[\mathit{false}]$. We do not discuss this minor point in the construction further, and leave the obvious details to the reader.

Let $\mathcal{S}$ be the balloon obtained by discarding from $\mathcal{B}$ all states not in σ. More precisely, the bag of $\mathcal{S}$ is the set of states that occur in the cyclic part of σ, and the string of $\mathcal{S}$ is the remaining states of σ, ordered in the order of their occurrence. The bag of $\mathcal{S}$ may be a subset of the bag of $\mathcal{B}$. It may happen, however, that the bag of $\mathcal{S}$ consists of a single state s from the string of $\mathcal{B}$ (or s and $\hat{s}$); this is in the case where either s is halting or s is the last state to satisfy R_s.

Next we show that $\neg F^*$ still holds in $\mathcal{S}$. We strengthen the claim, to apply to every state and every subformula of $\neg F^*$. However, it is not claimed that all of the theorems compiled in X still hold at each state in $\mathcal{S}$; this claim is not needed.

Proposition 3 *If G^* is a subformula of $\neg F^*$ and s is a state in $\mathcal{S}$, then G^* holds at s in $\mathcal{S}$ if and only if G^* holds at s in $\mathcal{B}$. In particular, all linearizations of $\mathcal{S}$ satisfy $\neg F^*$.*

Proof The proof proceeds by induction on the structure of the subformula of $\neg F^*$. As is common in proofs of this sort, the only important argument is that subformulas of the form $\Diamond G^*$ that hold at s in $\mathcal{B}$ also hold at s in $\mathcal{S}$. (Intuitively, this is because $\mathcal{S}$ is a "subballoon" of $\mathcal{B}$.) The argument that subformulas of the form $\Box G^*$ that hold at s in $\mathcal{S}$ also hold at s in $\mathcal{B}$ is exactly dual. All other arguments are trivial.

Within the main induction, we perform an auxiliary induction on the distance from the state considered to $\mathcal{S}$'s bag.

As a base case, we prove the claim for the states in $\mathcal{S}$'s bag. There are several subcases:

- If $\mathcal{S}$'s bag is a singleton $\{s\}$, then s must satisfy $[\mathit{false}]^*$, by construction of σ. Since $\vdash [\mathit{false}] \Rightarrow G \equiv \Diamond G$, if $\Diamond G^*$ holds at s in $\mathcal{B}$ then G^* holds at s in $\mathcal{B}$. By induction hypothesis, G^* holds at s in $\mathcal{S}$, and by temporal reasoning $\Diamond G^*$ holds at s in $\mathcal{S}$.

- If $\mathcal{S}$'s bag is a pair $\{s, \hat{s}\}$, then it must be that s is not a halting state, and that R_s holds in no state after s in $\mathcal{B}$'s string (if any) and in no state in $\mathcal{B}$'s bag other than s. Furthermore, $O_s = R_s$. Therefore, s satisfies $R_s \wedge \Box(R_s \Rightarrow [R'_s])^*$ in $\mathcal{B}$, and hence also $\Box R_s$, since $\vdash R_s \wedge \Box(R_s \Rightarrow [R'_s]) \Rightarrow \Box R_s$. Assume that s satisfies $\Diamond G^*$ in $\mathcal{B}$. By temporal reasoning, s satisfies $\Diamond(R_s \wedge G^*)$ in $\mathcal{B}$. But s is the last state in $\mathcal{B}$ that satisfies R_s, and so it must be that s satisfies G^*. By induction hypothesis, G^* holds at s in $\mathcal{S}$ as well, and by temporal reasoning $\Diamond G^*$ holds at s in $\mathcal{S}$.

- Otherwise, $\mathcal{S}$'s bag is a subset of $\mathcal{B}$'s bag, and not a singleton. In particular, there is no halting state in the bag. Let R be the disjunction of all O_t for t in $\mathcal{S}$'s bag. By construction of σ, all states in $\mathcal{B}$'s bag that satisfy R are also in $\mathcal{S}$'s bag—the point being that σ makes the biggest cycle possible. In $\mathcal{B}$, it must be that each of the states in $\mathcal{S}$'s bag satisfies $\Box(R \Rightarrow [R'])^*$. This formula simply says that all states in $\mathcal{B}$'s bag that are also in $\mathcal{S}$'s bag must be followed by another state in $\mathcal{S}$'s bag. Moreover, we have that $\vdash R \wedge \Box(R \Rightarrow [R']) \Rightarrow \Box R$. This yields that each state s in $\mathcal{S}$'s bag satisfies $\Box R$ in $\mathcal{B}$'s bag. Let $\Diamond G^*$ hold at s in $\mathcal{B}$'s bag. Then, by temporal reasoning, $\Diamond(R \wedge G^*)$ holds at s in $\mathcal{B}$'s bag. In other words, G^* holds at some state t in $\mathcal{B}$, and t satisfies R. Since t satisfies R, it must also be in $\mathcal{S}$'s bag. Thus, by induction hypothesis, t also satisfies G^* in $\mathcal{S}$, and hence s satisfies $\Diamond G^*$ in $\mathcal{S}$.

Next we consider the states in $\mathcal{S}$'s string. We assume the claim has been proved for all states at distance no bigger than n from $\mathcal{S}$'s bag; we consider the state s at distance $n+1$. Suppose that s satisfies $\Diamond G^*$ in $\mathcal{B}$ and that s is in $\mathcal{S}$'s string. Since $\vdash [R'_s] \wedge \Diamond G \Rightarrow G \vee \Diamond(R_s \wedge \Diamond G)$, either G^* must hold at s, or $\Diamond G^*$ must hold at the state that follows s, in $\mathcal{B}$. In the former case, the complexity of the formula considered has decreased. In the latter case, the complexity of the formula considered has remained the same but we are closer to $\mathcal{S}$'s bag, since s is a state in $\mathcal{S}$'s string and t follows s. In either case, the induction hypothesis immediately yields the desired result.

Since all linearizations of $\mathcal{B}$ satisfy $\neg F^*$, we conclude that all linearizations of $\mathcal{S}$ satisfy $\neg F^*$. ∎

The final step in our proof is constructing an interpretation $(\mathbf{S}, I)$ and checking that $(\mathbf{S}, I, \sigma)$ satisfies $\neg F$.

We take $\mathbf{S}$ to be the set of states in $\mathcal{B}$ that occur in σ.

Each proposition symbol P_i that occurs in F has a value at each state in the balloon. We take $I_\rho(P_i)$ to be the set of states of σ where this value is true. Similarly, if A_i is an action symbol in

F, then the proposition symbol $[A_i]^*$ has a value at each state in the balloon (because it occurs in $(X \Rightarrow F)^*$). We take $I_\alpha(A_i)$ to be the set of pairs of states (s,t) such that $[A_i]^*$ is true in s. Two slight oddities should be noticed here. The first one is that we are interpreting an action symbol much as a proposition symbol (the second component of the pair, t, plays no role). The second one is that if $[\mathit{false}]^*$ holds in s then $(s,t) \in I_\alpha(A_i)$, somewhat arbitrarily.

All remaining proposition symbols and action symbols can be interpreted at will, as they do not affect the meaning of F.

In order to show that $(\mathbf{S}, I, \sigma) \models \neg F$, we prove a stronger proposition:

Proposition 4 *If G is a subformula of $\neg F$ and s is a state in σ, then G holds at s in σ if and only if G^* holds at s in $\mathcal{S}$. In particular, σ satisfies $\neg F$.*

Proof The proof is by induction on the structure of G. The only nontrivial case is for formulas of the form $[A]$. In this case, we consider separately the subcases where $[\mathit{false}]^*$ holds at s in $\mathcal{S}$ and where it does not. In both subcases, we use that some of the theorems compiled in X hold at s in $\mathcal{S}$; this is true because the theorems hold at s in $\mathcal{B}$ and because they are free of $\Box$ and $\Diamond$.

If $[\mathit{false}]^*$ holds at s, then $[A]^*$ holds at s for every A of interest, since $\vdash [\mathit{false}] \Rightarrow [A]$. Also, if $[\mathit{false}]^*$ holds at s, the construction of σ yields that σ loops at s, thus $[\mathit{false}]$ holds at s in σ; therefore, $[A]$ holds at s in σ for every A, by definition of the semantics of $[\,]$. Thus, $[A]^*$ holds at s in $\mathcal{S}$, and $[A]$ holds at s in σ.

If $[\mathit{false}]^*$ does not hold at s, then we can use the axioms to decompose A into a boolean combination of proposition symbols, primed proposition symbols, and action symbols. More precisely, consider the following primitive-recursive function d:

$$\begin{aligned}
d([P_i]) &\triangleq [P_i] \\
d([P_i']) &\triangleq [P_i'] \\
d([A_i]) &\triangleq [A_i] \\
d([\mathit{false}]) &\triangleq \mathit{false} \\
d([\neg A]) &\triangleq \neg d([A]) \\
d([A \wedge B]) &\triangleq d([A]) \wedge d([B]) \\
d([(\neg P)']) &\triangleq d([\neg P']) \\
d([(P \wedge Q)']) &\triangleq d([P' \wedge Q'])
\end{aligned}$$

One can derive by induction that $[A]$ and $d([A])$ are provably equivalent for every A:

$$\vdash \neg[\mathit{false}] \Rightarrow [A] \equiv d([A])$$

and this is also valid, of course:

$$\models \neg[\mathit{false}] \Rightarrow [A] \equiv d([A])$$

Therefore, it suffices to consider $[A]$ in the cases where A is *false*, a proposition symbol, a primed proposition symbol, or an action symbol:

- $A = \mathit{false}$: We have that s satisfies $\neg[\mathit{false}]^*$ in $\mathcal{S}$. By construction, σ does not fall into a loop in s (though it may loop between s and $\hat{s}$). Therefore, s satisfies $\neg[\mathit{false}]$ in σ.

- $A = P_i$: Since $\vdash \neg[\mathit{false}] \Rightarrow [P_i] \equiv P_i$, we have that s satisfies $[P_i]^*$ in $\mathcal{S}$ if and only if s satisfies P_i in $\mathcal{S}$. Since $\models \neg[\mathit{false}] \Rightarrow [P_i] \equiv P_i$, similarly, s satisfies $[P_i]$ in σ if and only if s satisfies P_i in σ. Finally, the definition of I_ρ yields that s satisfies P_i in $\mathcal{S}$ exactly when s satisfies P_i in σ.

- $A = P_i'$: If s satisfies $[P_i']^*$ in $\mathcal{S}$, then P_i is one of the conjuncts in the assignment R_s. In the construction of σ, the state immediately after s must satisfy R_s, and hence P_i. Therefore, the semantics yields that s satisfies $[P_i']$ in σ. Conversely, suppose that s does not satisfy $[P_i']^*$ in $\mathcal{S}$; then $\neg P_i$ is one of the conjuncts in the assignment R_s. In the construction of σ, the state immediately after s must satisfy R_s, and hence $\neg P_i$. Therefore, the semantics yields that s satisfies $[(\neg P_i)']$ and not $[P_i']$ in σ.

- $A = A_i$: The definition of I_α is designed to make this case trivial.

We derive that $(\mathbf{S}, I, \sigma) \models \neg F$, as a special case. ∎

This concludes the completeness proof. It follows from the proof that PTLA possesses a finite model property, and hence it is decidable. In fact, it seems likely that the validity problem for PTLA is decidable in polynomial space.

5 Conclusions

We have presented a complete proof system for a propositional temporal logic of actions, PTLA. Lamport has considered extensions of the basic temporal logic of actions, and it seems worthwhile to search for axiomatizations of some of them as well.

The simplest extension consists in adding formulas of the form $[A]_{P_0,\ldots,P_n}$ to the logic; this yields the general temporal logic of actions. Roughly, $[A]_{P_0,\ldots,P_n}$ says that the action A will take place the next time that one of $P_0, \ldots, P_n$ changes value. A further extension consists in adding existential quantification over propositions, for hiding internal state. These extensions make it possible to formulate proofs that a program implements another program, and lead to a simple compositional semantics for concurrent systems. As the logic becomes more powerful, however, it becomes more difficult to choose appropriate proof principles. A complete axiomatization might help in this choice.

Acknowledgements

Leslie Lamport encouraged this work, and helped in describing his logic and in defining PTLA. Rajeev Alur, Cynthia Hibbard, and Jim Saxe suggested improvements in the exposition.

References

[Gol87] Robert Goldblatt. *Logics of Time and Computation*. Number 7 in CSLI Lecture Notes. CSLI, Stanford, California, 1987.

[GPSS80] D. Gabbay, A. Pnueli, S. Shelah, and Y. Stavi. On the temporal analysis of fairness. In *Seventh Annual ACM Symposium on Principles of Programming Languages*, pages 163–173. ACM, January 1980.

[HC68] G. E. Hughes and M. J. Cresswell. *An Introduction to Modal Logic.* Methuen Inc., New York, 1968.

[HM85] Matthew Hennessy and Robin Milner. Algebraic laws for nondeterminism and concurrency. *Journal of the ACM*, 32(1):137–161, January 1985.

[INM84] INMOS. *Occam Programming Manual.* Prentice-Hall, Inc., Englewood Cliffs, New Jersey, 1984.

[Lam89] Leslie Lamport. A simple approach to specifying concurrent systems. *Communications of the ACM*, 32(1):32–45, January 1989.

[Lam90] Leslie Lamport. A temporal logic of actions. Research Report SRC57, Digital Equipment Corporation, Systems Research Center, April 1990.

[Pnu77] A. Pnueli. The temporal logic of programs. In *Proceedings of the 18th Symposium on the Foundations of Computer Science.* IEEE, November 1977.

[Pnu81] A. Pnueli. The temporal semantics of concurrent programs. *Theoretical Computer Science*, 13:45–60, 1981.

[Pra76] Vaughn R. Pratt. Semantical considerations on Floyd-Hoare logic. In *17th Symposium on Foundations of Computer Science*, pages 109–121. IEEE, October 1976.

[SC85] A. P. Sistla and E. M. Clarke. The complexity of propositional linear temporal logic. *Journal of the ACM*, 32(3):733–749, July 1985.

Convergence of Iteration Systems

(Extended Abstract)

Anish ARORA [1,2] *Paul ATTIE* [1,2] *Michael EVANGELIST* [1]
Mohamed GOUDA [1,2]

1. Microelectronics and Computer Technology Corporation, Austin
2. Department of Computer Sciences, The University of Texas at Austin

Abstract

An iteration system is a set of assignment statements whose computation proceeds in steps: at each step, an arbitrary subset of the statements is executed in parallel. The set of statements thus executed may differ at each step; however, it is required that each statement is executed infinitely often along the computation. The convergence of such systems (to a fixed point) is typically verified by showing that the value of a given variant function is decreased by each step that causes a state change. Such a proof requires an exponential number of cases (in the number of assignment statements) to be considered. In this paper, we present alternative methods for verifying the convergence of iteration systems. In most of these methods, upto a linear number of cases need to be considered.

1 Introduction

Iteration systems are a useful abstraction for computational, physical and biological systems that involve "truly concurrent" events. In computing science, they can be used to represent self-stabilizing programs, neural networks, transition systems and array processors. This wide applicability derives from the simplicity and generality of the formalism.

nformally, an iteration system is defined by a finite set of variables, V. Associated with each ariable is a function called its *update function.* The computation of the system proceeds in teps. At each step, the variables in an arbitrary subset of V are updated by assigning each ariable the value obtained from applying its update function to the current system state. The et of variables thus updated may differ at each step; however, it is required that each variable n V be updated infinitely often.

n allowing an arbitrary subset of variables to be updated at each step, our formalism admits a arge number of widely varying computations. These range from the sequential computations in vhich exactly one variable is updated at every step to the parallel computation in which each ariable is updated at every step. By comparison, traditional semantics admit computations of esser variety. For example, interleaving requires one enabled event to be executed at each step see, for instance, the work on CSP [Hoa], UNITY [CM] and I/O Automata [Lyn]), whereas naximal parallelism requires that all enabled events are executed at every step (see, for instance, he work on systolic arrays [KL] and cellular automata [Wol]).

property of interest in iteration systems is convergence. This property is useful in studying the elf-stabilization of distributed programs (cf. [Dij1], [BGM], [BGW], [Dij] and [GE]), convergence f iterative methods in numerical analysis (cf. [Rob] and [BT]), and self-organization in neural etworks (cf. [Koh] and [Arb]). Informally, an iteration system is called convergent if on starting n an arbitrary state the system is guaranteed to reach a fixed point; that is, a state in which no pdate can cause a state change. The standard method for verifying that an iteration system convergent is to exhibit a variant function (cf. [Gri]) whose value is bounded from below nd is decreased by each step that causes a state change. Since any subset of the variables can e updated in a step, the number of cases that need to be considered are $2^n - 1$, where n is he number of variables in the system. In this paper, we discuss new methods for verifying the onvergence of iteration systems. Nearly all these methods require upto n cases to be considered.

he rest of this paper is organized as follows. In Section 2, we formally define iteration systems nd their dependency graphs. (The dependency graph of an iteration system captures the depends on" relation between the variables in the system.) In Section 3, we identify two classes f iteration systems, namely those whose dependency graphs are acyclic or self-looping, and resent a theorem that establishes efficient proof obligations for verifying the convergence of hese two classes. This theorem is then extended to general, deterministic iteration systems in

Section 4. In Section 5, we show that, with minor modifications, our results continue to hold in nondeterministic iteration systems. Concluding remarks are in Section 6.

2 Iteration Systems

An *iteration system*, I, is defined by the pair (V, F), where

- V is a finite, nonempty set of variables. Each variable v in V has a predefined domain Q_v. Let Q denote the cartesian product of the domains of all variables in V.
- F is a set of "update" functions with exactly one function f_v associated with each variable v in V, where f_v is a mapping from Q to Q_v.

A *state* q of I is an element of Q. We adopt the notation q_v to denote the value of variable v in state q. A state q is called a *fixed point* of I iff for each variable v in V, $f_v(q) = q_v$.

A *step* of I is defined to be a subset of V; informally, a step identifies those variables that are updated when the step is executed. A *round* of I is a minimal, finite sequence of steps with the property that each variable in V is an element of at least one step in the round. A *computation* of I is an infinite sequence of rounds. Notice that since each variable is updated at least once in every round, each variable is updated infinitely often in every computation.

The *application* of a finite sequence of steps S to a state q, denoted $S \circ q$, is the state q' defined inductively as follows:

- if S is empty, then $q' = q$
- if S is a single step, then for every variable v in V,

$$q'_v = \begin{cases} f_v(q) & \text{, if } v \in S \\ q_v & \text{, otherwise} \end{cases}$$

- if S is the concatenation of two sequences $S = S'; S''$, then $q' = S'' \circ (S' \circ q)$.

A computation C is called *convergent* iff for every state q, there exists a finite prefix, S, of C such that $S \circ q$ is a fixed point of I. An iteration system is called *convergent* iff all of its computations are convergent.

As shown in the following examples of iteration systems, it is convenient to represent an iteration system by a set of assignment statements, one for each variable. Each statement has the form ariable⟩ := ⟨*corresponding update function*⟩. We will later prove each of these iteration systems be convergent.

xample 1: (Greatest Common Divisor)

et x, y and z be variables that range over the natural numbers. Then, the three assignment atements

$x := \underline{\text{if}}\ x > y\ \underline{\text{then}}\ x - y\ \underline{\text{else}}\ x$

$y := \underline{\text{if}}\ x < y\ \underline{\text{then}}\ y - x\ \underline{\text{else}}\ y$

$z := \underline{\text{if}}\ x = y\ \underline{\text{then}}\ 0\ \underline{\text{else}}\ z + 1$

efine a convergent iteration system. At fixed point, the value of x is the greatest common visor of the initial values of x and y, and $y = x$ and $z = 0$. □

xample 2: (Minimum of a bag)

et x be an integer array of size n. Then, the following assignment statements:

$x[1] := x[1]$

$x[2] := \min(x[2], x[1])$

⋮

$x[n] := \min(x[n], x[n-1])$

efine a convergent iteration system. At fixed point, the value of each $x[i]$ is the minimum of e initial values in the sub-array $x[1], x[2], \ldots, x[i]$. □

xample 3: (Shortest Path)

onsider the directed graph

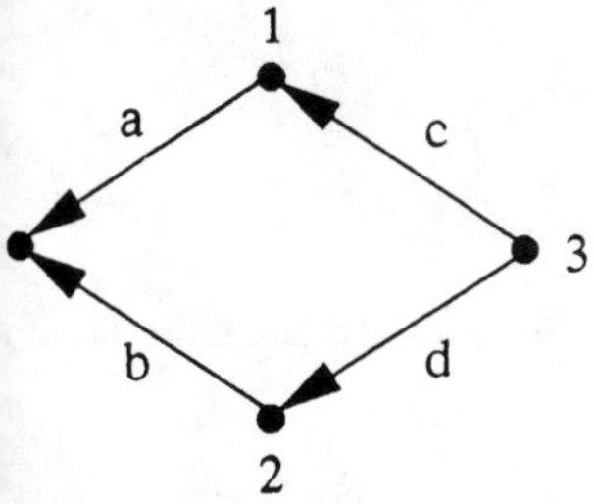

has four nodes 0-3, and four edges. Each edge is labeled with a non-negative integer constant noting its length. Associated with each node i is a variable $v[i]$, of type *record*, that has two teger components, $first[i]$ and $second[i]$. The assignment statements

$v[0] := (0, 0)$

$v[1] := (a, 0)$

$v[2] := (b, 0)$

$v[3] :=$ <u>if</u> $first[1] + c \leq first[2] + d$ <u>then</u> $(first[1] + c, 1)$ <u>else</u> $(first[2] + d, 2)$

define a convergent iteration system. At fixed point, each $first[i]$ is the length of the shortest path from node i to node 0, and $second[i]$ is the nearest neighbor to node i along this path. Extending the above system for an arbitrary directed graph is straightforward. □

The objective of this paper is to identify proof obligations that are sufficient to establish the convergence of iteration systems. Towards this end, the following two definitions will prove useful shortly.

Let v and w be variables in V. We say v *depends on* w iff there exist two states q and q' of I such that q and q' differ only in their value of w and $f_v(q) \neq f_v(q')$. Informally, v *depends on* w iff a change in the value of w can cause a change in the value assigned to v by its update function f_v.

The *dependency graph* of I is a directed graph whose nodes correspond to the variables in V and whose directed edges correspond to the *depends on* relation; that is, the set of nodes of the dependency graph is $\{n_v | v \in V\}$ and its set of directed edges is $\{(n_v, n_w) | v \in V,\ w \in V, \text{ and } v \textit{ depends on } w\}$.

The dependency graphs for the iteration systems in Examples 1, 2 and 3 are as follows:

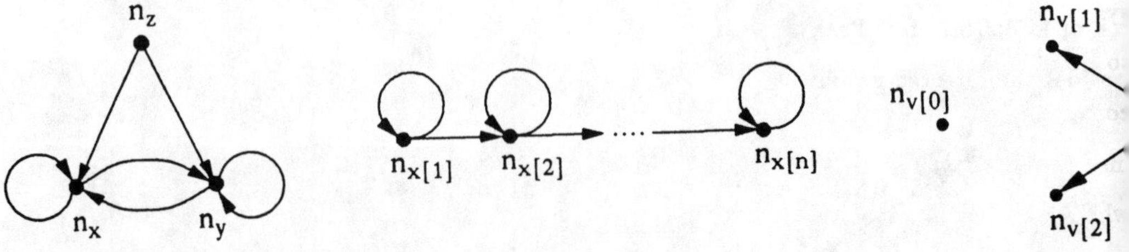

Henceforth, we shall use 'variable' and 'node' interchangeably when referring to the dependency graph of an iteration system.

3 Convergence of Non-Cyclic Systems

An iteration system is called *acyclic* iff its dependency graph is acyclic. It is called ***self-looping*** iff its dependency graph has one or more cycles, and all its cycles are self-loops.

In this section we state a fundamental theorem concerning the convergence of acyclic and self-looping iteration systems. The implications of this theorem are discussed subsequently.

Theorem 1:

If an iteration system is

(a) acyclic, then it is convergent, and if it is

(b) self-looping and has one convergent computation, then it is convergent. □

The examples in Section 2 illustrate Theorem 1. For instance, the iteration system of Example 3 is acyclic; thus, each of its computations is convergent by Theorem 1(a). The iteration system of Example 2 is self-looping, and any computation where the first step updates $x[1]$, the second updates $x[2]$, and so on is convergent; thus, each computation of the system is convergent by Theorem 1(b).

As mentioned in the introduction, verifying the convergence of an iteration system is generally accomplished by exhibiting a variant function whose value is bounded from below and is decreased by each step that causes a state change. Such a proof requires $2^n - 1$ cases to be considered, where n is the number of variables in the system. In contrast, Theorem 1 shows that verifying the convergence of acyclic systems requires no such case analysis.

The theorem also states that the convergence of all computations of a self-looping iteration system can be established from the convergence of a computation of choice. One possibility is to choose this computation to be the one in which each variable is updated at every step. The convergence of this computation can then be proved by the variant function method which, in this instance, needs only one case to be considered. Another possibility is to choose computations in which exactly one variable is updated at each step; in this instance, the variant function method requires n cases.

Theorem 1 cannot be made to apply to all iteration systems. Consider, for example, the iteration system defined by the two assignment statements

$$x := y$$

$$y := x.$$

Although this system has many computations that are convergent (for example, all computations where exactly one variable is updated at the first step), it also has a computation that is not convergent (for example, the computation where both x and y are updated at each step).

Thus, unlike Theorem 1, one cannot establish that this system is convergent by exhibiting one convergent computation. Similar examples have been presented in [Dij] and [Rob].

In fact, it is straightforward to show that Theorem 1 cannot be made to apply to any class of iteration systems that properly includes acyclic and self-looping systems. The proof for this follows from a construction that exhibits, for each directed graph G that has a cycle of two or more nodes, an iteration system I such that the dependency graph of I is G, and I has both convergent and non-convergent computations.

The following lemma states that for any cyclic iteration system I there is a self-looping system which captures a subset of the computations of I and, thereby, is a possible implementation for I. This shows that the class of self-looping systems is rich.

Lemma 2:

For each iteration system I that is neither acyclic nor self-looping, there exists a self-looping iteration system I' that satisfies the following two conditions:

- There is a one-to-one correspondence between the states of I and those of I'.
- Every computation of I' is a computation of I. □

System I' in Lemma 2 is self-looping; hence, its convergence can be established by Theorem 1(b). (The convergence of I', however, does not necessarily imply the convergence of the original system I.) Consider, for instance, the iteration system in Example 1. This system is neither acyclic nor self-looping because its dependency graph has a maximally strongly connected component consisting of variables x and y. By replacing these two variables by one variable with two components, also called x and y for convenience, we obtain the following implementation of the system:

$$(x, y) := (\underline{\text{if}}\ x > y\ \underline{\text{then}}\ x - y\ \underline{\text{else}}\ x\ ,\ \underline{\text{if}}\ x < y\ \underline{\text{then}}\ y - x\ \underline{\text{else}}\ y\ )$$

$$z := \underline{\text{if}}\ x = y\ \underline{\text{then}}\ 0\ \underline{\text{else}}\ z + 1.$$

As this system is self-looping, its convergence can be established by Theorem 1(b).

4 Convergence of Cyclic Iteration Systems

In this section, we generalize our analysis for the convergence of acyclic and self-looping systems to the convergence of general iteration systems. Our starting point is to note the basic characteristic of an iteration system that is neither acyclic nor self-looping, namely the existence of at east one maximally strongly connected component in its dependency graph that consists of two or more nodes. For convenience, we call a maximally strongly connected component that has wo or more nodes a *district.*

et D be a district in the dependency graph of an iteration system, I. The *iteration system associated with* D is the iteration system (V_D, F_D) that satisfies the following two conditions:

- The set of variables, V_D, is the set of all variables in D together with each variable that is not in D but some variable in D depends on it.
- The set of update functions, F_D, is defined as follows. The update function for a variable in D is the same as its update function in I, whereas the update function for a variable in $V_D \setminus D$ is the identity function for that variable.

or instance, the iteration system in Example 1 has one district whose associated iteration ystem can be defined by the two assignment statements

$x :=$ if $x > y$ then $x - y$ else x

$y :=$ if $x < y$ then $y - x$ else y.

n iteration system is called *district-convergent* iff the iteration system associated with each district in the dependency graph of the system is convergent. Since acyclic and self-looping systems o not have any districts in their dependency graphs, they are trivially district-convergent.

n iteration system is called *0-cyclic* iff its dependency graph has no maximally strongly connected component that consists of a single node with a self-loop; otherwise, the iteration system called *1-cyclic.* Note that each iteration system is either 0-cyclic or 1-cyclic; in particular, cyclic systems are 0-cyclic whereas self-looping ones are 1-cyclic.

he following theorem generalizes Theorem 1.

heorem 3

a district-convergent iteration system is

(a) 0-cyclic, it is convergent, and if it is

(b) 1-cyclic and has one convergent computation, then it is convergent. □

the remainder of this section we identify two proof obligations which are sufficient to establish he convergence of an iteration system that is associated with a district. These obligations consists of exhibiting either a variant function for each node in a selected set of nodes in the district (Lemma 4), or a single variant function for the whole district (Lemma 5).

The intuition underlying Lemma 4 is to "break" each cycle in the district by ensuring that some distinguished variable on the cycle will eventually reach a stable value. This is achieved by exhibiting a variant function for the distinguished variable whose value decreases each time the value of the variable is changed by an update. Once every distinguished variable in the district becomes stable (i.e. has a fixed value), the iteration system associated with the district starts to behave like an acyclic system and, so, eventually reaches a fixed point.

A more general approach to solving the same problem is to exhibit a variant function for all the variables in the district. The value of this function is decreased by each step that causes a state change. See Lemma 5 below.

In what follows, let

D be a district in some iteration system I,
(V_D, F_D) be the iteration system associated with D,
Q_D be the set of states of (V_D, F_D), and
f_v be the update function of a variable v in (V_D, F_D), and
$V_{D'}$ be the set of variables in D', a subgraph of D, and
N be an arbitrary set that is well-founded under some relation $<$.

Lemma 4: *(Local Variant)*

If each directed cycle in D has a variable v and a variant function $\# : Q_v \to N$ such that for each state q in Q_D,

$$\#(f_v(q)) < \#(q_v) \quad \vee \quad (f_v(q) = q_v)$$

then the iteration system (V_D, F_D) is convergent. □

Lemma 5: *(Global Variant)*

If there is a variant function $\# : Q_D \to N$ such that for each state q in Q_D, and for each strongly connected component D' in D,

$$\#(V_{D'} \circ q) < \#(q) \quad \vee \quad (V_{D'} \circ q = q)$$

then the iteration system (V_D, F_D) is convergent. □

As an example, both Lemma 4 and Lemma 5 can be used to show that the iteration system associated with the district in Example 1 is convergent. In using Lemma 4, let the variant unctions for both x and y be their respective identity functions. In using Lemma 5, let the global variant function be the sum of x and y. In either case, the convergence of the entire ystem in Example 1 can now be established by Theorem 3.

5 Convergence of Nondeterministic Systems

So far, the definition of an iteration system associates exactly one (deterministic) update function vith each variable. We now extend this definition to allow each variable to be updated by more han one update function. More specifically, we associate with each variable v a finite, non-mpty set of update functions F_v. At each step in which v is updated, one of the functions in F_v is chosen to update v. This choice is arbitrary except for the requirement that each update unction in F_v is chosen infinitely often in every computation. (Note that this is possible because each variable is updated infinitely often in every computation.)

In the next example, we represent a nondeterministic iteration system by a set of assignment tatements (with choice), one for each variable. If $F_v = \{f, g, \ldots, h\}$ then the assignment tatement that updates v has the form:

$$v := f \mid g \mid \ldots \mid h$$

Example 4. (Nondeterministic Shortest Path)
We exhibit a nondeterministic iteration system for the directed graph in Example 3. The non-determinism makes it possible to reduce the 'atomicity' of the update functions. In fact, every update function refers uniquely to one edge in the graph, as follows:

$$v[0] := (0, 0)$$

$$v[1] := (a, 0)$$

$$v[2] := (b, 0)$$

$$\begin{aligned} v[3] := &\ \underline{\text{if}}\ \ first[1] + c \leq first[3] \vee second[3] = 1\ \underline{\text{then}}\ (first[1] + c, 1)\ \underline{\text{else}}\ v[3] \\ \mid &\ \underline{\text{if}}\ \ first[2] + d \leq first[3] \vee second[3] = 2\ \underline{\text{then}}\ (first[2] + d, 2)\ \underline{\text{else}}\ v[3] \end{aligned}$$

This system is convergent to some fixed point, and when it is at a fixed point, each $first[i]$ is the length of the shortest path from node i to node 0, and each $second[i]$ is the nearest neighbor o node i along this path. □

We now redefine four concepts that were introduced earlier in order to accommodate the extension to nondeterminism.

- A state q of an iteration system is a *fixed point* iff for each variable v and each update function f_v in F_v, $f_v(q) = q_v$.
- A computation is said to be *convergent* iff for each state q and for each choice of update functions in the computation there exists a finite prefix S of the computation such that $S \circ q$ is a fixed point.
- The *depends on* relation is redefined as follows: variable v *depends on* variable w iff there exist two states q and q' such that q and q' differ only in their value of w and $f_v(q) \neq f_v(q')$ for some f_v in F_v.
- We augment the notion of the *iteration system* (V_D, F_D) *associated with* a district D in the dependency graph of an iteration system I. With each variable v in V_D that is also in D, we now associate its set of update functions in I, i.e. F_v. The set of update functions for every variable in V_D but not in D is defined to be the set that contains only the identity function for that variable.

The results presented in Sections 3 and 4 apply, with minor modifications, to nondeterministic iteration systems. Stated informally, modifications are needed on two accounts. Firstly, unlike acyclic deterministic systems, acyclic nondeterministic systems are not necessarily convergent. (Consider, for instance, the system $x := 0 \mid 1$.) However, as we show in the full paper [AAEG], to establish convergence of an acyclic nondeterministic system it is sufficient to exhibit that the system has a fixed point. (The convergence of the iteration system in Example 4 follows from this result.) Secondly, to account for nondeterministic updates, the requirement associated with variant functions (see lemmas 4 and 5) is modified. The modified obligation requires that, for *each possible choice* of update function(s), the value assigned by the variant function decreases each time the value of the variable(s) in question is changed by an update.

The details of the modified results and their proofs are relegated to the full paper.

6 Conclusions

We have defined a very general model of computation that exhibits true concurrency, and have considered convergence as a typical property of systems expressed in this model. We established everal results that reduce the proof burden involved in establishing convergence.

Vhen analyzing concurrent systems, convergence can be used to model termination. Since some rogress (that is, eventuality) properties of a concurrent system can be reduced to the termi-ation of a derived system, our techniques are applicable in verifying progress properties. (See, or example, [GFMR] where the eventual enabledness of an action is reduced to termination.)

'here are several issues that need to be investigated in extending this work. Other than identi-ying more general sufficiency conditions and determining their scope, a comparison of the 'rate' f convergence of various types of computations is needed. Methods regarding other properties, uch as safety, also need to be studied.

References

Arb] M.A. Arbib, *Brains, Machines and Mathematics*, Springer-Verlag, 1987.

AAEG] A. Arora, P. Attie, M. Evangelist and M. Gouda, "Convergence of Iteration Systems", submitted for publication in *Distributed Computing* (Special Issue on Self-Stabilization).

3GM] J.E. Burns, M.G. Gouda, and R.E. Miller, "On relaxing interleaving assumptions", *Proceedings of the MCC Workshop on Self-Stabilizing Systems*, MCC Technical Report #STP-379-89.

3GW] G.M. Brown, M.G. Gouda, and C.-L. Wu, "Token Systems that Self-Stablize", *IEEE Transactions on Computers 38(6)*, pp. 845-852, 1989.

3T] D.P. Bertsekas and J.N. Tsitsiklis, *Parallel and Distributed Computation*, Prentice-Hall, 1989.

CM] K.M. Chandy and J. Misra, *Parallel Program Design: A Foundation*, Addison-Wesley Publishing, 1988.

)ij] E.W. Dijkstra, EWD306 "The Solution to a Cyclic Relaxation Problem", 1973. Reprinted in *Selected Writings on Computing: A Personal Perspective*, Springer-Verlag, pp. 34-35, 1982.

)ij1] E.W. Dijkstra,"Self-stabilizing Systems in Spite of Distributed Control", *Communications of the ACM 17(11)*, pp. 643-644, 1973.

[Gri] D.Gries, *The Science of Programming*, Springer-Verlag, 1981.

[GE] M.G. Gouda and M. Evangelist, "Convergence Response Tradeoffs in Concurrent Systems", *MCC Technical Report #STP-124-89*; also submitted to *ACM TOPLAS*.

[GFMR] O. Grumberg, N. Francez, J.A. Makovsky and W.P. deRoever, "A proof rule for fair termination of guarded commands", *Information and Control 66*, pp. 83-102, 1985.

[Har] F. Harary, *Graph Theory*, Addison-Wesley Publishing, 1972.

[Hoa] C.A.R. Hoare, *Communicating Sequential Processes*, Prentice-Hall International, 1985.

[Koh] T. Kohonen, *Self-Organization and Associative Memory*, Springer-Verlag, 1984.

[KL] H.T. Kung and C.E. Leiserson, "Systolic Arrays (for VLSI)," in *Sparse Matrix Proc.*, 1978.

[Lyn] N.A. Lynch, "I/O Automata: A Model for Discrete Event Systems," *Proc. of 22nd Annual Conference on Information Sciences and Systems*, 1988.

[Rob] F. Robert, *Discrete Iterations - A Metric Study*, Springer-Verlag, 1986.

[Wol] S. Wolfram, *Theory and Applications of Cellular Automata, Advanced Series on Complex Systems*, Vol.1, World Scientific Publishing, 1986.

Process Algebra with a Zero Object

J.C.M. Baeten
Department of Software Technology, Centre for Mathematics and Computer Science
P.O.Box 4079, 1009 AB Amsterdam, The Netherlands

J.A. Bergstra
Programming Research Group, University of Amsterdam
P.O.Box 41882, 1009 DB Amsterdam, The Netherlands

Department of Philosophy, State University of Utrecht
Heidelberglaan 2, 3584 CS Utrecht, The Netherlands

The object 0 acts as a zero for both sum and multiplication in process algebra. The constant δ, representing deadlock or inaction, is only a left zero for multiplication. We will call 0 predictable failure.

1980 Mathematics Subject Classification (1985 revision): 68Q45, 68Q55, 68Q65, 68Q50.
1987 CR Categories: F.4.3, D.2.10, D.3.1, D.3.3.
Key words & Phrases: process algebra, zero, deadlock, inaction, failure.
Note: Partial support received by ESPRIT basic research action 3006, CONCUR, and by RACE contract 1046, SPECS. This document does not necessarily reflect the views of the SPECS consortium.
Note: this article is a revision of BAETEN & BERGSTRA [90], but leaving out the material in section 6.

1. INTRODUCTION

The object 0 acts as a 0 for both sum and multiplication in process algebra. The constant δ representing deadlock or inaction is only a left zero for multiplication. The purpose of this paper is to indtroduce a constant 0 in process algebra and discuss its properties.

We will call 0 *predictable failure*. A predictable failure differs from deadlock (inaction) in the sense that a system will actively try to avoid it. The axioms for 0 incorporate the intention of a system to avoid failure whenever possible. The axioms for δ (in particular $x \neq 0 \Rightarrow x + \delta = x$) incorporate the intention of a process to make progress if it can.

In fact 0 stands for a truly empty process, its execution is simply inconceivable. The process 0 also occurs in PONSE & DE VRIES [89] (but is called δ there!).

A process specification involving 0 or renaming into 0 is not executable. It must be implemented, which means that one has to provide an equivalent (or better) specification not involving 0 or renaming into 0. Due to this observation 0 is a high level feature that plays a role in system design and specification, rather than in implementation.

2. AXIOMATIZATION

2.1. BASIC PROCESS ALGEBRA WITH ZERO

We start out from the theory of Basic Process Algebra as described in BERGSTRA & KLOP [84a, 85, 86]. For a recent survey article see BERGSTRA & KLOP [89].

We have a set of **atomic actions**, A. Each atomic action is a constant of the sort P, the sort of processes that the theory is about. Then, we have two binary operators on P: + is **alternative composition** or sum, and · is **sequential composition** or product. For this signature, we have the first five axioms in table 1 below (A1-5), constituting BPA. In table 1, x,y,z are arbitrary elements of P.

Then we add the **zero** constant, obeying the next three axioms (Z1-3). Usually, we also have the constant δ in the theory, called **deadlock** or **inaction**. The two axioms for inaction need conditions, in order to avoid clashes with the zero axioms ($A6^0$, $A7^0$). The conditions use a predicate on processes ≠ 0, that determines whether or not a term can be proved equal to the zero process. This predicate is axiomatized in the last four axioms of table 1. There, we have $a \in A$. We put $A_\delta = A \cup \{\delta\}$, $A_{0\delta} = A \cup \{0,\delta\}$, $BPA_0 = BPA + Z1\text{-}3$, $BPA_{0\delta} = BPA_0 + A6^0, A7^0 + NZ1\text{-}4$.

Of all operators, + will bind the weakest, and · the strongest. We often leave out the · sign.

$x + y = y + x$	A1
$x + (y + z) = (x + y) + z$	A2
$x + x = x$	A3
$(x + y) \cdot z = x \cdot z + y \cdot z$	A4
$(x \cdot y) \cdot z = x \cdot (y \cdot z)$	A5
$x \cdot 0 = 0$	Z1
$x + 0 = x$	Z2
$0 \cdot x = 0$	Z3
$x \neq 0 \Rightarrow x + \delta = x$	$A6^0$
$x \neq 0 \Rightarrow \delta \cdot x = \delta$	$A7^0$
$\delta \neq 0$	NZ1
$a \neq 0$	NZ2
$x \neq 0, y \neq 0 \Rightarrow x \cdot y \neq 0$	NZ3
$x \neq 0 \Rightarrow x + y \neq 0$	NZ4

TABLE 1. $BPA_{0\delta}$

The difference between 0 and δ requires further comments:

i. δ is an inactive process, if a system reduces to δ it deadlocks. Of course the mere occurrence of δ in a process expression (like $a + b(\delta + c)$) need not indicate a deadlock.

Compare this to the expression $e = 2 + 0 \cdot (5 + 8)$; in no way the occurrence of 0 in e implies that e vanishes.
ii. $x + \delta = x$ is a progress rule: the system eventually discovers that δ is no option and proceeds with x (if possible).
iii. $x \cdot 0 = 0$ is a liveness rule: the system must, after completion of x, eventually execute 0, i.e. assert false. This form of liveness is not required for δ.
iv. $0 + \delta = \delta$ because δ is 'better than' 0.
v. $0 \cdot x = 0$ and $\delta \cdot x = \delta$ are explained by the non-executability of 0, resp. δ. Note that we have in particular $\delta \cdot 0 = 0$. Unfortunately, the explanation of this in philosophical terms is shallow.

2.2. PROJECTION

We can extend the signature given above with the **projection operators** π_n (for $n \geq 1$). The axioms are straightforward adaptations of the usual ones (see BERGSTRA & KLOP [84a]). In table 2, $a \in A$.

$\pi_n(0) = 0$
$\pi_n(\delta) = \delta$
$\pi_n(a) = a$
$x \neq 0 \Rightarrow \pi_1(ax) = a$
$\pi_{n+1}(ax) = a \cdot \pi_n(x)$
$\pi_n(x + y) = \pi_n(x) + \pi_n(y)$

TABLE 2. Projection

2.3 INFINITARY RULES

The axioms given above constitute a complete theory for finite processes, processes represented by closed terms, i.e. equality on finite processes (in a graph model to be presented further on) coincides with derivability from the axioms. When we are dealing with infinite processes however, processes specified by means of recursive equations, we need additional proof principles. In this paper, we will discuss 4 such principles.

First, we consider the *Approximation Induction Principle* (AIP). AIP can be maintained in the present setting as in BERGSTRA & KLOP [86]. Roughly, this proof rule states that two processes should be identified if all their projections are equal. More explicitly:

$$\text{for all } n \quad \pi_n(x) = \pi_n(y) \quad \Rightarrow \quad x = y.$$

This rule is valid for finitely branching processes only (a process is finitely branching if it has a representation as a finitely branching graph or tree, see further on).

Next, we consider two principles that deal with solutions of recursive equations. The *Recursive Definition Principle* (RDP) states that every recursive specification has at least one solution, and the *Recursive Specification Principle* (RSP) states that every guarded recursive specification has at most one solution. Here, guarded roughly means that every variable occurring in the right-hand side of an equation must be preceded by an atomic action. For more details and formal definitions, see BERGSTRA & KLOP [86].

Now let us consider these rules in the present setting. Notice that the guarded equation $x = a \cdot x$ has two solutions: 0 and a^{ω} (the process that indefinitely performs a). It follows that the principle RSP has to be relaxed: every guarded system of recursion equations not containing an occurrence of 0 has at most one solution different from 0. Thus more formally, RSP^0 states:

Let E be a guarded recursive specification with k process names such that none of the equations of E involves either 0 or a renaming into 0 (see below). Such E is called **0-free**.
Let $X = (x_1, ..., x_k)$ and $Y = (y_1, ..., y_k)$ be process vectors of length k.
Then $x_1 \neq 0$ & ... & $x_k \neq 0$ & $y_1 \neq 0$ & ... & $y_k \neq 0$ & $X = E(X), Y = E(Y)$
implies $X = Y$.

Besides RSP^0 there is also RDP^0 which states that:

every 0-free guarded system of recursion equations possesses at least one solution vector consisting of processes different from 0.

The last infinitary proof rule that we will consider is the *Limit Rule*. We describe this rule in 4.4.

2.4 RENAMING INTO ZERO

0_δ is an operator that substitutes 0 for δ. We will call this operator **deadlock prevention**. Let $a \in A$.

$0_\delta(0) = 0$
$0_\delta(\delta) = 0$
$0_\delta(a) = a$
$0_\delta(ax) = a \cdot 0_\delta(x)$
$0_\delta(x + y) = 0_\delta(x) + 0_\delta(y)$

TABLE 3. Deadlock prevention

An example: $0\delta(ab + c(de\delta + a\delta)) = ab$.

Note that the equation $0_\delta(x \cdot y) = 0_\delta(x) \cdot 0_\delta(y)$ leads to a problem as follows: Let $x = a^{\omega}$ (i.e. the unique solution different from 0 of the equation $z = a \cdot z$) and $y = \delta$, then $x \cdot y = x$ because of AIP, hence $x = 0_\delta(x) = 0_\delta(x \cdot y) = 0_\delta(x) \cdot 0_\delta(y) = 0_\delta(x) \cdot 0 = 0$, which contradicts the assumptions on x.

Also note that it is incorrect to rename an atomic action a into 0 by an operator 0_a, for that leads to $0 = 0_a(a) = 0_a(a + \delta) = 0_a(a) + 0_a(\delta) = 0 + \delta = \delta$.

3. SEMANTICS

3.1 TRANSITION RULES

We define a **structured operational semantics** as in VAN GLABBEEK [87], on congruence classes of process expressions, i.e. a $\xrightarrow{a}$ relation holds between two terms iff they are provably equal to the format in the rules in table 4 below. Likewise, a term satisfies

$\xrightarrow{a} \surd$ iff it can be written in the form $a + x$. For closed terms, the rules now determine an **action graph**. On action graphs, we will then define a notion of **bisimulation** as in BERGSTRA & KLOP [86]. In this definition, we have to pay special attention to the separate status of the process 0. Note that we have to define the rules on congruence classes of process expressions, since we need the predicate $\neq 0$ in the definition. We cannot define action rules on terms in the format of GROOTE & VAANDRAGER [89] or GROOTE [90].

$$a + x \xrightarrow{a} \surd$$
$$x \neq 0 \;\Rightarrow\; a \cdot x + y \xrightarrow{a} x$$

TABLE 4. Action rules

We can also add rules in order to deal with recursive equations. In that case, however, the transition system may become undecidable because $x \neq 0$ is in general not decidable. To see this notice that for every recursively enumerable set $W_e \subseteq \mathbb{N}$ with recursive index e and every $n \in \mathbb{N}$ a guarded recursive specification over ACP can be uniformly computed with solution $p(e,n)$ such that:

$$n \in W_e \Leftrightarrow \text{for some } k \;\; p(e,n) = i^k \cdot \delta$$
$$n \notin W_e \Leftrightarrow p(e,n) = i^\omega.$$

This construction can be found in BERGSTRA & KLOP [84b]. It follows that:

$$n \in W_e \Leftrightarrow 0_\delta(p(e,n)) = 0,$$

hence it is undecidable whether $X = 0$ for recursively specified X in general.

3.2 GRAPH MODEL

We can also define a graph model directly. Let $\mathbb{T}$ be the set of all rooted labeled **trees**, where all edges are labeled with elements of A, and all endpoints labeled with an element of $\{\surd, \delta, 0\}$. The interpretation of the constants and operators of $BPA_{0\delta}$ is straightforward:

- $[\![\delta]\!]$ is the one-node graph labeled with δ, $[\![0]\!]$ is the one-node graph labeled with 0;
- $[\![a]\!]$ is the two-node graph with one edge labeled a, and the endpoint labeled $\surd$;
- if g,h are in $\mathbb{T}$, then $g + h$ is obtained by identifying the roots of g and h. If one graph is $[\![0]\!]$, the result is just the other graph. If one graph is $[\![\delta]\!]$, the result is also the other graph, unless that graph is $[\![0]\!]$ (in which case it is $[\![\delta]\!]$);
- if g,h are in $\mathbb{T}$, then $g \cdot h$ is obtained by first making a copy of h for each $\surd$-endpoint of g. Then we remove the $\surd$-label, and identify the endpoint with the root of its copy.

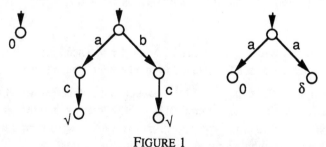

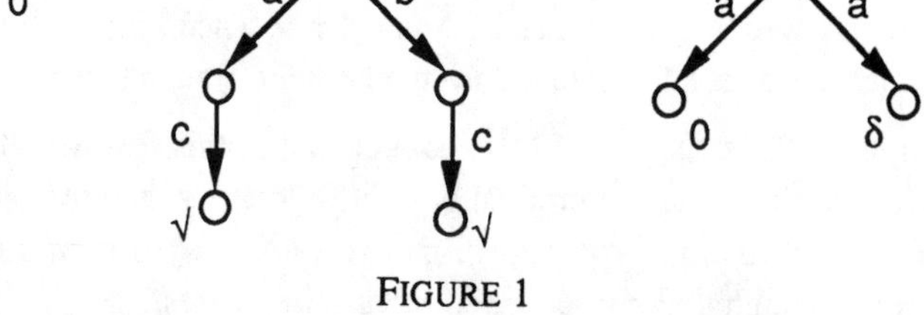

FIGURE 1

We see some examples in fig. 1 above. We see the trees that represent the terms 0, $(a + b)\cdot c$, $a\cdot\delta + a\cdot 0$. Recursively specified processes will in general have infinite trees.

If g is a tree in $\mathbb{T}$, and s a node in g, then we call s a **zero node** of g if every maximal path in g from s must end in a 0-labeled point. Now we can give the definition of bisimulation on these trees.

3.2.1 DEFINITION

Let $g,h \in \mathbb{T}$. We say g,h are **bisimilar**, $g \leftrightarrow h$, if there exists a relation R (called a **bisimulation**) on nodes of g and h, such that

1. the domain of R consists of all non-zero nodes of g;
2. the range of R consists of all non-zero nodes of h;
3. either g,h are both the zero graph, or the roots of g,h are related;
4. if R(s,t) and $s \xrightarrow{a} s'$ and s' non-zero, then there is a non-zero t' such that $t \xrightarrow{a} t'$ and R(s',t');
5. if R(s,t) and $t \xrightarrow{a} t'$ and t' non-zero, then there is a non-zero s' such that $s \xrightarrow{a} s'$ and R(s',t');
6. if R(s,t) and s,t are endpoints, then they have the same label.

3.2.2 PROPOSITION

Bisimulation is a congruence on $\mathbb{T}$.

In order to prove that $\mathbb{T}/\leftrightarrow$ is a good model for the theory $BPA_{0\delta}$, we first need a couple of lemmas.

3.2.3 LEMMA

Let t be a closed term over $BPA_{0\delta}$ such that $BPA_{0\delta} \nvdash t \neq 0$. Then $BPA_{0\delta} \vdash t = 0$.

PROOF: By induction on the structure of t. If t is a constant, this follows immediately from the axioms in table 1. If t is a sum, say $t = t_1 + t_2$, and $BPA_{0\delta} \nvdash t \neq 0$, it follows from axiom NZ4 that $BPA_{0\delta} \nvdash t_1 \neq 0$ and $BPA_{0\delta} \nvdash t_2 \neq 0$. By induction hypothesis, $BPA_{0\delta} \vdash t_1 = 0$ and $BPA_{0\delta} \vdash t_2 = 0$ and consequently $BPA_{0\delta} \vdash t = 0$. Finally, if t is a product, say $t = t_1 \cdot t_2$, and $BPA_{0\delta} \nvdash t \neq 0$, it follows from axiom NZ3 that either $BPA_{0\delta} \nvdash t_1 \neq 0$ or $BPA_{0\delta} \nvdash t_2 \neq 0$. By induction hypothesis, we obtain either $BPA_{0\delta} \vdash t_1 = 0$ or $BPA_{0\delta} \vdash t_2 = 0$ In the first case, use Z3 and in the second case Z1 to obtain $BPA_{0\delta} \vdash t = 0$.

3.2.4 LEMMA

Let t be a closed term over $BPA_{0\delta}$ such that $BPA_{0\delta} \vdash t \neq 0$. Then t can be written without 0, i.e. there is a closed term s over $BPA_{0\delta}$ that does not contain 0, such that $BPA_{0\delta} \vdash t = s$.

PROOF: By induction on the structure of t. If t is a constant, this follows directly from the axioms in table 1. If t has the form of a product $t_1 \cdot t_2$, it follows by the previous lemma that $t_1 \neq 0$ and $t_2 \neq 0$. Then use induction hypothesis. Finally, if t has the form of a sum $t_1 + t_2$, it is by the previous lemma enough to consider the following four cases.

<u>Case 1:</u> $t_1 \neq 0$, $t_2 \neq 0$. Use induction hypothesis.

Case 2: $t_1 = 0$, $t_2 = 0$. It follows that $t = 0$, contradiction.
Case 3: $t_1 = 0$, $t_2 \neq 0$. It follows that $t = t_2$, and we can use the induction hypothesis for t_2.
Case 4: $t_1 \neq 0$, $t_2 = 0$. Just like case 3.

3.2.3 THEOREM

$BPA_{0\delta}$ is a sound and complete axiomatization of the model $\mathbb{T}/\leftrightarrow$ (for closed terms).

PROOF: First note that the graph model has a substructure consisting of the processes $\neq 0$. That structure is a model of BPA_δ, the theory without zero, and with axioms A1-7 (no conditions on A6 and A7). Thus there is only one process in which 0 features in an essential way and that is 0 itself.

Now soundness can be proved by inspection of the model. To prove completeness, consider two closed terms t,s over $BPA_{0\delta}$ and suppose $\mathbb{T}/\leftrightarrow \vdash t=s$. We use case distinction.
Case 1: $t \neq 0$, $s \neq 0$. By the previous lemma, t and s can be written as t^*, s^* without 0. By soundness $\mathbb{T}/\leftrightarrow \vdash t^* = s^*$. By the completeness theorem for BPA_δ, $BPA_\delta \vdash t^* = s^*$.
Case 2: $t = 0$, $s = 0$. Immediate.
Case 3: $t = 0$, $s \neq 0$. Write s as s^* without 0. The definition of $\leftrightarrow$ shows that $t \leftrightarrow s^*$ cannot hold, contradiction.
Case 4: $t \neq 0$, $s = 0$. Just like case 3.

4. EXTENSIONS

4.1 RENAMING

Now we will look at renamings of atomic actions into atomic actions or δ. (Renaming into zero was discussed in 2.5.) Let $f: A_\delta \rightarrow A_\delta$ be any function that keeps δ fixed, i.e. $f(\delta) = \delta$. Then the **renaming operator** ρ_f is defined by the axioms in table 5, where $a \in A_\delta$.

$\rho_f(0) = 0$
$\rho_f(a) = f(a)$
$\rho_f(x + y) = \rho_f(x) + \rho_f(y)$
$\rho_f(x \cdot y) = \rho_f(x) \cdot \rho_f(y)$

TABLE 5. Renaming

It is straightforward to define the renaming operators on the graph model of 3.2. A very useful example of a renaming operator is the **encapsulation operator** ∂_H (for $H \subseteq A$) that is based on the function g given by:

$$g(a) = \delta \quad \text{if } a \in H \qquad \text{and } g(a) = a \quad \text{if } a \notin H.$$

The composition of an encapsulation operator with the deadlock prevention operator $(0_\delta \circ \partial_H)$ will prevent any action of H from occurring. We will see applications of this in the sequel.

4.2 PARALLEL COMPOSITION

We can extend the theory $BPA_{0\delta}$ with parallel composition as in the theory ACP of BERGSTRA & KLOP [84a]. In order to axiomatize the parallel operator $\|$ (**merge**), we need two auxiliary operators $\mathbin{|\!\!\lfloor}$ **(left-merge)** and $|$ **(communication merge)**. The theory is parametrized by a **communication function** $|$, a binary function on the set of constants $A_{0\delta}$ that satisfies conditions C1-4 in table 8 below. Moreover, we have the encapsulation operator of 4.1, that is used to block communications with the outside. The theory ACP_0 consists of $BPA_{0\delta}$ plus the axioms in table 6 below. In table 6, $a,b,c \in A_{0\delta}$.

$a \mid b = b \mid a$		C1
$(a \mid b) \mid c = a \mid (b \mid c)$		C2
$a \neq 0 \Rightarrow \delta \mid a = \delta$		C3
$0 \mid a = 0$		C4
$x \parallel y = x \mathbin{\mid\!\!\lfloor} y + y \mathbin{\mid\!\!\lfloor} x + x \mid y$		CM1
$a \mathbin{\mid\!\!\lfloor} x = a \cdot x$		CM2
$ax \mathbin{\mid\!\!\lfloor} y = a \cdot (x \parallel y)$		CM3
$(x + y) \mathbin{\mid\!\!\lfloor} z = x \mathbin{\mid\!\!\lfloor} z + y \mathbin{\mid\!\!\lfloor} z$		CM4
$ax \mid b = (a \mid b) \cdot x$		CM5
$a \mid bx = (a \mid b) \cdot x$		CM6
$ax \mid by = (a \mid b) \cdot (x \parallel y)$		CM7
$(x + y) \mid z = x \mid z + y \mid z$		CM8
$x \mid (y + z) = x \mid y + x \mid z$		CM9
$\partial_H(0) = 0$		D0
$\partial_H(a) = a$	if $a \notin H$	D1
$\partial_H(a) = \delta$	if $a \in H$	D2
$\partial_H(x + y) = \partial_H(x) + \partial_H(y)$		D3
$\partial_H(x \cdot y) = \partial_H(x) \cdot \partial_H(y)$		D4

TABLE 6. ACP_0

On the graph model, we can define parallel composition as follows:

- the node set of graph $g \parallel h$ is the cartesian product of the node sets of g and h;
- there is an edge $(s,t) \xrightarrow{a} (s',t)$ iff there is an edge $s \xrightarrow{a} s'$ in g;

likewise, there is an edge $(s,t) \xrightarrow{a} (s,t')$ iff there is an edge $t \xrightarrow{a} t'$ in h;
and there is an edge $(s,t) \xrightarrow{a} (s',t')$ iff there are edges $s \xrightarrow{b} s'$ in g and $t \xrightarrow{c} t'$ in h with $b \mid c = a$;

- an endpoint (s,t) has a 0-label iff either s or t has a 0-label;

an endpoint (s,t) has a δ-label if one has a δ-label, and the other a δ or $\surd$-label;
an endpoint (s,t) has a $\surd$-label if both s and t have a $\surd$-label (all labels in non-endpoints are dropped).

It can be proved that with this definition, bisimulation is also a congruence for parallel composition, and all axioms of ACP_0 hold in the graph model.

4.3 DISCUSSION

Note that we can derive from ACP_0 that for all finite closed terms x we have

$0 \| x = 0 \mathbin{⌊\!⌊} x = x \mathbin{⌊\!⌊} 0 = x \mid 0 = 0 \mid x = 0.$

Now let us consider this for recursively defined processes. Look at the process B that is the non-zero solution of $B = b{\cdot}B$. We calculate: $0 \| B = 0 \mathbin{⌊\!⌊} B + B \mathbin{⌊\!⌊} 0 + 0 \mid B = 0{\cdot}B + bB \mathbin{⌊\!⌊} 0 + bB \mid 0 = 0 + b{\cdot}(B \| 0) + (b \mid 0){\cdot}B = b{\cdot}(B \| 0) = b{\cdot}b{\cdot}(0 \| B)$.

We also have $B = b{\cdot}b{\cdot}B$, so by RSP^0 we have either $0 \| B = B$ or $0 \| B = 0$.

Both options are consistent. Our choice is to put $0 \| B = 0$ and in general $0 \| x = 0$. We can motivate this if we use the *Limit Rule* of BAETEN & BERGSTRA [88]. We describe this rule next.

4.4 LIMIT RULE

Let $FCPE_0$ be the class of finite closed process expressions over ACP_0. Let $p(x_1,...,x_n) = q(x_1,...,x_n)$ be an equation over ACP0. The limit rule (LR) is as follows:

$$\frac{\text{for all } t_1,...,t_n \in FCPE_0 \quad p(t_1,...,t_n) = q(t_1,...,t_n)}{p(x_1,...,x_n) = q(x_1,...,x_n)} \quad \text{LR.}$$

The identity $0 \| x = 0$ follows from LR, because, as remarked above, $0 \| t = 0$ holds for all $t \in FCPE_0$.

4.5 PROJECTION AXIOMS

The limit rule has other applications of equal importance. Consider the projection operator π_n. The following hold for the projection operators: for $x,y \in FCPE_0$

$$\pi_n(x \| y) = \pi_n(\pi_n(x) \| \pi_n(y))$$
$$\pi_n(x{\cdot}y) = \pi_n(x){\cdot}\pi_n(y)$$
$$\pi_n(\partial_H(x)) = \partial_H(\pi_n(x))$$
$$\pi_n(x \mathbin{⌊\!⌊} y) = \pi_n(\pi_n(x) \mathbin{⌊\!⌊} \pi_n(y))$$
$$\pi_n(x \mid y) = \pi_n(\pi_n(x) \mid \pi_n(y)).$$

We call this set of equations EP. It follows from the limit rule that these identities are valid for all processes. Now consider once more the process B from 4.3. Then $\pi_n(0 \| B) = \pi_n(\pi_n(0) \| \pi_n(B)) = \pi_n(0 \| b^n) = 0$. This holds for all n and so by AIP $0 \| B = 0$. More generally, using AIP and EP we can prove that an identity $p(x_1,...,x_n) = q(x_1,...,x_n)$ holds for all recursively specified processes as soon as it holds for all finite processes. The proof proceeds just like in BAETEN & VAN GLABBEEK [87].

Summarizing the discussion, we have that our model satisfied LR, AIP, EP. The logical relationships are: LR $\vdash$ EP, and AIP + EP $\vdash$ LR for recursively specified processes only.

4.6 STATE OPERATOR

We can add a **state operator** λ to the theory along the same lines as in BAETEN & BERGSTRA [88]. The process $\lambda_s(x)$ represents the process x in state s. The state operator is parametrized by two functions **action** and **effect**. We write $a(s)$ for action(a,s) and $s(a)$ for effect(a,s) (a an action, s a state). When an action a is to be executed, $a(s)$ gives the resulting action, and $s(a)$ the resulting state. We will not allow that $a(s) = 0$, further one must assume that $s(0) = s$ and $0(s) = 0$, and $s(\delta) = s$, $\delta(s) = \delta$. Then the state operator works just as well as in the case of ACP. The axioms are displayed in table 7. We have $a \in A_\delta$. It is straightforward to define the state operator on the graph model of 3.2.

$\lambda_s(0) = 0$
$\lambda_s(a) = a(s)$
$\lambda_s(a \cdot x) = a(s) \cdot \lambda_{s(a)}(x)$
$\lambda_s(x + y) = \lambda_s(x) + \lambda_s(y)$

TABLE 7. State operator

4.7. PRIORITIES

In examples in section 5 we will also make use of the **priority operator** of BAETEN, BERGSTRA & KLOP [87]. This operator gives some actions priority over others in a sum context. An auxiliary operator $\lhd$ (**unless**) is needed to give a finite axiomatization. We assume that a partial ordering < is given on A (so 0 and δ are not ordered). Table 8 gives axioms on top of the axioms of ACP_0. We have $a,b \in A_\delta$.

$a \lhd b = a$	if *not* $a<b$
$a \lhd b = \delta$	if $a<b$
$0 \lhd x = 0$	
$x \lhd 0 = x$	
$z \neq 0 \Rightarrow x \lhd yz = x \lhd y$	
$x \lhd (y + z) = (x \lhd y) \lhd z$	
$xy \lhd z = (x \lhd z)y$	
$(x + y) \lhd z = x \lhd z + y \lhd z$	
$\theta(0) = 0$	
$\theta(a) = a$	
$\theta(xy) = \theta(x) \cdot \theta(y)$	
$\theta(x + y) = \theta(x) \lhd y + \theta(y) \lhd x$	

TABLE 8. Priority operator

The priority operator can be defined on the graph model of 3.2 similarly as in BAETEN, BERGSTRA & KLOP [87]: we prune away every branch that splits off at a node where there is a 'brother' edge with higher priority, that leads to a non-zero node.

4.8. WEAVING

In advance of an explanation of how to apply failure prediction in the design of (toy) control systems we will introduce a parallel composition operator that differs from the ACP merge. This operator is called **weaving**, because on trace sets it corresponds exactly to the weaving operator of trace theory, see REM [87]. It is denoted with $x \parallel_B y$ and has in failure semantics the same meaning as the corresponding operator of TCSP, see HOARE [85] from which the notation is taken. Our axioms explain it in terms of bisimulation semantics and therefore in terms of many other abstract semantic models. In table 9 we give an axiomatization on top of

BPA_δ, so *not* considering the extra constant 0. We have $B \subseteq A$, $a,b \in A_\delta$. We can add 0 by putting $0 \mathbin{⌊\!⌊}_B x = x \mathbin{⌊\!⌊}_B 0 = 0 \mid_B x = x \mid_B 0 = 0$.

$x \parallel_B y = x \mathbin{⌊\!⌊}_B y + y \mathbin{⌊\!⌊}_B x + x \mid_B y$	
$a \mathbin{⌊\!⌊}_B x = a \cdot x$	if $a \notin B$
$a \mathbin{⌊\!⌊}_B x = \delta$	if $a \in B$
$(a \cdot x) \mathbin{⌊\!⌊}_B y = a \cdot (x \parallel_B y)$	if $a \notin B$
$(a \cdot x) \mathbin{⌊\!⌊}_B y = \delta$	if $a \in B$
$(x + y) \mathbin{⌊\!⌊}_B z = x \mathbin{⌊\!⌊}_B z + y \mathbin{⌊\!⌊}_B z$	
$a \mid_B b = \delta$	if $a \notin B$ or $a \neq b$
$a \mid_B a = a$	if $a \in B$
$(a \cdot x) \mid_B b = (a \mid_B b) \cdot x$	
$a \mid_B (b \cdot x) = (a \mid_B b) \cdot x$	
$(a \cdot x) \mid_B (b \cdot y) = (a \mid_B b) \cdot (x \parallel_B y)$	
$(x + y) \mid_B z = (x \mid_B z) + (y \mid_B z)$	
$x \mid_B (y + z) = (x \mid_B z) + (y \mid_B z)$	

TABLE 9. Weaving

Weaving is a parallel composition that uses action sharing: the actions named in the subscript B must occur in a shared fashion for both x and y simultaneously. The above equations describe weaving on the bisimulation model. It is possible to describe weaving in terms of the merge of ACP. Then it is necessary to introduce copies of the atomic actions, so let for every $b \in B$, b^c be a new copy different from all other actions in x and y and let c be a renaming function that renames every $b \in B$ to b^c and leaves all other atoms unchanged. As a communication function we have $b^c \mid b^c = b$, all other communications are trivial. Then the following identity holds for all finite closed process expressions:

$$x \parallel_B y = \partial_{Bc}(\rho_c(x) \parallel \rho_c(y)).$$

The reason to have weaving in addition to $\parallel$ of ACP is that in many cases the shared action communication mechanism is quite pleasant and one would prefer not to be burdened with its encoding in terms of the merge operator.

5. APPLICATIONS

5.1. SYSTEMS CONTROL

In order to apply failure prediction we start from a system S that may be operated with actions from a set B, a set of *buttons*. For simplicity we assume that S is perpetual (does not terminate). Every now and then an error e may occur ($e \notin B$). A controller allows the use of S. The functionality of this controller is as follows:

$$C = \sum_{b \in B} instr(b) \cdot H(b) \cdot C,$$

where H(b) is a handler for the instruction b (instr(b) is an atomic action, H(b) need not be). H(b) may or may not perform the action b, meant as an instruction for S. We will choose the following equation for the handler:

$$H(b) = b \cdot done(b) + not(b)$$

The action not(b) denotes a signal from the controller that b may not be performed, the action done(b) is a controller signal indicating that b has successfully been performed. Both these actions are supposed not to occur in any other system component. Thus the external alphabet of the controller is $instr(B) \cup done(B) \cup not(B)$ and none of these actions is supposed to occur in S.

The handler uses a simulation program SIM that simulates the action b as an instruction for S. If this simulation reveals a problem (a predictable failure) then b is not enforced on S, otherwise it will be. We have:

$$SIM = 0_{\delta} \circ \partial_{\{e\}}(S).$$

Let < be the partial ordering of atomic actions that imposes not(b) < b for all $b \in B$ and no other relations. Then the controller together with the simulated system work as follows:

$$C\text{-}SIM = \theta_{<}(C \parallel_B SIM)$$

The system C-SIM allows not(b) if it will not allow b. C-SIM allows b if after b, S can proceed with at least one infinite trace of actions not involving e. Of course, C-SIM must be implemented in a way that does not use the constant 0 or the 'real' system S.

Now, finally, the controller together with the system S is given by:

$$C\text{-}S = C\text{-}SIM \parallel_B S.$$

Due to the nature of the weaving operator, the occurring ternary communication can be described in a very compact way.

5.2. SPECIFYING A PATH THROUGH A COMBINATORIAL EXPLOSION

It is well-known that any NP-complete problem can be solved non-deterministically in polynomial time. Essentially, this is done by non-deterministically guessing a value at each step. We formalize this as follows. Let P be a computable predicate on sequences of length k of natural numbers in the range {1,...,n}. The set of states is $S = \{\langle j,\sigma\rangle : 1 \le j \le k, \sigma$ a sequence of length j from $\{1,...,n\}\}$. We have atomic actions guess(i) (for $1 \le i \le n$). The action and effect functions are as follows:

- $guess(i)(\langle j,\sigma\rangle) = skip$ if $j < k$
- $guess(i)(\langle k,\sigma\rangle) = exit$ if $P(\sigma)$
- $guess(i)(\langle k,\sigma\rangle) = \delta$ if $\neg P(\sigma)$
- $\langle j,\sigma\rangle(guess(i)) = \langle j+1, \sigma * i\rangle$ if $j < k$
- $\langle k,\sigma\rangle(guess(i)) = \langle k, \sigma\rangle$
- all other actions a are inert, i.e. a(s) = a and s(a) = s for all states s.

Now define the process Q,R by:

$$Q = \lambda_{\langle 0,\varepsilon\rangle}\Big(\big(\sum_{i=1}^{n} guess(i)\big)^{k+1}\Big), \quad R = 0_{\delta}(Q).$$

R will equal 0 iff no sequence σ with P(σ) exists; otherwise, a sequence of length k will be accepted by R such that P holds.

5.3. TRAFFIC LIGHT

Let P be a point that travels on a one dimensional two way infinite discrete grid (i.e. the integers). At each moment in time the coordinates of the point are an integer pair (p, v) where p is the position on the grid and v is an integer denoting the velocity of P: if v = -3 this means that in one unit of time (say a second) P moves from p to p - 3. There are three actions for P and one of these is performed each second:

st	remain in the same state (keep the same speed in the same direction).
la	accelerate left: $v \to v - 1$,
ra	accelerate right: $v \to v + 1$.

Thus $P = (st + la + ra) \cdot tick \cdot P$ where tick marks the progress of a clock.

At the same time, there is a traffic light at position 10 on the grid. Every 3 seconds this light changes its colour, from green to red and back again:

$$TL = green \cdot tick \cdot tick \cdot tick \cdot red \cdot tick \cdot tick \cdot tick \cdot TL.$$

Here tick marks the progress of the same clock as for the moving object P.

We require the communication $tick \mid tick = t$. The composition of object and traffic light is $\partial_{\{tick\}}(P \parallel TL)$. The next step is that we have a state operator with triples consisting of an integer pair and a colour as states. The functions action and effect work as follows (p,v integers, c a colour):

effect	action
(p, v, c)(st) = (p, v, c)	st(p, v, c) = st
(p, v, c)(la) = (p, v-1, c)	la(p, v, c) = la
(p, v, c)(ra) = (p, v+1, c)	ra(p, v, c) = ra
(p, v, c)(red) = (p, v , red)	red(p, v, c) = red
(p, v, c)(green) = (p, v , green)	green(p, v, c) = green
(p, v, c)(t) = (p + v, v, c)	t(p, v, c) = δ if c = red & v>0 & $p \le 10 \le p+v$
	t(p, v, c) = t otherwise.

Thus, for instance the second line of this table says that if action la is performed in state (p,v,c), then we see the action la occurring, and the resulting state is (p,v-1,c).

The process $PTL = \lambda_{(0,\,0,\,green)}(\partial_{\{tick\}}(P \parallel TL))$ describes P starting in the position (0, 0) with the constraint that a deadlock occurs if P crosses the traffic light from left to right if it is green.

Next the process $PTLC = 0_\delta(PTL)$ describes P under the constraint that it will never cross the traffic light in red state from left to right. (C denotes correct functioning of the PTL combination).

Using the operator 0_δ it becomes possible to view all possible ways of correct behaviour as a process itself. Notice that if we view st, la, ra as control options for an agent that controls P then controlling P in the context PTL leaves the controlling agent all freedom of action (choice from st, la, ra at any moment). In contrast to this, the freedom of control in the context PTLC is limited.

We give examples of applying 0_δ in various states of PTL. For instance:

$0_\delta(\lambda_{(2,\,5,\,green)}(\partial_{\{tick\}}(P \parallel red \cdot tick \cdot tick \cdot tick \cdot TL))) = 0$, but for no v we have

$0_\delta(\lambda_{(11,\, v,\, red)}(\partial_{\{tick\}}(P \parallel TL))) = 0.$

Of course this is just a toy example but one may imagine a more complex control system for which disastrous events have to be avoided. Then the freedom of a controlling agent has to be limited in order to avoid problems. Using the operation 0_δ it becomes possible to specify a control system that disallows actions that must inevitably lead to a problematic stage (i.e. 0). Of course the implementation of such a control system is quite a different matter. Already in the simple case with moving point and traffic light above, a specification of PTLC without the use of 0 is not so straightforward.

Using 0, one may cut down a process graph to correct (failure free) process executions only.

In terms of a control system as described in 5.1 we get the following:

$B = \{la, ra, st\}$
$S = P$
$SIM = PTLC$
$C\text{-}SIM = \theta_<(C \parallel_B SIM)$
$C\text{-}S = C \parallel_B P.$

6. CONCLUDING REMARKS

6.1. RELATION BETWEEN ACP_0 AND ACP

ACP_0 is a generalization of ACP. The mechanism of generalization can be compared to the case in which one takes the positive rational numbers which combine a multiplicative group structure and an additive semi-group and adds 0 to it. One adds a single object and several laws become invalid. (Let us assume that one has defined $p / 0$ as 1 in order to avoid partial functions.)

6.2 EFFECTIVE COMPUTABILITY

The reason not to have 0 as a member of the core system ACP is that it is not effectively computable. That is to say that if we have a finite guarded recursive specification of a process X over ACP_0, it may be impossible to compute its finite projections π_n in a uniform way. The central axiom systems BPA, PA, ACP and its extensions in concrete process algebra all have the property that finite projections of finitely recursively specified processes can be determined in a uniform mechanical way. This simply means that ACP and its extensions in concrete process algebra can be viewed as an executable programming language. This is why we propose not to consider 0 a part of concrete process algebra (just as the empty step ε and the silent step τ are not part of concrete process algebra).

6.3 IMPLEMENTATION

Implementation of a recursive ACP0 specification first of all involves an elimination of 0. Now it must be noticed that interesting use of 0 happens just in those cases where elimination of 0 is possible but at a very high cost. In the examples 5.1, 5.2 and 5.3 this elimination is possible if the state operator is allowed. Elimination of 0 is also possible if in addition to ACP abstraction (τ_I) may be used.

In all of these examples it is not known to us whether an equivalent specification in ACP can be given (i.e. whether the state operator or abstraction operator are necessary strengthenings for an elimination of 0 and 0_δ).

6.4 RELATED WORK

In MILNER [89], a process 0 is introduced that replaces the constant NIL of CCS of MILNER [80]. This is just a notational matter and does not introduce semantic modifications as such. Nevertheless the notation differs from ours considerably in the sense that Milner's 0 definitely corresponds to our constant δ and not to our constant 0. Similarly the constant STOP of TCSP of OLDEROG & HOARE [86] corresponds to δ and not to (our) 0. We use δ because that makes the notation consistent with other papers about ACP (e.g. BERGSTRA & KLOP [89]). Because 0 is more truly a zero in process algebra than δ we preferred not to adapt our notation to the notation of Milner.

Of course Milner's restriction must be compared to our encapsulation operator and not to a substitution of (our) 0 for some actions. Thus $x / \{a, b\}$ in the notation of MILNER [80] corresponds to $\partial_{\{a, b\}}(x)$ in the case of ACP.

ACKNOWLEDGEMENTS

Hans Mulder and Sjouke Mauw have contributed to this paper through several critical remarks including the view that an undisputable zero in ACP must satisfy the law $x \cdot 0 = 0$ and not just $a \cdot 0 = 0$ for atomic actions a. In addition, the second author acknowledges extended discussions with C.A.R. Hoare about various aspects of this paper. We thank the referees for their careful review and many helpful comments.

REFERENCES

J.C.M. BAETEN & J.A. BERGSTRA [88], *Global renaming operators in concrete process algebra,* I&C 78, 1988, pp. 205-245.

J.C.M. BAETEN & J.A. BERGSTRA [90], *Process algebra with zero object and non-determinacy,* report P9002, Programming Research Group, University of Amsterdam 1990.

J.C.M. BAETEN & R.J. VAN GLABBEEK [87], *Merge and termination in process algebra,* in: Proc. 7th FST&TCS, Pune (K.V. Nori, ed.), Springer LNCS 287, 1987, pp. 153-172.

J.C.M. BAETEN, J.A. BERGSTRA & J.W. KLOP [86], *Syntax and defining equations for an interrupt mechanism in process algebra,* Fund. Inf. IX, 1986, pp. 127-168.

J.A. BERGSTRA & J.W. KLOP [84a], *Process algebra for synchronous communication,* I&C 60, 1984, pp. 109-137.

J.A. BERGSTRA & J.W. KLOP [84b], *The algebra of recursively defined processes and the algebra of regular processes,* in: Proc. 11th ICALP, Antwerpen (J. Paredaens, ed.), Springer LNCS 172, 1984, pp. 82-95.

J.A. BERGSTRA & J.W. KLOP [85], *Algebra of communicating processes with abstraction,* TCS 37, 1985, pp. 77-121.

J.A. BERGSTRA & J.W. KLOP [86], *Process algebra: specification and verification in bisimulation semantics,* in: Math. & Comp. Sci. II (M. Hazewinkel, J.K. Lenstra & L.G.L.T. Meertens, eds.), CWI Monograph 4, North-Holland, Amsterdam, 1986, pp. 61-94.

J.A. BERGSTRA & J.W. KLOP [89], *Process theory based on bisimulation semantics,* in: Linear Time, Branching Time and Partial Order in Logics and Models for Concurrency (J.W. de Bakker, W.-P. de Roever & G. Rozenberg, eds.), Springer LNCS 354, 1989, pp. 50-122.

R.J. VAN GLABBEEK [87], *Bounded nondeterminism and the approximation induction principle in process algebra,* in: Proc. STACS 87 (F.J. Brandenburg, G. Vidal-Naquet & M. Wirsing, eds.), Springer LNCS 247, 1987, pp. 336-347.

J.F. GROOTE & F.W. VAANDRAGER [89], *Structured operational semantics and bisimulation as a congruence,* extended abstract in: Proc. ICALP 89, Stresa (G. Ausiello, M. Dezani-Ciancaglini & S. Ronchi Della Rocca, eds.), Springer LNCS 372, 1989, pp. 423-438. Full version to appear in I&C.

J.F. GROOTE [90], *Transition system specifications with negative premises,* report CS-R8950, Centre for Math. & Comp. Sci. 1990. To appear in Proc. CONCUR'90, Springer LNCS.

C.A.R. HOARE [85], *Communicating sequential processes,* Prentice Hall International, 1985.

R. MILNER [80], *A calculus for communicating systems,* Springer LNCS 92, 1980.

R. MILNER [89], *Communication and concurrency,* Prentice Hall International, 1989.

E.-R. OLDEROG & C.A.R. HOARE [86], *Specification-oriented semantics for communicating processes,* Acta Informatica 23, 1986, pp. 9-66.

A. PONSE & F.-J. DE VRIES [89], *Strong completeness for Hoare logics of recursive processes: an infinitary approach,* report CS-R8957, Centre for Math. & Comp. Sci., Amsterdam 1989.

M. REM [87], *Trace theory and systolic computations,* in: Proc. PARLE Vol. I (J.W. de Bakker, A.J. Nijman & P.C. Treleaven, eds.), Springer LNCS 258, 1987, pp. 14-33.

On the Asynchronous Nature of Communication in Concurrent Logic Languages: a Fully Abstract Model based on Sequences *

Frank S. de Boer[1] Catuscia Palamidessi[2]

[1]Technische Universiteit Eindhoven
P.O. Box 513, 5600 MB Eindhoven, The Netherlands

[2]Dipartimento di Informatica, Università di Pisa
Corso Italia 40, 56125 Pisa, Italy

Abstract

The main contribution of this paper is to show that the nature of the communication mechanism of concurrent logic languages is essentially different from imperative concurrent languages. We show this by defining a compositional model based on sequences of input-output substitutions. This is to be contrasted with the compositionality in languages like CCS and TCSP, which requires more complicated structures, like trees and failure sets. Moreover, we prove that this model is fully abstract, namely that the information encoded by these sequences is necessary.

Regarding fully abstractness, our observation criterium consists of all the possible finite results, namely the computed answer substitution together with the termination mode (success, failure, or suspension). The operations we consider are parallel composition of goals and disjoint union of programs. We define a compositional operational semantics delivering sequences of input-output substitutions. Starting from this we obtain a fully abstract denotational semantics by requiring some closure conditions on sequences, that essentially model the monotonic nature of communication in concurrent logic languages. The correctness of this model is proved by refining the operational semantics in order to embody these closure conditions.

Key words and phrases: operational semantics, denotational semantics, concurrent logic languages, substitutions, sequences, compositionality fully abstractness.
1985 Mathematics Subject Classification: 68Q55, 68Q10.
1987 Computing Reviews Categories: D.1.3, D.3.1, F.1.2, F.3.2.

*Part of this work was supported by the ESPRIT BRA projects "Integration" and "SPEC".

1 Introduction

Compositionality is considered one of the most desirable characteristics of a formal semantics, since it provides a foundation of program verification and modular design. The difficulty in obtaining this property depends upon the *operators* for composing programs, the behaviour we want to describe (*observables*), and the degree of *abstractness* we want to reach. A compositional model is called *fully abstract* (with respect to some operators and observables) if it identifies programs that behave in the same way under all the possible contexts. A fully abstract model can be considered to be *the* semantics of a language: all the other compositional semantics can be reduced to it by abstracting from the redundant information. Fully abstractness is important, for instance, for deciding correctness of program transformation techniques. If a fully abstract model distinguishes the transformed program from the original one then the transformation is not correct (in the sense that it does not preserve the same behaviour under composition).

In the field of logic languages there are basically two operators for composing programs: the conjunction of goals and the union of clauses. The observables are usually related to the finite result: success, failure, and computed answer substitutions. For Concurrent Logic Languages compositionality has been studied mainly with respect to the conjunction of goals, whilst union of clauses has been considered only in simple cases (union of *disjoint* programs [GCLS88], and union of *nicely intersecting* programs [GMS89]). This is rather natural since in a concurrent framework the main operation is the parallel composition of processes. On the other hand, the class of observables has to be enriched by *suspension* (or *deadlock*).

The main problem of compositionality in concurrent languages is the description of deadlock behaviour. For languages like CCS and TCSP it is well-known that sequences are not sufficient, and that, on the other hand, trees contain too much control information to be fully abstract. In order to abstract from redundant branching information encoded by the tree structure different approaches have been proposed. The most well known consist of defining an appropriate equivalence relation on trees (see, for instance, bisimulation [Mil80]), and of grouping the *branching information* in *failure sets* [BHR84]. In general, failure set semantics is more abstract than bisimulation and it is proved to be fully abstract in the case of TCSP.

Until now, with respect to compositionality, Concurrent Logic Languages have been regarded just as a particular case of the classic paradigms. Therefore, the problem has been approached by the standard methods. De Bakker and Kok ([dBK88], [Kok88]) and De Boer et. al. ([dBKPR89a], [dBKPR89b]) use tree-like structures labeled with functions on substitutions. More simple tree-like structures, labeled by constraints, are used by Gabbrielli and Levi ([GL90]) and by Saraswat and Rinard ([SR89]). In [GCLS88] and in [GMS89] the authors approach the problem of fully abstractness by refining the failure set semantics of TCSP.

Let us try to argue why we think that Concurrent Logic Languages require a completely different approach. The communication mechanism in Concurrent Logic Languages is based on the *production* and *consumption* of bindings (substitutions) on shared variables. We can translate a CCS process by interpreting the action a as the production of a binding on a variable x_a and the complementary action $\bar{a}$ (in CCS parallel processes synchronize on complementary actions) as the consumption of this binding. The main difference is that the behaviour of complementary actions is not symmetrical as it is in CCS and in TCSP. Indeed, the production of a binding can always proceed, whilst the consumption has to wait. In other terms, the communication mechanism of concurrent logic languages is intrinsically *asynchronous*. The following example shows that this leads to an essentially different deadlock behaviour [1].

Example 1.1 *Let* $p_1 ::= \bar{a}\bar{b} + \bar{a}\bar{c} + \bar{a}\bar{d}$ *and* $p_2 ::= \bar{a}\bar{b} + \bar{a}(\bar{c} + \bar{d})$. *The failure set semantics distinguishes these two processes. Indeed, in case of synchronous communication, they behave differently under the context* $p ::= a(b+c)$. *The process* p_1 *can deadlock, by choosing the third branch, whilst* p_2 *cannot. In the case of asynchronous communication however, both processes have the same deadlock behaviour. The process* p_2 *can now deadlock by choosing the second branch, because* p *can independently decide to produce* b *(after* a*). In the formalism of concurrent logic languages, this example can be translated as follows (we use here the syntax of [GMS89].* $ask(t = u)$ *represents the consumption of a substitution satisfying the equation* $t = u$, *whilst* $tell(t = u)$ *represents the production of the most general substitution satisfying* $t = u$, *and* $|$ *is the commit operator). Figure 1 illustrates this example.*

$$\begin{array}{l}\{\ p_1(x_a, x_b, x_c, x_d) : -ask(x_a = a) \mid ask(x_b = b) \\ \quad p_1(x_a, x_b, x_c, x_d) : -ask(x_a = a) \mid ask(x_c = c) \\ \quad p_1(x_a, x_b, x_c, x_d) : -ask(x_a = a) \mid ask(x_d = d)\ \ \}\end{array}$$

$$\begin{array}{l}\{\ p_2(x_a, x_b, x_c, x_d) : -ask(x_a = a) \mid ask(x_b = b) \\ \quad p_2(x_a, x_b, x_c, x_d) : -ask(x_a = a) \mid p_3(x_c, x_d) \\ \quad p_3(x_c, x_d) : -ask(x_c = c) \mid \\ \quad p_3(x_c, x_d) : -ask(x_d = d) \mid \qquad\qquad\qquad \}\end{array}$$

$$\begin{array}{l}\{\ p(x_a, x_b, x_c, x_d) : -tell(x_a = a) \mid p'(x_b, x_c) \\ \quad p'(x_b, x_c) : -tell(x_b = b) \mid \\ \quad p'(x_b, x_c) : -tell(x_c = c) \mid \qquad\qquad \}\end{array}$$

This example indicates that in the asynchronous case the failure set semantics (at least as it is defined in [BHR84]) is not abstract enough. The essence of this redundancy relies in the following observation:

Example 1.2 *In the asynchronous case,* $p_1 ::= a(b + c)$ *is equivalent to* $p_2 ::= ab + ac$ *under every context. This is due to the fact that the choice present in* p_1 *does not depend upon the environment. After the production of* a, p_1 *can proceed to produce either* b *or* c *in the same way as* p_2.

This example may induce one to believe that simple sequences of bindings are sufficient for obtaining compositionality. However this is not true in general, because of the different behaviour of the complementary (consuming) case. Consider the following example:

[1] The term *deadlock* is used here with its classical meaning in the theory of concurrency. In concurrent logic programming, this kind of *deadlock* can correspond both to *failure* or to *suspension* (whilst, in the current terminology, it is associated only to *suspension*). In example 1.1 it correspond to *suspension*.

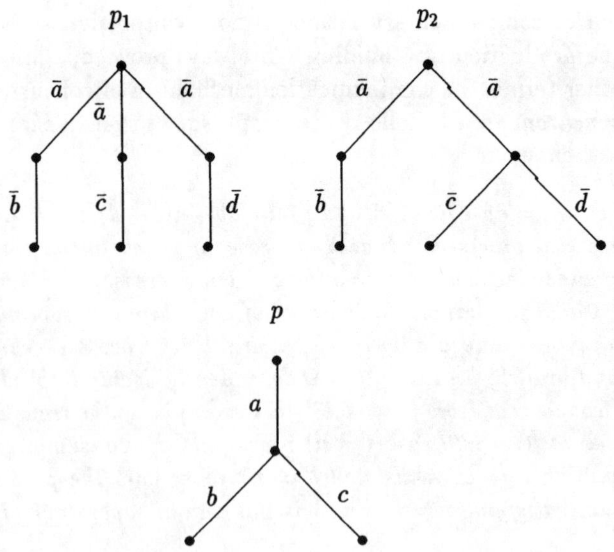

Figure 1: In logic programming p_1 and p_2 cannot be distinguished by p

Example 1.3 *The process* $p_1 ::= \bar{a}(\bar{b} + \bar{c})$ *is not equivalent to* $p_2 ::= \bar{a}\bar{b} + \bar{a}\bar{c}$. *They ar distinguished by the context* $p := ab$ *(*p_2 *can deadlock whilst* p_1 *cannot). This is due to th fact that the choice present in* p_1 *does depend upon the environment. This dependency is however, of a different nature from the synchronous case. Here it is related only to th past behaviour of the environment, i.e., to the bindings that have already been produced.*

This last remark indicates a possible way to solve the problem. Given a sequenc of substitutions representing the computation of a process with respect to an arbitrar environment, we add the information about who is the producer of each substitution the process or its environment. If the substitution obtained from such a sequence doe not provide the process with the necessary information to proceed then it will deadloc *assuming* that the environment does not produce any bindings anymore. The compositio of different processes then simply amounts to verifying that the assumptions made by on process about its environment are indeed validated by the other processes.

We will show that we can describe compositionally the behaviour of a process by mean of a Plotkin-like transition system, labeled by input-output substitutions. The behaviou of the possible environments are modeled by transitions labeled with input substitutions This kind of transition step does not occur in the usual transition system for CCS, an it allows here to obtain directly a compositional operational semantics based on (sets of sequences. These sequences are essentially different from the *scenarios* of [Sar85], wher input substitutions correspond to assumptions about the environment which are necessar for the process to proceed. As a consequence, compositionality is there obtained only fo the success set. The input-output sequences we use have been introduced in [GCLS88] a one component of the domain of the denotational semantics, the other ingredient bein the suspension set. Because of what is stated above, this first component would hav already been sufficient to define there a compositional (hence denotational) semantics.

The language described in [GCLS88] contains non-monotonic test predicates. However the real intricacies of the asynchronous and declarative nature of communication in logic languages come to surface in the monotonic case. Here even the sequences contain too much information about the particular order of production of bindings, information that cannot be sensed by monotonic contexts. This is mainly due to the fact that monotonic contexts cannot be specified to ask (only) a specific amount of bindings, they can always proceed when more bindings are provided. We will define therefore a refinement of the transition system, based on some additional steps that allow to abstract from the particular order of productions by buffering them in a kind of *active* store. Active is meant here in the sense that it can produce bindings itself. The monotonic case has also been investigated in [GMS89]. However the model presented in that paper is based on a refinement of failure set semantics, via some equivalence relation based on simulation of sequences.

In our paper we give a fully abstract semantics for a class of Flat Concurrent Logic Languages. This class will be denoted by HC($\mathcal{A},\mathcal{T}$), where HC stands for Horn Clauses and $\mathcal{A},\mathcal{T}$ are parameters which denote the set of primitives that can occur in the guards. We consider the following possibilities: $\mathcal{A} = \emptyset$ or $\mathcal{A} = Ask$, where

$$Ask = \{ask(E) \mid E \textit{ is a set of equations in the Herbrand universe}\}$$

and $\mathcal{T} = \emptyset$ or $\mathcal{T} = Tell$, where

$$Tell = \{tell(E) \mid E \textit{ is a set of equations in the Herbrand universe}\}.$$

This class includes, for instance, Oc [Hir87] (HC($\emptyset,\emptyset$)), Flat GHC [Ued88] (HC($Ask,\emptyset$)), ccH [GL90] and the language of [GMS89](HC(Ask, $Tell$)), that is a refined version of Flat CP [Sha89]. HC($\mathcal{A},\mathcal{T}$) can be seen as a particular instance of the cc/Herbrand framework [SR89]. For example, HC($Ask,\emptyset$) corresponds to Eventual Herbrand and HC(Ask, $Tell$) corresponds to atomic Herbrand.

We are mainly concerned with the above three instances of HC($\mathcal{A},\mathcal{T}$). Observe that HC($\emptyset,\emptyset$) is included in HC($Ask,\emptyset$) and HC($Ask,\emptyset$) is included in HC(Ask, $Tell$). A model that is compositional for a language is compositional also for the sublanguages, but not vice-versa. On the other side, with respect to fully abstractness the situation is just the reverse. Therefore each case must be considered separately. However we show that the fully abstract compositional model is the same for all these languages.

The plan of the paper is the following: In the next section we introduce some basic notions. In section 3 we give the syntax and an informal operational semantics of the class of languages we study. Then we present a compositional semantics based on a Plotkin-like labeled transition system. In the last section we present a refinement of this semantics which is fully abstract. Due to lack of space we omit the proofs which can be found in the full paper.

2 Preliminaries

In this section we briefly recall some basic notions and results about substitutions and unification. We will use (mainly) the terminology of [Apt88], [Llo87], [LMM88], and [Ede85], to which we refer for a more detailed presentation of these topics.

2.1 Substitutions

Let *Var* be a non empty set of *variables*, with typical elements $x, y, z, \ldots$. Let *Cons* be a set of *constructors*, with typical elements $a, b, c, \ldots$ (constructors with 0 arguments, or *constants*), and $f, g, h, \ldots$ (constructors with one or more arguments). We assume the presence of at least one constant.

The set *Term*, with typical elements $t, u, \ldots$, is the set of all the *terms* built on *Var* and on *Cons*. Examples of terms are $f(a), f(x), g(f(a), f(x)), \ldots$. The set of variables occurring in t is denoted by $var(t)$.

The set of *substitutions*, *Subst*, with typical elements $\vartheta, \sigma, \ldots$, consists of all the mappings ϑ from *Var* into *Term* such that the domain of ϑ, $dom(\vartheta) = \{x \in Var \mid \vartheta(x) \neq x\}$, is finite. We will use the set-theoretic notation $\{x/t \mid x \in dom(\vartheta) \wedge \vartheta(x) \equiv t\}$ to represent ϑ.

The *application* $t\vartheta$ of ϑ to t is defined as the term obtained by replacing each variable x in t by $\vartheta(x)$. The set $ran(\vartheta)$ (range of ϑ) is defined as $\bigcup_{x \in dom(\vartheta)} var(\vartheta(x))$. The substitution ϑ is *ground* iff $ran(\vartheta) = \emptyset$.

The *composition* $\vartheta\vartheta'$ of ϑ and ϑ' is defined by $(\vartheta\vartheta')(x) \equiv (\vartheta(x))\vartheta'$. We recall that the composition is associative, the empty substitution ϵ is the neutral element, and, for each term t, $t(\vartheta\vartheta') \equiv (t\vartheta)\vartheta'$ holds.

The restriction $\vartheta_{|V}$ of ϑ to a set of variables V is the substitution $\vartheta_{|V}(x) = \vartheta(x)$ for $x \in V$ and $\vartheta_{|V}(x) = x$ otherwise. We will abbreviate $\vartheta_{|var(\bar{A})}$ to $\vartheta_{|\bar{A}}$.

2.2 Unification

The notion of *unification* can be given, equivalently, either with respect to a set of sets of terms [Ede85, Llo87], or with respect to a set of equations [LMM88, Apt88]. We choose the second approach.

An *equation* is an expression of the form $t = u$, where t and u are terms. The set of sets of equations, with typical element E, will be denoted by *Eqn*. A set of equations E is *unifiable* iff there exists ϑ such that for all $t = u$ in E, $t\vartheta \equiv u\vartheta$ holds, where $\equiv$ is the syntactic identity on terms. ϑ is then called an *unifier* of E. The set of all the unifiers of E will be denoted by $unif(E)$. The set of all the *most general unifiers* of E is given by $mgu(E) = \{\vartheta \in unif(E) \mid \forall \vartheta' \in unif(E).\ \vartheta \leq \vartheta'\}$. It is well known that all the most general unifiers of a set E are equivalent, and this explains why, in the literature, we often find the terminology "*the* mgu of E".

Given a substitution ϑ the associated set of equations will be denoted by $\mathcal{E}(\vartheta)$.

2.3 Constraints on the Herbrand Universe

A constraint system is essentially a system of partial information that supports the notion of consistency and entailment. The notion of *constraint* was introduced in Logic Programming by Jaffar and Lassez [JL87]. Maher [Mah87] suggested the use of constraints to model logically the synchronization and communication mechanisms of concurrent logic languages. We restrict here to a particular class of constraints: the *ask* and *tell* constraints on the Herbrand universe [Sar89]. They are of the form $ask(E)$ and $tell(E)$ respectively where E is a set of equations. A constraint of the form $ask(E)$ or $tell(E)$ is *solved* by a substitution ϑ if $\vartheta = mgu(E)$.

We use the following notations: $Ask = \{ask(E) \mid E \in Eqn\}$ and $Tell = \{tell(E) \mid E \in Eqn\}$.

3 The class HC($\mathcal{A}, \mathcal{T}$). Syntax and informal operational semantics

The (parametric) sets $\mathcal{A}$ and $\mathcal{T}$ specify the set of *primitives* used in the guards. We restrict here to the following cases: $\mathcal{A} = \emptyset$ or $\mathcal{A} = Ask$, and $\mathcal{T} = \emptyset$ or $\mathcal{T} = Tell$.

Let *Pred* be a set of *predicates*, with typical elements $p, q, r, \ldots$. The set *Atom*, with typical elements $A, B, H \ldots$, is the set of all the *atomic formulas* built on *Pred* and on *Term*, and of the primitives of the form $tell(E)$. In the following, we use the abbreviations $\bar{A}, \bar{x}$ etc. to denote conjunctions of atoms, sequences of variables etc.

A *clause* C of HC($\mathcal{A}, \mathcal{T}$) is a formula of the form $H \leftarrow g_1{:}g_2 \mid \bar{B}$, where $g_1 \in \mathcal{A}$ and $g_2 \in \mathcal{T}$. The atom H is the head, $g_1{:}g_2$ is the guard and $\bar{B}$, a multiset of atoms, is the body of the clause. A *goal statement* G is a multiset of atoms $\bar{A}$. The empty goal statement is denoted by $\Box$. A HC($\mathcal{A}, \mathcal{T}$) program P is a (finite) set of clauses. An atom A in a goal G is seen as an (AND-) process. If A is of the form $tell(E)$ then:

- if E is unifiable then A *succeeds* producing the substitution $\vartheta = mgu(E)$, and the remaining goal $G \setminus A$ is instantiated by ϑ.
- if E is not unifiable then A and the whole goal *fails*

If A is an atomic formula, its computation proceeds by looking for a candidate clause in W.

Definition 3.1 *Let A be an atomic formula and let $C \equiv H \leftarrow g_1{:}g_2 \mid \bar{B}$ be a clause in P, renamed in order to have no variables in common with A. Then the clause C*

- is candidate *if*
  1. $\exists mgu(H = A) = \vartheta$,
  2. $\vartheta_{|A} = \epsilon$
  3. $\exists \vartheta_1$ *such that* $g_1\vartheta$ *is solved by* ϑ_1,
  4. $\vartheta_{1|A} = \epsilon$,
  5. $\exists \vartheta_2$ *such that* $g_2\sigma$ *is solved by* ϑ_2, *where* $\sigma = \vartheta\vartheta_1$.
- suspends *if (1) is satisfied but (2) is not satisfied, or (1), (2) and (3) are satisfied but (4) is not satisfied.*
- fails *in all the other cases.*

If there are candidate clauses, then the computation of A *commits* to one of them (i.e. no backtracking will take place), A is replaced by $\bar{B}\sigma\vartheta_2$, and the rest of the goal is instantiated by ϑ_2. If all the clauses for A fail, then A and the whole goal fail. If no clauses are candidate and at least one clause suspends, then A *suspends*. A can be *resumed* when its arguments get bound by other processes in the goal. If all the processes in the goal suspend, then the goal suspends.

4 A compositional operational semantics for HC($\mathcal{A}$, $\mathcal{T}$

To define the meaning of a goal in terms of its subgoals, we describe the behaviour of a goal as a sequence of *interactions* with its environment. Interactions are modeled as *input/output* substitutions. An input substitution is provided by the environment, whereas an output substitution is produced by the goal itself.

Definition 4.1

- *The set of input substitutions is* $Subst_I = \{\vartheta^I \mid \vartheta \in Subst\}$.
- *The set of output substitutions is* $Subst_O = \{\vartheta^O \mid \vartheta \in Subst\}$.
- *The set of input/output substitutions, with typical element* ϑ^ℓ, *is* $Subst_{IO} = Subst_I \cup Subst_O$.

The operational semantics we define is based on a transition system T labeled on $Subst_{IO}$. The *configurations* of T are pairs consisting of a goal and an infinite set of fresh variables (representing the variables still available for the renaming mechanism). To obtain a compositional operational semantics we need a compositional renaming mechanism. We propose the following (formal) solution. Let $\mathcal{P}_i(Var)$ be the set of all the infinite sets of variables. We assume the existence of a *partitioning* function $Part : \mathcal{P}_i(Var) \rightarrow \mathcal{P}_i(Var) \times \mathcal{P}_i(Var)$ such that

$$\forall V \in \mathcal{P}_i(Var)\ [\langle V_1, V_2\rangle = Part(V) \Rightarrow (V_1 \cup V_2 \subseteq V \ \wedge\ V_1 \cap V_2 = \emptyset)]$$

In this way we can split the "available variables" among the subgoals so to avoid clash of variables among the subcomputations.

Table 1 gives the rules for T describing the "successful" computation steps. We call them *computation rules*.

Table 1: The Transition System T. Computation Rules

C1	$\langle A, V\rangle \xrightarrow{\vartheta_2^O} \langle \bar{B}\sigma\vartheta_2, W\rangle$ **if** $\exists C \in P_V$ [C is candidate] $\wedge\ W = V \setminus var(P_V)$
C2	$\langle tell(E), V\rangle \xrightarrow{\vartheta^O} \langle \Box, V\rangle$ **where** $\vartheta = mgu(E)$
C3	$\langle A, V\rangle \xrightarrow{\vartheta^I} \langle A\vartheta, V\rangle$ **where** A is either an atomic formula or $tell(E)$
C4	$\dfrac{\langle \bar{A}_1, V_1\rangle \xrightarrow{\vartheta^I} \langle \bar{B}_1, W_1\rangle \quad \langle \bar{A}_2, V_2\rangle \xrightarrow{\vartheta^\ell} \langle \bar{B}_2, W_2\rangle}{\langle \bar{A}_1, \bar{A}_2, V\rangle \xrightarrow{\vartheta^\ell} \langle \bar{B}_1, \bar{B}_2, W_1 \cup W_2\rangle}$ **where** $\langle V_1, V_2\rangle = Part(V)$

The rule **C1** describes the normal atomic reduction step. The symbols $C, \bar{B}, \sigma$, and ϑ_2 are the ones of definition 3.1, whilst P_V denotes the program P renamed with the variables in V.

The rule **C2** describes the output of the substitution satisfying an atom of the form *ell*(E). Note that in the case of HC$(\mathcal{A}, \emptyset)$ this is the only rule that allows to produce indings on the (shared) variables of the goal.

The rule **C3** describes the input of a substitution produced by the external environ-nent. Note that ϑ is any substitution, namely it represent a *free assumption* on the nvironment. In other words, it does not depend upon what the goal needs to proceed, nd whether or not it will fail after this input. This is the point in which our transition ystem essentially differs from the semantics described in [Sar85] and in [GMS89], and his is why our sequences contain information enough to achieve compositionality.

Finally, **C4** describes the transition of a goal in terms of the transitions of the subgoals. Note that this is the rule that allows to check that the input assumptions really correspond o the outputs done by the environment.

Table 2 and table 3 illustrate the rules for failure and suspension respectively. We need o introduce in our configurations the symbols *fail* and *susp*, with the obvious meaning.

Table 2: The Transition System T. Failure Rules

F1 $\langle A, V\rangle \xrightarrow{\epsilon^O} \langle fail, V\rangle$ **if** $\forall C \in P_V$ [C fails]

F2 $\langle tell(E), V\rangle \xrightarrow{\epsilon^O} \langle fail, V\rangle$ **if** $\nexists mgu(E)$

F3 $$\frac{\langle A, V\rangle \xrightarrow{\epsilon^O} \langle fail, V\rangle}{\langle A, \bar{B}, V \xrightarrow{\epsilon^O} \langle fail, V\rangle}$$ **where** A is either an atomic formula or $tell(E)$

Table 3: The Transition System T. Suspension Rules

S1 $\langle A, V\rangle \xrightarrow{\epsilon^O} \langle susp, V\rangle$ **if** $\exists C \in P_V$ [C suspends] $\wedge$ $\forall C \in P_V$ [C is not candidate]

S2 $$\frac{\langle \bar{A}_1, V_1\rangle \xrightarrow{\epsilon^O} \langle susp, V_1\rangle \quad \langle \bar{A}_2, V_2\rangle \xrightarrow{\epsilon^O} \langle susp, V_2\rangle}{\langle \bar{A}_1, \bar{A}_2, V\rangle \xrightarrow{\epsilon^O} \langle susp, V_1 \cup V_2\rangle}$$ **where** $\langle V_1, V_2\rangle = Part(V)$

The operational semantics $\mathcal{O}$ based on this transition system T delivers sets of se-uences s of input/output substitutions, ended by a *termination mode*. We denote the et of these sequences as $Seq = Subst_{IO}{}^*.\{ss, ff, dd, \perp\}$. The set $Subst_{IO}{}^*$, with typical el-ment c, denotes the sequences of substitutions generated during the computation, whilst the symbols *ss*, *ff*, and *dd* represent the possible ways in which a process can terminate: *success*, *failure* and *suspension* (or *deadlock*) respectively. A sequence ending in $\perp$ will be interpreted as an *unfinished* sequence. The symbol α will denote an element ranging over the set $\{ss, ff, dd, \perp\}$.

We need first to introduce a technical notion: the *restriction operator*. Intuitively, this operator restricts the substitutions in the sequences generated by a goal A to only those bindings which affect the variables of A. This is not really necessary to achieve compositionality here, but it is a step towards fully abstractness, since, intuitively, we must identify all those processes that only differ for the assignments to the local variables (no contexts can distinguish them). Moreover, if V is the set of local variables available for the computation of A, we must delete all those sequences that are of the form $c.\vartheta^I.s$ such that $var(\vartheta)$ contains a variable in V that has not yet been introduced by c.

Definition 4.2 (Restriction operator) *Let $\bar{A}$ be a goal and let $V \in \mathcal{P}_i(Var)$. Then $r_{\bar{A},V} : \mathcal{P}(Seq) \to \mathcal{P}(Seq)$ is defined by*

$$r_{\bar{A},V}(S) \quad = \quad \{Restrict_{\bar{A}}(s) \mid s \in S \ \wedge \ [s = c.\vartheta^I.s' \Rightarrow var(\vartheta) \cap V \subseteq var(\bar{A}\vartheta_c)]\}$$

where ϑ_c denotes the substitution obtained by composing all the elements of c: $\vartheta_\lambda = \epsilon$ and $\vartheta_{\sigma.c} = \sigma\vartheta_c$ ($var(\vartheta)$ denotes the set of variables occurring in ϑ)

and

$$\begin{array}{lcl} Restrict_{\bar{A}}(\alpha) & = & \alpha \\ Restrict_{\bar{A}}(\vartheta^I.s) & = & \vartheta^I.Restrict_{\bar{A}}(s) \\ Restrict_{\bar{A}}(\vartheta^O.s) & = & \left\{ \begin{array}{ll} \vartheta^O_{|\bar{A}}.Restrict_{\bar{A}\vartheta}(s) & if\ \vartheta_{|\bar{A}} \neq \epsilon \\ Restrict_{\bar{A}}(s) & otherwise \end{array} \right. \end{array}$$

We can now define the operational semantics.

Definition 4.3 (Operational Semantics) *The operational semantics $\mathcal{O} : Goal \to [\mathcal{P}_i($ $Seq]$ is given by*

$$\begin{array}{lcl} \mathcal{O}[\![\Box]\!](V) & = & \{ss, \perp\} \\ \mathcal{O}[\![fail]\!](V) & = & \{ff, \perp\} \\ \mathcal{O}[\![susp]\!](V) & = & \{dd, \perp\} \\ \mathcal{O}[\![\bar{A}]\!](V) & = & r_{\bar{A},V}(\{\vartheta^\ell.\mathcal{O}[\![\bar{B}]\!](W) \mid \langle \bar{A}, V\rangle \xrightarrow{\vartheta^\ell} \langle \bar{B}, W\rangle\} \cup \{\perp\}) \end{array}$$

Note that this definition is recursive. We consider $\mathcal{O}$ as the least fixpoint of the transformation associated with its defining equations. The continuity of this transformation is basically defined with respect to the domain $\mathcal{P}_\perp(Seq)$, the set of subsets of *Seq* containing $\perp$, ordered by set inclusion.

From this operational semantics we obtain our observation criterium as follows:

Definition 4.4 (Observables)

$$Obs[\![\bar{A}]\!] = Result_{\bar{A}}(\mathcal{O}[\![\bar{A}]\!](V))$$

where V is the set $Var \setminus var(A)$, and

$$Result_{\bar{A}}(S) = \{\langle \vartheta_{c|\bar{A}}, \alpha\rangle \mid c.\alpha \in S \ \wedge \ c \in Subst^*_O \ \wedge \ \alpha \in \{ss, ff, dd\}\}.$$

In the following, when $V = Var \setminus var(A)$, we will write simply $\mathcal{O}[\![\bar{A}]\!]$ instead of $\mathcal{O}[\![\bar{A}]\!](V)$.

Note that in this definition we pick up the sequences entirely composed by output ubstitutions. This amounts to require that each substitution we observe has been really roduced. It is easy to see that this notion of observables, based on T, correspond to he set of *finite results* (computed answeer substitution, termination mode) that can be lerived by the classical operational semantics described in section 3.

To show the compositionality of the operational semantics we define the *parallel composition* $\|$. This operator, first introduced in [GCLS88], allows to combine sequences of nput/output substitutions that are equal at each point, apart from the labels, so modeling the interaction of a process with its environment. It corresponds to apply iteratively he rule **C4** of the transition system T.

Definition 4.5 (Parallel composition operator) *The* partial *operator* $\|$: $Seq \times Seq \rightarrow Seq$ *is defined by*

- $s_1 \parallel s_2 = s_2 \parallel s_1$
- $s \parallel ss = s$
- $\bot \parallel ff = ff$
- $dd \parallel dd = dd$
- $\bot \parallel \bot = \bot$
- $\vartheta^I.s_1 \parallel \vartheta^\ell.s_2 = \vartheta^\ell.(s_1 \parallel s_2)$

The natural extension of $\|$ *on sets will be still denoted by* $\|$.

The following result shows the compositionality of our operational semantics with espect to goal conjunction. The compositionality with respect to the union of disjoint rograms (or nicely intersecting programs) is obviously given by the set union.

Theorem 4.6 (Compositionality of $\mathcal{O}$)

$$\mathcal{O}[\![\bar{A}_1, \bar{A}_1]\!](V) = \mathcal{O}[\![\bar{A}_1]\!](V_1) \parallel \mathcal{O}[\![\bar{A}_1]\!](V_2), \quad \textit{where } \langle V_1, V_2 \rangle = Part(V).$$

Note that $\mathcal{O}$ is also a *denotational semantics*, since it is compositional and it is defined s the fixpoint of an (higher order) operator.

5 Fully abstract semantics for the languages of the class HC($\mathcal{A}, T$)

The operational semantics defined in the previous section is not fully abstract, this is due mainly to the fact that it enforces synchronization between the producer and the consumer. This synchronization consists in that a substitution produced by an atom has to be consumed at the same time by the other atoms of the goal. To abstract from this phenomenon we will define some closure conditions on sets of sequences of *elementary* substitutions. The fully abstract denotational semantics $\mathcal{D}$ then will be based on the transition system T plus these closure conditions.

We first introduce the notion of an elementary substitution.

Definition 5.1 *We denote by ESubst the set of all the elementary substitutions, namely the ones of the form $\{x/t\}$. Moreover we define*

- $ESubst_I = \{\vartheta^I \mid \vartheta \in ESubst\}$
- $ESubst_O = \{\vartheta^O \mid \vartheta \in ESubst\}$
- $ESubst_{IO} = ESubst_I \cup ESubst_O$
- $ESeq = ESubst^*_{IO}.\{ss, ff, dd, \bot\}$

To decompose a substitution into elementary ones we define

Definition 5.2 *The function* decomposition, $Dec : Subst_{IO} \to ESubst^*_{IO}$, *is defined as follows*

$$Dec(\{x_1/t_1, \ldots, x_n/t_n\}^\ell) = \{x_1/t_1\}^\ell \ldots \{x_n/t_n\}^\ell$$

We next give the definition of the closure conditions.

Definition 5.3 *Given an initial goal $\bar{A}$, a set of local variables V, we define $Closure_{\bar{A},V}(S)$ to be the minimal set that contains S and that is closed with respect to the conditions* **P1-P6** *of table 4.*

In this table 4 $\vartheta_1[\vartheta_2]$ denotes the substitution $\{x/t\vartheta_2 : x/t \in \vartheta_1\}$. With respect to the rules **P3** and **P4** we require that $dom(mgu(\mathcal{E}(\vartheta_2))) \subseteq ran(\vartheta_1)$. The first condition expresses that there is no direction involved in the communication of substitutions, that is, a substitution produced can also be consumed. The condition **P2** expresses that a input substitution received after the production of a substitution can also be received before. The following condition states that a input substitution preceded by a output substitution can be produced when the environment provides the instantiated output substitution. On the other hand the condition **P4** expresses that when the environment provides less information than the process is able to produce the additional information then can be produced by the process itself. Condition **P5** states that successive input (output) substitutions can be replaced by a equivalent sequence of substitutions. This equivalence is defined in such a way that local variables can be added and deleted.

Here is the definition of the denotational semantics $\mathcal{D}$.

Definition 5.4 (The denotational semantics)

$$\begin{array}{lcl}
\mathcal{D}[\![\Box]\!](V) & = & \{ss, \bot\} \\
\mathcal{D}[\![fail]\!](V) & = & \{ff, \bot\} \\
\mathcal{D}[\![susp]\!](V) & = & \{dd, \bot\} \\
\mathcal{D}[\![A]\!](V) & = & r_{A,V}(Closure_{A,V}(\{Dec(\vartheta^\ell).\mathcal{D}[\![\bar{B}]\!](W) \mid \langle A, V\rangle \xrightarrow{\vartheta^\ell} \langle \bar{B}, W\rangle\} \cup \{\bot\} \\
\mathcal{D}[\![\bar{A}_1, \bar{A}_2]\!](V) & = & r_{\bar{A}_1,\bar{A}_2,V}(Closure_{\bar{A}_1,\bar{A}_2,V}(\mathcal{D}[\![\bar{A}_1]\!](V_1) \parallel \mathcal{D}[\![\bar{A}_2]\!](V_2))), \\
 & & where\ \langle V_1, V_2\rangle = Part(V)
\end{array}$$

Table 4: The closure conditions with respect to an initial goal $\bar{A}$ and set of local variables V

P1 $c_1.\vartheta^O.s \in S \Rightarrow c_1.\vartheta^I.s \in S$

P2 $c_1.\vartheta_1^O.\vartheta_2[\vartheta_1]^I.s \in S \Rightarrow c_1.\vartheta_2^I.\vartheta_1[\vartheta_2]^O.s \in S$

P3 $c_1.\vartheta_1^O.Dec(\vartheta_2^I).s \in S \Rightarrow c_1.\vartheta_1[mgu(\mathcal{E}(\vartheta_2))]^I.Dec(\vartheta_2^O).s \in S$

P4 $c_1.\vartheta_1[mgu(\mathcal{E}(\vartheta_2))]^O.Dec(\vartheta_2^I).s \in S \Rightarrow c_1.\vartheta_1^I.Dec(\vartheta_2^O).s \in S$

P5 $c = c_1.\vartheta_1^\ell \ldots \vartheta_m^\ell.s \in S \Rightarrow c_1.\sigma_1^\ell \ldots \sigma_n^\ell.s \in S$,
where $(\vartheta_1 \cdots \vartheta_m)_{Var \setminus Z_1} = (\sigma_1 \cdots \sigma_n)_{Var \setminus Z_2}$,
with $Z_1 \cap var(c_1) = \emptyset$, $Z_2 \cap var(c) = \emptyset$ and
for $\ell = I$: Z_1 and Z_2 have no variables in common with V or $\bar{A}$,
for $\ell = O$: $Z_1, Z_2 \subseteq V$

P6 $c.\alpha \in S \Rightarrow c.\perp \in S$ where $\alpha \in \{ss, f\!f, dd\}$

We consider $\mathcal{D}$ as the least fixed point of the transformation (continous with respect to set inclusion) associated with its defining equations. The first question is now the correctness of $\mathcal{D}$, i.e., whether it distinguishes *enough.* In other words, if two processes that have the same denotational semantics give also the same observables. We show this by defining a transition system T' (see table 5), which adds to T some rules that model the closure conditions. Essentially T' describes the asynchronous communication by the use of a *store of bindings.* Substitutions produced are communicated to this store. The consumption of a substitution then consists of retrieving this substitution from the store. The store is modeled by adding to a goal constructs of the form *store*(E), which is to be interpreted as that the equation E is present in the store.

In T' we modify rule **C1** so that the output substitution of the guard is added to the store. In the rule **C3** A can now also be of the form *store*(E). It is important to note that the failure rules remain the same, as a consequence inconsistencies in the store do not lead to failure. Moreover we add the rules **C5**-**C9**. The rule **C5** allows to add information about the terms occurring in the goal to the store. The rule **C6** models the communication of a set of equations to the store. The rule **C7** describes how a store can resolve itself into an equivalent one. The rule **C8** describes the retrieval of some information from the store. The last rule allows for the retrieval of some information from the store by a *single* atom of a goal. Without this rule all the atoms in a goal are forced by rule **C4** to consume at each time the same amount of information from the store, thus introducing unnecessary synchronization. The rule **C9** may lead to additional failures, but these occur as inconsistencies in the store, and are as such not visible in the final result.

Table 5: The Transition System T'. Additional computation Rules

C1 $\langle A, V\rangle \stackrel{\epsilon O}{\rightarrow} \langle \bar{B}\sigma, store(\mathcal{E}(\vartheta_2)), W\rangle$ **if** $\exists C \in P_V[C$ is candidate$] \wedge W = V \setminus var(P_V)$

C5 $\langle p(t), V\rangle \stackrel{\epsilon O}{\rightarrow} \langle p(z), store(z = t), V \setminus \{z\}\rangle$ **where** $z \in V$

C6 $\langle tell(E), V\rangle \xrightarrow{\epsilon O} \langle store(E), V\rangle$

C7 $\langle store(E_1), \ldots store(E_m), V\rangle \xrightarrow{\epsilon O} \langle store(E'_1), \ldots store(E'_n), V \setminus Z\rangle$ **if** $[(E_1 \wedge \ldots E_m) \Leftrightarrow \exists Z\, (E'_1 \wedge \ldots E'_n \wedge Z \subseteq V$

C8 $\langle store(x = t), V\rangle \xrightarrow{\vartheta O} \langle \square, V\rangle$ **if** $\vartheta = mgu(x = t)$

C9 $$\frac{\langle \bar{A}, V\rangle \xrightarrow{\vartheta O} \langle \bar{B}, W\rangle}{\langle \bar{A}, V\rangle \xrightarrow{\epsilon O} \langle store(\mathcal{E}(\vartheta)), \bar{B}, W\rangle}$$

Given this transition system T' we prove that, for the operational semantics $\mathcal{O}'$ based on T', the following properties hold:

$$Dec(\mathcal{O}[\![\bar{A}]\!](V)) \subseteq \mathcal{D}[\![\bar{A}]\!](V) \subseteq \mathcal{O}'[\![\bar{A}]\!](V) \text{ and } Result(\mathcal{O}[\![\bar{A}]\!](V)) = Result(\mathcal{O}'[\![\bar{A}]\!](V))$$

Therefore we obtain (by monotonicity of the operator *Result*)

Theorem 5.5 (Correctness of $\mathcal{D}$)

$$Result(\mathcal{D}[\![\bar{A}]\!]) = Obs[\![\bar{A}]\!]$$

Here $\mathcal{D}[\![\bar{A}]\!]$ abbreviates $\mathcal{D}[\![\bar{A}]\!](V)$, where $V = Var \setminus var(\bar{A})$.

To formulate the fully abstractness of $\mathcal{D}$ we introduce the notion of an *initialized* program $P; \bar{A}$. We will write $\mathcal{D}[\![P; \bar{A}]\!]$ to make explicit that we consider $\mathcal{D}[\![\bar{A}]\!]$, the meaning of the goal $\bar{A}$, with respect to the program P.

Theorem 5.6 (Fully abstractness of $\mathcal{D}$) *For arbitrary initialized programs $P_1; \bar{A}_1$ and $P_2; \bar{A}_2$ if $\mathcal{D}[\![P_1; \bar{A}_1]\!] \neq \mathcal{D}[\![P_2; \bar{A}_2]\!]$ then there exists a $P; \bar{A} \in$ HC$(\emptyset, \emptyset)$,* a distinguishing context, *with no predicates in common with $P_1; \bar{A}_1$ and $P_2; \bar{A}_2$, such that*

$$Obs(P \cup P_1; \bar{A}, \bar{A}_1) \neq Obs(P \cup P_2; \bar{A}, \bar{A}_2).$$

6 Conclusion

We have studied in this paper the asynchronous nature of the communication in Concurrent Logic Languages. We have shown that to obtain a fully abstract semantics for these languages a quite different approach is required than for the imperative concurrent languages like CCS. One of the main differences consist in the description of deadlock behaviour. In Concurrent Logic Languages deadlock depends upon the *past* behaviour of a process, whereas in languages like CCS deadlock essentially depends upon the *current* state of the system as described by the failure sets.

A future research topic consist of generalizing the result to Concurrent Constraint Programming Languages. Another interesting line of research is to define a framework to study asynchronous communication in a more abstract setting.

References

[Apt88] K.R. Apt. Introduction to logic programming (revised and extended version. Technical Report CS-R8826, Centre for Mathematics and Computer Science, Amsterdam, 1988. To appear as a chapter in *Handbook of Theoretical Computer Science*, North-Holland (J. van Leeuwen, editor).

[BHR84] S.D. Brookes, C.A.R. Hoare, and W. Roscoe. A theory of communicating sequential processes. *JACM*, 31:499–560, 1984.

[dBK88] J.W. de Bakker and J.N. Kok. Uniform abstraction, atomicity and contractions in the comparative semantics of concurrent prolog. In *Proc. Fifth Generation Computer Systems*, pages 347–355, Tokyo, Japan, 1988. Extended Abstract, full version available as CWI report CS-8834. To appear on Theoretical Computer Science.

[dBKPR89a] F.S. de Boer, J.N. Kok, C. Palamidessi, and J.J.M.M. Rutten. Control flow versus logic: a denotational and a declarative model for guarded horn clauses. In *Proc. of the Symposium on Mathematical Foundations of Computer Science*, LNCS, pages 165–176, 1989.

[dBKPR89b] F.S. de Boer, J.N. Kok, C. Palamidessi, and J.J.M.M. Rutten. Semantic models for a version of parlog. In G. Levi and M. Martelli, editors, *Proc. of the Sixt International Conference on Logic Programming*, pages 621–636, Lisboa, 1989. MIT Press. Extended version to appear in Theoretical Computer Science.

[Ede85] E. Eder. Properties of substitutions and unifications. *Journal Symbolic Computation*, 1:31–46, 1985.

[GCLS88] R. Gerth, M. Codish, Y. Lichtenstein, and E. Shapiro. Fully abstract denotational semantics for concurrent prolog. In *Proc. of the Third IEEE Symposium on Logic In Computer Science*, pages 320–335, 1988.

[GL90] M. Gabbrielli and G. Levi. An unfolding reactive semantics for concurrent constraint programming. Technical Report TR ../90, Dipartimento di Informatica, Pisa, 1990.

[GMS89] H. Gaifman, M.J. Maher, and E. Shapiro. Rehactive behaviour semantics for concurrent constraint logic languages. In *Proc. of the North American Conference on Logic Programming*, 1989.

[Hir87] M. Hirata. Parallel list processing language oc and its self-description. *Computing Software*, 4(3):41–64, 1987. In Japanese.

[JL87] J. Jaffar and J.-L. Lassez. Constraint logic programming. In *Proc. ACM Symp. on Principles of Programming Languages*, pages 111–119, 1987.

[Kok88] J.N. Kok. A compositional semantics for concurrent prolog. In R. Cori and M. Wirsing, editors, *Proc. 5th Theoretical Aspects of Computer Science*, number 294 in LNCS, pages 373–388. Springer Verlag, 1988.

[Llo87] J.W. Lloyd. *Foundations of Logic Programming.* Springer Verlag, 1987. Second edition.

[LMM88] J.-L. Lassez, M.J. Maher, and K. Marriot. Unification revisited. In J. Minker, editor, *Foundations of deductive databases and logic programming*, Los Altos, 1988. Morgan Kaufmann.

[Mah87] M. J. Maher. Logic semantics for a class of committed choice programs. In J.-L. Lassez, editor, *Proc. of the Fourth Int. Conference on Logic Programming*, pages 877–893, Melbourne, 1987. MTI Press.

[Mil80] R. Milner. *A Calculus of Communicating Systems.* Number 92 in LNCS. Springer Verlag, New York, 1980.

[Sar85] V.A. Saraswat. Partial correctness semantics for cp(↓, |, &). In *Proc. of the Conf. on Foundations of Software Computing and Theoretical Computer Science*, number 206 in LNCS, pages 347–368, 1985.

[Sar89] V.A. Saraswat. *Concurrent Constraint Programming Languages.* PhD thesis, january 1989. To be published by MTI Press.

[Sha89] E.Y. Shapiro. The family of concurrent logic languages. *ACM Computing Surveys*, 21(3):412–510, 1989.

[SR89] V.A. Saraswat and M. Rinard. Concurrent constraint programming. Technical report, Carnegie-Mellon University, 1989.

[Ued88] K. Ueda. Guarded horn clauses, a parallel logic programming language with the concept of a guard. In M. Nivat and K. Fuchi, editors, *Programming of Future Generation Computers*, pages 441–456, Amsterdam, 1988. North Holland.

Verifying Temporal Properties of Processes

Julian Bradfield, Colin Stirling
Department of Computer Science
University of Edinburgh
The King's Buildings
EDINBURGH
U.K. EH9 3JZ
email: jcb@lfcs.ed.ac.uk, cps@lfcs.ed.ac.uk

Motivation

Iany interesting concurrent systems have infinite state spaces: examples include concurrent while rograms; Petri Nets; CCS (or CSP) processes with value passing. All of these examples can be iterpreted operationally as infinite labelled transition systems, structures of the form $(\mathcal{P}, \{\stackrel{a}{\longrightarrow} :$ $\in L\})$ where $\mathcal{P}$ is a set of points (states, markings, processes) and $\stackrel{a}{\longrightarrow}$ the appropriate binary ansition relation on $\mathcal{P}$ for each label a (action, set of events) belonging to the family L.

A very rich temporal logic for expressing properties of such transition systems is a slight extension f the modal mu-calculus [9, 6] where the modalities are indexed by families of labels instead of idividual labels. The question we address in this paper is: can model checking techniques, as itroduced in [3] be extended from finite to infinite state spaces? (Pragmatically, this means moving om automated to computer-aided verification techniques.) We provide an affirmative answer by resenting a *sound and complete* tableau system for proving temporal properties of states (processes markings) in *arbitrary* infinite transition system models. The tableau system extends local model lecking techniques as presented in [4, 7, 11, 14]. The delicate aspect is showing that a point (or set points) has, or lacks, a least fixed point property (a *liveness* property). The tableau proof system data independent and therefore generalizes standard methods commonly used in program logics uch as Hoare logics). The verification technique is illustrated on examples drawn from CCS [8]. e [1] for the application of the method to Petri Nets and [12] for its application to concurrent while ograms.

Section 2 provides examples of CCS processes and their properties. In section 3 the syntax and mantics of the (slightly extended) modal mu-calculus are described. The tableau proof system is esented in section 4, and finally we briefly examine applications in section 5.

Processes

ocesses of CCS determine labelled transition systems under their behavioural (transitional) seman- s. In this case $\stackrel{a}{\longrightarrow}$ is the usual CCS derivation relation with $E \stackrel{a}{\longrightarrow} F$ representing that process may become F by performing a. When L is the set of all actions, and $K \subseteq L$, we write $\stackrel{K}{\longrightarrow}$ for $_{\in K} \stackrel{a}{\longrightarrow}$. Moreover we write $\stackrel{a}{\Longrightarrow}$ for the observable derivation, defined as usual by

$$E \stackrel{\epsilon}{\Longrightarrow} F \quad \text{if} \quad E = E_0 \stackrel{\tau}{\longrightarrow} E_1 \stackrel{\tau}{\longrightarrow} \cdots \stackrel{\tau}{\longrightarrow} E_n = F \quad (n \geq 0)$$

$$E \stackrel{a}{\Longrightarrow} F \quad \text{if} \quad E \stackrel{\epsilon}{\Longrightarrow} E' \text{ and } E' \stackrel{a}{\longrightarrow} F' \text{ and } F' \stackrel{\epsilon}{\Longrightarrow} F$$

e now present four example processes, and some liveness and safety properties that may be ascribed them.

First, we have a simple finite state system, representing a road crossing a railway; unlike most crossings, this one keeps the barriers down except when a car actually approaches and tries to cross.

$$\begin{array}{lcl} Rail & \stackrel{\text{def}}{=} & train.green.tcross.\overline{red}.Rail \\ Road & \stackrel{\text{def}}{=} & car.up.ccross.\overline{down}.Road \\ Signal & \stackrel{\text{def}}{=} & \overline{green}.red.Signal + \overline{up}.down.Signal \\ Crossing & \stackrel{\text{def}}{=} & (Road \mid Rail \mid Signal)\backslash\{green, red, up, down\} \end{array}$$

Here, *train* and *car* represent the approach of a train and car respectively, *green* is the receipt of a green signal by the train, *tcross* is the train crossing and $\overline{red}$ automatically sets the lights red, *up* is the gates opening for the car, and *ccross* is the car crossing and $\overline{down}$ closes the gates. A crucial safety property of this system is that it is never possible for a train and a car both to be able to cross. A very desirable liveness property (which the crossing fails to have) is that whenever a car or train approaches, it eventually crosses. However, it does have this property if we assume that the signal is fair.

For a simple infinite state system, consider a process *Ticker* whose observable behaviour is to perform a finite number of *tick*s and then stop.

$$\begin{array}{lcll} Up_i & \stackrel{\text{def}}{=} & \tau.Up_{i+1} + Down_i & (i \geq 0) \\ Down_i & \stackrel{\text{def}}{=} & tick.Down_{i-1} & (i \geq 1) \\ Down_0 & \stackrel{\text{def}}{=} & \mathbf{0} & \\ Ticker & \stackrel{\text{def}}{=} & Up_0 & \end{array}$$

This process has the feature that it may diverge, that is, perform τs indefinitely. So the expression of "*Ticker* performs a finite number of ticks and then stops" must account for this; and we should also like to express the stronger property that "*Ticker* may diverge initially, but once it starts ticking, it does nothing else until it stops".

A more complex example (employing value passing) is a slot machine SM_n, with three components—IO, which handles taking and paying out money, B_n a bank holding n pounds, and D, the wheel-spinning decision component.

$$\begin{array}{lcl} IO & \stackrel{\text{def}}{=} & slot.\overline{bank}.(lost.\overline{loss}.IO + release(y).\overline{win}(y).IO) \\ B_n & \stackrel{\text{def}}{=} & bank.\overline{max}(n+1).left(y).B_y \\ D & \stackrel{\text{def}}{=} & max(z).\big(\overline{lost}.\overline{left}(z).D + \sum_{1\leq y\leq z} \overline{release}(y).\overline{left}(z-y).D\big) \\ SM_n & \stackrel{\text{def}}{=} & (IO \mid B_n \mid D)\backslash K \end{array}$$

where $K = \{\, bank, max(v), left(v), release(v) : v \in \mathbf{N}\}$. Again, we have a safety property, that the bank never has a negative amount of money, and a weak liveness property, that it is possible to win a million pounds infinitely often.

Finally, we have a complex example illustrating some of the issues raised by networks of similar processes. Consider a set of n processes placed on a grid of relay stations which allow them to communicate. Suppose that each process wishes to establish a dedicated and private communication link with each other process: to do this, each pair of processes requires a chain of dedicated relay stations between them. Since each relay can only communicate with its four immediate neighbours, some method is needed to achieve this. We shall consider one of the simplest: each process takes control of the $n-1$ (say) stations initially assigned to it, and instructs them to take control of a random neighbour; communication is established when two such chains owned by different processes meet head on, as in the picture:

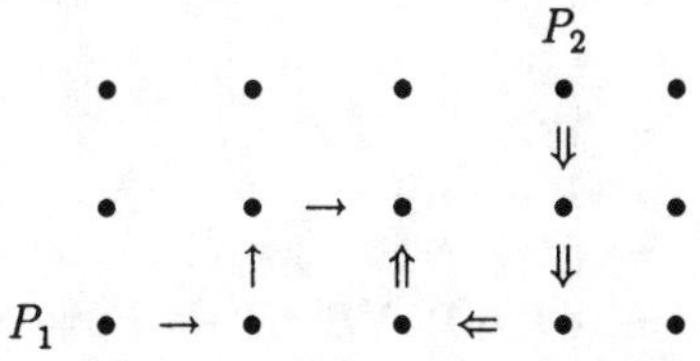

We shall suppose that each process is assigned a line of $n-1$ adjacent stations.

A crude model of such a system is given by letting $\{C_{ij}\}$ be an array of cells, each defined by

$$\begin{aligned} C_{ij} &\stackrel{\text{def}}{=} alloc_{ij}(k).(\overline{alloc}_{i\,j+1}(k).\mathbf{0} + \overline{alloc}_{i+1\,j}(k).\mathbf{0} + \overline{alloc}_{i\,j-1}(k).\mathbf{0} + \overline{alloc}_{i-1\,j}(k).\mathbf{0} \\ &\quad + alloc_{ij}(k').\text{if } k = k' \text{ then } \mathbf{0} \text{ else } \overline{line}(\min(k,k'), \max(k.k'))) \end{aligned}$$

That is, a cell waits to be allocated to some process k, and then tries to allocate its neighbours to k, and also listens to see if a different process k' wants this cell. If it allocates a neighbour, the cell plays no further part; if it is contacted by process k', it signals the successful establishment of a line from k to k'.

Suppose that process P_k is initially assigned cells $C_{i_k j_k+1}, \ldots, C_{i_k j_k+n-1}$: P_k is then just

$$P_k \stackrel{\text{def}}{=} |_{1 \le l < n} \overline{alloc}_{i_k\, j_k+l}(k).\mathbf{0}$$

The system is then the parallel composition of all cells and processes, restricted on the *alloc* actions. One would like to be able to say that every pair of processes connects, but that is clearly too much to expect of our simple random algorithm, so we shall instead consider the weak liveness property that it is possible for all pairs to eventually connect, and we shall show that this holds only if there are fewer than five processes.

3 The modal mu-calculus

The temporal logic used is (a slight extension of) the modal mu-calculus. The syntax for formulae Φ is as follows:

$$\Phi ::= Z \mid \neg\Phi \mid \Phi_1 \wedge \Phi_2 \mid [K]\Phi \mid \nu Z.\Phi$$

where Z ranges over a denumerable family of propositional variables, and K over *subsets* of a label set L. A restriction on $\nu Z.\Phi$ is that each free occurrence of Z in Φ should lie within the scope of an even number of negations. Derived operators are defined in the familiar way: $\mathrm{tt} \stackrel{\text{def}}{=} \nu Z.Z$; $\mathrm{ff} \stackrel{\text{def}}{=} \neg\mathrm{tt}$; $\Phi \vee \Psi \stackrel{\text{def}}{=} \neg(\neg\Phi \wedge \neg\Psi)$; $\langle K\rangle\Phi \stackrel{\text{def}}{=} \neg[K]\neg\Phi$; and $\mu Z.\Phi \stackrel{\text{def}}{=} \neg\nu Z.\neg\Phi[Z := \neg Z]$ where $\neg\Phi[Z := \neg Z]$ is the result of substituting $\neg Z$ for each free occurrence of Z. Further useful abbreviations (which also apply to the $\langle K\rangle$ modalities) are:

$$\begin{aligned} [-K]\Phi &\stackrel{\text{def}}{=} [L-K]\Phi \\ [a_1, \ldots, a_n]\Phi &\stackrel{\text{def}}{=} [\{a_1, \ldots, a_n\}]\Phi \\ [-]\Phi &\stackrel{\text{def}}{=} [L]\Phi \end{aligned}$$

Formulae of this logic are interpreted on labelled transition systems $\mathcal{T} = (\mathcal{P}, \{\stackrel{a}{\longrightarrow} : a \in L\})$. A model is a pair $(\mathcal{T}, \mathcal{V})$ where $\mathcal{T}$ is a transition system and $\mathcal{V}$ a valuation assigning sets of processes to propositional variables: $\mathcal{V}(Z) \subseteq \mathcal{P}$. We assume the customary updating notation: $\mathcal{V}[\mathcal{E}/Z]$ is the valuation $\mathcal{V}'$ which agrees with $\mathcal{V}$ except that $\mathcal{V}'(Z) = \mathcal{E}$. Finally the set of processes of $\mathcal{T}$ having the property Φ in the model $(\mathcal{T}, \mathcal{V})$ is inductively defined as $\|\Phi\|_{\mathcal{V}}^{\mathcal{T}}$ (where for ease of notation we drop the index $\mathcal{T}$ which is assumed to be fixed):

$$\begin{array}{lcl} \|Z\|_{\mathcal{V}} & = & \mathcal{V}(Z) \qquad (Z \in \text{Var}) \\ \|\neg\Phi\|_{\mathcal{V}} & = & \mathcal{P} - \|\Phi\|_{\mathcal{V}} \\ \|\Phi_1 \wedge \Phi_2\|_{\mathcal{V}} & = & \|\Phi_1\|_{\mathcal{V}} \cap \|\Phi_2\|_{\mathcal{V}} \\ \|[K]\Phi\|_{\mathcal{V}} & = & \{P \in \mathcal{P} : \forall P'. P \xrightarrow{K} P' \Rightarrow P' \in \|\Phi\|_{\mathcal{V}}\} \\ \|\nu Z.\Phi\|_{\mathcal{V}} & = & \bigcup\{\mathcal{E} \subseteq \mathcal{P} : \|\Phi\|_{\mathcal{V}[\mathcal{E}/Z]} \supseteq \mathcal{E}\} \end{array}$$

The expected clause for the derived operator $\mu Z.\Phi$ is:

$$\|\mu Z.\Phi\|_{\mathcal{V}} = \bigcap\{\mathcal{E} \subseteq \mathcal{P} : \|\Phi\|_{\mathcal{V}[\mathcal{E}/Z]} \subseteq \mathcal{E}\}$$

CCS processes are interpreted operationally as transition systems. The modal mu-calculus is a natural extension of Hennessy-Milner logic [5], offering a succinct temporal logic for processes. Moreover its set of closed formulas (those without free variables) also characterizes bisimulation equivalence [10]. The examples below illustrate the benefit of extending the modal mu-calculus to encompass modalities with sets of labels. In the general case it allows one to express temporal properties of processes engaged in value passing in a very simple and succinct fashion. Moreover, it furnishes a direct relationship with standard temporal logics such as CTL and LT, where labels are ignored, via the modalities $[-]$ and $\langle - \rangle$.

Assume any mu-calculus model for the *Crossing*. The safety property that it is never possible for both a train and a car both to be able to cross is expressed by the formula

$$\nu Z.([\mathit{tcross}]\mathrm{ff} \vee [\mathit{ccross}]\mathrm{ff}) \wedge [-]Z$$

More generally, $\nu Z.\Phi \wedge [-]Z$ expresses that Φ is invariantly true. The desirable liveness property that whenever a car approaches then eventually it crosses only holds if we assume that the signal is fair. So let Q be a propositional variable which is true when the crossing is in any state where *Road* has the form $\mathit{up}.\mathit{ccross}.\overline{\mathit{down}}.\mathit{Road}$ and let R hold when it is in a state where *Rail* has the form $\mathit{green}.\mathit{tcross}.\overline{\mathit{red}}.\mathit{Rail}$. ($Q$ and R can be thought of as probes in the sense of [13].) The liveness property is now: for any run if $\neg Q$ is true infinitely often and $\neg R$ is also true infinitely often then whenever a car approaches eventually it crosses. This is expressed by the following formula Φ employing embedded fixed point subformulae:

$$\begin{array}{lcl} \Phi & = & \nu Y.[\mathit{car}]\Phi_1 \wedge [-]Y \\ \Phi_1 & = & \mu X.\nu Y_1.(Q \vee [-\mathit{ccross}](\nu Y_2.(R \vee X) \wedge [-\mathit{ccross}]Y_2)) \wedge [-\mathit{ccross}]Y_1 \end{array}$$

In general for arbitrary models, the property that eventually action a *happens* (as opposed to *may* happen) is given by the formula $\mu Z.\langle - \rangle \mathrm{tt} \wedge [-a]Z$.

Another example CCS process from the previous section is *Ticker*. Its *observable* behaviour is to perform a finite number of ticks and then to stop. This description relates to its behaviour with respect to the $\stackrel{a}{\Longrightarrow}$ transitions. So it is useful to introduce the abbreviation

$$[\![\,]\!]\Phi \stackrel{\text{def}}{=} \nu Z.\Phi \wedge [\tau]Z$$

(where Z does not occur free in Φ). In a model process E has the property $[\![\,]\!]\Phi$ just in case every process F such that $E \stackrel{\epsilon}{\Longrightarrow} F$ has the property Φ. Then when $\tau \notin K$ the modality $[\![K]\!]$ is defined as $[\![\,]\!][K][\![\,]\!]$: consequently, $[\![K]\!]$ relates to $\stackrel{a}{\Longrightarrow}$ as $[K]$ relates to $\stackrel{a}{\longrightarrow}$. The property above of its observable behaviour is expressed as:

$$\mu Z.[\![\mathit{tick}]\!]Z$$

The ticker may also diverge initially, that is it may perform τs indefinitely. Divergence is given by the formula $\nu Z.\langle \tau \rangle Z$.

In the case of the slot machines SM_n, assume a model whose valuation assigns to the propositional variables Q_i the set $\{ SM_j : j < i \}$ for each $i \geq 0$. Now an important safety property is that SM_n never has a negative amount of money, so never pays out more than it contains:

$$\nu Z. \neg Q_0 \wedge [-] Z$$

A useful weak liveness property is that it may pay out a million (pounds) infinitely often, expressible as

$$\nu Y. \mu Z. (\langle \overline{win}(10^6) \rangle Y) \vee \langle - \rangle Z$$

The weak liveness property for the process network is

$$\mu Z. \Psi \vee \langle - \rangle Z$$

where $\Psi \stackrel{\text{def}}{=} \bigwedge_{1 \leq j < i \leq n} \langle \overline{line}(j,i) \rangle \mathrm{tt}$. This formula is stronger than necessary: all we need is that all $\overline{line}(k,k')$ actions should happen eventually, in any order; although this can be expressed in the mu-calculus, the resulting formula is large; see section 5 for further discussion.

These examples show that we are interested in whether a particular process (state or point) or set of processes $\mathcal{E}$ has a temporal property Φ. But the meaning of Φ is defined in terms of *all* processes having it as a property. Clearly, in general *computing* the set $\|\Phi\|_{\mathcal{V}}^{\mathcal{T}}$ is not only unnecessarily cumbersome (if one is just interested in whether $\mathcal{E} \subseteq \|\Phi\|_{\mathcal{V}}^{\mathcal{T}}$), but also is not really feasible either for large formulae involving numerous embeddings of the fixed point operators, or for very large (let alone infinite) transition systems. Discovering fixed point sets in general is not easy, and is therefore liable to lead to errors. Instead we would like simpler (and consequently safer) methods for verifying that processes have, or fail to have, temporal properties.

Here a temporal property checker is developed as a tableau system in the style of [11] but extended to cope with arbitrary *infinite* transition system models. The rules turn out to be straightforward, but the successful termination condition for least fixed point subformulae is subtle.

4 The tableau system

Assume a fixed mu-calculus model $(\mathcal{T}, \mathcal{V})$ where $\mathcal{T}$ may have infinite size. We wish to show that some process, or more generally some set of processes $\mathcal{E}$ has the property Φ, that $\mathcal{E} \subseteq \|\Phi\|_{\mathcal{V}}^{\mathcal{T}}$. We present a tableau system for proving this, built on sequents of the form $\mathcal{E} \vdash_{\Delta} \Phi$ where Δ is a *definition list* which keeps track of unrolling fix-point formulae. A definition is given as $U = \Psi$ where U is a propositional constant and Ψ a fixed point formula. A finite sequence Δ of definitions $(U_1 = \Psi_1, \ldots, U_n = \Psi_n)$ has the two properties that each U_i is distinct, and secondly that each Ψ_i can only mention propositional constants belonging to the set $\{U_1, \ldots, U_{i-1}\}$.[1] Lists of definitions can be extended: if U is not declared in Δ and Ψ only mentions constants declared in Δ, then $\Delta \cdot U = \Psi$ is the definition list that results from appending $U = \Psi$ to Δ. When Δ is a definition list and U is declared to be Ψ in Δ, then $\Delta(U)$ is Ψ.

The rules of the tableau below are inverse natural deduction type rules. The premise sequent is the goal to be achieved (that $\mathcal{E} \subseteq \|\Phi_{\Delta}\|_{\mathcal{V}}^{\mathcal{T}}$) while the consequents are the subgoals. The rules are presented only for formulae in positive form (where all negations are moved inwards by using the dual operators $\vee$, $\langle K \rangle$ and $\mu Y.$). We assume that σ ranges over ν or μ. In the Diamond rule we

[1] We assume the interpretation of formulae relative to definition lists as in [11] by in effect treating constants as variables: if Δ is $(U_1 = \Psi_1, \ldots, U_n = \Psi_n)$ then $\|\Phi_{\Delta}\|_{\mathcal{V}}^{\mathcal{T}} = \|\Phi\|_{\mathcal{V}_n}^{\mathcal{T}}$ where $\mathcal{V}_0 = \mathcal{V}$ and $\mathcal{V}_{i+1} = \mathcal{V}_i[\|\Psi_{i+1}\|_{\mathcal{V}_i}^{\mathcal{T}} / U_{i+1}]$.

assume that f is a function from the set $\mathcal{E}$ to the set $f(\mathcal{E})$ such that for all $E \in \mathcal{E}$, $E \xrightarrow{K} f(E)$:

And	$\dfrac{\mathcal{E} \vdash_\Delta \Phi_1 \wedge \Phi_2}{\mathcal{E} \vdash_\Delta \Phi_1 \qquad \mathcal{E} \vdash_\Delta \Phi_2}$	
Or	$\dfrac{\mathcal{E} \vdash_\Delta \Phi_1 \vee \Phi_2}{\mathcal{E}_1 \vdash_\Delta \Phi_1 \qquad \mathcal{E}_2 \vdash_\Delta \Phi_2}$	$\mathcal{E} = \mathcal{E}_1 \cup \mathcal{E}_2$
Box	$\dfrac{\mathcal{E} \vdash_\Delta [K]\Phi}{\mathcal{E}' \vdash_\Delta \Phi}$	$\mathcal{E}' = \{ P' : \exists P \in \mathcal{E} \,.\, P \xrightarrow{K} P' \}$
Diamond	$\dfrac{\mathcal{E} \vdash_\Delta \langle K \rangle\Phi}{f(\mathcal{E}) \vdash_\Delta \Phi}$	
Constant Introduction	$\dfrac{\mathcal{E} \vdash_\Delta \sigma Z.\Phi}{\mathcal{E} \vdash_{\Delta'} U}$	$\Delta' = \Delta \cdot (U = \sigma Z.\Phi)$
Constant Unfolding	$\dfrac{\mathcal{E} \vdash_\Delta U}{\mathcal{E} \vdash_\Delta \Phi[Z := U]}$	$\Delta(U) = \sigma Z.\Phi$
Thinning	$\dfrac{\mathcal{E} \vdash_\Delta \Phi}{\mathcal{E}' \vdash_\Delta \Phi}$	$\mathcal{E}' \supseteq \mathcal{E}$

To test if every process in $\mathcal{E}$ has the property Φ (relative to Δ) one has to achieve the goal $\mathcal{E} \vdash_\Delta \Phi$ by building a tableau, a proof tree whose root is labelled with this initial sequent. Sequents labelling the immediate successors of a node are determined by an application of one of the rules. The boolean, Box and Diamond rules are straightforward. New constants are introduced when fixed point formulae are met, and then these are unfolded. The Thinning rule allows us to enlarge the set being checked, and is essential for completeness of the system (and need only be applied to sequents with fixed point formulae or constants).

An essential missing ingredient is when a node in a proof tree counts as a leaf. We assume that the rules above only apply to nodes that are not terminal. A node $\mathbf{n}$ labelled by the sequent $\mathcal{F} \vdash_\Delta \Psi$ is *terminal* if one of the following conditions holds:

(i) $\mathcal{F} = \emptyset$

(ii) $\Psi = Z$ or $\Psi = \neg Z$

(iii) $\Psi = \langle K \rangle \Phi$ and $\exists F \in \mathcal{F} \,.\, \mathrm{not}(F \xrightarrow{K})$

(iv) $\Psi = U$ and $\Delta(U) = \sigma Z.\Phi$ and there is a node above $\mathbf{n}$ in the proof tree labelled $\mathcal{E} \vdash_{\Delta'} U$ with $\mathcal{E} \supseteq \mathcal{F}$

A nodel fulfilling condition (iv) is called a σ-terminal. A node obeying conditions (i) or (iv) when $\sigma = \nu$ is called a *successful* terminal, whereas a node only fulfilling (iii) is *unsuccessful*. In the case of (ii) success depends on the valuation $\mathcal{V}$ of the model: if $\mathcal{F} \subseteq \mathcal{V}(Z)$ when Ψ is Z, or $\mathcal{F} \subseteq \mathcal{P} - \mathcal{V}(Z)$ when Ψ is $\neg Z$ then it is successful, otherwise it is unsuccessful. The definition of a successful μ-terminal, a node complying with (iv) when $\sigma = \mu$, is more complex and requires a little notation.

Suppose a node $\mathbf{n}$ is labelled by $\mathcal{E} \vdash_\Delta \Phi$ and $\mathbf{n}'$ is an immediate successor of $\mathbf{n}$ labelled $\mathcal{E}' \vdash_{\Delta'} \Phi'$. We say that $E' \in \mathcal{E}'$ at $\mathbf{n}'$ is a *dependant* of $E \in \mathcal{E}$ at $\mathbf{n}$ if

- the rule applied to $\mathbf{n}$ is (And), (Or), (Constant Introduction), (Constant Unfolding) or (Thinning), and $E = E'$, or
- the rule is (Box) and $E \xrightarrow{K} E'$, or
- the rule is (Diamond), and $E' = f(E)$.

Assume that the *companion* of a σ-terminal is that node above it which makes it a terminal. Next we define a *trail* to F at a μ-terminal $\mathbf{n}$ from E at its companion $\mathbf{m}$ to be a sequence of pairs of nodes and processes $(\mathbf{n}_1, E_1), \ldots, (\mathbf{n}_k, E_k)$ with $(\mathbf{n}_1, E_1) = (\mathbf{m}, E)$ and $(\mathbf{n}_k, E_k) = (\mathbf{n}, F)$, such that for all i with $1 \leq i < k$ either

(i) E_{i+1} at $\mathbf{n}_{i+1}$ is a dependant of E_i at $\mathbf{n}_i$, or

(ii) $\mathbf{n}_i$ is the immediate predecessor of a σ-terminal node $\mathbf{n}'$ (where $\mathbf{n}' \neq \mathbf{n}$) whose companion is $\mathbf{n}_j$ for some $j \leq i$, and $\mathbf{n}_{i+1} = \mathbf{n}_j$ and E_{i+1} at $\mathbf{n}'$ is a dependant of E_i at $\mathbf{n}_i$.

Then each companion node $\mathbf{n}$ of a μ-terminal induces a pre-order $\sqsupset_{\mathbf{n}}$ by $E \sqsupset_{\mathbf{n}} F$ if there is a trail to F at $\mathbf{n}$ from E at its companion. A μ-terminal node $\mathbf{n}$ is successful if the ordering $\sqsubset_{\mathbf{n}}$ induced by its companion $\mathbf{n}$ is well-founded, that is, there is no infinite descending chain

$$E_0 \sqsupset_{\mathbf{n}} E_1 \sqsupset_{\mathbf{n}} \cdots$$

Finally we say that a tableau is *successful* if it is finite and all its leaves are successful terminals. The tableau technique is both *sound* and *complete* for arbitrary (infinite) transition systems.

Theorem. *$\mathcal{E} \vdash_\Delta \Phi$ is the root of some successful tableau on $(\mathcal{T}, \mathcal{V})$ iff $\mathcal{E} \subseteq \|\Phi_\Delta\|^{\mathcal{T}}_{\mathcal{V}}$*

A full proof of this theorem is given in [2]. The proof is very intricate and directly extends the techniques used in [11]. An interesting question is if the elegant semantic account of local model checking presented in [4, 14] can be extended to cover infinite models.

This tableau method can therefore be used to verify arbitrary temporal properties of arbitrary concurrent systems on arbitrary data (such as Petri Nets, CCS and CSP processes, and concurrent while programs). As the technique is data independent it generalizes standard methods used in program logics which employ induction or appeal to reasoning on well founded structures. Despite its generality we now show that it is applicable to proving properties of processes.

5 Examples

In this section we demonstrate the tableau system on the CCS examples of section 2. Recall the level crossing. Its safety property was expressed as $\Phi = \nu Z.([tcross]\mathrm{ff} \lor [ccross]\mathrm{ff}) \land [-]Z$. The successful tableau for this property of the crossing (where () is the empty definition list) is:

$$\dfrac{\{\mathit{Crossing}\} \vdash_{()} \Phi}{\dfrac{\dfrac{\mathcal{E} \vdash_{()} \Phi}{\dfrac{\mathcal{E} \vdash_\Delta U}{\dfrac{\mathcal{E} \vdash_\Delta ([tcross]\mathrm{ff} \lor [ccross]\mathrm{ff}) \land [-]U}{\dfrac{\mathcal{E} \vdash_\Delta [tcross]\mathrm{ff} \lor [ccross]\mathrm{ff}}{\dfrac{\mathcal{E}_1 \vdash_\Delta [tcross]\mathrm{ff}}{\varnothing \vdash_\Delta \mathrm{ff}} \quad \dfrac{\mathcal{E}_2 \vdash_\Delta [ccross]\mathrm{ff}}{\varnothing \vdash_\Delta \mathrm{ff}}} \qquad \dfrac{\mathcal{E} \vdash_\Delta [-]U}{\mathcal{E} \vdash_\Delta U}}}} \; {\scriptstyle \Delta=(U=\Phi)}}{}}$$

where $\mathcal{E}_1$ is all derivatives of *Crossing* such that the *Rail* component is not in the state $tcross.\overline{red}.Rail$, and $\mathcal{E}_2$ is all derivatives such that the *Road* component is not in the state $ccross.\overline{down}.Road$, and $\mathcal{E} = \mathcal{E}_1 \cup \mathcal{E}_2$. By brute force, one checks that $\mathcal{E}$ is closed under $\stackrel{-}{\longrightarrow}$, thus giving the success of the terminal $\mathcal{E} \vdash_\Delta U$.

Recall the fair liveness property for the crossing given by the formula Φ:

$$\begin{aligned} \Phi &= \nu Z.[car]\Phi_1 \land [-]Z \\ \Phi_1 &= \mu X.\nu Y_1.(Q \lor [-ccross](\nu Y_2.(R \lor X) \land [-ccross]Y_2)) \land [-ccross]Y_1 \end{aligned}$$

where Q holds when the crossing is in any state where *Road* can by itself perform *up* and R holds in any state where *Rail* can by itself perform *green*. A proof that the crossing has this property is given by the following tableau where $[-cc]$ abbreviates $[-ccross]$:

$$\{Crossing\} \vdash_{()} \Phi$$
$$\mathcal{E} \vdash_{()} \Phi$$
$$\mathcal{E} \vdash_{\Delta_1} W$$
$$\mathcal{E} \vdash_{\Delta_1} [car]\Phi_1 \wedge [-]W$$
$$\mathcal{E} \vdash_{\Delta_1} [car]\Phi_1 \qquad\qquad \mathcal{E} \vdash_{\Delta_1} [-]W$$
$$\mathcal{E}' \vdash_{\Delta_1} \Phi_1 \qquad\qquad \mathcal{E} \vdash_{\Delta_1} W$$
$$\mathcal{E}_1 \vdash_{\Delta_1} \Phi_1$$
$$\mathcal{E}_1 \vdash_{\Delta_2} U$$
$$\mathcal{E}_1 \vdash_{\Delta_2} \nu Y_1.(Q \vee [-cc](\nu Y_2.(R \vee U) \wedge [-cc]Y_2)) \wedge [-cc]Y_1$$
$$\mathcal{E}_1 \vdash_{\Delta_3} V_1$$
$$\mathcal{E}_1 \vdash_{\Delta_3} (Q \vee [-cc](\nu Y_2.(R \vee U) \wedge [-cc]Y_2)) \wedge [-cc]V_1$$
$$\mathcal{E}_1 \vdash_{\Delta_3} Q \vee [-cc]\nu Y_2.(R \vee U) \wedge [-cc]Y_2 \qquad\qquad \mathcal{E}_1 \vdash_{\Delta_3} [-cc]V_1$$
$$\mathcal{E}_1 - \mathcal{E}_2 \vdash_{\Delta_3} Q \qquad \mathcal{E}_2 \vdash_{\Delta_3} [-cc]\nu Y_2.(R \vee U) \wedge [-cc]Y_2 \qquad \mathcal{E}_1' \vdash_{\Delta_3} V_1$$
$$\mathcal{E}_3 \vdash_{\Delta_3} \nu Y_2.(R \vee U) \wedge [-cc]Y_2$$
$$\mathcal{E}_3 \vdash_{\Delta_4} V_2$$
$$\mathcal{E}_3 \vdash_{\Delta_4} (R \vee U) \wedge [-cc]V_2$$
$$\mathcal{E}_3 \vdash_{\Delta_4} R \vee U \qquad\qquad \mathcal{E}_3 \vdash_{\Delta_4} [-cc]V_2$$
$$\mathcal{E}_3 \vdash_{\Delta_4} R \qquad \emptyset \vdash_{\Delta_4} U \qquad\qquad \emptyset \vdash_{\Delta_4} V_2$$

The definition lists appealed to in this tableau are as follows:

$$\begin{aligned}
\Delta_1 &= (W = \Phi)\\
\Delta_2 &= \Delta_1 \cdot U = \Phi_1\\
\Delta_3 &= \Delta_2 \cdot V_1 = \nu Y_1.(Q \vee [-ccross](\nu Y_2.(R \vee U) \wedge [-ccross]Y_2)) \wedge [-ccross]Y_1\\
\Delta_4 &= \Delta_3 \cdot V_2 = \nu Y_2.(R \vee U) \wedge [-ccross]Y_2
\end{aligned}$$

The set $\mathcal{E}$ consists of all possible states of the crossing. $\mathcal{E}'$ contains all the configurations in which *car* has just happened, and this is then thinned to $\mathcal{E}_1$ by closing under derivation by non-*ccross* actions; $\mathcal{E}_1'$ is the set of non-*ccross* derivatives of $\mathcal{E}_1$ (and is therefore contained in $\mathcal{E}_1$); $\mathcal{E}_2$ is that subset of $\mathcal{E}_1$ failing the property Q; and $\mathcal{E}_3$ consists of the appropriate successors of $\mathcal{E}_2$. The success of this tableau is easily established: the inclusion conditions hold, and $\sqsubset$ relation for the least fixed point is well-founded because it is empty!

A simple demonstration of how the technique works for infinite state spaces is given by showing that the *Ticker* from section 2 has the property $\mu Z.[tick]Z$, that its observable behaviour is just to perform a *finite* number of ticks.

$$\{\,Ticker\,\} \vdash_{()} \mu Z.[tick]Z$$
$$\{\,Up_i, Down_j : i \geq 0, j \geq 0\,\} \vdash_{()} \mu Z.[tick]Z \qquad \Delta{=}(U{=}\mu Z.[tick]Z)$$
$${}_1\{\,Up_i, Down_j : i \geq 0, j \geq 0\,\} \vdash_{\Delta} U$$
$${}_2\{\,Up_i, Down_j : i \geq 0, j \geq 0\,\} \vdash_{\Delta} [tick]U$$
$${}_3\{\,Down_j : j \geq 0\,\} \vdash_{\Delta} U$$

For this tableau to be successful, we have to show that the μ-terminal leaf node (3) is successful. It clearly satisfies the inclusion condition. Recall that we have to show that $\sqsubset_{(1)}$ is well-founded, and that in this case $E \sqsupset_{(1)} F$ if there is a trail from E at (1) (the companion of (3)) to F at (3). Since there are no other μ-terminals, this is true exactly if there is a finite sequence of processes E_i $(0 \le i \le n,\ n \ge 0)$ such that $E = E_0 \stackrel{tick}{\Longrightarrow} E_1 \stackrel{tick}{\Longrightarrow} \cdots \stackrel{tick}{\Longrightarrow} E_n = F$. But since it can be seen from the process definitions that $E \stackrel{tick}{\Longrightarrow} F$ just in case either $E = Up_i$ and $F = Down_j$ for $i, j \ge 0$, or $E = Down_i$ and $F = Down_{i-1}$ for $i \ge 1$, (which can be proved formally by induction on i and j), $\sqsubset_{(1)}$ is well-founded.

We noted that *Ticker* satisfies the stronger property $[\![\,]\!]\mu Z.[tick](Z \wedge [-tick]\mathrm{ff})$, i.e. that although it may diverge initially, once it starts ticking, it does nothing else until it stops. We can prove this by using the system with $\stackrel{K}{\longrightarrow}$, and expanding the abbreviated modality. The formula becomes

$$\nu Y.(\mu Z.[tick](Z \wedge [-tick]\mathrm{ff})) \wedge [\tau]Y$$

with the following tableau, where $\Psi = \mu Z.[tick](Z \wedge [-tick]\mathrm{ff})$:

$$\dfrac{\{Ticker\} \vdash_{()} \nu Y.\Psi \wedge [\tau]Y}{\dfrac{\{Up_i, Down_j\} \vdash_{()} \nu Y.\Psi \wedge [\tau]Y}{\dfrac{\{Up_i, Down_j\} \vdash_{\Delta} U}{\dfrac{\{Up_i, Down_j\} \vdash_{\Delta} \Psi \wedge [\tau]U}{\dfrac{\{Up_i, Down_j\} \vdash_{\Delta} \Psi}{\dfrac{1\,\{Up_i, Down_j\} \vdash_{\Delta'} V}{\dfrac{2\,\{Up_i, Down_j\} \vdash_{\Delta'} [tick](V \wedge [-tick]\mathrm{ff})}{\dfrac{3\,\{Down_j : j \ge 0\} \vdash_{\Delta'} V \wedge [-tick]\mathrm{ff}}{4\,\{Down_j : j \ge 0\} \vdash_{\Delta'} V \qquad \dfrac{\{Down_j : j \ge 0\} \vdash_{\Delta'} [-tick]\mathrm{ff}}{\varnothing \vdash_{\Delta'} \mathrm{ff}}}}}}\;{\scriptstyle \Delta' = \Delta \cdot V = \Psi} \qquad \dfrac{\{Up_i, Down_j\} \vdash_{\Delta} [\tau]U}{\{Up_i : i \ge 1\} \vdash_{\Delta} U}}}\;{\scriptstyle \Delta = (U = \nu Y.\Psi \wedge [\tau]Y)}}}$$

Most rule applications here are straightforward. Again, to show that (4) is a successful μ-terminal, note that there is a trail from E at (1) to F at (4) just in case $E = E_0 \stackrel{tick}{\longrightarrow} \cdots \stackrel{tick}{\longrightarrow} E_n = F$ $(n \ge 1)$, and since $E \stackrel{tick}{\longrightarrow} F$ iff $E = Down_i$ and $F = Down_{i-1}$, $\sqsubset_{(1)}$ is well-founded.

The slot machine is infinite state, like *Ticker*, but also has a relatively large state space modulo an integer parameter, so as with the *Crossing* we shall leave it to the enthusiastic reader to verify our statements about the derivatives. The safety property, that the machine never pays out more than it has in its bank, has the following tableau where Q_0 indicates that the slot machine owes money:

$$\dfrac{\{SM_n : n \ge 0\} \vdash_{()} \nu Z.\neg Q_0 \wedge [-]Z}{\dfrac{\mathcal{E} \vdash_{()} \nu Z.\neg Q_0 \wedge [-]Z}{\dfrac{\mathcal{E} \vdash_{\Delta} U}{\dfrac{\mathcal{E} \vdash_{\Delta} \neg Q_0 \wedge [-]U}{\mathcal{E} \vdash_{\Delta} \neg Q_0 \qquad \dfrac{\mathcal{E} \vdash_{\Delta} [-]U}{\mathcal{E} \vdash_{\Delta} U}}}\;{\scriptstyle \Delta = (U = \nu Z.\neg Q_0 \wedge [-]Z)}}}$$

As in previous cases we first use the (Thinning) rule to enlarge the set of slot machines to the set $\mathcal{E}$ of derivatives of $\{SM_n : n \ge 0\}$: it is merely tedious to check that $\mathcal{E}$ can be partitioned into nine subsets (according to the state of each component: SM_n, $(bank.(lost.\overline{loss}.IO + release(y).\overline{win}(y).IO) \mid B_n \mid D)\backslash K$, etc.), each parametrized by an integer variable (or in some cases, two), and that this set is closed under $\stackrel{-}{\longrightarrow}$ and does not include any SM_j for $j < 0$.

The verification that the slot machine has the weak liveness property that a million pounds can be won infinitely often is given by the following successful tableau:

$$
\begin{array}{c}
\{\, SM_n : n \geq 0 \,\} \vdash_{()} \nu Y.\mu Z.\langle \overline{win}(10^6)\rangle Y \vee \langle - \rangle Z \\
\hline
\mathcal{E} \vdash_{()} \nu Y.\mu Z.\langle \overline{win}(10^6)\rangle Y \vee \langle - \rangle Z \\
\hline
\mathcal{E} \vdash_{\Delta} U \quad \scriptstyle \Delta=(U=\nu Y.\mu Z.\langle \overline{win}(10^6)\rangle Y \vee \langle - \rangle Z) \\
\hline
\mathcal{E} \vdash_{\Delta} \mu Z.\langle \overline{win}(10^6)\rangle U \vee \langle - \rangle Z \\
\hline
{}_1\,\mathcal{E} \vdash_{\Delta'} V \quad \scriptstyle \Delta'=\Delta\cdot V=\mu Z.\langle \overline{win}(10^6)\rangle U \vee \langle - \rangle Z \\
\hline
{}_2\,\mathcal{E} \vdash_{\Delta'} \langle \overline{win}(10^6)\rangle U \vee \langle - \rangle V \\
\hline
\begin{array}{c}
\mathcal{E}_1 \vdash_{\Delta'} \langle \overline{win}(10^6)\rangle U \\
\hline
\mathcal{E}_1' \vdash_{\Delta'} U
\end{array}
\qquad
\begin{array}{c}
{}_3\,\mathcal{E}_2 \vdash_{\Delta'} \langle - \rangle V \\
\hline
{}_4\,\mathcal{E}_2' \vdash_{\Delta'} V
\end{array}
\end{array}
$$

Some explanation is called for here, but again, the details are omitted. As before, $\mathcal{E}$ is the set of all derivatives. The vital rules in this tableau are the disjunction at node (1), where $\mathcal{E}_1$ is exactly those processes capable of performing a $\overline{win}(10^6)$ action, and $\mathcal{E}_2$ is the remainder; and the (Diamond) rule at node (3), where f is defined to ensure that $\mathcal{E}_1$ is eventually reached: for process with less than 10^6 in the bank, f chooses events leading towards loss, so as to increase the amount in the bank; and for processes with more than 10^6, f chooses to $\overline{release}(10^6)$. The formal proof requires partitioning $\mathcal{E}_2$ into several classes, each parametrized by an integer n, and showing that while $n < 10^6$, n is strictly increasing over a cycle through the classes; then when $n = 10^6$, f selects a successor that is not in $\mathcal{E}_2$, and so a chain $E_0 \xrightarrow{-} \cdots$ through nodes (1), (2), (3), (4) terminates, and therefore $\sqsubset_{(1)}$ is well-founded.

Finally, we consider the network of processes and relay stations. Recall that a weak liveness property is $\mu Z.\Psi \vee \langle - \rangle Z$ where Ψ is the formula $\bigwedge_{1 \leq j < i \leq n} \langle \overline{line}(j,i) \rangle \mathtt{tt}$ A successful tableau for this has the form:

$$
\begin{array}{c}
\{\text{initial config}\} \vdash_{()} \mu Z.\Psi \vee \langle - \rangle Z \\
\hline
\mathcal{E}_1 \vdash_{()} \mu Z.\Psi \vee \langle - \rangle Z \\
\hline
\mathcal{E}_1 \vdash_{\Delta} U \\
\hline
\mathcal{E}_1 \vdash_{\Delta} \Psi \vee \langle - \rangle U \\
\hline
\begin{array}{c}
\mathcal{E}_2 \vdash_{\Delta} \Psi \\
\hline
\begin{array}{c}
\mathcal{E}_2 \vdash_{\Delta} \langle \overline{line}(1,2) \rangle \mathtt{tt} \\
\hline
\mathcal{F}_{1,2} \vdash_{\Delta} \mathtt{tt}
\end{array}
\quad \cdots \quad
\begin{array}{c}
\mathcal{E}_2 \vdash_{\Delta} \langle \overline{line}(n-1,n) \rangle \mathtt{tt} \\
\hline
\mathcal{F}_{n-1,n} \vdash_{\Delta} \mathtt{tt}
\end{array}
\end{array}
\qquad
\begin{array}{c}
\mathcal{E}_3 \vdash_{\Delta} \langle - \rangle U \\
\hline
\mathcal{E}_4 \vdash_{\Delta} U
\end{array}
\end{array}
$$

We describe, fairly informally, the construction of a successful tableau. Choose a cell owned by P_1 and a cell owned by P_2. Draw a line on the grid from one cell to the other. Starting from the initial configuration, select P_1 to fire the $\overline{alloc}$ action for the chosen cell, and then similarly for P_2; then continue choosing $\overline{alloc}$ actions to allocate the next cell on the line along from P_1. When the line has been followed to the P_2 cell, that cell will be in the state $\overline{line}(1,2).\mathbf{0}$.

Repeat for the other pairs of processes, drawing each new line so that it does not cross any earlier line. When all pairs have been connected in this manner, the resulting configuration satisfies Ψ, and forms the sole member of the set $\mathcal{E}_2$; all the intervening configurations form $\mathcal{E}_3$, and $\mathcal{E}_4$ is $\mathcal{E}_3$ less the start state.

Now suppose a successful tableau exists. Follow the $\sqsupset$ relation from the initial state until it terminates, and each time through the diamond rule, draw a line on the grid corresponding to the action chosen: if the action involves the $\overline{alloc}_{i\,j+1}$ action of C_{ij}, draw a line from (i,j) to $(i,j+1)$. The result of this procedure is to draw disjoint lines between each pair of processes, for a $\overline{line}(k,k')$ action is possible only if a line from a P_k cell meets a line from a P_k' cell.

Now, since we stipulated that the cells owned by a given process be adjacent cells on a line, they may be collapsed to a point to give the complete graph on n vertices; since the lines do not cross, the graph is planar; and therefore no successful tableau exists if $n \geq 5$.

This system can be elaborated in many ways. For instance consider a persistent system that reconfigures after, say, a power failure. In this case, one would want to specify that after a $\overline{reset}$ action (which we assume causes all components to revert to their initial state) Ψ is achieved again, and that this cycle continues. Further, if we adopt the less restrictive requirement that Ψ should be that all $\overline{line}$ actions occur in any order, we have an example of a fairly complex cylic property. These properties can be expressed in the mu-calculus; however, the formulae become quite long. A simple cyclic property is that every run involves the sequence of labels $(ab)^{\omega}$. This is expressed by the following formula:

$$\nu Z.[-a]\mathrm{ff} \wedge \langle - \rangle \mathrm{tt} \wedge [-]([-b]\mathrm{ff} \wedge \langle - \rangle \mathrm{tt} \wedge [-]Z)$$

This illustrates the desirability of providing the user with a suitably chosen set of formula macros to express common properties succintly and, if feasible, to give derived tableau rules for such macros. This is one topic for further investigation: another important topic is the use of compositional reasoning within the system.

References

[1] J. Bradfield, 'Proving temporal properties of Petri Nets', *Proc. 11th Int. Conf. on Theory and Application of Petri Nets*, 1990.

[2] J. Bradfield, C. Stirling, 'Local model checking for infinite state spaces', Technical Report LFCS-90-115, University of Edinburgh 1990.

[3] E. M. Clarke, E. A. Emerson, A. P. Sistla, 'Automatic verification of finite-state concurrent systems using temporal logic specifications: a practical approach' *Tenth ACM Symposium on Principles of Programming Languages.* ACM, Austin, Texas, 1983.

[4] R. Cleaveland, 'Tableau-based model checking in the Propositional Mu-Calculus', Technical Report, University of Sussex 1988. (To appear in *Acta Informatica*)

[5] M. Hennessy, R. Milner, 'Algebraic laws for nondeterminism and concurrency', *J. Assoc. Comput. Mach.* 32, 137-161, 1985.

[6] D. Kozen, 'Results on the propositional mu-calculus', *Theoretical Computer Science* **27**, 333–354, 1983.

[7] K. Larsen, 'Proof systems for Hennessy-Milner logic with recursion', *Proc. CAAP* 1988. (To appear in *Inf. and Comp.*)

[8] R. Milner, *Communication and Concurrency*, Prentice-Hall 1989.

[9] V. Pratt, 'A decidable μ-calculus', *Proc. 22nd. FOCS*, 421–27, 1981.

.0] C. Stirling, 'Temporal logics for CCS', LNCS 354, 660-672, 1989.

.1] C. Stirling, D. J. Walker, 'Local model checking in the modal mu-calculus', *TAPSOFT 89* (369–382). LNCS 351, 1989. (To appear in *Theoretical Computer Science*)

2] C. Stirling, D. J. Walker, 'A general tableau technique for verifying temporal properties of concurrent programs', *Proc. Int. BCS-FACS Workshop on Semantics For Concurrency*, 1990.

3] D. J. Walker, 'Automated analysis of mutual exclusion algoritms using CCS', *Formal Aspects of Computing* 1, 273-292, 1989.

4] G. Winskel, 'Model checking the modal **nu-Calculus**', *Proc. ICALP* 1989.

Testing Equivalences and Fully Abstract Models for Probabilistic Processes

Ivan Christoff
Department of Computer Systems, Uppsala University
S-75120, Uppsala, Sweden

Abstract

We present a framework in which the observable behavior of probabilistic processes is distinguished through testing. Probabilistic transition systems are used to model the operational behavior of processes. The observable behavior of processes is studied in terms of probabilities for successful interaction with tests. Based on these probabilities three equivalences are defined. We define three denotational models, and show that each model contains exactly the necessary information for verification of one of the equivalences.

1 Introduction

Prompted by a growing trend towards distributed concurrent computing, much work has been devoted during the last decade to develop theories in which the observable behavior of communicating concurrent processes can be studied. Several such theories, commonly referred to as *process algebras*, have emerged, e.g. CCS by Milner [Mi 80], CSP by Hoare [Ho 85], and ACP by Bergstra and Klop [BK 84].

The work presented in this paper falls in the area of process theory, and is related to research dealing with models and relations for studying the observable behavior of: (i) non-probabilistic processes through testing [DH 84, Ab 87, Ph 87, He 88], and (ii) probabilistic processes [LS 89, GJS 90, GSST 90].

From the research addressing non-probabilistic processes, we have adopted the approach to use testing for distinguishing the observable behavior of processes. Of the work referenced in (i), our work is closest related to the work of De Nicola and Hennessy [DH 84]. The main difference is that the semantics of our testing notion are, in the terminology of Pnueli [Pn 85], "barbed semantics" (i.e. the outcome of the interaction of a process with a test implicitly depends on the observable events that can be performed by the process but are not offered in the test), while the semantics of their testing notion are "broom semantics" (i.e. the outcome of the interaction of a process with a test, only depends on the combinations of observable events offered in the last step of the test that can be performed by the process).

The work referenced in (ii) includes: a process model, testing notion, relations and logics, for probabilistic processes by Larsen and Skou [LS 89], an algebra for probabilistic processes (PCCS), by Giacalone, Jou and Smolka [GJS 90], and a classification and inter-relation of models for probabilistic processes by van Glabbeek et al [GSST 90]. Our work is closest related to the work of Larsen and Skou, through the approach to use testing to distinguish the observable behavior of probabilistic processes. The main differences between our work and theirs, concern process model and relations. In the terminology of van Glabbeek et al, we use a "generative model" (i.e. a model in which the probabilities in processes are normalized according to the events offered by the environment), while they use a "reactive model" (i.e. a model in which processes do not allow a choice between simultaneously offered events). Their relations for probabilistic processes are based on the notion of bisimulation, while ours are of test equivalence type.

The work presented in this paper is based on results from [Ch 89], where we introduced a testing notion, which was used to define relations for distinguishing the observable behavior of probabilistic processes. The main results of this paper are based on these relations (three partial orders and the corresponding equivalences). We have defined three denotational models, and shown that the models are *fully abstract* with respect to the equivalences, i.e. each model contains exactly the necessary information for verification of one of the equivalences.

The organization of this paper is: the process model and testing notion, are defined in section 2. The testing notion is used in section 3, to define three partial orders, and the corresponding equivalences, for probabilistic processes. An alternative view of tests and three denotational models, are given in section 4. We show that each of the denotational models is fully abstract with respect to one of the equivalences defined in section 3. A summary of results, and plans for future work, are given in section 5.

2 Operational Models

In this section we present operational models for: probabilistic processes, tests and testing (i.e. probabilistic processes interaction with tests). The testing notion will be used in section 3, to define relations between probabilistic processes. The operational models defined in this section are based on Plotkin's notion of labeled transition systems [Pl 81].

2.1 Processes

We shall use labeled transition systems, in which probabilities are assigned to transitions, to model the operational behavior of probabilistic processes.

Definition 2.1 A finite labeled *probabilistic transition system* (PTS) is a triple:

$$(S, E, \pi),$$

where
- S is a finite *set of states*, ranged over by s, s', $s1$, $s2$, etc.
- $E=L\cup\{\tau\}$ is a finite *set of events*, ranged over by e, e', etc., where L denotes a *set of observable events*, ranged over by a, a', etc., and τ denotes an *unobservable event.*
- $\pi : S \times E \times S \rightarrow [0,1]$, is a *transition probability function.* Let $S_{\text{can-act}}=\{s\in S \mid \exists e\in E, \exists s'\in S, \pi(s,e,s')>0\}$. A requirement for π is:

$$\text{(R1)} \qquad \forall s\in S_{\text{can-act}}: \qquad \Sigma_{e\in E, s'\in S}\,\pi(s,e,s')=1$$

The function π gives the probabilities for performing events at different states. For example, $\pi(s,e,s')$ denotes the probability for performing event e at state s, and moving to state s'. Requirement (R1) disallows 'idling' at states at which events can be performed.

We shall use the following notation:

$s \xrightarrow{e}_{p} s'$	to denote that $\pi(s,e,s') = p$,
$s \xrightarrow{e} s'$	to denote that $\pi(s,e,s') > 0$,
$s \xrightarrow{e}$	to denote that $\exists s'$: $\pi(s,e,s') > 0$,
$s \not\xrightarrow{e}$	to denote that $\neg\exists s'$: $\pi(s,e,s') > 0$,
$s \xrightarrow{\tau^n} s'$	to denote $s \xrightarrow{\tau} s_1 \xrightarrow{\tau} \ldots\ s_{n-1} \xrightarrow{\tau} s'$ (where $n\geq 0$),
$s \xrightarrow{\tau^n a}$	to denote $s \xrightarrow{\tau^n} s' \xrightarrow{a}$.

In addition, we shall use L^* to denote the *set of strings of observable events* (ranged over by σ, σ', etc.), and ε to denote the *empty string*. Juxtaposition will be used to denote concatenation of strings, e.g. the concatenation of the strings a and σ is denoted $a\sigma$. We consider the strings εa and a equal.

Definition 2.2 Let (S,E,π) be a PTS. A function for computing the states reached after performing a string in a PTS, $\mathit{After}: 2^S \times L^* \rightarrow 2^S$, is defined for all $s \in S, S_1, S_2 \subseteq S$ and $a, \sigma \in L^*$, inductively as:

1) $\mathit{After}(\{\},\sigma) = \{\}$
2) $\mathit{After}(\{s\},\varepsilon) = \{s' \in S \mid s \xrightarrow{\tau^n} s'\}$
3) $\mathit{After}(\{s\},a) = \bigcup_{s' \in S'} \mathit{After}(\{s'\},\varepsilon)$, where $S' = \{s'' \in S \mid s \xrightarrow{\tau^n a} s''\}$
4) $\mathit{After}(S_1 \cup S_2, a) = \mathit{After}(S_1,a) \cup \mathit{After}(S_2,a)$
5) $\mathit{After}(S_1, a\sigma) = \mathit{After}(\mathit{After}(S_1,a),\sigma)$

Intuitively, $\mathit{After}(\{s\},\sigma)$ defines the set of states reachable by τ's, after σ is performed from s.

Definition 2.3 (Probabilistic process) Let (S,E,π) be a PTS. We interpret each $s \in S$ as a *probabilistic process*, defined by (S_s, E, π_s), where: s is the initial state, $S_s = \{s' \in S \mid \exists \sigma \in L^*, s' \in \mathit{After}(\{s\},\sigma)\}$ defines the set of states reachable from s, and $\pi_s: S_s \times E \times S_s \rightarrow [0,1]$, defined for all $s', s'' \in S_s$ and $e \in E$ as $\pi_s(s',e,s'') = \pi(s',e,s'')$, characterizes the operational behavior of s.

Our graphical representation for probabilistic processes is illustrated in figure 2.1.

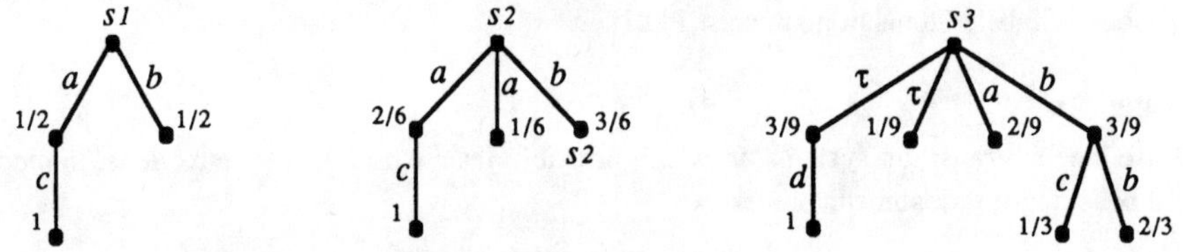

Figure 2.1 Probabilistic processes.

The processes are represented in a tree-like style, in which 'loops' are represented by a reference to a node, e.g. consider the 'loop' corresponding to the transition $(s2,b,s2)$ in $s2$.

So far, we have not discussed the impact of different environments on the ability of probabilistic processes to perform events. We have tacitly assumed, that all the events that can be performed by a process are provided by the environment. In the next subsection, we shall define the type of environments with which probabilistic processes can interact.

2.2 Tests

The environment of a probabilistic process is defined operationally, using the notion of labeled transition systems. We shall use the terms test and environment as synonyms.

Definition 2.4 A finite labeled *transition system* (TS) is a triple:

$$(T, L, \rightarrow),$$

where
- T is a finite *set of states*, ranged over by t, t', $t1$, etc.
- L is a finite *set of observable events*, ranged over by a, a', etc.
- $\rightarrow \subseteq T \times L \times T$, is a *transition relation*. A requirement for $\rightarrow$ is:

$$\text{(R2)} \qquad \forall t,t',t'' \in T,\ \forall a \in L:\ (t,a,t'),(t,a,t'') \in \rightarrow \quad \Rightarrow \quad t' = t''$$

The transition relation $\rightarrow$ represents the set of possible transitions for all events in L, at all states in T. For example, $(t,a,t')\in\rightarrow$ denotes the transition for performing event a at state t, and moving to state t'. Requirement (R2) disallows nondeterminism.

We shall use the following notation: $t\xrightarrow{a}t'$ to denote that $(t,a,t')\in\rightarrow$, $t\xrightarrow{a}$ to denote that $\exists t'\in T$: $(t,a,t')\in\rightarrow$, and $t\not\xrightarrow{a}$ to denote that $\neg\exists t'\in T$: $(t,a,t')\in\rightarrow$.

Definition 2.5 (Test) Let $(T,L,\rightarrow)$ be a TS. We interpret each $t\in T$ as a *test*, defined by $(T_t,L,\rightarrow_t)$, where: t is the initial state, $T_t=\{t'\in T \mid \exists\sigma\in L^*, t'\in \mathcal{After}(\{t\},\sigma)\}$ defines the set of states reachable from t, and $\rightarrow_t\subseteq T_t\times L\times T_t$ characterizes the operational behavior of t.

Definition 2.6 (Sequential tests) Let $(T,L,\rightarrow)$ be an arbitrary TS. We define the *set of sequential tests* in T, as:

$$T_{seq}=\{t\in T \mid \forall t',t'',t'''\in T_t, \forall a,a'\in L, (t',a,t''),(t',a',t''')\in\rightarrow \;\Rightarrow\; (a=a' \wedge t''=t''')\}$$

Intuitively, $T_{seq}\subseteq T$ denotes the set of tests which have at most one transition at each state. We shall refer to each $t\in T_{seq}$ as a *sequential test.*

Our graphical representation of tests is illustrated in figure 2.2.

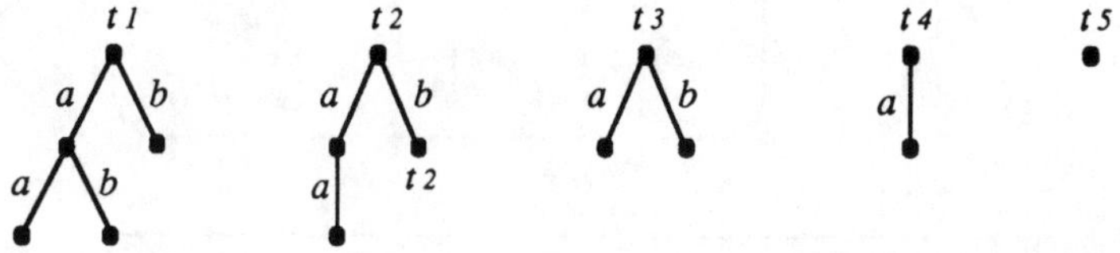

Figure 2.2 Tests.

The tests *t4* and *t5* are examples of sequential tests. Note that the graphical representation of tests, is very similar to the one of probabilistic processes.

2.3 Testing Probabilistic Processes

We shall present a testing notion, which defines how the behavior of a probabilistic process is restricted depending on the observable events that can be performed by the environment (i.e. a test is regarded as a specific environment).

Definition 2.7 (Testing system) Let $\mathcal{S}=(S,E,\pi)$ and $\mathcal{T}=(T,L,\rightarrow)$, be an arbitrary PTS and TS, respectively. We assume that $\mathcal{S}$ and $\mathcal{T}$ are constructed from the same set of observable events, i.e. $L=E-\{\tau\}$. A *testing system*, $\mathcal{TS}(\mathcal{S},\mathcal{T})$, is a quadruple:

$$(S,\, T,\, E,\, \pi'),$$

where $\pi' : (S\times T)\times E\times(S\times T)\rightarrow[0,1]$, is a *transition probability function*, defined by the following four rules:

1) $s\xrightarrow{\tau}s' \quad\Rightarrow\quad \pi'((s,t),\tau,(s',t)) = \pi(s,\tau,s') \,/\, \sum_{e\in \mathcal{Ev}(s,t), s'\in S}\pi(s,e,s')$

2) $s\xrightarrow{a}s',\ t\xrightarrow{a}t' \quad\Rightarrow\quad \pi'((s,t),a,(s',t')) = \pi(s,a,s') \,/\, \sum_{e\in \mathcal{Ev}(s,t), s'\in S}\pi(s,e,s')$

3) $s\xrightarrow{a}s',\ t\not\xrightarrow{a} \quad\Rightarrow\quad \pi'((s,t),a,(s',t)) = 0$

4) $s\not\xrightarrow{a} \quad\Rightarrow\quad \pi'((s,t),a,(s,t)) = 0$

where $\mathcal{Ev}(s,t)=\{a\in L \mid s\xrightarrow{a}, t\xrightarrow{a}\}\cup\{\tau \mid s\xrightarrow{\tau}\}$, is the set of events that can be performed at (s,t).

Note that the values of π', for the transitions possible at a state (s,t), are calculated from the values of π for the corresponding transitions at s. Since some of the observable transitions at s may not be possible at (s,t), due to lack of such transitions at t, the values of π and π' for corresponding transitions will in general be different. Note also that π' meets requirement (R1) (from definition 2.1), which means that a testing system can be regarded as a probabilistic process.

For testing systems, we shall adopt the notation for transitions from PTSs. For example, $(s,t)\xrightarrow{a}$ denotes that there exists $(s',t')\in S\times T$, such that $\pi'((s,t),a,(s',t'))>0$, while $(s,t)\xrightarrow{\tau}(s',t)$ denotes that $\pi'((s,t),\tau,(s',t)) > 0$, etc.

Let (S,T,E,π') be a testing system. We interpret each $(s,t)\in S\times T$ as a testing system, defined by $(S_s,T_t,E,\pi'_{(s,t)})$, where: (s,t) is the initial state, $S_s \times T_t \subseteq S \times T$ defines the set of states for (s,t), and $\pi'_{(s,t)}$: $(S_s\times T_t) \times E \times (S_s\times T_t) \to [0,1]$, defined for all $(s',t'),(s'',t'')\in S_s\times T_t$ and $e\in E$ as $\pi'_{(s,t)}((s',t'),e,(s'',t''))=\pi'((s',t'),e,(s'',t''))$, characterizes the operational behavior of (s,t).

Our graphical representation for testing systems is illustrated in figure 2.3.

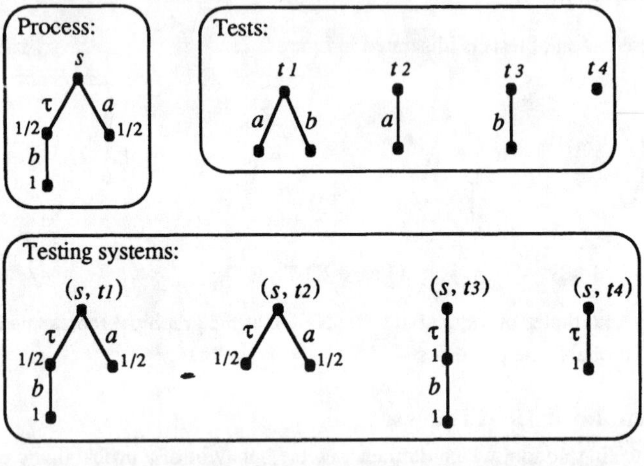

Figure 2.3 A probabilistic process, tests, and the corresponding testing systems.

Each testing system (s,ti), for $i\in\{1,2,3,4\}$, represents all possible interactions of s with ti. Consequently, (s,ti) models the behavior of s, when s is confined to an environment modeled by ti.

3 Three Equivalences

In this section we shall use the testing notion from section 2 to define relations for probabilistic processes. Processes are distinguished through two types of probabilities in testing systems: probabilities for performing strings of observable events, and probabilities for performing single observable events. We shall refer to the two different perspectives as *the sequential view*, and *the stepwise view*, respectively. The different views lead to alternative definitions of the relations. We shall show that the alternative definitions of the relations coincide.

3.1 The Sequential View

We shall present the sequential view of probabilities for performing strings of observable events in testing systems, and define functions with which such probabilities can be computed. The functions will be used to define three partial orders, and the corresponding equivalences.

To illustrate the sequential view we give an example. Consider the testing system shown in figure 3.1.

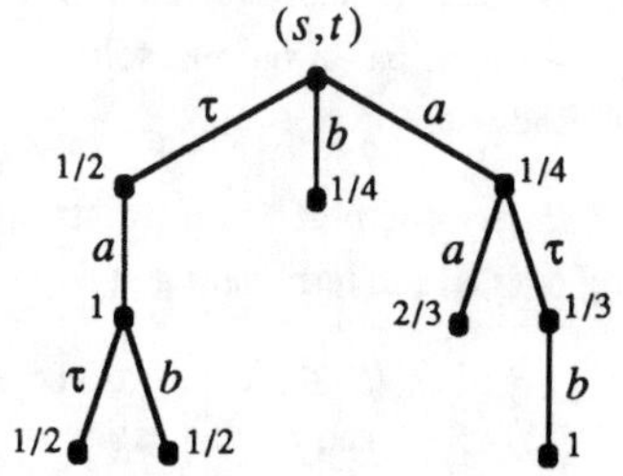

Figure 3.1 A testing system.

If we want to compute "the probability for performing the string *ab* from state (s,t)", the probabilities for all *ab* strings which can be performed starting from state (s,t), must be computed and added. These probabilities can be computed in a step-by-step manner. For our example, we compute first the probabilities for performing the *a* events (which are 1/2 and 1/4, respectively), and from the states reached immediately after the *a* events are performed, we compute the probabilities for performing *b* (which are 1/2 and 1/3, respectively). To compute "the probability for performing the string *ab* from state (s,t)", we multiply the probabilities for performing the events *a* and *b* (to obtain the probabilities for the two *ab* strings), and sum the resulting probabilities, i.e. $(1/2)\cdot(1/2) + (1/4)\cdot(1/3) = 1/3$.

To define a function, with which probabilities for strings of observable events in testing systems can be computed in the above illustrated step-by-step manner, we need to exactly define what we consider to be one step.

Definition 3.1 (Just-after states) Let $(S,T,L\cup\{\tau\},\pi')$ be testing system. A function for computing the states reached in a testing system immediately after a string of observable events is performed, $\mathcal{J}ust\text{-}after: 2^{S\times T}\times L^* \rightarrow 2^{S\times T}$, is defined for all $(s,t)\in S\times T$, $S',S''\subseteq S\times T$ and $a,\sigma\in L^*$, inductively as:

1) $\mathcal{J}ust\text{-}after(\{\},\sigma) = \{\}$
2) $\mathcal{J}ust\text{-}after(S',\varepsilon) = S'$
3) $\mathcal{J}ust\text{-}after(\{(s,t)\},a) = \{(s',t')\in S\times T \mid (s,t)\xrightarrow{\tau^n a}(s',t')\}$
4) $\mathcal{J}ust\text{-}after(S'\cup S'',a) = \mathcal{J}ust\text{-}after(S',a) \cup \mathcal{J}ust\text{-}after(S'',a)$
5) $\mathcal{J}ust\text{-}after(S',a\sigma) = \mathcal{J}ust\text{-}after(\mathcal{J}ust\text{-}after(S',a),\sigma)$

Intuitively, $\mathcal{J}ust\text{-}after(\{(s,t)\},\sigma)$ defines the set of states reached immediately after the string of observable events σ is performed from (s,t), i.e. no τ's are performed after the last observable event in σ is performed. We can now define a function which computes the probability for performing a string σ, starting at a state (s,t) and terminating at a specific state just-after σ.

Definition 3.2 (String probability) Let $(S,T,L\cup\{\tau\},\pi')$ be a testing system. A function, $\mathcal{P}': (S\times T)\times L^* \times (S\times T) \rightarrow [0,1]$, is defined for all $(s,t)\in S\times T$ and $a,\sigma\in L^*$, inductively as:

1) $\mathcal{P}'((s,t),\varepsilon,(s,t)) = 1$
2) $\mathcal{P}'((s,t),a,(s',t')) =$ if $(s',t')\notin \mathcal{J}ust\text{-}after(\{(s,t)\},a)$ then 0
   else $\sum_{(s',t)\in S'} \pi'((s,t),\tau,(s',t))\cdot\mathcal{P}'((s',t),a,(s',t')) + \pi'((s,t),a,(s',t'))$
3) $\mathcal{P}'((s,t),a\sigma,(s',t')) = \sum_{(s'',t'')\in \mathcal{J}ust\text{-}after(\{(s,t)\},a)} \mathcal{P}'((s,t),a,(s'',t''))\cdot\mathcal{P}'((s'',t''),\sigma,(s',t'))$

where $S'=\{(s',t)\in S\times T \mid (s,t)\xrightarrow{\tau}(s',t)\}$.

Intuitively, $\mathcal{P}'((s,t),\sigma,(s',t'))$ is the probability for performing σ, starting at (s,t) and terminating at (s',t'). Note from (2), that in general a linear equation system needs to be solved in order to compute $\mathcal{P}'((s,t),a,(s',t'))$. The linear equation system has a unique solution, since it consists of equally many linearly independent equations and unknowns.

We can now define the *total probability* for performing a string σ starting at (s,t), as the sum of probabilities for performing σ starting at (s,t), and terminating at any state just-after σ.

Definition 3.3 (Total string probability) Let $(S,T,L\cup\{\tau\},\pi')$ be a testing system. A function, $\mathcal{P}$: $(S \times T) \times L^* \rightarrow [0,1]$, is defined for all $(s,t)\in S\times T$ and $\sigma\in L^*$, as:

$$\mathcal{P}((s,t),\sigma) = \Sigma_{(s',t')\in \mathcal{J}ust\text{-}after(\{(s,t)\},\sigma)}\, \mathcal{P}'((s,t),\sigma,(s',t'))$$

Intuitively, $\mathcal{P}((s,t),\sigma)$ is the probability for performing the string of observable events σ, starting at (s,t). We shall give an example of how $\mathcal{P}((s,t),\sigma)$ can be computed. Consider the testing system shown in figure 3.2.

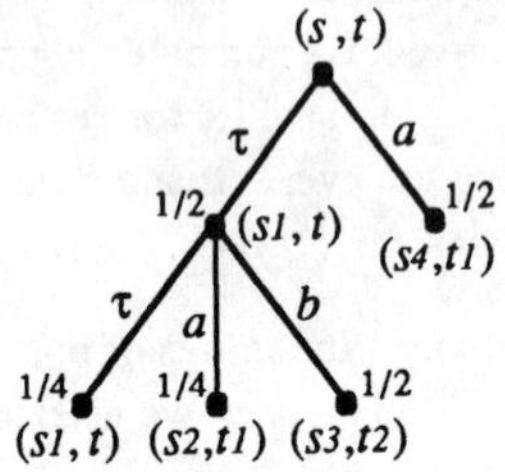

Figure 3.2 Computing $\mathcal{P}((s,t),a)$.

According to definition 3.3, to compute $\mathcal{P}((s,t),a)$ we need to compute $\mathcal{P}'((s,t),a,(s2,t1))$ and $\mathcal{P}'((s,t),a,(s4,t1))$. We use rule (2) in definition 3.2, which gives us two linear equation systems:

$$\text{(i)} \quad \begin{cases} \mathcal{P}'((s,t),a,(s2,t1)) = \pi'((s,t),\tau,(s1,t))\cdot\mathcal{P}'((s1,t),a,(s2,t1)) + \pi'((s,t),a,(s2,t1)) \\ \mathcal{P}'((s1,t),a,(s2,t1)) = \pi'((s1,t),\tau,(s1,t))\cdot\mathcal{P}'((s1,t),a,(s2,t1)) + \pi'((s1,t),a,(s2,t1)) \end{cases}$$

$$\text{(ii)} \quad \begin{cases} \mathcal{P}'((s,t),a,(s4,t1)) = \pi'((s,t),\tau,(s1,t))\cdot\mathcal{P}'((s1,t),a,(s4,t1)) + \pi'((s,t),a,(s4,t1)) \\ \mathcal{P}'((s1,t),a,(s4,t1)) = 0 \end{cases}$$

Solving equation systems (i) and (ii), we get: $\mathcal{P}'((s,t),a,(s2,t1))=1/6$ and $\mathcal{P}'((s,t),a,(s4,t1))=1/2$. Consequently $\mathcal{P}((s,t),a) = 2/3$.

We shall now use the function $\mathcal{P}$ to define three partial orders for probabilistic processes.

Definition 3.4 (Partial orders) Let $(S,T,L\cup\{\tau\},\pi')$ be a testing system. Three partial orders are defined for all probabilistic processes $s,s'\in S$,

1) $s \leq_{tr} s' \iff \forall t\in T_{seq}, \forall\sigma\in L^*: \quad \mathcal{P}((s,t),\sigma) \leq \mathcal{P}((s',t),\sigma)$

2) $s \leq_{wte} s' \iff \forall t\in T, \forall\sigma\in L^*: \; \Sigma_{a\in L}\mathcal{P}((s,t),\sigma a)/\mathcal{P}((s,t),\sigma) \leq \Sigma_{a\in L}\mathcal{P}((s',t),\sigma a)/\mathcal{P}((s',t),\sigma)$

3) $s \leq_{ste} s' \iff \forall t\in T, \forall\sigma\in L^*: \quad \mathcal{P}((s,t),\sigma) \leq \mathcal{P}((s',t),\sigma)$

The following observation can be made about the strength of the partial orders.

Proposition 3.1 (Strength of partial orders) Let $(S, T, L\cup\{\tau\}, \pi')$ be a testing system. For all probabilistic processes $s,s'\in S$,

$$(s \leq_{\text{tr}} s') \;\Leftarrow\; (s \leq_{\text{wte}} s') \;\Leftarrow\; (s \leq_{\text{ste}} s')$$

Proof It is obvious that $(s \leq_{\text{wte}} s') \Leftarrow (s \leq_{\text{ste}} s')$, since $(s \leq_{\text{ste}} s')$ requires that for all strings σ and all tests t, $\mathcal{P}((s,t),\sigma) \leq \mathcal{P}((s',t),\sigma)$, while $(s \leq_{\text{wte}} s')$ only puts requirements on sums of probabilities for strings of observable events. The claim $(s \leq_{\text{tr}} s') \Leftarrow (s \leq_{\text{wte}} s')$ can be verified by observing that $(s \leq_{\text{tr}} s')$ is a special case of $(s \leq_{\text{wte}} s')$, namely when only sequential tests are considered. □

Definition 3.5 (Equivalences) Let $(S,T,L\cup\{\tau\},\pi')$ be a testing system. Three equivalences: *probabilistic trace equivalence* ($=_{\text{tr}}$), *weak probabilistic test equivalence* ($=_{\text{wte}}$), and *strong probabilistic test equivalence* ($=_{\text{ste}}$), are defined for all $s,s'\in S$ and $i\in\{\text{tr},\text{wte},\text{ste}\}$:

$$s =_i s' \quad\Leftrightarrow\quad (s \leq_i s') \wedge (s' \leq_i s)$$

Intuitively, $s=_{\text{tr}}s'$ means that s and s' have the same deadlock properties for interactions with sequential tests, $s=_{\text{wte}}s'$ means that after performing any string of observable events σ, s and s' have the same deadlock properties for interactions with any test, while $s=_{\text{ste}}s'$ means that s and s' have the same deadlock properties for interactions with any test, and the same 'successful terminations' (i.e. probabilities for performing specific strings of observable events).

Proposition 3.2 (Strength of equivalences) Let $(S,T,L\cup\{\tau\},\pi')$ be a testing system. For all probabilistic processes $s,s'\in S$,

$$(s =_{\text{tr}} s') \;\Leftarrow\; (s =_{\text{wte}} s') \;\Leftarrow\; (s =_{\text{ste}} s')$$

Proof Follows directly from definition 3.5 and proposition 3.1. □

Examples that illustrate the equivalences are given in figure 3.3.

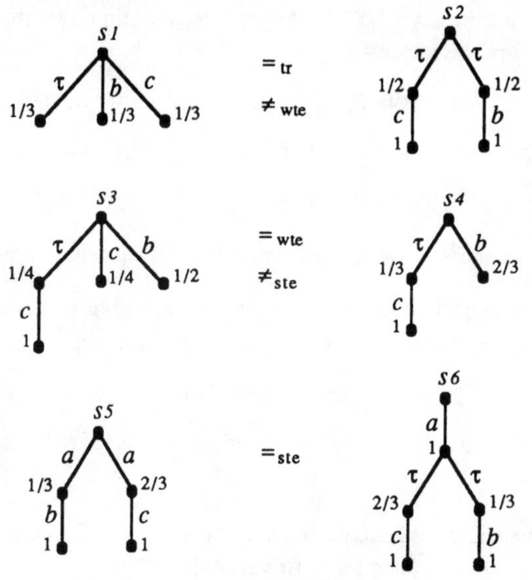

Figure 3.3 The equivalences.

The test for which $s1 \neq_{\text{wte}} s2$ and $s3 \neq_{\text{ste}} s4$, is: b or c. For that test, the sums of probabilities for $s1$ and

$s2$ are 2/3 and 1, respectively, while the probabilities for performing c for $s3$ and $s4$ are 1/2 and 1/3, respectively.

3.2 The Stepwise View

We shall present the stepwise view of probabilities for strings of observable events in testing systems, and define functions with which such probabilities can be computed. The functions will be used to give an alternative definition of the relations from subsection 3.1. We shall show that the two definitions of the relations coincide. The main reason for introducing the stepwise view, and the corresponding alternative definitions for the relations, is the proof of the full abstraction results in section 4.

To illustrate the stepwise view we shall give an example. Consider figure 3.1. Instead of computing the probability for performing the string ab from state (s,t), we may wish to compute "the probability for performing b from the just-after a states". We can compute that probability, by dividing the probability for performing ab from (s,t), with the probability for performing a from (s,t). For our example, "the probability for performing b from the just-after a states", is 4/9. We shall now define a function with which such probabilities can be computed.

Definition 3.6 Let $(S,T,L\cup\{\tau\},\pi')$ be a testing system. A function for computing the probability for performing an observable event just-after a string of observable events is performed, $\mathcal{R}: (S \times T) \times L^* \to [0,1]$, is defined for all $(s,t)\in S \times T$ and $a,\sigma \in L^*$, as:

1) $\mathcal{R}((s,t),\varepsilon) = 1$

2) $\mathcal{R}((s,t),\sigma a) = \mathcal{P}((s,t),\sigma a) / \mathcal{P}((s,t),\sigma)$

Intuitively, $\mathcal{R}((s,t),\sigma a)$ computes the probability for performing a, from the states reached just-after σ is performed from (s,t). We shall use the function $\mathcal{R}$ to give alternative definitions for the partial orders, and the corresponding equivalences, defined in subsection 3.1.

Definition 3.7 (Partial orders) Let $(S,T,L\cup\{\tau\},\pi')$ be a testing system. Three partial orders are defined for all probabilistic processes $s,s'\in S$, as:

1) $s <<_{\mathrm{tr}} s' \quad \Leftrightarrow \quad \forall t\in T_{\mathrm{seq}}, \forall\sigma\in L^*: \quad \mathcal{R}((s,t),\sigma) \le \mathcal{R}((s',t),\sigma)$

2) $s <<_{\mathrm{wte}} s' \quad \Leftrightarrow \quad \forall t\in T, \forall\sigma\in L^*: \quad \Sigma_{a\in L}\mathcal{R}((s,t),\sigma a) \le \Sigma_{a\in L}\mathcal{R}((s',t),\sigma a)$

3) $s <<_{\mathrm{ste}} s' \quad \Leftrightarrow \quad \forall t\in T, \forall\sigma\in L^*: \quad \mathcal{R}((s,t),\sigma) \le \mathcal{R}((s',t),\sigma)$

Using the above defined partial orders we can now define the corresponding equivalences.

Definition 3.8 (Equivalences) Let $(S,T,L\cup\{\tau\},\pi')$ be a testing system. Three equivalences are defined for all probabilistic processes $s,s'\in S$ and $\mathrm{i}\in\{\mathrm{tr},\mathrm{wte},\mathrm{ste}\}$, as:

$$s \equiv_{\mathrm{i}} s' \quad \Leftrightarrow \quad (s <<_{\mathrm{i}} s') \wedge (s' <<_{\mathrm{i}} s)$$

We shall give a theorem that relates the partial order definitions based on the sequential view, and on the stepwise view.

Theorem 3.1 (Alternative characterization of equivalences) Let $(S,T,L\cup\{\tau\},\pi')$ be a testing system. For all probabilistic processes $s,s'\in S$ and for $\mathrm{i}\in\{\mathrm{tr},\mathrm{wte},\mathrm{ste}\}$,

$$s =_{\mathrm{i}} s' \quad \Leftrightarrow \quad s \equiv_{\mathrm{i}} s'$$

Proof See appendix.

We have shown that for i∈{tr,wte,ste} an alternative characterization of $=_i$ can be given in terms of $\equiv_i$. In section 4 we shall use the intuition behind $\equiv_i$ to construct three denotational models for probabilistic processes.

4 Fully Abstract Models

We present an alternative notion for tests, which is used to define three denotational models for probabilistic processes. Each of these models is shown to be fully abstract with respect to one of the equivalences defined in section 3 (i.e. each denotational model contains exactly the necessary information for verification of one equivalence).

4.1 Offerings

An alternative notion for tests, called *offerings*, and functions for computing probabilities for strings of observable events for probabilistic processes interacting with offerings, are defined. These functions are used for the definition of the denotational models.

Definition 4.1 (Offerings) Let L be a set of observable events. The *set of offerings* for L is defined as: $O=2^L-\{\}$. To range over O we use L', L'', etc.

We shall use O^* to denote the *set of strings of offerings* (ranged over by o, o', etc.), and λ to denote the *empty string*. Juxtaposition will be used to denote concatenation, e.g. the concatenation of the strings L' and o is denoted L'o. We consider the strings $\lambda L'$ and L' equal.

Definition 4.2 Let $(S,L\cup\{\tau\},\pi)$ be a PTS. A function for computing probabilities for strings of observable events, when probabilistic processes interact with offerings, $Q'\colon S\times O^*\times L^*\times S\rightarrow[0,1]$, is defined for all $s,s'\in S$, $L',o\in O^*$ and $a,\sigma\in L^*$, inductively as:

1) $Q'(s,\lambda,a,s')=0$

2) $Q'(s,o,\varepsilon,s)=1$

3) $Q'(s,L',a,s')=$ if $s'\notin \mathit{Just\text{-}after}(\{s\},a)$ then 0

$$\text{else } \Sigma_{s''\in S'}\pi_{L'}(s,\tau,s'')\cdot Q'(s'',L',a,s')+\pi_{L'}(s,a,s')$$

4) $Q'(s,L'o,a\sigma,s')=\Sigma_{s''\in \mathit{Just\text{-}after}(\{s\},a)}Q'(s,L',a,s'')\cdot Q'(s'',o,\sigma,s')$

where $S'=\{s'''\in S\mid s\xrightarrow{\tau}s'''\}$, and for all $e\in E$, $\pi_{L'}(s,e,s')=\pi(s,e,s')/\Sigma_{e'\in L'\cup\{\tau\},s'\in S}\pi(s,e',s')$.

Intuitively, $Q'(s,o,\sigma,s')$ is the probability for performing the string σ, starting at s and terminating at s', where o defines the events offered when σ is performed. For example, $Q'(s,L'L'',ab,s')$ is the probability for performing the string ab, starting at s and terminating at s', where the events in L' are offered when a is performed, while the events in L'' are offered when b is performed. Note from rule (3) that in general a linear equation system has to be solved in order to compute $Q'(s,o,\sigma,s')$.

Definition 4.3 Let $(S,L\cup\{\tau\},\pi)$ be a PTS. A function for computing the total probability for performing a string of observable events, $Q\colon S\times O^*\times L^*\rightarrow[0,1]$, is defined for all $s\in S$, $o\in O^*$ and $\sigma\in L^*$, as:

$$Q(s,o,\sigma)=\Sigma_{s'\in \mathit{Just\text{-}after}(\{s\},\sigma)}Q'(s,o,\sigma,s')$$

Intuitively, $Q(s,o,\sigma)$ is the probability for performing the string σ starting at s, under the restriction that only events in o can be performed.

We shall use the string function | |, on strings of events and strings of offerings. | | has its usual meaning, i.e. $|\sigma|$ and $|o|$ denotes the length of σ and o respectively, with $|\varepsilon|=|\lambda|=0$. For example, $|L'L''|=2$.

Definition 4.4 Let $(S, L\cup\{\tau\}, \pi)$ be a PTS. A function for computing probabilities for observable events from the states reached just-after a string σ is performed in a PTS, $\mu : S \times O^* \times L^* \rightarrow [0,1]$, is defined for all $s\in S$, L',$o\in O^*$ and $a,\sigma\in L^*$, as:

1) $\mu(s,\lambda,\varepsilon) = 1$
2) $|o| < |\sigma| : \mu(s,o,\sigma) = 0$
3) $|o| > |\sigma| : \mu(s,o,\sigma) = \mu(s,o',\sigma)$ where $|o'| = |\sigma|$, and $o'o''=o$
4) $|o| = |\sigma| : \mu(s,oL',\sigma a) = Q(s,oL',\sigma a) / Q(s,o,\sigma)$

Intuitively, $\mu(s,oL',\sigma a)$ computes the probability for performing a, from the states reached immediately after σ is performed from s, under the restriction that only events in L' can be performed.

Definition 4.5 Let L^* be a set of strings. We define a function that transforms a string of events to a string of offerings, $Sets : L^* \rightarrow O^*$, for all $\sigma\in L^*$, inductively as: (1) $Sets(\varepsilon)=\lambda$, (2) $Sets(a)=\{a\}$ and (3) $Sets(a\sigma)=Sets(a)Sets(\sigma)$, where juxtaposition denotes concatenation.

Intuitively, $Sets(\sigma)$ transforms a string of events to a string of offerings, where each offering contains only one event from σ. For example, $Sets(abc) = \{a\}\{b\}\{c\}$.

4.2 Three Denotational Models

We shall use the μ function to define denotational models for probabilistic processes.

Definition 4.6 (Denotational models) Let $(S, L\cup\{\tau\}, \pi)$ be a PTS. Three denotational models: *probabilistic trace result systems* ($\mathcal{RS}_{tr}$), *weak probabilistic test result systems* ($\mathcal{RS}_{wte}$), and *strong probabilistic test result systems* ($\mathcal{RS}_{ste}$), are defined for all $s\in S$, and $i\in\{tr,wte,ste\}$, as:

$$\mathcal{RS}_i[\![s]\!] = (L, \mu_i),$$

where μ_i is a *probability function*, with

- $\mu_{tr} : L^* \rightarrow [0,1]$ defined as: $\mu_{tr}(\sigma) = \mu(s,Sets(\sigma),\sigma)$
- $\mu_{wte} : O^* \times L^* \rightarrow [0,1]$ defined as: $\mu_{wte}(oL',\sigma a) = \Sigma_{a\in L}\mu(s,oL',\sigma a)$
- $\mu_{ste} : O^* \times L^* \rightarrow [0,1]$ defined as: $\mu_{ste}(o,\sigma) = \mu(s,o,\sigma)$

As a shorthand notation for $\mathcal{RS}_i[\![s]\!]$ we shall use $[\![s]\!]_i$. Two denotations are considered equivalent if and only if the corresponding probability functions are equivalent. For example, let $[\![s]\!]_{tr}=(L,\mu_{tr})$ and $[\![s']\!]_{tr}= (L, \mu'_{tr})$, then:

$$[\![s]\!]_{tr} = [\![s']\!]_{tr} \quad \Leftrightarrow \quad \forall\sigma\in L^*: \quad \mu_{tr}(\sigma) = \mu'_{tr}(\sigma)$$

Our intention with the $\mathcal{RS}_{tr}$, $\mathcal{RS}_{wte}$ and $\mathcal{RS}_{ste}$ models, was to define interpretations for probabilistic processes which contain exactly the necessary information for establishing probabilistic trace equivalence, probabilistic test equivalence and strong probabilistic test equivalence, respectively. The finest level of distinction provided by our testing notion, is reflected by the $\mathcal{RS}_{ste}$ model.

Proposition 4.1 (Relation between the denotational models) Let $(S, L\cup\{\tau\}, \pi)$ be a PTS. For all probabilistic processes $s,s'\in S$,

$$[\![s]\!]_{ste} = [\![s']\!]_{ste} \quad \Rightarrow \quad [\![s]\!]_{wte} = [\![s']\!]_{wte} \quad \Rightarrow \quad [\![s]\!]_{tr} = [\![s']\!]_{tr}$$

Proof Obvious from the definitions of the functions: μ_{ste}, μ_{wte}, and μ_{tr}. □

4.3 Full Abstraction Theorem

We shall define full abstraction in our setting, and show that each of the denotational models is fully abstract with respect to one of the equivalences defined in section 3.

Full abstraction is often shown in relation to a language, e.g. see [HP 79, BCL 85]. However, the concept is not only relevant in a language context, and can be established for models without the use of a language, e.g. see [Jo 89]. We shall give a generic definition of full abstraction in the style of [HP 79, Jo 89], and interpret it in our setting.

We define the *context of a process*, $\mathcal{C}[\cdot]$, as an environment with which the process can interact. The dot can be regarded as a 'place holder'. When a probabilistic process s is considered in a context, the place holder $\cdot$ is replaced with s. In our setting a context is interpreted as a test.

Definition 4.7 (Full abstraction) Let $\mathcal{D}_i$ and $\mathcal{O}$ be models for probabilistic processes, and $=_i$, be an equivalence for processes in $\mathcal{O}$. The model $\mathcal{D}_i$ is considered *fully abstract* with respect to $=_i$, if for all probabilistic processes s and s'

$$\text{for all contexts } \mathcal{C}[\cdot]: \quad \mathcal{O}(\mathcal{C}[s]) =_i \mathcal{O}(\mathcal{C}[s']) \quad \Leftrightarrow \quad \mathcal{D}_i(s) = \mathcal{D}_i(s')$$

In our setting, $i \in \{tr, wte, ste\}$. For a specific test $t \in \boldsymbol{T}$: $\mathcal{C}_t[s]=(s,t)$, while $\mathcal{O}$ refers to the operational characterization for probabilistic processes in terms of the function $\mathcal{P}$. For probabilistic trace equivalence, and strong probabilistic test equivalence, $\mathcal{O}(\mathcal{C}_t[s])$ corresponds to $\forall \sigma \in \boldsymbol{L}^*$: $\mathcal{P}((s,t),\sigma)$ (the distinction between the equivalences is by the type of tests used), while for weak probabilistic test equivalence $\mathcal{O}(\mathcal{C}_t[s])$ corresponds to $\forall \sigma \in \boldsymbol{L}^*$: $\Sigma_{a \in L} \mathcal{P}((s,t),\sigma a)/\mathcal{P}((s,t),\sigma)$. $\mathcal{D}_i(s)$ corresponds to $[\![s]\!]_i$.

We shall now phrase the main result of the paper, by claiming full abstraction of the denotational models with respect to the equivalences defined for the operational model in section 3.

Theorem 4.1 (Full abstraction for the denotational models) Let $(\boldsymbol{S}, \boldsymbol{L} \cup \{\tau\}, \pi)$ be a PTS. For any two probabilistic processes $s,s' \in \boldsymbol{S}$ and $i \in \{tr, wte, ste\}$,

$$s =_i s' \quad \Leftrightarrow \quad [\![s]\!]_i = [\![s']\!]_i$$

Proof See appendix.

In [Ch 90], we have defined representations for the denotational models in terms of labeled trees.

5 Summary of Results

We have presented an operational model, and a testing notion for probabilistic processes. The observable behavior of probabilistic processes is distinguished through probabilities for processes successful interaction with tests. Using the testing notion, three partial orders and the corresponding equivalences ($=_{tr}$, $=_{wte}$, and $=_{ste}$), were defined.

To better reflect the nature of the equivalences, we defined three denotational models: $\mathcal{RS}_{tr}$, $\mathcal{RS}_{wte}$, and $\mathcal{RS}_{ste}$. We showed that each of the models contains exactly the necessary information for verification of one of the equivalences. Based on the denotational models, algorithms for verification of the relations have been developed and implemented [Ak 89, Ch 89].

We have currently defined a language with composition for probabilistic processes [Ch 90], in which we plan to give axiomatic characterization of the equivalences. An interesting topic for future work would be, following the line of work by Larsen and Skou [LS 89], to define a logic for our process model, and relate the different notions of equivalence between probabilistic processes to sets of formulas that must be satisfied by the processes in order for the processes to be equivalent.

Acknowledgements

I would like to thank: Linda Christoff, Bengt Jonsson, Kim G. Larsen, and John Tucker, for reading, discussing and providing helpful comments on drafts of this manuscript. This work has been supported by the Swedish National Board for Technical Development.

References

[Ab 87] S. ABRAMSKY. Observation equivalence as a testing equivalence. *Theor. Comp. Sci.* **53**, pp 225-241, 1987.

[Ak 89] N. AKINAGA. Decision Procedures for Relations on Probabilistic Transition Systems. MSc thesis, Department of Computer Systems, Uppsala University, Uppsala, Sweden, 1989. Available as report DoCS 89/19.

[BK 84] J.A. BERGSTRA, J.W. KLOP. Process algebra for synchronous communication. *Information and Control* **60**, pp 109-137, 1984.

[BCL 85] G. BERRY, P.-L. CURIEN, J.-J. LÉVY. Full abstraction for sequential languages: the state of the art. *Algebraic Methods in Semantics*, eds. M. Nivat and J.C. Reynolds, Cambridge University Press, pp 89-132, 1985.

[Ch 89] I. CHRISTOFF. Distinguishing probabilistic processes through testing. Manuscript presented at *Nordic Workshop on Program Correctness*, Uppsala, Sweden, 1989.

[Ch 90] I. CHRISTOFF. Testing Equivalences for Probabilistic Processes. PhD thesis, Department of Computer Systems, Uppsala University, Uppsala, Sweden, 1990. In preparation.

[DH 84] R. DE NICOLA, M. HENNESSY. Testing equivalences for processes. *Theor. Comp. Sci.* **34**, pp 83-133, 1984.

[GJS 90] A. GIACALONE, C.-C. JOU, S.A. SMOLKA. Algebraic reasoning for probabilistic concurrent systems. *Proc. Working Conf. on Programming Concepts and Methods (IFIP TC2)*, Sea of Galilee, Israel, 1990.

[GSST 90] R.v. GLABBEEK, S.A. SMOLKA, B. STEFFEN, C. TOFTS. Reactive, generative, and stratified models of probabilistic processes. *Proc. 5th IEEE Symp. on Logic in Computer Science*, Philadelphia, Pennsylvania, 1990.

[He 88] M. HENNESSY. *Algebraic Theory of Processes*, MIT Press, 1988.

[HP 79] M. HENNESSY, G. PLOTKIN. Full abstraction for a simple parallel programming language. *Proc. 8th Symp. on Mathematical Foundations of Computer Science, Lecture Notes in Computer Science* **74**, pp 108-120, 1979.

[Ho 85] C.A.R. HOARE. *Communicating Sequential Processes*. Prentice-Hall, 1985.

[Jo 89] B. JONSSON. A fully abstract trace model for dataflow networks. *Proc. 16th ACM Symp. on Principles of Programming Languages*, pp 155-165, 1989.

[LS 89] K.G. LARSEN, A. SKOU. Bisimulation through probabilistic testing. *Proc. 16th ACM Symp. on Principles of Programming Languages*, pp 344-352, 1989.

[Mi 80] R. MILNER. A Calculus of Communicating Systems. *Lecture Notes in Computer Science* **92**, 1980.

[Ph 87] I. PHILLIPS. Refusal testing. *Theor. Comp. Sci.* **50**, pp 241-284, 1987.

[Pl 81] G. PLOTKIN. A Structural Approach to Operational Semantics. Technical Report DAIMI FN-19, Computer Science Department, Aarhus University, Aarhus, Denmark, 1981.

[Pn 85] A. PNUELI. Linear and branching structures in the semantics and logics of reactive systems. *Proc. 12th Intl. Coll. on Automata, Languages and Programming, Lecture Notes in Computer Science* **194**, pp 15-32, 1985.

Appendix: Proofs

We give proof outlines for theorems 3.1 and 4.1. Theorem 3.1, the alternative characterization theorem, is used for the proof of theorem 4.1, the full abstraction theorem.

Proof Outline for the Alternative Characterization Theorem

The main idea behind the proof of the alternative characterization theorem is to relate the functions $\mathcal{P}$ and $\mathcal{R}$ to each other, and use that result to show that for i∈{tr,wte,ste} the equivalences $=_i$ and $\equiv_i$ coincide.

Definition 1 Let L be a set of observable events. A function for computing the substrings of a string of events, $\mathit{Prefix}: L^* \rightarrow 2^{L^*}$, is defined for all $a,\sigma \in L^*$, inductively as: (1) $\mathit{Prefix}(\varepsilon)=\{\varepsilon\}$ and (2) $\mathit{Prefix}(\sigma a)=\mathit{Prefix}(\sigma)\cup\{\sigma a\}$.

Lemma 1 The functions $\mathcal{P}$ and $\mathcal{R}$ are related in the following manner:

$$\forall t\in \boldsymbol{T}, \forall \sigma\in \boldsymbol{L}^*: \qquad \mathcal{P}((s,t),\sigma) = \prod_{\sigma'\in Prefix(\sigma)} \mathcal{R}((s,t),\sigma')$$

Proof The proof is by induction on the length of σ. It is trivial to establish that the lemma holds for $|\sigma|=0$. We assume that the lemma holds for all $\sigma\in \boldsymbol{L}^*$ such that $|\sigma| \leq |\sigma_1|$, and that there exists $\sigma_1 a\in \boldsymbol{L}^*$. Then for all $t\in \boldsymbol{T}$:

$$\prod_{\sigma'\in Prefix(\sigma_1 a)} \mathcal{R}((s,t),\sigma') = \{\text{by the definition of } Prefix\} = (\prod_{\sigma'\in Prefix(\sigma_1)} \mathcal{R}((s,t),\sigma'))\cdot \mathcal{R}((s,t),\sigma_1 a) =$$

$$= \{\text{by the inductive hypothesis}\} = \mathcal{P}((s,t),\sigma_1)\cdot \mathcal{R}((s,t),\sigma_1 a) = \{\text{by the definition of } \mathcal{R}\} = \mathcal{P}((s,t),\sigma_1 a)$$

By the induction axiom the lemma holds for all $\sigma\in \boldsymbol{L}^*$. □

Proof of Theorem 3.1 (Alternative characterization of equivalences) Consider theorem 3.1 for i=ste. Using the definitions of $=_{ste}$ and $\equiv_{ste}$, and lemma 1, we can rephrase the theorem as:

$$\forall t\in \boldsymbol{T}, \forall \sigma\in \boldsymbol{L}^*: \quad \prod_{\sigma'\in Prefix(\sigma)} \mathcal{R}((s,t),\sigma') = \prod_{\sigma'\in Prefix(\sigma)} \mathcal{R}((s',t),\sigma') \quad \Leftrightarrow \quad \mathcal{R}((s,t),\sigma) = \mathcal{R}((s',t),\sigma)$$

We shall consider both directions. (⇐) Obvious. (⇒) We shall show that

$$(1) \quad \forall t\in \boldsymbol{T}, \forall \sigma\in \boldsymbol{L}^*: \quad \prod_{\sigma'\in Prefix(\sigma)} \mathcal{R}((s,t),\sigma') = \prod_{\sigma'\in Prefix(\sigma)} \mathcal{R}((s',t),\sigma') \quad \Rightarrow \quad \mathcal{R}((s,t),\sigma) = \mathcal{R}((s',t),\sigma)$$

The proof is by induction on the length of σ. It is trivial to establish that (1) holds for $|\sigma|=0$. We assume that (1) holds for all $\sigma\in \boldsymbol{L}^*$ such that $|\sigma| \leq |\sigma_1|$, and that there exists $\sigma_1 a\in \boldsymbol{L}^*$. Then for all $t\in \boldsymbol{T}$:

$$\prod_{\sigma'\in Prefix(\sigma_1 a)} \mathcal{R}((s,t),\sigma') = \prod_{\sigma'\in Prefix(\sigma_1 a)} \mathcal{R}((s',t),\sigma')$$

{by the definition of $Prefix$} ⇔

$$(\prod_{\sigma'\in Prefix(\sigma_1)} \mathcal{R}((s,t),\sigma'))\cdot \mathcal{R}((s,t),\sigma_1 a) = (\prod_{\sigma'\in Prefix(\sigma_1)} \mathcal{R}((s',t),\sigma'))\cdot \mathcal{R}((s',t),\sigma_1 a)$$

{by the induction hypothesis} ⇒

$$\mathcal{R}((s,t),\sigma_1 a) = \mathcal{R}((s',t),\sigma_1 a)$$

By the induction axiom theorem 3.1 for i=ste holds for all $\sigma\in \boldsymbol{L}^*$. The proof of theorem 3.1 for i=tr is done in exactly the same manner, but with restriction on the types of tests used, i.e. $t\in \boldsymbol{T}_{seq}$.

It is trivial to prove theorem 3.1 for i=wte. The definitions of $=_{wte}$ and of $\equiv_{wte}$ coincide, if we use the definition of $\mathcal{R}$ in terms of $\mathcal{P}$, in the definition of $\equiv_{wte}$. □

Proof Outline for the Full Abstraction Theorem

The main idea behind the proof of the full abstraction theorem is to use the alternative characterization of the equivalences in terms of $\mathcal{R}$, and show that it coincides with the definition of the denotational models in terms of μ. We shall first examine the relation between tests and offerings.

Definition 2 Let $(\boldsymbol{T},\boldsymbol{L},\rightarrow)$ be a TS. A function for generation of strings of offerings from tests, $Gen\text{-}o$: $\boldsymbol{T}\times \boldsymbol{L}^* \rightarrow \boldsymbol{O}^*$, is defined for all $t\in \boldsymbol{T}$ and $a,\sigma\in \boldsymbol{L}^*$, inductively as:

1) $Gen\text{-}o(t,\varepsilon)=\lambda$,

2) $Gen\text{-}o(t,a\sigma)=$ if $Can(a,t)=\{\}$ then λ else $\{a'\in L \mid t\xrightarrow{a'}\}\bullet Gen\text{-}o(t',\sigma)$

where $Can(a,t)=\{t'\in \boldsymbol{T} \mid t\xrightarrow{a} t'\}$, and (•) denotes concatenation.

Intuitively, $Gen\text{-}o(t,\sigma)$ generates a string of offerings o, in which the elements are selected from t with the string σ. For example, the strings of offerings that can be generated from a test t are: $\bigcup_{\sigma\in L^*}\{Gen\text{-}o(t,\sigma)\}$.

Lemma 2 The functions $\mathcal{P}'$ and $\mathcal{Q}'$ are equal, i.e.

$$\forall t \in T, \forall \sigma \in L^*: \quad \mathcal{P}'((s,t),\sigma,(s',t')) = \mathcal{Q}'(s, Gen\text{-}o(t,\sigma),\sigma,s')$$

Proof Compare definitions 3.2 and 4.2. Note especially that the normalization of probabilities for testing systems (see definition 2.7), coincides with the normalization of probabilities with respect to offerings used in definition 4.2. □

Corollary 1 The functions $\mathcal{P}$ and $\mathcal{Q}$ are equal, i.e.

$$\forall t \in T, \forall \sigma \in L^*: \quad \mathcal{P}((s,t),\sigma) = \mathcal{Q}(s, Gen\text{-}o(t,\sigma),\sigma)$$

Proof Follows directly from definitions 3.3 and 4.3, and lemma 2. □

Corollary 2 The functions $\mathcal{R}$ and μ are equal, i.e.

$$\forall t \in T, \forall \sigma \in L^*: \quad \mathcal{R}((s,t),\sigma) = \mu(s, Gen\text{-}o(t,\sigma),\sigma)$$

Proof Follows directly from definitions 3.6 and 4.4, and corollary 1. □

Proof for Theorem 4.1 (Full abstraction theorem) We shall first consider the theorem for i=ste. To prove the theorem, theorem 3.1 is used to replace $=_{ste}$ with $\equiv_{ste}$. We need to show that:

$$\forall t \in T, \forall \sigma \in L^*: \qquad \mathcal{R}((s,t),\sigma) = \mathcal{R}((s',t),\sigma) \quad \Leftrightarrow \quad \mu_{ste}(Gen\text{-}o(t,\sigma),\sigma) = \mu'_{ste}(Gen\text{-}o(t,\sigma),\sigma)$$

Using the definition of μ_{ste} (in terms μ), and corollary 2, we can establish that the theorem holds for i=ste.

Similarly, to prove the theorem for i∈{tr,wte}, theorem 3.1 is used to replace $=_{tr}$ with $\equiv_{tr}$, and $=_{wte}$ with $\equiv_{wte}$, respectively. Using the definitions of μ_{tr} and of μ_{wte} (in terms μ), and corollary 2, we can establish that the theorem holds for i∈{tr,wte}. □

A Preorder for Partial Process Specifications

Rance Cleaveland
Department of Computer Science
Box 8206
North Carolina State University
Raleigh, NC 27695-8206
USA

Bernhard Steffen
Department of Computer Science
Aarhus University
Ny Munkegade 116
DK 8000 Aarhus C
DENMARK

Abstract

This paper presents a behavioral preorder for relating *partial* process specifications to implementations and establishes that it is "adequate" in the following sense: *one* specification may be used to characterize *all* implementations of a network component that are correct for any network context exhibiting a particular interface. This property makes the preorder particularly suitable for reasoning compositionally about networks of processes. The paper also gives a sound and complete axiomatization for finite processes of the largest precongruence contained in this new preorder.

Introduction

increasingly popular approach to reasoning about concurrent systems involves the use of a *behavioral order* to relate specifications and implementations. In this framework, a specification is a process t may be only *partially* defined; an implementation meets such a specification if it is larger than latter in the preorder (the intuition being that a correct implementation provides "at least" the avior dictated by the specification). These specifications afford implementors more freedom in cting appropriate implementation strategies than the more usual equivalence-based techniques do, this fact can be exploited when one wishes to specify a component that is to be used in a particular *work context*. Such contexts often impose constraints on the states its components may enter; one reflect this in a specification by allowing unreachable states to be undefined, since any elaboration hese states will not affect the overall system. This enables a form of *compositional* verification, e once an appropriate partial specification has been developed for a component of a system, one st only verify an implementation with respect to this specification—the remainder of the system is mportant. To carry this approach further, rather than defining a particular context in which the ired component is to execute, one may instead fix the *interface* that the context is guaranteed to vide. Then, provided that a given context supplies the stipulated interface, any implementation of partial specification will be guaranteed to behave in the desired fashion. This approach affords a her degree of modularity in program development, since one may continue to refine the context well as the component; as long as the context continues to provide the stipulated interface, any lementation of the partial specification will behave in the desired fashion.

The goal of this paper is to develop a preorder that is tailored for such context-dependent vertion. To this end, we introduce the notion of *adequacy for partial specification* to characterize orders that are "complete" for this kind of reasoning. If a preorder has this property, then all ect implementations of a component for any totally defined system with a given interface behavior be specified by means of a single partial specification. The *specification preorder* is then develd by modifying the well-known *divergence preorder* [1, 11, 14, 15, 16], so that the resulting relation enjoys this property. We also give a sound and complete axiomatization for the largest precongruenc contained in this new preorder for finite processes.

The remainder of the paper is organized as follows. The next section briefly describes Milner' Calculus of Communicating Systems (CCS) [10, 12] and presents a behavioral equivalence, as well a the divergence preorder. Section 3 then describes the new preorder and establishes that, in contrast t the divergence preorder, it is adequate for partial specification. Section 4 gives a sound and complet axiomatization for the preorder, and the final section contains our conclusions and directions for futur work.

2 Processes, Equivalences and the Divergence Preorder

In CCS processes are built from a collection of *atomic actions*, or *communications*, using *proce constructors*. The actions are given in terms of a set Λ of labels, or *ports*, and a distinguished interna or *silent*, action τ. Let $\overline{\Lambda} = \{\overline{\lambda} \mid \lambda \in \Lambda\}$ be the set of *complements* of Λ [12]; then the set *Act* actions is defined as

$$Act = \Lambda \cup \overline{\Lambda} \cup \{\tau\}$$

In the remainder of the paper we let λ range over Λ and $a, b, \ldots$ range over *Act*. Intuitively, an actio $\lambda \in \Lambda$ can be thought of as the consumption of an input on port λ, while $\overline{\lambda}$ represents the offerin of an output on port λ; τ represents an internal, or unobservable, computation step. We extend th complementation operator to $\Lambda \cup \overline{\Lambda}$ by defining $\overline{\overline{\lambda}} = \lambda$.

Let *Var* be a set of process variables ranged over by $X, Y, \ldots$, let $L \subseteq \Lambda$, and let $f \in Act \to Act$ b a *relabeling*: $f(\tau) = \tau$ and $f(\overline{a}) = \overline{f(a)}$. The syntax of processes is given by the following grammar.

$$P \ ::= \ X \mid nil \mid \bot \mid a.P \mid P + P \mid P|P \mid P\backslash L \mid P\langle L\rangle \mid P[f] \mid \mu X.P$$

The operator μ binds the free occurrences of X in p in the usual sense in process $\mu X.p$, and the proce $p[q/X]$ gotten by substituting p for the free occurrences of X in p is also defined in the standard fashio We also impose an additional syntactic constraint: in $\mu X.p$, any free occurrences of X in p must b *guarded*, i.e. must occur in subterms of the form $a.p'$ for some a and p'. Let *Proc* be the set of tern generated by the grammar and satisfying the guardedness condition. In what follows we shall u $p, q, \ldots$ to range over *Proc*.

Processes have the following intuitive meaning. The processes *nil* and $\bot$ represent the terminate and undefined processes, respectively, while $a.p$ denotes a process that engages in an a action befo behaving like p. The operator $+$ represents external choice, $|$ parallel composition, $[f]$ relabeling, an $\backslash L$ restriction. The process $p\langle L\rangle$ "exports" only the actions whose labels come from L; all other actio are made internal. Finally, $\mu X.p$ is a recursively defined process that is a distinguished solution to th equation $X = p$. We shall often give such processes as a series of equations.

The formal semantics of CCS may given as an *extended labeled transition system* [15, 1 $\langle Proc, Act, \rightarrow, \uparrow\rangle$. The relation $\rightarrow \subseteq Proc \times Act \times Proc$ is the *transition relation* describing the actio processes may perform and the effect on processes of performing them, while $\uparrow$, the *partiality predicat* indicates which processes are underdefined. We shall write $p \xrightarrow{a} q$ and $p \uparrow$ in lieu of $\langle p, a, q\rangle \in \rightarrow$ ar $p \in \uparrow$. The definitions are given in Figure 1. The definition of $\rightarrow$ may be extended to sequences $s \in Ac$ in the obvious fashion. It should also be noted that $\langle L\rangle$ may be derived from the other operators; th process $p\langle L\rangle$ has the same transitions as $(p|\mu X. \sum_{a \in (\Lambda - L)} a.X)\backslash(\Lambda - L)$.

The $\rightarrow$ relation does not distinguish between observable and unobservable actions; in order reflect that τ is internal, and hence not visible, we define the *weak* transition relation, $\Rightarrow$, the *glob partiality* (or *underspecification*) predicate, $\Uparrow$, and the *local partiality* relation, $\Uparrow \subseteq Proc \times Act$, follows.

Definition 2.1 *Let $s = a_1 a_2 \ldots a_n$ be a sequence of visible actions.*

1. $\stackrel{s}{\Rightarrow} \; = \; (\xrightarrow{\tau})^* \circ \xrightarrow{a_1} \circ (\xrightarrow{\tau})^* \circ \xrightarrow{a_2} \circ \cdots \circ (\xrightarrow{\tau})^* \circ \xrightarrow{a_n} \circ (\xrightarrow{\tau})^*$, *where $\circ$ denotes relational compositio Notice that for the empty sequence ϵ, $\stackrel{\epsilon}{\Rightarrow} \; = (\xrightarrow{\tau})^*$.*

→ is the least relation satisfying the following.

$$
\begin{array}{rcl}
a.p \xrightarrow{a} p & & \\
p \xrightarrow{a} p' & \Rightarrow & p+q \xrightarrow{a} p',\ q+p \xrightarrow{a} p',\ p|q \xrightarrow{a} p'|q,\ q|p \xrightarrow{a} q|p',\ p[f] \xrightarrow{f(a)} p'[f] \\
p \xrightarrow{a} p', q \xrightarrow{\overline{a}} q' & \Rightarrow & p|q \xrightarrow{\tau} p'|q' \\
p \xrightarrow{a} p', a \notin (L \cup \overline{L}) & \Rightarrow & p\backslash L \xrightarrow{a} p\backslash L \\
p \xrightarrow{a} p', a \in (L \cup \overline{L}) & \Rightarrow & p\langle L\rangle \xrightarrow{a} p'\langle L\rangle \\
p \xrightarrow{a} p', a \notin (L \cup \overline{L}) & \Rightarrow & p\langle L\rangle \xrightarrow{\tau} p'\langle L\rangle \\
p[\mu X.p/X] \xrightarrow{a} p' & \Rightarrow & \mu X.p \xrightarrow{a} p'
\end{array}
$$

is the complement of ↓, which is the set defined by:

$$
\begin{array}{rcl}
nil\downarrow,\ X\downarrow,\ a.p\downarrow & & \\
p\downarrow,\ q\downarrow & \Rightarrow & p+q\downarrow,\ p|q\downarrow \\
p\downarrow & \Rightarrow & p\backslash L\downarrow,\ p[f]\downarrow \\
p[\mu X.p/X]\downarrow & \Rightarrow & \mu X.p\downarrow
\end{array}
$$

Figure 1: The definition of the transition relation and partiality predicate.

2. $p \Uparrow$ *if there is a* p' *such that* $p \stackrel{s}{\Rightarrow} p'$ *and* $p' \uparrow$.

3. $p \Uparrow a$ *if* $p \Uparrow$ *or if there is a* p' *such that* $p \stackrel{a}{\Rightarrow} p'$ *and* $p' \Uparrow$.

/e shall write $p \Downarrow$ (or equivalently $p \Downarrow \epsilon$) and $p \Downarrow a$ in place of $\neg(p \Uparrow)$ and $\neg(p \Uparrow a)$. Intuitively, $p \Uparrow a$ ieans that p may be triggered by means of an a action into reaching a partially defined state. A rocess satisfying $\Uparrow a$ (or $\Uparrow$) is called *a-partial* (or *globally partial*). We shall also say that a process is *tally defined* if for every s and p' with $p \stackrel{s}{\Rightarrow} p'$, $p' \Downarrow$, and *partially defined* otherwise.

Our definition of $\Uparrow$ is slightly different from the one proposed by Walker [16]; he also defined $p \Uparrow$ to ld if p was capable of an infinite internal computation, because his intent was to capture a notion of *vergence* rather than underdefinedness. It should be noted that all our results hold when Walker's edicate is substituted for ours.

We now define the operational equivalence that we use in the paper. *Specification equivalence* is a fined version of the standard *observational equivalence* of CCS that is given in terms of *specification simulations.*

efinition 2.2 *A relation* $R \subseteq Proc \times Proc$ *is a* specification bisimulation *if* pRq *implies that:*

1. $p \Uparrow$ *if and only if* $q \Uparrow$.

2. *for all* $\alpha \in (Act - \{\tau\}) \cup \{\epsilon\}$,

   (a) $p \stackrel{\alpha}{\Rightarrow} p'$ *implies* $\exists q'.\ q \stackrel{\alpha}{\Rightarrow} q' \wedge p'Rq'$, *and*

   (b) $q \stackrel{\alpha}{\Rightarrow} q'$ *implies* $\exists p'.\ p \stackrel{\alpha}{\Rightarrow} p' \wedge p'Rq'$.

ecification equivalence, $\approx^s$, is defined as $\bigcup\{ R \mid R$ is a specification bisimulation$\}$, which is the rgest specification bisimulation; thus $p \approx^s q$ exactly when there is a specification bisimulation R th pRq. It is readily shown that $\approx^s$ is an equivalence relation and that $\approx^s$ is preserved by all the ocess constructors except $+$. There is a well-known method for remedying this, which we shall not rsue in this paper. Observation equivalence, $\approx$, is like $\approx^s$; it is given in terms of *bisimulations*, which e specification bisimulations with condition 1 omitted. Clearly, $\approx^s$ is finer than $\approx$.

The *divergence preorder*, $\sqsubseteq$, may be used to determine when one process is "more specified" than other. It does so by using the local partiality relation to relax selectively the bisimulation conditions. is relation is given in terms of *prebisimulations.*

Definition 2.3 *A relation $R \subseteq Proc \times Proc$ is a* prebisimulation *if pRq implies that the following hol for all $\alpha \in (Act - \{\tau\}) \cup \{\epsilon\}$.*

1. $p \stackrel{\alpha}{\Rightarrow} p' \Rightarrow \exists q'.\, q \stackrel{\alpha}{\Rightarrow} q' \wedge p'Rq'$, *and*
2. $p \Downarrow \alpha \Rightarrow [q \Downarrow \alpha \wedge (q \stackrel{\alpha}{\Rightarrow} q' \Rightarrow \exists p'.\, p \stackrel{\alpha}{\Rightarrow} p' \wedge p'Rq')]$.

Then $\underset{\approx}{\sqsubset} = \bigcup\{ R \mid R \text{ is a prebisimulation} \}$ is the largest prebisimulation; it is also a preorder, and i preserved by all the process constructors except $+$. Let $\simeq$ be the equivalence relation associated wit $\underset{\approx}{\sqsubset}$, i.e. $p \simeq q$ if and only if $p \underset{\approx}{\sqsubset} q$ and $q \underset{\approx}{\sqsubset} p$.

It is well-known that $\approx$ and $\underset{\approx}{\sqsubset}$, the two main verification relations in CCS, induce slightly differen semantics on processes, as the next result shows.

Theorem 2.4 *There exist processes p_1, p_2, q_1 and q_2 such that:*

1. $p_1 \approx q_1$ *and* $p_1 \not\simeq q_1$, *and*
2. $p_2 \not\approx q_2$ *and* $p_2 \simeq q_2$.

However, it does turn out that $\approx^s$ is a "common denominator" of both $\simeq$ and $\approx$.

Theorem 2.5 *If $p \approx^s q$ then $p \simeq q$ and $p \approx q$.*

In the remainder of the paper we will focus our attention on what we call *network preorders* which are preorders that preserve the $|$ and $\backslash L$ operators and which coincide with $\approx^s$ for totall defined processes. The divergence preorder is a network preorder, as are the preorders defined i [7, 8, 9, 14, 16].

3 Adequacy, and the Specification Preorder

In this section we illustrate a shortcoming of the divergence preorder as an specification-implementatio relation that affects its applicability to compositional reasoning. We then introduce a new preorde and show that it overcomes this limitation.

As we mentioned in the introduction, one way of using preorders to reason about processes exploit the fact that systems may restrict the states their components can enter. Therefore, when specifyin such components, one need not specify them completely, but may instead give partial specification in which states that the component may not enter are marked as undefined. In [4], the notion c *adequacy* was introduced to formalize this: a (partial) specification q is *adequate* for context $(p|.)\backslash$ if, for every q' with $q \underset{\approx}{\sqsubset} q'$, $(p|q)\backslash L \approx^s (p|q')\backslash L$. If q is adequate for $(p|.)\backslash L$, then any implementatio of q may be used in place of q without disturbing the behavior of the overall system. This suggests proof technique for establishing the equivalence of $(p|q)\backslash L$ and $(p|q')\backslash L$: first determine whether q i adequate for $(p|.)\backslash L$, and if it is then show that $q \underset{\approx}{\sqsubset} q'$.

The definition of adequacy may be relativized with respect to the preorder used. If $\leq$ is a preorde a process q is $\leq$-adequate for $(p|.)\backslash L$ if for every q' with $q \leq q'$, $(p|q)\backslash L \approx^s (p|q')\backslash L$. When th particular preorder is clear from context, we shall omit explicit reference to it. The notion of adequac may also be extended to *classes* of contexts in the obvious fashion: if C is a class of contexts, then is adequate for C if it is adequate for each context in C.

We are interested in classes of contexts that may be characterized by *interface behaviors*, i.e. by th behavior that contexts provide along the internal communication ports shared with the missing com ponent. Intuitively, such a component cannot distinguish between contexts sharing the same interfac behavior because the communication patterns they supply to the component are indistinguishabl For the purposes of modular program development, it should also be the case that the adequacy c a component depends only on the communication patterns provided by a context, since if this is th case one may alter the context without disturbing the behavior of the missing component, provide the interface behavior of the context remains the same.

For our purposes, an *interface behavior* is a pair of the form $\langle L, I\rangle$, where $L \subseteq \Lambda$ and I is a process hose actions come from the set $L \cup \overline{L} \cup \{\tau\}$. A context $(p|.)\backslash L$ *satisfies* $\langle L, I\rangle$, written $(p|.)\backslash L \models \langle L, I\rangle$, $p\langle L\rangle \approx^{\bullet} I$—in other words, if the behavior of p that is visible along the ports L is the same as I. Ve shall abuse terminology somewhat in identifying an interface behavior with the class of contexts aat satisfy it; in particular, we shall refer to the adequacy of a partial specification with respect to n interface behavior.

When q is adequate for $\langle L, I\rangle$, the following (compositional) line of reasoning is sound: to establish aat $(p|q)\backslash L \approx^{\bullet} (p|q')\backslash L$, it is sufficient to show that $(p|.)\backslash L \models \langle L, I\rangle$ and $q \leq q'$. We are now going elaborate on this kind of reasoning.

efinition 3.1 *A process r is a* minimal cover *for an interface specification $\langle L, I\rangle$ and a process q if ie following holds:*

1. $\forall\, (p|.)\backslash L \models \langle L, I\rangle.\ (p|r)\backslash L \approx^{\bullet} (p|q)\backslash L$, *and*

2. $\forall r' \in$ *Proc.* $(\ \forall (p|.)\backslash L \models \langle L, I\rangle.\ (p|r')\backslash L \approx^{\bullet} (p|q)\backslash L\ )$ implies $r \leq r'$.

ituitively, a process r is a minimal cover for an interface behavior $\langle L, I\rangle$ and a process q if it may be sed interchangeably with q in any context with this interface behavior, and also specifies *any* other rocess that may be used interchangeably with q in *all* these contexts. Note that this does *not* necesarily imply that r is adequate for $\langle L, I\rangle$, only that every "correct" implementation is an elaboration r. However, if the preorder being used to relate component specifications and implementations is a etwork preorder, and if $(I|q)\backslash L$ is *totally defined*, then it does turn out that r is adequate. This is a onsequence of the following theorem.

heorem 3.2 *Suppose that $\leq$ is a network preorder, let q and $\langle L, I\rangle$ be such that $(I|q)\backslash L$ is totally efined, and let r be a minimal cover for $\langle L, I\rangle$ and q. Then the following holds.*

$$(\ \forall (p|.)\backslash L \models \langle L, I\rangle.\ (p|r')\backslash L \approx^{\bullet} (p|q)\backslash L\ )\ \textit{if and only if}\ \ r \leq r'.$$

hus, for networks whose interface behavior prohibits a component entering a globally partial state, inimal covers may be seen as *minimally adequate specifications* of their components; they are the eakest possible specifications that guarantee correctness for all the contexts with this interface beavior. This motivates the following requirement for a suitable specification-implementation relation:

efinition 3.3 *A preorder $\leq$ is* adequate for partial specification *if for every interface specification $\langle L, I\rangle$ and every process q, there exists a minimal cover.*

In general, the fact that r is a minimal cover for q and $\langle L, I\rangle$ does not guarantee that it is a minimal over for each $(p|.)\backslash L \models \langle L, I\rangle$, with minimal cover defined analogously for particular contexts. That for particular $\leq$ and $(p|.)\backslash L \models \langle L, I\rangle$ there may be r' such that $(p|q)\backslash L \approx^{\bullet} (p|r')\backslash L$ and yet $\not\leq r'$. This phenomenon is due to the fact that in some instances, a component need not cater for very communication offered by the context in order to ensure correctness of the over-all system. For stance, the context may offer several choices of communication to the component, any of which will ad to the same behavior of the resulting system; so long as the component responds to one of these ommunications appropriately, the system will be correct. This excludes the existence of a minimal over for this particular context, unless one introduces a disjunctive mechanism as has been done in]. However, for any interface behavior $\langle L, I\rangle$ there are contexts for which the minimal cover for the terface behavior is the minimal cover for the context as well.

efinition 3.4 *Let $\leq$ be a network preorder, $\langle L, I\rangle$ be an interface behavior, and q be a process such at there exists a minimal cover r for q and $\langle L, I\rangle$. Then a context $(p|.)\backslash L$ is* distinctive *for q and $\langle L, I\rangle$ iff r is also a minimal cover for $(p|.)\backslash L$.*

Let $S(p)$ be the set of reachable *states* of p, i.e. $S(p) = \{p' \mid \exists s \in (Act - \{\tau\})^*.\ p \stackrel{s}{\Rightarrow} p'\}$. Then the next theorem gives a useful sufficient condition for determining when a context is distinctive.

Theorem 3.5 *Given any network preorder, a context* $(p|.)\backslash L$ *is distinctive for* q *and* $\langle L, I\rangle$ *if for no* $p' \in S(p)$ *and* $L' \subseteq L$ *such that* $p'\backslash L' \not\approx^s p'$ *it is the case that* $(p'\backslash L')\langle(\Lambda - L)\rangle \approx^s p'\langle(\Lambda - L)\rangle$.

Intuitively, this theorem says if it is the case that no communication action that p is eventually capable of may be omitted without affecting the external (i.e. non-communication) behavior of p, then $(p|.)\backslash L$ is distinctive. In fact, in a language without hiding where the synchronized actions remain unchanged, any process would be distinctive. In our setting, distinctive contexts may be obtained from an interface behavior $\langle L, I\rangle$ by inserting unique "probes" (actions not in $L \cup \overline{L}$) after every action in $L \cup \overline{L}$ in I. This transformation recovers the information about internal communications that was lost without changing the interface behavior.

It turns out that $\sqsubseteq_{\approx}$ is *not* adequate for partial specification. The reason for this is best illustrated by an example. Consider the following partially defined buffer.

$$\begin{aligned} \text{PBUFF1} &= in.\text{PBUFF1}' \\ \text{PBUFF1}' &= \overline{out}.\text{PBUFF1} + in.\perp \end{aligned}$$

It is the case that PBUFF1 is adequate for the interface specification $\langle L, I\rangle$ given by:

$$\begin{aligned} L &= \{in, out\} \\ I &= \overline{in}.out.I \end{aligned}$$

Now consider the context $(p|.)\backslash\{in, out\}$, where p is given below; note that $(p|.)\backslash L \models \langle L, I\rangle$. Moreover $(p|.)\backslash L$ is distinctive for PBUFF1 and $\langle L, I\rangle$, since $\overline{receive}$ is not possible until after both in and $\overline{out}$ have occurred.

$$\begin{aligned} p &= (S|R)\backslash ack \\ S &= send.\overline{in}.ack.S \\ R &= out.\overline{receive}.\overline{ack}.R \end{aligned}$$

Here p consists of a sender, S, and a receiver, R. The sender uses the buffer to send a message to R and then awaits the receipt of an acknowledgement. The receiver awaits the appearance of a message from the buffer and then sends an acknowledgement.

Now consider the totally defined one-place buffer BUFF1 defined as follows.

$$\text{BUFF1} = in.\overline{out}.\text{BUFF1}$$

We have that

$$(p|\text{PBUFF1})\backslash\{in, out\} \approx^s (p|\text{BUFF1})\backslash\{in, out\}$$

However, neither PBUFF1 $\not\sqsubseteq_{\approx}$ BUFF1 nor BUFF1 $\not\sqsubseteq_{\approx}$ PBUFF1 holds. In fact, it turns out that there is no process r with

$$(p|r)\backslash\{in, out\} \approx^s (p|\text{BUFF1})\backslash\{in, out\}$$

that is less than PBUFF1 and BUFF1 in the divergence preorder. Thus $\sqsubseteq_{\approx}$ cannot be adequate for partial specification. If it were, then $\langle L, I\rangle$ would have a minimal cover r, and since $(p|.)\backslash L$ is distinctive for PBUFF1 and this interface behavior, r would also be a minimal cover for this context and therefore cause a contradiction.

Intuitively, the reason why $\sqsubseteq_{\approx}$ is not adequate for partial specification is that, whenever $p \sqsubseteq_{\approx} q$, then q must match every a-transition p is capable of, even when the only a-transition available to p leads to an undefined state. The preorder we now present treats a-partiality "catastrophically" in the sense that, if a state is a-partial, then *none* of its a-transitions need be matched by a state that is considered larger in the preorder. This relation is defined as follows.

efinition 3.6 *The* specification preorder, $\lesssim$, *is defined as the largest relation* $\subseteq$ *Proc* $\times$ *Proc such at the following holds for all* $\alpha \in (Act - \{\tau\}) \cup \{\epsilon\}$ such that $p \Downarrow \alpha$.

1. $q \Downarrow \alpha$, *and*

2. $p \overset{a}{\Rightarrow} p' \Rightarrow \exists q'.\ q \overset{a}{\Rightarrow} q' \wedge p'Rq'$, *and*

3. $q \overset{a}{\Rightarrow} q' \Rightarrow \exists p'.\ p \overset{a}{\Rightarrow} p' \wedge p'Rq'$.

his preorder may be readily seen to have the following properties.

heorem 3.7 *Let* $p, q \in Proc$ *with* $p \sqsubseteq q$. *Then* $p \lesssim q$.

orollary 3.8 *Suppose* $p \approx^s q$. *Then* $p \lesssim q$.

he central result of this section, however, is:

heorem 3.9 (Adequacy for Specification)
he specification preorder $\lesssim$ *is adequate for partial specification.*

heorem 4.1 shows that $\lesssim$ is preserved by $|$ and $\backslash L$, and as $\lesssim$ clearly coincides with $\approx^s$ on totally defined ocesses, $\lesssim$ is a network preorder. Therefore, Theorem 3.2 may be applied, thereby establishing the istence of minimally $\lesssim$-adequate specifications.

Returning to the previous example, it turns that PBUFF1 $\lesssim$ BUFF1. The minimal cover for BUFF1 and $(p|.)\backslash\{in, out\}$ is MBUFF1, which is given as follows.

$$\begin{array}{rcl} \text{MBUFF1} & = & in.\text{MBUFF1}' + out.\perp + \overline{in}.\perp \\ \text{MBUFF1}' & = & \overline{out}.\text{MBUFF1} + \overline{in}.\perp + out.\perp \end{array}$$

hen p and q are finite-state, it is also possible to construct *automatically* a minimal cover for $(p|.)\backslash L$ d q. This can be done by adapting a technique presented in [5].

Up to now we have characterized contexts on the basis of their interface behaviour. It turns out at this degree of detail is unnecessary, because component processes also cannot distinguish contexts at provide the same sequences, or *traces*, of communication at the interface. In order to make this plicit, let $\mathcal{L}(p)$, the *language* or set of possible traces of p, be defined as follows.

$$\mathcal{L}(p) = \{\, s \in (Act - \{\tau\})^* \mid \exists p'.\ p \overset{s}{\Rightarrow} p' \,\}$$

ien Theorem 3.2 and can be strengthened as follows.

heorem 3.10 *Let* $\leq$ *be a network preorder,* q *a component process and* $\langle L, I\rangle$ *an interface behaviour ch that* $(I|q)\backslash L$ *is totally defined. Furthermore, let* p *be distinctive for* q *and* $\langle L, I\rangle$, *and* r *be a inimal cover for* q *and* $\langle L, I\rangle$. *Then the following characterizations of a process* r' *are equivalent:*

1. $r \leq r'$

2. $(p|r')\backslash L \approx^s (p|q)\backslash L$

3. $\forall (p'|.)\backslash L \models \langle L, I\rangle.\ (p'|r')\backslash L \approx^s (p'|q)\backslash L$

4. $\forall p'.\ \mathcal{L}(p'\langle L\rangle) \subseteq \mathcal{L}(I)$ *implies* $(p'|r')\backslash L \approx^s (p'|q)\backslash L$

The point of this theorem is the fourth characterization, which says that the knowledge of the language of possible communications provided at the interface is as good as the knowledge of the precise interface behaviour, or equivalently, of the particular distinctive context, for our purposes. In particular, having developed a correct implementation of a component for a distinctive context, one may subsequently change the context and still preserve the correctness of the component, as long as the language of possible communications does not increase. Thus our notion of adequacy supports any semantic treatment of context that is finer than interface trace containment and coarser than interface behavioural equivalence. For example, it also caters for the *simulation-based* approach proposed by [13].

The observation that sets of traces and partially defined processes semantically reflect the dual nature of the concepts of context and component is also inherent in [4, 5]. In the former, modal formulas are used to describe the conditions a context must abide by for a partial specification to be adequate, while in the latter (prefix-closed) sets of *traces* are used to approximate the interface a component may expect from a context. As the modal formulas in [4] have a characterization in terms of prefix-closed sets of traces, both papers specify contexts by means of their language of possible communications.

4 A Sound and Complete Axiomatization

This section is devoted to a sound and complete (inequational) axiomatization for $\lesssim$ for *finite* terms (i.e. terms not involving recursion). To obtain such an axiomatizations, it is desirable that $\lesssim$ be a *precongruence*, so that the substitution of larger terms for smaller ones is guaranteed to result in a larger overall term. Unfortunately, $\lesssim$ is not a precongruence because of well-known complications that result from the interplay between τ and $+$. However, we do have the following.

Theorem 4.1 *Let $p \lesssim q$. Then the following are true.*

1. $a.p \lesssim a.q$
2. $p|r \lesssim q|r$
3. $p \backslash L \lesssim q \backslash L$
4. $p[f] \lesssim q[f]$

We fix the problem with $+$ in the usual way. Let $\lesssim^C$ be the largest precongruence contained in $\lesssim$; it is defined as follows: $p \lesssim^C q$ if for all contexts $C[]$, $C[p] \lesssim C[q]$. We then axiomatize $\lesssim^C$.

In this section we need not consider the $\langle L \rangle$ operator, because, as mentioned above, it is a derived operator. Thus all the laws for $\langle L \rangle$ can be obtained from the laws for the other operators. Now, Figure 2 contains the sound and complete axiomatization for $\underset{\sim}{\sqsubset}^C$, the largest congruence contained in $\underset{\sim}{\sqsubset}$ developed by Walker in [16]. Here $p = q$ is short-hand for $p \leq q \wedge q \leq p$; the usual rules involving the properties of a precongruence (substitutivity, transitivity, etc.) are assumed. In axiom INT, the notation $\{+\perp\}$ signifies that $\perp$ may be a summand of the indicated terms. Moreover, it is a summand of the right-hand side of the axiom if and only if it is a summand of either of the terms on the left-hand side.

These rules are also sound for $\lesssim^C$, since it is strictly coarser than $\underset{\sim}{\sqsubset}^C$. In order to develop a complete axiomatization, then, we need to add more axioms. Figure 3 contains the necessary additions. Axiom $\perp 3$ implies that a-transitions leading to the undefined state $\perp$ need not be matched by higher processes, while Axioms $\perp 4$ and $\perp 5$ reflect the fact that a-partial is catastrophic. Note that since $(x+\perp) \Uparrow a$ for any a it follows that $+$ is *strict*. It should also be noted that Axiom $\perp 2$ may now b

$$
\begin{array}{lrcl}
\text{A1} & x + x & = & x \\
\text{A2} & x + y & = & y + x \\
\text{A3} & x + (y + z) & = & (x + y) + z \\
\text{A4} & x + nil & = & x
\end{array}
$$

$$
\begin{array}{lrcl}
\text{INT} & \multicolumn{3}{l}{\text{Let } p, q \text{ denote } \sum \alpha_i.p_i\{+ \perp\},\ \sum \beta_j.q_j\{+ \perp\}, \text{ respectively. Then}} \\
 & p|q & = & \sum \alpha_i.(p_i|q) + \sum \beta_j.(p|q_j) + \sum_{\alpha_i = \overline{\beta_j} \in A} \tau.(p_i|q_j)\{+ \perp\}
\end{array}
$$

$$
\begin{array}{lrcl}
\text{RES1} & nil \backslash L & = & nil \\
\text{RES2} & \perp \backslash L & = & \perp \\
\text{RES3} & (\alpha.x)\backslash L & = & \begin{cases} nil & \text{if } \alpha \in L \cup \overline{L} \\ \alpha.(x\backslash\lambda) & \text{otherwise} \end{cases} \\
\text{RES4} & (x + y)\backslash L & = & (x\backslash L) + (y\backslash L)
\end{array}
$$

$$
\begin{array}{lrcl}
\text{REL1} & nil\,[f] & = & nil \\
\text{REL2} & \perp [f] & = & \perp \\
\text{REL3} & (\alpha.x)[f] & = & f(\alpha).(x[f]) \\
\text{REL4} & (x + y)[f] & = & x[f] + y[f]
\end{array}
$$

$$
\begin{array}{lrcl}
\tau 1 & a.\tau.x & = & a.x \\
\tau 2 & x + \tau.x & = & \tau.x \\
\tau 3 & a.(x + \tau.y) & = & a.(x + \tau.y) + a.y
\end{array}
$$

$$
\begin{array}{lrcl}
\perp 1 & \perp & \leq & x \\
\perp 2 & \tau.(x + \perp) & \leq & x + \perp
\end{array}
$$

Figure 2: The sound and complete axiomatization of $\mathrel{\sqsubset\!\!\!\sim}^C$ for finite processes.

$$
\begin{array}{lrcl}
\perp 3 & x + a.\perp & \leq & x \\
\perp 4 & x + \perp & = & \perp \\
\perp 5 & a.x + a.\perp & = & a.\perp
\end{array}
$$

Figure 3: Additional axioms for $\lesssim^C$.

rived from the other axioms, as the following line of reasoning indicates.

$$
\begin{array}{rcll}
\tau.(x + \perp) & = & \tau.\perp & \perp 4 \\
 & = & \perp + \tau.\perp & \tau 2 \\
 & = & \perp & \perp 4 \\
 & \leq & x + \perp & \perp 1
\end{array}
$$

Let A be the axioms in Figures 2 and 3, and define $p \leq_A q$ to hold if p may be proved less than or ual to q using A. Then we have:

heorem 4.2 (Soundness and Completeness)
, $q \in Proc.$ $p \lesssim q$ iff $p \leq_A q$.

Conclusions and Directions for Future Research

this paper we have presented the *specification preorder*, a specification-implementation relation that tailored to support *context-dependent verification.* Its main feature is that it is *adequate for partial specification*; in contrast to the *divergence preorder* of [16], it allows *all* correct implementations of network component that are correct for any context of a given *interface behavior* to be characterize by a single specification. We have also given a sound and complete axiomatization of the larges precongruence contained in the specification preorder for finite processes.

The adequacy property of the specification preorder makes it particularly suitable for composi tional proof techniques like the ones presented in [4, 5], which both are elaborations of the technique prosed in [7, 16]. The first paper develops proof rules for establishing semantic equivalence betwee processes, while the second proposes a method for compositionally minimizing the state space of complex system. It should also be noted that the verification of the specification preorder can easil be automated for finite-state systems. The Concurrency Workbench [2, 3], for example, contains a algorithm for preorder checking that can easily be adapted to compute the specification preorder.

The design of the specification preorder was guided by the desire for simplicity: the resulting relatio should arise from straightforward intuitions, afford a clean theoretical treatment, and permit as man implementations as possible. It is our opinion that the specification preorder with its catastrophi interpretation of a-partiality satisfies these criteria. Nevertheless, other choices are also possibl For example, another possibility would require that a-transitions to convergent states from a-partia states must be matched; this amounts to adopting a "noncatastrophic" view of a-partiality. A thir possibility would involve introducing constants into the language that enable one to indicate when state is a-partial. In the framework studied in this paper, a state can be a-partial and incapable c an a-transition only when it is globally partial. If one introduces these constants (which have bee presented in another context in [14, 15]), then it becomes possible to define states that are not globall partial, are incapable of an a-transition, and yet are a-partial. It appears that both of these refine treatments would yield preorders that are adequate for partial specification, and there may well b applications where such refinements are advantageous for a specification language. However, they ar unnecessarily complicated for our purpose.

Larsen and Thomsen have also considered preorders that do not require every a-transition in specification to be matched by a specification [8]. Their relations rely on the introduction of two kind of transitions: *may* transitions, which the specification permits implementations to match, and *mu* transitions, which the specification requires implementations to match. They also require that ever must transition must also be a may transition. This preorder is much finer than ours, since the onl "may" transitions that are not also "must" transitions in our framework are those leading to globall divergent states. Although the additional expressive power is useful in some settings (for example, i the general framework for equation solving given in [9]), for our purpose it is unnecessary.

Acknowledgements

We would like to thank Kim Larsen, H. Qin and Frits Vaandrager for their insightful comments on a earlier version of this paper.

References

[1] Abramsky, S. "Observation Equivalence as a Testing Equivalence." *Theoretical Computer Scienc* v. 53, 1987, pp. 225–241.

[2] Cleaveland, R., J. Parrow and B.U. Steffen. "A Semantics-Based Tool for the Verification Finite-State Systems." In *Proceedings of the Ninth IFIP Symposium on Protocol Specificatio Testing and Verification*, June 1989. To be published by North-Holland.

[3] Cleaveland, R., J. Parrow and B.U. Steffen. "The Concurrency Workbench." In *Proceedings of th Workshop on Automatic Verification Methods for Finite-State Systems*, Lecture Notes in Con puter Science 407. Springer-Verlag, Berlin, 1989.

] Cleaveland, R. and B. Steffen. "When is 'Partial' Adequate? A Logic-Based Proof Technique for Partial Specifications." In *Proceedings of the Fifth Annual Symposium on Logic in Computer Science*, pp. 108–117. Computer Society Press, Los Alamitos, 1990.

] Graf, S. and B. Steffen. "Using Interface Specifications for Compositional Reduction." To appear in *Proceedings of the Workshop on Computer-Aided Verification.*

] Larsen, K.G., and R. Milner. "Verifying a Protocol Using Relativized Bisimulation." In *Proceedings of ICALP 87*, Lecture Notes in Computer Science 267. Springer-Verlag, Berlin.

] Larsen, K.G., and B. Thomsen. "Compositional Proofs by Partial Specification of Processes." Technical Report 87-20, University of Aalborg, July 1987.

] Larsen, K.G. and B. Thomsen. "A Modal Process Logic." In *Proceedings of the Third Annual Symposium on Logic in Computer Science*, pp. 203–210. Computer Society Press, Washington DC, 1988.

] Larsen, K. and Xinxin, L. "Equation Solving through an Operational Semantics of Context." In *Proceedings of the Fifth Annual Symposium on Logic in Computer Science*, pp. 108–117. Computer Society Press, Los Alamitos, 1990.

Milner, R. *A Calculus of Communicating Systems.* Lecture Notes in Computer Science 92. Springer-Verlag, Berlin, 1980.

Milner, R. "A Modal Characterization of Observable Machine Behaviour." Lecture Notes in Computer Science 112, pp. 25-34. Springer-Verlag, Berlin, 1981.

Milner, R. *Communication and Concurrency.* Prentice-Hall, New York, 1989.

Shurek, G. and O. Grumberg. "The Modular Framework of Computer-Aided Verification: Motivation, Solutions and Evaluation Criteria." To appear in *Proceedings of the Workshop on Computer-Aided Verification.*

Steffen, B.U. "Characteristic Formulae for CCS with Divergence." In *Proceedings of Eleventh International Colloquium on Automata, Languages and Programming*, 1989.

Stirling, C. "Modal Logics for Communicating Systems." *Theoretical Computer Science*, v. 49, 1987, pp. 311-347.

Walker, D. "Bisimulations and Divergence." In *Proceedings of the Third Annual Symposium on Logic in Computer Science*, pp. 186-192. Computer Society Press, Washington DC, 1988.

Back and Forth Bisimulations

Rocco De Nicola
Istituto di Elaborazione dell' Informazione, C.N.R.
Via S. Maria 46, I-56126 Pisa, Italy

Ugo Montanari
Dipartimento di Informatica, Università di Pisa
Corso Italia 40, I-56125 Pisa, Italy

Frits Vaandrager
Centre for Mathematics and Computer Science
P.O. Box 4079, 1009 AB Amsterdam, The Netherlands

This paper is concerned with bisimulation relations which do not only require related agents to simulate each others behavior in the direction of the arrows, but also to simulate each other when going back in history. First it is demonstrated that the back and forth variant of strong bisimulation leads to the same equivalence as the ordinary notion of strong bisimulation. Then it is shown that the back and forth variant of Milner's observation equivalence is different from (and finer than) observation equivalence. In fact we prove that it coincides with the branching bisimulation equivalence of Van Glabbeek & Weijland. Also the back and forth variants of branching, η and delay bisimulation lead to branching bisimulation equivalence. The notion of back and forth bisimulation moreover leads to characterizations of branching bisimulation in terms of abstraction homomorphisms and in terms of Hennessy-Milner logic with backward modalities. In our view these results support the claim that branching bisimulation is a natural and important notion.

1. Introduction

The notion of bisimulation relation has been introduced by Park [18]. It leads to an equivalence on labelled transition systems which, in case image finiteness is assumed, coincides with the strong equivalence of Milner [12]. The great importance and usefulness of bisimulations in the theory of concurrent systems is evident: Mathematically, bisimulation is a very pleasant notion. It is closely related to the non-well-founded sets of Aczel [1] and leads to a natural first behavioral abstraction from transition systems. Algebraically, in the setting of CCS-like languages, bisimulations lead to elegant and simple laws [10]. Moreover, bisimulation equivalence has a beautiful characterization in terms of Hennessy-Milner logic [10]. Bisimulations are also important from a practical point of view because with the algorithm of Paige & Tarjan [17], bisimulation equivalence on finite state automata can be decided extremely fast in $O(m \log n)$ time (where m is the number of transitions and n the number of states).

In this paper we introduce the concept of back and forth bisimulations. In a back and forth bisimulation the agents can not only simulate each others behavior in the direction of the arrows but also when going backward in their history. In general, given the definition of any bisimulation, one can define a corresponding back and forth version of it. We want to explore the relationships between bisimulations and their back and forth variants.

Back and forth bisimulations are interesting for several reasons. First of all, it is always

First and second authors where supported by ESPRIT project 3011 (CEDISYS). The research of the third author was supported by RACE project no. 1046, Specification and Programming Environment for Communication Software (SPECS) and by ESPRIT project no. 3006 (CONCUR).

intriguing to see what are the consequences of small modifications of important definitions. More important is the connection with temporal and modal logics. These logics give rise to equivalences on transition systems and Kripke structures (two states are equivalent iff they satisfy the same formulas) and it appears to be very useful to give operational characterizations of these equivalences. A well known result in this area is that the equivalence induced by Hennessy-Milner logic (HML) formulas coincides with (strong) bisimulation equivalence [10]. Other results of this kind are for instance reported in [4]: two bisimulation-based equivalences over Kripke structures are related to two variants of CTL* [8]. In the world of temporal and modal logic, there has been a lot of interest in past-tense operators (see for instance [11, 20]). If one is looking for operational characterizations of the equivalences induced by logics with a past-tense operator, it seems natural to consider back and forth bisimulations.

In Section 2.1 of this paper, we demonstrate that the back and forth variant of strong bisimulation leads to the same equivalence as the ordinary notion of strong bisimulation. This results clarifies an earlier result of HENNESSY & STIRLING [11]. They showed that in the context of classic labelled transition systems, the extension of HML with a reverse operator does not lead to any increase in discriminating power. HENNESSY & STIRLING [11] did not consider abstraction of silent actions. In Section 2.2, we show that the back and forth variant of the (weak) observation equivalence of MILNER [15] is different from (and finer than) observation equivalence. In fact we prove that it coincides with the branching bisimulation equivalence of VAN GLABBEEK & WEIJLAND [9]. We will play the same game with other equivalences and prove that the back and forth versions of branching bisimulation, the *η-bisimulation* of BAETEN & VAN GLABBEEK [3] and the *delay bisimulation* of [21] (first introduced by MILNER [13] under the somewhat confusing name observational equivalence), all lead to branching bisimulation equivalence. Hence, branching bisimulation equivalence arises as a kind of 'fixed point' of the back and forth operation. This result supports the claim that branching bisimulation is a natural and important notion.

In Section 3, we study the relationships of back and forth bisimulation with abstraction homomorphisms. Abstraction homomorphisms have been introduced for the first time in [6], for labelled event structures. The tight relation of abstraction homomorphisms with bisimulation has been discussed in [5], where it is also proved that, under some rather restrictive conditions, given any transition system $\mathcal{A}$, there is always a unique minimal homomorphic system. Both important properties have been extended in [16] to partial ordering labelings, and they are proved to hold under significantly milder conditions. Similar results (but apparently applicable only to transition systems without τ) have been generalized to saturating quasi-homomorphisms of Φ algebras in [2] (following [19]). In the present paper, the main result about abstraction homomorphisms is the evidence of their flexibility in expressing several notions of equivalence: we show that observation equivalence can be characterized in terms of abstraction homomorphims that preserve successors, whereas branching bisimulation corresponds with abstraction homomorphisms which preserve both successors and predecessors.

The final section presents a logical characterization of branching bisimulation equivalence in terms of *back and forth logic*, an extension of Hennessy-Milner logic with backward modalities. In order to establish this logical characterization, we essentially use the result of Section 2.2 which says that branching bisimulation and weak back and forth bisimulation coincide. The logical characterization with backward modalities is also reported in [7]. In addition that paper presents two other logics for branching bisimulation equivalence. The first is an extension of HML with a kind of 'until' operators, which turn out to be definable in terms of the modalities of back and forth logic. The second is CTL* without the next-time operator interpreted over all paths, not just the maximal ones.

2. Operational characterizations

2.1. Strong bisimulation.
We start the technical part of this paper with a discussion of a simple, but fundamental case: the back and forth variant of strong bisimulation. Strong bisimulation equivalence, like all other equivalences in this paper, will be defined on the states of *labelled transition systems (LTS's)*. Below we recall the definitions of a LTS and strong bisimulation.

2.1.1. Definition. A *labelled transition system* (or *LTS*) is a triple $(S, A, \longrightarrow)$ where:
- S is a set of *states*,
- A is a set of *actions*; the *silent action* τ is not in A; we write $A_\tau = A \cup \{\tau\}$;
- $\longrightarrow \subseteq S \times A_\tau \times S$ is the *transition relation*; an element $(r, \alpha, s) \in \longrightarrow$ is called a *transition* and is usually written as $r \xrightarrow{\alpha} s$.

We let $r,s,..$ range over S; $a,b,..$ over A and α, β over A_τ.

2.1.2. Definition. Let $\mathscr{Q} = (S, A, \longrightarrow)$ be a LTS. A relation $R \subseteq S \times S$ is called a *(strong) bisimulation* if it is symmetric and satisfies: if $r\,R\,s$ and $r \xrightarrow{\alpha} r'$, then there exists an s' such that $s \xrightarrow{\alpha} s'$ and $r'\,R\,s'$.
Two states r,s are *(strongly) bisimilar*, abbreviated $\mathscr{Q}: r \leftrightarrow s$ or $r \leftrightarrow s$, if there exists a strong bisimulation relating r and s.

It is well known (and easy to see) that the arbitrary union of strong bisimulation relations is again a strong bisimulation; $\leftrightarrow$ is the maximal strong bisimulation on S. Moreover $\leftrightarrow$ is an equivalence relation.

2.1.3. Back and forth bisimulation on states. Back and forth bisimulations do not only require that related states can simulate each other in a *forward* direction but also also that they can simulate each others behavior in *backward* direction. The definition which comes to mind first is obtained by adding to the definition of a bisimulation relation the condition[1]:
if $r\,R\,s$ and $r' \xrightarrow{\alpha} r$, then there exists an s' such that $s' \xrightarrow{\alpha} s$ and $r'\,R\,s'$.
The resulting equivalence distinguishes the states r and s in Figure 1 below. From r it is possible to do a, then b, and then go back with a, which is not possible from s.

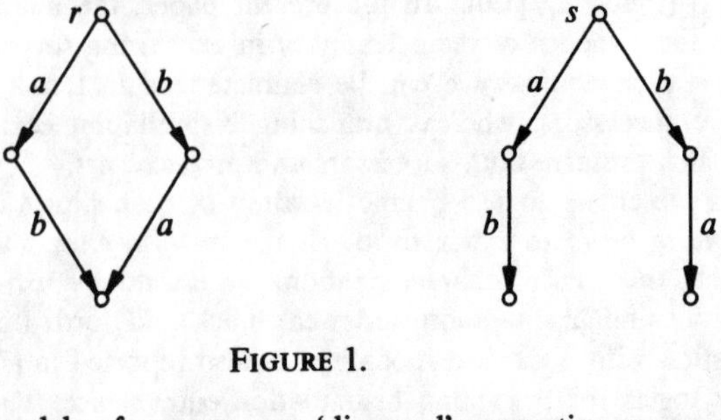

Figure 1.

In some non-interleaved models of concurrency 'diamond' properties are used to express concurrency of events. In these models the states r and s of Figure 1 are distinguished since the behavior from r corresponds to a pair a and b of concurrent events, whereas s describes a system which either does an a causally followed by a b, or a b causally followed by an a. Back and forth bisimulations cannot capture the intuitions behind these models: if we replace in Figure 1 the labels b by a, then there exists a back and forth bisimulation between r and s, even though

1. This generalization arose from discussion of the first author with Colin Stirling at Edinburgh

from a true concurrency point of view these states are different. At this moment, we do not see how the above back and forth variant of bisimulation equivalence could be useful. Probably everything would work if autoconcurrency is not permitted.

Therefore, in this paper a different type of back and forth bisimulations will be studied: it is possible to move back from a state, but only along the path which represents the *history* that brought one to this state. This means that a bisimulation is no longer a relation between states, but is instead a relation between histories. Below, after some preliminary definitions, we define the new notion of back and forth bisimulation.

2.1.4. DEFINITION. For a set K, the notation K^* denotes the set of finite sequences of elements of K. We denote concatenation of sequences by juxtaposition. With λ we denote the empty sequence and with $|\sigma|$ the length of sequence σ.

2.1.5. DEFINITION. Let $\mathcal{Q}=(S,A,\rightarrow)$ be a LTS. A sequence $(s_0,\alpha_1,s_1)\cdots(s_{n-1},\alpha_n,s_n)\in\rightarrow^*$ is called a *path* from s_0. A *run* (also called *history* or *computation*) from $s\in S$ is a pair (s,π), where π is a path from s. We write $run_{\mathcal{Q}}(s)$, or just $run(s)$ for the set of runs from s and $run_{\mathcal{Q}}$ for the set of runs in $\mathcal{Q}$. We let π,.. range over paths and ρ,σ,θ,.. over runs.

2.1.6. DEFINITION. Let $\rho=(s_0,\pi)\in run(s_0)$ with $\pi=(s_0,\alpha_1,s_1)\cdots(s_{n-1},\alpha_n,s_n)$.
- $first(\rho) = s_0$,
- $path(\rho) = \pi$,
- $last(\rho) = s_n$,
- $states(\rho) = s_0 s_1 \cdots s_n$,
- $label(\rho) = \alpha_1 \cdots \alpha_n$,
- concatenation of runs is denoted by juxtaposition, the result $\sigma\sigma'$ is defined iff $last(\sigma)=first(\sigma')$, the operation is associative and empty runs behave as (left and right) identities,
- $\sigma\xrightarrow{\alpha}\sigma'$ if for some run $\theta=(s,(s,\alpha,s'))$: $\sigma'=\sigma\theta$.

2.1.7. DEFINITION. Let $\mathcal{Q}=(S,A,\rightarrow)$ be a LTS. Two states $r,s\in S$ are *strongly back and forth bisimilar*, abbreviated $\mathcal{Q}:r\leftrightarrow_{bf}s$ or $r\leftrightarrow_{bf}s$, if there exists a symmetric relation $R\subseteq run_{\mathcal{Q}}\times run_{\mathcal{Q}}$, called a *strong back and forth bisimulation*, satisfying
1. $(r,\lambda)\,R\,(s,\lambda)$;
2. if $\rho\ R\ \sigma$ and $\rho\xrightarrow{\alpha}\rho'$, then there exists a σ' such that $\sigma\xrightarrow{\alpha}\sigma'$ and $\rho' R\sigma'$;
3. if $\rho\ R\ \sigma$ and $\rho'\xrightarrow{\alpha}\rho$, then there exists a σ' such that $\sigma'\xrightarrow{\alpha}\sigma$ and $\rho' R\sigma'$.

The following proposition tells us that, when all actions are visible, the possibility to go back 'in ones own history' does not result in any additional distinguishing power: the resulting equivalence is the same as the 'forward only' strong bisimulation equivalence.

2.1.8. PROPOSITION. *Let $\mathcal{Q}=(S,A,\rightarrow)$ be a LTS. Then for all $r,s\in S$:*

$$\mathcal{Q}:r\leftrightarrow_{bf}s \quad\Leftrightarrow\quad \mathcal{Q}:r\leftrightarrow s.$$

PROOF. "$\Leftarrow$" Suppose $r\leftrightarrow s$. Let ct be the mapping that associates to each path π in $\mathcal{Q}$ its *colored trace*, i.e. the sequence which is obtained from π by replacing each state by its bisimulation equivalence class (this terminology is borrowed from [9]). So

$$ct((s_0,\alpha_1,s_1)...(s_{n-1},\alpha_n,s_n)) = (s_0/_{\leftrightarrow},\alpha_1,s_1/_{\leftrightarrow})...(s_{n-1}/_{\leftrightarrow},\alpha_n,s_n/_{\leftrightarrow}).$$

Define relation R by:

$$R = \{(\rho,\sigma),(\sigma,\rho)\,|\,\rho\in run(r),\ \sigma\in run(s)\ \&\ ct(path(\rho))=ct(path(\sigma))\}.$$

It is straightforward to check that R is a back and forth bisimulation between r and s.

"⇒" Suppose $r \leftrightarrow_{bf} s$. Let R be a back and forth bisimulation between r and s. Define

$$R' = \{(last(\rho), last(\sigma)) \mid \rho R \sigma\}.$$

Again, it is straightforward to check that R is a strong bisimulation between r and s. □

2.2. Weak bisimulation

Weak bisimulation equivalence or observation equivalence is a variant of strong bisimulation equivalence that has been proposed by MILNER [14], to take into account the 'invisible' nature of the silent step τ. In this section we will see that, in contrast to the case of strong bisimulation, weak bisimulation differs from its back and forth variant.

2.2.1. DEFINITION. Let $(S,A,\rightarrow)$ be a LTS. Let $\epsilon \notin A_\tau$. Define $\overset{\epsilon}{\Rightarrow}$ as the transitive and reflexive closure of $\overset{\tau}{\rightarrow}$. So $r \overset{\epsilon}{\Rightarrow} s$ says that there exists a path from r to s consisting of zero or more τ-transitions. Further we define for $a \in A$:

$$r \overset{a}{\Rightarrow} s \quad \Leftrightarrow \quad \exists r', s' : r \overset{\epsilon}{\Rightarrow} r' \overset{a}{\rightarrow} s' \overset{\epsilon}{\Rightarrow} s.$$

We let $k,l,..$ range over $A_\epsilon = A \cup \{\epsilon\}$.

2.2.2. DEFINITION. Let $(S,A,\rightarrow)$ be a LTS. A relation $R \subseteq S \times S$ is called a *weak bisimulation* if it is symmetric and satisfies: if $r R s$ and $r \overset{k}{\Rightarrow} r'$, then there exists an s' such that $s \overset{k}{\Rightarrow} s'$ and $r' R s'$.

Two states r,s are *weakly bisimilar* or *observation equivalent*, abbreviated $\mathcal{Q}: r \leftrightarrow_\tau s$ or $r \leftrightarrow_\tau s$, if there exists a weak bisimulation relating r and s.

Again, it is easy to see that the arbitrary union of weak bisimulation relations is a weak bisimulation; $\leftrightarrow_\tau$ is the maximal weak bisimulation on S. Moreover $\leftrightarrow_\tau$ is an equivalence relation. Since any strong bisimulation is also a weak bisimulation, we have $\leftrightarrow \subseteq \leftrightarrow_\tau$.

There is an obvious way to generalize the relations $\overset{k}{\Rightarrow}$ to runs:

2.2.3. DEFINITION. Let $\mathcal{Q}=(S,A,\rightarrow)$ be a LTS. Define for $\rho, \sigma \in run_{\mathcal{Q}}$:

$$\rho \overset{\epsilon}{\Rightarrow} \sigma \quad \Leftrightarrow \quad \exists n \geqslant 0 \; \exists \rho_0, ..., \rho_n : \rho = \rho_0, \; \rho_n = \sigma \; \& \; [\forall 0 \leqslant i < n : \rho_i \overset{\tau}{\rightarrow} \rho_{i+1}].$$

$$\rho \overset{a}{\Rightarrow} \sigma \quad \Leftrightarrow \quad \exists \rho', \sigma' : \rho \overset{\epsilon}{\Rightarrow} \rho' \overset{a}{\rightarrow} \sigma' \overset{\epsilon}{\Rightarrow} \sigma.$$

Now consider the following 'weak' variant of the back and forth bisimulation:

2.2.4. DEFINITION. Let $\mathcal{Q}=(S,A,\rightarrow)$ be a LTS. Two states $r,s \in S$ are *weakly back and forth bisimilar*, abbreviated $\mathcal{Q}: r \leftrightarrow_{\tau bf} s$ or $r \leftrightarrow_{\tau bf} s$, if there exists a symmetric relation $R \subseteq run_{\mathcal{Q}} \times run_{\mathcal{Q}}$, called a *weak back and forth bisimulation*, satisfying

1. $(r,\lambda) R (s,\lambda)$;
2. if $\rho R \sigma$ and $\rho \overset{k}{\Rightarrow} \rho'$, then there exists a σ' such that $\sigma \overset{k}{\Rightarrow} \sigma'$ and $\rho' R \sigma'$;
3. if $\rho R \sigma$ and $\rho' \overset{k}{\Rightarrow} \rho$, then there exists a σ' such that $\sigma' \overset{k}{\Rightarrow} \sigma$ and $\rho' R \sigma'$.

Interestingly, weak bisimulation equivalence and weak back and forth bisimulation equivalence are different. In Figures 2 and 3 below, two counterexamples are presented. The states p and q in Figure 2 are not weak back and forth bisimulation equivalent because there exists no weak back and forth bisimulation relating the corresponding empty runs: from q it is possible to do an a in such a way that always after going back with an a there is a possibility of doing b. This behavior is not possible from p. The counterexample of Figure 3 is similar.

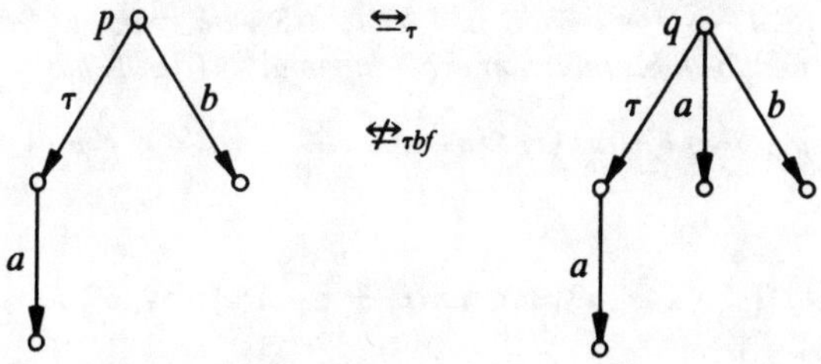

FIGURE 2.

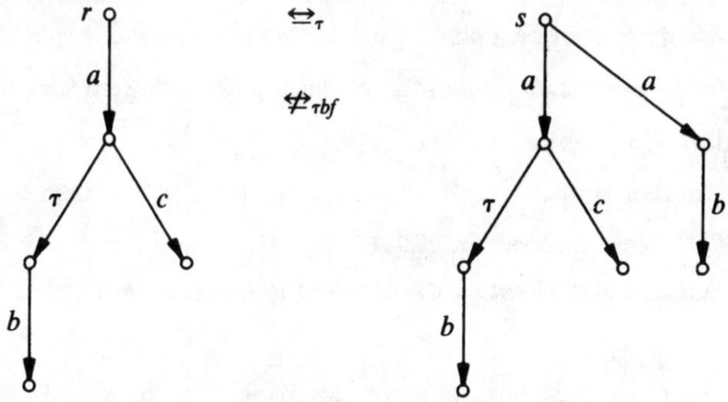

FIGURE 3.

Since any strong back and forth bisimulation is also a weak back and forth bisimulation we have $\leftrightarrow_{bf} \subseteq \leftrightarrow_{\tau bf}$ and hence, by Proposition 2.1.8, $\leftrightarrow \subseteq \leftrightarrow_{\tau bf}$. The example of Figure 4 shows that this inclusion is strict.

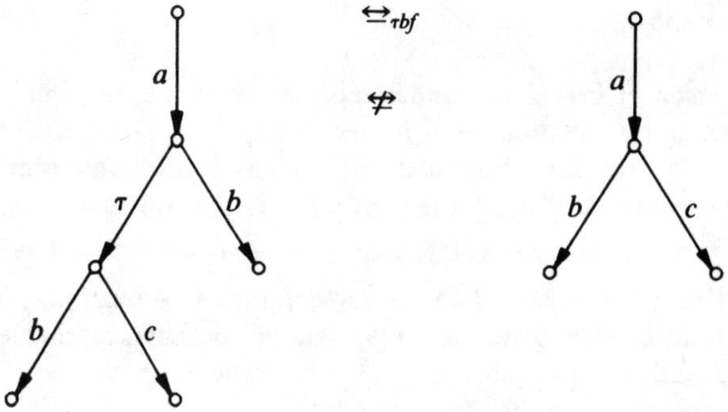

FIGURE 4.

We conclude this section with a technical lemma which will be needed to relate $\leftrightarrow_{\tau bf}$ with branching bisimulation.

2.2.5. LEMMA. *Let $(S,A,\rightarrow)$ be a LTS and let $r,s \in S$ with $r \leftrightarrow_{\tau bf} s$. Let $R \subseteq run_{\mathcal{Q}} \times run_{\mathcal{Q}}$ be the maximal weak back and forth bisimulation between r and s. Then R has the following X-property:*

$$\forall \rho,\rho' \in run(r)\ \forall \sigma,\sigma' \in run(s) : [\rho \stackrel{\epsilon}{\Rightarrow} \rho',\ \sigma \stackrel{\epsilon}{\Rightarrow} \sigma',\ \rho R \sigma' \ \&\ \rho' R \sigma] \Rightarrow \rho' R \sigma'.$$

PROOF. Define relation R' by:

$$R' = R \cup \{(\rho',\sigma'), (\sigma',\rho') \mid \exists \rho \in run(r)\ \exists \sigma \in run(s) : \rho \stackrel{\epsilon}{\Rightarrow} \rho',\ \sigma \stackrel{\epsilon}{\Rightarrow} \sigma',\ \rho R \sigma' \ \&\ \rho' R \sigma\}$$

We prove that R' is a weak back and forth bisimulation. Since R is the maximal back and forth bisimulation and $R \subseteq R'$ by construction, $R = R'$. Thus R has the X-property.
Clearly R' is symmetric. Moreover $(r,\lambda) R' (s,\lambda)$ because $(r,\lambda) R (s,\lambda)$. Suppose $\rho' R' \sigma'$ with $\rho' \in run(r)$ and $\sigma' \in run(s)$. If $\rho' R \sigma'$ then the back and forth conditions 2 and 3 are trivially fulfilled. Otherwise there must be a ρ and a σ such that: $\rho \stackrel{\epsilon}{\Rightarrow} \rho'$, $\sigma \stackrel{\epsilon}{\Rightarrow} \sigma'$, $\rho R \sigma'$ and $\rho' R \sigma$.

We check transfer property 2. Suppose $\rho' \stackrel{k}{\Rightarrow} \rho''$. Then $\rho \stackrel{k}{\Rightarrow} \rho''$. Since $\rho R \sigma'$, there exists a σ'' such that $\sigma' \stackrel{k}{\Rightarrow} \sigma''$ and $\rho'' R' \sigma''$.

Next we check transfer property 3. If $\rho'' \stackrel{k}{\Rightarrow} \rho'$, then, since $\rho' R \sigma$, there exists a σ'' such that $\sigma'' \stackrel{k}{\Rightarrow} \sigma$ and $\rho'' R' \sigma''$. Now observe that $\sigma'' \stackrel{k}{\Rightarrow} \sigma'$.
The remaining case that $\sigma' R' \rho'$ with $\sigma' \in run(s)$ and $\rho' \in run(r)$ is symmetric. □

2.3. Branching bisimulation

In this section we prove the main result of this paper: the back and forth variant of weak bisimulation equivalence coincides with the branching bisimulation of VAN GLABBEEK & WEIJLAND [9].

2.3.1. DEFINITION. Let $\mathcal{Q} = (S,A,\rightarrow)$ be a LTS. A relation $R \subseteq S \times S$ is called a *branching bisimulation* if it is symmetric and satisfies: if $r R s$ and $r \stackrel{\alpha}{\longrightarrow} r'$, then either $\alpha = \tau$ and $r' R s$, or there exist s_1, s' such that $s \stackrel{\epsilon}{\Rightarrow} s_1 \stackrel{\alpha}{\longrightarrow} s'$, $r R s_1$ and $r' R s'$.

Two states r,s are *branching bisimilar*, abbreviated $\mathcal{Q}: r \leftrightarrow_b s$ or $r \leftrightarrow_b s$, if there exists a branching bisimulation relating r and s.

Again the arbitrary union of branching bisimulation relations is a branching bisimulation; $\leftrightarrow_b$ is the maximal branching bisimulation on S. Moreover $\leftrightarrow_b$ is an equivalence relation. Since any strong bisimulation is a branching bisimulation and any branching bisimulation is a weak bisimulation, we have $\leftrightarrow\ \subseteq\ \leftrightarrow_b\ \subseteq\ \leftrightarrow_\tau$. It is worth noting that we could have strengthened the above definition by requiring *all* intermediate states in $s \stackrel{\epsilon}{\Rightarrow} s_1$ to be related with r. The following lemma implies that this would have led to the same equivalence relation. Moreover, we could have also asked, as in the original definition of [9], that all the states reachable from s' via silent sequences be related with s; again, by simple considerations, it can be concluded that the equivalence we would obtain would be the same.

2.3.2. LEMMA (cf. Lemma 1.3 of [9]). *Let $\mathcal{A}=(S,A,\rightarrow)$ be a LTS, let $n>0$ and let $(r_0,\tau,r_1)...(r_{n-1},\tau,r_n)$ be a path in $\mathcal{A}$ with $r_0 \leftrightarrow_b r_n$. Then for all $0 \leq i \leq n$: $r_0 \leftrightarrow_b r_i$.*

PROOF. Define for $i>0$:

$$R_0 = \leftrightarrow_b$$

$$R_i = R_{i-1} \cup \{(r,r'), (r',r) \mid \exists r'' : r \overset{\epsilon}{\Rightarrow} r' \overset{\tau}{\longrightarrow} r'' \;\&\; r \, R_{i-1} \, r''\}$$

$$R_\omega = \bigcup_{i<\omega} R_i$$

First we show that R_ω has the property that we want to prove for $\leftrightarrow_b$. Let for some $n>0$, $(r_0,\tau,r_1) \cdots (r_{n-1},\tau,r_n)$ be a path with $r_0 \, R_\omega \, r_n$. By induction on n we prove that for all $0 \leq i \leq n$: $r_0 \, R_\omega \, r_i$.

If $n=1$ the statement is trivially correct.

Now take $n>1$. Since $r_0 \, R_\omega \, r_n$, there exists an $m<\omega$ with $r_0 \, R_m \, r_n$. By definition of R_{m+1}: $r_0 \, R_{m+1} \, r_{n-1}$. Thus $r_0 \, R_\omega \, r_{n-1}$ and, by induction hypothesis, $r_0 \, R_\omega \, r_i$ for all $0 \leq i \leq n$.

Next we will prove with induction that, for every $n<\omega$, R_n is a branching bisimulation. Thus R_ω is a branching bisimulation and $R_\omega \subseteq \leftrightarrow_b$. But by construction $\leftrightarrow_b \subseteq R_\omega$. Hence $\leftrightarrow_b = R_\omega$ and we have proved the lemma.

R_0 is a branching bisimulation because $\leftrightarrow_b$ is.

Suppose that, for certain $n>0$, R_{n-1} is a branching bisimulation. We prove that R_n is a branching bisimulation too. By construction R_n is symmetric. Suppose $r \, R_n \, r'$ and $r \overset{\alpha}{\longrightarrow} s$. If $r \, R_{n-1} \, r'$, then the transfer property is trivially fulfilled. In the other case there are two possibilities:

1. For some r'': $r \overset{\epsilon}{\Rightarrow} r' \overset{\tau}{\longrightarrow} r''$ and $r \, R_{n-1} \, r''$. Using $r \, R_{n-1} \, r''$, a first possibility is that $\alpha=\tau$ and $s \, R_{n-1} \, r''$. But this means that $r' \overset{\epsilon}{\Rightarrow} r' \overset{\tau}{\longrightarrow} r''$ with $r \, R_n \, r'$ and $s \, R_n \, r''$. Otherwise there are r_1, r_2 such that $r'' \overset{\epsilon}{\Rightarrow} r_1 \overset{\alpha}{\longrightarrow} r_2$, $r \, R_{n-1} \, r_1$ and $s \, R_{n-1} \, r_2$. But then $r' \overset{\epsilon}{\Rightarrow} r_1 \overset{\alpha}{\longrightarrow} r_2$, $r \, R_n \, r_1$ and $s \, R_n \, r_2$.
2. For some r'': $r' \overset{\epsilon}{\Rightarrow} r \overset{\tau}{\longrightarrow} r''$ and $r' \, R_{n-1} \, r''$. Then $r' \overset{\epsilon}{\Rightarrow} r \overset{\alpha}{\longrightarrow} s$, $r \, R_n \, r$ and $s \, R_n \, s$. □

2.3.3. THEOREM. *Let $\mathcal{A}=(S,A,\rightarrow)$ be a LTS. Then for all $r,s \in S$:*

$$\mathcal{A}: r \leftrightarrow_{\tau bf} s \quad \Leftrightarrow \quad \mathcal{A}: r \leftrightarrow_b s.$$

PROOF. "$\Leftarrow$" Suppose $r \leftrightarrow_b s$. Let *cct* be the mapping that associates to each path π in $\mathcal{A}$ its *concrete colored trace*, i.e. the sequence which is obtained from π by replacing each state by its branching bisimulation equivalence class. So

$$cct((s_0,\alpha_1,s_1) \cdots (s_{n-1},\alpha_n,s_n)) = (s_0/_{\leftrightarrow_b},\alpha_1,s_1/_{\leftrightarrow_b}) \cdots (s_{n-1}/_{\leftrightarrow_b},\alpha_n,s_n/_{\leftrightarrow_b}).$$

Let *ct* be the mapping that associates to each path π in $\mathcal{A}$ its *(abstract) colored trace*, i.e. the sequence which is obtained from $cct(\pi)$ by removing all elements (C,τ,C) from the sequence. Define relation R by

$$R = \{(\rho,\sigma),(\sigma,\rho) \mid \rho \in run(r),\ \sigma \in run(s) \;\&\; ct(path(\rho))=ct(path(\sigma))\}.$$

Using Lemma 2.3.2, it is straightforward to check that R is a weak back and forth bisimulation between r and s.

"$\Rightarrow$" Suppose $r \leftrightarrow_{\tau bf} s$. Let $R \subseteq run_{\mathcal{A}} \times run_{\mathcal{A}}$ be the maximal weak back and forth bisimulation between r and s. Define

$$R' = \{(last(\rho), last(\sigma)) \mid \rho \, R \, \sigma\}.$$

Clearly $r \, R' \, s$. We show that R' is a branching bisimulation.

R' is symmetric because R is.

Suppose $r_0 R' s_0$. Then there are ρ, σ with $\rho R \sigma$, $last(\rho)=r_0$ and $last(\sigma)=s_0$. Suppose that $r_0 \xrightarrow{\alpha} r'$. Let $\rho'=\rho\,(r_0,(r_0,\alpha,r'))$. In the proof of the transfer property we distinguish between two cases.

1. $\alpha \neq \tau$. Since $\rho \overset{\alpha}{\Rightarrow} \rho'$ and $\rho R \sigma$, there exist σ_1, σ_2, σ' such that $\sigma \overset{\epsilon}{\Rightarrow} \sigma_1 \xrightarrow{\alpha} \sigma_2 \overset{\epsilon}{\Rightarrow} \sigma'$ and $\rho' R \sigma'$. Since $\sigma_2 \overset{\epsilon}{\Rightarrow} \sigma'$, there exists a $\hat{\rho}$ such that $\hat{\rho} \overset{\epsilon}{\Rightarrow} \rho'$ and $\hat{\rho} R \sigma_2$. But since the last transition of ρ' has label α, $\hat{\rho}=\rho'$ so that $\rho' R \sigma_2$. Because $\sigma_1 \overset{\alpha}{\Rightarrow} \sigma_2$, there exists a $\bar{\rho}$ such that $\bar{\rho} \overset{\epsilon}{\Rightarrow} \rho \xrightarrow{\alpha} \rho'$ and $\bar{\rho} R \sigma_1$. Now use that R has the X-property to obtain $\rho R \sigma_1$. But this gives us the transfer property:

$$s_0 \overset{\epsilon}{\Rightarrow} end(\sigma_1) \xrightarrow{\alpha} end(\sigma_2)\,,\quad r_0 R' end(\sigma_1) \text{ and } r' R' end(\sigma_2).$$

2. $\alpha=\tau$. Since $\rho \overset{\epsilon}{\Rightarrow} \rho'$ and $\rho R \sigma$, there is an $n \geqslant 0$ and there are σ_i for $0 \leqslant i \leqslant n$ such that $\sigma_0=\sigma$, for $0<i\leqslant n$: $\sigma_{i-1} \xrightarrow{\tau} \sigma_i$, and $\rho' R \sigma_n$.
If $n=0$ then $r' R' s_0$ and we have proved the transfer property. If $n>0$ then we can go back with an ϵ-move from σ_n to σ_{n-1}. A first possibility is that ρ' can simulate this step by doing nothing: $\rho' R \sigma_{n-1}$. If this is the case then either $n=1$ and we are ready, or we can go back one more ϵ-step from σ_{n-1} to σ_{n-2}. Repeating this, we either find $\rho' R \sigma_0$, in which case we have proved the transfer property for branching bisimulation since $r' R' s_0$, or, for some $m>0$ with $\rho' R \sigma_m$, we have that a backward step to σ_{m-1} is simulated by a backward step $\hat{\rho} \overset{\epsilon}{\Rightarrow} \rho \xrightarrow{\tau} \rho'$ with $\hat{\rho} R \sigma_{m-1}$. In this case we use the X-property (Lemma 2.2.5) to obtain $\rho R \sigma_{m-1}$. This gives us the transfer property for branching bisimulation since:

$$s_0 \overset{\epsilon}{\Rightarrow} end(\sigma_{m-1}) \xrightarrow{\tau} end(\sigma_m),\quad r_0 R' end(\sigma_{m-1}) \text{ and } r' R' end(\sigma_m).$$ □

Now we have shown that the back and forth variant of weak bisimulation coincides with branching bisimulation, it becomes natural to consider the back and forth variant of branching bisimulation. Let the symbol $\leftrightarrows_{bbf}$ denote this equivalence. One can easily prove the following

2.3.4. **Theorem.** *Let $\mathcal{Q}=(S,A,\rightarrow)$ be a LTS. Then for all $r,s \in S$:*

$$\mathcal{Q}{:}r \leftrightarrows_{bbf} s \;\Leftrightarrow\; \mathcal{Q}{:}r \leftrightarrows_b s.$$

Proof. Similar to but much easier than the proof of Theorem 2.3.3. □

As a corollary of Theorem 2.3.3 and Theorem 2.3.4 one can show that also the back and forth versions of some equivalences in between $\leftrightarrows_\tau$ and $\leftrightarrows_b$ coincide with $\leftrightarrows_b$. Consider the following two definitions:

2.3.5. **Definition.** Let $\mathcal{Q}=(S,A,\rightarrow)$ be a LTS. A relation $R \subseteq S \times S$ is called a *η-bisimulation* if it is symmetric and satisfies: if $r R s$ and $r \xrightarrow{\alpha} r'$, then either $\alpha=\tau$ and $r' R s$, or there exist s_1,s_2,s' such that $s \overset{\epsilon}{\Rightarrow} s_1 \xrightarrow{\alpha} s_2 \overset{\epsilon}{\Rightarrow} s'$, $r R s_1$ and $r' R s'$.

Two states r,s are *η-bisimilar*, abbreviated $\mathcal{Q}{:}r \leftrightarrows_\eta s$ or $r \leftrightarrows_\eta s$, if there exists an η-bisimulation relating r and s.

2.3.6. **Definition.** Let $\mathcal{Q}=(S,A,\rightarrow)$ be a LTS. A relation $R \subseteq S \times S$ is called a *delay bisimulation* if it is symmetric and satisfies: if $r R s$ and $r \xrightarrow{\alpha} r'$, then either $\alpha=\tau$ and $r' R s$, or there exist s_1,s' such that $s \overset{\epsilon}{\Rightarrow} s_1 \xrightarrow{\alpha} s'$ and $r' R s'$.

Two states r,s are *delay bisimilar*, abbreviated $\mathcal{Q}{:}r \leftrightarrows_d s$ or $r \leftrightarrows_d s$, if there exists a delay bisimulation relating r and s.

The notion of *η-bisimulation* was first introduced by Baeten & Van Glabbeek [3] as a finer version of observation equivalence. Delay bisimulations are used by WEIJLAND [21]. A variant of delay bisimulation - only differing in the treatment of divergence - first appeared in MILNER [13], also under the name observational equivalence. From the definitions it follows right away that $\leftrightarrow_b \subseteq \leftrightarrow_\eta$, $\leftrightarrow_b \subseteq \leftrightarrow_d$, $\leftrightarrow_\eta \subseteq \leftrightarrow_\tau$ and $\leftrightarrow_d \subseteq \leftrightarrow_\tau$. The example of Figure 2 can be used to show that the second and third inclusion are strict. The example of Figure 3 illustrates the strictness of the other two inclusions.

2.3.7. COROLLARY. *Let $\leftrightarrow_{\eta bf}$ and $\leftrightarrow_{dbf}$ denote the back and forth variants of η- and delay bisimulation respectively. Let $\mathcal{A}=(S,A,\longrightarrow)$ be a LTS. Then for all $r,s \in S$:*

$$\mathcal{A}: r \leftrightarrow_{\eta bf} s \quad \Leftrightarrow \quad \mathcal{A}: r \leftrightarrow_b s \quad \Leftrightarrow \quad \mathcal{A}: r \leftrightarrow_{dbf} s.$$

PROOF. Suppose $r \leftrightarrow_{\eta bf} s$. Then there exists a back and forth η-bisimulation relating r and s. But any back and forth η-bisimulation is also a weak back and forth bisimulation. Therefore $r \leftrightarrow_{\tau bf} s$. But this implies $r \leftrightarrow_b s$ by Theorem 2.3.3.
Suppose $r \leftrightarrow_b s$. Then $r \leftrightarrow_{bbf} s$ by Theorem 2.3.4. So there exists a back and forth branching bisimulation between r and s. But any back and forth branching bisimulation is also a back and forth η-bisimulation. Therefore $r \leftrightarrow_{\eta bf} s$.
The remaining two implications can be proved in the same way. □

3. ALGEBRAIC CHARACTERIZATION

In this section we show that observation equivalence can be characterized in terms of abstraction homomorphims that preserve successors, whereas branching bisimulation corresponds with abstraction homomorphisms which preserve both successors and predecessors.

3.1. DEFINITION. A *category of labelled computations* (CLAC) $\mathcal{C} = (S, C, src, trg, ;, id, A, o)$ is defined as follows:

- $(S, C, src, trg, ;, id)$ is a category, i.e. S is a set of objects called *states*; C is a set of arrows called *computations*; $src, trg : C \rightarrow S$ are functions associating to every computation its *source* and *target*; the binary operation $;: C \times C \rightarrow C$ of *concatenation* is partial: $\rho=\rho';\rho''$ is defined iff $trg(\rho')=src(\rho'')$, with $src(\rho)=src(\rho')$ and $trg(\rho)=trg(\rho'')$; concatenation has an identity $id(s)$ (both left and right) for each state s and is associative;
- A is a set of *actions*; the silent action τ is not in A;
- $o : C \rightarrow A^*$ is a *labeling* function which respects concatenation, i.e. $o(\rho;\rho')=o(\rho)\, o(\rho')$.

On computations we define a prefix preorder: we have $\rho_1 \leqslant \rho_2$ iff there is a computation σ with $\rho_2=\rho_1;\sigma$. We let $succ(\rho)=\{\sigma \mid \rho \leqslant \sigma\}$ and $pred(\rho)=\{\sigma \mid \sigma \leqslant \rho\}$.

3.2. DEFINITION. Let $\mathcal{A}=(S,A,\longrightarrow)$ be a LTS. Its *associated* CLAC is $\mathcal{C}(\mathcal{A}) = (S, C, src, trg, ;, id, A, o)$ where C is the set $run_{\mathcal{A}}$ of runs of $\mathcal{A}$, $src(\rho)=first(\rho)$, $trg(\rho)=last(\rho)$, operation ';' is run concatenation, $id(s)$ is the empty run from s, and $o(\rho)$ is obtained from $label(\rho)$ by removing all τ's.

3.3. DEFINITION. Let $\mathcal{C} = (S, C, src, trg, ;, id, A, o)$ and $\mathcal{C}' = (S', C', src', trg', ;', id', A', o')$ be CLAC's. A pair of *surjective* functions $k=\langle f,g\rangle$, $f: C \rightarrow C'$ and $g : S \rightarrow S'$ is a *forward abstraction homomorphism*, and we write $\mathcal{C} \xrightarrow{k}_{fw} \mathcal{C}'$, iff:

i) $g(src(\rho)) = src'(f(\rho))$ and $g(trg(\rho)) = trg'(f(\rho))$,
ii) $f(\rho;\sigma) = f(\rho) ;' f(\sigma)$ and $f(id(s)) = id'(g(s))$,
iii) $o(\rho) = o'(f(\rho))$ and
iv) $f(succ(\rho)) = succ'(f(\rho))$, where f is extended to sets.

In words, a forward abstraction homomorphism must respect sources, targets, concatenations, identities, observation and successors. A pair $k=\langle f,g\rangle$ is called a *back and forth abstraction homomorphism*, and we write $\mathcal{C} \xrightarrow{k}_{bf} \mathcal{C}'$, iff, besides (i)-(iv), it satisfies also:

v) $f(pred(\rho)) = pred'(f(\rho))$,

i.e. if it respects predecessors.

3.4. DEFINITION. Let $\mathcal{A}=(S,A,\longrightarrow)$ be a LTS. A pair of states $r,s \in S$ is *forward preserving*, and we write $r \leftrightarrow_{fp} s$, iff there exist $\mathcal{C}$ and $k=<f,g>$ with $\mathcal{C}(\mathcal{A}) \xrightarrow{k}_{fw} \mathcal{C}$ and $g(r)=g(s)$. Similarly, a pair of states $r,s \in S$ is *back and forth preserving*, and we write $r \leftrightarrow_{bfp} s$, iff there exist $\mathcal{C}$ and $k=<f,g>$ with $\mathcal{C}(\mathcal{A}) \xrightarrow{k}_{bf} \mathcal{C}$ and $g(r)=g(s)$.

The following theorem says that forward and back and forth abstraction homomorphisms correspond to weak observation equivalence and branching bisimulation respectively.

3.5. THEOREM. *Let $\mathcal{A}=(S,A,\longrightarrow)$ be a LTS. Then, given $r,s \in S$, we have*
a) $r \leftrightarrow_{fp} s$ *iff* $r \leftrightarrow_{\tau} s$;
b) $r \leftrightarrow_{bfp} s$ *iff* $r \leftrightarrow_{\tau bf} s$.
PROOF. We sketch the proof of (a). The proof of (b) is similar. Define a one-to-one correspondence between weak bisimulation equivalences and forward abstraction homomorphisms. When going from homomorphisms $<f,g>$ to bisimulations $\sim$, the construction is straightforward, assuming $s \sim s'$ iff $g(s)=g(s')$. In the other direction it is convenient first to extend $\sim$ to computations letting $\rho \sim \sigma$ iff $src(\rho) \sim src(\sigma)$, $trg(\rho)=trg(\sigma)$ and $o(\rho)=o(\sigma)$, and then to take $f(\rho)=\rho/_{\sim}$ and $g(s)=s/_{\sim}$.
The result follows since if a pair belongs to a bisimulation, it also belongs to a bisimulation equivalence. □

3.6. REMARK. CLAC's and abstraction homomorphisms have been introduced here for a labelling function o ranging over A^*. However, exactly the same definition (for both the forward and the back and forth case) could be given in terms of a different observation function o_{po}, yielding for instance a partial ordering observation $o_{po}(\rho)$ of a computation ρ. This allows us to develop an algebraic theory of observation equivalence, in both its traditional version and its branching variant, also for true concurrency.

4. LOGICAL CHARACTERIZATION

Theorem 2.3.3 suggests a logical characterization of branching bisimulation which is a variant of the Hennessy-Milner logic [10]. The characterization relies on a well-known theorem from HENNESSY & MILNER [10], which we recall first.

4.1. DEFINITION. Let A be a given alphabet of symbols. The set $HML(A)$, often abbreviated to HML, of *Hennessy-Milner logic formulas over A* is given by the following grammar:

$$\phi ::= T \mid \phi \wedge \phi \mid \neg\phi \mid \langle\alpha\rangle\phi.$$

Let $\mathcal{A}=(S,A,\longrightarrow)$ be a LTS. The *satisfaction relation* $\models \subseteq S \times HML$ is the least relation such that:
- $s \models T$ for all $s \in S$,
- $s \models \phi \wedge \psi$ iff $s \models \phi$ and $s \models \psi$,
- $s \models \neg\phi$ iff not $s \models \phi$,
- $s \models \langle\alpha\rangle\phi$ iff for some $t \in S$: $s \xrightarrow{\alpha} t$ and $t \models \phi$.

The following notations are standard:
- F stands for $\neg T$,
- $\phi \vee \psi$ stands for $\neg(\neg\phi \wedge \neg\psi)$,
- $[\alpha]\phi$ stands for $\neg\langle\alpha\rangle\neg\phi$.

Let $\mathcal{L}$ be a set of logical formulas with an associated satisfaction relation. Two states $r,s \in S$ are *$\mathcal{L}$-equivalent*, abbreviated $\mathcal{A}{:}r \sim_{\mathcal{L}} s$ or $r \sim_{\mathcal{L}} s$, if for all formulas ϕ in $\mathcal{L}$: $s \models \phi \iff t \models \phi$.

Transition system $\mathcal{A}$ is called *image finite* if for all $s \in S$ and $a \in A$ the set $\{t \mid s \xrightarrow{a} t\}$ is finite.

4.2. THEOREM ([10]). *Let $\mathcal{A}=(S,A,\longrightarrow)$ be an image finite LTS. Then for all $r,s \in S$:*

$$r \leftrightarroweq s \quad \Leftrightarrow \quad r \sim_{HML} s.$$

4.3. DEFINITION. Let A be a given alphabet and let k range over A_ϵ. The set $BFL(A)$ (or just BFL) of *back and forth logic formulas over A* is defined by the following grammar:

$$\phi ::= T \mid \phi \wedge \phi \mid \neg \phi \mid \langle k \rangle \phi \mid \langle \leftarrow k \rangle \phi.$$

Let $\mathcal{A}=(S,A,\longrightarrow)$ be a LTS. The *satisfaction relation* $\models \subseteq run_{\mathcal{A}} \times BFL$ is the least relation such that:

- $\rho \models T$ for all $\rho \in run_{\mathcal{A}}$,
- $\rho \models \phi \wedge \psi$ iff $\rho \models \phi$ and $\rho \models \psi$,
- $\rho \models \neg \phi$ iff not $\rho \models \phi$,
- $\rho \models \langle k \rangle \phi$ iff for some run ρ': $\rho \overset{k}{\Rightarrow} \rho'$ and $\rho' \models \phi$,
- $\rho \models \langle \leftarrow k \rangle \phi$ iff for some run ρ': $\rho' \overset{k}{\Rightarrow} \rho$ and $\rho' \models \phi$.

The *satisfaction relation* $\models \subseteq S \times BFL$ is defined by: $s \models \phi$ iff $(s,\lambda) \models \phi$.

4.4. THEOREM. *Let $\mathcal{A}=(S,A,\longrightarrow)$ be a LTS with an image finite double arrow relation (i.e. for all $r \in S$ and $k \in A_\epsilon$ the set $\{s \mid r \overset{k}{\Rightarrow} s\}$ is finite). Then for all $r,s \in S$:*

$$r \leftrightarroweq_b s \quad \Leftrightarrow \quad r \sim_{BFL} s.$$

PROOF. Let $r,s \in S$ with

$$\mathcal{A}: r \leftrightarroweq_b s. \tag{1}$$

By Theorem 2.3.3, this is equivalent to:

$$\mathcal{A}: r \leftrightarroweq_{\tau bf} s. \tag{2}$$

Consider the LTS $bf(\mathcal{A})$, which is obtained by replacing the single step transitions between states in S with the corresponding many step forward and backward arrows between paths in $\mathcal{A}$. More precisely, we define:

$$bf(\mathcal{A}) = (run_{\mathcal{A}},\ A_{bf},\ \longrightarrow_{bf}),$$

where:

$$A_{bf} = A_\epsilon \cup \{-k \mid k \in A_\epsilon\}$$

and for $\rho, \rho' \in run_{\mathcal{A}}$ and $k \in A_\epsilon$:

$$\rho \overset{k}{\longrightarrow}_{bf} \rho' \quad \Leftrightarrow \quad \rho \overset{k}{\Rightarrow} \rho', \text{ and}$$

$$\rho \overset{-k}{\longrightarrow}_{bf} \rho' \quad \Leftrightarrow \quad \rho' \overset{k}{\Rightarrow} \rho.$$

Observe that any weak back and forth bisimulation on $\mathcal{A}$ is a strong bisimulation on $bf(\mathcal{A})$ and vice versa. Thus, (2) is equivalent to:

$$bf(\mathcal{A}): (r,\lambda) \leftrightarroweq (s,\lambda). \tag{3}$$

Because the double arrow relation in $\mathcal{A}$ is image finite and because we only consider finite runs, it easily follows that $bf(\mathcal{A})$ is image finite. This means that we can use Theorem 4.2 to obtain that (3) is equivalent to:

$$bf(\mathcal{A}): (r,\lambda) \sim_{HML(A_{bf})} (s,\lambda). \tag{4}$$

Let f be the bijective mapping that associates to each Hennessy-Milner formula in $HML(A_{bf})$ a back and forth formula in $BFL(A)$ by replacing each $"-"$ by a $"\leftarrow"$. Let $\rho \in run_{\mathcal{Q}}$ and $\phi \in HML(A_{bf})$. A simple inductive argument gives that $\rho \models \phi$ relative to $bf(\mathcal{Q})$ iff $\rho \models f(\phi)$ relative to $\mathcal{Q}$. Therefore we may conclude that (4) is equivalent to:

$$\mathcal{Q}: r \sim_{BFL(A)} s. \tag{5}$$

The theorem now follows from the equivalence of (1)-(5). □

4.5. EXAMPLE. Let p,q be as in Figure 2, and r,s as in Figure 3. Let $[k] = \neg\langle k\rangle\neg$ and $[\leftarrow k] = \neg\langle\leftarrow k\rangle\neg$.
If $\phi = \langle a\rangle[\leftarrow a]\langle b\rangle T$ then $q \models \phi$ while $p \not\models \phi$.
If $\phi' = [a][b]\langle\leftarrow b\rangle\langle c\rangle T$ then $r \models \phi'$ while $s \not\models \phi'$.

REFERENCES

[1] P. ACZEL (1988): *Non-well-founded sets*, CSLI Lecture Notes No.14, Stanford University.

[2] A. ARNOLD & A. DICKY (1989): *An algebraic characterization of transition system equivalences*. Information and Computation 82, pp. 198-229.

[3] J.C.M. BAETEN & R.J. VAN GLABBEEK (1987): *Another look at abstraction in process algebra*. In: Proceedings ICALP 87, Karlsruhe (Th. Ottman, ed.), LNCS 267, Springer-Verlag, pp. 84-94.

[4] M.C. BROWNE, E.M. CLARKE & O. GRUMBERG (1988): *Characterizing finite Kripke structures in propositional temporal logic*. Theoretical Computer Science 59(1,2), pp. 115-131.

[5] I. CASTELLANI (1987): *Bisimulations and abstraction homomorphisms*. Journal of Computer and System Sciences 34, pp. 210-235.

[6] I. CASTELLANI, P. FRANCESCHI & U. MONTANARI (1983): *Labeled event structures: a model for observable concurrency*. In: Proceedings IFIP TC2 Working Conference on Formal Description of Programming Concepts - II, Garmisch (D. Bjørner, ed.), North-Holland, pp. 383-400.

[7] R. DE NICOLA & F.W. VAANDRAGER (1990): *Three logics for branching bisimulation (extended abstract)*. In: Proceedings 5th Annual Symposium on Logic in Computer Science (LICS 90), Philadelphia, USA, IEEE Computer Society Press, Los Alamitos, CA, pp. 118-129, full version to appear as CWI Report CS-R9012.

[8] E.A. EMERSON & J.Y. HALPERN (1986): *'Sometimes' and 'Not Never' revisited: on branching time versus linear time temporal logic*. JACM 33(1), pp. 151-178.

[9] R.J. VAN GLABBEEK & W.P. WEIJLAND (1989): *Branching time and abstraction in bisimulation semantics (extended abstract)*. In: Information Processing 89 (G.X. Ritter, ed.), Elsevier Science Publishers B.V. (North Holland), pp. 613-618.

[10] M. HENNESSY & R. MILNER (1985): *Algebraic laws for nondeterminism and concurrency*. JACM 32(1), pp. 137-161.

[11] M. HENNESSY & C. STIRLING (1985): *The power of the future perfect in program logics*. Information and Control 67, pp. 23-52.

[12] R. MILNER (1980): *A Calculus of Communicating Systems*, LNCS 92, Springer-Verlag.

[13] R. MILNER (1981): *Modal characterisation of observable machine behaviour*. In: Proceedings CAAP 81 (G. Astesiano & C. Bohm, eds.), LNCS 112, Springer-Verlag, pp. 25-34.

[14] R. MILNER (1983): *Calculi for synchrony and asynchrony*. Theoretical Computer Science 25, pp. 267-310.

[15] R. MILNER (1989): *Communication and concurrency*, Prentice-Hall International.

[16] U. MONTANARI & M. SGAMMA (1989): *Canonical representatives for observational equivalence classes*. In: Resolution Of Equations In Algebraic Structures, Vol. I, Algebraic Techniques (H. Aït-Kaci & M. Nivat, eds.), Academic Press, pp. 293-319.

[17] R. PAIGE & R. TARJAN (1987): *Three partition refinement algorithms*. SIAM Journal on Computing 16(6), pp. 973-989.

[18] D.M.R. PARK (1981): *Concurrency and automata on infinite sequences*. In: Proceedings 5th

GI Conference (P. Deussen, ed.), LNCS 104, Springer-Verlag, pp. 167-183.

[19] J. SIFAKIS (1984): *Property-preserving homomorphisms of transition systems*. In: Proceedings Logics of Programs, 1983 (E. Clarke & D. Kozen, eds.), LNCS 164, Springer-Verlag, pp. 458-473.

[20] C. STIRLING (1990): *Modal and temporal logics*. In: Handbook of Logic in Computer Science, Vol I (S. Abramsky, ed.), to appear.

[21] W.P. WEIJLAND (1989): *Synchrony and asynchrony in process algebra*. Ph.D. Thesis, University of Amsterdam.

Reduction and Design of Well-behaved Concurrent Systems[1]

Jörg Desel
Institut für Informatik der TU München
Arcisstr. 21, D-8000 München 2

Abstract

It is shown that each live and safe free-choice system without frozen token can be reduced either to a live and safe marked T-graph (marked graph) or to a live and safe marked P-graph (state machine). The four proposed reduction rules are purely local and preserve the behavioural properties in both directions. Hence the method can be used for both, effective analysis and correct design.

The class of systems which can be reduced to marked P-graphs (T-graphs, respectively) can be characterized without using the reduction rules by their P- and T-components. The two classes are not disjoint; systems in the intersection of the classes can be reduced to a unique systems with only two elements.

1 Introduction

Studying transformations and reductions of models for concurrent systems we were led to distinguish *distributed systems* and *parallel devices*.

Often a distributed system is a collection of subsystems which are connected by means of communication links. Concurrency is an inherent phenomenon of distributed systems since the subsystems work mutually independent as long as they do not wait for a communication. The design of well-behaved distributed systems is a severe problem. The analysis with respect to liveness, fairness or safety properties is known to be a highly complex task.

Concurrency plays a completely different role in parallel devices. The core idea of parallel devices is the parallelization of independent actions to gain a speed-up of computations. In contrast to distributed systems the presence of concurrency in parallel devices does not cause synchronization problems such as danger of deadlocks.

Both aspects of concurrency can appear together when parts of distributed systems behave like parallel devices.

Distributed systems should have the capability to react differently on different inputs. Hence usually we find the notion of alternatives or choice. The selection of one alternative (i.e., a concrete input) may have global consequences. As with concurrency, a specific kind of alternatives can be distinguished where each choice has only local effects.

We shall concentrate on concurrent systems which are well-behaved in a sense to be defined. The behaviour restriction includes aspects of liveness, safety and fairness. It turns out that local aspects of concurrency and alternatives of well-behaved systems can be identified locally.

[1]Work supported partly by the Esprit Basic Research Action No. 3148: DEMON.

We propose reduction rules which can be seen as a grammar for well-behaving local parts of distributed systems. These rules include concepts such as:

ction → action; action

ction → PARBEGIN action PAREND

ction → CHOICEBEGIN action CHOICEEND

The exhaustive application of the rules leads to the abstraction of the internal behaviour of ocal devices. Moreover, subparts of the systems which might be distributed but behave like ocal devices can be reduced without effecting the well-behavedness.

The (four) rules proposed in this paper are purely local: Substructures which allow the application of a rule are limited in size and can easily be identified. The respective reductions are ery simple. The main result is that maximally reduced well-behaved systems either exhibit o concurrency or have no alternatives. In other words:

n well-behaved systems either concurrency or choice is purely local.

The result has two consequences: the analysis of concurrent systems (w.r.t. well-behavedness) an be reduced to the analysis of systems without concurrency or without choice which is much impler as the general case. The design of well-behaved concurrent systems can always be done n two steps: First the global topology of concurrency or of choice has to be designed. Then he grammar rules can be applied to get an arbitrary well-behaved system.

ormally, we use Petri-Nets for modelling concurrent systems and we consider the class of *live nd safe free choice systems without frozen token* to be the class of *well-behaved systems.*

With these notions the results read as follows:

ach well behaved system can be either reduced to a well-behaved S-graph (which exhibits hoice but no concurrency) or to a well-behaved T-graph (which exhibits concurrency but no hoice). We shall give a characterization of well-behaved systems which can be reduced to S-raphs (T-graphs, respectively) without using reductions. These characterization results give ew insight to the general structure of well-behaved systems. The intersection of these two lasses is not empty; well-formed nets without global aspects of concurrency or choice can be educed to a net with only two elements.

he paper is organized as follows: In section 2 we introduce the used model for well-behaved ystems and the reduction rules. We define live and safe free choice systems without frozen okens and well-formed nets as their structural counterparts. The reduction rules and some of heir properties are presented as well. For proving the main results in section 4 we need some eneral properties of the considered class of systems which are given in section 3. Section 5 oncludes the paper with consequences and related work.

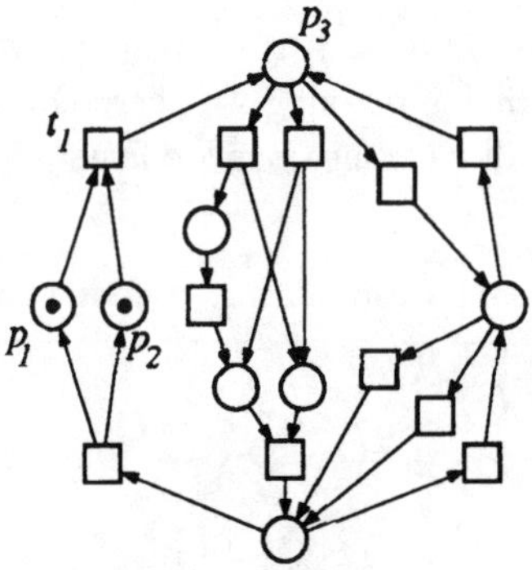

igure 1

2 The Model and the Reduction Rules

We shall use *Petri-Nets* as a formal model of well-behaved systems (for formal definitions of the used concepts see also [BF 87]). Consider the example of figure 1.

The *places* of the net (denoted by circles) denote local conditions. The *transitions* (squares) represent actions. A state of the system is given by the set of those conditions which hold at the state (denoted by tokens on the respective places). States are also called *markings.* A transition can *occur* when all its preconditions hold. Hence in the example the transition t_1 can occur. A transition consumes all tokens of its preconditions and produces tokens on all postconditions when it occurs. So, by the occurrence of t_1, p_1 and p_2 loose their tokens while p_3 gets a token.

Formally, a *net* is a triple $N = (P_N, T_N, F_N)$ where P_N and T_N are disjoint sets of places and transitions and $F_N \subseteq ((P_N \times T_N) \cup (T_N \times P_N))$ is the *flow relation* (denoted by directed arcs).

A marking M of N is a subset of P_N.

A net N together with a marking M is called *system.*

The (sequential) behaviour of a system is given by its set of finite or infinite *runs* $M_0t_1M_1t_2M_2\ldots$ where $M_0 = M$ and for each $i > 0$, t_i transforms the marking M_{i-1} to the marking M_i.

Each marking M_i appearing in a run is called *reachable marking.*

Abusing language, we also call a run of (N, M') a run of (N, M) if M' is a reachable marking of (N, M) (i.e., suffixes of runs are runs).

We call a system (N, M) (modelled by a net N together with a marking M of N) *well-behaved free-choice system* (or just *well-behaved system*) if it has the following properties:

Connectedness and Finiteness

The net N – seen as a graph – is connected and has only finitely many elements.

Independence of Alternatives

Alternatives are modelled by forwardly branched places. Independence of alternatives means that whenever two distinct transitions have a common input place, then both have only this input place. This property is called *free choice property.* Corresponding *free choice nets* do not include the following substructure:

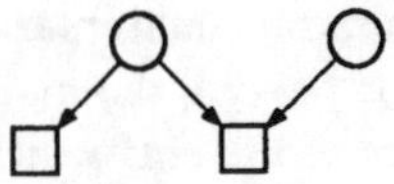

Liveness

Liveness means lack of partial deadlocks. A transition t is live if it can always get enabled again. Formally, for each reachable marking M' there is a run starting with M' which includes t. A system is live if all its transitions are live.

Safety

Safety means impossibility of contact situations. A contact situation (w.r.t. transition t) looks as follows:

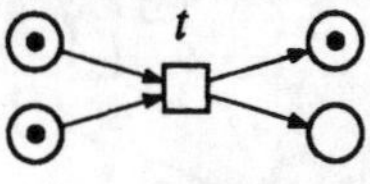

t is enabled but some of its output places which are not input places as well are already marked. In safe systems for each reachable marking there are no contact situations.[2]

Token Fairness

The fairness notion we use here excludes the following situation:

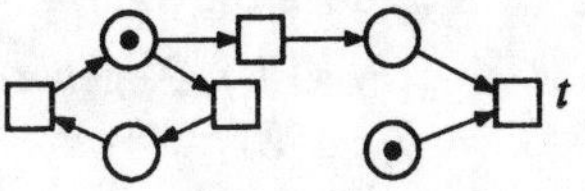

Some of the preconditions of t hold but t might never get enabled.
The usual name of this fairness property is *lack of frozen token*. It requires that each infinite run starting with an arbitrary reachable marking M' consumes all tokens of M'. More precisely, for each proper subset M'' of M' we require that there is no infinite run which starts with M''.

All together we consider the class of *live and safe free choice systems without frozen token* to be the class of *well-behaved systems.* This class has been introduced in [GT 84] and it is closey related to well-behaved bipolar synchronization schemes (also introduced in [GT 84]). In particular, each well-behaved bipolar synchronization scheme can be transformed to a well-behaved system while the converse direction is still open (see section 5). The system depicted in figure 1 is a well-behaved system.

A core issue of net theory is the deduction of behavioural properties of systems from structural properties of the underlying nets and vice versa. Hence we are also interested in the class of nets (without markings) which can be marked in such a way that the marked net is a well-behaved system. We shall call this class *well-formed nets.* One of the advantages of well-formed nets is that they can be characterized without any marking considerations [BD 89]. Obviously all well-formed nets are connected finite free choice nets since these properties of well-behaved systems are purely structural. So we concentrate on connected finite free choice nets in the sequel.

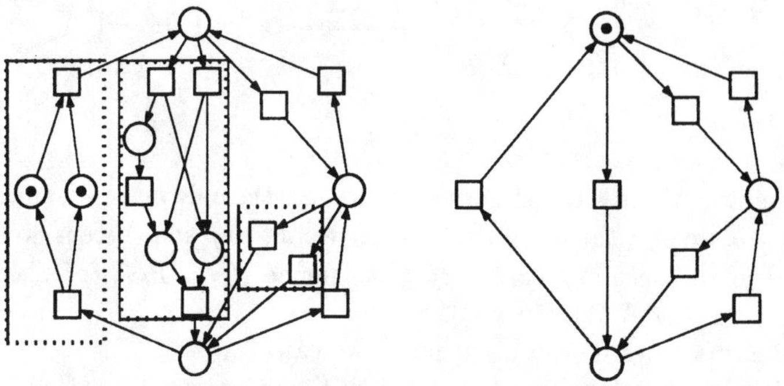

Figure 2

In figure 2 we show on the left side the system already known from figure 1. The areas depicted by means of dotted lines are candidates for abstractions since the internal concurrency and

[2]In contrast to ***Elementary Net Systems*** [Th 86] we do not consider marked loops as contact situations

choice of these parts do not effect the respective environments. In fact, the reduction mechanism we propose abstracts from the internal behaviour of these parts and leads to the system shown on the right side of figure 2.

Before defining the reduction rules we introduce the following convenient notations:
Let $N = (P_N, T_N, F_N)$ be a net. For $X \subseteq P_N \cup T_N$, X generates a *subnet* N' of N as follows: $P_{N'} = P_N \cap X$, $T_{N'} = T_N \cap X$ and $F_{N'} = F_N \cap (X \times X)$. We do not distinguish the set X and the subnet generated by X. In particular, the *set* $N = P_N \cup T_N$ generates the *net* N.
For $x \in N$, the *preset* ${}^\bullet x$ is defined as ${}^\bullet x = \{y \in N \mid (y, x) \in F_N\}$ and the *postset* $x^\bullet$ is defined as $x^\bullet = \{y \in N \mid (x, y) \in F_N\}$. For $X \subseteq N$, ${}^\bullet X = \bigcup_{x \in X} {}^\bullet x$ and $X^\bullet = \bigcup_{x \in X} x^\bullet$.
Dealing with nets and subnets the notion of pre- and postsets might be ambiguous. We shall always use it with respect to the highest net level (which is called N throughout this paper).

Let (N, M) be a free-choice system.

P-Reduction

If N has two distinct places p and p' such that ${}^\bullet p = {}^\bullet p'$ (they have the same input transitions), $p^\bullet = p'^\bullet$ (they have the same output transitions) and $p \in M$ iff $p' \in M$ then they are redundant in a strong sense. A P-reduction removes one of these places and yields the resulting system $(N \setminus \{p'\}, M \setminus \{p'\})$.

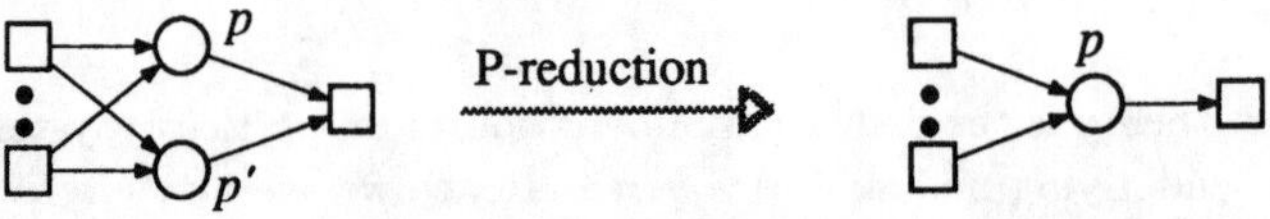

T-Reduction

Similarly, if N has two distinct transitions t and t' such that ${}^\bullet t = {}^\bullet t'$ and $t^\bullet = t'^\bullet$ then both transitions perform the same marking transformation and we can apply a *T-reduction* . The resulting system is $(N \setminus \{t'\}, M)$.

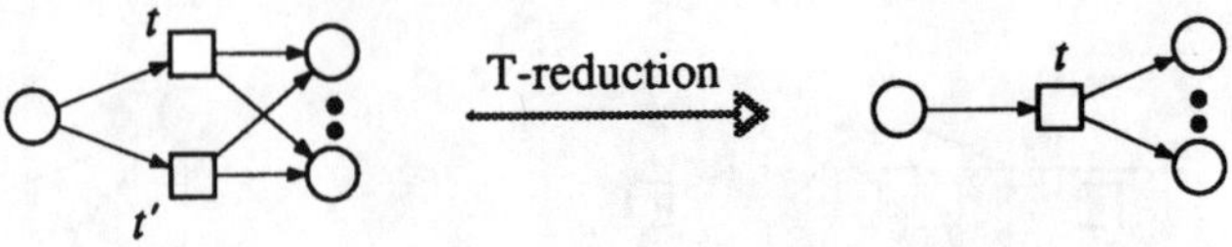

F-Reduction

If N has a place p and a transition t such that t is the only output transition of p, p is the only input place of t, there is no arc leading from t to p and there is no contact situation w.r.t. t then we can apply an *F-reduction.* Define $P_{N'} = P_N \setminus \{p\}$, $T_{N'} = T_N \setminus \{t\}$ and $F_{N'} = (F_N \cap (N' \times N')) \cup ({}^\bullet p \times t^\bullet)$.
Concerning markings we can distinguish two cases:
$p \notin M$: then the marking is not changed by the reduction (i.e., $M' = M$).
$p \in M$: t is enabled by M since p is the only input place of t. Before applying the reduction, we transform the marking by the occurrence of t (i.e., $M' = (M \setminus \{p\}) \cup t^\bullet$).

The F-reduction yields the system (N', M') (where $N' = (P_{N'}, T_{N'}, F_{N'})$).

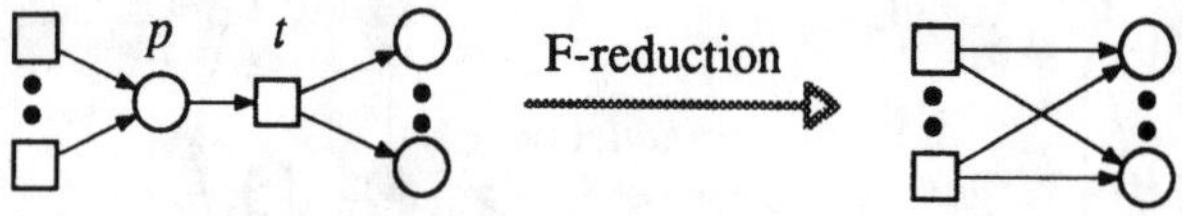

A-Reduction

If N has transitions t, t' such that ${}^\bullet t \subseteq t'^\bullet$, $|{}^\bullet t| > 1$ and for no $p \in {}^\bullet t$ holds ${}^\bullet p = \{t'\}$ then we can apply an *A-reduction*. The resulting net is $(P_N, T_N, (F_N \setminus (\{t'\} \times {}^\bullet t)) \cup (\{t'\} \times t^\bullet))$, i.e. we remove the arcs from t' to ${}^\bullet t$ and add arcs from t' to $t^\bullet$.

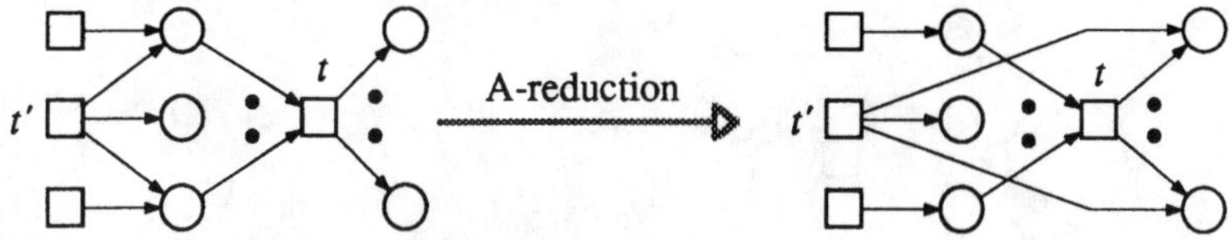

Markings are not changed by A-reductions.
In contrast to the other rules a net is not transformed to a smaller net by A-reductions. Hence the name 'reduction' needs a justification which we shall give later.

We state two of the most important properties of reductions already in this section but delay the proofs to section 3 since we have to introduce some necessary tools before.

In case of well-behaved systems the conditions for applying a reduction rule only depend on structural properties. We say *a net N is reducible* w.r.t. P-, T-, F- or A-reduction if (N, M) is reducible for any marking M. *N is reduced to N'* if (N, M) is reduced to (N', M') for any M and M'.

Lemma 2.1

A well-behaved system can be reduced by means of a P-, T-, F- or A-reduction if and only if the corresponding well-formed net can be reduced.

Lemma 2.1 is an important prerequisite for our further proceeding. We only have to consider reductions of well-formed nets and know that there exist unique corresponding reductions of well-behaved systems. However, the converse does not hold: if N is reduced to the net N' and M' is a marking of N' then there is not necessarily a *unique* marking M such that (N, M) is reduced to (N', M'). Ambiguities only occur in case of F-reductions. If N is reduced to N' by means of an F-reduction w.r.t. a place p and a transition t and $t^\bullet \subseteq M'$ then we can choose either the marking $M = M'$ or the marking $M = (M' \setminus t^\bullet) \cup \{p\}$.

Lemma 2.2

Let (N, M) be a free choice system which can be reduced by a P-, T-, F- or A-reduction and let (N', M') be the resulting (free choice) system.
(N, M) is a well-behaved system iff (N', M') is well-behaved.
N is well-formed if and only if N' is well-formed.

Now let us return to our leading example. Figure 3 shows a maximal reduction sequence which, in fact, ends with the system already known from figure 2. This systems exhibits no concurrency.

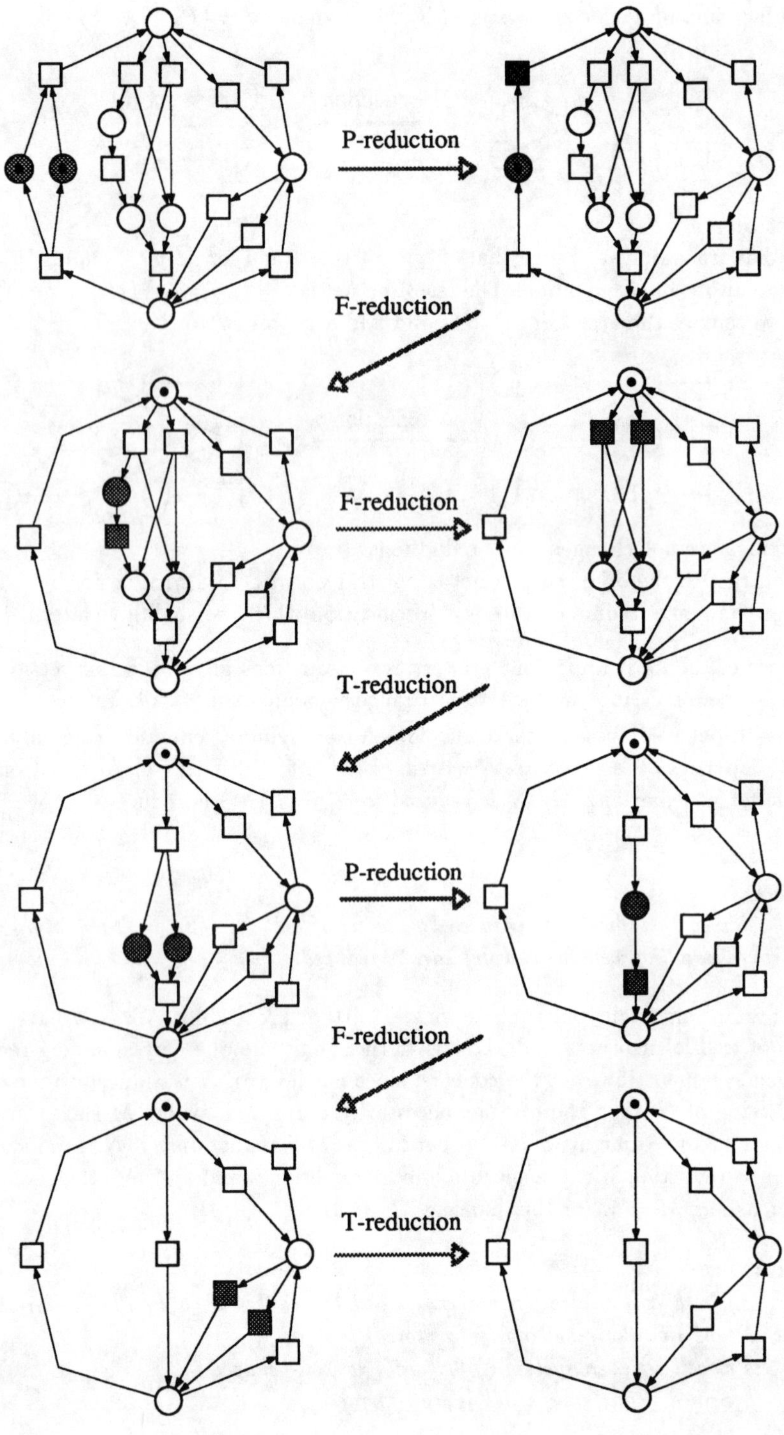
P-reduction
F-reduction
F-reduction
T-reduction
P-reduction
F-reduction
T-reduction

Figure 3

3 Properties of Well-formed Nets

The first property of well-formed nets we shall need is *strong connectedness.* In contrast to (weak) connectedness, being strongly connected requires that for each pair (x, y) of elements we find a (directed) path which starts with x and ends with y. This property holds for all connected live and safe systems [BD 89]. It is used to prove lemma 2.1:

Proof of lemma 2.1:

Let (N, M) be well-behaved. We have to show: If N can be reduced then the conditions concerning the marking M holds. Such conditions only appear with P- and F-reductions. Assume that we can apply a P-reduction w.r.t. places p and p'. Since N is strongly connected there is a transition $t \in p^{\bullet}(= p'^{\bullet})$. By the liveness property, t can get enabled and hence we can reach a marking $\widehat{M}$ with $\{p, p'\} \subseteq \widehat{M}$. Since by each transition occurrence either the marking of p as well as the marking of p' is changed or both remain unchanged we get that $p \in M$ if and only if $p' \in M$.
For F-reductions we required that there is no contact-situation w.r.t. the transition t. This property holds for (N, M) by definition since (N, M) is safe. ■

After reducing a net once we can reduce it again if possible, and so on. With each P-, T- or F-reduction step the size of the net, i.e., the number of its elements, decreases. This does not hold for A-reductions. Nevertheless, for strongly connected nets we can apply A-reductions only finitely often.

Lemma 3.1
Each strongly connected net can only finitely often get reduced.

Proof: We only have to consider A-reductions. Clearly, A-reductions preserve strong connectedness. Each strongly connected net is covered by simple cycles. The total number of cycles which cover a net is preserved but some cycles get smaller while no cycle gets larger by an A-reduction. ■

Hence, for each strongly connecetd net N we find a net N' which is obtained by maximally applying P-, T-, F- and A-reductions. We call a net *reduced* if we can not apply a reduction. We say a net is P- (T-, F-, A-)-reduced if no P- (T-, F-, A-)-reduction can be applied.
Simple examples of well-formed nets are strongly connected *P-graphs* and *T-graphs.*
In P-graphs each transition has exactly one input place and exactly one output place. For strongly connected P-graphs, each marking with exactly one element obviously defines a well-behaved system. Thus strongly connected P-graphs are well-formed nets.
T-graphs are nets dual to P-graphs in a sense to be defined later. Places of T-graphs have exactly one input- and exactly one output transition. Each strongly connected T-graph can be marked lively and safely [CHEP 71]. It can be seen easily that live and safe marked T-graphs are well-behaved systems. Hence strongly connected T-graphs are well-formed nets too.
With these notions and lemma 2.1 the result to be proven reads as follows:
Each reduced well-formed net is either a P-graph or to a T-graph.

Each strongly connected net is covered by (simple) cycles. Now consider a strongly connected P-graph and one of its cycles. The subnet generated by the elements of this cycle is a strongly connected T-graph with the additional property that for each transition its pre- and postset

belongs to the T-graph as well. We call such a substructure *T-component.* Hence each strongly connected P-graph is covered by T-components and, similarly, each strongly connected T-graph is covered by *P-components* (which are defined like T-components but with places and transitions exchanged). Moreover, strongly connecetd P-graphs are covered by P-components (namely by itself) and the same holds for T-graphs and T-components. The generalization of these properties to arbitrary well-formed nets is due to Hack [Ha 72]: *each well-formed net is covered by P-components and by T-components.*
Examples of P- and T-components of our example net are given in figure 4.

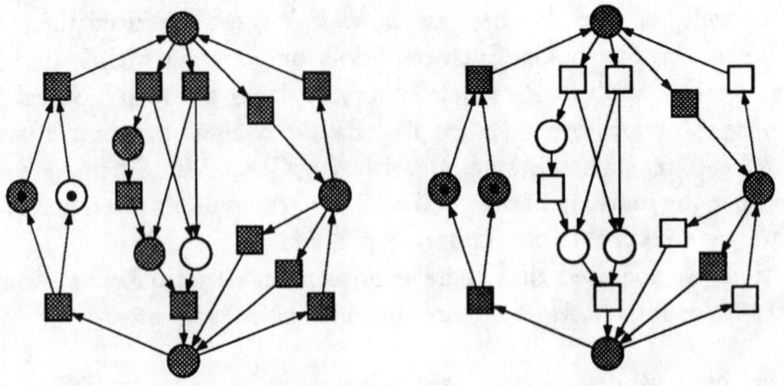

A P-components is depicted A T-components is depicted

Figure 4
In the following lemma systems instead of nets are considered. Its proof is also in [Ha 72].

Lemma 3.2
Let (N, M) be a live free choice system. It is safe iff there is for each $x \in N$ a P-component X with $x \in X$ and $|X \cap M| = 1$ (and hence $|X \cap M'| = 1$ for each reachable marking M')

Another structural property, which is related to the fairness property is proven in [BD 89]:

Lemma 3.3
A live and safe free choice system (N, M) is fair (in our sense) iff each P-component of N has a nonempty intersection with each T-component of N.

Before proving lemma 2.2 we clarify how P- and T-components are transformed by A-reductions

Lemma 3.4
Let N be a free choice net and let an A-reduction transform N to N' (with ${}^\bullet t \subseteq t'^\bullet$).
$X \subseteq N$ is a P-component of N iff X is a P-component of N'.
For $X \subseteq N, t' \notin X$, X is a T-component of N iff X is a T-component of N'.
For $X \subseteq N, t' \in X$, $X \cup \{t\} \cup {}^\bullet t$ is a T-component of N iff $X \setminus (\{t\} \cup {}^\bullet t)$ is a T-component of N'.

Proof: Straightforward verification.

Proof of lemma 2.2:
The propositions obviously hold for P-, T- and F-reductions.

So it remains to consider A-reductions (w.r.t. ${}^{\bullet}t \subseteq t'^{\bullet}$).
Assume (N, M) is well-behaved. By lemma 3.2 N is covered by P-components which carry one token each and and by lemma 3.3 each P-component intersects each T-component. Since each intersection of a P- and a T-component includes a cycle we get the very same results for (N', M) with lemma 3.4 . So, again with lemma 3.2 and lemma 3.3, it remains to show that (N', M) is live.
We get for each run of (N, M) a run of (N', M) by removing the respective first occurrences of t (and its subsequent markings) after occurrences of t'. So we only have to show that the liveness of t is preserved (i.e. that not all occurrences of t are removed): By the definition of A-reductions each place $p \in {}^{\bullet}t$ has more than one input transition in N. Hence each such p has at least one input transition in N'. These transitions are live. Since t is the only transition in N' which removes tokens of ${}^{\bullet}t$ by the free choice property, t is live in N' as well.
Assume now that (N', M') is well-behaved. With similar reasoning as above we only have to prove the liveness of (N, M). We can substitute for each run of (N', M') each occurrence of t' by subsequent occurrences of t' and t (and the intermediate marking). Hence the liveness of (N, M) follows directly from the liveness of (N', M').

As a consequence we get that N is well-formed if and only if N' is well-formed. ■

.s interchanging places and transitions transforms T-graphs to P-graphs and vice versa we find n according duality concept for well-formed nets. However, at least the free choice property violated if we do not convert the arc directions at the same time. Hence the net *dual* to a et $N = (P_N, T_N, F_N)$ is defined as $N^{-d} = (T_N, P_N, F_N^{-1})$.

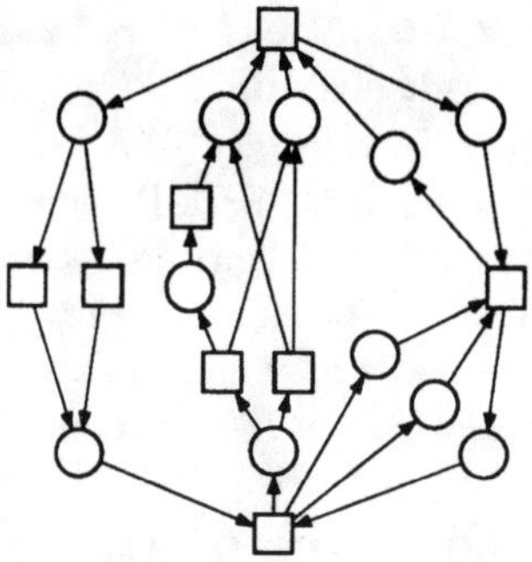

The dual net of the net of figure 1.
igure 5

ack proved [Ha 72] for live and safe free choice systems:

emma 3.5
If (N, M) is a live and safe free choice system then we find a marking M^{-d} of N^{-d} such that (N^{-d}, M^{-d}) is a live and safe free choice system as well.

Corollary 3.6
A net N is well-formed iff N^{-d} is well-formed.

Proof: It remains to show that N^{-d} behaves fair for a live and safe marking. Since each T-component of N^{-d} is a P-component of N and each P-component of N^{-d} is a T-component of N this follows with lemma 3.3. ▸

We use the concept of P- and T-components to show in the following lemma that for well-formed nets N the third premiss for the application of an F-reduction is only necessary if $|N| = 2$. Hence, if we can assume that $|N| > 2$, we can apply an F-reduction whenever we find a place p and a transition t such that $p^\bullet = \{t\}$, ${}^\bullet t = \{p\}$ and there is no contact w.r.t. t.

Lemma 3.7
Let N be a well-formed net. If $p \in P_N$ and $t \in T_N$ such that $p^\bullet = \{t\}$, ${}^\bullet t = \{p\}$ and $(t,p) \in F_N$ then $N = (\{p\}, \{t\}, \{(p,t),(t,p)\})$.

Proof: Consider a P-component X of N with $t \in X$. Since p is the only input place of t we get $p \in X$. Since p is an output place of t, X includes no other output places of t and hence $t^\bullet = \{p\}$. Dually, ${}^\bullet p = \{t\}$. The result follows since N is strongly connected. ▸

Similarly it can be shown that the third premiss for A-reduction is not necessary whenever the considered net is P-reduced and well-formed:

Lemma 3.8
Let N be a P-reduced well-formed net. If t and t' are transitions of N with ${}^\bullet t \subseteq t'^\bullet$ and $|{}^\bullet t| > 1$ then $t \neq t'$ and for each $p \in {}^\bullet t$ holds $|{}^\bullet p| > 1$.

Proof: Assume $t = t'$. Then ${}^\bullet t \subseteq t^\bullet$. Let $p \in {}^\bullet t$. Each P-component with p contains t as well and hence no other place than p in $t^\bullet$. Thus no P-component contains a place p' in $t^\bullet \setminus {}^\bullet t$. Therefore ${}^\bullet t = t^\bullet$. Since $|{}^\bullet t| > 1$ no place in ${}^\bullet t$ is forward branched. So each T-component including a place $p \in {}^\bullet t$ only includes t and p. In particular for all places $p \in {}^\bullet t$ holds ${}^\bullet p = p^\bullet = \{t\}$. Thus, since N is P-reduced, there is only one such place and $|{}^\bullet t| = 1$ – contradicting the hypothesis.

If for any place p holds ${}^\bullet p = \{t'\}$ and $p^\bullet = \{t\}$ then each T-component which contains t also contains t'. This implies that for all places $p' \in {}^\bullet t$ holds ${}^\bullet p' = \{t'\}$. Hence, since N is P-reduced, there is only one such place – contradicting the hypothesis. ▸

4 Main Results

Theorem 4.1
Each reduced well-formed net N is either a P-graph or a T-graph.

Proof: Define $Q = \{p \in P_N \mid |p^\bullet| > 1\}$. The proof is divided into the following steps:

(a) For each $p \in Q$ we find a $p' \in p^{\bullet\bullet} \cap Q$.

(b) For each $p \in Q$, the marked net $(N, \{p\})$ is well-behaved.

(c) For each $t \in Q^\bullet$ we find a $p' \in t^\bullet \cap Q$.

(d) $Q^{\bullet\bullet} \subseteq Q$.

(e) N is a P-graph or a T-graph.

(a)
Assume the contrary.
Then we find a $p \in Q$ such that all $p' \in p^{\bullet\bullet}$ have only one output transition. Since N is F-reduced, all these output transitions have more than one input place. So the setting looks as follows:

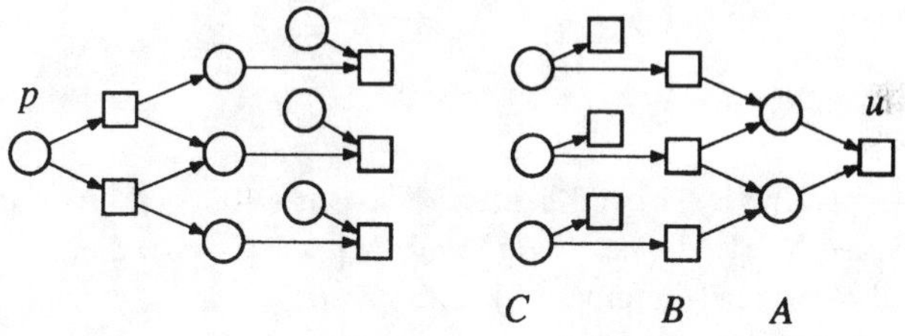

On the right side we depicted the same substructure of the (well-formed) dual net N^{-d}. Set $u = p$ (u is a transition of N^{-d}. We use a new symbol to avoid confusion.) Define $A = {}^\bullet u$, $B = {}^\bullet A$ and $C = {}^\bullet B$ (preset-notation w.r.t. N^{-d}). By the assumption $|A| > 1$ and all places of C have at least two output transitions.
If for all $t \in B$ holds $t^\bullet = A$ we could apply a P-reduction and remove one element of A. This corresponds to a T-reduction of N – contradicting the assumption that N is T-reduced. Hence we find a $t \in B$ which is an input transition of some, but not of all places of A.
Let M be a marking of N^{-d} such that (N^{-d}, M) is well-behaved. By the above and since (N^{-d}, M) is live, we can reach a marking M' which includes some but not all elements of A. Starting with M', we play the token game with the following strategy: For each reached marking we choose the occurrence of an enabled transtion t such that the set $|t^\bullet \cap A|$ is minimal.
By the fairness property we will eventually be forced to enable u since there are already some tokens in A which can only get removed by u. So we eventually reach a marking which includes A. Let $t \in B$ be the transition which occurrence leads to such a marking. By the free choice property and since all places of C are forward branched ${}^\bullet t = \{v\}$ for a place $v \in C$. Let v' be a place in $t^\bullet \cap A$. Since we followed our strategy, all other choices of transitions in $v^\bullet$ also lead to markings which include A. Hence $v^\bullet \subseteq {}^\bullet v'$.
Now consider the net N again. v and v' are transitions of N with ${}^\bullet v \subseteq v'^\bullet$ and $|{}^\bullet v| > 1$ – contradicting the assumption that N is A- and P-reduced (lemma 3.8).

(b)
Let $p \in Q$ and consider a marking M with $p \in M$ such that (N, M) is well-behaved.

Such a marking exists since each transition in $p^\bullet$ can be enabled in well-behaved systems. By the free choice property M enables all transitions of $p^\bullet$. Using (a) we find a $t \in p^\bullet$ and a $p' \in Q \cap t^\bullet$. So the occurrence of t leads to the marking $(M \setminus \{p\}) \cup t^\bullet$ In particular $p' \in M'$. We continue by choosing a $t' \in p'^\bullet$ such that $t'^\bullet \cap Q \neq \emptyset$. Repeating this strategy arbitrarily leads to an infinite system run where all marked places of $M \setminus \{p\}$ carry frozen tokens. Hence, by the fairness assumption, we conclude that $M = \{p\}$.

(c)
Assume the contrary. Then we find a $p \in Q$ and a $t \in p^\bullet$ such that $t^\bullet \cap Q = \emptyset$. Hence we find for each place $p' \in t^\bullet$ a transition t' such that $p'^\bullet = \{t'\}$ and $|^\bullet t'| > 1$ since N is F-reduced.

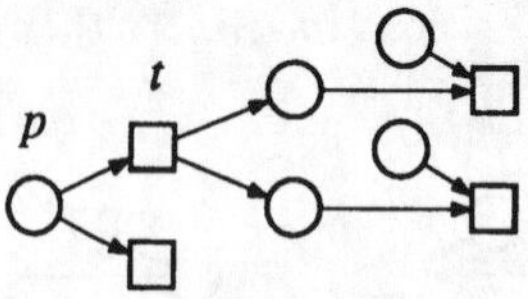

$\{p\}$ is a live marking by (b) and hence $t^\bullet$ is live as well. So there is at least one transition $t' \in t^{\bullet\bullet}$ enabled by $t^\bullet$. Hence $^\bullet t' \subseteq t^\bullet$ and $|^\bullet t'| > 1$ – contradicting the assumption that N is A- and P-reduced (lemma 3.8).

(d)
For each $t \in Q^\bullet$ we find a $p' \in t^\bullet \cap Q$ by (c). Since $(N, \{p'\})$ is well-behaved by (a), there is no reachable marking which properly includes $\{p'\}$ by the fairness property. Hence $t^\bullet = \{p'\}$.

(e)
Assume that N is not a T-graph. Then N has at least two distinct T-components and Q is not empty. By (d) $Q^{\bullet\bullet} \subseteq Q$ and hence, since N is strongly connected, $Q = P_N$. By the free-choice property no transition of N is backward branched. So there is only one P-component of N and N is hence a P-graph. ■

Together with lemma 2.2 we get the following corollary:

Corollary 4.2

Let (N, M) be a strongly connected free choice system and let (N', M') be the system obtained by maximally reducing (N, M).
(N, M) is well-behaved if and only if (N', M') is either a well-behaved marked P-graph or a well-behaved marked T-graph.

Since we reduce each well-formed net to either a P-graph or a T-graph we can distinguish two classes of well-formed nets. These classes turn out to be not disjoint. We can characterize the classes by means of the P- and T-components. Well-formed nets which are reducible to P-graphs have the property that all its P-components have a common element. By duality, for nets of the other class the T-components have a nonempty intersection. These properties are robust w.r.t. reduction:

Lemma 4.3

Let N be a well-formed net which can be reduced to N'. Then the P-components (T-components, respectively) of N have a nonempty intersection iff the same property holds for N'.

Proof: The only non-trivial case is the intersection of all T-components for A-reductions. If, with the names like in lemma 3.4, an $x \in \{t\} \cup {}^\bullet t$ belongs to all T-components of N, then t belongs to all T-components since the places in ${}^\bullet t$ are not forward branched and hence the set $t^\bullet$ is included in all T-components as well. Thus $t^\bullet$ is included in all T-components of N'. ■

If N is a well-formed T-graph then it has only one T-component and in particular its T-components have a nonempty intersection. Hence each well-formed net which can be reduced to a T-graph has this property. Dually, if N can be reduced to a well-formed P-graph then its P-components have a nonempty intersection. So for each well-formed net one of the properties holds. Conversely, whenever the T-components of a reduced P-graph do not have a common intersection then obviously the same holds for all well-formed nets which can be reduced to that P-graph. The same implication holds for reduced T-graphs with an empty intersection of P-components by duality. So we only have to treat the case that the P-components as well as the T-components of a well-formed net intersect. We show that those nets can be reduced to cycles of two elements (a place and a transition) which is a P-graph as well as a T-graph.

Theorem 4.4

Let N be a well-formed net.
N can be reduced to a T-graph (P-graph) iff the T-components (P-components, respectively) of N have a nonempty common intersection.
N can be reduced to a cycle with two elements iff the P-components as well as the T-components of N have a nonempty common intersection.

Proof: With the above considerations it remains to treat the case that the T-components of N have a nonempty intersection and the P-components of N have a nonempty intersection. By theorem 4.1 N can be reduced to a P-graph or to a T-graph. Assume w.l.o.g. that N can be reduced to a P-graph N'. It can be seen easily that the T-components of N' are just simple cycles since neither their places nor their transitions are branched. These cycles have a common intersection, i.e. there is a place p which is included in all cycles of N'. Consider a maximal (w.r.t. the number of elements) cycle of N'. We claim that this cycle (and hence all cycles) has only two elements. Otherwise we find arcs (p', t) and (t, p) of N' where p is included in all cycles and $p' \neq p$. p' is forward branched since N' is an F-reduced P-graph. Let $t' \in p'^\bullet$, $t' \neq t$. Since each cycle through t' includes p and a maximal cycle leads through p', t and p, $t'^\bullet = \{p\}$. Hence N' is not T-reduced w.r.t. t and t' – contradicting the hypothesis.
So N' has only one place p and for all transitions t of N' holds: ${}^\bullet t = t^\bullet = \{p\}$. Since N' is P-reduced there is only one such transition and hence $|N'| = 2$. ■

The two classes can behaviourly be interpreted as follows:

If the T-components of the underlying net of a well-behaved system have a common intersection then each run which reproduces a marking 'passes through' that element (it is shown in [BD 89] that each reproducing run uses all transitions of at least one T-component). Since the number of different markings is finite, the same holds for infinite runs. Hence, if the common element is a transition we know that this transition plays the role of a bottleneck of the system. Without the occurrence of this transition there are only finite runs. Analogously, if the common element is a place then this place has to change its marking during each reproducing run of the system.

If the P-components of a well-formed net N have a common intersection then this intersection clearly includes at least one transition t. The marking $^\bullet t$ is a live (and safe) marking of N (lemma 3.2). If the intersection includes a place p then $\{p\}$ marks N lively.

5 Consequences and Discussion

Corollary 4.2 proves that each well-behaved system (and only well-behaved systems) can be reduced to well-behaved marked P-graphs or well-behaved marked T-graphs. Hence we obtain an analysis tool to decide well-behavedness of systems which only needs criteria for well-behavedness of marked P-graphs and marked T-graphs. These are very simple in case of P-graphs (strong connectedness and exactly one token). In case of T-graphs sufficient conditions for well-behavedness are also much simpler then in case of arbitrary systems [CHEP 71, GT 84]. Using the converse direction our result can be seen as a design tool for the correct construction of arbitrary well-behaved systems.

Other consequences of our result are:

Well-behaved systems and well-formed nets possess a lot of additional properties. Most of them require lengthy proofs. By means of reduction we have the possibility to show the respective properties for P- and T-graphs and generalize them by induction.

For well-behaved bipolar synchronization schemes a set of reduction rules is known [GT 84]. These reductions include global rules with respect to the marking transformation, i.e., global considerations are necessary to obtain a well-behaved marking. They reduce each well-behaved scheme to a unique minimal one which has only one node. For each well-behaved bipolar synchronization scheme there is a corresponding well-behaved free-choice system [GT 84] while the converse is still an open problem. Our results on reductions of well-behaved free choice systems solve this problem. We show in a subsequent paper that for each well-behaved free-choice system we find a translation into bipolar synchronization schemes [De 90]. In particular rules corresponding to our reduction rules work for bipolar synchronization schemes as well. For minimal well-behaved systems (T-graphs or P-graphs) a translation to well-behaved schemes can easily be given. Hence we get the general correspondence by induction on the number of possible reduction steps.

There are several papers dealing with local transformations and reductions of nets (including [Va 79], [AN 83], [SM 83], [Be 87]). A comparison of our work with other approaches yields two main differences:

The class of systems we consider is quite restricted. Usually rules are applied to arbitrary classes of marked nets. Therefore very sophisticated rules have been proposed. The rules we presented here are partly just trivial cases of known reductions. On the other hand these approaches do not lead to completeness results. Interrelations between classes of systems (which are not just defined by means of the reduction rules) such that each system of one class can be reduced to a system of the other (simpler) one have not been emphasized.

Another approach of analyzing systems means of reductions can be found in [ES 90]. In this paper it is shown how live and *bounded* free choice can be reduced. Compared with our pproach, the class of systems under consideration is larger (no fairness required) and there only one minimal net (a loop with two nodes). Safety of systems is not preserved by the eductions and the rules are non-lokal, i.e., the application of a rule can depend on an arbitrary rge part of the system.

or a class of nets (named regular nets) it is shown in [DG 84] that all nets of this class can e reduced to simple loops by non-local reduction rules.

he following aspect of our approach is not satisfactory yet and needs further investigations: t is not known whether A-reductions are necessary. In fact, we did not find any example of well-behaved system which could not be reduced by P-, T- and F-reductions. One starting oint in proving that A-reductions are unnecessary could be the observation that, instead of -reductions, the dual of A-reductions could be used as well (in contrast to the other reduction ules, we did not use a dual counterpart of A-reductions).

References

An 83] Andre, C.: Structural Transformations Giving B-equivalent PT-Nets. In: Pagnoni, A, ozenberg, G. (ed.): Informatik-Fachberichte No. 66: Application and Theory of Petri Nets. p14-28, Springer-Verlag (1983).

BF 87] Best, E., Fernandez, C.: Notations and Terminology on Petri Net Theory. Second lition. Arbeitspapiere der GMD No. 195 (1987).

BD 89] Best, E., Desel, J.: Partial Order Behaviour and Structure of Petri Nets. Arbeitspaiere der GMD No. 373 (1989). Accepted for publication in Formal Aspects of Computing.

Be 87] Berthelot, G.: Transformations and Decompositions of Nets. LNCS 254: Petri Nets, entral Models and Their Properties, pp. 359-376 (1987).

DG 84] Datta, A., Ghosh, S.: Synthesis of a Class of Deadlock-Free Petri Nets. Journal of e ACM, Vol. 31, No. 3, pp.486-506 (1984).

De 90] Desel, J.: Live and Safe Free-choice Systems without Frozen Token are Well-behaved ipolar Synchronization Schemes. Forthcoming paper (1990).

CHEP 71] Commoner, F., Holt, A.W., Even, S., Pnueli, A.: Marked Directed Graphs. urnal of Computer and System Science 5, pp. 511-523 (1971).

ES 90] Esparza, J; Silva, M.: Top-Down Synthesis of Live and Bounded Free Choice Nets. 1th International Conference on Application and Theory of Petri Nets, Paris (1990).

GT 84] Genrich, H.J., Thiagarajan, P.S.: A Theory of Bipolar Synchronization Schemes. CS Vol. 30, pp.241-318 (1984).

Ha 72] Hack, M.: Analysis of Production Schemata by Petri Nets. TR-94, MIT-MAC (1972).

SM 83] Suzuki, I.; Murata, T.: A Method for Stepwise Refinement and Abstraction of Petri ets. Journal of Computer and System Sciences, Vol. 27, pp. 51-76 (1983).

Th 87] Thiagarajan, P.S.: Elementary Net Systems. In: Brauer, W., Reisig, W., Rozenberg, . (ed): LNCS 254: Petri Nets, Central Models and Their Properties, pp. 26-59 (1987).

Va 79] Valette, R.: Analysis of Petri Nets by Stepwise Refinements. Journal of Computer d System Sciences, Vol. 18, pp. 35-46 (1979).

Synthesis Rules for Petri Nets, and How they Lead to New Results *

Javier Esparza
Institut für Informatik
Universität Hildesheim
Samelsonplatz 1 D-3200-Hildesheim

Abstract

Three kits of rules for top-down synthesis of Petri nets are introduced. The properties and expressive power of the kits are compared. They are then used to characterise the class of structurally live Free Choice nets by means of the rank of the incidence matrix.

Introduction

It is a well learnt lesson that systems, even not very large ones, can be successfully built only by means of disciplined design. A particular implementation of this discipline is the top-down paradigm: systems are designed through stepwise refinement of a simple initial system. Correctness is proved using induction: the initial system is shown to meet the specification, and the refinements are shown to preserve it (i.e if a system in the refinement sequence meets the specification, so does its succesor). This process is well understood for sequential systems, but not so much for concurrent ones. Petri nets provide a formal framework where this problem can be addressed.

In a previous paper [3], some research along this line was carried out. Two requirements which are part of the specification of many systems were selected: absence of global or partial deadlocks and absence of overflows in finite stores. In the Petri net formalism they correspond to the properties of liveness and boundedness. On the other hand, also the class of systems was restricted to the ones modelled by means of Free Choice nets, a class of nets which permits to represent both concurrency and nondeterminism, but constraints the interplay between both. The goal was to provide a sound and complete kit of refinements rules, i.e a kit of rules which preserves liveness and boundedness, and allows one to synthetise every live and bounded Free Choice net. It turned out that two refinement rules sufficed.

The technical details in [3] possibly hid the simplicity of the final result. The first goal of this paper is to overcome this problem: the kit of rules is presented here again, hopefully

*This work was partially supported by the DEMON Esprit Basic Research Action 3148

in a readable form, and is situated in the context of other new kits, whose properties and relationships are considered. The second goal is to show how synthesis procedures can lead to a deeper understanding of the class of nets they synthetise: a simple algebraic characterisation of structural liveness is given for the class of Free Choice nets that can be decomposed into State Machines (Free Choice nets in which concurrency is due to synchronous communication between sequential processes by rendez-vous). More precisely, it is shown that structural liveness (the existence of a marking for which the net is live) can be decided calculating the rank of the incidence matrix of the net. From this result, Hack's duality theorem and a polynomial algorithm for deciding liveness can be easily derived.

The paper is organised as follows: section 1 contains basic definitions. In section 2, three sets of refinement rules for top-down synthesis of nets are described, the last one being the one introduced in [3] and mentioned above, and their properties discussed (this part can be considered, up to a certaint extent, a survey). The algebraic characterisation is proved in section 3. Finally, section 4 shows the two consequences mentioned above.

1 Basic definitions

N denotes the set $\mathbf{N} = \{0, 1, 2, \ldots\}$. **Z** is the set of integers $\mathbf{Z} = \{\ldots -2, -1, 0, 1, 2, \ldots\}$. **Q** is the set of rational numbers. The cardinality of a set X is denoted $|X|$. A *class* is a set whose elements are also sets.

1.1 Nets

A *net* is a triple $N = (S, T, F)$ where

- $S \cap T = \emptyset$
- $F \subseteq (S \times T) \cup (T \times S)$.

The elements of S, T and F are called *places*, *transitions* and *arcs*, respectively. Places and transitions are called generically *nodes*. We assume that S and T are totally ordered, and denote $n = |S|$, $m = |T|$. n and m inherit the subscripts of the net they refer to. The *Pre-set* ${}^\bullet x$ of $x \in S \cup T$ is given by ${}^\bullet x = \{y \in S \cup T \mid (y, x) \in F\}$. The *Post-set* $x^\bullet$ of $x \in S \cup T$ is given by $x^\bullet = \{y \in S \cup T \mid (x, y) \in F\}$. We also define for $X \subseteq S \cup T$

$${}^\bullet X = \bigcup_{x \in X} {}^\bullet x \qquad X^\bullet = \bigcup_{x \in X} x^\bullet$$

If a node x belongs to more than one net, and there is ambiguity about which net the dot $^\bullet$ refers to, the name of the net is added as a subscript, unless it is N: $({}^\bullet x)_{N'}$ denotes the Pre-set of x in N'.

The operation $\cup$ on sets is extended to nets in the natural way. Given two nets $N_1 = (S_1, T_1, F_1)$, $N_2 = (S_2, T_2, F_2)$ we define

$$N_1 \cup N_2 = (S_1 \cup S_2, T_1 \cup T_2, F_1 \cup F_2)$$

The intersection of nets is defined analogously. Notice that both the union and the intersection of two nets is a net.
W denotes the characteristic function of F. The matrix $C = ||c_{ij}||, 1 \leq i \leq n, 1 \leq j \leq m$ with

$$c_{ij} = W(t_j, s_i) - W(s_i, t_j)$$

is called the *incidence matrix* of N. That is, to each place of the net corresponds a row of the incidence matrix. Risking confusion, we denote this row with the name of the place. The rank of a matrix A (i.e the maximal number of linearly independent rows) is denoted by $r(A)$.
A *path* of a net $N = (S,T,F)$ is an alternating sequence $(x_1, f_1, x_2, \ldots, f_{r-1}, x_r)$ of elements of $S \cup T$ and F such that $\forall i, 1 \leq i \leq r-1$: $f_i = (x_i, x_{i+1})$. A path with arcs in a certain set $F' \subseteq F$ is called an F'–path.
Given a net $N = (S,T,F)$, the net $N^{-d} = (T, S, F^{-1})$ is called the *reverse-dual* net of N. It is easy to see that if C is the incidence matrix of N, then $-C^T$ is the incidence matrix of N^{-d}.
A net $N = (S,T,F)$ is called an S*–graph* iff $\forall t \in T$: $|{}^\bullet t| = 1 = |t^\bullet|$. N is called a T*–graph* iff $\forall s \in S$: $|{}^\bullet s| = 1 = |s^\bullet|$. N is called *free choice* iff $\forall s \in S : |s^\bullet| > 1 \Rightarrow {}^\bullet(s^\bullet) = \{s\}$.
A net $N' = (S', T', F')$ is a *subnet* of N, denoted $N' \subseteq N$, iff

$$S' \subseteq S \quad T' \subseteq T \quad F' = F \cap ((S' \times T') \cup (T' \times S'))$$

A place $s' \in S'$ is a *way-in* (*way-out*) place of $N' \subseteq N$ iff ${}^\bullet s' \not\subseteq T'$ ($s'^\bullet \not\subseteq T'$). $N' \subseteq N$ is an S*–component* (T*–component*) of N iff N' is a strongly connected S–graph (T–graph) and $T' = {}^\bullet S' \cap S'^\bullet$ ($S' = {}^\bullet T' \cup T'^\bullet$).
N is said to be *State Machine Decomposable* (*Marked Graph Decomposable*) iff there is a set $\{N_1, \ldots, N_r\}$ of S–components (T–components) of N such that

$$N = \bigcup_{i=1}^{r} N_i$$

State Machine Decomposable will be shortened to SMD.

1.2 Place/transition nets or Petri nets

A function $M: S \to \mathbf{N}$ is called a *marking*. Markings are also represented in vector form: the ith component of the vector corresponds to $M(s_i)$.
A *Place/Transition net* or *Petri net*, is a pair (N, M_0) where N is a net and M_0 is a marking called *initial marking*. A transition $t \in T$ is *enabled* at a marking M iff $\forall s \in {}^\bullet t$: $M(s) > 0$. If $t \in T$ is enabled at a marking M then t may *occur* yielding a new marking M' given by

$$\forall s \in S: \quad M'(s) = M(s) - W(s,t) + W(t,s)$$

$M[t\rangle M'$ denotes the fact that M' is reached from M by the occurrence of t.

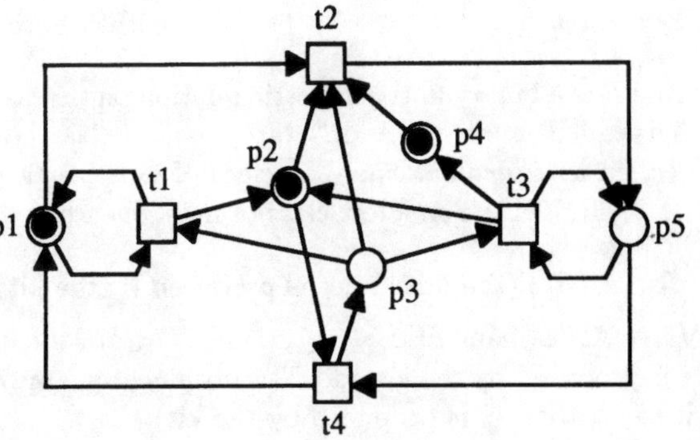

Figure 1: A live and bounded Petri net whose underlying net is not structurally bounded

A sequence of transitions, $\sigma = t_1 t_2 \ldots t_r$, is a *transition sequence of* (N, M_0) iff there exists a sequence $M_0 t_1 M_1 t_2 M_2 \ldots t_r M_r$ such that $\forall i, 1 \leq i \leq r: M_{i-1}[\, t_i \rangle M_i$. The marking M_n is said to be *reachable* from M_0 by the occurrence of σ. The set of reachable markings of (N, M_0) is denoted by $R(N, M_0)$.

A Petri net (N, M_0) is *k-bounded* iff $\forall s \in S, \forall M \in R(N, M_0): M(s) \leq k$. (N, M_0) is *bounded* iff $\exists k \in \mathbf{N}: (N, M_0)$ is k-bounded. A net N is *structurally bounded* iff $\forall M_0 \; \exists k \in \mathbf{N}: (N, M_0)$ is bounded.

A transition $t \in T$ is *live* in (N, M_0) iff $\forall M \in R(N, M_0) \; \exists M' \in R(N, M): M'$ enables t. (N, M_0) is *live* iff all $t \in T$ are live. N is *structurally live* iff $\exists M_0: (N, M_0)$ is live. Structurally live and structurally bounded is shortened to SL&SB.

It follows from the definition that if N is $SL\&SB$, then there is a marking M_0 that makes (N, M_0) live and bounded. But the converse is not true. The Petri net of figure 1, taken from [8], is an example. It is live and bounded for the given marking, but unbounded for the marking (11110) (firing $t_2 t_3 t_3 t_4$ the marking $(11120) > (11110)$ is reached). Hence, the underlying net is not structurally bounded.

2 Refinement kits

Top-down synthesis of nets is performed starting from a very simple net to which refinement rules are applied stepwise. Given a class of rules (a *kit*), the nets that can be produced applying a finite number of times the elements of the kit is the class of nets generated by these rules. In this section we first formalise these concepts. Then we introduce three different refinement kits, together with their properties. The seed of the synthesis procedure is called *initial net.*

Definition 2.1 *Initial net*

The net $N_0 = (\{s\}, \{t\}, \{(s,t), (t,s)\})$ is called *initial net.* ■ 2.1

A refinement rule allows us to transform a net into another one, which under some criterion is considered more complex than the old one. This is adequately represented by means of an antisymmetric binary relation.

Definition 2.2 *Refinement rules*

A refinement rule R is a binary antisymmetric relation on the class of nets $\mathcal{N}$. Given $(N, \hat{N}) \in R$, N is called *source net* and $\hat{N}$ *target net*. A class $\{R_1, \ldots, R_a\}$ of refinement rules is called a *refinement kit*. The class of nets *produced* by $\{R_1, \ldots, R_a\}$, denoted $\mathcal{N}(R_1, \ldots, R_a)$, is the smallest class of nets given by:

1. $N_0 \in \mathcal{N}(R_1, \ldots, R_a)$ (the initial net is produced by the kit)
2. If $N \in \mathcal{N}(R_1, \ldots, R_a)$ and $\exists i, 1 \leq i \leq a : (N, \hat{N}) \in R_i$, then $\hat{N} \in \mathcal{N}(R_1, \ldots, R_a)$ (i.e. if N is produced by the kit and $\hat{N}$ is obtained by applying one of the rules of the kit to N, then $\hat{N}$ is produced by the kit).

A sequence of nets (N_i), $0 \leq i \leq r$, where $r \in \mathbf{N}$, such that N_0 is the initial net and

$$\forall i,\ \ 0 \leq i \leq (r-1) : (N_i, N_{i+1}) \in R_j,\ \ 1 \leq j \leq a$$

is called a *synthesis sequence* of N_r in $\mathcal{N}(R_1, \ldots, R_a)$. ■ 2.2

We can now introduce the three kits we deal with in this paper. Each of them is composed by two rules.

The SMD kit The name of this kit is due to the fact that it produces all SMD nets (as well as others, but we are not interested in this). The other kits will be obtained taking subrelations of the rules of this kit, and therefore will generate smaller classes of nets.

The two rules R_1, R_2 can be explained very easily in a graphical way. R_1 consists of the addition of a new place to a net, with the condition that the new place must have at least one input arc and one output arc. Figure 2 shows an example, where the new place and its input and output arcs are printed in boldface.

R_2 consists of the substitution of a place by a connected S-graph. Two possible substitutions of the place $\hat{s}$ of figure 2 are shown in figure 3 (the S–graphs are printed in boldface). Nevertheless, we impose two further conditions: one on the substituted place s and other on the S–graph $N' = (S', T', F')$ substituting it.

First, s must satisfy $\forall t \in {}^\bullet s\colon |t^\bullet| > 1$ and $\forall t \in s^\bullet\colon |{}^\bullet t| > 1$, except maybe at the very beginning, when refining the only place of the initial net.

Second, every place of N' must be contained in an F'–path starting at a way-in place and ending at a way-out place of N'.

The substitution on the left of figure 3 is a legal one, since it satisfies both conditions. The substitution on the right does not satisfy the second. s'_1 is the only way-in and the only way-out place of the S-graph. Nevertheless, no path of the S–graph starting and ending at s'_1 contains s'_2.

This second condition is included for the following reason: it is easy to see that if the source net is strongly connected, the substitutions leading to a strongly connected target net are exactly the ones that satisfy it.

The formal definitions of the two rules are given next.

Let $N = (S, T, F)$, $\hat{N} = (\hat{S}, \hat{T}, \hat{F})$ be two nets.

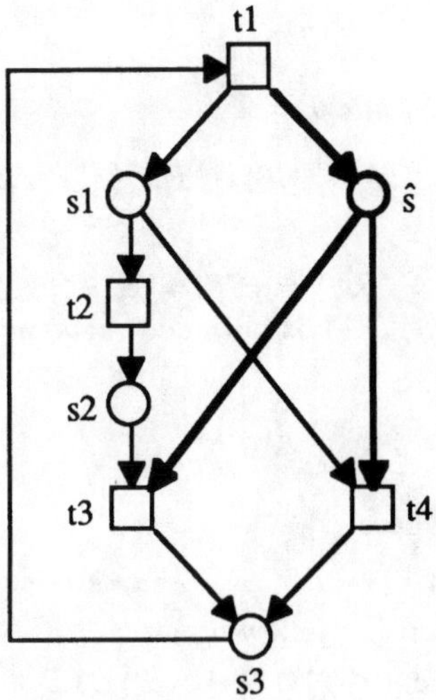

Figure 2: The refinement rule R_1

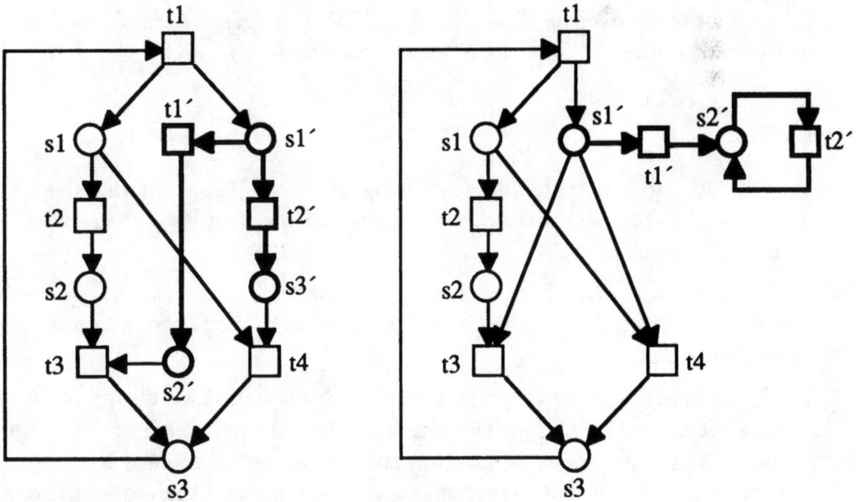

Figure 3: Two possible substitutions of the place $\hat{s}$ of figure 2 by an S–graph. Only the left one is allowed by rule R_2

Rule 1 $(N, \hat{N}) \in R_1$ *iff:*

1. $\hat{S} = S \cup \hat{s}$, *where* $\hat{s} \notin S$, *and* $\hat{T} = T$

2. $\hat{F} = F \cup F_{\hat{s}}$, *where* $F_{\hat{s}} \subseteq (\{\hat{s}\} \times T) \cup (T \times \{\hat{s}\})$ *and* $({}^\bullet\hat{s})_{\hat{N}} \neq \emptyset \neq (\hat{s}^\bullet)_{\hat{N}}$ ■ R1

Definition 2.3 Let $N = (S, T, F)$, $N' = (S', T', F')$ be two nets, and $s \in S$. It is said that the net $\hat{N} = (\hat{S}, \hat{T}, \hat{F})$ is obtained *replacing* s *by* N' *in* N iff

1. $\hat{S} = (S - \{s\}) \cup S'$
2. $\hat{T} = T \cup T'$
3. $\hat{F} = \bar{F} \cup F' \cup F_s$, where
   - $\bar{F} = F \cap ((\hat{S} \times \hat{T}) \cup (\hat{T} \times \hat{S}))$ *(arcs remaining in* N *after removing* s*)*
   - F_s is obtained in the following way:

     for each arc $f \in F \cap (\{s\} \times T) \cup (T \times \{s\})$: select arbitrarily a place $s' \in S'$, replace s by s' and add the resulting arc to F_s
     (arcs replacing the input and output arcs of s*)*

■ 2.3

Rule 2 $(N, \hat{N}) \in R_2$ *iff there exist* $s \in S$ *and an S-graph* $N' = (S', T', F') \subseteq \hat{N}$ *satisfying:*

1. $N = N_0 \vee (\forall t \in {}^\bullet s\colon |t^\bullet| > 1 \wedge \forall t \in s^\bullet\colon |{}^\bullet t| > 1)$

2. $\hat{N}$ *is obtained replacing* s *by* N' *in* N

3. $\forall s' \in S'$: *there exists an* F'*-path* $(s'_{in}, \ldots, s', \ldots, s'_{out})$, *where* s'_{in} *and* s'_{out} *are a way-in and a way-out place of* N', *respectively.* ■ R2

Remark 2.4

1. It is easy to see that, when applying R_2, $s = \sum_{s' \in S'} s'$ (where, as was announced in the past section, we identify a place with its corresponding row in the incidence matrix).
2. The nets of $\mathcal{N}(R_1, R_2)$ are strongly connected. This can be proved inductively: the initial net is strongly connected and the two rules preserve strong connectedness.
3. Notice that R_2 always operates on the places added by means of R_1, except when it is applied at N_0. The reason is that places not added by R_1 (and not the initial place) must come from the refinement of a place by means of R_2. Then they belong to a subnet that is an S–graph, and therefore have one single input transition or one single output transition in the net. But in this case they do not satisfy condition 1 for the application of R_2.

■ 2.4

Now we state the property that gives name to the kit.

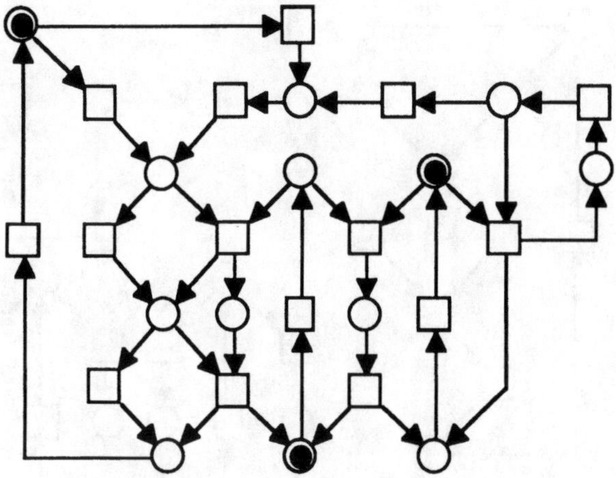

Figure 4: An SMD Petri net

Proposition 2.5

Let N be an SMD net. Then $N \in \mathcal{N}(R_1, R_2)$.

Proof: (sketch). We give here an outline of the synthesis procedure. Assume that N is connected (otherwise we generate its connected components separately). Let $\mathcal{C} = \{N_1, \ldots, N_r\}$ be a set of S–components of N that cover it, with the following property:

$$\forall i,\ 1 \leq i \leq r-1 \colon N_{i+1} \cap (\bigcup_{j=1}^{i} N_j) \neq \emptyset$$

It is not difficult to see that such a set exists. The procedure synthesises first N_1, then $N_1 \cup N_2$, $N_1 \cup N_2 \cup N_3$ and so on.

Since N_1 is an strongly connected S–graph, it can be obtained applying $R2$ to the initial net N_0.

Consider now the net $N' = (\bigcup_{j=1}^{i} N_j)$. N_{i+1} has a part in common with this net, plus one or more several "private" connected subnets. For each of these subnets, we use $R1$ to add a new place to N', which is connected to N' in the same way than the corresponding subnet. Then these places are substituted by the subnets themselves using $R2$. The net so obtained is $N_{i+1} \cup (\bigcup_{j=1}^{i} N_j)$ ■ 2.5

We illustrate this construction by means of an example. Consider the SMD net of figure 4. The set of S–components described above is shown in figure 5. The steps of the synthesis are described in figure 6. After the first application of $R2$, N_1 has been generated. We add then a place, which is expanded to yield $N_1 \cup N_2$. Similarly, the final net $N = N_1 \cup N_2 \cup N_3$ is produced.

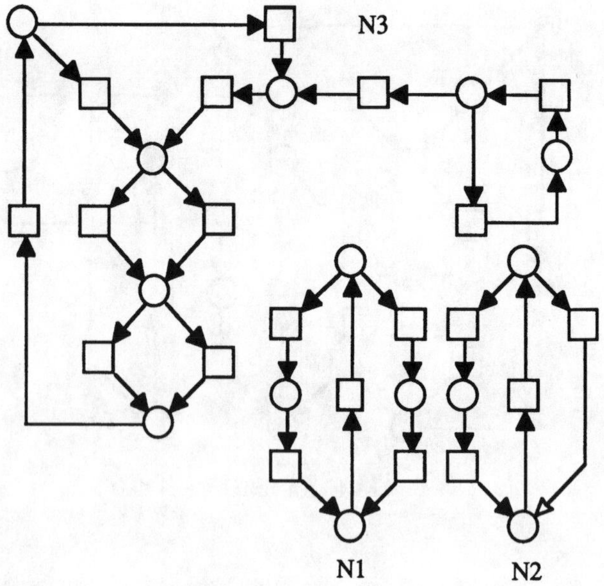

Figure 5: A set of S–components that cover the net of figure 4

The SL&SB kit The kit we introduce now is more interesting, because it produces only SL&SB nets (this is one of the results of [3]). Its two rules are modifications of the rules R_1 and R_2 of the SMD kit.
Let $N = (S,T,F)$, $\hat{N} = (\hat{S},\hat{T},\hat{F})$ be two nets.

Rule 3 $(N,\hat{N}) \in R_3$ *iff:*

1. $(N,\hat{N}) \in R_1$ *(i.e. $\hat{N}$ is obtained adding a place to N)*
2. *The new place $\hat{s}$ is a linear combination of places of S, i.e. $\hat{s} = \sum_{i=1}^{n} \lambda_i s_i$.* ■ R3

The place $\hat{s}$ added to the net of figure 2 satisfies the second requirement. It is not difficult to see that $\hat{s} = s_1 + s_2$.

Rule 4 $(N,\hat{N}) \in R_4$ *iff the following conditions hold:*

1. $(N,\hat{N}) \in R_2$
2. *The S-graph $N' = (S',T',F') \subseteq N$ described in Rule R_2 satifies that $\forall s' \in S'$, $\forall$ way-out place $s'_{out} \in S'$: there exists an F'-path $(s',\ldots,s'_{out})$.* ■ R4

Due to condition 2, every place of N' can receive tokens (through the way-in places it is connected to), which can afterwards reach *any* of the way-out places. The idea lying behind this construction is that, if tokens are needed to fire one of the output transitions of a certain way-out place, it should be always possible for the tokens of N' to reach that place. Notice that the refinement performed with the net of figure 2 as source net and the net of the left of figure 3 as target net does not satisfy this condition: there is no F'-path from s'_3 to the way-out place s'_2.
It is immediate from the definition that $\mathcal{N}(R_3,R_4) \subseteq \mathcal{N}(R_1,R_2)$, but the reverse is not true.

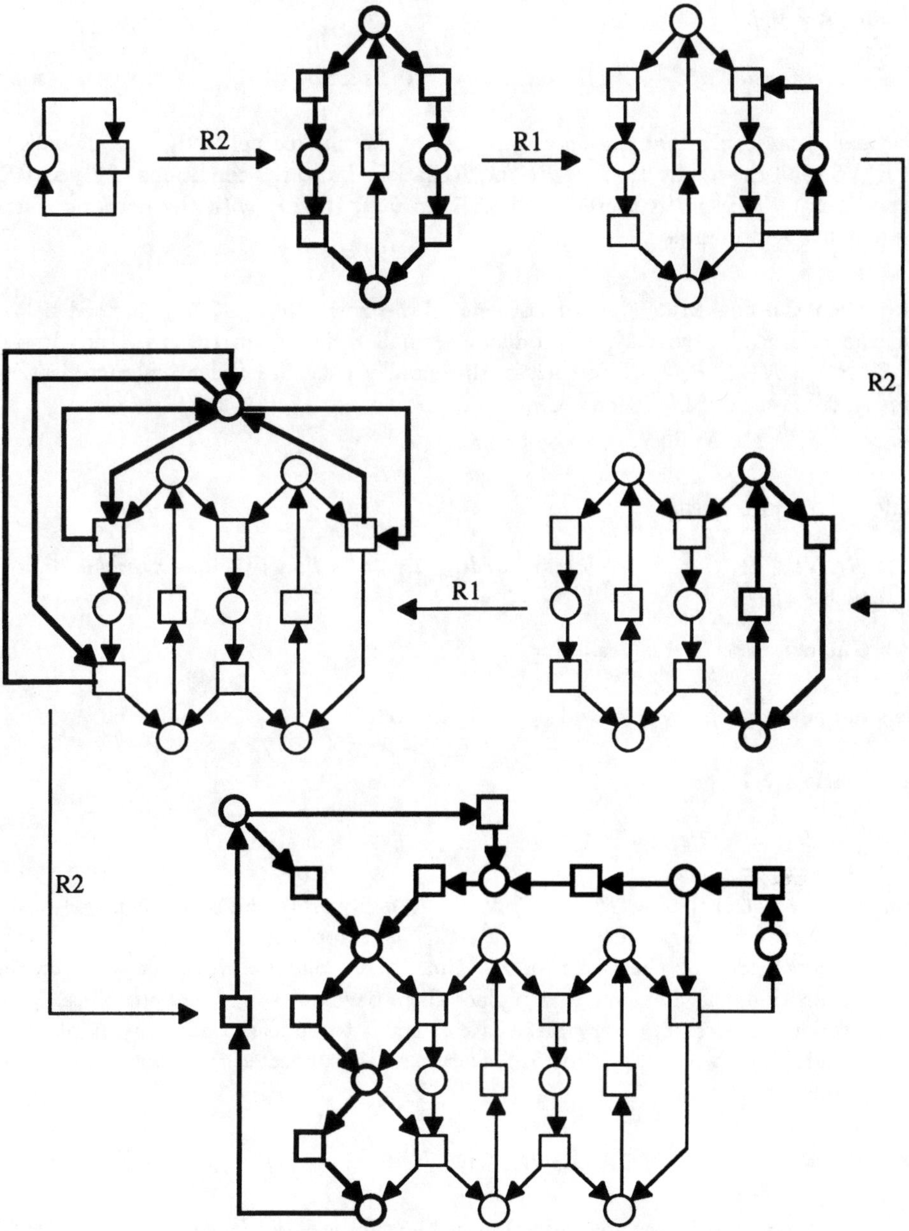

Figure 6: Synthesis of the net of figure 4

Theorem 2.6 *[3]*

Let N be a net. If $N \in \mathcal{N}(R_3, R_4)$, then N is SL&SB. ■ 2.6

The reader can check that the underlying net of the marked net in figure 4 belongs to $\mathcal{N}(R_3, R_4)$ (all the applications of R_1 and R_2 satisfied also the conditions of R_3 and R_4 respectively). This net is therefore SL&SB. In fact, the net with the marking of the figure is live and bounded.

The Free Choice kit This kit consists of the new rule $R_5 \subseteq R_3$, defined below, and the rule R_4. Obviously, it produces a smaller class than the previous kit (i.e. $\mathcal{N}(R_5, R_4) \subseteq \mathcal{N}(R_3, R_4)$). Nevertheless, this smaller class has a clear characterisation: namely, it is that of SL&SB Free Choice nets.
Let $N = (S, T, F)$, $\hat{N} = (\hat{S}, \hat{T}, \hat{F})$ be two nets.

Rule 5 $(N, \hat{N}) \in R_5$ *iff:*

1. *$(N, \hat{N}) \in R_3$ (i.e. $\hat{N}$ is obtained adding a place $\hat{s}$ that is a linear combination of places of N)*
2. *The new place $\hat{s}$ satisfies $|\hat{s}^\bullet| = 1$.* ■ R5

We state now a property that will be used later on.

Proposition 2.7

$\mathcal{N}(R_5, R_4) = \mathcal{N}(R_5, R_2)$.

Proof: It is obvious that $\mathcal{N}(R_5, R_4) \subseteq \mathcal{N}(R_5, R_2)$. To prove the other inclusion notice that, as was mentioned in remark 2.4.3, R_2 is applied, except at N_0, only to the places added by the action of R_5. Since these places have exactly one output transition, the S-graphs that replace them have one single way-out place. Then condition 4 of R_2 (every place is connected to at least one way-out place) is equivalent to condition 2 of R_4 (every place is connected to *all* way-out places). This means that

$$(N \in \mathcal{N}(R_5, R_2) \wedge (N, \hat{N}) \in R_2) \Rightarrow (N, \hat{N}) \in R_4$$

The inclusion follows by induction on the synthesis sequences. ■ 2.7

Theorem 2.8 *[3]*

Let N be an SL&SB Free Choice net. Then $N \in \mathcal{N}(R_5, R_4)$. ■ 2.8

Using classical theory of Free Choice nets, this result can be somewhat extended.

Proposition 2.9 *[7] [1]*

The four following classes of nets are identical:

1. $\mathcal{N}(R_5, R_4)$
2. *The class of SL&SB Free Choice nets*
3. *The class of structurally live SMD Free Choice nets*
4. *The class of Free Choice nets that can be endowed with a live and bounded marking.* ■ 2.9

In [5], a kit of eigt rules was given for the synthesis of well behaved Bipolar Schemes. These models are included in the class of live and 1-bounded Free Choice nets, and can therefore be generated by the two-rule Free Choice kit, once the rules are slightly modified to deal with markings. Nevertheless the (modified) Free Choice kit does not guarantee 1-boundedness.

We would like to finish the section with some comments about the locality of the rules of the second and third kit. A rule is said to be *local* if, loosely speaking, the conditions for its application can be determined to hold or not by examining a small environment of a node of the graph. Local rules are easier to handle than non-local ones. In the three kits, the even-numbered rules are local, because the conditions can be checked examining only the sustituted place and its input and output transitions. The odd-numbered rules are possibly easier to state, but more difficult to deal with, since they are non-local: the required linear combination could involve places of the system situated apart ones from the others. This problem was also present in the kit of [5] and, in fact, it is probably inherent to the problem. It is a widespread conjecture, though so far no proof has been given, that it is not possible to generate all and only strongly connected graphs (and therefore all and only strongly connected nets) using local rules only. Since a well known result of net theory states that live and bounded nets are strongly connected [1], we should not expect be able to synthesise live and bounded nets using local rules only.

3 An algebraic characterisation of live SMD-FC nets

We prove in this section the result mentioned in the introduction: structural liveness of SMD-Free Choice nets can be characterised by means of the rank of the incidence matrix (we prove it for $\mathcal{N}(R_5, R_4)$, which by theorem 2.8 is the same class). More precisely, the theorem is stated as follows:

Theorem 3.1

Let $N = (P, T, F)$ be a net, C its incidence matrix and $a = |F \cap (S \times T)|$ (i.e. a is the number of arcs of N leading from a place to a transition).

Then $N \in \mathcal{N}(R_5, R_4)$ iff N is SMD and $r(C) = n + m - a - 1$. ■ 3.1

The 'if' part of this characterisation was conjectured by Silva [9], and the 'only if' part by Campos and Chiola [2].

The proof requires some previous definitions and lemmata. We introduce the function $\mathcal{F} : \mathcal{N} \longrightarrow \mathbf{Z}$ given by $\mathcal{F}(N) = r(C) - (n + m - a - 1)$. In a first step we will show that, if N is an SMD net, then $\mathcal{F}(N) \geq 0$. Then we will prove that $\mathcal{F}(N) = 0$ iff $N \in \mathcal{N}(R_5, R_4)$.

Let us start by showing some useful relationships between n, m and a, very easy to check.

Proposition 3.2

Let $N = (S, T, F)$ be a strongly connected net. Then:

1. *$a \geq n$, and $a = n$ if N is a Marked Graph*
2. *$a \geq m$, and $a = m$ if N is a State Machine*

Proof: Follows easily from the definitions of State Machine and Marked Graph.

■ 3.2

Lemma 3.3

Let N be a strongly connected net and $(N, \hat{N}) \in R_1$. Then:

1. *$\mathcal{F}(\hat{N}) - \mathcal{F}(N) \geq 0$*
2. *$\mathcal{F}(\hat{N}) - \mathcal{F}(N) = 0 \Leftrightarrow (N, \hat{N}) \in R_5$.*

Proof: $\hat{N}$ is obtained by adding a place $\hat{s}$ to N such that $|{}^\bullet\hat{s}| \neq 0 \neq |\hat{s}^\bullet|$. This means

$$\hat{n} = n + 1 \qquad \hat{m} = m \qquad \hat{a} = a + |\hat{s}^\bullet| > a$$

It follows:

$$\mathcal{F}(\hat{N}) - \mathcal{F}(N) = r(\hat{C}) - r(C) - 1 + |\hat{s}^\bullet| \geq 0$$

Since $r(\hat{C}) - r(C) \leq 1$, the equality holds iff $(r(\hat{C}) = r(C) \wedge |\hat{s}^\bullet| = 1)$. But in this case $\hat{s}$ is linear combination of places of S, and by definition $(N, \hat{N}) \in R_5$.

■ 3.3

Lemma 3.4

Let N be a strongly connected net and $(N, \hat{N}) \in R_2$. Then $\mathcal{F}(\hat{N}) - \mathcal{F}(N) = 0$.

Proof: $\hat{N}$ is obtained by replacing a place $s \in S$ by an S-graph $N' = (S', T', F')$. We claim that $r(\hat{C}) - r(C) = n' - 1$. Let us prove first that if the claim is true then the result follows. It is not difficult to see that

$$\hat{n} = n + n' - 1 \qquad \hat{m} = m + a' \qquad \hat{a} = a + a'$$

and therefore

$$\mathcal{F}(\hat{N}) - \mathcal{F}(N) = r(\hat{C}) - r(C) - (n' - 1) - a' + a' = 0$$

Now we prove the claim.

(a) $r(\hat{C}) - r(C) \leq n' - 1$. This is easy, because $\hat{C}$ has $(n' - 1)$ rows more than C (the row corresponding to s is removed and the n' rows corresponding to S' added). It is clear that the rank cannot grow more than the difference between the number of rows of C and $\hat{C}$.

(b) $r(\hat{C}) - r(C) \geq n' - 1$. Consider the matrix $\hat{C}_s$, obtained by adding to $\hat{C}$ the row of C corresponding to the place s. By the first part of remark 2.4, s is a linear combination of places of S'. Therefore $r(\hat{C}_s) = r(\hat{C})$.
Let V_1 be a set of linearly independent rows of C with $|V_1| = r(C)$. Let V_1' be a set of rows of $\hat{C}$, corresponding to $S_1' \subseteq S'$, with $|V_1'| = n' - 1$. We show that the vectors of $V_1 \cup V_1'$ are linearly independent.
As N' is an S-graph, $\exists s' \in S_1' : (({s'}^{\bullet})^{\bullet} \not\subseteq S_1' \vee {}^{\bullet}({}^{\bullet}s') \not\subseteq S_1')$ (for instance, in the net on the left of figure 3, if we remove s_1', then t_1' and t_2' have no input place. Similarly with s_2' and s_3'). Then $\exists t' \in T' : ({}^{\bullet}t' \cup t'^{\bullet}) \cap S_1' = s'$.
It follows that $\hat{c}_s(s', t') \neq 0$ and $\forall s \in S_1 \cup S_1', s \neq s' : \hat{c}_s(s', t') = 0$. Then clearly the row corresponding to s' is not a linear combination of the other vectors of $V_1 \cup V_1'$, and hence $V_1 \cup V_1'$ is a set of linearly independent vectors. In consequence

$$r(\hat{C}) = r(\hat{C}_s) \geq |V_1 \cup V_1'| = n' - 1 + r(C)$$

and the claim is proved.

■ 3.4

Using these two lemmas, it is now easy to prove the following result.

Proposition 3.5

Let (N_i), $0 \leq i \leq r$ be a synthesis sequence in $\mathcal{N}(R_1, R_2)$. Then $\mathcal{F}$ is non-decreasing and non-negative on (N_i)

Proof: $\mathcal{F}$ is non-decreasing by lemmata 3.4 and 3.3. It is non-negative because $\mathcal{F}(N_0) = 0$ and it is non-decreasing. ■ 3.5

Finally, we are well prepared to prove theorem 3.1.

Proof of theorem 3.1

Proof: ($\Rightarrow$): Let $(N_i), 1 \leq i \leq r$ be a synthesis sequence of N in $\mathcal{N}(R_3, R_5)$. By lemmas 3.4 and 3.3

$$\forall i,\ 1 \leq i \leq (r-1) : \mathcal{F}(N_{i+1}) = \mathcal{F}(N_i)$$

Since $\mathcal{F}(N_0) = 0$, we have $\mathcal{F}(N) = 0$, and the result follows.

($\Leftarrow$): As N is an SMD net, $N \in \mathcal{N}(R_1, R_2)$ by proposition 2.5. Let $(N_i), 1 \leq i \leq r$ be a synthesis sequence of N in $\mathcal{N}(R_1, R_2)$. Since $\mathcal{F}$ is non-decreasing, we have

$$\forall i, 1 \leq i \leq (r-1) : \mathcal{F}(N_{i+1}) = \mathcal{F}(N_i)$$

Then, by lemma 3.3, $N \in \mathcal{N}(R_5, R_2)$. By proposition 2.7, $N \in \mathcal{N}(R_5, R_4)$.

■ 3.1

4 Consequences

We show in this section two results that can be derived from theorem 3.1. The first one is a new proof of Hack's duality theorem [7]. This result was proved by M. Silva [9], assuming that theorem 3.1 was true.

Theorem 4.1

Let N be a Free Choice net. N is SL&SB iff the reverse dual net of N is SL&SB.

Proof: Assume N is SL&SB, and let N^{-d} be the reverse dual of N. We show that N^{-d} satisfies the conditions of theorem 3.1. First, it is easy to see that N^{-d} is also Free Choice. A result of [7] ensures that N is Marked Graph Decomposable. As the reverse-dual of a T-component is an S-component, it follows that N^{-d} is SMD.

Moreover, by theorem 3.1, $\mathcal{F}(N) = n + m - 1 - a$. Places of N correspond to transitions of N^{-d}, and viceversa. Arcs leading from places to transitions are transformed into themselves (because the arcs are reversed!). Therefore $\mathcal{F}(N^{-d}) = m + n - 1 - a$.

As N^{-d} satisfies all the conditions of theorem 3.1, the result holds. ■ 4.1

We can obtain from this theorem the following corollary:

Corollary 4.2

Let N be an FC net. N is SL&SB iff N is Marked Graph Decomposable and $r(C) = n + m - 1 - a$.

Proof: (⇒): $N^{-d} = (T, P, F^{-1})$ is SL&SB by theorem 4.1. Then, by theorem 3.1, N^{-d} is SMD and $r(-C^T) = m + n - 1 - a$. It follows that N is SMD and $r(C) = n + m - 1 - a$.

(⇐): N^{-d} is SMD and satisfies the equation. Therefore, by theorem 3.1, N^{-d} is SL&SB. By theorem 4.1, N is SL&SB as well. ■ 4.2

It can be proved that a structurally live SMD-FC net is live iff all its S-components are marked at M_0 [7]. Using this property, the following result is obtained in [3].

Proposition 4.3

Let (N, M_0) be an SMD-FC Petri net such that N is structurally live. Then (N, M_0) is live iff the following Linear Programming problem

$$\begin{array}{ll} \textit{minimise} & 0^T \cdot Y \\ \textit{subject to} & Y^T \cdot C = 0 \\ & Y^T \cdot M_0 = 0 \\ & Y \geq 0 \end{array}$$

has no solution. ■ 4.3

Notice that the optimisation function vanishes. This means that an optimal solution exists if and only if there exists a vector $Y \in \mathbf{Q}^n$ satisfying the constraints.
This results leads immediately to the following theorem:

Theorem 4.4

Liveness of marked SMD-FC nets is decidable in polynomial time.

Proof: Obvious, since liveness can be decided calculating the rank of the incidence matrix and solving a Linear Programming problem, problems for which polynomial algorithms exist. ■ 4.4

Conclusions

We have introduced and commented in this paper several kits of refinement rules for performing top-down synthesis of nets. Many results (in fact, the more difficult ones) were obtained in a previous paper. Here, free of the technical details, we have tried to give the kits a clear organisation, point out their relationships and present them in their simplest form. Our purpose was to convince the reader that small kits (in fact, containing two elements) of easy-to-describe rules suffice to produce all the nets of non-trivial subclasses enjoying desirable properties. Moreover, we wanted to show that these kits are powerful tools for deepening our knowledge on the classes of nets they generate. We hope to have achieved this by presenting a particular result: SMD-FC nets are structurally live if and only if an equation relating the rank of the incidence matrix to the number of places, transitions and arcs of the net holds. This theorem allowed us also to derive Hack's duality theorem, and to prove that liveness of SMD-FC nets is a polynomial problem.
Two interesting questions remain open, both concerning the class $\mathcal{N}(R_3, R_4)$: Are there simple characterisations of this class, first in terms of meaningful conditions on the structure of the nets (such as the Free Choice property is for $\mathcal{N}(R_5, R_4)$), and second in purely algebraic terms? We conjecture that the answer to the first question is "yes", and to the second "no". Nevertheless, we also believe that it should be posible to find some "quasi-algebraic" characterisation of this class.

References

[1] Best, E.: Structure Theory of Petri Nets: the Free Choice Hiatus, *Advanced Course on Petri Nets, Bad Honnef*, Lecture Notes on Computer Science 254-255, Springer (1987).

[2] Campos, J; Chiola, G.: Private communication (1989).

[3] Esparza, J.; Silva, M.: Top-down Synthesis of Live & Bounded Free Choice Nets. To appear in: *Proceedings of the XI International Conference on Applications and Theory of Petri Nets*, Paris (June 1990).

[4] Esparza, J.: Structure Theory of Free Choice nets. Ph.D. Thesis, University of Zaragoza (June 1990).

[5] Genrich, H.; Thiagarajan, P.S.: A theory of bipolar synchronisation schemes, *Theoretical Computer Science*, 30, pp. 241-318 (1984).

[6] Jones, N.; Landweber, L.; Lien, Y.: Complexity of some Problems in Petri Nets, *Theoretical Computer Science* Vol. 4, pp. 277-299 (1977).

[7] Hack, M.H.T.: Analysis of Production Schemata by Petri Nets. *TR-94 MIT*, Boston (1972, corrected June 1974).

[8] Reisig, W.: Petri Nets. An Introduction. *EATCS monographs on Theoretical Computer Science*, Vol. 4 (1985).

[9] Silva, M.: Private communication (1989).

'he Need for Headers: An Impossibility Result for Communication over Unreliable Channels

Alan Fekete
Software Systems Research Group
Department of Computer Science
University of Sydney
NSW 2006
AUSTRALIA

Nancy Lynch
MIT Lab for Computer Science
545 Technology Square
Cambridge, MA 02139
U.S.A.

Abstract

It is proved that any protocol that constructs a reliable data link service using an unreliable physical channel service necessarily includes in the packets some header information that enables the protocol to treat different packets differently. The physical channel considered is permitted to lose, but not reorder or duplicate packets. The formal framework used for the proof is the I/O automaton model.

Introduction

obably the most common use for formal models of concurrent programming is to verify that otocols meet their specifications. A less common, but nonetheless important use for such models to prove impossibility results, showing that no protocol can possibly solve a particular problem. e believe that it is important to the coherence of our research field that the same formal models used as the foundation for both kinds of activities.

Some of the features of a formal model that are needed to support impossibility results are e same as those needed for verification; for instance, both kinds of models must allow separate scription of the problem to be solved and the allowable implementations. However, some other tures may be different. A useful model for proving impossibility results should include a notion "fair" or "admissible" execution, which describes exactly when system components are required continue taking steps; otherwise, it will be impossible to guarantee that the system satisfies y liveness properties. Also, a model for proving impossibility results needs to designate which tivities are under the control of the protocol and which are under the control of the environment; herwise, a protocol might "solve" a problem in a trivial way, simply by preventing some of the uts from occurring. A more extensive discussion of these features can be found in [12].

We believe that the I/O automaton model [10, 11] can serve as a reasonable basis for both rifying concurrent algorithms and for proving impossibility results. The model has already been ed in verifying a wide range of algorithms (see, for example, [10, 3, 4, 5, 7, 14, 16, 17]). In s paper, we use it to prove an impossibility result, namely, that any protocol that implements eliable data link service by using an unreliable physical channel service necessarily includes in

e authors were supported in part by the National Science Foundation under grant CCR-86-11442, by the Office Naval Research under contract N00014-85-K-0168 and by the Defense Advanced Research Projects Agency un- contracts N00014-83-K-0125 and N00014-89-J-1988. The first author was supported in part by the Research undation for Information Technology of the University of Sydney.

the packets some header information that enables the protocol to treat different packets different We believe that the main interest of this work is not so much in the result itself (no one has e suggested using a protocol without header information) but rather in the way the model is used show nonexistence of a protocol with certain properties.

There has recently been a lot of research in the distributed computing theory research co munity on the problem of constructing a reliable message transmission service using an underlyi unreliable packet transmission service (see [9, 1, 18, 13], for example). Most of this work has dressed the case where the physical channel is especially unreliable, in that it can lose packets a also deliver packets out of order. In these cases the natural protocol, due to Stenning ([15]) pla each message in a packet with a sequence number as header, and repeatedly sends the packet ur its receipt has been acknowledged. The difficulty with this protocol is that the sequence numb increase without bound, and the papers mentioned above explore the possibility of using a fix size header. By contrast, in this paper we consider using a FIFO (but possibly lossy) physi channel. There are many protocols known for this situation, most being variants on the Altern ing Bit protocol [2], in which packets and acknowledgments contain a single bit header. We sh that this header is needed, in that there is no protocol that solves the same problem without usi some header to distinguish between packets. A key modeling issue is how to define "headers" in arbitrary protocol, without assuming a particular structure (such as [sequence-number,messag for the packets. The definitions we use are adapted (and simplified) from those in [9, 6].

The rest of the paper is organized as follows. In Section 2, we show how we model physi channels, and also show the existence of a "universal" physical channel, which exhibits all behaviors that can arise in any physical channel. In Section 3, we give the specification for corr data link behavior. In Section 4, we define data link protocols and show what it means for them implement correct data link behavior using two unidirectional physical channels. In Section 5 prove a preliminary impossibility result, namely, we show the impossibility of implementing a d link service using *identical* packets in each direction. This result contains most of the complex of our main result, but avoids the issue of how to model the headers. In Section 6, we discuss h to define the headers used by an arbitrary protocol, and invoke our preliminary result to obt the main theorem. A summary of the results we need about the I/O automaton model is given Appendix A, and the construction of a universal physical channel is given in Appendix B.

2 The Physical Layer

The physical layer is the lowest layer in the OSI Reference Model hierarchy, and is implemen directly in terms of the physical transmission media. A standard interface to the physical la permits implementation of the higher layers independently of the transmission media. In a typi setting, a physical layer interacts with higher layers at two endpoints, a "transmitting station" a "receiving station". The physical layer receives messages called "packets" from the higher la at the transmitting station, and delivers some of the packets to the higher layer at the receiv station. The physical layer can lose packets. While it is also possible for packets to be corrupted the transmission medium, we assume that the physical layer masks such corrupted packets us error-detecting codes. Thus, the only faulty behavior we consider is loss of packets.

In this section, we give a specification for physical layer behavior; in particular, we specif channel that ensures FIFO delivery of packets. It is convenient to parameterize the specification b channel name n and by an alphabet P of legal *packets.* The specification will be given (as usual in I/O automaton model) by a pair, $PL^{n,P}$, consisting of an action signature, $sig(PL^{n,P})$, describ the interface the layer provides, and a set of sequences of actions, $scheds(PL^{n,P})$, describing allowed interactions. $PL^{n,P}$ has the action signature given formally as follows.

ɔut actions:
send_pkt$^n(p), p \in P$
ıtput actions:
rec_pkt$^n(p), p \in P$

The *send_pkt*$^n(p)$ action represents the sending of packet p on the physical channel by the ınsmitting station, and the *rec_pkt*$^n(p)$ represents the receipt of packet p by the receiving station. e will refer to these actions as ***physical layer actions*** (for n and P). In order to define the set sequences, *scheds*$(PL^{n,P})$, we define first a collection of properties, reflecting the operation a "good" physical channel. The properties are defined with respect to $\beta = \pi_1\pi_2\ldots$ a (finite or inite) sequence of physical layer actions, and a ***correspondence relation***, a binary relation between e *send_pkt*^{n} events and the *rec_pkt*^{n} events in β. The correspondence relation is intended to model e association that can be set up between the event modeling the sending of a packet and the ent modeling the receipt of the same packet. Complications are caused by the fact that the same ta might be sent repeatedly, and so the sending of two such identical packets is modeled by two currences of the same action *send_pkt*$^n(p)$. The first property gives basic requirements on the rrespondence.

L1) 1. If an event π_i = *rec_pkt*$^n(p)$ corresponds to an event π_j = *send_pkt*$^n(q)$, then $p = q$, and also $j < i$, that is, the event π_j precedes π_i in β.
2. Each *rec_pkt*^{n} event corresponds to exactly one *send_pkt*^{n} event.
3. Each *send_pkt*^{n} event corresponds to *at most* one *rec_pkt*^{n} event.

We next define the FIFO property. It says that those packets that are delivered have their ceive_pkt* events occurring in the same order as their *send_pkt* events. Note that (PL2) may be ue even if a packet is delivered and some packet sent earlier is not delivered: there can be gaps the sequence of delivered packets representing lost packets.

L2) (FIFO) Suppose that the event π_i = *send_pkt*$^n(p)$ in β corresponds to the event π_j = *rec_pkt*$^n(p)$, and π_k = *send_pkt*$^n(p')$ corresponds to π_l = *rec_pkt*$^n(p')$. Then $i < k$ if and only if $j < l$.

So far, the properties listed have been safety properties, that is, when they hold for a sequence y also hold for any prefix of that sequence. The final property is a liveness property. It says at if repeated send events occur, then eventually some packet is delivered.

L3) If infinitely many *send_pkt*^{n} actions occur in β, then infinitely many *rec_pkt*^{n} actions occur in β.

Now we define *scheds*$(PL^{n,P})$ to be the set of those sequences β of physical layer actions for ich there exists a correspondence such that (PL1), (PL2), and (PL3) are all satisfied for β and at correspondence. A ***FIFO physical channel*** for n and P is any I/O automaton whose external nature is *sig*$(PL^{n,P})$ and whose fair behaviors are all in the set *scheds*$(PL^{n,P})$.

We close this section by defining "universal FIFO physical channels", that is, automata whose r behaviors are exactly the set of sequences permitted by the specification above. Namely, we that an I/O automaton is a ***universal physical channel*** for n and P if it is a FIFO physical annel for n and P and the set of its fair behaviors is exactly the set *scheds*$(PL^{n,P})$. We give the ıstruction of a universal FIFO physical channel in Appendix B, but for most of this paper all at we will need is the fact that one exists:

Lemma 2.1 *For any n and P, there exists a universal FIFO physical channel for n and P.*

The following lemma will allow us to argue in Section 5 that certain sequences of actions whi we will construct by extending behaviors of a universal FIFO physical channel are themselves fa behaviors of the channel. The intuition behind this result is that after any history, every packet n yet delivered might be lost, and then subsequent activity of the channel would be as if it restart in an initial state.

Lemma 2.2 *Let C be a universal physical channel for n and P. Let β be a finite behavior of and γ any fair behavior of C. Then $\beta\gamma$ is a fair behavior of C.*

3 The Data Link Layer

The data link layer is the second lowest layer in the hierarchy, and is implemented using the servic of the physical layer. Generally, it is implemented in terms of two physical channels, one in eac direction. It provides a reliable one-hop message delivery service, which can in turn be used by tl next higher layer.

In this section, we specify a weak form of data link behavior. We again assume that there a two endpoints, a "transmitting station" and a "receiving station". The data link layer receiv messages from the higher layer at the transmitting station, and delivers them at the receivin station. The data link layer guarantees that every message that is sent is eventually received. Tl order of the messages need not be preserved, however. We give a parameterized specification f data link layer behavior[1] denoted DL^M, where M is an alphabet of legal *messages.* The interfa is described by $sig(DL^M)$, which is given formally as follows.

Input actions:
    $send\_msg(m)$, $m \in M$
Output actions:
    $rec\_msg(m)$, $m \in M$

The $send\_msg(m)$ action represents the sending of message m on the data link by the transmi ting station, and the $rec\_msg(m)$ represents the receipt of message m by the receiving station. W will refer to these actions as *data link layer actions.* In order to define the set of allowed interactio $scheds(DL^M)$, we again define a collection of auxiliary properties. They are defined for a sequen $\beta = \pi_1\pi_2\ldots$ of data link layer actions and a correspondence relationship, a binary relation betwee the *send_msg* events and the *rec_msg* events in β. The first property is analogous to (PL1) a gives elementary requirements on the correspondence.

(DL1)
1. If an event $\pi_i = rec\_msg(m)$ corresponds to an event $\pi_j = send\_msg(n)$, then $m = n$ and also $j < i$, that is, the event π_j precedes π_i in β.
2. Each $rec\_msg(m)$ event corresponds to exactly one $send\_msg(m)$ event.
3. Each $send\_msg(m)$ event corresponds to *at most* one $rec\_msg(m)$ event.

The remaining property is the data link layer liveness property. It says that all messages th are sent are eventually delivered. This property expresses the reliability of the message deliver guaranteed by the data link layer.

---

[1] A stronger form of data link behavior, including a FIFO message delivery property, is described in [9]. Whi the specification in this paper is less interesting for describing properties of a useful data link layer, it is adequa for proving our impossibility result. A similar impossibility result for the stronger data link specification followi immediately from ours.

L2) If π is a *send_msg*(m) event occurring in β, then there is a *rec_msg*(m) event in β corresponding to π.

Note that (DL1) and (DL2) together imply that there is exactly one *rec_msg*(m) event corre- onding to each *send_msg*(m) event. Now we define *scheds*(DL^M) to be the set of sequences β of ta link layer actions for which there exists a correspondence relation such that (DL1) and (DL2) satisfied for β and the correspondence relation.

We have the following immediate consequence of the definition:

mma 3.1 *If β is in scheds(DL^M) then the number of send_msg events in β is equal to the mber of rec_msg events in β.*

Data Link Implementation

this section, we define a "data link protocol", which is intended to be used to implement the ta link layer using the services provided by the physical layer. A data link protocol consists two automata, one at the transmitting station and one at the receiving station. These au- mata communicate with each other using two physical channels, one in each direction. They also mmunicate with the outside world, through the data link layer actions we defined in Section 3.

Let t and r again be names (for the transmitting and receiving station respectively). Let M, P^{tr} d P^{rt} be alphabets (of messages, forward packets and backwards packets, respectively). Then *transmitting automaton* for (t, r) and (M, P^{tr}, P^{rt}) is any I/O automaton having the following ternal action signature.

put actions:
- *send_msg*(m), $m \in M$
- *rec_pkt*$^{rt}(p)$, $p \in P^{rt}$

itput actions:
- *send_pkt*$^{tr}(p)$, $p \in P^{tr}$

addition, there can be any number of internal actions. That is, a transmitting automaton receives quests from the environment of the data link layer to send messages to the receiving station r. It nds packets to r over the physical channel to r. It also receives packets over the physical channel om r. Similarly, a *receiving automaton* for (t, r) and (M, P^{tr}, P^{rt}) is any I/O automaton having e following external signature.

put actions:
- *rec_pkt*$^{tr}(p)$, $p \in P^{tr}$

utput actions:
- *send_pkt*$^{rt}(p)$, $p \in P^{rt}$
- *rec_msg*(m), $m \in M$

gain, there can also be any number of internal actions. That is, a receiving automaton receives ckets over the physical channel from t. It sends packets to t over the physical channel to t, and delivers messages to the environment of the data link layer. A *data link protocol* for (t, r) and $I, P^{tr}, P^{rt})$ is a pair (A^t, A^r), where A^t is a transmitting automaton for (t, r) and (M, P^{tr}, P^{rt}), d A^r is a receiving automaton for (t, r) and (M, P^{tr}, P^{rt}). (Often we will omit mention of the ation names and the alphabets, if these are clear from context.)

Now we are ready to define correctness of data link protocols. Informally, we say that a da link protocol is "correct" provided that when it is composed with any "correct physical laye (i.e. a pair of FIFO physical channels from t to r and from r to t, respectively), the resulti system yields correct data link layer behavior. This reflects the fundamental idea of layering, th the implementation of one layer should not depend on the details of the implementation of oth layers, so that each layer can be implemented and maintained independently. Formally, suppo (A^t, A^r) is a data link protocol for (t, r) and (M, P^{tr}, P^{rt}). We say that (A^t, A^r) is *correct* provid that the following is true. For all C^{tr} and C^{rt} such that C^{tr} is a FIFO physical channel for and P^{tr}, and C^{rt} is a FIFO physical channel for rt and P^{rt}, and for every sequence β that is t projection on the actions of data link layer actions of a fair behavior of the composition of A^t, A C^{tr} and C^{rt}, it is the case that β is in *scheds*(DL^M).

The definition of correctness for a data link protocol requires us to examine its behavior wh combined with any possible FIFO physical channels. However, examining the definition sho that we are able to prove the impossibility of a correct protocol satisfying certain requirements merely demonstrating that no such protocol works when combined with a specific pair of physic channels. In fact we will do our impossibility proofs by considering a system constructed with a arbitrarily chosen universal FIFO physical channels.[2]

If $A = (A^t, A^r)$ is a data link protocol for (t, r) and (M, P^{tr}, P^{rt}), C^{tr} is a universal FIF physical channel for tr and P^{tr}, and C^{rt} is a universal FIFO physical channel for rt and P^{rt}, th we denote by $D(A)$ an automaton that is the composition of A^t, A^r, C^{tr} and C^{rt}. Then by virt of Theorem A.4, we have the following result:

Lemma 4.1 *A data link protocol A is correct if and only if for every fair behavior β of $D(A)$, t projection of β on the data link layer actions is an element of scheds(DL^M).*

Corollary 4.2 *Suppose that A is a correct data link protocol. Then in every fair behavior of $D(A$ the number of send_msg events is equal to the number of rec_msg events.*

Proof: This is immediate from Lemma 4.1 and Lemma 3.1. ∎

5 Impossibility of Having All Packets Identical

We show here the impossibility of constructing a correct data link protocol that uses only a sing type of packet (and so needs no header) to transmit a sequence of identical messages. This resu seems weaker than the result we want (that it is impossible to construct a correct data link protoc where all packets contain the same header) but in Section 6 we will show that in fact the desire result follows from this. By making this simplification we postpone some of the difficult modelin issues, and allow the reader to see the style of impossibility proof in a simpler setting. The techniqu of measuring the number of headers by the size of the packet alphabet when the message alphab has size one, was used earlier without formal proof of a reduction result in [13].

The proof takes the following form: we assume for the purpose of obtaining a contradictio that A is a correct data link protocol in which, in each direction, all packets are identical. The Corollary 4.2 implies that, in every fair behavior of $D(A)$, the number of *send_msg* events is equ to the number of *rec_msg* events. Then in Lemmas 5.1 - 5.3, we deduce a series of three prelimina facts about the states of the end stations during executions of $D(A)$, by showing that the failu of one of these facts, coupled with the previously derived facts, would enable us to construct a fa

[2] We do not use the fact in this paper, but we note that since the universal FIFO physical channels have a possible behaviors of any FIFO physical channels, we can prove that a particular protocol is correct by analyzing in a system with just universal FIFO physical channels.

havior in which the number of *send_msg* events is unequal to the number of *rec_msg* events. nally, we use these facts in Lemma 5.4 to construct two fair behaviors of $D(A)$ with identical ojections at the receiving automaton, but in one of which two messages are sent while in the other ly one message is sent. Since the projections at the receiver are equal, the two executions contain e same number of *rec_msg* events. Thus one of them will have the number of *send_msg* events equal to the number of *rec_msg* events. This yields a contradiction to the original assumption at the protocol was correct.

1 Preliminary Lemmas

this subsection, we assume that A is a correct data link protocol for (t, r) and (M, P^{tr}, P^{rt}) with $^{tr}| = |P^{rt}| = 1$. We also assume that each of A^t and A^r is deterministic, that is, it has only one tial state, at most one locally controlled action is enabled in each state, and at most one new ate can be reached by applying an action in a state. As we will see later, this involves no loss of nerality. Given a state of an end station (i.e., A^t or A^r) in such a protocol A, there is a *unique* r execution fragment of that automaton that commences with the given state and includes no put actions. (This execution corresponds to running the automaton from the given state in such way that it receives no inputs, for as long as it can keep taking steps.) We will say that the given ate is *quiescing* if this fragment contains only finitely many *send_pkt* events.

Our first lemma says that from any state in any execution, if the transmitting automaton is n without receiving any inputs (that is, with no *send_msg* or *rec_pkt*^{rt} events) then it must send ly finitely many packets to the receiver.

mma 5.1 *If α is a finite execution of $D(A)$, then A^t is quiescing in the final state of α.*

oof: The idea of the proof is as follows. If A^t sends infinitely many packets with no response en A^r has no hope of determining how many *send_msg* events have happened. In particular, we ow that A^r cannot tell the difference between the situation in which one additional *send_msg* ent occurs after the given finite execution and the situation in which no such events occur.

More precisely, suppose that A^t is not quiescing in the final state of α. Let $\beta = beh(\alpha)$. Then nsider the behavior $\beta send\_msg(m)$ where m is some arbitrary message in the message alphabet A. This behavior has an extension that is fair and contains no extra *send_msg* events (by eorem A.1). By Corollary 4.2, there is a finite prefix of this extension, say $\beta send\_msg(m)\beta'$ ich contains as many *rec_msg* events as there are *send_msg* events in $\beta send\_msg(m)$, namely, e more than the number of *send_msg* events in α.

Let k be the number of *rec_pkt*^{tr} events in β'. Since A^t is not quiescing in the final state of α, ere is a finite behavior of A^t that is an extension of $\beta|A^t$, say $(\beta|A^t)\gamma$, where γ consists of exactly *send_pkt*^{tr} events (but no *send_msg* or *rec_pkt*^{rt} events).

Now we consider the sequence $\beta\gamma(\beta'|A^r)$ of actions of $D(A)$. We show that this sequence a (not necessarily fair) behavior of $D(A)$, by showing that its projection on each of the four mponents of the system is a behavior of that component.

1. The projection on A^t is $(\beta|A^t)\gamma$, which is a behavior of A^t by construction.

2. The projection on A^r is $(\beta\beta')|A^r$ (since γ involves only actions of A^t). This is a behavior of A^r since it is equal to $(\beta send\_msg(m)\beta')|A^r$.

3. The projection on C^{tr} is $(\beta|C^{tr})(\gamma|C^{tr})((\beta'|A^r)|C^{tr})$. Since $\gamma|C^{tr}$ is a sequence of k *send_pkt*$^{tr}(p)$ events and $(\beta'|A^r)|C^{tr}$ is a sequence of k *rec_pkt*$^{tr}(p)$ events, where p is the unique element of the packet alphabet P^{tr}, it follows from the universality of C^{tr} that $(\gamma|C^{tr})((\beta'|A^r)|C^{tr})$ is a fair behavior of C^{tr}. Then Lemma 2.2 implies that $(\beta|C^{tr})(\gamma|C^{tr})((\beta'|A^r)|C^{tr})$ is a behavior of C^{tr}.

4. The projection on C^{rt} is $(\beta|C^{rt})((\beta'|A^r)|C^{rt})$ (since γ involves no actions of C^{rt}) and this a behavior of C^{rt} since $(\beta'|A^r)|C^{rt}$ consists only of $send\_pkt^{rt}$ events, which are inputs to C (and I/O automata are input-enabled).

Since its projection on each component is a behavior of that component, the sequence $\beta\gamma(\beta'|A$ is a behavior of $D(A)$, by Theorem A.4.

Now consider the two behaviors $\beta\gamma(\beta'|A^r)$ and $(\beta)send\_msg(m)\beta'$. They both have the sar projection on A^r and hence contain the same number of *rec_msg* events. By the argument at t beginning of this proof, this number is exactly one more than the number of *send_msg* events β. On the other hand, the number of *send_msg* events in $\beta\gamma(\beta'|A^r)$ is the same as the number *send_msg* events in $\beta\gamma$ (the two sequences having the same projection on A^t), and this is the san as the number of *send_msg* events in β. Thus, the number of *send_msg* events in the behavi $\beta\gamma(\beta'|A^r)$ is one fewer than the number of *rec_msg* events in the same behavior $\beta\gamma(\beta'|A^r)$.

Now when we consider a fair extension of $\beta\gamma(\beta'|A^r)$ that contains no further *send_msg* event as given by Theorem A.1, we find that it contains more *rec_msg* events than *send_msg* event This contradicts Corollary 4.2.

Our second lemma says that from any state in any execution, if the receiving automaton is r without receiving any inputs, then it must send infinitely many packets to the transmitter.

Lemma 5.2 *If α is a finite execution of $D(A)$, then A^r is not quiescing in the final state of α.*

Proof: Suppose the contrary: that A^r is quiescing in the final state. Once again, we reach contradiction by constructing two fair behaviors of $D(A)$ with the same number of *rec_msg* even but different numbers of *send_msg* events.

Let α_1 be the fair execution fragment of A^t containing no inputs and starting from the sta of A^t at the end of α. Similarly let α_2 be the fair execution fragment of A^r containing no inpu and starting from the state of A^r at the end of α. By definition, the projection of $\alpha\alpha_1$ on A^t is fair execution of A^t, and likewise the projection of $\alpha\alpha_2$ on A^r is a fair execution of A^r. By Lemn 5.1, α_1 contains only finitely many $send\_pkt^{tr}$ events, and by assumption α_2 contains only finite many $send\_pkt^{rt}$ events. Let γ be any sequence of actions formed by interleaving the sequenc $beh(\alpha_1)$ and $beh(\alpha_2)$.

We claim that $beh(\alpha)\gamma$ is a fair behavior of $D(A)$. We show this by showing that its projecti on each of the four components of the system is a fair behavior of that component.

1. The projection on A^t is just $beh((\alpha|A^t)\alpha_1)$, which is a fair behavior of A^t by the definition α_1.

2. The projection on A^r is $(beh(\alpha|A^r)\alpha_2)$ which is a fair behavior of A^r by the definition of c

3. The projection on C^{tr} is just the projection of beh(α) on that channel followed by a fini number of $send\_pkt^{tr}$ events. This is a fair behavior of C^{tr} by the universality of C^{tr} a Lemma 2.2.

4. Similarly the projection on C^{rt} is a fair behavior of C^{rt}.

Since its projection on each component is a fair behavior of that component, Theorem A implies that $beh(\alpha)\gamma$ is a fair behavior of $D(A)$. By Corollary 4.2 the number of *rec_msg* events $beh(\alpha)\gamma$ equals the number of *send_msg* events in the same sequence.

Now let m be an arbitrary element of the message alphabet. We construct another fair behavi $beh(\alpha)send\_msg(m)\gamma'$, which contains the same number of *rec_msg* events as in $beh(\alpha)\gamma$, b contains one more *send_msg* event, which yields a contradiction.

Let α_3 be the fair execution fragment of A^t containing no inputs and starting from the state A^t at the end of α*send_msg*(m). By Lemma 5.1, α_3 contains only finitely many *send_pkt*^{tr} nts. Now we consider the sequence of actions γ' formed by interleaving (in any fashion) the uences $beh(\alpha_3)$ and $beh(\alpha_2)$. Just as above, $beh(\alpha)$*send_msg*$(m)\gamma'$ is a fair behavior of $D(A)$, l so by Corollary 4.2, the number of *rec_msg* events in $beh(\alpha)$*send_msg*$(m)\gamma'$ is equal to the nber of *send_msg* events. However, since $beh(\alpha)$*send_msg*$(m)\gamma'|A^r$ and $beh(\alpha)\gamma|A^r$ are both al to $beh(\alpha|A^r)beh(\alpha_2)$, we see that the number of *rec_msg* events in $beh(\alpha)$*send_msg*$(m)\gamma'$ is same as the number of *rec_msg* events in $beh(\alpha)\gamma$. By the equalities proved above, $beh(\alpha)\gamma$ l $beh(\alpha)$*send_msg*$(m)\gamma'$ contain the same number of *send_msg* events, which is false. This is a tradiction. ■

The third lemma further characterizes the behavior of a correct data link protocol by showing t the transmitter must both send and receive infinitely many packets.

mma 5.3 *If α is a fair execution of $D(A)$ that contains a finite nonzero number of send_msg ions, then $\alpha|A^t$ contains infinitely many send_pkt^{tr} actions and infinitely many rec_pkt^{rt} actions.*

oof: We show that every other possibility leads to a contradiction.

1. $\alpha|A^t$ contains infinitely many *send_pkt*^{tr} actions and finitely many *rec_pkt*^{rt} actions. Then there is a suffix of $\alpha|A^t$ that contains no input actions (neither *send_msg* nor *rec_pkt*^{rt} actions) but contains infinitely many *send_pkt*^{tr} actions. The state of A^t at the start of this suffix must be not quiescing, which contradicts Lemma 5.1.

2. $\alpha|A^t$ contains finitely many *send_pkt*^{tr} actions and finitely many *rec_pkt*^{rt} actions. Then $\alpha|A^r$ contains finitely many *rec_pkt*^{tr} actions (since the channel C^{tr} delivers at most as many packets as were sent) and contains finitely many *send_pkt*^{rt} actions (since a fair execution of C^{rt} would contain an infinite number of *rec_pkt*^{rt} events if it contained an infinite number of *send_pkt*^{rt} events). Thus there is a suffix of $\alpha|A^r$ that contains no input events and only a finite number of *send_pkt*^{rt} events. The state of A^r at the start of this suffix is quiescing, which contradicts Lemma 5.2.

3. $\alpha|A^t$ contains finitely many *send_pkt*^{tr} actions and infinitely many *rec_pkt*^{rt} actions. First consider the maximal execution of A^r starting from the initial state of A^r and containing no input actions. This execution is a fair execution of A^r. Let β be the behavior of this execution. By Lemma 5.2, β contains infinitely many *send_pkt*^{rt} actions. Let γ be the sequence of actions obtained by interleaving β and $beh(\alpha|A^t)$ in such a way that for each i the i-th *rec_pkt*^{rt} action is immediately preceded by the i-th *send_pkt*^{rt} action. We claim that γ is a fair behavior of $D(A)$. Its projections on A^t and A^r are fair behaviors by construction. Its projection on C^{rt} is just *send_pkt*$^{rt}(p)$*rec_pkt*$^{rt}(p)$*send_pkt*$^{rt}(p)$*rec_pkt*$^{rt}(p)\ldots$ (where $P^{rt} = \{p\}$) which is a fair behavior since (PL1)-(PL3) are satisfied using the obvious correspondence between each *rec_pkt*^{rt} event and the immediately preceding *send_pkt*^{rt} event. Finally, its projection on C^{tr} is a fair behavior since it consists of the sending of a finite number of packets and the delivery of none (as β contains no inputs to A^r, in particular no *rec_pkt*^{tr} actions) and this clearly satisfies (PL1)-(PL3).

   We observe that β (and hence γ) cannot contain any *rec_msg* action. (Otherwise, take the prefix of β up to and including the first *rec_msg* event, regard it as a behavior of $D(A)$ where all the actions take place at A^r, and extend it to a fair behavior of $D(A)$ which contains no inputs to $D(A)$, that is, no *send_msg* actions. This contradicts Corollary 4.2.) However, γ contains a nonzero number of *send_msg* events (the same ones as in α). Thus γ is a fair behavior of $D(A)$ which does not contain the same number of *rec_msg* events as of *send_msg* events, contradicting Corollary 4.2.

5.2 The Theorem

We now use the facts proved in the previous subsection to construct two executions that lo identical to the receiver, but have different numbers of messages sent at the transmitting end.

Lemma 5.4 *Suppose that A is a correct data link protocol for (t, r) and (M, P^{tr}, P^{rt}) with $|P^{tr}|$ $|P^{rt}| = 1$, and suppose that each of A^t and A^r is deterministic.*

Let m be an arbitrary element of the message alphabet. Let β_1 and β_2 be two fair behavic of $D(A)$ such that β_1 begins with send_msg(m) and contains no other send_msg event, and begins with send_msg(m)send_msg(m) and contains no other send_msg event. Then there ex fair behaviors $\hat{\beta}_1$ and $\hat{\beta}_2$ of $D(A)$ such that $\hat{\beta}_j|A^t = \beta_j|A^t$ for $j = 1, 2$, and such that $\hat{\beta}_1|A^r = \hat{\beta}_2|A$

Proof: Applying Lemma 5.3 to the behavior β_1, we see that we may express $\beta_1|A^t$ as an infini sequence

$$send\_msg(m)\beta_1^1\gamma_1^1\beta_1^2\gamma_1^2\beta_1^3\ldots$$

where each β_1^i consists only of $send\_pkt^{tr}$ actions, each γ_1^i consists only of rec_pkt^{rt} actions, a where each β_1^i (except possibly β_1^1) and each γ_1^i contains a finite, nonzero number of $send\_pkt^{tr}$ rec_pkt^{rt} events. [3] Similarly, we may express $\beta_2|A^t$ as an infinite sequence

$$send\_msg(m)send\_msg(m)\beta_2^1\gamma_2^1\beta_2^2\gamma_2^2\beta_2^3\ldots$$

where each β_2^i consists only of $send\_pkt^{tr}$ actions, each γ_2^i consists only of rec_pkt^{rt} actions, a where each β_2^i (except possibly β_2^1) and each γ_2^i contains a finite, nonzero number of $send\_pkt^{tr}$ rec_pkt^{rt} events. Let a_j^i denote the number of rec_pkt^{rt} events in γ_j^i, for $j = 1, 2$.

We next claim that there exists an infinite fair behavior of A^r of the following form:

$$\eta^1 rec\_pkt^{tr}(p)\eta^2 rec\_pkt^{tr}(p)\eta^3 rec\_pkt^{tr}(p)\ldots$$

(where p is the unique element of P^{tr}), such that each η^i is a finite sequence containing exact $\max(a_1^i, a_2^i)$ $send\_pkt^{rt}$ events and no input events of A^r. In order to prove this claim, we first sho inductively on i, that $\eta^1 rec\_pkt^{tr}(p)\eta^2 rec\_pkt^{tr}(p)\eta^3 rec\_pkt^{tr}(p)\ldots\eta^i rec\_pkt^{tr}(p)$ is a behavior A^r; this follows because any state of A^r after $\eta^1 rec\_pkt^{tr}(p)\eta^2 rec\_pkt^{tr}(p)\eta^3 rec\_pkt^{tr}(p)$. $\eta^i rec\_pkt^{tr}(p)$ is not quiescent according to Lemma 5.2, and $rec\_pkt^{tr}(p)$ is an input action of We then observe that since A^r is deterministic and the given infinite sequence contains infinite many locally controlled events of A^r, any execution with this behavior is a fair execution, so t behavior is a fair behavior.

Now we define $\hat{\beta}_1$ to be the infinite sequence

$$send\_msg(m)\beta_1^1\eta^1\gamma_1^1\beta_1^2 rec\_pkt^{tr}(p)\ldots\eta^{i-1}\gamma_1^{i-1}\beta_1^i rec\_pkt^{tr}(p)\ldots$$

and similarly define $\hat{\beta}_2$ to be the infinite sequence

$$send\_msg(m)send\_msg(m)\beta_2^1\eta^1\gamma_2^1\beta_2^2 rec\_pkt^{tr}(p)\ldots\eta^{i-1}\gamma_2^{i-1}\beta_2^i rec\_pkt^{tr}(p)\ldots$$

We claim that these sequences have all the required properties.

First we see that $\hat{\beta}_1|A^t = send\_msg(m)\beta_1^1\gamma_1^1\beta_1^2\ldots\gamma_1^{i-1}\beta_1^i\ldots$, since η^i and $rec\_pkt^{tr}(p)$ co sist entirely of actions of A^r. Thus $\hat{\beta}_1|A^t = \beta_1|A^t$. Similarly, $\hat{\beta}_2|A^t = \beta_2|A^t$. Also, $\hat{\beta}_1|A^r$ $\eta^1 rec\_pkt^{tr}(p)\ldots\eta^{i-1}rec\_pkt^{tr}(p)\ldots$ since $send\_msg(m)$, β_1^i and γ_1^i consist entirely of events of Also, $\hat{\beta}_2|A^r = \eta^1 rec\_pkt^{tr}(p)\ldots\eta^{i-1}rec\_pkt^{tr}(p)\ldots$. Thus, $\hat{\beta}_1|A^r = \hat{\beta}_2|A^r$.

We show that $\hat{\beta}_1$ is a fair behavior of $D(A)$ by examining its projection on each component.

---

[3] The exception is due to the fact that we do not know whether the first packet sent by A^t precedes or follows first packet received by A^t.

1. The projection on A^t is equal to $\beta_1|A^t$ as we observed above, which is a fair behavior of A^t by Theorem A.3 since β_1 is a fair behavior of $D(A)$.

2. The projection on A^r is equal to $\eta^1 rec\_pkt^{tr}(p) \ldots \eta^{i-1} rec\_pkt^{tr}(p) \ldots$ as observed above, which was shown in the earlier claim to be a fair behavior of A^r.

3. The projection on C^{tr} is equal to $\beta_1^1 \beta_1^2 rec\_pkt^{tr}(p) \ldots \beta_1^i rec\_pkt^{tr}(p) \ldots$, i.e., a finite (possibly zero length) sequence of $send\_pkt^{tr}(p)$ events followed by an infinite sequence of segments, each consisting of a finite nonzero number of $send\_pkt^{tr}(p)$ events, followed by one $rec\_pkt^{tr}(p)$ event; this is because none of $send\_msg(m)$, η^i or γ_1^i contains any $send\_pkt^{tr}$ or rec_pkt^{tr} events. It is clear that a correspondence can be found for which this satisfies (PL1)-(PL3). Since C^{tr} is a universal FIFO physical channel, this is a fair behavior of C^{tr}.

4. The projection on C^{rt} is equal to $\eta^1 \gamma_1^1 \ldots \eta^{i-1} \gamma_1^{i-1} \ldots$, i.e., a sequence of $\max(a_1^1, a_2^1)$ $send\_pkt^{rt}(p')$ events followed by a_1^1 $rec\_pkt^{rt}(p')$ events, then $\max(a_1^2, a_2^2)$ $send\_pkt^{rt}(p')$ events followed by a_1^2 $rec\_pkt^{rt}(p')$ events, and so on, where p' is the unique element of P^{rt}. Since each a_1^i is nonzero but finite, it is clear that a correspondence can be found for which this sequence satisfies (PL1)-(PL3), and so this is a fair behavior of C^{rt}.

Now by Theorem A.4 this shows that $\hat{\beta}_1$ is a fair behavior of $D(A)$. By exactly similar argu-ents, $\hat{\beta}_2$ is a fair behavior of $D(A)$, which completes the proof. ∎

We can now put the pieces together to prove our impossibility result for packet alphabets of ;e 1.

heorem 5.5 *There is no correct data link protocol for (t, r) and (M, P^{tr}, P^{rt}) with $|P^{tr}| = |P^{rt}| =$*

roof: Suppose for the purpose of reaching a contradiction that A is a correct data link protocol th $|P^{tr}| = |P^{rt}| = 1$.

First, we deal with the potential nondeterminism of the end-stations. By Theorem A.2, there a deterministic automaton B^t (respectively, B^r) with fair behaviors that are a subset of the ir behaviors of A^t (respectively, A^r). Let $B = (B^t, B^r)$, which is also a data link protocol with $^{tr}| = |P^{rt}| = 1$. Now by Theorems A.3 and A.4, $fairbehs(D(B)) \subseteq fairbehs(D(A))$, and so B correct (using Lemma 4.1 and the fact that A is correct).

Now, let m be an arbitrary element of the message alphabet. By Theorem A.1, there are ir behaviors β_1 and β_2 of $D(B)$ such that β_1 begins with $send\_msg(m)$ and contains no other *nd_msg* event, and β_2 begins with $send\_msg(m) send\_msg(m)$ and contains no other *send_msg* ent. Consider the fair behaviors $\hat{\beta}_1$ and $\hat{\beta}_2$ whose existence is shown in Lemma 5.4. Since the otocol B is correct, each $\hat{\beta}_j$ satisfies Corollary 4.2 (that is, the number of *rec_msg* events in $\hat{\beta}_j$ uals the number of *send_msg* events in $\hat{\beta}_j$). Now the number of *send_msg* events in $\hat{\beta}_j$ is just the umber of *send_msg* events in $\hat{\beta}_j|B^t$ which equals the number of *send_msg* events in $\beta_j|B^t$ by the operties of $\hat{\beta}_j$. Since $\beta_j|B^t$ contains j *send_msg* events, we deduce that $\hat{\beta}_j$ contains j *rec_msg* ents, contradicting the fact that $\hat{\beta}_1|B^r$ and $\hat{\beta}_2|B^r$ are equal, and so contain the same number of *c_msg* events. ∎

Impossibility of Having No Headers

ost data link protocols in the literature use a finite packet alphabet in each direction, since packets e required to be of limited size. However, it is normally the case that the packets are treated as ving two separate parts: a header (which determines what is to be done with the packet) and an encapsulated message (treated as an uninterpreted bit string). Indeed, one can envisage protocc that allow packets of unbounded size because the included messages may have unbounded size, a yet use only a fixed size of header (and thus a finite number of headers). Here we sketch one way which one can model the existence of headers in a protocol, without assuming that the packets a necessarily structured explicitly with two parts, one a control field and the other an uninterpret message. We then show, using a reduction, how our impossibility result for identical packets impli a corresponding impossibility result for the case of infinite packet alphabets without headers.

We model the "headers" used by a protocol as follows. Let $A = (A^t, A^r)$ be a data link protoc for (t, r) and (M, P^{tr}, P^{rt}). Let $\equiv$ be an equivalence relation on the domain $M \cup P^{tr} \cup P^{rt}$ $states(A^t) \cup states(A^r) \cup acts(A^t) \cup acts(A^r)$. Then $\equiv$ is said to be a *header relation* for A provid that the following conditions hold.

1. $\equiv$ only relates elements of the same kind, i.e., elements of M, or P^{tr}, or $states(A^t)$, etc. Als a start state cannot be related to a non-start state. Moreover, if $a \equiv a'$ for two actions and a', then a and a' are identical except possibly for a difference in their message or pack parameter. Further, every pair a and a' of locally controlled events of A^t (respectively, of A such that $a \equiv a'$, a and a' are in the same class of $part(A^t)$ (respectively, of $part(A^r)$).
2. For each pair m, m' of messages in M, $send\_msg(m) \equiv send\_msg(m')$ if and only if $m \equiv m$ and $rec\_msg(m) \equiv rec\_msg(m')$ if and only if $m \equiv m'$.
3. For each pair p,p' of packets in P^{tr}, $send\_pkt^{tr}(p) \equiv send\_pkt^{tr}(p')$ if and only if $p \equiv p'$, a $rec\_pkt^{tr}(p) \equiv rec\_pkt^{tr}(p')$ if and only if $p \equiv p'$.
4. For each pair p,p' of packets in P^{rt}, $send\_pkt^{rt}(p) \equiv send\_pkt^{rt}(p')$ if and only if $p \equiv p'$, a $rec\_pkt^{rt}(p) \equiv rec\_pkt^{rt}(p')$ if and only if $p \equiv p'$.
5. For every two states q and q' of A^t (respectively, of A^r) with $q \equiv q'$, if action a is enabled q then there is an action a' with $a \equiv a'$, such that a' is enabled in q'.
6. For every two states q and q' of A^t (respectively, of A^r) and every two actions a and a' of (respectively, of A^r) such that $q \equiv q'$ and $a \equiv a'$, if r is a state such that (q, a, r) is a step A^t (respectively, of A^r) and action a' is enabled in state q', then there exists a state r' su that $r \equiv r'$ and (q', a', r') is a step of A^t (respectively, of A^r).

For a data link protocol A for (t, r) and (M, P^{tr}, P^{rt}) with a header relation $\equiv$, we define the s $headers(A, t, r, \equiv)$ to be the set of equivalence classes of packets in P^{tr}. Similarly $headers(A, r, t, \equiv$ is the set of equivalence classes of packets in P^{rt}. We think of each equivalence class of packets being those (in one direction) with the same pattern of bits in the header. Informally, the way packet is processed must depend only on the header – for example, if receiving a packet takes t protocol to a state where release of a message to the higher layer is possible, then receiving a other packet containing the same header will also take the protocol to a state where release of message to the higher layer is possible (however, it may be a different message that is released We note that for a data link protocol A, the diagonal relation, where each message, action etc. equivalent only to itself, is a header relation for A. We say that A has *no header under* $\equiv$ if each $headers(A, t, r, \equiv)$ and $headers(A, r, t, \equiv)$ is a singleton set, that is, all packets in P^{tr} are relate by $\equiv$, as are all packets in P^{rt}. We say that A has *no header* if there exists a header relation $\equiv$ f A such that A has no header under $\equiv$.

In order to prove that headers are necessary for a data link protocol, we show how to redu the question of the existence of a protocol with sets of header equivalence classes of a given siz to the question of the existence of a protocol using packet alphabets of that size. This will allo us to show that there is no correct data link protocol that has no header using our earlier resu that there is no correct data link protocol with packet alphabets of size one. The intuition behir is reduction comes from the case where packets have the simple form (header, message) and the otocol works uniformly over all message alphabets. In this case, if the protocol is applied to a essage alphabet of size one, the packet alphabet will be identical to the set of possible headers. Of urse, the proof must be more complicated than this, since we do not assume any simple structure packets.

eorem 6.1 *Suppose* $A = (A^t, A^r)$ *is a correct data link protocol for* (t, r) *and* (M, P^{tr}, P^{rt})*. If is a header relation for A such that* $|headers(A, t, r, \equiv)| = h_1$ *and* $|headers(A, r, t, \equiv)| = h_2$*, then ere are alphabets* M', Q^{tr} *and* Q^{rt} *with* $|M'| = 1$, $|Q^{tr}| = h_1$ *and* $|Q^{rt}| = h_2$ *and a correct data k protocol* $B = (B^t, B^r)$ *for* (t, r) *and* (M', Q^{tr}, Q^{rt})*.*

oof: Choose m to be an arbitrary element of M and put $M' = \{m\}$, $Q^{tr} = headers(A, t, r, \equiv)$ d $Q^{rt} = headers(A, r, t, \equiv)$. [4] These alphabets clearly have the correct cardinalities. Now let B^t the transmitting automaton for (t, r) and (M', Q^{tr}, Q^{rt}) defined as follows. The input actions B^t are $send\_msg(m)$ and $rec\_pkt^{rt}(p')$ where p' is an element of Q^{rt}, the output actions are $nd\_pkt^{tr}(p')$ where p' is an element of Q^{tr}, and the internal actions are the internal actions of . We say that an action π' of B^t *represents* an action π of A exactly when one of the following nditions holds:

- π' is either $send\_msg(m)$ or an internal action of A^t and $\pi = \pi'$
- π' is $send\_pkt^{tr}(p')$ and $\pi = send\_pkt^{tr}(p)$ for some p which is an element of p'
- π' is $rec\_pkt^{rt}(p')$ and $\pi = rec\_pkt^{rt}(p)$ for some p which is an element of p'.

e states and start states of B^t are the same as those of A^t. The transition relation of B^t includes $, \pi', s)$ exactly when there exists some π that is represented by π' for which $(s', \pi, s) \in steps(A^t)$. e partition $part(B^t)$ relates locally controlled actions π_1' and π_2' exactly when $part(A^t)$ relates ne (and hence all) pairs π_1 and π_2 such that π_1 is represented by π_1' and π_2 is represented by π_2'.

Similarly, let B^r be the receiving automaton for (t, r) and (M', Q^{tr}, Q^{rt}) defined as follows. e input actions of B^r are $rec\_pkt^{tr}(p')$ where p' is an element of Q^{tr}, the output actions are $c\_msg(m)$ and $send\_pkt^{rt}(p')$ where p' is an element of Q^{rt}, and the internal actions are the ernal actions of A^r. We say that an action π of A^r *is represented by* an action π' of B^r exactly en one of the following conditions holds:

- π' is either $rec\_msg(m)$ or an internal action of A^r and $\pi = \pi'$
- π' is $send\_pkt^{rt}(p')$ and $\pi = send\_pkt^{rt}(p)$ for some p which is an element of p'
- π' is $rec\_pkt^{tr}(p')$ and $\pi = rec\_pkt^{tr}(p)$ for some p which is an element of p'.

e states and start states of B^r are the same as those of A^r. The transition relation of B^r includes $, \pi', s)$ exactly when there exists some π that is represented by π' for which $(s', \pi, s) \in steps(A^r)$. e partition $part(B^r)$ relates locally controlled actions π_1' and π_2' exactly when $part(A^r)$ relates ne (and hence all) pairs π_1 and π_2 such that π_1 is represented by π_1' and π_2 is represented by π_2'.

It is easy to check that (B^t, B^r) is a data link protocol for (t, r) and (M', Q^{tr}, Q^{rt}). We claim at it is correct, proving the theorem.

To prove the claim, it suffices to take an arbitrary fair behavior of $D(B)$ and produce a corre- ondence for its projection on the data link layer actions such that (DL1) and (DL2) are satisfied. r this we use the specific universal FIFO physical channels $C^{tr,P^{tr}}$ and $C^{rt,P^{rt}}$ whose construc- n is given in Appendix B, rather than arbitrary ones as we have done previously. The universal

[4] Thus, each packet name in the alphabet Q^{tr} is a set of packet names from the alphabet P^{tr}.

channel used clearly does not affect the set of behaviors of $D(A)$, but it will make it easier to rela executions of different systems, since the construction we give has no internal actions and wor uniformly for different packet alphabets.

Thus we consider an arbitrary fair execution $s'_0, \pi'_1, s'_1, \pi'_2, s'_2, \ldots$ of $D(B)$. From this we c construct an execution $s_0, \pi_1, s_1, \pi_2, s_2, \ldots$ of $D(A)$ such that π_i is represented by π'_i for each i, t state of A^t (respectively, of A^r) in s_i is the same as the state of B^t (respectively, of B^r) in s'_i, a the state of $C^{tr,P^{tr}}$ (respectively, of $C^{rt,P^{rt}}$) in s_i is related to the state of $C^{tr,Q^{tr}}$ (respectively, $C^{rt,Q^{rt}}$) in s'_i in the natural way: the values for the variables S, $counter_1$ and $counter_2$ are the sar in $C^{tr,P^{tr}}$ (respectively, in $C^{rt,P^{rt}}$) as in $C^{tr,Q^{tr}}$ (respectively, in $C^{rt,Q^{rt}}$), and for each n the value $packet(n)$ in $C^{tr,P^{tr}}$ (respectively, in $C^{rt,P^{rt}}$) is one element of its value in $C^{tr,Q^{tr}}$ (respectively, $C^{rt,Q^{rt}}$) except when both values are undefined.[5] This execution is in fact a fair execution of $D(A$ as is seen by observing that no action $rec\_msg(m')$ for $m' \neq m$ is enabled in any state s_i (using t correctness of A and the fact that no action π_j is $send\_msg(m')$), and that therefore if a local controlled action of $D(A)$ is enabled in s_i then it is represented by a locally controlled action of $D($ that is enabled in s'_i. Since this execution is fair, its behavior has projection on the data link lay actions that is an element of $scheds(DL^M)$. However the two executions have identical projectio on the data link layer actions (the actions differ only for *send_pkt* and *receive_pkt* events, whi are not included in the projection). Thus for $s'_0, \pi'_1, s'_1, \pi'_2, s'_2, \ldots$, its projection on the data li layer actions has a correspondence between *send_msg* and *rec_msg* events that satisfies (DL1) a (DL2). Thus by Theorem 4.1, B is correct.

Theorem 6.2 *There is no correct data link protocol that has no header.*

Proof: Immediate from Theorems 5.5 and 6.1.

7 Conclusions

In this paper, we have proved the impossibility of constructing a reliable data link service usi an unreliable physical channel service, where the packets sent over the physical channels carry header information. This result was proved using the I/O automaton formal model for concurre systems.

Some remarks are in order about the style of the proof. The proof is carried out entirely the "semantic level", i.e., in terms of set-theoretic concepts such as states and sequences of actio Many formal models (including I/O automata) come equipped with a syntactic component design to facilitate protocol description and verification. It appears that such syntactic structure is of litt value (in fact, may get in the way) for carrying out impossibility proofs. We also note that t proof (as all other impossibility proofs so far discovered) is designed to be understood by peop not by machines. Because of their complexity, we see little evidence that interesting impossibili proofs can (or should) be automated. This is in contrast to the situation for proofs of protoc correctness.

The proof is not an simple as one might expect from the simple and natural statement of t result. It would be desirable to polish and simplify the proof further. But in carrying out such simplification, one must be careful not to gloss over any of the genuinely subtle issues (e.g., t treatment of fairness or the control of actions) that arise in the proof.

It remains to consider how the same proof can be carried out, or the same result can be prov in other ways, using other popular formal models for concurrency. We suggest this as a challen for users of those other models. It seems to us that key issues that must be addressed in doing t are those of fairness and control of actions.

[5] The inductive construction of the execution $s_0, \pi_1, s_1, \ldots$ is straightforward.

Finally, we wish to reiterate our belief that the theory of concurrency should be developed in vay that uses the same underlying formal model as a basis for both verification work and for possibility proofs (and more broadly, for other combinatorial and complexity-theoretic results).

:knowledgements

: thank Yishay Mansour for his help in our initial discussions about this work, and also Steve nzio and Isaac Saias for their useful suggestions for improving the presentation.

eferences

] Attiya, H., Fischer, M., Wang, D.-W., and Zuck, L., "Reliable Communication Using Unreliable Channels", manuscript.

] Bartlett, K., Scantlebury, R., and Wilkinson, P., "A Note on Reliable Full-Duplex Transmission over Half-Duplex Links" *Communications of the ACM*, 12(5):260-261, May 1969.

] Bloom, B., "Constructing Two-Writer Atomic Registers" *Proceedings of 6th ACM Symposium on Principles of Distributed Computing*, pp. 249-259, August 1987.

] Chou, C.-T., and Gafni, E., "Understanding and Verifying Distributed Algorithms Using Stratified Decomposition" *Proceedings of 7th ACM Symposium on Principles of Distributed Computing*, pp. 44-65, August 1988.

] Fekete, A., Lynch, N., and Shrira, L., "A Modular Proof of Correctness for a Network Synchronizer" *Proceedings of the 2nd International Workshop on Distributed Algorithms*, Amsterdam, Netherlands, July 1987, (J. van Leeuwen, ed), pp. 219-256. Lecture Notes in Computer Science 312, Springer-Verlag.

] Fekete, A., Lynch, N. A., Mansour, Y. and Spinelli, J., "The Data Link Layer: The Impossibility of Implementation in Face of Crashes", Technical Memo, TM-355b, Laboratory for Computer Science, Massachusetts Institute of Technology.

] Fekete, A, Lynch, N., Merritt, M., and Weihl, W., "Commutativity-Based Locking for Nested Transactions" to appear in JCSS.

] Lynch, N., and Goldman, K., "Distributed Algorithms" Research Seminar Series MIT/LCS/RSS-5, Laboratory for Computer Science, Massachusetts Institute of Technology, Cambridge, MA, 1989.

] Lynch, N. A., Mansour, Y. and Fekete, A., "Data Link Layer: Two Impossibility Results," *Proceedings of 7th ACM Symposium on Principles of Distributed Computing*, pp. 149-170, August 1988.

] Lynch N. A. and Tuttle M. R., "Hierarchical Correctness Proofs for Distributed Algorithms," *Proceedings of the 6th ACM Symposium on Principles of Distributed Computing*, pp. 137-151, August 1987.

] Lynch, N., and Tuttle, M., "An Introduction to Input/Output Automata" *CWI Quarterly*, 2(3):219-246, September 1989.

] Lynch, N., "A Hundred Impossibility Proofs for Distributed Computing", *Proceedings of 8th ACM Symposium on Principles of Distributed Computing*, pp. 1-28, August 1989.

[13] Mansour, Y., and Schieber, B., "The Intractability of Bounded Protocols for non-FIFO Cha nels" *Proceedings of 8th ACM Symposium on Principles of Distributed Computing*, pp. 59-7 August 1989.

[14] Nipkow, T., "Proof Transformations for Equational Theories" *Proceedings of 5th Annual IE Symposium on Logic in Computer Science*, pp. 278-288, June 1990.

[15] Stenning, N., "A Data Transfer Protocol" *Computer Networks*, 1:99-110, 1976.

[16] Troxel, G., "A Hierarchical Proof of an Algorithm for Deadlock Recovery in a System usi Remote Procedure Calls" MS thesis, Massachusetts Institute of Technology, Department Electrical Engineering and Computer Science Cambridge, MA. January, 1990.

[17] Welch, J., Lamport, L., and Lynch, N., "A Lattice-Structured Proof of a Minimum Spanni tree Algorithm" *Proceedings of 7th ACM Symposium on Principles of Distributed Computi* pp. 28-43, August 1988.

[18] Wang, D.-W., and Zuck, L., "Tight Bounds for the Sequence Transmission Problem", *Pr ceedings of 8th ACM Symposium on Principles of Distributed Computing*, pp. 73-84, Augu 1989.

A The I/O Automaton Model

The *input/output automaton* model was defined in [10] as a tool for modeling concurrent a distributed systems. We refer the reader to [10] and to the expository paper [11] for a comple development of the model, plus motivation and examples. Here, we mention two extensions to t definitions of [10] and [11], and also provide a summary of the theorems about I/O automata th are needed for our results.

We extend the definition of fairness to execution fragments, not just executions. Namely, execution fragment α of an automaton A is said to be *fair* if the following condition holds for ea class C of $part(A)$: if α is finite, then no action of C is enabled in the final state of α, while i is infinite, then either α contains infinitely many events from C, or else α contains infinitely ma occurrences of states in which no action of C is enabled. Thus, a fair execution fragment gives "f turns" to each class of $part(A)$.

An I/O automaton is said to be *deterministic* if it has only one initial state, at most one loca controlled action is enabled in each state, and at most one new state can be reached by applyi an action in a state.

The following theorem says that no matter what has happened in any finite execution, and matter what inputs continue to arrive from the environment, an automaton can continue to ta steps to give a fair execution.

Theorem A.1 *Let A be an I/O automaton and let γ be a sequence of input actions of A. Supp that β is a finite schedule of A. Then there exists a fair schedule β' of A such that β' is an extens of β and $\beta'|in(A) = (\beta|in(A))\gamma$. Moreover, the same is true for behaviors in place of schedules.*

The following theorem shows that there is no loss of generality in restricting attention to terministic solutions to specifications.

Theorem A.2 *If A is an I/O automaton, then there is a deterministic I/O automaton B s that $fairbehs(B) \subseteq fairbehs(A)$.*

The remaining theorems relate executions, schedules and behaviors of a composition to those the automata being composed.

heorem A.3 *Let $\{A_i\}_{i\in I}$ be a strongly compatible collection of automata, and let $A = \Pi_{i\in I}A_i$. $\alpha \in execs(A)$ then $\alpha|A_i \in execs(A_i)$ for all $i \in I$. Moreover, the same result holds for fairexecs, heds, fairscheds, behs and fairbehs in place of execs.*

heorem A.4 *Let $\{A_i\}_{i\in I}$ be a strongly compatible collection of automata, and let $A = \Pi_{i\in I}A_i$. t β be a sequence of actions in acts(A). If $\beta|A_i \in scheds(A_i)$ for all $i \in I$, then $\beta \in scheds(A)$. oreover, the same result holds for fairscheds, behs and fairbehs in place of scheds.*

The Universal FIFO Physical Channels

re we show how to construct the universal FIFO physical channels whose existence we claimed Lemma 2.1.

First, we define a set S of ordered pairs (i,j) of positive integers to be a *delivery set* provided at it satisfies the following two conditions: for each positive integer j, S includes a unique element $j)$, and for each positive integer i, it includes at most one element (i,j). We say that a delivery is *monotone* provided there are no pairs (i_1,j_1) and (i_2,j_2) in S with $i_1 < i_2$ and $j_1 \geq j_2$.

The state of the physical channel $C^{n,P}$ has two counters, $counter_1$ and $counter_2$, an infinite notone delivery set S of pairs of positive integers, and a partial mapping *packet* from the set positive integers to P. The counter $counter_1$ represents the number of *send_pkt* actions, and $unter_2$ represents the number of *receive_pkt* actions, that have occurred so far. The set S termines which packets are delivered, and in what order – it contains pairs (i,j) that correlate j-th *receive_pkt* event with the i-th *send_pkt* event. Thus the restrictions in the definition of lelivery set correspond to the conditions (PL1) and (PL2). The mapping *packet* associates with integer i the packet that was sent in the i-th *send_pkt* event. Initially $counter_1$ and $counter_2$ zero and *packet* is undefined everywhere. In a particular execution, the set S is initialized to arbitrary monotone delivery set and remains fixed.

The transition relation for the automaton $C^{n,P}$ consists of all triples (s',π,s) described by the lowing code.[6]

$nd\_pkt^n(p)$
fect: $counter_1 \leftarrow counter_1 + 1$
$\quad packet(counter_1) \leftarrow p$

$c\_pkt^n(p)$
econdition: $packet(i) = p$ and $(i, counter_2 + 1) \in S$, for some i
fect: $counter_2 \leftarrow counter_2 + 1$

The partition puts all the output actions of $C^{n,P}$ (that is, all the $rec\_pkt^n(p)$ actions, for all $\in P$) in a single class.

mma B.1 *The automaton $C^{n,P}$ is a universal FIFO physical channel.*

[6] This style of describing I/O automata by giving preconditions (that is, conditions on s') and effects (that is, peratives to be executed sequentially to transform s' to give s) is used in [11]. It is not fundamental to the model, is rather a notational convenience for describing sets of triples.

A Temporal Approach to Algebraic Specifications *

Yulin Feng † Junbo Liu

FB3 Mathematik und Informatik
Universität Bremen
Postfach 330 440, D-2800 Bremen 33
F.R. Germany

Usenet: {feng,liu}%informatik.Uni-Bremen.de

Abstract. **This paper is contributed to make connections between models for algebraic and temporal specifications. It brings a different viewpoint for classical algebras and algebraic specifications. Every algebra we concern here is finitely generated and associated with an implicit transition structure. The operators in the algebra may be partially defined. The class of algebras could be used as Kripke semantic models to interpret the temporals, so that we can do temporal reasoning about system behaviours such as safety and liveness properties. The unification of notions in algebraic and temporal specifications has many advantages for system developments. We may use a formal temporal deduction system to verify some dynamic properties from premises of algebraic specifications; or a temporal requirement specification may be used to develop systems in the style of top-down refinements. The notion of C-algebras has been crucial all along this work. In this paper we present the concepts, definitions and some basic theorems on C-algebras. Moreover, there exists a minimally defined algebra which is the initial one for each partially defined specification. The example of a lift controller is finally used to illustrate how to reason about the temporal behaviours of an algebraic specification.**

1. Introduction

Every approach to algebraic specifications is based on some logical systems to make descriptions really precise and to allow computers to perform interpretations. The classical approach to algebraic specifications is based on equational logic, which has been proved very successful for specifying systems in abstraction. However, a major drawback of this approach has been its limited capacity to deal with systems involving concurrence. A large number of approaches (cf. [1,2,4,11,13,16-18]) has been taken so far to extend traditional algebraic framework to cope with this problem. The value of temporal logic for specifying concurrent systems is also well appreciated (cf. [5,8,15,19]). In order to reason about dynamic behaviours of systems, it seems useful to combine techniques from both algebraic and temporal approaches.

---

* This work is partly supported by the Commission of the European Communities under the ESPRIT Programme, Contract #390, PROSPECTRA Project.

† On leave from University of Science & Technology of China, Hefei, Anhui 230026, China.

t is essential that the temporal and algebraic specifications are based on different semantic models. To combine classical algebraic specifications with techniques in temporal ogic, it must be clarified how these models can be unified. Our purpose in this paper is ry to answer the question and to integrate algebraic specifications into a temporal framework, so that we can do temporal reasoning about their behaviour both with respect to afety and liveness properties.

Some aspects characterizing our approach are listed below.

- We take a different view of algebraic specifications. Each object in an algebra is onsidered as a temporal entity that evolves by some update operators.

- We consider hierarchically defined specifications. Each signature can only define ne new sort, while other sorts involved are considered to be predefined.

- As the notion of partial operators is of practical importance for system specifications, we consider a class of partial algebras as the semantics of concerning specifications with a set of positive conditional equations. A definition of existential equality is sed to interpret equality for partially defined terms.

- Algebraic specifications are integrated into a temporal framework. The logic used ere is first order branching time temporal logic. The Kripke semantics of the logic is defined with respect to a transition structure. The class of algebras we concern here is alled C-algebras. For a partially defined specification, there exists a unique (up to isomorphism) C-algebra which is initial in the class of all of C-algebras satisfying equations n the specification.

- Emphasis is put on the abstract data type of which the system evolves. It is essential that the specification is property-oriented, only to define the necessary properties n system states, while to leave rooms for different implementations. It may be implemented in a system either in a sequential, or in a concurrent system. From a methodological viewpoint, the unification of algebraic and temporal specifications may bring the advantages for system development in two aspects. We may use a formal temporal deduction system to verify some dynamic conclusions from premises of algebraic specifications; n the other side, a temporal requirement specification may be used to develop systems n the style of top-down refinements.

he authors consider the paper as a first approach to combine the different approaches lgebraic and temporal) for system specification. The layout of the paper is as follows. ection 2 is a brief introduction to temporal logic. Section 3 is the main part of the paper. is devoted to the concepts, definitions and some theorems on C-algebras. To illustrate e temporal reasoning from an algebraic specification, we give the lift example in section . We in the last section conclude with some remarks and indicate some problems and e further research directions.

. Temporal Logic

emporal logic was originally developed by philosophers for reasoning about the ordering of events in time without introducing time explicitly. As the proper sequencing of events is crucial to the correctness of systems, temporal logic has most successfully been used in computer science, especially for specifying concurrent systems.

Based on different semantic models, we distinguish between linear-time and branch-time temporal logic. The logic used in this paper is a kind of first-order temporal logic in branch time, which we call Computation Tree Logic(CTL for short).

The CTL language is an extension of first order language by adding some of generic modal operators: A(all paths), E(some path), X(next) and U(until). Moreover, we assume the set of variables to be partitioned into two subsets which we call the set of global and local variables, respectively. Terms in CTL language are defined in the same way as that in first order logic. The unique form of atomic formulas is t1 = t2, where t1, t2 are terms.

In defining well formed CTL formulas, we distinguish between state and path formulas. The state formulas are interpreted in states from a Kripke structure; in contrast, the path formulas are interpreted over linear sequences of states through paths.

State formulas

- Every atomic formula t1 = t2 is a state formula, where t1, t2 are terms.
- If f1 and f2 are state formulas, so are f1 & f2 and ~f1.
- If g is a path formula, then Ag and Eg are state formulas.

Path formulas

- If f1 and f2 are state formulas, then Xf1 and U[f1,f2] are path formulas.

The Kripke semantics of CTL language is defined with respect to a model

$$M = (S, I, R, W)$$

where S is a non-empty set of state and R is a relation on S. We assume R total on S. This assumption does not restrict the strength of results in this paper. Each state in S is a local variable valuation with respect to the domain, i.e. an assignment of an element of |S| to every local variable. W is the set of path with respect to S and R. Typically for instance W may be constructed by actions from processes in a concurrent system. The number of states in S may be finite or infinite.

From a given state s_0, a path from the state s_0 in W is an infinite sequence of states

$$\sigma: s_0 s_1 \dots s_i s_{i+1} \dots$$

such that for every $i \geq 0$, $(s_i, s_{i+1}) \in R$. In the following we use σ^i to denote the suffix of σ starting at s_i, σ_i the ith state on the path, counting from zero. The component W in the model is a set of all of possible paths on S with respect to R.

Finally, I is a global assignment with respect to S, i.e. an assignment of the domain of S to every global variable; an assignment of a function I(N) to every function symbol N.

or a given model M, the term evaluation could be recursively defined by

$M^{(I,si)}(x) = I(x)$ for each global variable x

$M^{(I,si)}(y) = si(y)$ for each local variable y

$M^{(I,si)}(N(t1, ..., tn)) = I(N)(M^{(I,si)}(t1), ..., M^{(I,si)}(tn))$

Ve use the standard notation to indicate that a formula valid in a model. By M,s |= f we ean that the formula f holds at the state s of the model. Similarly, if g is a path formula, 1, σ |= g means that g holds along the path σ in the model. Let f be a state formula. hen we mean by M |= f that f is valid in M, i.e. for all of s $\in$ S, M,s |= f. For a class C of odels, f is valid in C if M |= f, for every model M in C. Assuming that f1 and f2 are ate formulas and g1, g2 are path formulas, the relation |= can be defined inductively as llows.

or state formulas:

M,s |= t1=t2	iff	$M^{(I,s)}(t1) = M^{(I,s)}(t2)$
M,s |= ~f	iff	not M,s |= f
M,s |= f1$\wedge$f2	iff	M,s |= f1 and M,s |= f2
M,s |= Ag	iff	for all paths σ starting from s such that M,σ |= g
M,s |= Eg	iff	for some path σ starting from s such that M,σ |= g

or path formulas:

M,σ |= Xf	iff	M,σ_1 |= f
M,σ |= U[f1,f2]	iff	there exists k ≥ 0 such that M,σ_k |= f2 and for all $0 \leq j < k$, M,σ_j |= f1

/e take over all conventions and definitions concerning further logical operators, paren- eses and abbreviations. For example, we define

AF f $\leftrightarrow$ AU[true, f] -- f inevitably holds

EF f $\leftrightarrow$ EU[true, f] -- f potentially holds

AG f $\leftrightarrow$ ~EF ~f -- f holds globally

EG f $\leftrightarrow$ ~AF ~f -- f holds on all states along some path

n axiomatic deduction system of CTL has been established and used for verification of TL formulas in various application ([7], [8]).

. C-Algebra

is a different point of view for algebras and algebraic specifications that we distinguish gebraic constraints between two levels: at the level of generation and at the level of mporal evolution of attributes. Each object in an algebra is considered as a temporal entity that evolves by some update operators, while preserving their abstract identity of sort . The formalism presented here extends the algebraic framework to deal with temporal behaviours, without losing the advantages of the conventional algebraic approach.

3.1 Signature and Algebra

A signature SIG=(S,Σ) consists of a family S of sorts and a family Σ of operator names together with their arity definitions. As we consider hierarchically defined specifications, each signature can only define a new sort, while other sorts involved are considered to be predefined. When we define the new sort and new operators in a signature, we consider those sorts and operators in predefined signatures as primitives.

The operators in a signature are divided into three groups: origin operators stand for initial objects; update operators stand for changes in objects, and finally, attribute operators stand for properties of objects which are used to determine the partiality of update operators. Syntactically,

let s_0 be the new sort and $s_1, s_2, ..., s_n, s_{n+1}$ are predefined sorts, every origin or update operator is in the form: $N\ s_1\ s_2\ ...\ s_n$, which means in classical algebraic signature that $N: s_1 \times s_2 \times ... \times s_n \times s_0 \to s_0$ for origin operators and for update operators.

The attribute operators are in the form

$$N \to s_1$$

which means classically that

$$N: s_0 \to s_1.$$

The definitions of terms in SIG=(S,Σ) are as usual in conventional algebraic specifications. It is evident that any attribute operators can only be applied after origin or update operators.

For example we have a signature of stack as

```
stack = alphabet +
    sort      stack
    origin    empty
    gopns     push alphabet
              pop
    popns     top → alphabet
```

Let s be a variable over stack, then for every $x \in$ alphabet,

(push x): empty
pop: (push x): s
top: (push x): pop: s

are well formed terms in $T_{S,\Sigma}(X)$, X={x,s}.

otice: We use the Currying denotation and put : between operators to emphasize our oncern of the main sort whenever we write terms in this paper.

he intended meaning of symbol names in a signature can be denoted by algebras. An al-ebra consists of a family of sets and related operators and usually is heterogeneous or any-sorted.

efinition 3.1.1 (SIG-algebra) a SIG-Algebra consists of

(i) a sort indexed family of sets, i.e. $\{ A_s \mid s \in S\}$,

(ii) functions for operators in Σ . Let an operator N be with arity $s_1 \times s_2 \times \cdots \times s_n \rightarrow$ s_{n+1} in the classical sense, then we have in the algebra a function N^A: $As_1 \times A\, s_2 \times \cdots$ $As_n \rightarrow As_{n+1}$, where N^A may be a partial function.

efinition 3.1.2 (C-algebra) SIG-algebra A is called C-algebra, if it is associated with relation $R \subseteq A_s \times A_s$ where s is the main sort in S and R is the union $\cup R_N^{As}$ for all ori-in and update operators in Σ such that for each operator $N\ s_1\ s_2 \cdots s_n$ and the relation $R_N^{As} \subseteq A_s \times A_s$, $(a_1,a_2) \in R_N^{As}$ if and only if $a_1, a_2 \in A_s$ and $N^A(a_1, t_1, t_2, \cdots, t_n) = a_2$ or some $t_1 \in As_1, t_2 \in As_2, \cdots, t_n \in As_n$.

It is easy to see that a C-algebra with the defined relation R can be considered as a ansition structure used to interpret semantic of temporal modal operators. From a tem-oral point of view, each finitely generated object in the algebra is considered as a tem-oral entity evolving by update operators. The occurrence of update operators is the only ause of change in the state of object. The state of object at each instant depends on the ace of all update operators that have already happened in its life time. The attribute op-rators are used to check the attribute properties of objects. Therefore, C-algebras can e used as semantic models of CTL. In this way we bring algebras into a unified temporal amework, so that we can do temporal reasoning about their behavior both with respect safety and liveness properties.

.2 Initial Semantics

o characterize operator behaviours in partial algebra, we have to concern the defined-ess of terms. We introduce a special semantic predicate *D* to express definedness. et $T_{S,\Sigma}$ be the ground term algebra on SIG(S,Σ). For a given C-algebra A, $A \models D(t)$, t $T_{S,\Sigma}$ iff the interpretation of t in A is defined , therefore, term t has $eval_A(t)$ in A s its interpretation.

here are two possibilities to interpret equations on partially defined terms.

- Strong equality

$A \models (t1 = t2)$ iff $(A \models \sim D(t1)\ \&\ A \models \sim D(t2))$ or
$(A \models D(t1)\ \&\ A \models D(t2)\ \&\ t1^A = t2^A)$

- Existential equality

$A \models (t1 = t2)$ iff $A \models D(t1)\ \&\ A \models D(t2)\ \&\ t1^A = t2^A$

Both interpretations are essentially equivalent in their expressive power. To avoid tha the tedious definedness predicate clutter the specifications, we prefer to use the existential equality.

Also there are different kinds of homomorphism for partial algebras defined in the literature. The definition of homomorphism used here for C-algebras is total. Let A, B be two algebras on SIG =(S, Σ), φ is a homomorphism between A and B, i.e. $\varphi = \{ \varphi_s : As \rightarrow Bs\}$ if for all operator N with arity $s_1 \times s_2 \times \cdots \times s_n \rightarrow s$ in the signature and all $t_1 \in As_1, t_2 \in As_2, \cdots, t_n \in As_n$, then

$N^A (t_1, t_2, \cdots, t_n)$ defined => $N^B (\varphi(t_1), \varphi(t_2), \cdots, \varphi(t_n))$ defined and

$$\varphi (N^A(t_1, t_2, \cdots, t_n)) = N^B(\varphi(t_1), \varphi(t_2), \cdots, \varphi(t_n))$$

Definition 3.2.1 (Syntax of Specification) A partially defined specification PSPEC(S,Σ, E) consists of a signature SIG=(S,Σ) and a set E of positive conditional equations, in which every conditional equation is in the form

$$\wedge_{i=0}^{m} (pi = qi) \rightarrow t1 = t2$$

where pi, qi, t1, t2 are terms, $\wedge_{i=0}^{m}\ (pi = qi)$ $(m \geq 0)$ is the conditional part of the equation.

Moreover, for each specification a set DE of axioms and deduction rules can be established to determine definedness of ground terms as follows:

(i) for each equation of E in the form $\wedge_{i=0}^{m}\ (pi = qi) \rightarrow t1 = t2$ and an arbitrary substitution $\sigma: X \rightarrow T_{S,\Sigma}$, we have

$$\wedge_{i=0}^{m} (D(pi\sigma)\ \&\ D(qi\sigma)\ \&\ pi\sigma = qi\sigma) \vdash D(t1\sigma)\ \&\ D(t2\sigma)\ \&\ t1\sigma = t2\sigma$$

In particular for t1 = t2 in E, we have $D(t1\sigma)\ \&\ D(t2\sigma)\ \&\ t1\sigma = t2\sigma$.

(ii) for each term $t \in T_{S,\Sigma}$, $D(t) \vdash D(t')$ where t' is a subterm of t.

Given a partial specification PSPEC(S,Σ,E) we use **Palg** (PSPEC) to denote the all C-algebras that satisfy E:

Palg (PSPEC) = {A : A is C-algebra on (S,Σ) and A |= E}

ertainly not all C-algebras are equally interesting to our approach; we will concentrate n initial algebra in the sense of "no junk" and "no confusion", and, moreover, partiality ere motivates another requirement, that is minimal definedness with respect to initial al- ebraic semantics. In the following we construct such an algebra and prove it to be the itial object in **Palg** (PSPEC).

efinition 3.2.2 (Partial Term Algebra) Let $PT_{S,\Sigma}$ be a partial term algebra on (S, Σ) efined as follows:

(i) for each sort s in S we define $PT_s = \{ t : t \in T_{s,\Sigma}$ and **DE** |- D(t) $\}$.

(ii) for each operator N with arity $s_1 \times s_2 \times \cdots \times s_n \to s$ in the signature and all

$_1 \in PTs_1, t_2 \in PTs_2, \cdots, t_n \in PTs_n$, if DE |- $N(t_1, t_2, \cdots, t_n)$, then

$$N^{PT}(t_1, t_2, \cdots, t_n) = N(t_1, t_2, \cdots, t_n)$$

emma 3.2.3 $PT_{S,\Sigma}$ is a subalgebra of $T_{S,\Sigma}$.

he proof is obvious by definition of subalgebra (in the partial sense).

efinition 3.2.4 (Induced Congruence Relation) For a given set of equations E, the rela- on $\equiv_E$ over ground terms is inductively defined on $PT_{S,\Sigma}$ as follows: Assume that σ is n arbitrary substitution : $X \to PT_{S,\Sigma}$

(1) for t1 = t2 in E, we have $t1\sigma \equiv_E t2\sigma$

(2) for $\wedge_{i=0}^{m}$ (pi = qi) → t1 = t2 in E, if for all i, $pi\ \sigma \equiv_E qi\ \sigma$ then $t1\sigma \equiv_E t2\sigma$

(3) $t \equiv_E t$

(4) if $t1 \equiv_E t2$ then $t2 \equiv_E t1$

(5) if $t1 \equiv_E t2$ and $t2 \equiv_E t3$ then $t1 \equiv_E t3$

(6) for all operator N with arity $s_1 \times s_2 \times \cdots \times s_n \to s$ on (S,Σ) and all $t1 \in PTs_1$, $\in PTs_2, \cdots, tn \in PTs_n$, if $t1 \equiv_E t1' \ldots tn \equiv_E tn'$ and **DE** |- D(N(t1,...tn)), then (t1,...tn) $\equiv_E$ N(t1',...tn')

efine the quotient algebra on $PT_{S,\Sigma}$ w.r.t. $\equiv_E$, let

$$QPT_{S,\Sigma} = PT_{S,\Sigma} / \equiv_E ,$$

e call $QPT_{S,\Sigma}$ the partial quotient algebra w.r.t. PSPEC.

emma 3.2.5 The partial quotient algebra $QPT_{S,\Sigma}$ is a PSPEC-algebra, i.e.

$$QPT_{S,\Sigma} \in \textbf{Palg}\ (\text{PSPEC})$$

Proof. We need to show that $QPT_{S,\Sigma}$ satisfies all equations in E . Assume an equation in E of the form

$$\wedge_{i=0}^{m} (pi = qi) \rightarrow t1 = t2.$$

For simplicity, let the substitution $\sigma : X \rightarrow QPT_{S,\Sigma}$ have a general form [u] / s , where s is the main sort and $u \in PT_{S,\Sigma}$ and assume for all i , pi σ = qi σ , i.e. pi([u]) = qi([u]). So pi(u) $\equiv_E$ qi(u) . By the definition of $\equiv_E$, t1(u) $\equiv_E$ t2(u). Hence t1([u]) = t2([u]), i.e. t1σ = t2σ .

#

From the definitions and lemmas above, we obtain our intended algebra which is the initial object in **Palg** (PSPEC). We have the following theorem.

Theorem 3.2.6 $QPT_{S,\Sigma}$ is initial in **Palg** (PSPEC), i.e. for every A $\in$ **Palg** (PSPEC) , there is a unique homomorphism $\varphi : QPT_{S,\Sigma} \rightarrow A$.

Proof. To show $QPT_{S,\Sigma}$ is initial in **Palg** (PSPEC) , for A $\in$ **Palg** (PSPEC) and $t \in PT_{S,\Sigma}$ let

$$\varphi([t]) = eval_A(t) .$$

Since DE |- D(t) , so A |= D(t), t is defined in A and $eval_A(t)$ is some object t^A in A. For all t' $\in$ [t], i.e. t $\equiv_E$ t' . By the definition of evaluation of terms, we conclude that $eval_A(t) = eval_A(t')$. Hence φ is well defined.

To see that φ is a homomorphism from $QPT_{S,\Sigma}$ to A , for each operator N with arity $s_1 \times s_2 \times \cdots \times s_n \rightarrow s$ and $ti \in PT_{S,\Sigma}$ with the sort si , if N^Q ([t1], [t2], ..., [tn]) is defined , then we have

$$\begin{aligned} \varphi (N^Q([t1], [t2], ..., [tn]) &= \varphi ([N(t1, t2, ..., tn)]) \\ &= eval_A(N(t1, t2, ..., tn)) \\ &= N^A(eval_A(t1), ..., eval_A(tn)) \\ &= N^A(\varphi(t1), \varphi(t2), ..., \varphi(tn)). \end{aligned}$$

So φ is a homomorphism .

Finally we come to prove the uniqueness of φ . Assume there is another homomorphism $h : QPT_{S,\Sigma} \rightarrow A$, by structural induction in a similar way as that in classical algebraic approach it is easy to show that $\varphi = h$. #

The initial semantics of PSPEC is then defined as all algebras isomorphic to initial alge-a, i.e.

$$\textbf{Init}\ (\textbf{Palg}\ (\text{PSPEC})) = \{A : A \in \textbf{Palg}\ (\text{PSPEC})\ ,\ A \cong QPT_{S,\Sigma}\}$$

efinition 3.2.7 (Semantics of Specification) Let PSPEC(S, Σ, E) be a specification, then e define its semantic as

$$\textbf{Sem}(\text{PSPEC}) = \textbf{Init}\ (\textbf{Palg}\ (\text{PSPEC}))$$

As we explained before, the specification of C-algebra has a special form in the sense at each time there is only one mainly concerned sort. To clarify this idea on the specifi- ation level we use the concept of hierarchy.

efinition 3.2.8 (Hierarchical Specifications) Let HT = (S, Σ, E, HT') be a given specifi- ation with HT' = PSPEC(S', Σ', E'). HT is called hierarchical specification if

(i) HT is partially complete with respect to HT', i.e. for all $t \in T_{S,\Sigma}$ with sort s ∈ ',

HT |- D(t) => exist $t' \in T_{S',\Sigma'}$ and HT |- t' = t

(ii) HT is hierarchical consistent, i.e. for all $t, t' \in T_{S',\Sigma'}$

HT |- t = t' => HT' |- t = t'

orollary 3.2.9 Let HT be a hierarchical specification. Then there exists an initial C-al- ebra for HT.

.3 Temporal Reasoning for C-algebras

or a given signature SIG=(S,Σ) , $L_{S,\Sigma}$ is a CTL language with all symbols of variables, unctions and their arity definitions being the same as that in the signature. Therefore ach conditional equation in PSPEC(S,Σ,E) is a well formed temporal formula, it is in ct a state formula in $L_{S,\Sigma}$. Any equation provable from E in equational logic is valid all C-algebras of **Palg** (PSPEC). Moreover, as an temporal entity relation is assigned C-algebras, it is hopeful that we can conclude more than purely equational reasoning. rom the premises E, further temporal properties, especially liveness properties, can be roved in a formal system of temporal logic when they are expressed in temporal formu- s. Generally we have

heorem 3.3.1 Given a specification PSPEC(S,Σ,E) and a temporal formula f in $L_{S,\Sigma}$, if is provable from E in a formal system of temporal logic, then f is valid in all C-algebras **Palg** (PSPEC).

y the consistency of the formal proving system in temporal logic, the conclusion is evi- ent.

heorem 3.3.2 Given a specification PSPEC(S,Σ,E), an equation is provable from E in quational logic, if and only if it is provable from E in a formal system of temporal logic.

Proof. Let e be an equation in E. If e is provable from E in a formal proving system c temporal logic, by Th. 3.3.1, the equation e is valid in all C-algebras. According to equa tional theory, it is provable from E in equational logic.

#

It is a natural question to ask whether a general temporal formula f is provable from E i a temporal calculus provided that f is valid in all C-algebras in **Palg** (PSPEC). Thi problem is still open.

4. The Lift Example

To illustrate how to express about temporal behaviours of an algebraic specification, her we give a non trivial example, a system controller for a passenger lift. The lift controller i specified by an algebraic specification. All essential aspects of a lift behaviour, except th case of time delay and emergency, are specified in the specification. We assume that th lift system is intended to serve a building with an arbitrary but fixed number of floors. Th lift has one button for each floor, and each floor has separate buttons to request ascend ing and descending the lift. Also, in the lift there is a button to open or to re-open it door. The lift movements are responses to user demands, a passenger can tell the sys tem that he wants it to visit a certain level by pressing a corresponding button and the lif will do so sooner or later. Various strategies can be imagined how to fulfil the ridin wishes, however, some of basic requirements should be guaranteed.

We assume some predefined specifications **bool, dir, level** and **level_set** etc. The ne sort we are going to define is **state** that characterizes the system states generated b the lift movements. In a reality, a lift operator can only perform in some states that satis fy pre-conditions of the operator. In other words, these operators are partial operators All of possible system states with partially defined operators constitutes a partial alge bra.

We now in the following give a specification to specify **state** in the lift system. Every C algebra which satisfied with the specification is also satisfied with those temporal proper ties derived from the specification in a unified framework of temporal logic. We expec the given specification readable and understandable, there is no need for more explana tions here.

state = bool + dir + level + level_set +

sort state
origin start
gopns move_up_step
move_down_step
get_calling level_set level_set level_set
popns direction → dir /* dir = {up, down} */
current_floor → level
ups → level_set /* the levels asked for up */
downs → level_set /* the levels asked for down */

eqns

direction: start = up
current_floor: start = 0
ups: start = empty
downs: start = empty

for s ∈ state,
s1∈ state :: dierction:s = up and ups:s =/= epmty
or direction:s = down and ups:s =/= epmty and downs:s = empty,
s2∈ state :: dierction:s = up and ups:s =/= epmty
or direction:s = down and ups:s =/= epmty and downs:s = empty,
x ∈ level , uu,vv,ww ∈ level_set ==>

direction: get_calling(uu,vv,ww):s = direction : s
current_floor: get_calling(uu,vv,ww):s = current_floor: s
ups:get_calling(uu,vv,ww):s = ups:s ∪(filter (>current_floor:s) uu ∪vv∪ww)
downs: get_calling(uu,vv,ww): = downs:s ∪(filter (current_floor:s>) uu ∪vv∪ww)

direction: move_up_step:s1 = up
direction: move_down_step:s2 = down
current_floor: move_up_step:s1 = current_floor:s + 1
current_floor: move_down_step:s2 = current_floor:s - 1
ups: move_up_step:s1= sub(ups:s1, current_floor:s + 1)
ups: move_down_step:s2= empty
downs: move_down_step:s2 = sub(downs:s1, current_floor:s - 1)
downs: move_up_step:s1 = empty

It is natural that each lift controller must satisfy the following requirements:

- If a passenger ask to serve at a certain level, the lift will be eventually arrive in that level.

- To optimize the lift movements, the lift changes of its directions only if in the old direction there is no any requests.

These requirements could be expressed by temporal formulas in $L_{S,\Sigma}$ such that

(i) x in (ups:s ∪ downs:s) →
AF(current_floor:s = x)

(ii) direction:s = up →
AU[direction:s = up, (ups:s) = empty]

direction:s = down →
AU[direction:s = down, (downs:s) = empty]

These formulas could be proved from **eqns** of the lift specification in a temporal deduction
system, according to Th. 2.2.1, they are valid in all C-algebras in Alg(state). Every lift

control system will be satisfiable with these requirements, so long as it satisfies with all of equations in PSPEC(**state**).

We omit all proof here, for more precise treatment please refer to [7] [8] .

#

5. Concluding Remarks

The notion of C-algebra has been crucial all along this work. It brings a different point of view for algebraic specifications. In fact, when we consider an abstract data type in a programming language, what we concern is the class of data which are reachable from a primitive data in the type through applications of admissible operators. It is evident that such a class of data is finitely generated and has an implicit transition structure. In this sense, the notion of C-algebras is natural and reasonable . It can be used as Kripke semantic models to interpret the temporals. In this way we integrate algebraic specification into a temporal framework, so that we can do temporal reasoning about system behaviours both in safety and liveness properties.

We have presented the concepts, definitions and some basic theorems on C-algebras As the notion of partial operators in algebras is of practical importance for system specifications, a class of partial algebras has been considered. We have shown that there exists a minimally defined algebra, i.e. initial algebra (unique to isomorphism) for every partially defined algebraic specification.

This paper is contributed to make connections between different models for algebraic and temporal specifications. A non trivial example is used here to illustrate how to express about the temporal behaviour of an algebraic specification. The unification of notions in algebraic and temporal specifications will bring many advantages for system development We may use a formal temporal deduction system to verify some dynamic properties from premises of algebraic specifications; or the temporal requirement specification may be used to develop system in the style of top-down refinements.

However, there are more problems which need to be addressed and clarified further. For example,

- The connection between the temporal calculus and term rewriting.
- The trace theory for the unified framework of algebraic and temporal specifications.
- The methodological considerations for system development by combining techniques of algebraic and temporal specifications.

All of these problems appeal for future research in this area.

Acknowledgment

We are indebted to B. Krieg-Brückner for encouragements and helps in the research ork. There is a constant interaction with the PROSPECTRA group at Bremen universi- : B. Hoffman, B. Gersdorf, J. Holten, S. Kahrs, D. Plump, R. Seifert and Z. Qian et al , e wish to thank all of them.

eferences

1] J. Bergstra and J. Klop, Algebra of communicating processes with abstraction. TCS 7, 77-121, 1985.

2] J. Bergstra and J. Klop, Process Theory Based on Bisimulation Semantics, Report 8824 Univ. of Amsterdam, Holland, 1988.

3] M. Broy and M. Wirsing, Partial abstract types, Acta Informatica 18, 47-64, 1982.

4] M. Broy, Requirement and Design Specification for Distributed Systems, in F.Vogt d., Proc. Concurrency 88, LNCS 335, Springer, 1988.

5] E. Clarke, E. Emerson and A. Sistla, Automatic verification of finite-state concurrent stems using temporal logic specifications, ACM Trans. on Prog. Languages and Sys- ms, 8(2), 244-263, 1986.

6] H. Ehring, B. Mahr, Fundamentals of algebraic specification 1: Equations and initial emantics, EATCS Monographs on TCS, Springer, 1985.

7] Y. Feng, H. Lin and C.S.Tang, A proof system for temporal logic programs, Computer esearch and Development (in Chinese), 22(10), 1985.

8] Y. Feng, X. Zhao and D. Guo, Modelling and verification of concurrent systems, J. of omputers, 13(1) (in Chinese), 1990.

9] C.A.R. Hoare, Communicating sequential processes, Comm. ACM 21(3), 1978.

10] S. Kaplan, Conditional rewrite rules, TCS 33, 175-193, 1984.

11] S. Kaplan, Algebraic specificaton of concurrent systems, TCS 69, 69-115, 1989.

12] H.Kreowski, Partial algebras flow from algebraic specifications, 14th ICALP proc., NCS267, 521-530, 1987.

13] B. Krieg-Brückner, The PROSPECTRA methodology of program development, in J. alewski ed., Proc. IFIP/IFAC Working Conf. on Hardware and Software for Real Time rocess Control, 1988.

14] F. Kroeger, Temporal Logic of Programs, EATCS Monographs on TCS, Springer, 987.

15] L. Lamport, While waiting for the millennium: formal specification and verification of oncurrent systems now, in F. Vogt ed., Proc. Concurrency 88, LNCS 335, Springer, 1988.

16] K. Larsen and L. Xinxin, Compositionality through an operational semantics of con- xts. Tech. Report R89-13, Aalborg Univ. Center, Denmark, 1989.

17] R. Milner, Calculus of Communicating Systems, LNCS 92, Springer, 1980.

18] R. Milner, Lecture on a calculus for communicating systems, Seminar on Concurren- , Carnegie Mellon Univ., LNCS 197, Springer, 1984.

19] A. Pnueli, Linear and branching structures in the semantics of logics and reactive stems, in Proc. 12th ICALP, LNCS 194, Springer, 1985.

20] A. Sernadas, J. Fiadeiro, C. Sernadas and H.Ehrich, Abstract object types: a tempo- l perspective, in B. Banieqbal, H. Barringer and A. Pnueli eds, Proc. Temporal logic in ecifications, LNCS 398, Springer, 1989.

21] C. Zhou and C.A.R. Hoare, Partial corectness of communicating processes, in Proc. nd Intl. Conf. on Distributed Comput. Systems, Paris, 1981.

Superimposition for Interacting Processes

Nissim Francez* Ira R. Forman
Microelectronics and Computer Technology Corp., Austin, Texas

Abstract. The paper defines an *operator* for superimposition, in contrast to previous tran formational views. It does so in the context of the *multiparty interaction* as the primitive f synchronization and interprocess communication in distributed programs. The operator is give a (structured) operational semantics, and some aspects of its methodological importance a discussed. The papers distills the essentials of superimposition, in contrast to other approach where its role is mixed with other programming aspects. A simple example of its use is provide Large design examples and more details are presented elsewhere.

1 Introduction

The notion of *superimposition* is already widely used informally, but only recently has receive more recognition and formal treatment ([BF88], [Kat87], [Kat89] and [CM88]). Superimpos tion allows a separation of concerns paradigm where one imposes an additional *control* activit (referred to as a *regulator*) achieving one concern on top of some *basic* activity, which achiev another concern. This paper presents an adaptation of superimposition to the context of coord nated distributed programming using *multiparty interactions* as the interprocess communicatio and synchronization mechanism. The presentation is in terms of a language called *Interactin Processes (IP)*. To make the paper self-contained, we include a brief section presenting the (co of) the language and its semantics, even though this core was already presented in [Fra89] an [AFG90]. Details of the full language can be found in [FF90a], and a full account of the languag and its accompanying design methodology will appear in [FF90b].

Our treatment attempts to capture the bare essentials of superimposition. We focus o superimposition as a language construct representing an *operator*, in contrast to previous pr sentations by others that are transformational in nature. A more extensive comparison wit other definitions of superimposition appears below, together with some classification of its var ous conceptions.

Following [BF88], we list some special cases of the construct, all of which were coined a "superimposition" by their proposers.

Distributed detection: Since [Fra80], [DS80] the detection of distributed termination, an later of arbitrary *stable* properties of global states [CL85], became an extensively studie problem. It actually constitutes one of the first applications of superimposition. Th concern of the superimposed regulator is to detect the termination of the underlying bas program without interfering with it.

Handshaking: This is the main problem in the asynchronous implementation of language with synchronous communication used as guards, as in CSP [Hoa78]. In a more genera form it also occurs in *coordination algorithms* implementing the synchronization require

*On sabbatical from Computer Science Dept., Technion, Haifa, Israel

for multiparty interactions. The first algorithm for achieving handshaking was presented in [Sch78], and several others followed (see [BS83] for a survey). The superimposition here consists of communicating components known as *pollers*, whose task is to single out a *matching pair* of communication requests among the basic components.

Converting traversal to election: The problem of *symmetry breaking* in a distributed system, known also as the *leader election* problem, has been extensively studied. In [KKM85] it is shown that every graph-traversal algorithm can be converted into an election algorithm by superimposition. The principle is to merge control messages and basic messages. The latter are used to direct and moderate the former which carry the real information.

Converting traversal to garbage-collection: It is shown in [TTvL86] that distributed garbage-collection algorithms can be systematically generated by superimposing termination detection algorithms on *graph marking* algorithms.

Communication-closed layering: In [EF82] the notion of a *communication-closed layer* in a distributed program is introduced. A decomposition of a distributed program into such layers has the property that no inter-layer communications occur during the execution. Thus, every execution of such a program is equivalent to one where all processes are synchronized at layer boundaries. The latter are much simpler to reason about and to verify. As suggested in [EF82] and [SF86], a fragment of a distributed program can be *made* a layer by the superimposition of synchronization components. Such layers are important in the design of phase-based distributed algorithms.

Following the proposal of [BF88], we propose to have superimposition as an *explicit* language construct. A similar approach was adopted in *UNITY* [CM88] and is also advocated in [Kat87], [Kat89] and [KFE90]. There are two views of how this is to be achieved. One view is a *transformational* view, where superimposition is viewed as a program transformation, transforming the program that achieves the basic concern into a program achieving the combined concern. In *UNITY* very modest transformations are considered. In [Kat87] and [Kat89] the transformation involves the program specifications[1] (controlling the degree of invasiveness as explained below), the grammar of the programming language, instantiation of generic units and elaborate parametrization and binding rules. While the above are useful methodological components, they are not directly associated with superimposition and should be treated separately.

The other view is to regard superimposition as an *operator* having an independent semantic definition. To this view [BF88] alludes. We adhere to this view and define a simple form for the operator that is sufficiently expressive from the methodological point of view. In particular, we make strong usage of the fact that interactions in *IP* have names, which may be used for the association needed to regulate the basic program. While bearing strong resemblance to concurrent composition, superimposition is a different operator. In particular, it is a *non-commutative* operator, as expected for a support of a layering methodology of distributed programming.

Recent research revealed three different conceptions of superimposition, ordered by the degree of "influence" the superimposition has over the basic computation. For programs in normal form[2], these three classes[3] can be characterized also by the kind of semantic properties of the basic computation preserved by superimposition [KJ89].

[1] Still, the main reason to include a reference to the specification in the superimposition declaration of [Kat89] is in connection to correctness arguments, with which we do not deal here.

[2] Or *flat*, where *all* interaction parts are guards of one top-level iteration.

[3] The names of these three classes are ours.

Spectative superimposition: This is the weakest superimposition proposed and has been adopted in *UNITY*. According to this conception, the regulator can modify its own variables in terms of the basic variables and have additional computations. However, the regulator can neither affect the control of the basic program nor modify basic variables. When viewed as a transformation, the guards of the basic program may not be modified. This approach preserves both *safety* and *liveness* properties of flat *IP* programs.

Regulative superimposition: According to this view, the superimposition can restrict the computations of the basic computation by either delaying certain operations or even forbidding them [BF88]. When viewed as a transformation, it allows also for *guard strengthening*. According to this approach, *safety* properties of flat programs are preserved, but *liveness* properties need not be preserved.

Invasive superimposition: This is the most liberal proposal, by [Kat87] [Kat89], which allows also for basic variable updates by the regulator. Invasive superimpositions preserve neither safety nor liveness properties.

The *spectative* version is too weak for many applications (e.g., termination enforcement of Table 2 below cannot be expressed by it). While invasive superimposition is convenient in some programming tasks (see [Kat89]), there do not exist at this time any known ways to constrain invasive superimposition to preserve correctness properties. For our purposes, we chose *regulative superimposition* for *IP*, because its preservation of safety properties.[4]

2 The IP core language

In this section we present the core of the mini-language, *IP (Interacting Processes)*. Its main feature is the usage of multiparty interactions as *guards*. *IP* generalizes both Dijkstra's original guarded commands language [Dij75], which has only boolean guards, and *CSP* [Hoa78], which has synchronous binary communication as guards.

A program $P :: [P_1 \parallel \cdots \parallel P_n]$ (alternatively[5] denoted also as $[\underset{i=1,n}{\parallel} P_i]$), consists of a *concurrent composition* of $n \geq 1$ (fixed n) *processes*, having *disjoint* local states (i.e., no shared variables). A *process* $P_i, 1 \leq i \leq n$, consists of a statement S, called its *body*, S taking one of the following forms:

- Dummy statement - *skip*: A statement with no effect on the state.
- Assignment statement - $x := e$: The variable x and the expression e (from the underlying expression language) are local to P_i. Assignments have their usual meaning of state transformation.
- Interaction statement - $a[\bar{v} := \bar{e}]$: Here a is[6] the *interaction name* and $a[\bar{v} := \bar{e}]$ is an *interaction part*, containing an optional parallel assignment. The process P_i is a *participant* of the interaction a. All variables in $\bar{v}$ are local to P_i and pairwise different. The expressions $\bar{e}$ may involve variables *not local* to P_i (belonging to other participants of that interaction). An interaction a is *readied* by a process P_i, if control of P_i has reached a point where executing a is one of the possible continuations. An interaction a is *enabled* only when *all* its participants have readied it. Thus, an interaction synchronizes all its participants, with

[4] There is a case where this is not quite true; the case is described in Section 3.

[5] When convenient, mnemonic process names are used, e.g., [*Producer* ||*Consumer* ||*Buffer*].

[6] See below the meaning of an *absent* interaction name.

all the parts executed in parallel. The assignments in an interaction body are the means for *interprocess communication*, in addition to synchronization. When an interaction starts, a temporary *combined state* is formed, allowing participating processes to access variables in the local states of other participants, by means of the computed expressions. Thus, this is a *symmetric* interprocess communication construct, involving an arbitrary number of participating processes. Synchronous *send* - *receive* constructs (*handshaking, rendezvous*) are special cases of interactions.

Upon termination of a local interaction part a participating process resumes its local thread of control. The two formal semantics below differ on the issue of *post-synchronization.* If the body $\bar{v} := \bar{e}$ is empty for some participating process, the effect of the interaction on that process is pure synchronization. A process may have several interaction parts with the same interaction name, allowing it to execute different assignments in different executions of the interaction.

An assignment statement $x := e$ can be seen as an abbreviation to an *anonymous* one-party interaction $\llbracket x := e \rrbracket$, containing the assignment within its (only) part. By having nameless interactions, one enforces purely local actions. Similarly, the *skip* statement can be seen as an abbreviation to $\llbracket\,\rrbracket$, an empty anonymous one-party interaction.

- *Sequential composition* - $S_1; S_2$: S_1 is executed; if and when it terminates, S_2 is executed.
- *Nondeterministic selection* - $[\,[]_{k=1}^{m} B_k \& a_k \llbracket \bar{v}_k := \bar{e}_k \rrbracket \rightarrow S_k]$:
Here $B_k \& a_k \llbracket \bar{v}_k := \bar{e}_k \rrbracket$ is a *guard*, having two components: B_k is a local boolean expression , and $a_k \llbracket \bar{v}_k := \bar{e}_k \rrbracket$ is an interaction (part) guard. S_k is any statement. When a nondeterministic selection statement is evaluated in some state, the k'th guard is *open* if B_k is true in that state. The interaction a_k is *readied* by P_i at that state; the guard is *enabled* if it is open and the interaction a_k is enabled. Executing the statement involves the following steps. First, an evaluation of all boolean parts to determine the collection of open guards. If this collection is empty the statement *fails.* Otherwise an enabled guard is passed (simultaneously with the execution of all the other matching bodies in the other participating parties) and then S_k is executed. In case no open guard is enabled, execution is *blocked* until some open guard is enabled (possibly forever).
- *Nondeterministic iteration* - $*[\,[]_{k=1}^{m} B_k \& a_k \llbracket \bar{v}_k := \bar{e}_k \rrbracket \rightarrow S_k]$: Similar to the nondeterministic selection, but the whole procedure is repeated after each execution of a guarded command, and execution terminates once none of the guards are open. At this point, no restrictions are imposed on consecutive choices during the execution of a nondeterministic iteration. *Fairness* assumptions for *IP* are considered in [AFK88], [AFG90] and [FF90b]. Two fairness assumptions are briefly reviewed below.

The *participants* of an interaction a, are *all* the processes having a reference to the interaction ame a anywhere in their program. The collection of participants of interaction a is denoted by $_a$. Because interaction names are not storable at this point, $\mathcal{P}_a$ is *syntactically* determinable. hus, an *IP* process can be conceived as having a *behavior*, consisting of a set of sequences interactions (possibly tree-structured, if so desired). Labeled interactions are the *visible* terface of a process. The statement S in the body of a process definition is a description of e way the process "brings about" any required sequence of its behavior, in coordination with her participants of the interactions in that behavior. If one would like to "tailor" a specific ecification method for *IP* programs, it might be in terms of such behaviors.

Note that at this point nested concurrency is excluded by the above description.

2.1 Formal semantics

In this subsection we present a formal definitions of the operational semantics, based on Plotkin' *sos - structured operational semantics* definition method[Plo83]. We define two different semantics for IP: *serialized* and *overlapping* (compare with a similar distinction in [BK88] an [GFK84]). The former is better suited for correctness-proofs, while the latter captures bette the behavior induced by typical implementations.

Serialized Semantics. The central characteristic of this semantics is that local actions an interactions take place one at a time. A *configuration* for IP is defined as $C = \langle [\mathop{\|}\limits_{i=1,n} S_i], \sigma$ and consists of a concurrent syntactic continuation and a *global* state. Here S_i is the rest of th program that process P_i has still to execute. The *empty* continuation E satisfies $E; S = S$; $E =$ S for any S.

A configuration $\langle [E \parallel \cdots \parallel E], \sigma \rangle$ is a *terminal* configuration. For a state σ, we use the usua notions of a *variant* $\sigma[c/x]$ and $\sigma[\bar{c}/\bar{x}]$, obtained from σ by changing the value of x to c ($\bar{x}$ to $\bar{c}$ respectively) and preserving the values of all other variables. We use $\sigma(e)$ to denote the valu of an expression e in a state σ.

We now define the *(serialized) transition* relation '$\rightarrow$' among configurations.

$$\langle [S_1 \parallel \ldots \parallel S_i \parallel \ldots \parallel S_n], \sigma \rangle \rightarrow \langle [S_1 \parallel \ldots \parallel E \parallel \ldots \parallel S_n], \sigma \rangle \quad (1$$

for any $1 \leq i \leq n$, iff $S_i = skip$, or $S_i = *[\|_{j=1}^{n_i} B_j \& a_j [\![\bar{v}_j := \bar{e}_j]\!] \rightarrow T_j]$ and $\neg \vee_{j=1}^{n_i} B_j$ holds in σ This transition means that loop exit and dummy statements are atomic.

$$\langle [S_1 \parallel \ldots \parallel S_i \parallel \ldots \parallel S_n], \sigma \rangle \rightarrow \langle [S_1 \parallel \ldots \parallel E \parallel \ldots \parallel S_n], \sigma[\sigma(e)/x] \rangle \quad (2$$

for any $1 \leq i \leq n$, iff $S_i = (x := e)$. The meaning of this transition is that local assignments ar atomic computation steps. Recall that processes do not share any variables.

$$\langle [S_1 \| \ldots \| S_{i_1-1} \| S_{i_1} \| S_{i_1+1} \| \ldots \| S_{i_k-1} \| S_{i_k} \| S_{i_k+1} \| \ldots \| S_n], \sigma \rangle \rightarrow \quad (3$$
$$\langle [S_1 \| \cdots \| S_{i_1-1} \| S'_{i_1} \| S_{i_1+1} \| \cdots \| S_{i_k-1} \| S'_{i_k} \| S_{i_k+1} \| \cdots \| S_n], \sigma' \rangle$$

iff the following holds: There is an interaction a with a set of participants $\mathcal{P}_a = \{i_1, \ldots, i_k\}$ (fo some $1 \leq k \leq n$), and for every $i \in \mathcal{P}_a$ one of the following conditions holds:

1. $S_i = a[\![\bar{v}_i := \bar{e}_i]\!]$ and $S'_i = E$
2. $S_i = [\|_{j=1}^{n_i} B_j \& a_j [\![\bar{v}_j := \bar{e}_j]\!] \rightarrow T_j]$ and there exists some $j, 1 \leq j \leq n_i$, s.t. B_j holds in σ $a_j = a$ and $S'_i = T_j$
3. $S_i = *[\|_{j=1}^{n_i} B_j \& a_j [\![\bar{v}_j := \bar{e}_j]\!] \rightarrow T_j]$ and there exists some $j, 1 \leq j \leq n_i$, s.t. B_j holds in σ $a_j = a$, $S'_i = T_j; S_i$.

Finally, for all these cases, $\sigma' = \sigma[\sigma(\bar{e})/\bar{v}]$, with $\bar{v} = \uplus_{i \in \mathcal{P}_a} \bar{v}_i$, $\bar{e} = \uplus_{i \in \mathcal{P}_a} \bar{e}_i$. Here '$\uplus$' denote *ordered union*, where the variables and expressions are ordered by increasing process index.

This transition captures the atomicity of a multiparty interaction and defines its execution Note that in the syntactic continuation, all the participating processes advance in their program and all update their state in accordance with the interaction body. Thus, participant post-synchronization is implied by the serialized semantics. The nondeterminism embodied ir selections and repetitions is induced by the meta-level nondeterminism of the many possible ways an appropriate transition rule can be applied.

For any $1 \leq i \leq n$, if

$$\langle [S_1 \parallel \ldots \parallel S_i \parallel \ldots \parallel S_n], \sigma \rangle \rightarrow \langle [S'_1 \| \cdots \| S'_i \| \cdots \| S'_n], \sigma' \rangle \quad (4$$

hen

$$\langle [S_1;\ T_1\| \cdots \|S_i; T_i\| \cdots \|S_n;\ T_n], \sigma\rangle \rightarrow \langle [S'_1;\ T_1\| \cdots \|S'_i; T_i\| \cdots \|S'_n;\ T_n], \sigma'\rangle$$

This rule captures the meaning of sequential composition, by allowing single steps to be arried out in a context where there is a syntactic continuation, which is preserved by the transtion. Recall the equalities satisfied by empty continuations, by which a terminated sequential mponent "disappears" from the configuration.

For this semantics, we define the following notions.

omputation: A *(serialized) computation* π of P on σ is a maximal (finite or infinite) sequence of configurations $C_i, i \geq 0$, such that:

1. $C_0 = \langle P, \sigma\rangle$.
2. For all $i \geq 0$, if C_i is not the last configuration in π, then $C_i \rightarrow C_{i+1}$.

We denote by $\mathcal{C}_P(\sigma)$ the set of all computations of P starting in configuration $\langle P,\ \sigma\rangle$, and by $\mathcal{C}_P \stackrel{df.}{=} \bigcup_{\sigma} \mathcal{C}_P(\sigma)$ the set of all computations of P.

ermination: The computation π *terminates* iff it is finite and its last configuration is terminal.

eadlock: The computation π *deadlocks* iff it is finite and its last configuration is *not* terminal.

eadiness: An interaction a is *readied* by $P_i \in \mathcal{P}_a$ in a configuration C iff S_i has one of the forms 1. - 3. in clause (3) of the definition of '$\rightarrow$'. We denote this property by $ready_i(a)$. The *readiness set* of process P_i in some configuration C is the collection of all interactions readied by P_i in C, namely $\mathcal{R}_i \stackrel{df.}{=} \{a \mid ready_i(a)\}$.

nabledness: An interaction a is *enabled* in a configuration C iff it is readied by all its participants, i.e., $\wedge_{P_i \in \mathcal{P}_a} ready_i(a)$ holds in C. In particular, the enabledness condition for an interaction is the conjunction of all the *local* readiness conditions of its parts over all its participants. Thereby, each participant has some local control over the global event of the occurrence of an interaction.

onflict: Two enabled interactions a_1 and a_2 are in *conflict* in a configuration C iff both are enabled in C and they have non-disjoint set of participants, i.e., $\mathcal{P}_{a_1} \cap \mathcal{P}_{a_2} \neq \emptyset$. The *conflict set* $\mathcal{C}$ in a configuration C is the collection of all pairwise conflicting interactions in C.

ndependence: Two interactions a_1 and a_2 are *independent* iff $\mathcal{P}_{a_1} \cap \mathcal{P}_{a_2} = \emptyset$., i.e., they have no common participant.

Overlapping semantics. The main characteristic of this semantics is that it is *not* an terleaving of actions. Rather, actions have (finite, but unspecified) *duration*, so that an action n start while another action is in progress (obviously, a non-conflicting one). This is still presented as an interleaving of atomic *action fragments*, which are of a finer grain of atomicity en that of program actions.

In order to allow the treatment of this semantics within the same transitional framework, e augment configurations with an additional hidden state, consisting of a boolean array (of ze n, the number of processes), called the *readiness* state and denoted by ρ. The meaning of $i] = true, 1 \leq i \leq n$, in a configuration C, is that process P_i is currently free to engage in n action (either local or with other participants); otherwise, P_i is *engaged* in some action. In r semantics, the "real" local-state transformation is instantaneous and resets the readiness bit the acting process(es). However, a participating process can not engage in any other action (including a local one), until another transition, setting the readiness bit, has taken place. Also in this semantics participants of an interaction are synchronized upon entrance, but are not synchronized upon exit of the interaction. We denote the resulting transition relation by '$\longrightarrow$'. Its defining clauses are similar to the ones in the serialized case, with the following changes.

1. A interaction can only take place if the readiness bits of all participants are *true*.

2. In the resulting configuration, the readiness bits of all participants are equal to *false*.

3. The following transition is added

$$\langle [S_1 \| \cdots \| S_n], \sigma, (\rho[1], \ldots, \rho[i-1], false, \rho[i+1], \ldots, \rho[n]) \rangle \longrightarrow \qquad (5)$$
$$\langle [S_1 \| \cdots \| S_n], \sigma, (\rho[1], \ldots, \rho[i-1], true, \rho[i+1], \ldots, \rho[n]) \rangle$$

   for every $1 \leq i \leq n$. This transition marks the completion of P_i's part in the interaction P_i is executing.

An *overlapping* computation π of P is defined similarly to the serialized case, except that the transition '$\longrightarrow$' replaces '$\rightarrow$', and the following conditions are added:

1. In the initial configuration $\rho[i] = true$ for all $1 \leq i \leq n$.

2. In a terminal configuration $\rho[i] = true$ for all $1 \leq i \leq n$.

3. In an infinite computation, $\rho[i] = true$ infinitely-often, for every $1 \leq i \leq n$ (i.e., every action terminates within a finite time).

The modified definition of readiness, accommodating the readiness bit, is presented below.

Readiness: Process P_i *readies* an interaction a in a configuration C iff P_i has one of the three forms of (3), in addition to $\rho[i] = true$.

This modification induces similar modifications of the definitions of enabledness, conflict and independence, all taking into account the readiness bits.

Programming examples may be found in the above listed references for definitions of *IP*.

Fairness. There are two basic requirements which fit the current context, being natural generalizations of fairness in either nondeterministic sequential choice or fairness for binary rendezvous. For lack of space, we do not elaborate more on this issue here.

Weak interaction fairness: Every interaction continuously enabled is eventually executed.

Strong interaction fairness: Every interaction infinitely often enabled is eventually executed.

3 Superimposition in IP

We present a simple example to introduce the concept in this specific context. In [FF90b] we present the methodological principles of superimposition and larger examples.

Suppose we wish to find a solution to the system of equations

$$x = f(x,y), \; y = g(x,y)$$

There are two concerns of interest: convergence and termination. Let us assume (with appropriate constraints on f, g and on the domain of x and y) that convergence is achievable via

$Q_1 : x : T := x0;\ *[\ px[\ x := f(x,y)\ ] \rightarrow skip\ []\ py[\ ] \rightarrow skip\ ]$

$\|$

$Q_2 : y : T = y0;\ *[\ py[\ y := g(x,y)\ ] \rightarrow skip\ []\ px[\ ] \rightarrow skip\ ]$

Table 1: Basic computation for an iterative solution to (x,y) = (f(x,y),g(x,y))

$P_1 :: xold : T := x\infty, \{\ x0 \neq x\infty\ \}\ xdone : \textbf{boolean} := false;$
$\quad *[\ \neg\ xdone \wedge xold = x\ \&\ stop[\ ] \rightarrow xdone := true$
$\quad []\ \neg\ xdone\ \&\ px[\ xold := x\ ] \rightarrow skip$
$\quad []\ \neg\ xdone\ \&\ py[\ ] \rightarrow skip\ ]$

$\|$

$P_2 :: yold : T := y\infty, \{\ y0 \neq y\infty\ \}\ ydone : \textbf{boolean} := false;$
$\quad *[\ \neg\ ydone \wedge yold = y\ \&\ stop[\ ] \rightarrow ydone := true$
$\quad []\ \neg\ ydone\ \&\ py[\ yold := y\ ] \rightarrow skip$
$\quad []\ \neg\ ydone\ \&\ px[\ ] \rightarrow skip\ ]$

Table 2: Regulator for termination enforcement

haotic) iterations of applications of f and g, starting for arbitrary initial values. In other ords, assume that the concurrent composition in Table 1 does converge to a fixed point of the stem. We wish to superimpose a computation that discovers when the fixed point is reached.

Table 2 shows a regulator that can be superimposed over the basic convergence computation. he indices[7] of the processes indicate which processes of the regulator correspond to which ocesses of the basic computation. A regulator looks much like a concurrent composition of ocesses except that some variables are unbound (e.g., x and y); these variables are expected belong to the basic computation and can be referenced by the regulator but not be assigned by it. A regulator may also use the names of interactions of the basic computation; this has e effect of making such a process of the regulator a party to the basic interaction – as a result regulator process can control the readiness of an interaction part based on whether or not the gulator's guard for the interaction is open. A regulator can have its own interactions (e.g., *op*), which are used to communicate between its processes.

The *exact* meaning of superimposing a regulator is defined below. Here, we informally scribe the behavior of combining a regulator and a basic computation in Table 2.

First, Q_1 and P_1 form a *single* process R_1 (similarly Q_2 and P_2 form a single process R_2). rocess R_i (for $i = 1, 2$) is a participant in both $px[...]$ and $py[...]$. The guard for $px[...]$ in is the conjunction of both the guard for $px[...]$ in Q_i and the guard for $px[...]$ in P_i. In is way the regulating process can do more than just prevent interactions from occurring, it n terminate iterations in the basic process.

In the extended interactions $px[...]$ and $py[...]$, the regulating processes P_1 and P_2 have riables *oldx* and *oldy* to retain the previous values of x and y, respectively. Convergence is tected by noticing a non-modifying update, i.e., $x = oldx \wedge y = oldy$.

Next note that the P_i share an "added" interaction $stop[...]$, not present in the basic com- itation. Enabledness of this interaction is determined only by the regulating processes. Again ese "added" interactions are interactions of R_i and are interleaved in such a way that no ocess executes more than one interaction at a time. As long as neither coordinate has reached s fixed point value, neither part of $stop[...]$ is readied. Suppose x reaches a value such that $= oldx$ (i.e., $x = f(x, y)$). At this point, P_1 readies $stop[...]$, but keeps on readying $px[...]$

[7] In the case where processes are named instead of indexed, regulating processes are associated with basic ocesses of the same name. More flexible association schemes are also possible.

P_1 :: *xcount* : **integer** := 0, *xdone* : **boolean** := *false*;
$*[\neg\, xdone \wedge xcount = max$ & *stop*[ ] → *xdone* := *true*
▯ ¬ *xdone* & *px*[ ] → *xcount* := *xcount* + 1
▯ ¬ *xdone* & *py*[ ] → *skip*]
‖
P_2 :: *ycount* : **integer** := 0, *ydone* : **boolean** := *false*;
$*[\neg\, ydone \wedge ycount = max$ & *stop*[ ] → *ydone* := *true*
▯ ¬ *ydone* & *py*[ ] → *ycount* := *ycount* + 1
▯ ¬ *ydone* & *px*[ ] → *skip*]

Table 3: Regulator to limit the number of iterations

in case this is not yet a fixed point state (i.e., $oldy \neq y$ implying $y \neq g(x,y)$). However, once the fixed point state is reached, both participants R_i, $i = 1,2$ ready *stop*[...], which becomes enabled. By assuming strong interaction fairness[8], the *stop*[...] interaction is eventually executed, after which $xdone = ydone = true$ holds. This terminates R_i, because neither iteration statement in P_i or Q_i has any open guards (which causes loop exit). This is a case where the regulator has forced the termination of an infinite loop in the basic computation. Thus the combined program properly terminates after a fixed point is reached (a simple argument may be used to show that no premature termination is possible).

In this example, no infinite execution of "added" interactions is possible (due to the *xdone* and *ydone* guards in the P_i). In general, repeated execution of "added" interactions might be possible and again starvation of the basic computation by the regulator is excluded by fairness.

Next we show (in Table 3) the proper separations of concerns induced by superimposition by showing another regulator, which terminates the basic iterative computation by limiting the number of iterations[9]. This is useful in case that the fixed point is only reachable in the limit and some approximation to it is needed. The way this combination of regulator and basic computation behaves should be clear from the previous description. The two regulators are similar in structure, differing in the way the *stop*[...] interaction is eventually enabled.

3.1 Definition of superimposition for IP

This section defines superimposition in terms of several semantic requirements, accompanied by an operational semantics satisfying them. To simplify the presentation, we assume that the superimposed computation has exactly the same number of processes as that in the basic computation

First, we distinguish between *closed* and *open IP* programs. In a closed program, every variable mentioned by some process is local to some (possibly different) process. On the other hand, an open program may refer to *free* variables, not belonging to any of its processes. An open program $P :: [\underset{i=1,n}{\|} P_i]$ is a *regulator* for a closed program $Q :: [\underset{i=1,n}{\|} Q_i]$ iff every variable free in P_i is bound in $Q_i, i = 1,n$. Note that each process of a regulator will also be referred to as a regulator when the context is clear the process of the regulator is meant. A regulator P for Q may use some of the same interaction labels as Q; these are called *shared* interactions. The regulator may also have its own interactions that do not appear in the basic program, called *added* interactions. Also note that, except in degenerate cases where P and Q are *disjoint*, if P

[8]When *stop*[...] is enabled, each execution of either *px*[...] or *py*[...] disables *stop*[...]. This means that weak interaction fairness is not strong enough to ensure termination of the computation.

[9]Here *max* is assumed to be a constant. When extending superimposition to parametrized abstractions (*teams* in the terminology of [FF90b] and [FF90a]), it may become a team parameter.

a regulator for Q then Q is *not* a regulator for P.

We define the *superimposition* R of P over Q, denoted by $R :: \frac{P}{Q}$, whenever P is a regulator Q, the latter referred to as the *basic* program. This notation should suggest the association at P "rides over" Q. We write $R :: \frac{P}{Q}$ to mean $\underset{i=1,n}{\|} R_i :: \frac{P_i}{Q_i}$ overloading the superimposition mbol. In the general case of $\frac{\underset{i=1,m}{\|} P_i}{\underset{j=1,n}{\|} Q_j}$ where $n < m$, P_k for $n < k \leq m$ are additional regulator ocesses that can use only its own variables. If $n > m$, the Q_k for $m < k \leq n$ are unregulated. ius, the resulting program R can be viewed as $[\underset{i=1,n}{\|} R_i]$. Obviously, R should be built over the me communication network[10].

We require that the *local* superimposition $\frac{P_i}{Q_i}$, $i = 1, n$, adhere to the following requirements.

cal-state access: P_i can read [11] the local state of Q_i, $i = 1, n$, but not vice versa.

This is the main source of asymmetry of this construct, reflecting the non-commutativity of e operator. It is satisfied due to the definition of a regulator and the closedness of the basic ogram. In *IP*, the local-state access ability manifests itself in two forms. First, guards in the gulator can refer to basic variables. Second, assignments to local variables in the regulator may e terms involving basic variables. The interleaving rules that follow guarantees the absence of y "collisions" in variable access.

cal interleaving: For each i, $i = 1, n$, P_i and Q_i interleave their non-shared interactions.

We now relate to P's ability to control Q's interactions. This ability has two aspects:

- Temporary delay of some interactions.
- Termination enforcement, preventing any further interaction.

These aspects are handled by the next two requirements. Regulators may share interaction els with the basic computation; this has the effect of making the regulator a partner in the eraction. In this manner the superimposition can delay (or block) an interaction of the basic nposition by not opening its own guard for that interaction (as *Terminate* does in Figure Processes of a regulator may interact independently of the basic computation by means non-shared interaction names, for "private" purposes. This is crucial in deadlock detection orithms, where the regulator does its "real" work *after* the basic computation has deadlocked.

teraction Participation: For $i = 1, n$, P_i participates in the interactions shared with Q_i, which are readied by R_i only if readied by *both* P_i and Q_i.

The multiparty interaction allows the synchronization of the basic computation and the ulator. In this manner the regulator can strengthen the guard of an interaction to regulate its ecution. Whenever the regulator's guard for a shared interaction becomes false, this interaction not take place even if Q readies all of its parts for that interaction. Thus, it is temporarily ayed, until some added interactions in P take place, turning that guard true again. This tes care of the first aspect of regulation mentioned above. Note that the basic computation Q 'unaware" of delays generated by superimposition because there is no mechanism to discover y an interaction is not enabled.

[0]Bougé [Bou85] termed this criterion *generic*, that is, components R_i and R_j should interact only if Q_i and interact, which requires no new links in the communications network.

[1]For readers familiar with [BF88], we do not assume here asynchronous access to basic state components.

$P_i :: *[B_1\&a[...] \rightarrow \cdots \,[]\, B_2\&b[...] \rightarrow \cdots]$
$Q_i :: *[C_1\&a[...] \rightarrow \cdots []C_2\&c[...] \rightarrow \cdots]$

Table 4: Exemplifying the Joint Termination Convention

$R_i' :: *[B_1 \wedge C\&a[...] \rightarrow \cdots \,[]\, B_2\&b[...] \rightarrow \cdots \,[]\, C_2\&c[...] \rightarrow \cdots]$

Table 5: An equivalent process

The combination of local interleaving and local state access must be interpreted properly - the problems of asynchronous access are non-existent in IP because P_i and Q_i interleave their interactions. Also the guard of an interaction must be true when the interaction is executed. Therefore, if Q_i executes an interaction, the guards of P_i must be re-evaluated before P_i can be considered ready for an interaction part, because Q_i may have changed the values of variables referenced in P_i.

Once the process R_i is formed by superimposing P_i on Q_i, the question of when R_i terminates arises. This is answered by the next requirement.

Joint Termination Convention: R_i terminates if and when both P_i and Q_i jointly terminate. P_i and Q_i terminate when there is no open guard. Note that for the guard of a shared interaction to be open, the guarding condition for both P_i and Q_i must hold.

P_i can delay the basic computation by participating in some Q_i interactions but not opening the guards until P_i wants Q_i to progress. P_i can also enforce termination of the basic computation by participating in all Q_i interactions but not opening the guards after P_i has determined that Q_i should terminate. It is important to remember that in a superimposition neither the basic process nor the superimposed process terminate separately. One can only speak of the termination of the composite. Table 4 contains an example that helps us better understand the Joint Termination Convention. In this example, $a[...]$ is a shared interaction while $b[...]$ is an added one. In order for R_i to ready interaction $a[...]$, C_1 and B_1 must hold. Thus, when B_1 becomes false, the regulator P_i is *temporarily* delaying interaction $a[...]$. Suppose interaction $b[...]$ is executed, turning B_1 to true; this causes the guard of $a[...]$ to open. The process R_i terminates when no guard is open; this happens when $\neg(B_1 \wedge C_1) \wedge \neg B_2 \wedge \neg C_2$. Consequently, the composite process of Table 4 can be shown equivalent (in terms of the generated computations) to the one in Table 5.

The Joint Termination Convention is based on the principle that termination should be based on stable properties. To appreciate this point consider an alternative choice for the termination convention: R_i terminates when both P_i and Q_i have individually terminated. P_i (Q_i) terminates, if a configuration is reached in which all its guards are not open. Under this definition, P_i terminates because none of its guards hold. However, subsequent computation by Q_i might change the value of a guard in P_i. Thus P_i is terminating on a condition that is not stable; we believe that this choice would lead to error-prone language where programmers would find programs terminated prematurely (because our actual choice for the Joint Termination Convention is more natural) or programs fail to terminate (because the implementation scheduled interactions in an unexpected way).

Finally, a requirement is needed to ensure that both the basic computation and the regulator do make progress. For example, consider the superimposition in Table 2. To terminate the computation, one must ensure that the interaction *stop* is executed once it becomes enabled. This is not a trivial matter, because the interactions *px* and *py* continue to be enabled.

airness: Fairness must be strong enough to ensure that both the superimposition and the basic computation achieve their liveness properties in the presence of each other.

his does not prescribe a specific fairness condition. The condition will be dependent on what veness properties need to be achieved in a particular context. Termination enforcement as is one in both Tables 2 and 3 requires strong interaction fairness. Termination detection can e expressed with just weak interaction fairness, because after termination only the detecting teraction is enabled.

Superimposition has been defined for flat IP programs. In this restricted domain, a regulator eserves all safety properties because the regulator can only reduce the behavior of the basic mputation. There would be one exception to this when extending superimposition to the full re language.[12] Suppose one had a nonterminating loop where a safety property is violated by statement succeeding the loop. For example,

$$P_i :: x := 0; *[true\&a[\,] \to skip]; x := 1$$

ere $x = 0$ is an invariant (assuming the Joint Termination Convention is appropriately inter- eted for loop termination). However, if one superimposed a regulator that terminated the op the subsequent assignment would negate the invariant. This case cannot arise for flat IP ograms, because it the sequence were encode as a guarded alternative to the loop, there would no way for the regulator to make the guard true.

Note that if P vacuously regulates Q, i.e., P is closed and shares no interaction names, then $\frac{P}{Q}$ a *disjoint* version of the UNITY *union* operator. To obtain the non-disjoint union, either both and Q should be open or variables of the same name in both should be identified. Similarly, P vacuously regulates Q, this combination is the same as disjoint parallel composition.

2 Formal semantics

e next institute the above requirements by presenting a *structured* operational semantics ad- ring to them. A configuration now contains syntactic continuations of the form $[\ \|_{i=1,n} \frac{P_i}{Q_i}\ ]$, here $P ::\ [\ \|_{i=1,n} P_i\ ]$ is a regulator for $Q ::\ [\ \|_{i=1,n} Q_i\ ]$. Only the serialized version is presented. ne overlapping one can be obtained by analogous modification for the core language.

First, rule (1) is split to handle separately *skip* and loop exit (in this case termination, due the restriction to normal form). The *skip* rule is combined with the rule for local assignment.

$$If\ \langle\, [S_1 \,\|\, \ldots \,\|S_i\,\| \ldots \|S_n],\ \sigma\,\rangle {\rightarrow} \langle [\,S_1 \| \cdots \|E\| \cdots \|S_n],\ \sigma' \rangle \tag{6}$$

r any $1 \le i \le n$, where S_i is either *skip* or $x := e$, then

$$\langle [\,\tfrac{S_1}{Q_1} \| \cdots \| \tfrac{S_i}{Q_i} \| \cdots \| \tfrac{S_n}{Q_n}],\ \sigma \rangle {\rightarrow} \langle [\,\tfrac{S_1}{Q_1} \| \cdots \| \tfrac{E_i}{Q_i} \| \cdots \| \tfrac{S_n}{Q_n}],\ \sigma' \rangle$$

d also

$$\langle [\,\tfrac{P_1}{S_1} \| \cdots \| \tfrac{P_i}{S_i} \| \cdots \| \tfrac{P_n}{S_n}],\ \sigma \rangle {\rightarrow} \langle [\,\tfrac{P_1}{S_1} \| \cdots \| \tfrac{P_i}{E_i} \| \cdots \| \tfrac{P_n}{S_n}],\ \sigma' \rangle.$$

Thus, each component can execute a local action independently of the other.

$$If\ \langle\, [S_1 \,\|\, \ldots \,\|S_i\,\| \ldots \|S_n],\ \sigma\,\rangle {\rightarrow} \langle [\,S_1 \| \cdots \|E\| \cdots \|S_n],\ \sigma \rangle \tag{7}$$

any $1 \le i \le n$, where S_i is $*[\ \square_{j=1,n_i}\ B_j \&\ a_j[\ldots] \to \cdots]$ and $\neg \bigvee_{j=1,n_i} B_j$ holds, then

$$\langle [\,\tfrac{P_1}{S_1} \| \cdots \| \tfrac{P_i}{S_i} \| \cdots \| \tfrac{P_n}{S_n}],\ \sigma \rangle {\rightarrow} \langle [\,\tfrac{P_1}{S_1} \| \cdots \| \tfrac{P_i}{E_i} \| \cdots \| \tfrac{P_n}{S_n}],\ \sigma \rangle.$$

[2]We are indebted to Shmuel Katz for this example.

This rule expresses *local termination* of the basic program. Note that the regulator does no[t] have a corresponding rule, since its guards may refer to basic variables, to be still modified b[y] the basic program. Thus, falsity of these guards need not be a stable property.

$$\langle [\, R_1 \| \cdots \| \frac{*[\, \underset{j=1,n_i}{[]} \; B_j \& a_j [\![\ldots]\!] \rightarrow \cdots}{*[\, \underset{k=1,n_i{}'}{[]} \; C_j \& b_k [\![\ldots]\!] \rightarrow \cdots} \| \cdots \| R_n],\ \sigma\rangle \rightarrow \langle [\, R_1 \| \cdots \| \tfrac{E}{E} \| \cdots \| R_n],\ \sigma\rangle \qquad (8$$

for any $1 \leq i \leq n$, if the following condition applies.

- For a_j an added interaction in P, $\neg B_j$ holds.
- For b_k an added interaction in Q, $\neg C_k$ holds.
- For $b_j = a_j$ a shared interaction in P, Q, $\neg(C_j \wedge B_k)$ holds.

This rule embodies the *Joint Termination Convention*, the means of a regulator to *enforc[e]* termination of the basic computation. Suppose

$$\langle\, [S_1 \| \ldots \| S_{i_1-1} \| S_{i_1} \| S_{i_1+1} \| \ldots \| S_{i_k-1} \| S_{i_k} \| S_{i_k+1} \| \ldots \| S_n],\ \sigma\,\rangle \rightarrow \qquad (9)$$

$$\langle [S_1 \| \cdots \| S_{i_1-1} \| S'_{i_1} \| S_{i_1+1} \| \cdots \| S_{i_k-1} \| S'_{i_k} \| S_{i_k+1} \| \cdots \| S_n], \sigma' \rangle$$

where there is an interaction a with a set of participants $\mathcal{P}_a = \{i_1, \ldots, i_k\}$ (for some $1 \leq k \leq n$) and for every $i \in \mathcal{P}_a$, $S_i = *[[]_{j=1}^{n_i} B_j \& a_j [\![\bar{v}_j := \bar{e}_j]\!] \rightarrow T_j]$ and there exists some $j, 1 \leq j \leq n_i$, such that B_j holds in σ, $a_j = a$, $S'_i = T_j; S_i$, and $\sigma' = \sigma[\sigma(\bar{e})/\bar{v}]$, with $\bar{v} = \uplus_{i \in \mathcal{P}_a} \bar{v}_i$, $\bar{e} = \uplus_{i \in \mathcal{P}_a} \bar{e}_i$, where $\uplus$ is ordered union. Then,

1. $$\langle [\tfrac{S_1}{Q_1} \| \cdots \| \tfrac{S_{i_1-1}}{Q_{i_1-1}} \| \tfrac{S_{i_1}}{Q_{i_1}} \| \tfrac{S_{i_1+1}}{Q_{i_1+1}} \| \cdots \| \tfrac{S_{i_k-1}}{Q_{i_k-1}} \| \tfrac{S_{i_k}}{Q_{i_k}} \| \tfrac{S_{i_k+1}}{Q_{i_k+1}} \| \cdots \| \tfrac{S_n}{Q_n}], \sigma \rangle \rightarrow$$

   $$\langle [\tfrac{S_1}{Q_1} \| \cdots \| \tfrac{S_{i_1-1}}{Q_{i_1-1}} \| \tfrac{S'_{i_1}}{Q_{i_1}} \| \tfrac{S_{i_1+1}}{Q_{i_1+1}} \| \cdots \| \tfrac{S_{i_k-1}}{Q_{i_k-1}} \| \tfrac{S'_{i_k}}{Q_{i_k}} \| \tfrac{S_{i_k+1}}{Q_{i_k+1}} \| \cdots \| \tfrac{S_n}{Q_n}], \sigma' \rangle$$

   in case interaction a is an added interaction in P.

2. Analogous, if interaction a is local to Q.

This rule captures the interleaving of non-shared actions between the regulator and basic program. Next, controlling shared interactions is captured by a rule.

$$\langle [\tfrac{S_1}{T_1} \| \cdots \| \tfrac{S_{i_1-1}}{T_{i_1-1}} \| \tfrac{S_{i_1}}{T_{i-1}} \| \tfrac{S_{i_1+1}}{T_{i_1+1}} \| \cdots \| \tfrac{S_{i_k-1}}{T_{i_k-1}} \| \tfrac{S_{i_k}}{T_{i_k}} \| \tfrac{S_{i_k+1}}{T_{i_k+1}} \| \cdots \| \tfrac{S_n}{T_n}], \sigma' \rangle \rightarrow \qquad (10)$$

$$\langle [\tfrac{S_1}{T_1} \| \cdots \| \tfrac{S_{i_1-1}}{T_{i_1-1}} \| \tfrac{S'_{i_1}}{T'_{i-1}} \| \tfrac{S_{i_1+1}}{T_{i_1+1}} \| \cdots \| \tfrac{S_{i_k-1}}{T_{i_k-1}} \| \tfrac{S'_{i_k}}{T'_{i_k}} \| \tfrac{S_{i_k+1}}{T_{i_k+1}} \| \cdots \| \tfrac{S_n}{T_n}], \sigma' \rangle$$

where there is a *shared* interaction a with a set of participants $\mathcal{P}_a = \{i_1, \ldots, i_k\}$ (for some $1 \leq k \leq n$), and for every $i \in \mathcal{P}_a$, $S_i = *[[]_{j=1}^{n_i} B_j \& a_j [\![\bar{v}_j := \bar{e}_j]\!] \rightarrow G_j]$, $T_i = *[[]_{j=1}^{n'_i} C_j \& a'_j [\![\bar{v'}_j := \bar{e'}_j]\!] \rightarrow H_j]$ and there exists some $j, 1 \leq j \leq n_i$ and $k, 1 \leq k \leq n'_i$, such that $B_j \wedge C_k$ holds in σ, $a_j = a'_k$, $S'_i = G_j; S_i$, $T'_i = H_k; T_i$, and σ' defined similarly, capturing both basic and regulated variable updates in the joint interaction.

Finally, sequential composition is defined to allow separate transition at the basic and regulator level. We omit the formal rule.

Some language issues. Following our desire to stress concepts rather than syntax, this aper has not addressed several issues of language design. Rather than the simple name equiva- nce binding used here, more complex schemes of binding the superimposed processes to the pro- sses of the basic computation might be desired in programming language. In the full language F90b], there is an abstraction mechanism (called *team*), which has the suitable parametriza- on and genericity capabilities to allow more general ways of expressing superimposition. Thus, the example in Table 1, if the system of equations had three variables instead of two, and e functions represented, respectively, by three processes instead of two, a different regulator ould be needed, referring to the three guarded interactions. Thus, one would like to have a *neric regulator*, instances of which are superimposed on different basic programs which are tructurally similar". As already mentioned, we consider this instantiation of genericity to be dependent of superimposition because it has other applications as well. [Cai89] describes bind- g conventions that were adopted for superimposition in Estelle. Furthermore, a programming nguage might have operators for renaming processes or for hiding processes. Note, however, e difference of our approach to such "adjustments" from that of [Kat89]. We view the basic ogram as an unmodifiable, *given* module, and would adjust a generic regulator to fit the given asic program. According to [Kat89], the basic program "serves" the superimposition, and it as to be adjusted to fit a given regulator (in our terms).

Conclusions

uperimposition was presented here as an explicit language construct. It is important to notice at superimposition is radically different from using an algorithm as a *subroutine*, or even an vocation of an abstraction encapsulating that algorithm (by means of a concurrent enrolement some team in the terminology of [FF90b], [FF90a]). It is really a parallel composition oper- or of a special kind. However, in contrast to the standard concurrent composition operator, perimposition was until recently of a *semantic* nature, not explicit in the syntactic structure the resulting program. In this sense, it is similar to several other semantic structures of stributed programs not explicit in their syntax. For two other recent semantic decomposi- ons of distributed programs, the reader is referred to [SF86], where an arbitrary distributed ogram is conceived as a composition of *diffusing computations* (thereby yielding an efficient tection algorithm) and to [SD87], where a certain message-filtration technique yields a clear rrectness proof of a distributed spanning-tree construction [GHS83]. A very general treatment semantically-based compositions and decompositions can be found in [FFG90]. The notion constructing a distributed algorithm from building blocks is a promising methodology, much ore so than for sequential algorithms (e.g., see [Gaf86]). Superimposition is a central and tal tool in applying this methodology, and hence the importance of its study. Note that in ntrast to more traditional stepwise refinement, the superimposition operator supports better building-block methodology. For a *given* basic program, one can design several superimposi- ons, each augmenting the same basic program with different control enhancement (as shown our simple example). In particular, iterated superimposition is possible, due to equivalence the superimposition of normal form regulator and basic program to another normal form ogram. Examples may be found in [FF90b]. Orthogonally, one can design a *given* regulator, be superimposed on a variety of basic programs, augmenting each of them with the control hancement provided by that regulator. Distributed termination detectors are an excellent ample where this usage is manifested.

Finally, while general research in verifying properties of superimpositions is still in progress several researchers, the particularly simple form of the construct as fitting our context seems lend itself to simple proof-rules.

Acknowledgements The authors thank Paul Attie for a critical reading of an early version of this paper and Shmuel Katz for numerous discussions of superimposition, which clarified the relationship between our two approaches.

References

[AFG90] Paul C. Attie, Nissim Francez, and Orna Grumberg. Fairness and hyperfairness in multi-party interactions. In *17th ACM-POPL*, pages 292 – 305, San-Francisco, CA, January 17-19 1990.

[AFK88] Krzysztof R. Apt, Nissim Francez, and Shmuel M. Katz. Appraising fairness in distributed languages. *Distributed Computing*, 2:226 – 241, August 1988.

[BF88] Luc Bougé and Nissim Francez. A compositional approach to superimposition. In *15th ACM-POPL*, pages 240–249, San Diego, CA, January 13-15 1988.

[BK88] Ralph -J. R. Back and Reino Kurki-Suonio. Distributed cooperation with action systems. *TOPLAS*, 10(4):513–554, October 1988.

[Bou85] Luc Bougé. *Symmetry and generity for CSP distributed systems.* Technical Report, LITP, Univ. Paris, May 7 1985.

[BS83] G. N. Buckley and A. Silberschatz. An efficient implementation for the generalized input-output construct of csp. *ACM Trans. on Programming Languages and Systems*, 5(2):223–235, April 1983.

[Cai89] B. Caillaud. The superimposition of estelle programs: a tool for the implementation of observation and control algorithms. In *FORTE '89*, October 1989.

[CL85] K. Mani Chandy and Leslie Lamport. Distributed snapshots: determining global states of distributed systems. *ACM Transactions on Computer Systems*, 3(1):63–75, February 1985.

[CM88] K. Mani Chandy and Jayadev Misra. *Parallel program design: A foundation.* Addison-Wesley, 1988.

[Dij75] Edsger W. Dijkstra. Guarded commands, nondterminacy and, formal derivation of programs. *Communications of the ACM*, 18(8):453–457, August 1975.

[DS80] Edsger W. Dijkstra and C. S. Scholten. Termination detection for diffusing computations. *IPL*, 11(1):1–4, 1980.

[EF82] Tzilla Elrad and Nissim Francez. Decomposition of distributed programs into communication-closed layers. *SCP*, 2:155 – 173, 1982.

[FF90a] Nissim Francez and Ira R. Forman. Conflict propagation. In *IEEE International Conference on Computer Languages (ICCL'90)*, pages 155 – 168, New Orleans, LA, March 12-15 1990.

[FF90b] Nissim Francez and Ira R. Forman. *Interacting Processes: A Multiparty Approach to Coordinated Distributed Programming.* Forthcoming book, 1990.

[FFG90] Limor Fix, Nissim Francez, and Orna Grumberg. Semantics-driven decomposition for the verification of distributed programs. In *IFIP WG 2.2/2.3 working conference on programming concepts and methods*, Sea of Galilee, Israel, April 2 - 5 1990.

ra80] Nissim Francez. Distributed termination. *ACM Trans. on Programming Languages and Systems*, 2(1):42–55, January 1980.

ra89] Nissim Francez. Cooperative proofs for distributed programs with multi-party interactions. *IPL*, 32:235 –242, September 22 1989.

af86] E. Gafni. Perspectives on distributed network protocols: a case for building blocks. In *Proc. MILCOMM 86*, October 1986.

FK84] Orna Grumberg, Nissim Francez, and Shmuel M. Katz. Fair termination of communicating processes. In *Third ACM-PODC*, Vancouver, BC, Canada, August 1984.

HS83] R. G. Gallager, R. A. Humblet, and P. M. Spira. A distributed algorithm for minimum-weight spanning trees. *IEEE Trans. on Communications*, 31(6):756–762, 1983.

oa78] C. A. R. Hoare. Communicating sequential processes. *Communications of the ACM*, 21(8):666–677, August 1978.

at87] Shmuel M. Katz. *A superimposition control construct for distributed systems.* Technical Report, Microelectronics and Computer Technology Corp., August 1987.

at89] Shmuel M. Katz. A superimposition control construct for distributed systems. *Submitted to ACM-TOPLAS*, November 1989.

FE90] Shmuel M. Katz, Ira R. Forman, and W. Michael Evangelist. Language constructs for distributed systems. In *IFIP TC2 Working Conference on Programming Concepts and Methods*, Sea of Galilee, Israel, April 2-5 1990.

J89] Reino Kurki-Suonio and Hannu-Matti Jävinen. Action system approach to the specification and design of distributed systems. In *Proceedings Fifth Int'l Workshop on Software Specification and Design*, pages 34–40, May 19-20 1989.

KM85] E. Korach, S. Kutten, and S. Moran. A modular technique for the design of efficient distributed leader finding algorithms. In *Proc. 4th Annual ACM Symp. on Principles of Distributed Computing*, 1985.

lo83] Gordon D. Plotkin. An operational semantics for csp. In D. Bjorner, editor, *Formal description of programming concepts*, North-Holland, Amsterdam, 1983. IFIP TC.2 WG conference, Garmisch Partenkierchen.

ch78] J. S. Schwarz. *Distributed synchronization of communicating sequential processes.* Technical Report, Dept. Artificial Intelligence, Univ. Edinburgh, 1978.

D87] Frank A. Stomp and Willem P. De Roever. *A correctness proof of a distributed minimum-weight spanning tree algorithm.* Technical Report, Dept. Informatics, Nimegen Univ., April 1987.

F86] Nir Shavit and Nissim Francez. A new approach to detection of totally indicative stability. In *Proc. 13th ICALP*, July 1986.

TvL86] G. Tel, R. B. Tan, and J. van Leeuwen. *The derivation of graph marking algorithms from distributed termination algorithms.* Technical Report, Univ. Utrecht, August 1986.

An Implementation of a Translational Semantics for an Imperative Language

Lars-åke Fredlund
Bengt Jonsson
Joachim Parrow
Swedish Institute of Computer Science, Stockholm*

Abstract

We present a semantics for an imperative programming language, Lunsen, with constructs for concurrency and communication. The semantics is given through a translation into CCS. We have implemented this translation within the framework of the Concurrency Workbench, which is a tool for analysis of finite-state systems in CCS. The point of the translational semantics is that by imposing restrictions on Lunsen so that the semantics of a program is finite-state, we can analyze Lunsen programs automatically using the Concurrency Workbench. As an illustration we include an analysis of a mutual exclusion algorithm.

1 Introduction

Concurrent programs often exhibit complex behaviors, and it is therefore important to develo methods and tools for analyzing them rigorously. Many implementations have been developed fo automatic analysis of concurrent programs [CPS89, BdSV88, CES86, RRSV87, GLZ89]. Most o these tools are designed for simple models of programs, e.g. finite-state transition systems [CPS89 BdSV88, GLZ89]. However, many concurrent algorithms are naturally formulated in some imperativ programming language with constructs for concurrency. In order to analyze a program using the tool just mentioned, the program must first be translated manually into the appropriate model.

In this paper, we present an automated translation of a concurrent programming language with im perative features into CCS. The imperative language, Lunsen, is ALGOL-like and contains standar constructs for sequential programming — such as assignments, procedures, and arrays — as well a constructs for parallel execution of processes. Processes can communicate both via shared variable and through synchronous channels.

The formal semantics of Lunsen is defined through a translation into CCS [Mil89]. We have imple mented this translation within the framework of the the Concurrency Workbench (CWB). The poin of our implementation is that CWB can be used to analyze Lunsen programs. CWB is a versatil tool which can automatically decide e.g. whether two concurrent systems are equivalent, or whethe a system satisfies a property formulated in a modal logic.

The main source of inspiration for the formal semantics of Lunsen is the semantics of a sequentia language given by Robin Milner in Chapter 8 of his book [Mil89]. In order to make the programs effectively analyzable, we impose restrictions on the language Lunsen so that the semantics of a

*Address: SICS, Box 1263, S-164 28 Kista, SWEDEN.

ogram will be finite-state: this means e.g. that the types of variables must contain only finitely any elements, and that procedures cannot call each other recursively in an arbitrary manner.

here exist other automated tools for analyzing concurrent programs written in imperative lan-ages. EMC [CES86] is a tool for checking that a program satisfies a formula formulated in a anching time temporal logic. EMC has a preprocessor which accepts programs written in a sim-e CSP-like language. Xesar [RRSV87] is a tool for checking similar properties for communication otocols defined in an extension of PASCAL with facilities for communication. Auto [BdSV88] and AV [GLZ89] are tools for analysis of concurrent systems, which are related to the Concurrency orkbench. Other translations of imperative languages into CCS include a translation from Ada by ennessy and Li [HL83], a translation from CSP by Astesiano and Zucca [AZ81], and a translation om NIL by Smolka and Strom [SS86].

the next section, we define the syntax of Lunsen and give an informal semantics. The translation Lunsen to CCS goes via an intermediate language, Typed CCS, which is an extension of CCS at is presented in Section 3. The translation itself is presented in Section 4. Section 5 discusses me optimizations to the translation, and Section 6 illustrates the analysis by an example: a mutual clusion algorithm due to Peterson. Conclusions are found in Section 7.

Lunsen: Syntax and Informal Semantics

unsen is an imperative language belonging to the ALGOL family of strongly typed languages. his means that variables and imperative constructions such as while-loops are fundamental to the nguage. Lunsen does not include dynamic constructions such as pointers and creating of objects. e also place restrictions on procedure calling; these restrictions have the effect that Lunsen programs n be executed without a runtime stack or a heap area and ensure that programs in Lunsen will be *ite-state.*

unsen also contains non-sequential primitives. Commands are executed in parallel with the **par** mmand and may communicate either synchronously by sending messages over ports or via shared riables. Furthermore, nondeterministic choice can be expressed in the language.

unsen programs communicate with the outside world (the environment of programs) either by nding messages on ports that are visible to the environment or through global, or *visible*, variables the program which can be accessed by the environment.

he syntax for Lunsen programs is given in Table 1 using a dialect of BNF. Objects written inside anted brackets (*[ ]*) are optional, and we let the ¦ symbol denote alternatives (instead of grouping em on different lines). We presuppose a set of *identifiers* partitioned into *constants* ranged over I_c, *type identifiers* ranged over by I_t, *procedure identifiers* ranged over by I_p, *Lunsen variables* nged over by I_v, *array variables* ranged over by I_a, *program identifiers* ranged over by I_s, and *port entifiers* ranged over by I_m.

e will now give an informal description of the meaning of the Lunsen constructions.

- Programs

  A program consists of a declaration section and a command which invokes the execution of the program.

- Declarations

  Procedures, types, variables and ports are defined in a declaration section. The order between definitions is not significant.

- Types

S::=	**program** I_s; D C **endprog**;	Program
D::=	ϵ	Empty declaration
	type $I_t = T$;	Type definition
	var I_v:[G] $I_t = E$;	Variable declaration
	procedure I_p[(I_v:A I_t,...,I_v:A I_t)] D C **endproc**;	Procedure definition
	port I_m:[**visible**] I_t;	Port declaration
	D D	
A::=	**in** | **out** | **inout**	Parameter usage
G::=	**read** | **write** | **readwrite**	Visibility of variables
T::=	$\{I_c,\ldots,I_c\}$	Enumerated type
	ordered $\{I_c,\ldots,I_c\}$	Ordered type
	array $[I_t,\ldots,I_t]$ **of** I_t	Array type
	$I_t * \ldots * I_t$	Tuple type
E::=	I_c	Constant
	I_v	Variable
	$I_a[E,\ldots,E]$	Array expression
	$(E,\ldots,E)$	Tuple expression
	# I_c E	Tuple access
	if $E \rightarrow E$ |...| $E \rightarrow E$ [**else** E] **endif**	Deterministic choice
	$E = E$	Equality
	$E \leq E$	Less-than-or-equal
	succ E	Successor function
	pred E	Predecessor function
	not E | E **and** E | E **or** E	Boolean functions
C::=	**begin** C **end**	Compound statement
	skip	No action
	C ; C	Sequencing
	C **par** C	Parallel composition
	$I_v := E$	Assignment
	$I_a[E,\ldots,E] := E$	Array assignment
	I_p[$(E,\ldots,E)$]	Procedure call
	if $E \rightarrow C$ |...| $E \rightarrow C$ [**else** C] **endif**	Deterministic choice
	when $P \rightarrow C$ |...| $P \rightarrow C$ **endwhen**	Port synchronization choice
	while E **do** C **endwhile**	While loop
	P	Port command
P::=	τ	Invisible action
	I_m ! E	Send value
	I_m ? I_v	Receive value into variable

Table 1: The syntax of Lunsen

The only standard type in the language is **boolean**, defining the constants **true** and **false**. New types can be defined by enumerating the constants of the type, or by forming array types or tuple types. For example,

```
type int4 = {1,2,3,4};
```

defines a new type **int4** and also the predicate = over that type (e.g. $2 = 2$). By adding the keyword **ordered** before the enumerated set of constants, the relation $\leq$ will also be defined on that type as well as the functions **succ**(x) (the successor function) and **pred**(x) (the predecessor function) (in this case, applying **succ** to 4 generates an error message). As an example of an array type,

```
type arrint4 = array[int4] of boolean;
```

defines an array type of four elements, assuming the previous definition of **int4**. Each element of such an array is capable of storing one of the values **true** or **false**. An example of a definition of a tuple type is

```
type tup = int4 * int4;
```

- Variables

A variable is defined using the **var** declaration. A variable must be supplied with an initial value when it is declared. An example:

```
var a:int4 = 2;
```

The optional keywords **read**, **write** and **readwrite** determine how and if the variable is visible to a potential observer (user) of the program. If a variable is declared as **read**-able the observer can inquire of the value of that variable. If a variable is **write**-able the outside observer can modify the value of that variable. The **readwrite** keyword combines the effects of **read** and **write**.

Array variables are not full members of the language in that we place some restrictions on their usage. They cannot be passed as parameters to procedures, cannot be communicated via ports, nor can they be assigned to as a single entity. It is of course possible to assign values to elements of array variables, for example,

```
var a:arrint4 = {true, true, true, true};
a[1] := false
```

- Ports

Concurrent processes may use ports to synchronize their activities. A process sends a value on a port using the *!* operator and receives values into variables using the *?* operator. A process attempting communication on a port will halt its execution until another process is also ready for communication. As an example:

```
port p:int4;
var  x:int4;

p?x par p!2 par p!3
```

The three commands p?x, p!2, and p!3 represent three processes that execute in parallel. Each process is suspended until another process is able to communicate with it. Communication is binary. In an execution of the three commands, either the value 2 or the value 3 is assigned to the variable x.

If a port is declared to be **visible**, an observer of the program will be able to communicate with the program through that port. Otherwise the port is only accessible within the program (and in the scope of the port declaration).

- Procedures

  A procedure contains a declaration section and a command. A procedure accepts zero or more parameters. **in** parameters are used to supply values to the procedure. **out** parameters communicate results from the procedure back to the caller. **inout** parameters combine the effects of **in** and **out**.

  The semantics for procedure calls is as follows. First **in** and **inout** parameters are evaluated (*call-by-value* style) and temporary copies of **out** and **inout** parameters are created. Then the command in the procedure body is executed. After the execution of the procedure body the values in the temporary **inout** and **out** variables are copied back to the variables supplied in the actual procedure call command.

  In order to ensure that Lunsen programs are finite-state, the following (syntactic) restrictions on admissible procedure calls are enforced:

  1. A parent may always call its child.
  2. A child may never call its parent.
  3. A sibling may call another sibling as long as the call is made *tail-recursively*.
  4. A procedure may call itself as long as the call is made tail-recursively.

  Here *tail-recursive* means that no command can occur after the call command in the calling procedure, i.e. after the end of the execution of the called procedure, the calling procedure does not have to be resumed. Furthermore we demand that if an **inout** or **out** variable occurs at position i in the enumerated list of **inout** and **out** formal parameters to the procedure in which the tail-recursive call is made, then it should occur at the same place in the corresponding enumerated list of **inout** and **out** actual parameters to the called procedure in the tail-recursive call. We also require that these two enumerated lists have the same number of elements, i.e. no extra **inout** or **out** parameters are allowed in the formal parameter list of the tail-recursively called procedure. To illustrate:

```
procedure p;
  procedure p1;
    p;                Illegal! P1 may not call its parent
    p2                But may call its sibling tail-recursively
  endproc;

  procedure p2;
    p1;               Illegal! p2 may not call its sibling non-tail-recursively
    p2                But may call itself tail-recursively
  endproc;

  p1; p1              P may call p1 (a child) non tail-recursively as well as tail-recursively
endproc;
```

- Expressions

  An expression represents a value which can be passed as an **in** parameter to procedures and assigned to variables.

  The **if** expression evaluates its conditional expressions in sequential order. The value of the **if** expression is the value of the expression corresponding to the first true boolean expression. If no boolean expression evaluates to true and there exists an **else** clause, the value of the **if** expression is the value of the **else** expression. Otherwise the **if** expression aborts. The projection function $\#I_c\ E$ will return the component number I_c of the tuple expression E. Tuple components are numbered consecutively starting from 1.

- Commands

  Standard commands such as sequencing (;), assignment (:=) exists in Lunsen and have their usual meaning. Note that the execution of the assignment command is non-atomic: the evaluation of the right-hand side is separate from storing the result into the left-hand side.

  The **if** command is similar to the **if** expression. The execution of the **when** command is suspended until one of its communication events can take place; then the corresponding command is executed. The τ event will take place spontaneously, without having to wait for communication with another process. Thus the **when** command may introduce explicit non-determinism in a program. As an example:

```
port synch:boolean;

when
  tau -> p1
| synch!true -> p2
endwhen
```

The execution of the **when** command in the example can proceed in two ways: either through the spontaneous τ event, in which case p1 is executed, or by sending on the synch port, in which case p2 is executed; the last alternative requires that another process is ready to communicate on the synch port.

The **par** command creates two processes that execute in parallel. Given

```
program Pvar;
  type int4 = ordered {1,2,3,4};
  var v:int4;

  v := 1;
  v := succ(v) par v := succ(v)
endprog;
```

the value of v may become either 2 or 3, i.e. the execution of parallel commands is finely grained. This is due to the semantics of the assignment command, which is executed non-atomically.

CCS and Typed CCS

vo versions of CCS (Calculus of Concurrent Systems) are used in this paper. The first one is e *basic* calculus. The second version is called TCCS (Typed CCS), and is closely related to the *lue-passing calculus* in Chapter 2.8 of [Mil89]. TCCS extends basic CCS in that action prefixes are

S::=	$K\ E$ **where** $D;\ldots;D$	TCCS program
D::=	**type** $t = T$	Type definition
	port $p : t$	Port definition
	$K\ \tilde{v} : t \stackrel{\text{def}}{=} A$	Agent definition
T::=	$\{c,\ldots,c\}$	Enumerated type
	ordered $\{c,\ldots,c\}$	Ordered enumerated type
	$t * \ldots * t$	Tuple type
A::=	**nil**	The **nil** agent
	$K\ E$	Agent identifier expression
	$P.A$	Prefixing
	$A + A$	Choice
	$A \mid A$	Parallel composition
	$A\backslash\{p,\ldots,p\}$	Restriction
	$A[p/p,\ldots,p/p]$	Relabeling
	if E **then** A **else** A	If agent
E::=	c	Constant
	v	Variable
	$(E,\ldots,E)$	Tuple expression
	$\#c\ E$	Tuple access
	$E = E$	Equality
	$E \leq E$	Less than or equal
	not $E \mid E$ **and** $E \mid E$ **or** E	Boolean functions
P::=	τ	The τ action
	$p!E$	Send value E on port p
	$p?\tilde{v}$	Receive any value on port p
	$p?{=}E$	Receive value E on port p

Table 2: The syntax of TCCS

explicitly parameterized on data values. TCCS acts as an intermediary language in the translatio from Lunsen to CCS. This enables us to separate the concerns of flow control in Lunsen from concern related to typing and value-passing.

We first briefly review basic CCS. Let $A, B, \ldots$ range over *agents*, and let $a, b, \ldots$ range over por names. The complementary port of a is denoted by $\overline{a}$. Two agents can communicate if one of ther has a port named a and the other a port named $\overline{a}$. We extend the set of port names with the silen action τ to form the set of CCS actions. We let α range over the set of actions. The operators use in the basic calculus are: prefixing ($\alpha.A$), choice ($A + B$), parallel composition ($A \mid B$), restrictio ($A\backslash L$) on a set of ports L, and relabeling ($A[f]$) where f is a function that relabels the ports in A As usual we write $\sum_{i=1}^{n} A_i$ for $A_1 + \cdots + A_n$.

For TCCS we presuppose a set of *types* ranged over by t, a set of *agent identifiers* ranged over b K, a set of *port names* ranged over by p, q, a set of TCCS *variables* ranged over by v, and a set o *constant values* ranged over by c. We write $\tilde{v}$ for a (possibly empty) tuple of variables ($v_1, \ldots, v_n$ and similarly $\tilde{c}$ for a tuple of constants.

The syntax for TCCS is defined in Table 2 using the BNF dialect. There are two predefined types **boolean** with the two elements **true** and **false**, and **unit** with the only element (). This elemen may be omitted in expressions; for example $p!.A$ is short for $p!().A$.

TCCS variable v is *bound* in the input prefix $p?v.A$; more generally $p?\tilde{v}.A$ binds all variables in $\tilde{v}$. imilarly an agent definition $K\tilde{v} : t \stackrel{\text{def}}{=} A$ binds the variables $\tilde{v}$ in A. We only consider TCCS programs hich are well typed and where all variables occur bound. Thus, in a TCCS program each agent lentifier, port name, TCCS variable, and constant value is associated with a unique type as given in TCCS definition (the type of a variable is considered the same as the type of the port or identifier here it is bound). In the following we write $typeof(X)$ for the type associated with such an object r a tuple of such objects) X. We also write $A[\tilde{c}/\tilde{v}]$ to mean the TCCS agent gained by substituting ach free occurrence of v_i by c_i.

he meaning of TCCS is defined by a function $\mathcal{T}[\![C]\!]$, which maps a TCCS construction C into basic CS. In this definition we do not distinguish between a closed expression (an expression without ariables) and the constant value it denotes when the operators "=", "**and**" etc. are given the bvious interpretations. We further assume that for each TCCS port p and constant c of the same ype there is a basic CCS action p_c.

he first clause in the definition of $\mathcal{T}[\![]\!]$ is:

$$\mathcal{T}[\![K\ E \textbf{ where } D_1; \ldots; D_n]\!] = \mathcal{T}[\![K\ E]\!]$$

here the right hand side is to be interpreted with respect to the basic CCS agent identifier definitions troduced by $D_1, \ldots D_n$ as follows. TCCS type definitions and TCCS port definitions do not result any basic CCS identifier definitions. Each TCCS agent identifier definition $K\ \tilde{v} : t \stackrel{\text{def}}{=} A$ yields the t of CCS agent identifier definitions: $K_{\tilde{c}} \stackrel{\text{def}}{=} \mathcal{T}[\![A[\tilde{c}/\tilde{v}]]\!]$ forall $\tilde{c}$ of type t.

he translation of a TCCS agent is defined in Table 3. Note in particular the *determined* input onstruct $p?{=}E.A$. This results in a TCCS agent which can only accept a particular value (as etermined by E) on port p; such a construct turns out to be useful in defining the semantics of unsen arrays.

A	$\mathcal{T}[\![A]\!]$
nil	**nil**
$K\ E$	K_E
$\tau.A$	$\tau.\mathcal{T}[\![A]\!]$
$p?\tilde{v}.A$	$\sum_{typeof(\tilde{c})=typeof(p)} p_{\tilde{c}}.\mathcal{T}[\![A[\tilde{c}/\tilde{v}]]\!]$
$p!E.A$	$\overline{p_E}.\mathcal{T}[\![A]\!]$
$p?{=}E.A$	$p_E.\mathcal{T}[\![A]\!]$
$A_1 + A_2$	$\mathcal{T}[\![A_1]\!] + \mathcal{T}[\![A_2]\!]$
$A_1 \mid A_2$	$\mathcal{T}[\![A_1]\!] \mid \mathcal{T}[\![A_2]\!]$
$A \backslash L$	$\mathcal{T}[\![A]\!] \backslash \{p_c : p \in L, c \in typeof(p)\}$
$A[f]$	$\mathcal{T}[\![A]\!][f']$ where $f'(p_c) = f(p)_c$
if true then A_1 **else** A_2	$\mathcal{T}[\![A_1]\!]$
if false then A_1 **else** A_2	$\mathcal{T}[\![A_2]\!]$

Table 3: Translation of TCCS agents into CCS agents

or example, $\mathcal{T}[\![K(x : \textbf{boolean}) \stackrel{\text{def}}{=} p?y.q!y.r?{=}x.\textbf{nil}]\!]$ yields the set of basic CCS agents identifier def- itions

$$\begin{cases} K_{true} \stackrel{\text{def}}{=} p_{true}.\overline{q_{true}}.r_{true}.\textbf{nil} + p_{false}.\overline{q_{false}}.r_{true}.\textbf{nil} \\ K_{false} \stackrel{\text{def}}{=} p_{true}.\overline{q_{true}}.r_{false}.\textbf{nil} + p_{false}.\overline{q_{false}}.r_{false}.\textbf{nil} \end{cases}$$

4 A Formal Semantics for Lunsen

4.1 Combinators

When translating the Lunsen constructions into TCCS, we will use a number of basic combinators similar to the ones defined in Milner's book [Mil89]. The combinator *Before* will be used to model sequential composition of two agents; the first agent must signal that it has finished "running" by using the combinator *Done* before the other agent can start "running". The *Par* combinator is used to model two processes running in parallel.

Each expression will return its value by using the combinator *Result(value)*. Such a value can be bound to a TCCS variable in a process agent through the combinator *Into*, as in *expr Into*$(x)\,(ag)$; the TCCS variable x will here be bound to the value of *expr* in the TCCS process agent *ag*. The combinator *Into_l* is a variant of *Into*, allowing a list of variables to be bound to a list of expressions in an agent body. We use the syntax *hd* :: *tail* for a list consisting of the head *hd* and tail list *tl*. [] will denote the empty list. If we are certain that a list consists of a fixed number of elements, say e_1 and e_2, this is written as $[e_1, e_2]$. The combinators are defined in Table 4.

$$\begin{aligned}
Done &= done!.\mathbf{nil} \\
P\ Before\ Q &= (P[b/done] \,|\, b?.Q)\backslash\{b\} \\
P\ Par\ Q &= (P[d1/done] \,|\, Q[d2/done] \,|\, d1?.d2?.Done)\backslash\{d1, d2\} \\
Result(v) &= result!v.\mathbf{nil} \\
E\ Into(x)\,(A) &= (E[i/result] \,|\, i?x.A)\backslash\{i\} \\
\\
[]\ Into_l([])\,(A) &= A \\
E :: RestE\ Into_l(x :: Restx)\,(A) &= E\ Into(x)\,(RestE\ Into_l(Restx)\,(A))
\end{aligned}$$

Table 4: The basic combinators

4.2 Variables

A non-array variable v of type T is translated to the TCCS agent

$$Reg_v(y : T) \stackrel{\text{def}}{=} put_v?(x).Reg_v(x) + get_v!(y).Reg_v(y)$$

A value can be stored in the variable v by sending the new value on the port put_v. Reading of values from v is accomplished by receiving the current value from the port get_v. An array variable v of type **array** $[T_{index}]$ **of** T_{store} will be represented as the family of agents

$$\forall i \in T_{index} : Reg_{v\_i}(y : T_{store}) \stackrel{\text{def}}{=} get_v!(i, y).Reg_{v\_i}(y) + \sum_{x \in T_{store}} put_v?{=}(i, x).Reg_{v\_i}(x)$$

We will use $store(a)$ to denote the type of elements in an array a (T_{store} in the example), and $index(a)$ to denote the index type of the array (T_{index} in the example). So, reading the array element $a[2]$ is accomplished in the TCCS agent

$$\sum_{z \in store(a)} get_a?{=}(2, z).B$$

When we present the formal semantics of Lunsen below, we will for simplicity assume that no variables are of an array type (except in explicit array access expressions and commands). Each rule involving gisters should thus be extended with a case where the variable is an array, in which case a set of gister agents Reg_{v_i} should be used instead of the single agent Reg_v.

.3 Access and Restriction sorts

n access sort $\mathcal{ACC}$ for a declaration in Lunsen is the set of TCCS port names by which it it possible r other Lunsen declarations and commands to interact with that declaration. For a variable v, the cess sort will consist of put_v and get_v.

he restriction sort $\mathcal{RAC}$ for an declaration contains the port names of that declaration which should ot be accessible to an external observer of the program. It will be identical to $\mathcal{ACC}$ except when sible variables or ports occur in a program. The function $\mathcal{RAC}$ is presented in Table 5.

$$
\begin{aligned}
\mathcal{RAC}(\textbf{var } I_v{:}I_t = E;) &= \{put_{I_v}, get_{I_v}\}\\
\mathcal{RAC}(\textbf{var } I_v{:}\textbf{read } I_t = E;) &= \{put_{I_v}\}\\
\mathcal{RAC}(\textbf{var } I_v{:}\textbf{write } I_t = E;) &= \{get_{I_v}\}\\
\mathcal{RAC}(\textbf{var } I_v{:}\textbf{readwrite } I_t = E;) &= \emptyset\\
\mathcal{RAC}(\textbf{procedure } I_p[(I_v{:}A\ I_t,\ldots,I_v{:}A\ I_t)]\ D\ C\ \textbf{endproc};) &= \emptyset\\
\mathcal{RAC}(\textbf{port } I_m{:}\textbf{visible } I_t;) &= \emptyset\\
\mathcal{RAC}(\textbf{port } I_m{:}I_t;) &= \{I_m\}\\
\mathcal{RAC}(D_1\ D_2) &= \mathcal{RAC}(D_1)\bigcup\mathcal{RAC}(D_2)
\end{aligned}
$$

Table 5: The function $\mathcal{RAC}$ for computing restriction sorts

.4 The translation

'e define the translation $\mathcal{L}[\![]\!]$ from commands in Lunsen to TCCS agents in separate tables for eclarations, expressions, and commands. Constructions not generating any TCCS "code" are not sted in the tables. These for example include definitions of types. First we show the translation of e **program** statement:

$$\mathcal{L}[\![\textbf{program } I_s;\ D\ C\ \textbf{endprog};]\!] = I_s,\ \text{where } I_s \stackrel{\text{def}}{=} (\mathcal{L}[\![D]\!]\,|\,(\mathcal{L}[\![C]\!]\ \mathit{Before}\ \textbf{nil}))\backslash\mathcal{RAC}_{\mathcal{D}}$$

ranslation of declarations is listed in Table 6. In the rule for procedure translation **(1)**, $x_1^i,\ldots,x_n^i$ atch the subset of formal parameters to the procedure $(I_{v1}^i,\ldots,I_{vn}^i)$ that are **in** or **inout** paramers. Similarly, $x_1^o,\ldots,x_n^o$ match the subset of formal parameters to the procedure $(I_{v1}^o,\ldots,I_{vn}^o)$ that e **out** or **inout** parameters. The intuition behind the translation of a procedure is that the **in** and **out** actual parameter values $x_1^i,\ldots,x_n^i$ are supplied as parameters to the agent identifier I_p in the anslation of the call command. Temporary copies of all parameters are made and the actual values the parameters are stored in the registers $Reg_{I_{vi}}$. When the execution of the translated command $[\![C]\!]$ in the procedure has finished, the values of the temporary registers and returned as a TCCS ple expression through the use of the *Result* combinator.

he translation of Lunsen expressions into TCCS agents is shown in Table 7 and translation of mmands in Table 8.

the translation of a non-tail-recursive procedure call **(2)** in Table 8, we assume that **in** and **out** parameters are denoted by E_1^i through E_n^i and the **out** and **inout** parameters by E_1^o through E_n^o. Note that the actual parameters corresponding to **out** and **inout** formal parameters must be riables (I_v). In the translation we evaluate the **in** expressions, continue with the execution of the translated procedure I_p which will as previously described end its execution by returning the value of the "out" parameters. We then store back those values in the variables (I_v) passed as actual parameters to the procedure call.

The translation of the tail-recursive call (**3**) is more straightforward because we do not have to store back the results of the procedure call into the actual parameters.

Our semantics for scoping of procedures and procedure calling follow the informal rules presented in 2. Since the informal rules are nonproblematic we omit a formal specification here. Likewise, we do not present the function which determines if a procedure call is a tail-recursive one.

We will examplify the translation rules by informally discussing the result of applying them to the `Pvar` program in Section 2, in the introduction to Lunsen Commands. The resulting CCS specification will consist of two agents in parallel, the first representing the variable `v` in the form of a register agent, the second the execution flow in the program. The agent representing the execution flow starts by assigning `1` to `v`, then splits into two new agents composed by the CCS parallel operator. Each of these two agents will perform two atomic actions: first compute the value `succ(v)`, and then assign that value to `v`. Since the translation of the assignment command consists of two atomic actions, the resulting value assigned to `v` can be either 2 or 3:

- if the computation of `succ(v)` by the two agents are performed directly after each other, the result will be 2
- if one of these agents gets to both compute `succ(v)` and then store that value in `v` before the other agent computes `succ(v)`, the result will be 3

After both agents have performed their assignment to `v`, the execution of the program terminates.

5 Implementation details

As mentioned in the introduction, one aim is to use the output from the Lunsen compiler in analyses performed by the Concurrency Workbench. This means that care has to be taken to ensure that all generated agents are finite-state. In this section we briefly comment on some problems in this respect.

5.1 Problems

The basic Combinators from Section 4.1 may introduce agents which are syntactically non-finite state, although they are equivalent with finite state agents.

For example, the TCCS agent definition $A \stackrel{\text{def}}{=} p.Done\ Before\ A$ is translated into the CSS agent $A = (p.\overline{done}.nil[b/done]|b.A)\backslash\{b\}$. The expansion of this agent using the *expansion law* yields the agent definition $A = p.(nil[b/done]|A)\backslash\{b\}$. The CWB could here reduce the agent expression to $p.A$ using a rule equating $(nil[Z]|X)\backslash Y$ to $X\backslash Y$. Instead, CWB chooses the strategy of continuing to expand the $(nil[b/done]|A)\backslash\{b\}$ expression, resulting in the expression $p.(nil[b/done]|p.(nil[b/done]|A)\backslash\{b\})\backslash\{b\}$. Obviously, the expansion process will never terminate.

To avoid these problems the basic combinators are implemented in the following way:

- P *Before* Q is implemented as the agent obtained by substituting each occurrence of $done!.$**nil** in P with Q.
- E *Into*(x) (A) is implemented as: the agent obtained by substituting each occurrence of a $result!v.$**nil** expression in E with $A\{v/x\}$ (A where we substitute v for x).

$$\mathcal{L}[\![D_1\ D_2]\!] \;=\; \mathcal{L}[\![D_1]\!] \,|\, \mathcal{L}[\![D_2]\!]$$

$$\mathcal{L}[\![\mathbf{var}\ I_v{:}[G]\ I_t = E;]\!] \;=\; Reg_{I_v}(\mathcal{L}[\![E]\!]) \text{ where } E \text{ is a constant expression}$$

$$\mathcal{L}[\![\mathbf{procedure}\ I_p(I_{v1}{:}A\ I_{t1},\ldots,I_{vn}{:}A\ I_{tn});\ D\ C\ \mathbf{endproc};]\!] = \mathbf{nil}$$

We define the TCCS agent $I_p((x_1^i,\ldots,x_n^i)) \stackrel{\mathrm{def}}{=} (\underbrace{Reg_{I_{v1}}(\_) \mid \ldots \mid Reg_{I_{vn}}(\_)}_{\text{Temp. copies of parms.}} \mid$

$$\underbrace{put^i_{I_{v1}}!x^i_1.\ldots.put^i_{I_{vn}}!x^i_n}_{\text{In(out) parms.}} .\ \mathcal{L}[\![C]\!]\ Before \qquad (1)$$

$$\underbrace{get^o_{I_{v1}}?x^o_1.\ldots.get^o_{I_{vn}}?x^o_n}_{\text{(in)Out parms.}} .Result((x^o_1,\ldots,x^o_n)) \mid$$

$$\mathcal{L}[\![D]\!]) \backslash ACC_{\mathcal{D}} \bigcup ACC_{I_{v1}} \bigcup \ldots \bigcup ACC_{I_{vn}}$$

Table 6: Translation of declarations (D)

$$\mathcal{L}[\![I_c]\!] \;=\; Result(I_c)$$

$$\mathcal{L}[\![I_v]\!] \;=\; get_{I_v}?x.Result(x)$$

$$\mathcal{L}[\![I_a[E_1,\ldots,E_n]]\!] \;=\; [\mathcal{L}[\![E_1]\!],\ldots,\mathcal{L}[\![E_n]\!]]\ Into\_l([x_1,\ldots,x_n]) \; (\textstyle\sum_{y\in store(I_a)}(get_{I_a}?{=}((x_1,\ldots,x_n),y).Result(y)))$$

$$\mathcal{L}[\![(E_1,\ldots,E_n)]\!] \;=\; [\mathcal{L}[\![E_1]\!],\ldots,\mathcal{L}[\![E_n]\!]]\ Into\_l([x_1,\ldots,x_n])\ Result((x_1,\ldots,x_n))$$

$$\mathcal{L}[\![\#I_c\ E]\!] \;=\; \mathcal{L}[\![E]\!]\ Into(x)\,(\#\ I_c\ x)$$

$$\mathcal{L}[\![\mathbf{if}\ E_{b1} \rightarrow E_1 \mid\ldots\mid E_{bn} \rightarrow E_n\ \mathbf{endif}]\!] \;=\; \mathcal{L}[\![E_{b1}]\!]\ Into(x_1)\ (\mathbf{if}\ x_1\ \mathbf{then}\ \mathcal{L}[\![E_1]\!]\ \mathbf{else} \ldots \mathbf{else}\ \mathcal{L}[\![E_{bn}]\!]\ Into(x_n)\ (\mathbf{if}\ x_n\ \mathbf{then}\ \mathcal{L}[\![E_n]\!]\ \mathbf{else\ nil}))$$

$$\mathcal{L}[\![\mathbf{if}\ E_{b1} \rightarrow E_1 \mid\ldots\mid E_{bn} \rightarrow E_n\ \mathbf{else}\ E\ \mathbf{endif}]\!] \;=\; \mathcal{L}[\![E_{b1}]\!]\ Into(x_1)\ (\mathbf{if}\ x_1\ \mathbf{then}\ \mathcal{L}[\![E_1]\!]\ \mathbf{else} \ldots \mathbf{else}\ \mathcal{L}[\![E_{bn}]\!]\ Into(x_n)\ (\mathbf{if}\ x_n\ \mathbf{then}\ \mathcal{L}[\![E_n]\!]\ \mathbf{else}\ \mathcal{L}[\![E]\!]))$$

$$\mathcal{L}[\![E_1 = E_2]\!] \;=\; [\mathcal{L}[\![E_1]\!],\mathcal{L}[\![E_2]\!]]\ Into\_l([x_1,x_2])\,(Result(x_1 = x_2))$$

$$\mathcal{L}[\![E_1 \leq E_2]\!] \;=\; [\mathcal{L}[\![E_1]\!],\mathcal{L}[\![E_2]\!]]\ Into\_l([x_1,x_2])\,(Result(x_1 \leq x_2))$$

$$\mathcal{L}[\![\mathbf{succ}\ E]\!] \;=\; \mathcal{L}[\![E]\!]\ Into(x)\,(Result(\mathbf{succ}(x)))$$

$$\mathcal{L}[\![\mathbf{pred}\ E]\!] \;=\; \mathcal{L}[\![E]\!]\ Into(x)\,(Result(\mathbf{pred}(x)))$$

$$\mathcal{L}[\![\mathbf{not}\ E]\!] \;=\; \mathcal{L}[\![E]\!]\ Into(x)\,(Result(\mathbf{not}(x)))$$

$$\mathcal{L}[\![E_1\ \mathbf{and}\ E_2]\!] \;=\; \mathcal{L}[\![E_1]\!]\ Into(x_1)\,(\mathbf{if}\ x_1 = \mathbf{false\ then}\ Result(\mathbf{false})\ \mathbf{else}\ \mathcal{L}[\![E_2]\!])$$

$$\mathcal{L}[\![E_1\ \mathbf{or}\ E_2]\!] \;=\; \mathcal{L}[\![E_1]\!]\ Into(x_1)\,(\mathbf{if}\ x_1 = \mathbf{true\ then}\ Result(\mathbf{true})\ \mathbf{else}\ \mathcal{L}[\![E_2]\!])$$

Table 7: Translation of expressions (E)

$$\mathcal{L}[\![\textbf{begin } C \textbf{ end}]\!] = \mathcal{L}[\![C]\!]$$

$$\mathcal{L}[\![\textbf{skip}]\!] = \mathit{Done}$$

$$\mathcal{L}[\![C_1 \,;\, C_2]\!] = \mathcal{L}[\![C_1]\!]\ \mathit{Before}\ \mathcal{L}[\![C_2]\!]$$

$$\mathcal{L}[\![\mathrm{C}\ \textbf{par}\ C]\!] = \mathcal{L}[\![C_1]\!]\ \mathit{Par}\ \mathcal{L}[\![C_2]\!]$$

$$\mathcal{L}[\![I_v := E]\!] = \mathcal{L}[\![E]\!]\ \mathit{Into}(x)\,(\mathit{put}_{I_v}!x.\mathit{Done})$$

$$\begin{aligned}\mathcal{L}[\![I_a[E_1,\ldots,E_n] := E_f]\!] = {} & [\mathcal{L}[\![E_f]\!], \mathcal{L}[\![E_1]\!], \ldots, \mathcal{L}[\![E_n]\!]]\ \mathit{Into\_l}([x_f, x_1, \ldots, x_n]) \\ & (\mathit{put}_{I_a}!((x_1,\ldots,x_n), x_f).\mathit{Done})\end{aligned}$$

Non tail-recursive procedure call (**2**):

$$\begin{aligned}\mathcal{L}[\![I_p[(E_1,\ldots,E_n)]]\!] = {} & [\mathcal{L}[\![E^i_1]\!], \ldots, \mathcal{L}[\![E^i_n]\!]]\ \mathit{Into\_l}([x^i_1, \ldots, x^i_n]) \\ & I_p(x^i_1,\ldots,x^i_n)\ \mathit{Into}(y^o_1,\ldots,y^o_n)\ (\mathit{put}_{E^o_1}!y^o_1.\ldots.\mathit{put}_{E^o_n}!y^o_n.\mathit{Done})\end{aligned}$$

Tail recursive procedure call (**3**):

$$\mathcal{L}[\![I_p[(E_1,\ldots,E_n)]]\!] = [\mathcal{L}[\![E^i_1]\!], \ldots, \mathcal{L}[\![E^i_n]\!]]\ \mathit{Into\_l}([x^i_1, \ldots, x^i_n])\,(I_p(x^i_1,\ldots,x^i_n))$$

$$\begin{aligned}\mathcal{L}[\![\textbf{if } E_{b1} \rightarrow C_1 \mid \ldots \mid E_{bn} \rightarrow C_n \textbf{ endif}]\!] = {} & \mathcal{L}[\![E_{b1}]\!]\ \mathit{Into}(x_1)\ (\textbf{if } x_1 \textbf{ then } \mathcal{L}[\![C_1]\!] \textbf{ else } \ldots \\ & \textbf{else } \mathcal{L}[\![E_{bn}]\!]\ \mathit{Into}(x_n)\ (\textbf{if } x_n \textbf{ then } \mathcal{L}[\![C_n]\!] \textbf{ else nil}))\end{aligned}$$

$$\begin{aligned}\mathcal{L}[\![\textbf{if } E_{b1} \rightarrow C_1 \mid \ldots \mid E_{bn} \rightarrow C_n \textbf{ else } C \textbf{ endif}]\!] = {} & \mathcal{L}[\![E_{b1}]\!]\ \mathit{Into}(x_1)\ (\textbf{if } x_1 \textbf{ then } \mathcal{L}[\![C_1]\!] \textbf{ else } \ldots \\ & \textbf{else } \mathcal{L}[\![E_{bn}]\!]\ \mathit{Into}(x_n)\ (\textbf{if } x_n \textbf{ then } \mathcal{L}[\![E_n]\!] \textbf{ else } \mathcal{L}[\![C]\!]))\end{aligned}$$

$$\mathcal{L}[\![\textbf{when } P_1 \rightarrow C_1 \mid \ldots \mid P_n \rightarrow C_n \textbf{ endwhen}]\!] = \mathcal{L}[\![P_1]\!]\ \mathit{Before}\ \mathcal{L}[\![C_1]\!] + \ldots + \mathcal{L}[\![P_n]\!]\ \mathit{Before}\ \mathcal{L}[\![C_n]\!]$$

$$\begin{aligned}\mathcal{L}[\![\textbf{while } E \textbf{ do } C \textbf{ endwhile}]\!] = {} & W, \text{ where } W \stackrel{\text{def}}{=} \mathcal{L}[\![E]\!]\ \mathit{Into}(x) \\ & (\textbf{if } x \textbf{ then } \mathcal{L}[\![C]\!]\ \mathit{Before}\ W \textbf{ else } \mathit{Done})\end{aligned}$$

$$\mathcal{L}[\![I_m\ !\ E]\!] = \mathcal{L}[\![E]\!]\ \mathit{Into}(x)\,(I_m!x.\mathit{Done})$$

$$\mathcal{L}[\![\tau]\!] = \tau.\mathit{Done}$$

$$\mathcal{L}[\![I_m\ ?\ I_v]\!] = I_m?x.\mathit{put}_{I_v}!x.\mathit{Done}$$

Table 8: Translation of commands (C and P)

- *A Par B* is defined as $p_1!.p_2!p_1?.p_2?.Done$, where p_1 and p_2 are ports unique to the program. Two new agents are also added, $A' \stackrel{\text{def}}{=} p_1?.A$ *Before* $p_1!.A'$ and $B' \stackrel{\text{def}}{=} p_2?.B$ *Before* $p_2!.B'$.

n example: assuming $A \stackrel{\text{def}}{=} a.A + done!.\mathbf{nil}$ and $B \stackrel{\text{def}}{=} b.B$, then *A Before B* is equivalent to A' where $'$ is a new agent: $A' \stackrel{\text{def}}{=} a.A + b.B$.

he compiler furthermore "lifts up" all parallel compositions (and restrictions) into the main program gent (the one named in the program statement). In this process unique names for restricted objects re created. Thus the main program agent will appear:

$$Spec \stackrel{\text{def}}{=} (Ag_1 | \ldots | Ag_n) \backslash \{p_1, \ldots, p_n\}$$

nd the TCCS agents Ag_i will contain only + (choice), . (prefixing) and agent identifiers (K's).

.2 Optimizations

Ve also do some further optimizations for the sake of efficiency of the analysis of the generated CCS gents in the CWB:

- Lunsen variables which can be accessed by at most one concurrent process do not need associated Register agents (cf. Section 4.2). As an example,

```
procedure p;
  var v:boolean = false;
  q!v
endproc;
```

  The variable v in the example can only be used in the procedure p and since the procedure contains no **par** command, the variable cannot be accessed by multiple processes concurrently. Therefore we need not actually use a register agent for v but may encode it directly in the TCCS agent corresponding to the translation of procedure p.

- Some useful algebraic manipulations: simplify $A + A$ into A, $A + \mathbf{nil}$ into A and $X.\tau.Y$ into $X.Y$.

An Example

n this section we will present an example of a distributed algorithm which is normally, and perhaps ost clearly, formulated in an imperative language. Our example is an algorithm for mutual exclusion ue to Peterson [PS85]. The algorithm will be defined in Lunsen and formally verified with the oncurrency Workbench. In the following we assume the reader to have some familiarity with the odal logic supported by the Workbench. An introduction to this logic and its use for verifying utual exclusion algorithms can be found in a recent article by Walker [Wal89]; we will use the same lgorithm and correctness criteria.

irst we will formulate Peterson's mutual exclusion algorithm in Lunsen, assuming two concurrent rocesses competing to enter their critical sections. Then we will minimize the resulting CCS agents .r.t. observation equivalence using the Workbench. Finally we will determine whether the algorithm eets the safety and liveness demands. This will be done by checking if the minimized agents satisfy a air of propositions formulated in HML (Hennessy-Milner logic). These checks are done automatically y the Workbench.

In the formulation of the algorithm in Lunsen, we will use port communications to signal that process requests to enter, enters and leaves its critical section (*req*!i, *enter*!i and *exit*!i, where i i the name of one of the two concurrent processes). These are the only actions which will be visible t an observer of the program. Peterson's algorithm as formulated in Lunsen is presented in Figure 1 The state graph of the Lunsen program after translation into basic CCS, and minimization w.r.t observation equivalence by the Concurrency Workbench is displayed in Figure 2.

```
program Peterson;
  type int2 = {1, 2};
  type b_arr = array[int2] of boolean;

  var b:b_arr = {false, false};
  var k:int2;

  -- Ports visible to an observer
  port enter :visible int2;
  port exit  :visible int2;
  port req   :visible int2;

  procedure p(i:in int2, j:in int2);
    while true do
      b[i] := true; req!i;
      k := j;
      while (b[j] and (k = j)) do skip endwhile;
      enter!i;         -- Enter critical region
      exit!i;          -- Exit critical region
      b[i] := false
    endwhile
  endproc;

  p(1,2) par p(2,1) -- Start two processes running in parallel
endprog; -- Peterson
```

Figure 1: Peterson's algorithm

The mutual exclusion property we wish the algorithms to preserve can be formulated:

$$Mutex = \nu Z.(\neg(< exit!1 > true \wedge < exit!2 > true) \wedge [K]Z),$$

that is both processes should not be able to leave their critical sections at the same time, which implies that not both processes *are* in their critical section. $[K]Z$ in the previous formula is an abbreviation for $[K]Z \equiv \wedge_{a \in K}[a]Z$, where K is a set of actions. We will use

$$K = \{enter!1, enter!2, exit!1, exit!2, req!1, req!2\}$$

for verifying the mutual exclusion property. The liveness property can be formulated:

$$Live \equiv Live_1 \wedge Live_2$$

where

$$Live_i \equiv \nu Z.([req!i]\mu Y.(< exit!i > true \vee [K]Y) \wedge [K]Z)$$

This formula expresses that if a process i has requested execution of its critical section by *req*!i then there shall be no infinite path of actions not consisting of an *enter*!i action in the corresponding transition system. As pointed out by Walker this is just one interpretation of the liveness properties of the algorithm. When we check the safety property we find that $Peterson \models Mutex$ as hoped, i.e. two processes cannot be in their critical sections at the same time. We also find that it satisfies *Live*.

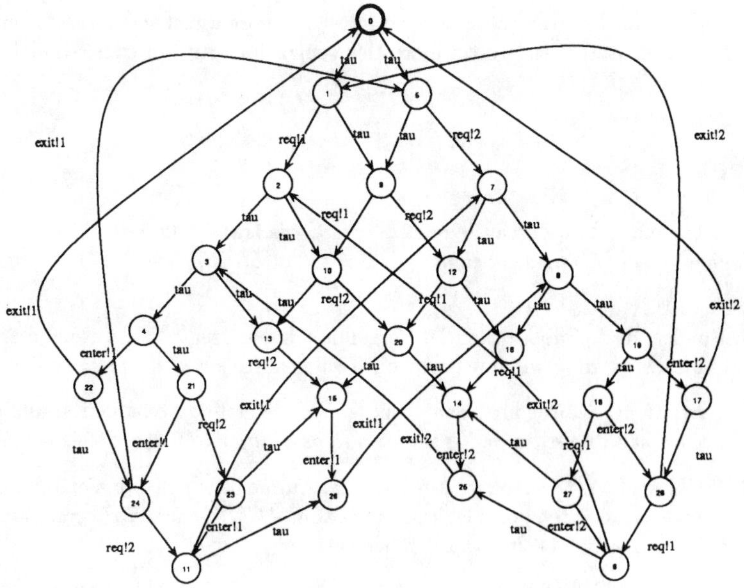

Figure 2: The graph of the minimized Peterson algorithm

7 Conclusions

We have presented an imperative language with constructs for sequential programming as well as constructs for parallel execution of processes, and an automated tool for analysis of concurrent programs written in the language. A few restrictions were imposed on the language to ensure that the semantics of a program is finite-state: e.g.: the types of variables must contain only finitely many elements, and that procedures cannot call each other recursively in an arbitrary manner. Many concurrent algorithms can be naturally described in this language. Our experience includes several other mutex algorithms (e.g. in [Lam86]) and two versions of the Alternating-Bit protocol [BSW69].

We have presented a formal semantics for the language, which represents the execution of a program on a multiprocessor with shared memory, without assuming that e.g. assignment statements are atomic. This means that the results of analysis are valid for a direct implementation of the algorithm.

The semantics was given through a translation to CCS. The efficiency of the resulting "code" is on par with the results of hand translations. The advantage of this is that CWB and associated systems can be used to carry out different forms of analysis. A disadvantage of the present implementation is that properties of programs must be formulated in terms of communication events and not in terms of predicates over the state of the program.

Desirable extensions of the language include a module concept to structure large programs and enable several instantiations of processes. This leads to problems with conflicting access sorts: a newly instantiated process needs to know on which ports it should communicate. One way to achieve this is to use CCS restriction and relabeling operators, but if these are used recursively, the resulting agents will not in general be finite state. Another way is to define the translation into the π-calculus [MPW89a, MPW89b] rather than into CCS (in the π-calculus an agent is explicitly parametrized on its free port names).

An interesting topic of further research is to give Lunsen a more direct semantics in terms of states and transitions between states, and to compare this semantics with the translational semantics in this paper.

References

[AZ81] E. Astesiano and E. Zucca. Semantics of CSP via translation into CCS. In *Mathematical Foundations of Computer Science*, volume 118 of *LNCS*, pages 172–182. Springer Verlag, 1981.

[BdSV88] G. Boudol, R. de Simone, and D. Vergamini. Experiment with Auto and Autograph on a simple case sliding window protocol. Technical Report 870, Inria, July 1988.

[BSW69] K. Bartlett, R. Scantlebury, and P. Wilkinson. A note on reliable full-duplex transmissions over half duplex lines. *Communications of the ACM*, 2(5):260–261, 1969.

[CES86] E.M. Clarke, E.A. Emerson, and A.P. Sistla. Automatic verification of finite-state concurrent systems using temporal logic specification. *ACM Trans. on Programming Languages and Systems*, 8(2):244–263, April 1986.

[CPS89] R. Cleaveland, J. Parrow, and B. Steffen. A semantics-based verification tool for finite-state systems. In *Protocol Specification, Testing, and Verification IX*, pages 287–302, 1989. North-Holland.

[GLZ89] J.C. Godskesen, K.G. Larsen, and M. Zeeberg. TAV users manual. In *Proc. Workshop on Automatic Verification Methods for Finite State Systems*, Grenoble, 1989.

[HL83] M. Hennessy and W. Li. Translating a subset of Ada into CCS. In D. Bjoerner, editor, *Formal Description of Programming Concepts II*, pages 227–249, Amsterdam, 1983 North-Holland.

[Lam86] L. Lamport. The mutual exclusion problem part II – statement and solutions. *Journal of the ACM*, 33(2), 1986.

[Mil89] R. Milner. *Communication and Concurrency.* Prentice Hall, 1989.

[MPW89a] R. Milner, J. Parrow, and D. Walker. A calculus of mobile processes, part I. Technical Report ECS-LFCS-89-85, Department of Computer Science, University of Edinburgh 1989.

[MPW89b] R. Milner, J. Parrow, and D. Walker. A calculus of mobile processes, part II. Technical Report ECS-LFCS-89-86, Department of Computer Science, University of Edinburgh 1989.

[PS85] J.L. Peterson and A. Silberschatz. *Operating System Concepts.* Addison-Wesley, 1985.

[RRSV87] J. Richier, C. Rodriguez, J. Sifakis, and J. Voiron. Verification in XESAR of the sliding window protocol. In *Protocol Specification, Testing, and Verification VII.* North-Holland, 1987.

[SS86] S.A. Smolka and R.E. Strom. A CCS semantics for NIL. In M. Wirsing, editor, *Formal Description of Programming Concepts III*, pages 347–368, Amsterdam, 1986. North-Holland.

[Wal89] D.J. Walker. Automated analysis of mutual exclusion algorithms using CCS. *Formal Aspects of Computing*, 1:273–292, 1989.

CCSR: A Calculus for Communicating Shared Resources *

Richard Gerber[†] Insup Lee[‡]

Department of Computer and Information Science

University of Pennsylvania

Philadelphia, PA 19104

Abstract

The timing behavior of a real-time system depends not only on delays due to process synchronization, but also on the availability of shared resources. Most current real-time models capture delays due to process synchronization; however, they abstract out resource-specific details by assuming idealistic operating environments. On the other hand, scheduling and resource allocation algorithms used for real-time systems ignore the effect of process synchronization except for simple precedence relations between processes. To bridge the gap between these two disciplines, we have developed a formalism called Communicating Shared Resources, or CSR. This paper presents the priority-based process algebra called the Calculus for Communicating Shared Resources (CCSR), which provides an equational characterization of the CSR language. The computation model of CCSR is resource-based in that multiple resources execute synchronously, while processes assigned to the same resource are interleaved according to their priorities. CCSR possesses a prioritized strong equivalence for terms based on strong bisimulation. The paper also describes a producer and consumer problem whose correct timing behavior depends on priority.

1 Introduction

The correctness of a real-time system depends not only on how concurrent processes interact, but also the time at which these interactions occur. In the tradition of untimed concurrency theory, however, formal models for time-dependent computation have treated the execution of processes abstractly, quite isolated from their operating environments. These environments often have a profound effect on the timing behavior of real-time systems, and cannot be ignored when reasoning about them.

To help bridge the gap between abstract computation models and implementation, we have developed a real-time formalism called *Communicating Shared Resources*, or CSR [4, 3]. CSR's

*This research was supported in part by ONR N000014-89-J-1131.

†Email: rich@linc.cis.upenn.edu

‡Email: lee@central.cis.upenn.edu

underlying computational model is *resource-based*, where a resource may be a processor, an Ethernet link, or any other constituent device in a real-time system. At any point in time, each resource has the capacity to execute an action consisting of only a single event or particle. However, a resource may host a set of many processes, and at every instant, any number of these processes may compete for its availability. "True" parallelism may take place only *between* resources; on a single resource, the actions of multiple processes must be interleaved. To arbitrate between competing events, CSR employs a priority-ordering among them.

Our priority semantics of CSR is based on the linear-history model [2], and its extension to real-time computing [9, 7]. While this semantics adequately captures the temporal properties of prioritized resource interaction, it does not easily lend itself to an equational characterization of the CSR language. It is for this reason that we have developed the Calculus for Communicating Shared Resources, or CCSR. Strongly influenced by SCCS [11, 13], CCSR is a process algebra that uses a synchronous form of concurrency, and possesses a term equivalence based on strong bisimilarity. Syntactically, CSR is a "richer" formalism, in that it contains many real-time language constructs such as timed interrupt-handlers, temporal scopes [10], and periodic processes. However, all of the CSR constructs can be formulated in CCSR, and further, CCSR provides the ability to perform equivalence proofs by syntactic manipulation.

This paper describes our resource-based view of concurrency, including the notion of strong prioritized equivalence based on *strong bisimulation* and the equational characteristics of the CCSR terms. To illustrate the effect of priority on computation, consider the following SCCS program fragment:

$$((a : P_1 \;+\; b : P_2) \times (\overline{a} : Q_1 \;+\; \overline{b} : Q_2))\upharpoonright(Act - \{a \cdot \overline{b}, b \cdot \overline{a}\})$$

where a and b are particles. In unprioritized SCCS, a strongly equivalent agent is:

$$1 : (P_1 \times Q_1)\upharpoonright(Act - \{a \cdot \overline{b}, b \cdot \overline{a}\}) \;+\; 1 : (P_2 \times Q_2)\upharpoonright(Act - \{a \cdot \overline{b}, b \cdot \overline{a}\})$$

That is, since the calculus has no underlying priority structure, the resulting term is nondeterministic. However, with the introduction of priority, the result can be much different. For example, assume that the priority of a is greater than that of b, or informally $a > b$, and $\overline{a} > \overline{b}$. In this case we would expect the first term of the summation to emerge; that is, the resulting unit action could not be treated as having no priority.

An interesting problem arises when priorities are "circular"; that is, when $a > b$ but $\overline{b} > \overline{a}$. Is the resulting term *inaction* or nondeterministic? If we view particles as "belonging" to system resources, there is no reason why such conflicts cannot arise. The problem is complicated further in SCCS, where *actions* consist of many such particles, each having its own priority. In order to assign composite priorities to actions, we clearly require some additional structure in our calculus.

Previous research has, with varying success, treated some issues of the priority problem. There has been a spate of effort directed toward defining models for concurrency based on "maximum parallelism" [15], in which if processes are ready to communicate, they *will* communicate. Thus maximum parallelism incorporates a very limited, bi-level priority scheme, where non-idle actions always take precedence over idle actions, and contention between non-idle actions is resolved nondeterministically. A priority scheme for CCS is treated in [1], in which particles can only communicate with inverses of the *same* priority; this avoids the multiple priority problem mentioned above. Further, it can be debated whether there is justification for priority in completely asynchronous contexts such as those defined by CCS.

The remainder of this paper is organized as follows. In Section 2, we briefly overview the language of CCSR. Then, in Section 3, we describe the resource-based action domain; in particular, we stress the priority function on actions, and the partitioning of particles according to resources constraints. The semantics of CCSR terms is defined in two steps: we define the unconstrained operational semantics for closed terms in Section 4, followed by the prioritized strong equivalence in Section 5. Section 6 presents an example of a real-time problem whose correct temporal behavior depends on priority.

2 The CCSR Language

The syntax of CCSR resembles, in large part, that of SCCS, and we wish to retain much of its flavor. The major differences occur in several places. First, the action domain contains sets of particles, does not form a monoid or a group. Second, our notion of communication is closer to that of CSP [6, 5], and is not performed with the use of inverse actions. In fact, there is no concept the inverse in our calculus, in that the basic combinator of actions is the union operation. Finally, and most importantly, our actions have priorities associated with them.

Let Σ be the set of all particles, or events, and let a, b, and c range over Σ. Let the letters A, B and C range $\mathcal{P}(\Sigma)$, and Γ range over $\mathcal{P}(\mathcal{P}(\Sigma))$. Let P range over the domain of terms, and let X range over the domain of term variables. As usual, we assume the existence of an infinite set of free term variables, FV. Also, we let the letter ϕ range over renaming functions on Σ; that is, $\phi \in \Sigma \rightarrow \Sigma$. We overload notation and extend such functions to sets in the usual way, where $\phi(A) = \{\phi(a) \mid a \in A\}$.

The following grammar defines the terms of CCSR:

$$P \;:=\; NIL \mid A : P \mid P + P \mid P\,{}_I\|_J P \mid [P]_I \mid P\backslash A \mid fix(X.P) \mid X$$

While we give formal semantics for these terms in subsequent sections, we briefly present some motivation for them here. The term NIL corresponds to $\mathbf{0}$ in SCCS – it can execute no action. The Action operator, "$A : P$", has the following behavior. At the first time unit, the action A is executed, proceeded by the term P. The Sum operator represents standard SCCS choice – either of the terms can be chosen to execute, subject to the constraints of the environment. The Conjunction operator $P\,{}_I\|_J P$ has two functions. It limits the resources that can be used by the two terms, and also forces synchronization between them. The Constrainment operator, $[P]_I$, denotes that the term P occupies *exactly* the resources represented in the index I. The Hiding operator $P\backslash A$ masks actions in P up to their priority, in that while the actions themselves are hidden, their priorities are still observable. The term $fix(X.P)$ denotes guarded recursion, allowing the specification of infinite behaviors.

The term X is a free variable that belongs to the infinite set FV. Any term in the calculus that contains a free variable is called *open*; a term that contains no free variables is considered *closed*. Staying within the terminology of CCS, we call closed terms *agents*.

3 A Resource-Based Action Domain

The considerations mentioned in section 1 have led us to construct a calculus with semantics based on resource constraints. In our execution model, we consider individual resources to be inherently sequential in nature. To put this in the parlance of SCCS, a single resource is capable of synchronously executing actions that consist, *at most*, of a single particle. Actions that consist of multiple particles must be formed by the synchronous execution of multiple resources.

This notion of execution leads to a natural partition of Σ into mutually disjoint subsets, each of which can be considered the set of particles available to a single resource. Letting $\mathcal{R}$ represent the index set of system resources, for some i in $\mathcal{R}$ we denote Σ_i as the collection of particles available to resource i. Also, since

$$\bigcup_{i \in \mathcal{R}} \Sigma_i = \Sigma$$

and

$$\forall i \in \mathcal{R}, \forall j \in \mathcal{R} \,.\, i \neq j,\ \Sigma_i \cap \Sigma_j = \emptyset,$$

the binary relation over $\mathcal{P}(\Sigma) \times \mathcal{P}(\Sigma)$ characterizing "belonging to the same set of resources" is an equivalence relation on actions.

As we have stated, a single resource is capable of executing actions that consist of at most one particle. We formalize this concept by defining, for each resource i, the domain of actions each is capable of performing:

$$\mathcal{D}_i = \{\{a\} \mid a \in \Sigma_i\} \cup \{\emptyset\}$$

We now can formally define the domain of actions $\mathcal{D}$:

$$\mathcal{D} = \{A \in p(\Sigma) \mid \forall i \in \mathcal{R},\ A \cap \Sigma_i \in \mathcal{D}_i\}$$

where $p(\Sigma)$ denotes the set of finite subsets of Σ. It is often convenient to map actions in $\mathcal{D}$ to the resource sets they inhabit. For a given action A, we use the notation $\mathcal{R}(A)$ to represent the resource set that executes particles in A:

$$\mathcal{R}(A) = \{i \in \mathcal{R} \mid \Sigma_i \cap A \neq \emptyset\}$$

It is important to briefly discuss the role of "$\emptyset$" in CCSR. What does it mean, for example, when for some $i \in \mathcal{R}$, $i \notin \mathcal{R}(A)$? It would be tempting to suggest that resource i is idling, but this is not necessarily true; resource i may not even be a member of the subsystem under observation. Thus, if $i \notin \mathcal{R}(A)$, we can only state that resource i is not contributing to the observed behavior. This notion is developed in the sequel.

3.1 Priorities

Each resource has a finite range of priorities at which its particles can execute. Letting mp_i be the maximum priority on resource i, we denote $PRI_i = [0, \ldots, mp_i] \subseteq \boldsymbol{N}$ as the set of priorities available to resource i.

Thus we can linearly order the particles in each Σ_i by a priority mapping $\pi_i \in \Sigma_i \rightarrow PRI_i$. Extending this ordering to $\mathcal{D}_i$, we construct a new mapping $\Pi_i \in \mathcal{D}_i \rightarrow PRI_i \cup \{\bot\}$ where for each A in $\mathcal{D}_i$,

$$\Pi_i(A) = \begin{cases} \pi_i(a) & \text{if } A = \{a\} \\ \bot & \text{otherwise} \end{cases}$$

and $\forall n \in PRI_i$, $\bot < n$. While technically unnecessary, we assign to the emptyset the undefined priority "$\bot$" to distinguish it from singleton actions.

Finally, we can define the partial ordering "$\leq_p$" that reflects our notion of priority over the domain $\mathcal{D}$. For all $A, B \in \mathcal{D}$,

$$A \leq_p B \quad \text{iff} \quad \forall i,\ \Pi_i(A \cap \Sigma_i) \leq \Pi_i(B \cap \Sigma_i)$$

3.2 Anonymous Execution Particles

In SCCS, the unit action "1" serves two distinctly different functions. One is to denote or idling, or a "busy waiting" condition. For example, the term $1 : (P \times Q)$ represents a process that idles for one time unit, and subsequently executes the term $P \times Q$. On the other hand, the unit can denote the combined actions of two communicating partners. For example, the term

$$(a : P) \times (\overline{a} : Q)$$

is strongly equivalent to $1 : (P \times Q)$, yet in the resource-based view on concurrency, it has a distinctly different meaning.

With the introduction of priority into the calculus, we cannot make such an equivalence between terms. For example, assume that that a and $\overline{a}$ are particulate actions, and that each particle has a nonzero priority on its respective processor. To allow the equivalence above would be contradictory to our execution model, in that two actions with nonzero priority could synchronize into an idle action. Priority mandates that there be a difference between time consumed by execution, and time consumed by a busy-wait condition. This restriction leads to the following constraint placed on the calculus: *At every time unit, each active resource in a system must contribute a minimum amount of observational information – the priority of particle being executed.* Thus priorities may be considered "lights" on a resource's control panel; whenever a particle is executed by the resource, the light corresponding to its priority is illuminated.

Note that a priority function π_i naturally partitions each Σ_i into equivalence classes. That is, for some $n \in PRI_i$, a particle a is in the class $[b]_i^n$ if and only if $\pi_i(a) = \pi_i(b) = n$. In CCSR, we use the symbol "τ_i^n" to represent a "canonical" particle from each class.

Now, let $a \in \Sigma_i$ be an arbitrary particle. Then there exists some τ_i^n in Σ such that $\pi_i(a) = \pi_i(\tau_i^n)$. Further, there is a unique renaming function ϕ_π such that $\phi_\pi(a) = \tau_i^n$. This renaming is unique up to particle priority, in that if $\pi_i(a) = \pi_i(b)$, but $a \neq b$, then

$$\phi_\pi(a) = \phi_\pi(b) = \tau_i^n$$

It follows that the τ_i^n are fixed-points of priority renaming; that is, $\phi_\pi(\tau_i^n) = \tau_i^n$.

Now we turn briefly to the topic of resource idling. As stated in section 3.1, when a resource contributes *no* particle to an action, it does not imply that the resource is in an idle state. Instead, the action merely remains unspecified with respect to that resource. In our semantics, idling is an *observable* behavior, corresponding to the τ action in CCS under strong bisimulation. When a resource i idles, it executes a particle τ_i^0; in the scenario portrayed above, the "light" corresponding to a priority of 0 is illuminated. Thus for every $i \in \mathcal{R}$, there is a τ_i^0 in Σ_i. This permits each resource to have the capacity to execute an idle particle. For a given set of resources $I \subseteq R$, the action composed of *all* of their idle particles is:

$$\mathcal{T}_I^0 = \{\tau_i^0 \mid i \in I\}$$

3.3 Synchronization

Resources synchronize through the use of *connection sets*, which can be thought of as the port connections between them. In a particulate calculus such as CCS, a particle "a" typically synchronizes with its inverse, or "$\overline{a}$," and the only *fully synchronized* action is "τ." This is a valid and even obvious approach, and certainly could be adapted for a priority-sensitive calculus. In CCSR, however, we take a more general approach. First, we wish to preserve the flavor of a fully synchronized action, without losing the ability to observe each of the action's constituent particles. (For example, when the CCS actions "a" and "$\overline{a}$" communicate, a degree of observability is lost, in that "τ" is fairly "generic.") Second, it is desirable to incorporate the SCCS expressibility of n-way communication within the structure of our prioritized action domain, $\mathcal{D}$.

Example 3.1 In CSP-type languages, the alphabet of particles is:

$$\{c_1!, c_1?, c_2!, c_2?, c_3!, c_3?, \ldots\}$$

where each c_i is considered a channel, $c_i!$ is interpreted as a write action, and $c_i?$ is interpreted as a read action. When a read and a write occur simultaneously on the same channel, the communication is considered successful. Thus, the connection sets in such languages are simply:

$$\{c_1!, c_1?\},\ \{c_2!, c_2?\},\ \{c_3!, c_3?\},\ \ldots$$

□

Akin to resources, connection sets partition Σ into mutually disjoint subsets of particles. We denote $\mathcal{C}$ as the index set of connection sets across the system, and for all $i \in \mathcal{C}$, C_i is a connection set. The connection sets must satisfy several properties:

1. The connection sets form a cover of Σ: $\bigcup_{i \in \mathcal{C}} C_i = \Sigma$.
2. The connection sets are mutually disjoint: $\forall i \in \mathcal{C}\, \forall j \in \mathcal{C}, i \neq j,\ C_i \cap C_j = \emptyset$.
3. The connection sets are constructed so that no particle of any Σ_i depends on synchronizing with another particle from Σ_i: $\forall i \in \mathcal{R}\, \forall j \in \mathcal{C},\ \Sigma_i \cap C_j \in \mathcal{D}_i$.
4. All "priority-canonical" particles belong to their own connection sets: $\forall a \in \Sigma\ \exists j \in \mathcal{C} \,.\, \{\phi_\pi(a)\} = C_j$.

The reason for condition 3 is apparent when we view connection sets in context of resource mapping. If the property did not hold, a processor would have to simultaneously execute two different particles for synchronization to occur. Such behavior violates our execution model, as each resource is sequential in nature.

Since every particle is a member of a *unique* connection set, we can construct a mapping from particles to the connection sets that contain them. We use the function *connections* to represent this. For any a in Σ, there is a distinct C_i such that $connections(a) = C_i$. For instance, in example 3.1, $connections(a?) = \{a?, a!\}$. We can naturally extend the *connections* function to sets of particles as follows:

$$Connections(A) = \bigcup_{a \in A} connections(a)$$

Definition 3.1 We say that a set is *fully synchronized* if it can be fully decomposed into a set of the connection sets (or it is empty). We use the predicate *sync* to represent this:

$$sync(A) \quad \text{iff} \quad Connections(A) = A$$

□

It is often convenient to be able to decompose a set $A \in \mathcal{D}$ into two parts: that which is fully synchronized, and that which is not. To do this, we make use of the following two definitions:

$$res(A) = \bigcup\{B \subseteq A \mid sync(B)\}$$
$$unres(A) = A - res(A)$$

Finally, we need not view actions in merely two ways, as being either synchronized or unsynchronized. An action can be synchronized with respect to a resource set I. This means that for an action A, all of the connections that can be made with particles on resources in I are made. This concept is particularly useful in defining the Conjunction operator. We let $\mathcal{D}_{\sigma(I)}$ denote the subdomain of $\mathcal{D}$ in which actions are synchronized with respect to I:

$$\mathcal{D}_{\sigma(I)} = \{A \in \mathcal{D} \mid A = Connections(A) \cap (\bigcup_{i \in I} \Sigma_i)\}$$

Of course if $sync(A)$ is true, then $A \in \mathcal{D}_{\sigma(\mathcal{R}(A))}$.

4 Semantics

In this section we give the *unconstrained* operational semantics for closed terms, in the style of [14]. By *unconstrained*, we mean that no priority structure is given to the domain. It is indeed possible to include our priority structure in a much more complicated set of transitional rules. The reason for this is that, when dealing with properties of priority, it is not sufficient to include in a rule's premise an action an agent *can* perform. In addition, it would be necessary to also include actions it *cannot* perform if higher priority actions are present. This highly complicates matters, and leads to infinite branching on finite terms. Thus we present an unconstrained version of our semantics, and we then refine our notion of priority with an equivalence relation based on it. In this we follow the path of [1] in their treatment of CCS priority, as well as [7] in their approach to maximum parallelism.

Let $\mathcal{E}$ represent the domain of closed terms. The labeled transition system $\langle \mathcal{E}, \rightarrow, \mathcal{D} \rangle$ is defined by the relation $\rightarrow \in \mathcal{E} \times \mathcal{D} \times \mathcal{E}$, whose members are denoted: "$P \xrightarrow{A} Q$". Throughout, we use the following notation. For a given set of resources $I \subseteq \mathcal{R}$, we let Σ_I represent the set $\bigcup_{i \in I} \Sigma_i$. Table 1 presents the unconstrained transition system. The Action, Sum and Recursion rules are straightforward, and similar to their counterparts in SCCS. The other operators, however, require special treatment.

Conjunction. The four side conditions are what makes the Conjunction operation different from a more general "product" combinator, such as that found in SCCS. First, the two resource sets I and J are mutually disjoint. This, combined with the next two side conditions, places a very strong constraint on the sorts that each term can execute. Not only are the particles in both of the sorts mutually disjoint, but they are drawn from completely different resources. This corresponds to our resource- oriented view of concurrency, in which the Conjunction operator merges the operations of two *different* subsystems.

The final side condition, "$A_1 \cup A_2 \in \mathcal{D}_{\sigma(I \cup J)}$", defines the essence our synchronization model: Assume that P can execute an action $A_P \subseteq \Sigma_I$. Likewise, assume that Q can execute an action $A_Q \subseteq \Sigma_J$. Then, if A_P and A_Q are to synchronize, they must be connected in the following sense:

- If any particle in $a \in A_P$ shares a connection set with some particle $b \in \Sigma_J$, then b *must* appear in A_Q.

Action : $A : P \xrightarrow{A} P$

SumL : $\dfrac{P \xrightarrow{A} P'}{P + Q \xrightarrow{A} P'}$ **SumR :** $\dfrac{Q \xrightarrow{A} Q'}{P + Q \xrightarrow{A} Q'}$

Conjunction :

$$\frac{P_1 \xrightarrow{A_1} P_1',\ P_2 \xrightarrow{A_2} P_2'}{P_1\,{}_I\|_J\,P_2 \xrightarrow{A_1 \cup A_2} P_1'\,{}_I\|_J\,P_2'} \quad (I \cap J = \emptyset,\ A_1 \subseteq \Sigma_I,\ A_2 \subseteq \Sigma_J,\ \mathcal{A}_1 \cup \mathcal{A}_2 \in \mathcal{D}_{\sigma(I \cup J)})$$

Hide : $\dfrac{P \xrightarrow{B} P'}{P \backslash A \xrightarrow{\phi_{hide(A)}(B)} P' \backslash A}$ $(sync(A),\ sync(A \cap B))$

Constrain : $\dfrac{P \xrightarrow{A} P'}{[P]_I \xrightarrow{A \cup (\mathcal{T}_I^0 - \mathcal{T}_{\mathcal{R}(A)}^0)} [P']_I}$ $(A \subseteq \Sigma_I)$

Recursion : $\dfrac{P \xrightarrow{A} P'}{fix(X.P) \xrightarrow{A} P'[(fix(X.P))/X]}$

Table 1: Unconstrained Transition System

- If any particle in $b \in A_Q$ shares a connection set with some particle $a \in \Sigma_I$, then a *must* appear in A_P.

Hiding. In CCSR we do not allow the general use of morphisms on actions. If we did, one could use it to reallocate a particle to a resource other than the one that "owns" it. Even a more restricted use of morphism, which only permitted functions that maintained the resource structure would still be too general – the connection set structure would then be violated. We are left with a very restricted use of morphism – one which reduces fully synchronized particles to their "anonymous" priority representation (see section 3.2). We call this operator Hiding, and denote it as $P \backslash A$.

Assume that $A \in \mathcal{D}$. We construct the function $\phi_{hide(A)}$ as follows. For all a in Σ,

$$\phi_{hide(A)}(a) = \begin{cases} \phi_\pi(a) & \text{if } a \in A \\ a & \text{otherwise} \end{cases}$$

Thus, all of the particles in A are reduced to their "canonical" priority representation.

Constrainment. The constrainment operator assigns terms to occupy *exactly* the resource set denoted by the index I. First, if the action A utilizes *more* than the resources in I, it is deleted. On the other hand, the particles in A utilize less than the set I, the action is augmented with the "idle" particles from each of the unused resources (see section 3.2).

Proposition 4.1 *All agents in $\mathcal{E}$ are well-defined, in that if $P \in \mathcal{E}$ and $P \xrightarrow{A} P'$, then $A \in \mathcal{D}$.*
The proof follows directly from the definition of the operators. □

5 Priority Equivalence

In our semantics, equivalence between processes is based on the concept of *strong bisimulation.*

Definition 5.1 *For a given transition system $\langle \mathcal{E}, \rightarrow, \mathcal{D} \rangle$, the symmetric relation $r \subseteq (\mathcal{E}, \mathcal{E})$ is a* strong bisimulation *if, for $(P,Q) \in r$ and $A \in \mathcal{D}$,*

1. *if $P \xrightarrow{A} P'$ then, for some Q', $Q \xrightarrow{A} Q'$ and $(P',Q') \in r$, and*
2. *if $Q \xrightarrow{A} Q'$ then, for some P', $P \xrightarrow{A} P'$ and $(P',Q') \in r$.* □

We let "$\sim$" denote *unconstrained strong equivalence*, or the largest such bisimulation with respect to the transition system $\langle \mathcal{E}, \rightarrow, \mathcal{D} \rangle$. As in [11, 12, 13], "$\sim$" exists, and is a congruence over the agents in $\mathcal{E}$.

In this section we define a new transitional system, $\langle \mathcal{E}, \rightarrow_\pi, \mathcal{D} \rangle$ grounded in our notion of priority. From this we derive a measure of *prioritized strong equivalence* based on strong bisimulations. Some care must be taken in this definition to ensure that it yields an equivalence with well-defined properties, properties that reflect a sound model of execution. To do this, we must find an adequate *preemption measure.* A preemption measure is a relation $\prec \in \mathcal{D} \times \mathcal{D}$ such that, for any $P \in \mathcal{E}$, $A, B \in \mathcal{D}$, when P may execute A and $B \prec A$, P will *never* execute B.

Definition 5.2 *For all $A \in \mathcal{D}$, $B \in \mathcal{D}$, $A \preceq B$ if and only if*

$$\mathcal{R}(A) = \mathcal{R}(B) \wedge unres(A) = unres(B) \wedge res(A) \leq_p res(B)$$

The relation "$\preceq$" defines a partial order over $\mathcal{D}$ and thus, we say $A \prec B$ if $A \preceq B$ and $B \not\preceq A$. □

Definition 5.3 *The labeled transition system* $\langle \mathcal{E}, \rightarrow_\pi, \mathcal{D}\rangle$ *is a relation* $\rightarrow_\pi \in \mathcal{E} \times \mathcal{D} \times \mathcal{E}$ *and is defined as follows:* $(P, A, P') \in \rightarrow_\pi$ *(or* $P \xrightarrow{A}_\pi P'$*) if:*

1. $P \xrightarrow{A} P'$, *and*

2. *For all* $A' \in \mathcal{D}$, $P'' \in \mathcal{E}$ *such that* $P \xrightarrow{A'} P''$, $A \not\prec A'$. □

The following result shows that "$\prec$" is progress-preserving, in that for a given transition $P \xrightarrow{A} P'$, either $P \xrightarrow{A} P'$ is itself executed under "$\rightarrow_\pi$", or some preempting transition $P \xrightarrow{A'} P''$, with $A \prec A'$, is executed under "$\rightarrow_\pi$".

Lemma 5.1 *If there is an* $A \in \mathcal{D}$ *and* $P, P' \in \mathcal{E}$ *such that* $P \xrightarrow{A} P'$, *then there exist* $A' \in \mathcal{D}$, $P'' \in \mathcal{E}$ *such that* $P \xrightarrow{A'}_\pi P''$ *with* $A \preceq A'$.

Proof: Assume that the conclusion is false. Then, setting $A_0 = A$ and inductively applying definition 5.3, we see that $\forall\, i \in \boldsymbol{N}$, $i > 0$, there exist $A_i \in \mathcal{D}$, $P'_i \in \mathcal{E}$ such that $P \xrightarrow{A_i} P'_i$ with $A_{i-1} \prec A_i$. So we have the infinite chain over $\mathcal{D}$: $A_0 \prec A_1 \prec A_2 \prec \ldots$, and by definition 5.2,

$$res(A_0) <_p res(A_1) <_p res(A_2) <_p \cdots$$

However, note that $\forall i, j \in \boldsymbol{N}$, $\mathcal{R}(A_i) = \mathcal{R}(A_j)$, and thus $\forall i, j \in \boldsymbol{N}$, $\mathcal{R}(res(A_i)) = \mathcal{R}(res(A_j))$. Further, since every $A \in \mathcal{D}$ is finite, $\mathcal{R}(res(A))$ is finite. Thus there are finitely many distinct priorities on sets using the resources in $\mathcal{R}(res(A))$: $\prod_{i \in \mathcal{R}(res(A))}(mp_i + 1)$ to be exact. So such infinite, strictly increasing chains cannot exist. □

We now define our notion of prioritized strong equivalence, "$\sim_\pi$".

Definition 5.4 *We denote "*$\sim_\pi$*" as the largest strong bisimulation over the transition system* $\langle \mathcal{E}, \rightarrow_\pi, \mathcal{D}\rangle$. □

Relying on the well-known theory found in [11, 12, 13], we state without proof that "$\sim_\pi$" exists, and that it is an equivalence relation over $\mathcal{E}$. The Appendix presents some of the equational characteristics of CCSR with respect to prioritized strong equivalence. Also, by the following theorem, we see that "$\sim_\pi$" forms a congruence over the operators.

Theorem 5.1 *Prioritized strong equivalence is a congruence over agents in* $\mathcal{E}$. *That is, for agents* P, Q *and* R *in* $\mathcal{E}$, A *in* $\mathcal{D}$, *such that* $P \sim_\pi Q$, *we have:*

$$\begin{array}{llll}
(1) & A : P \sim_\pi A : Q & & \\
(2a) & P + R \sim_\pi Q + R & (2b) & R + P \sim_\pi R + Q \\
(3a) & P\,{}_I\|_J R \sim_\pi Q\,{}_I\|_J R & (3b) & R\,{}_I\|_J P \sim_\pi R\,{}_I\|_J Q \\
(4) & P \backslash A \sim_\pi Q \backslash A & (5) & [P]_I \sim_\pi [Q]_I
\end{array}$$

We shall only present the proof for case (3a); the proofs for the other cases are similar. Before doing so, we require the following lemma.

Lemma 5.2 *Let* $P_1, P_2, P'_1, P'_2 \in \mathcal{E}$, *and* $A \in \mathcal{D}$. *If* $P_1\,{}_I\|_J P_2 \xrightarrow{A}_\pi P'_1\,{}_I\|_J P'_2$, *then* $P_1 \xrightarrow{A \cap \Sigma_I}_\pi P'_1$ *and* $P_2 \xrightarrow{A \cap \Sigma_J}_\pi P'_2$.

Proof: Denote $A_I = A \cap \Sigma_I$ and $A_J = A \cap \Sigma_J$. By definition 5.3, $P_1\,{}_I\|_J\,P_2 \xrightarrow{A} P_1'\,{}_I\|_J\,P_2'$. So by the transition rule for Conjunction, we have that $P_1 \xrightarrow{A_I} P_1'$ and $R \xrightarrow{A_J} R'$ with $A_I \cup A_J \in \mathcal{D}_{\sigma(I\cup J)}$. We now claim that $P_1 \xrightarrow{A_I}_\pi P_1'$ and $P_2 \xrightarrow{A_J}_\pi P_2'$.

To the contrary, assume it is false that $P_1 \xrightarrow{A_I}_\pi P_1'$. Then there is a $A_I' \in \mathcal{D}$ and $P_1'' \in \mathcal{E}$ such that $P_1 \xrightarrow{A_I'} P_1''$ with $A_I \prec A_I'$. By definition 5.2, $\mathcal{R}(A_I') = \mathcal{R}(A_I)$, and thus, $A_I' \subseteq \Sigma_I$. Since $unres(A_I') = unres(A_I)$ and $A_I \cup A_J \in \mathcal{D}_{\sigma(I\cup J)}$, we also have that $A_I' \cup A_J \in \mathcal{D}_{\sigma(I\cup J)}$. But these are exactly the side conditions required for $P_1\,{}_I\|_J\,P_2 \xrightarrow{A_I'\cup A_J} P_1''\,{}_I\|_J\,P_2''$. Now, since $res(A_I') >_p res(A_I)$, we have that

$$\begin{aligned} res(A_I \cup A_J) &= res(A_I) \cup res(A_J) \cup res(unres(A_I) \cup unres(A_J)) \\ &<_p res(A_I') \cup res(A_J) \cup res(unres(A_I) \cup unres(A_J)) \\ &= res(A_I') \cup res(A_J) \cup res(unres(A_I') \cup unres(A_J)) \\ &= res(A_I' \cup A_J) \end{aligned}$$

Also, $\mathcal{R}(A_I \cup A_J) = \mathcal{R}(A_I' \cup A_J)$ and $unres(A_I \cup A_J) = unres(A_I' \cup A_J)$, so $(A_I \cup A_J) \prec (A_I' \cup A_J)$. But this contradicts our original assumption. So, $P_1 \xrightarrow{A_I}_\pi P_1'$ and by a similar argument, $P_2 \xrightarrow{A_J}_\pi P_2'$. □

Proof of Theorem 5.1, (3a): Here we make use of the fact that "$\sim_\pi$" is the largest bisimulation with respect to $\langle \mathcal{E}, \rightarrow_\pi, \mathcal{D}\rangle$. Thus to prove that $P\,{}_I\|_J\,R \sim_\pi Q\,{}_I\|_J\,R$, it suffices to find any bisimulation r with respect to $\langle \mathcal{E}, \rightarrow_\pi, \mathcal{D}\rangle$ such that $(P\,{}_I\|_J\,R, Q\,{}_I\|_J\,R) \in r$, since $r \subseteq \sim_\pi$.

We claim that $r = \{(P\,{}_I\|_J\,R, Q\,{}_I\|_J\,R) \mid P \sim_\pi Q \wedge R \in \mathcal{E}\}$ is a strong bisimulation on $\langle \mathcal{E}, \rightarrow_\pi$ $\mathcal{D}\rangle$. By definition, $(P\,{}_I\|_J\,R, Q\,{}_I\|_J\,R)$ is in r. To prove that r satisfies property 1 of definition 5.1, assume there exist $P', R' \in \mathcal{E}$, $A \in \mathcal{D}$ such that

$$(\dagger) \qquad P\,{}_I\|_J\,R \xrightarrow{A}_\pi P'\,{}_I\|_J\,R'.$$

It suffices to show that for some Q', $Q\,{}_I\|_J\,R \xrightarrow{A}_\pi Q'\,{}_I\|_J\,R'$ and that $P' \sim_\pi Q'$. Let $A_I = A \cap \Sigma_I$ and $A_J = A \cap \Sigma_J$. By lemma 5.2 we have that $P \xrightarrow{A_I}_\pi P'$ and $R \xrightarrow{A_J}_\pi R'$.

Now because $P \sim_\pi Q$, we have that $Q \xrightarrow{A_I}_\pi Q'$, with $P' \sim_\pi Q'$. To finish showing that r enjoys property 1 of definition 5.1, we must prove that $Q\,{}_I\|_J\,R \xrightarrow{A}_\pi Q'\,{}_I\|_J\,R'$. Obviously part 1 of definition 5.3 is satisfied, so assume part 2 is violated. That is, assume there is some $A' \in \mathcal{D}$, $Q'', R'' \in \mathcal{E}$ such that $Q\,{}_I\|_J\,R \xrightarrow{A'} Q''\,{}_I\|_J\,R''$ with $A \prec A'$. Then by lemma 5.1, we know there exist some $A'' \in \mathcal{D}, Q''', R''' \in \mathcal{E}$ such that $Q\,{}_I\|_J\,R \xrightarrow{A''}_\pi Q'''\,{}_I\|_J\,R'''$ with $A' \preceq A''$, and hence $A \prec A''$.

Letting $A_I'' = A'' \cap \Sigma_I$, by lemma 5.2 have that $Q \xrightarrow{A_I''}_\pi Q'''$, and since $P \sim_\pi Q$, there is also some $P''' \in \mathcal{E}$ such that $P \xrightarrow{A_I''} P'''$. But this implies that $P\,{}_I\|_J\,R \xrightarrow{A''} P'''\,{}_I\|_J\,R'''$ with $A \prec A''$, again contradicting our assumption ($\dagger$). So $Q\,{}_I\|_J\,R \xrightarrow{A}_\pi Q'\,{}_I\|_J\,R'$ and the proof of property 1 is complete. By a symmetric argument, r satisfies property 2 in definition 5.1, and so r is a bisimulation. □

The next theorem shows that the strong equivalence defined by "$\sim_\pi$" is coarser than that defined by "$\sim$".

Theorem 5.2 *Let $P, Q \in \mathcal{E}$ and assume P is strongly equivalent to Q under the transition system $\langle \mathcal{E}, \rightarrow, \mathcal{D} \rangle$, (that is, $P \sim Q$). Then $P \sim_\pi Q$.*

Proof: We need only show that the relation "$\sim$" is a bisimulation on the transition system $\langle \mathcal{E}, \rightarrow_\pi, \mathcal{D} \rangle$. Assume $P \sim Q$, and let $P \xrightarrow{A}_\pi P'$. By definition 5.3, $P \xrightarrow{A} P'$. Since $P \sim Q$, there is some $Q' \in \mathcal{E}$ such that $Q \xrightarrow{A} Q'$ with $P' \sim Q'$. Thus we must prove that $Q \xrightarrow{A}_\pi Q'$. If this is false, there is some $A' \in \mathcal{D}$, $Q'' \in \mathcal{E}$ such that $Q \xrightarrow{A'} Q'$ and $A \prec A'$. But since $P \sim Q$, there is also some $P'' \in \mathcal{E}$ such that $P \xrightarrow{A'} P'$, which is a contradiction. Similarly, if $Q \xrightarrow{A}_\pi Q'$, then for some P', $P \xrightarrow{A}_\pi P'$ with $P' \sim Q'$. □

We now turn briefly to the subject of infinite terms. First we give the standard extension of "$\sim_\pi$" to terms with free variables.

Definition 5.5 *Let the set $\{X_1, \ldots, X_n\}$ include the free variable in the terms P and Q. Then $P \sim_\pi Q$ if, for all agents $P_1, \ldots, P_n$, $P[P_1/X_1, \ldots, P_n/X_n] \sim_\pi Q[P_1/X_1, \ldots, P_n/X_n]$.* □

With this definition we are able to show that "$\sim_\pi$" forms a congruence over recursive terms. For brevity we state the result without proof, which is performed by induction on transitional inference.

Theorem 5.3 *If $P \sim_\pi Q$, and at most X is free in P and Q, then $fix(X.P) \sim_\pi fix(X.Q)$.* □

6 An Example

In this section we present a simple example that illustrates the role of priority in CCSR. Our system is a time-critical Producer/Consumer problem that has two producers and one consumer; both producers possess real-time constraints that must be satisfied to ensure that they operate correctly. Our goal is to show that in some real-time applications, a system's correctness can hinge on the ability to implement priority.

First we introduce some notation that facilitates a concise specification of our system. For an action $A \in \mathcal{D}$ a term $P \in \mathcal{E}$, and a nonnegative integer t, let "$\delta_t A : P$" be the term that *must* execute the A action within t time units, but *may* execute $\emptyset$ up to that point. That is:

$$\delta_t A : P = \begin{cases} A : P & \text{if } t = 0 \\ (A : P) + (\emptyset : \delta_{t-1} A : P) & \text{otherwise} \end{cases}$$

Also, let "$A^t : P$" be the term that executes the action A for t time units before proceeding to P:

$$A^t : P = \begin{cases} A : P & \text{if } t = 1 \\ A : A^{t-1} : P & \text{otherwise} \end{cases}$$

The System is composed of three agents: *Consumer*, *Producer*$_1$ and *Producer*$_2$. Initially, *Producer*$_1$ can choose either to idle or to enter its "production" phase. In this phase, it "produces" for 1 time unit by executing the action $\{p_1\}$. Then it attempts to interrupt the *Consumer* by executing the action $\{int_1!\}$. However, if the interrupt is not accepted within 2 time units, *Producer*$_1$ deadlocks. If it is accepted, the agent has a latency of 2 time units before re-starting the loop:

$$Producer_1 = fix(X_{P_1}.\,(\emptyset : X_{P_1}) + (\{p_1\} : \delta_2\{int_1!\} : \emptyset^2 : X_{P_1}))$$

Producer$_2$ is exactly like *Producer*$_1$ except for one fact: it gives the *Consumer* 3 time units to accept its interrupt.

$$Producer_2 = fix(X_{P_2}.\ (\emptyset : X_{P_2})\ +\ (\{p_2\} : \delta_3\{int_2!\} : \emptyset^2 : X_{P_2})\ )$$

The *Consumer* waits for either *Producer*$_1$ or *Producer*$_2$ to interrupt. Once either interrupt is received, there is a digestion period of 2 time units, during which the action $\{c\}$ is executed in a critical section.

$$Consumer = fix(X_C.\ (0 : X_C)\ +\ (\{int_1?\} : \{c\}^2 : X_C)\ +\ (\{int_2?\} : \{c\}^2 : X_C)\ )$$

Let *Producer*$_1$ be hosted on resource 1, and let $\{p_1, int_1!\} \subseteq \Sigma_1$. Further, let $\pi_1(p_1) = 0$ and $\pi_1(int_1) = 0$, which makes *Producer*$_1$ a "passive" agent. Similarly, let *Producer*$_2$ be hosted on resource 2, and $\{p_2, int_2!\} \subseteq \Sigma_2$. Let $\pi_2(p_2) = 0$ and $\pi_2(int_2) = 0$.

Let *Consumer* be hosted on resource 3, with $\{c, int_1?, int_2?\} \subseteq \Sigma_3$; let $\pi_3(c) = 1$, $\pi_3(int_1?) = 2$, and $\pi_3(int_2?) = 1$. The only connection sets of importance are $C_1 = \{int_1!, int_1?\}$ and $C_2 = \{int_2!, int_2?\}$. All other particles are assumed to belong to their own connection sets. From this priority scheme, we have the property that if both interrupts "$int_1!$" and "$int_2!$" are raised simultaneously, the *Consumer* will handle "$int_1!$".

The entire system is posed as follows:

$$[\,(Producer_1\,{}_{\{1\}}\|_{\{2\}}\,Producer_2)\,{}_{\{1,2\}}\|_{\{3\}}\,Consumer\,]_{\{1,2,3\}}$$

We claim that this priority ordering is exactly the key to keeping the system deadlock-free. That is, it contains no proper derivatives that terminate in NIL. Assume that both interrupts "$int_1!$" and "$int_2!$" are raised simultaneously. In the priority-based system, "$int_1!$" is handled first, and there will be a delay of exactly 3 time units that "$int_2!$" is forced to wait. But *Producer*$_2$ *can* wait that long, and because *Producer*$_1$ cannot attempt to raise another interrupt for at least 4 time units, the system will remain safe. On the other hand, if "$int_2!$" had been serviced first, the system would not have been safe, as *Producer*$_1$ cannot wait for 3 time units to have "$int_1!$" serviced. In a semantics without priority structure (e.g., under the "$\longrightarrow$" transition system), the choice between the two interrupts would be nondeterministic. Thus, the system would not be deadlock-free.

7 Conclusion

In this paper we have presented a synchronous, priority-based process algebra called CCSR. Influenced by SCCS, this calculus gives an appropriate equational characterization of the CSR design language. The calculus, accompanied by a proof system, facilitates the syntactic manipulation of CCSR terms based on both resource configuration and priority ordering. Thus the formalism can be considered one step toward unifying abstract, real-time specifications with their resource-specific implementation environments.

We are currently incorporating a dynamic priority structure into the syntax and semantics of the CCSR model. Because deadline-driven scheduling can be formulated in terms dynamic priority, we will then be able to reason about the properties of real-time scheduling algorithms, and their efficacy in guaranteeing the deadlines of the processes with which they interact.

References

[1] R. Cleaveland and M. Hennessy. Priorities in Process Algebras. In *Proc. of IEEE Symposium on Logic in Computer Science*, 1988.

[2] N. Francez, D. Lehmann, and A. Pnueli. A Linear History Semantics for Distributed Programming. *Theoretical Computer Science*, 32:25–46, 1984.

[3] R. Gerber and I. Lee. Communicating Shared Resources: A Model for Distributed Real-Time Systems. In *Proc. 10th IEEE Real-Time Systems Symposium*, 1989.

[4] R. Gerber and I. Lee. The Formal Treatment of Priorities in Real-Time Computation. In *Proc. 6th IEEE Workshop on Real-Time Software and Operating Systems*, 1989.

[5] C.A.R. Hoare. Communicating sequential processes. *Communications of the ACM*, 21(8):666–676, August 1978.

[6] C.A.R. Hoare. *Communicating Sequential Processes.* Prentice-Hall, 1985.

[7] C. Huizing, R. Gerth, and W.P. de Roever. Full Abstraction of a Denotational Semantics for Real-time Concurrency. In *Proc.* 14th *ACM Symposium on Principles of Programming Languages*, pages 223–237, 1987.

[8] R. Janicki and P. Lauer. On the Semantics of Priority Systems. In *Proc. of Int. Conf. on Parallel Processing*, 1988.

[9] R. Koymans, R.K. Shyamasundar, W.P. de Roever, R. Gerth, and S. Arun-Kumar. Compositional Semantics for Real-Time Distributed Computing. In *Logic of Programs Workshop '85, LNCS 193*, 1985.

[10] I. Lee and V. Gehlot. Language Constructs for Distributed Real-Time Programming. In *IEEE Real-Time Systems Symposium*, 1985.

[11] R. Milner. Calculi for synchrony and asynchrony. *Theoretical Computer Science*, 25:267–310, 1983.

[12] R. Milner. *A Calculus for Communicating Systems.* LNCS 92, Springer-Verlag, 1980.

[13] R. Milner. *Communication and Concurrency.* Prentice-Hall, 1989.

[14] Gordon Plotkin. *A Structural Approach to Operational Semantics.* Technical Report DAIMI FN-19, Computer Science Dept., Aarhus University, 1981.

[15] A. Salwicki and T. Müldner. On the Algorithmic Properties of Concurrent Programs. In *Proceedings of Logic of Programs, LNCS 125*, 1981.

Appendix: Equational Characteristics of CCSR

(1) $P + NIL \sim_\pi P$

(2) $P + P \sim_\pi P$

(3) $P + Q \sim_\pi Q + P$

(4) $(P + Q) + R \sim_\pi P + (Q + R)$

(5) $(A : P) + (B : Q) \sim_\pi A : P \quad \text{if } A \prec B$

(6) $$(A : P)_I\|_J(B : Q) \sim_\pi \begin{cases} (A \cup B) : (P_I\|_J Q) & \text{if } A \subseteq \Sigma_I,\ B \subseteq \Sigma_J,\ A \cup B \in \mathcal{D}_{\sigma(I \cup J)} \\ NIL & \text{otherwise} \end{cases}$$

(7) $P_I\|_J NIL \sim_\pi NIL$

(8) $P_I\|_J Q \sim_\pi Q_J\|_I P$

(9) $(P_I\|_J Q)_{(I \cup J)}\|_K R \sim_\pi P_I\|_{(J \cup K)}(Q_J\|_K R)$

(10) $P_I\|_J(Q + R) \sim_\pi (P_I\|_J Q) + (P_I\|_J R)$

(11) $$(A : P)\backslash B \sim_\pi \begin{cases} \phi_{hide(B)}(A) : (P\backslash B) & \text{if } sync(A \cap B) \\ NIL & \text{otherwise} \end{cases}$$

(12) $(P + Q)\backslash B \sim_\pi P\backslash B + Q\backslash B$

(13) $NIL\backslash B \sim_\pi NIL$

(14) $$[[P]_I]_J \sim_\pi \begin{cases} [P]_J & \text{if } I \subseteq J \\ NIL & \text{otherwise} \end{cases}$$

(15) $[NIL]_I \sim_\pi NIL$

(16) $([P]_I)_I\|_J([Q]_J) \sim_\pi [P_I\|_J Q]_{I \cup J}$

(17) $[P]_I + [Q]_I \sim_\pi [P + Q]_I$

(18) $$[A : P]_I \sim_\pi \begin{cases} (A \cup (\mathcal{T}_I^0 - \mathcal{T}_{\mathcal{R}(A)}^0)) : [P]_I & \text{if } A \subseteq \Sigma_I \\ NIL & \text{otherwise} \end{cases}$$

The Linear Time - Branching Time Spectrum

(extended abstract)

R.J. van Glabbeek

Institut für Informatik der Technischen Universität
Arcisstraße 21, D-8000 München 2, Germany

In this paper eleven semantics in the linear time - branching time spectrum are presented in a uniform, model-independent way. Restricted to the domain of finitely branching, concrete, sequential processes, most semantics found in the literature that can be defined uniformly in terms of action relations coincide with one of these eleven. Several testing scenarios, motivating these semantics, are presented, phrased in terms of 'button pushing experiments' on generative and reactive machines. Finally nine of these semantics are applied to a simple language for finite, concrete, sequential, nondeterministic processes, and for each of them a complete axiomatization is provided.

Notes: The research reported in this paper has been initiated at the Centre for Mathematics and Computer Science (P.O. Box 4079, 1009 AB Amsterdam, The Netherlands), and finalized at the Technical University of Munich. It has been supported by Sonderforschungsbereich 342 of the TU München. Part of it was carried out in the preparation of a course Comparative Concurrency Semantics, given at the University of Amsterdam, spring 1988.

This is an extended abstract of Chapter I of my Ph.D Thesis *Comparative concurrency semantics and refinement of actions*, Free University, Amsterdam 1990. The full version also appeared as SFB-Bericht Nr. 342/../90 A, Institut für Informatik, Technische Universität München, 1990, and as Report CS-R9029, Centre for Mathematics and Computer Science, Amsterdam 1990.

INTRODUCTION

Process theory. A *process* is the behaviour of a system. The system can be a machine, an elementary particle, a communication protocol, a network of falling dominoes, a chess player, or any other system. Process theory is the study of processes. Two main activities of process theory are *modelling* and *verification*. Modelling is the activity of representing processes, mostly as elements of a mathematical domain or as expressions in a system description language. Verification is the activity of proving statements about processes, for instance that the actual behaviour of a system is equal to its intended behaviour. Of course, this is only possible if a criterion has been defined, determining whether or not two processes are equal, i.e. two systems behave similarly. Such a criterion constitutes the *semantics* of a process theory. (To be precise, it constitutes the semantics of the equality concept employed in a process theory.) Which aspects of the behaviour of a system are of importance to a certain user depends on the environment in which the system will be running, and on the interests of the particular user. Therefore it is not a task of process theory to find the 'true' semantics of processes, but rather to determine which process semantics is suitable for which applications.

Comparative concurrency semantics. This paper aims at the classification of process semantics.[1] The set of possible process semantics can be partially ordered by the relation 'makes strictly more identifications on processes than', thereby becoming a complete lattice[2]. Now the classification of some useful process semantics can be facilitated by drawing parts of this lattice and locating the positions of some interesting process semantics, found in the literature. Furthermore the ideas involved in the construction of these semantics can be unraveled and combined in new compositions, thereby creating an abundance of new process semantics. These semantics will, by their intermediate positions in the semantic lattice, shed light on the differences and similarities of the established ones. Sometimes they also turn out to be interesting

1. This field of research is called *comparative concurrency semantics*, a terminology first used by MEYER in [24].
2. The supremum of a set of process semantics is the semantics identifying two processes whenever they are identified by every semantics in this set.

their own right. Finally the semantic lattice serves as a map on which it can be indicated which emantics satisfy certain desirable properties, and are suited for a particular class of applications.

Most semantic notions encountered in contemporary process theory can be classified along four ifferent lines, corresponding with four different kinds of identifications. First there is the dichotomy of near time versus branching time: to what extent should one identify processes differing only in the ranching structure of their execution paths? Secondly there is the dichotomy of interleaving semantics ersus partial order semantics: to what extent should one identify processes differing only in the causal ependencies between their actions (while agreeing on the possible orders of execution)? Thirdly one ncounters different treatments of abstraction from internal actions in a process: to what extent should ne identify processes differing only in their internal or silent actions? And fourthly there are different pproaches to infinity: to what extent should one identify processes differing only in their infinite ehaviour? These considerations give rise to a four dimensional representation of the proposed semantic ttice.

However, at least three more dimensions can be distinguished. In this paper, stochastic and real-time spects of processes are completely neglected. Furthermore it deals with *uniform concurrency*[1] only. This eans that processes are studied, performing actions[2] $a,b,c,...$ which are not subject to further investigaons. So it remains unspecified if these actions are in fact assignments to variables or the falling of domoes or other actions. If also the options are considered of modelling (to a certain degree) the stochastic nd real-time aspects of processes and the operational behaviour of the elementary actions, three more arameters in the classification emerge.

rocess domains. In order to be able to reason about processes in a mathematical way, it is common ractice to represent processes as elements of a mathematical domain. Such a domain is called a *process omain*. The relation between the domain and the world of real processes is mostly stated informally. he semantics of a process theory can be modelled as an equivalence on a process domain, called a *emantic equivalence*. In the literature one finds among others:

graph domains, in which a process is represented as a *process graph*, or *state transition diagram*,
net domains, in which a process is represented as a (labelled) *Petri net*,
event structure domains, in which a process is represented as a (labelled) *event structure*,
explicit domains, in which a process is represented as a mathematically coded set of its properties,
projective limit domains, which are obtained as projective limits of series of finite term domains,
and *term domains*, in which a process is represented as a term in a system description language.

ction relations. Write $p \xrightarrow{a} q$ if the process p can evolve into the process q, while performing the action

The binary predicates $\xrightarrow{a}$ are called *action relations*. The semantic equivalences which are treated this paper will be defined entirely in terms of action relations. Hence these definitions apply to any rocess domain on which action relations are defined. Such a domain is called a *labelled transition sysm*. Furthermore they will be defined *uniformly* in terms of action relations, meaning that all actions are eated in the same way. For reasons of convenience, even the usual distinction between internal and ternal actions is dropped in this paper.

initely branching, concrete, sequential processes. Being a first step, this paper limits itself to a very simple ass of processes. First of all only *sequential* processes are investigated: processes capable of performing most one action at a time. Moreover attention is mainly restricted to *finitely branching* processes: rocesses having in each state only finitely many possible ways to proceed. A generalization to infinitely ranching processes can be found in the full version of this paper. Finally, instead of dropping the usual stinction between internal and external actions, one can equivalently maintain to study *concrete ocesses* in which no internal actions occur (and also no internal choices as in CSP [21]). For this sime class of processes, when considering only semantic equivalences that can be defined uniformly in rms of action relations, the announced semantic lattice collapses in six out of seven dimensions and overs only the *linear time - branching time* spectrum.

The term uniform concurrency is employed by DE BAKKER et al [5].
Strictly speaking processes do not perform actions, but systems do. However, for reasons of convenience, this paper somees uses the word process, when actually referring to a system of which the process is the behaviour.

Literature. In the literature on uniform concurrency 11 semantics can be found, which are uniformly definable in terms of action relations and different on the domain of finitely branching, sequential processes (see Figure 1).

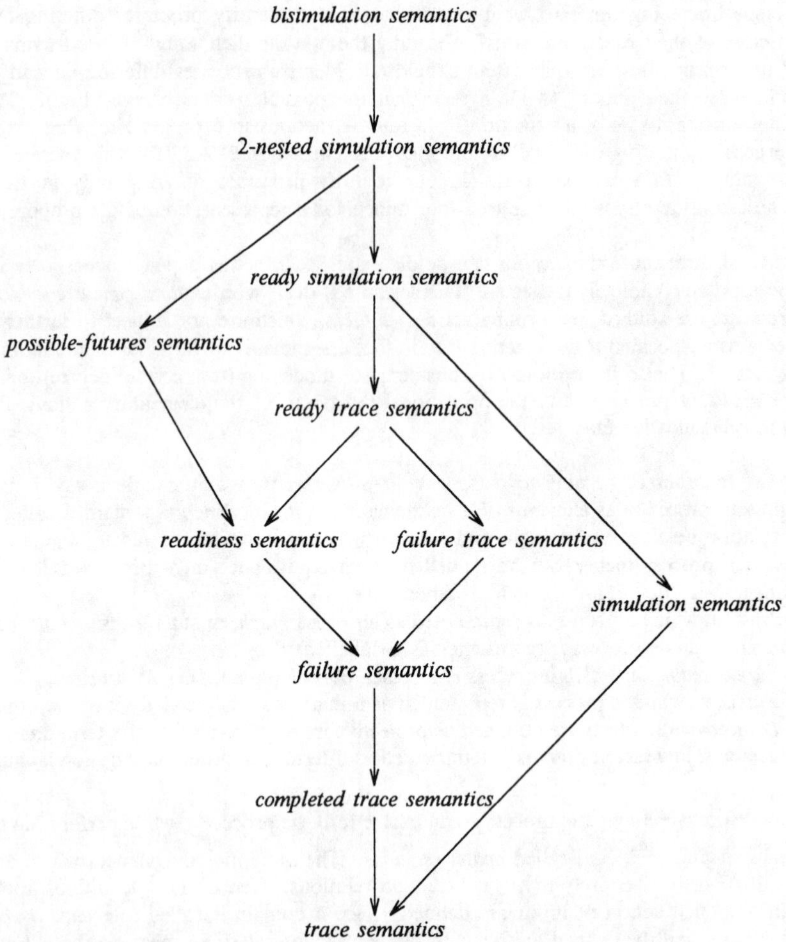

FIGURE 1. *The linear time - branching time spectrum*

The coarsest one (i.e. the semantics making the most identifications) is *trace semantics*, as presented in HOARE [20]. In trace semantics only *partial traces* are employed. The finest one (making less identifications than any of the others) is *bisimulation semantics*, as presented in MILNER [27]. Bisimulation semantics is the standard semantics for the system description language CCS (MILNER [25]). The notion of bisimulation was introduced in PARK [29]. Bisimulation equivalence is a refinement of *observational equivalence*, as introduced by HENNESSY & MILNER in [17]. On the domain of finitely branching, concrete, sequential processes, both equivalences coincide. Also the semantics of DE BAKKER & ZUCKER, presented in [6], coincides with bisimulation semantics on this domain. Then there are nine semantics in between. First of all a variant of trace semantics can be obtained by using *complete traces* besides (or instead of) partial ones. In this paper it is called *completed trace semantics*. *Failure semantics* is introduced in BROOKES, HOARE & ROSCOE [9], and used in the construction of a model for the system description language CSP (HOARE [19,21]). It is finer than completed trace semantics. The semantics based on *testing equivalences*, as developed in DE NICOLA & HENNESY [12], coincides with failure semantics on the omain of finitely branching, concrete, sequential processes, as do the semantics of KENNAWAY [22] and ARONDEAU [10]. This has been established in DE NICOLA [11]. In OLDEROG & HOARE [28] *readiness mantics* is presented, which is slightly finer than failure semantics. Between readiness and bisimulation mantics one finds *ready trace semantics*, as introduced independently in PNUELI [31] (there called *barbed mantics*), BAETEN, BERGSTRA & KLOP [4] and POMELLO [32] (under the name *exhibited behaviour seman-cs*). The natural completion of the square, suggested by failure, readiness and ready trace semantics elds *failure trace semantics*. For finitely branching processes this is the same as *refusal semantics*, intro-uced in PHILLIPS [30]. *Simulation equivalence*, based on the classical notion of *simulation* (see e.g. PARK 9]), is independent of the last five semantics. *Ready simulation semantics* was introduced in BLOOM, TRAIL & MEYER [8] under the name *GSOS trace congruence*. It is finer than ready trace as well as mulation equivalence. In LARSEN & SKOU [23] a more operational characterization of this equivalence as given under the name *⅔-bisimulation equivalence*. This characterization resembles the one used in is paper. Finally *2-nested simulation equivalence*, introduced in GROOTE & VAANDRAGER [15], is located etween ready simulation and bisimulation equivalence, and *possible-futures semantics*, as proposed in OUNDS & BROOKES [33], can be positioned between 2-nested simulation and readiness semantics. mong the semantics which are not definable in terms of action relations and thus fall outside the scope f this chapter, one finds semantics that take stochastic properties of processes into account, as in VAN LABBEEK, SMOLKA, STEFFEN & TOFTS [14] and semantics that make almost no identifications and are ardly used for system verification.

bout the contents. The first section of this paper introduces labelled transition systems and process aphs. A labelled transition system is any process domain that is equipped with action relations. The omain of *process graphs* or *state transition diagrams* is one of the most popular labelled transition sys-ems. In Section 2 all semantic equivalences mentioned above are defined on arbitrary labelled transition ystems. In particular these definitions apply to the domain of process graphs. Most of the equivalences an be motivated by the observable behaviour of processes, according to some testing scenario. (Two rocesses are equivalent if they allow the same set of possible observations, possibly in response on cer-in experiments.) I will try to capture these motivations in terms of *button pushing experiments* (cf. IILNER [25], pp. 10-12). Furthermore the semantics will be partially ordered by the relation 'makes at ast as many identifications as'. This yields the linear time - branching time spectrum. Counterexam-les are provided, showing that on the graph domain this ordering cannot be further expanded. Finally, Section 3, nine of the semantics are applied to a simple language for finite, concrete, sequential, non-eterministic processes, and for each of them a complete axiomatization is provided.

. LABELLED TRANSITION SYSTEMS AND PROCESS GRAPHS

1. Labelled transition systems. In this paper processes will be investigated, that are capable of perform-g actions from a given set *Act*. By an action any activity is understood that is considered as a concep-al entity on a chosen level of abstraction. Actions may be instantaneous or durational and are not quired to terminate, but in a finite time only finitely many actions can be carried out. Any activity of n investigated process should be part of some action $a \in Act$ performed by the process. Different activi-es that are indistinguishable on the chosen level of abstraction are interpreted as occurrences of the me action $a \in Act$.

A process is *sequential* if it can perform at most one action at the same time. In this paper only quential processes will be considered. A domain of sequential processes can often be conveniently presented as a labelled transition system. This is a domain **A** on which infix written binary predicates $\xrightarrow{a}$ are defined for each action $a \in Act$. The elements of **A** represent processes, and $p \xrightarrow{a} q$ means that can start performing the action a and after completion of this action reach a state where q is its maining behaviour. In a labelled transition system it may happen that $p \xrightarrow{a} q$ and $p \xrightarrow{b} r$ for ifferent actions a and b or different processes p and q. This phenomena is called *branching*. It need not e specified how the choice between the alternatives is made, or whether a probability distribution can be tached to it.

OTATION: For any alphabet Σ, let Σ^* be the set of *strings* over Σ. Write ϵ for the empty string, $\sigma\rho$ for the concatenation of σ and $\rho \in \Sigma^*$, and a for the string, consisting of the single symbol $a \in \Sigma$.

DEFINITION: A *labelled transition system* is a pair $(\mathbf{A}, \rightarrow)$ with $\mathbf{A}$ a class and $\rightarrow \subseteq \mathbf{A} \times Act \times \mathbf{A}$, such that for $p \in \mathbf{A}$ and $a \in Act$ the class $\{q \in \mathbf{A} \mid p \xrightarrow{\sigma} q\}$ is a set.

Let for the remainder of this paper $(\mathbf{A}, \rightarrow)$ be a labelled transition system, ranged over by $p,q,r,...$. Write $p \xrightarrow{a} q$ for $(p,a,q) \in \rightarrow$. The binary predicates $\xrightarrow{a}$ are called *action relations*.

DEFINITIONS (Remark that the following concepts are defined in terms of action relations only):

- The *generalized action relations* $\xrightarrow{\sigma}$ for $\sigma \in Act^*$ are defined inductively by:
  1. $p \xrightarrow{\epsilon} p$, for any process p.
  2. $(p,a,q) \in \rightarrow$ with $a \in Act$ implies $p \xrightarrow{a} q$ with $a \in Act^*$.
  3. $p \xrightarrow{\sigma} q \xrightarrow{\rho} r$ implies $p \xrightarrow{\sigma\rho} r$.

  In words: the generalized action relations $\xrightarrow{\sigma}$ are the reflexive and transitive closure of the ordinary action relations $\xrightarrow{a}$. $p \xrightarrow{\sigma} q$ means that p can evolve into q, while performing the sequence σ of actions. Remark that the overloading of the notion $p \xrightarrow{a} q$ is quite harmless.
- The set of *initial actions* of a process p is defined by: $I(p) = \{a \in Act \mid \exists q: p \xrightarrow{a} q\}$.
- A process $p \in \mathbf{A}$ is *finitely branching* if for each $q \in \mathbf{A}$ with $p \xrightarrow{\sigma} q$ for some $\sigma \in Act^*$, the set $\{(a,r) \mid q \xrightarrow{a} r,\ a \in Act,\ r \in \mathbf{A}\}$ is finite.

1.2. Process graphs.

DEFINITION: A *process graph* over a given alphabet *Act* is a rooted, directed graph whose edges are labelled by elements of *Act*. Formally, a process graph g is a triple $(\text{NODES}(g), \text{EDGES}(g), \text{ROOT}(g))$, where

- $\text{NODES}(g)$ is a set, of which the elements are called the *nodes* or *states* of g,
- $\text{ROOT}(g) \in \text{NODES}(g)$ is a special node: the *root* or *initial state* of g,
- and $\text{EDGES}(g) \subseteq \text{NODES}(g) \times Act \times \text{NODES}(g)$ is a set of triples (s,a,t) with $s,t \in \text{NODES}(g)$ and $a \in Act$: the *edges* or *transitions* of g.

If $e = (s,a,t) \in \text{EDGES}(g)$, one says that e *goes from* s *to* t. A (finite) *path* π in a process graph is an alternating sequence of nodes and edges, starting and ending with a node, such that each edge goes from the node before it to the node after it. If $\pi = s_0(s_0,a_1,s_1)s_1(s_1,a_2,s_2) \cdots (s_{n-1},a_n,s_n)s_n$, also denoted as $\pi: s_0 \xrightarrow{a_1} s_1 \xrightarrow{a_2} \cdots \xrightarrow{a_n} s_n$, one says that π *goes from* s_0 *to* s_n; it *starts* in s_0 and *ends* in $end(\pi) = s_n$. Let $\text{PATHS}(g)$ be the set of paths in g starting from the root. If s and t are nodes in a process graph then t *can be reached from* s if there is a path going from s to t. A process graph is said to be *connected* if all its nodes can be reached from the root; it is a *tree* if each node can be reached from the root by exactly one path. Let $\mathbf{G}$ be the domain of connected process graphs over a given alphabet *Act*.

DEFINITION: Let $g,h \in \mathbf{G}$. A *graph isomorphism* between g and h is a bijective function $f: \text{NODES}(g) \rightarrow \text{NODES}(h)$ satisfying

- $f(\text{ROOT}(g)) = \text{ROOT}(g)$ and
- $(s,a,t) \in \text{EDGES}(g) \Leftrightarrow (f(s),a,f(t)) \in \text{EDGES}(h)$.

Graphs g and h are *isomorphic*, notation $g \cong h$, if there exists a graph isomorphism between them. In this case g and h differ only in the identity of their nodes. Remark that graph isomorphism is an equivalence on $\mathbf{G}$.

Finitely branching connected process graphs can be pictured by using open dots (∘) to denote nodes, and labelled arrows to denote edges, as can be seen in Section 2. There is no need to mark the root of such a process graph if it can be recognized as the unique node without incoming edges, as is the case in all my examples. These pictures determine process graphs only up to graph isomorphism, but usually this suffices since it is virtually never needed to distinguish between isomorphic graphs.

EFINITION: For $g \in \mathbf{G}$ and $s \in$ NODES (g), let g_s be the process graph defined by
NODES $(g_s) = \{t \in$ NODES $(g) \mid$ *there is a path going from s to t*$\}$,
ROOT $(g_s) = s \in$ NODES (g_s),
and $(t,a,u) \in$ EDGES (g_s) iff $t,u \in$ NODES (g_s) and $(t,a,u) \in$ EDGES (g).

f course $g_s \in \mathbf{G}$. Remark that $g_{\text{ROOT}(g)} = g$. Now on $\mathbf{G}$ action relations $\xrightarrow{a}$ for $a \in Act$ are defined by $\xrightarrow{a} h$ iff (ROOT $(g), a, s) \in$ EDGES (g) and $h = g_s$. This makes $\mathbf{G}$ into a labelled transition system.

SEMANTIC EQUIVALENCES

1. Trace semantics. $\sigma \in Act^*$ is a *trace* of a process p, if there is a process q, such that $p \xrightarrow{\sigma} q$. Let $T(p)$ enote the set of traces of p. Two processes p and q are *trace equivalent* if $T(p) = T(q)$. In trace seman-cs (T) two processes are identified iff they are trace equivalent.

Trace semantics is based on the idea that two processes are to be identified if they allow the same set f observations, where an observation simply consists of a sequence of actions performed by the process n succession.

2. Completed trace semantics. $\sigma \in Act^*$ is a *complete trace* of a process p, if there is a process q, such that $\xrightarrow{\sigma} q$ and $I(q) = \varnothing$. Let $CT(p)$ denote the set of complete traces of p. Two processes p and q are *ompleted trace equivalent* if $T(p) = T(q)$ and $CT(p) = CT(q)$. In completed trace semantics (CT) two rocesses are identified iff they are completed trace equivalent.

Completed trace semantics can be explained with the following (rather trivial) *completed trace machine.*

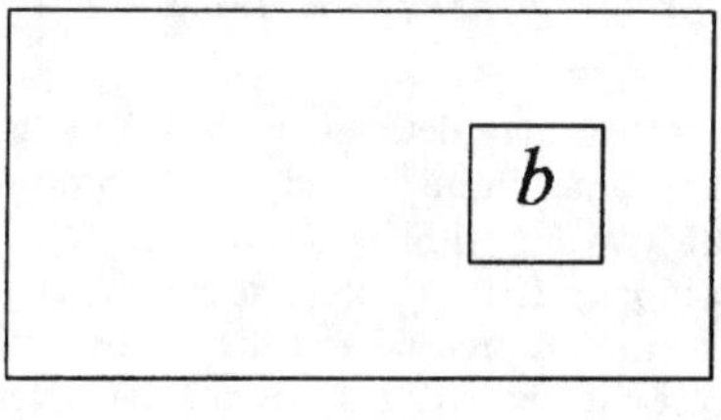

FIGURE 2. *The completed trace machine*

he process is modelled as a black box that contains as its interface to the outside world a display on hich the name of the action is shown that is currently carried out by the process. The process auto-omously choses an execution path that is consistent with its position in the labelled transition system $A, \rightarrow$). During this execution always an action name is visible on the display. As soon as no further ction can be carried out, the process reaches a state of deadlock and the display becomes empty. Now e existence of an observer is assumed that watches the display and records the sequence of actions isplayed during a run of the process, possibly followed by deadlock. It is assumed that an observation kes only a finite amount of time and may be terminated before the process stagnates. Two processes re identified if they allow the same set of observations in this sense.

The *trace machine* can be regarded as a simpler version of the completed trace machine, were the last ction name remains visible in the display if deadlock occurs (unless deadlock occurs in the beginning lready). On this machine traces can be recorded, but stagnation can not be detected, since in case of eadlock the observer may think that the last action is still continuing.

rite $\mathfrak{S} \preccurlyeq \mathfrak{T}$ if semantics $\mathfrak{S}$ makes at least as much identifications as semantics $\mathfrak{T}$. This is the case if the quivalence corresponding with $\mathfrak{S}$ is equal to or coarser than the one corresponding with $\mathfrak{T}$. Trivially $\preccurlyeq CT$ (as in Figure 1). The following counterexample shows that the reverse does not hold.

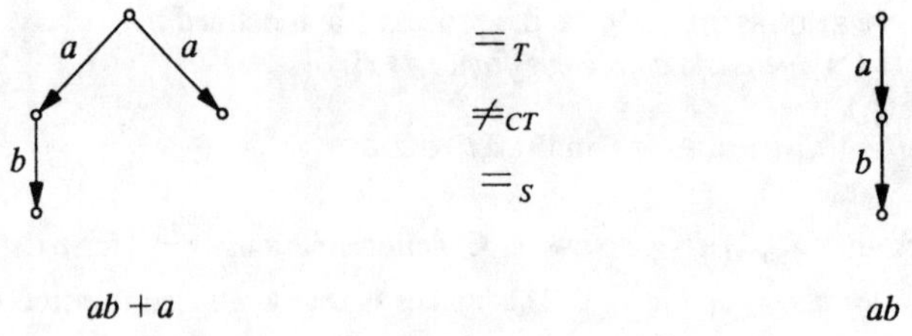

COUNTEREXAMPLE 1

2.3. Failure semantics. The *failure machine* contains as its interface to the outside world not only th display of the completed trace machine, but also a switch for each action $a \in Act$ (as in Figure 3).

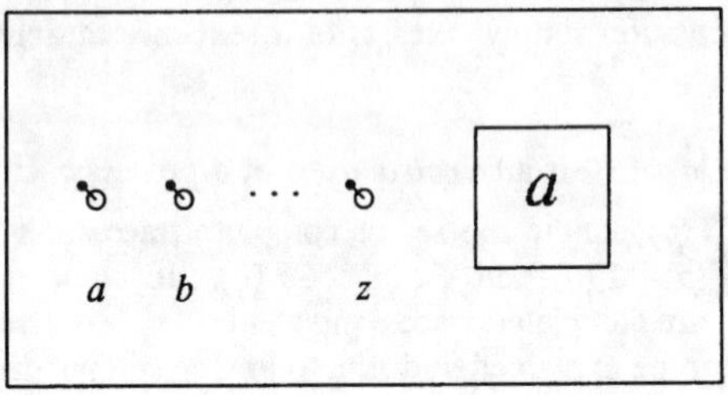

FIGURE 3. *The failure trace machine*

By means of these switches the observer may determine which actions are *free* and which are *blocked* This situation may be changed any time during a run of the process. As before, the process auto nomously choses an execution path that fits with its position in $(\mathbf{A}, \rightarrow)$, but this time the process may only start the execution of free actions. If the process reaches a state where all initial actions of it remaining behaviour are blocked, it can not proceed and the machine stagnates, which can be recognized from the empty display. In this case the observer may record that after a certain sequence of actions σ the set X of free actions is refused by the process. X is therefore called a *refusal set* and $<\sigma, X>$ a *failure pair*. The set of all failure pairs of a process is called its *failure set*, and constitutes its observable behaviour.

DEFINITION: $<\sigma, X> \in Act^* \times \mathcal{P}(Act)$ is a *failure pair* of a process p, if there is a process q, such that $p \xrightarrow{\sigma} q$ and $I(q) \cap X = \emptyset$. Let $F(p)$ denote the set of failure pairs of p. Two processes p and q are *failure equivalent* if $F(p) = F(q)$. In failure semantics (F) two processes are identified iff they are failure equivalent.

PROPOSITION 1: $CT \prec F$.
PROOF: For "$CT \preccurlyeq F$" it suffices to show that $CT(p)$ can be expressed in terms of $F(p)$:

$$CT(p) = \{\sigma \in Act^* \mid <\sigma, Act> \in F(p)\}.$$

"$CT \not\succcurlyeq F$" follows from Counterexample 2. □

This version of failure semantics is taken from HOARE [21]. In BROOKES, HOARE & ROSCOE [9], where failure semantics was introduced, the refusal sets are required to be finite. It is not difficult to see that for finitely branching processes the two versions yield the same failure equivalence. In fact this follows immediately from the following proposition, that says that, for finitely branching processes, the failure pairs with infinite refusal set are completely determined by the ones with finite refusal set.

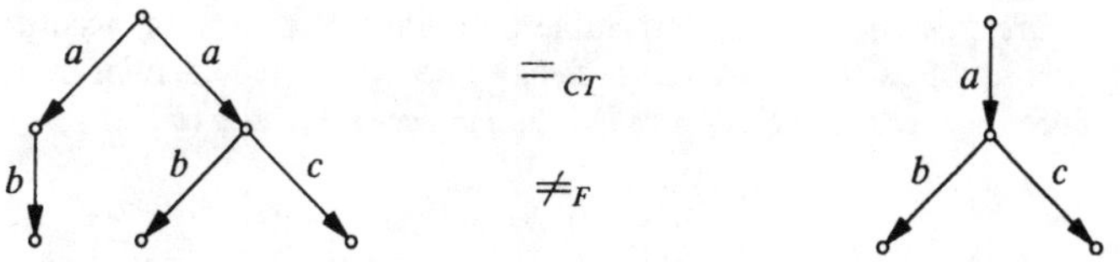

$ab + a(b+c)$ $\qquad\qquad$ $a(b+c)$

COUNTEREXAMPLE 2

ROPOSITION 2: Let $p \in \mathbf{A}$ and $\sigma \in T(p)$. Put $Cont(\sigma) = \{a \in Act \mid \sigma a \in T(p)\}$.
 Then, for $X \subseteq Act$, $<\sigma, X> \in F(p) \Leftrightarrow <\sigma, X \cap Cont(\sigma)> \in F(p)$.
. If p is finitely branching then $Cont(\sigma)$ is finite.
ROOF: Straightforward. □

DE NICOLA [11] several equivalences, that were proposed in KENNAWAY [22], DARONDEAU [10] and E NICOLA & HENNESY [12], are shown to coincide with failure semantics on the domain of finitely ranching transition systems without internal moves. For this purpose he uses the following alternative haracterization of failure equivalence.

EFINITION: Write p *after* σ *MUST* X if for each $q \in \mathbf{A}$ with $p \xrightarrow{\sigma} q$ there is an $r \in \mathbf{A}$ and $a \in X$ such that $\xrightarrow{a} r$. Put $p \simeq q$ if for all $\sigma \in Act^*$ and $X \subseteq Act$: p *after* σ *MUST* $X \Leftrightarrow q$ *after* σ *MUST* X.

ROPOSITION 3: Let $p, q \in \mathbf{A}$. Then $p \simeq q \Leftrightarrow F(p) = F(q)$.
ROOF: p *after* σ *MUST* $X \Leftrightarrow (\sigma, X) \notin F(p)$ [11]. □

HENNESSY [16], a model for nondeterministic behaviours is proposed in which a process is represented an *acceptance tree*. An acceptance tree of a finitely branching process p without internal moves or ternal nondeterminism can be represented as the set of all pairs $<\sigma, X> \in Act^* \times \mathcal{P}(Act)$ for which ere is a process q, such that $p \xrightarrow{\sigma} q$ and $X \subseteq I(q)$. It follows that for such processes *acceptance tree uivalence* coincides with failure equivalence.

4. *Failure trace semantics.* The *failure trace machine* has the same layout as the failure machine, but is oes not stagnate permanently if the process cannot proceed due to the circumstance that all actions it is repared to continue with are blocked by the observer. Instead it idles - recognizable from the empty isplay - until the observer changes its mind and allows one of the actions the process is ready to perrm. What can be observed are traces with idle periods in between, and for each such period the set of ctions that are not blocked by the observer. Such observations can be coded as sequences of members nd subsets of Act.

EXAMPLE: The sequence $\{a,b\}cdb\{b,c\}\{b,c,d\}a(Act)$ is the account of the following observation: At e beginning of the execution of the process p, only the actions a and b were allowed by the observer. pparently, these actions were not on the menu of p, for p started with an idle period. Suddenly the oserver canceled its veto on c, and this resulted in the execution of c, followed by d and b. Then again idle period occurred, this time when b and c were the actions not being blocked by the observer. fter a while the observer decided to allow d as well, but the process ignored this gesture and remained le. Only when the observer gave the green light for the action a, it happened immediately. Finally, the ocess became idle once more, but this time not even one action was blocked. This made the observer alize that a state of eternal stagnation had been reached, and disappointed he terminated the observaon.

A set $X \subseteq Act$, occurring in such a sequence, can be regarded as an offer from the environment, that is fused by the process. Therefore such a set is called a *refusal set*. The occurrence of a refusal set may interpreted as a 'failure' of the environment to create a situation in which the process can proceed ithout being disturbed. Hence a sequence over $Act \cup \mathcal{P}(Act)$, resulting from an observation of a process

p may be called a *failure trace* of p. The observable behaviour of a process, according to this testin scenario, is given by the set of its failure traces, its *failure trace set*. The semantics in which processes ar identified iff their failure trace sets coincide, is called *failure trace semantics* (*FT*).

DEFINITIONS:

- The *refusal relations* $\xrightarrow{X}$ for $X \subseteq Act$ are defined by: $p \xrightarrow{X} q$ iff $p = q$ and $I(p) \cap X = \varnothing$.
  $p \xrightarrow{X} q$ means that p can evolve into q, while being idle during a period in which X is the set c actions allowed by the environment.
- The *failure trace relations* $\xrightarrow{\sigma}$ for $\sigma \in (Act \cup \mathcal{P}(Act))^*$ are defined as the reflexive and transitive clo sure of both the action and the refusal relations. Again the overloading of notation is harmless.
- $\sigma \in (Act \cup \mathcal{P}(Act))^*$ is a *failure trace* of a process p, if there is a process q, such that $p \xrightarrow{\sigma} q$. Le $FT(p)$ denote the set of failure traces of p. Two processes p and q are *failure trace equivalent* i $FT(p) = FT(q)$.

PROPOSITION 4: $F \prec FT$.

PROOF: For "$F \preccurlyeq FT$" it suffices to show that $F(p)$ can be expressed in terms of $FT(p)$:

$$<\sigma, X> \in F(p) \Leftrightarrow \sigma X \in FT(p).$$

"$F \not\succcurlyeq FT$" follows from the following counterexample. □

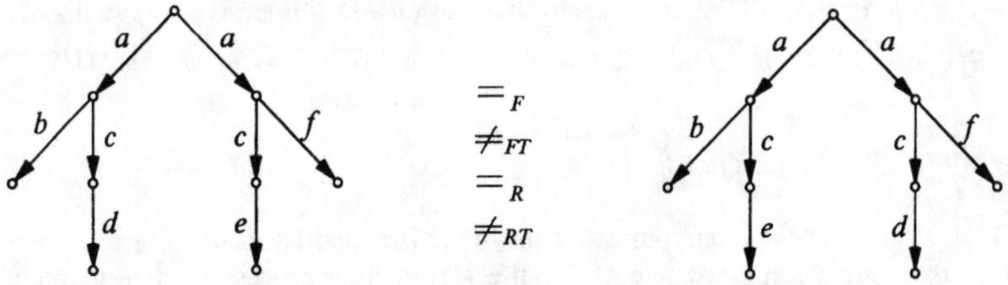

$a(b+cd)+a(f+ce)$ $\qquad$ $a(b+ce)+a(f+cd)$

COUNTEREXAMPLE 3

2.5. *Ready trace semantics.* The *Ready trace machine* is a variant of the failure trace machine that i equipped with a lamp for each action $a \in Act$.

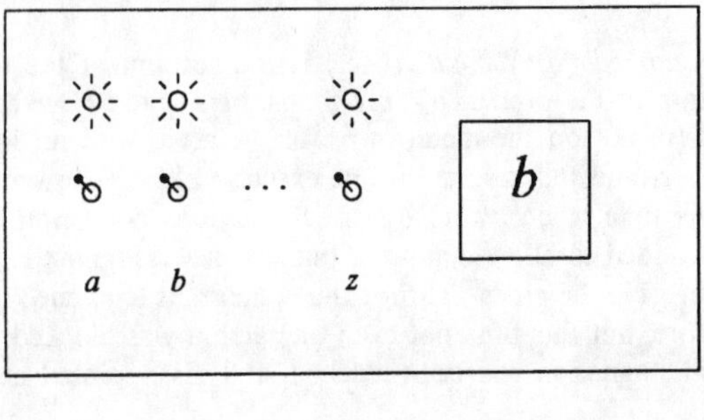

FIGURE 4. *The ready trace machine*

Each time the process idles, the lamps of all actions the process is ready to engage in are lit. Of course all these actions are blocked by the observer, otherwise the process wouldn't idle. Now the observer can e which actions could be released in order to let the process proceed. During the execution of an tion no lamps are lit. An observation now consists of a sequence of members and subsets of *Act*, the tions representing information obtained from the display, and the sets of actions representing informaon obtained from the lights. Such a sequence is called a *ready trace* of the process, and the subsets curring in a ready trace are referred to as *menus*. The information about the free and blocked actions now redundant. The set of all ready traces of a process is called its *ready trace set*, and constitutes its servable behaviour.

EFINITIONS:

The *ready trace relations* $\overset{\sigma}{\ast\!\!\rightarrow}$ for $\sigma \in (Act \cup \mathcal{P}(Act))^*$ are defined inductively by:

1. $p \overset{\varepsilon}{\ast\!\!\rightarrow} p$, for any process p.
2. $p \xrightarrow{a} q$ implies $p \overset{a}{\ast\!\!\rightarrow} q$.
3. $p \overset{X}{\ast\!\!\rightarrow} q$ with $X \subseteq Act$ whenever $p = q$ and $I(p) = X$.
4. $p \overset{\sigma}{\ast\!\!\rightarrow} q \overset{\rho}{\ast\!\!\rightarrow} r$ implies $p \overset{\sigma\rho}{\ast\!\!\rightarrow} r$.

The special arrow $\overset{\sigma}{\ast\!\!\rightarrow}$ had to be used, since further overloading of $\xrightarrow{\sigma}$ would cause confusion with the failure trace relations.

$\sigma \in (Act \cup \mathcal{P}(Act))^*$ is a *ready trace* of a process p, if there is a process q, such that $p \overset{\sigma}{\ast\!\!\rightarrow} q$. Let $RT(p)$ denote the set of ready traces of p. Two processes p and q are *ready trace equivalent* if $RT(p) = RT(q)$. In ready trace semantics (RT) two processes are identified iff they are ready trace equivalent.

BAETEN, BERGSTRA & KLOP [4], PNUELI [31] and POMELLO [32] ready trace semantics was defined ghtly differently. By the proposition below, their definition yields the same equivalence as mine.

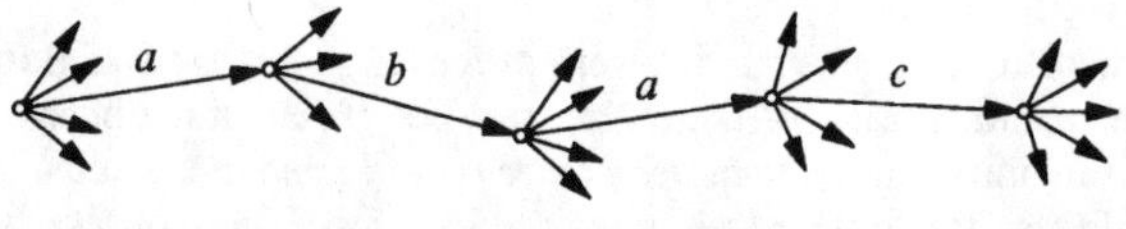

EFINITION: $X_0 a_1 X_1 a_2 \cdots a_n X_n \in \mathcal{P}(Act) \times (Act \times \mathcal{P}(Act))^*$ is a *normal ready trace* of a process p, if there e processes $p_1, \cdots, p_n$ such that $p \xrightarrow{a_1} p_1 \xrightarrow{a_2} \cdots \xrightarrow{a_n} p_n$ and $I(p_i) = X_i$ for $i = 1, \cdots, n$. Let $RT_N(p)$ note the set of normal ready traces of p. Two processes p and q are ready trace equivalent in the sense [4, 31, 32] if $RT_N(p) = RT_N(q)$.

ROPOSITION 5: Let $p, q \in \mathbb{A}$. Then $RT_N(p) = RT_N(q) \Leftrightarrow RT(p) = RT(q)$.
ROOF: The normal ready traces of a process are just the ready traces which are an alternating sequence sets and actions, and vice versa the set of all ready traces can be constructed form the set of normal ady traces by means of doubling and leaving out menus. □

ROPOSITION 6: $FT \prec RT$.
ROOF: For "$FT \preccurlyeq RT$" it suffices to show that $FT(p)$ can be expressed in terms of $RT(p)$:

$\sigma = \sigma_1 \sigma_2 \cdots \sigma_n \in FT(p)$ $(\sigma_i \in Act \cup \mathcal{P}(Act))$ $\Leftrightarrow$ $\exists \rho = \rho_1 \rho_2 \cdots \rho_n \in RT(p)$ $(\rho_i \in Act \cup \mathcal{P}(Act))$ such that for $i = 1, ..., n$ either $\sigma_i = \rho_i \in Act$ or $\sigma_i, \rho_i \subseteq Act$ and $\sigma_i \cap \rho_i = \varnothing$.

$T \not\succcurlyeq RT$" follows from Counterexample 4. □

5. *Readiness semantics.* The *readiness machine* has the same layout as the ready trace machine, but, like e failure machine, can not recover from an idle period. By means of the lights the menu of initial tions of the remaining behaviour of an idle process can be recorded, but this happens at most once ring an observation of a process, namely at the end. An observation either results in a trace of the ocess, or in a pair of a trace and a menu of actions by which the observation could have been extended the observer wouldn't have blocked them. Such a pair is called a *ready pair* of the process, and the set all ready pairs of a process is its *ready set*.

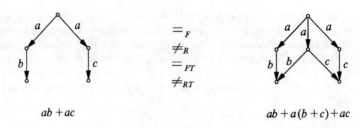

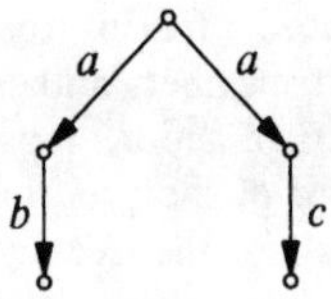

$=_F$
$\neq_R$
$=_{FT}$
$\neq_{RT}$

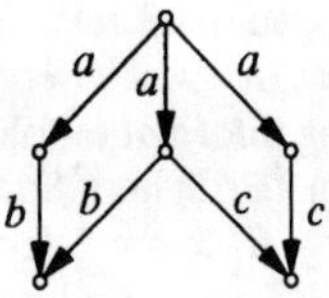

$ab+ac$ $\qquad\qquad$ $ab+a(b+c)+ac$

COUNTEREXAMPLE 4

DEFINITION: $<\sigma,X>\in Act^*\times\mathcal{P}(Act)$ is a *ready pair* of a process p, if there is a process q, such th $p\xrightarrow{\sigma}q$ and $I(q)=X$. Let $R(p)$ denote the set of ready pairs of p. Two processes p and q are *rea equivalent* if $R(p)=R(q)$. In readiness semantics (R) two processes are identified iff they are read equivalent.

PROPOSITION 7: *$F\prec R\prec RT$, but R and FT are independent.*
PROOF: For "$F\preccurlyeq R$" it suffices to show that $F(p)$ can be expressed in terms of $R(p)$:

$$<\sigma,X>\in F(p) \Leftrightarrow \exists Y\subseteq Act: <\sigma,Y>\in R(p)\ \&\ X\cap Y=\varnothing.$$

For "$R\preccurlyeq RT$" it suffices to show that $R(p)$ can be expressed in terms of $RT(p)$:

$$<\sigma,X>\in R(p) \Leftrightarrow \sigma X\in RT(p).$$

"$R\not\succcurlyeq FT$" (and hence "$R\not\succcurlyeq RT$") follows from Counterexample 3, and "$R\not\preccurlyeq FT$" (and hence "$R\not\preccurlyeq F$") fo lows from Counterexample 4. □

Two preliminary versions of readiness semantics were proposed in ROUNDS & BROOKES [33]. In *possibl futures semantics* (PF) the menu consists of the entire trace set of remaining behaviour of an idle proces instead of only the set of its initial actions; in *acceptance-refusal semantics* a menu may be any finite sub set of initial actions, while also the finite refusal sets of Subsection 2.3 are observable.

DEFINITION: $<\sigma,X>\in Act^*\times\mathcal{P}(Act^*)$ is a *possible-future* of a process p, if there is a process q, such tha $p\xrightarrow{\sigma}q$ and $T(q)=X$. Let $PF(p)$ denote the set of possible futures of p. Two processes p and q ar *possible-futures equivalent* if $PF(p)=PF(q)$.

DEFINITION: $<\sigma,X,Y>\in Act^*\times\mathcal{P}(Act)\times\mathcal{P}(Act)$ is a *acceptance-refusal triple* of a process p, if X and are finite and there is a process q, such that $p\xrightarrow{\sigma}q$, $X\subseteq I(q)$ and $Y\cap I(q)=\varnothing$. Let $AR(p)$ denote th set of acceptance-refusal triples of p. Two processes p and q are *acceptance-refusal equivalent* i $AR(p)=AR(q)$.

It is not difficult to see that for finitely branching processes acceptance-refusal equivalence coincides wit readiness equivalence: $<\sigma,X>$ is a ready pair of a process p iff p has an acceptance-refusal tripl $<\sigma,X,Y>$ with $X\cup Y=Cont(\sigma)$ (as defined in the proof of Proposition 2).

2.7. Infinite observations. All testing scenarios up till now assumed that an observation takes only a finit amount of time. However, they can be easily adapted in order to take infinite behaviours into account.

DEFINITION:

- For any alphabet Σ, let Σ^ω be the set of infinite sequences over Σ.
- $a_1a_2\cdots\in Act^\omega$ is an *infinite trace* of a process $p\in\mathbb{A}$, if there are processes $p_1,p_2,\cdots$ such tha $p\xrightarrow{a_1}p_1\xrightarrow{a_2}\cdots$. Let $T^\omega(p)$ denote the set of infinite traces of p.
- Two processes p and q are *infinitary trace equivalent* if $T(p)=T(q)$ and $T^\omega(p)=T^\omega(q)$.
- p and q are *infinitary completed trace equivalent* if $CT(p)=CT(q)$ and $T^\omega(p)=T^\omega(q)$. Note that i

this case also $T(p)=T(q)$.
p and q are *infinitary failure equivalent* if $F(p)=F(q)$ and $T^\omega(p)=T^\omega(q)$.
p and q are *infinitary ready equivalent* if $R(p)=R(q)$ and $T^\omega(p)=T^\omega(q)$.
Infinitary failure traces and infinitary ready traces $\sigma\in(Act\cup\mathcal{P}(Act))^\omega$ and the corresponding sets $FT^\omega(p)$ and $RT^\omega(p)$ are defined in the obvious way. Two processes p and q are *infinitary failure trace equivalent* if $FT^\omega(p)=FT^\omega(q)$, and likewise for infinitary ready trace equivalence.

/ith Königs lemma one easily proves that for finitely branching processes all infinitary equivalences incide with the corresponding finitary ones.

8. *Simulation semantics.* The testing scenario for finitary simulation semantics resembles that for trace mantics, but in addition the observer is, at any time during a run of the investigated process, capable f making arbitrary (but finitely) many copies of the process in its present state and observe them dependently. Thus an observation yields a tree rather than a sequence of actions. Such a tree can be ded as an expression in a simple modal language.

EFINITIONS:
The set $\mathcal{L}_S$ of *simulation formulas* over Act is defined inductively by:
1. $T\in\mathcal{L}_S$.
2. If $\phi,\psi\in\mathcal{L}_S$ then $\phi\wedge\psi\in\mathcal{L}_S$.
3. If $\phi\in\mathcal{L}_S$ and $a\in Act$ then $a\phi\in\mathcal{L}_S$.

The *satisfaction relation* $\models\subseteq\mathbf{A}\times\mathcal{L}_S$ is defined inductively by:
1. $p\models T$ for all $p\in\mathbf{A}$.
2. $p\models\phi\wedge\psi$ if $p\models\phi$ and $p\models\psi$.
3. $p\models a\phi$ if for some $q\in\mathbf{A}$: $p\xrightarrow{a}q$ and $q\models\phi$.

Let $S(p)$ denote the set of all simulation formula that are satisfied by the process p: $S(p)=\{\phi\in\mathcal{L}_S\mid p\models\phi\}$. Two processes p and q are *finitary simulation equivalent* if $S(p)=S(q)$.

he following concept of *simulation*, occurs frequently in the literature (see e.g. PARK [29]). The derived tion of *simulation equivalence* coincides with finitary simulation equivalence for finitely branching ocesses.

EFINITION: A *simulation* is a binary relation R on processes, satisfying, for $a\in Act$:
if pRq and $p\xrightarrow{a}p'$, then $\exists q'$: $q\xrightarrow{a}q'$ and $p'Rq'$.
ocess p *can be simulated by* q, notation $s\subsetneq t$, if there is a simulation R with pRq.
and q are *similar*, notation $p\leftrightarrows q$, if $p\subsetneq q$ and $q\subsetneq p$.

ROPOSITION 8: *Similarity is an equivalence on the domain of processes.*
ROOF: It has to be checked that $p\subsetneq p$, and $p\subsetneq q\ \&\ q\subsetneq r\ \Rightarrow\ p\subsetneq q$.
The identity relation is a simulation with pRp.
If R is a simulation with pRq and S is a simulation with qSr, then the relation $R\circ S$, defined by $x(R\circ S)z$ iff $\exists y$: $xRy\ \&\ ySz$, is a simulation with $p(R\circ S)r$. □

ence the relation will be called *simulation equivalence*.

ROPOSITION 9: Let $p,q\in\mathbf{A}$ be finitely branching processes. Then $p\leftrightarrows q\ \Leftrightarrow\ S(p)=S(q)$.
ROOF: See HENNESSY & MILNER [18]. □

ne testing scenario for simulation semantics differs from that for finitary simulation semantics, in that th the duration of observations and the amount of copies that can be made each time are not required be finite.

ROPOSITION 10: *Simulation semantics (S) is finer than trace semantics ($T\prec S$), but independent of the her semantics presented so far.*
ROOF: For "$T\preccurlyeq S$" it suffices to show that $T(p)$ can be expressed in terms of $S(p)$:

$$\sigma\in T(p)\ \Leftrightarrow\ \sigma T\in S(p)$$

"$S \not\succeq CT$" (and hence "$S \not\succeq RT$" etc.) follows from Counterexample 1, and "$S \not\preceq RT$" (and hence "$S \not\preceq T$" etc.) follows from Counterexample 5 below. □

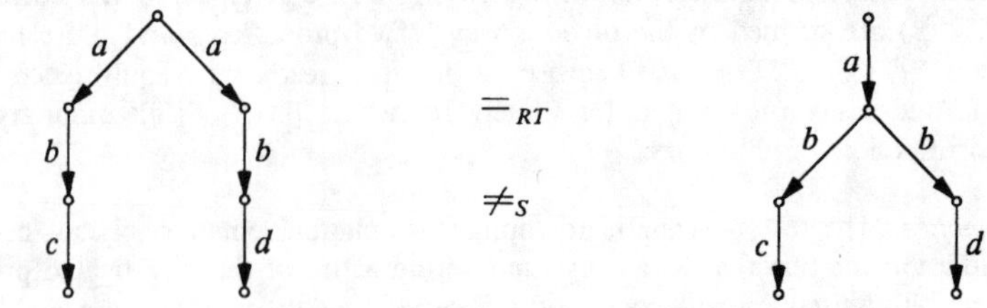

$abc + abd$ $\qquad\qquad$ $a(bc + bd)$

COUNTEREXAMPLE 5

2.9. Ready simulation semantics. Of course one can also combine the copying facility with any of the other testing scenarios. The observer can then plan experiments on one of the machines from the Subsections 2.2 to 2.6 together with a *duplicator*, an ingenious device by which one can duplicate the machine whenever and as often as one wants. In order to represent observations, the modal language from the previous subsection needs to be slightly extended.

DEFINITIONS:

- The *completed simulation formulas* and the corresponding satisfaction relation are defined by means of the extra clauses:
  4. $0 \in \mathcal{L}_{CS}$.
  4. $p \models 0$ if $I(p) = \varnothing$.
- For the *failure simulation formulas* one needs:
  4. If $X \subseteq Act$ then $X \in \mathcal{L}_{FS}$.
  4. $p \models X$ if $I(p) \cap X = \varnothing$.
- For the *ready simulation formulas*:
  4. If $X \subseteq Act$ then $X \in \mathcal{L}_{RS}$.
  4. $p \models X$ if $I(p) = X$.
- For the *failure trace simulation formulas*:
  4. If $\phi \in \mathcal{L}_{FTS}$ and $X \subseteq Act$ then $X\phi \in \mathcal{L}_{FTS}$.
  4. $p \models X\phi$ if $I(p) \cap X = \varnothing$ and $p \models \phi$.
- And for the *ready trace simulation formulas*:
  4. If $\phi \in \mathcal{L}_{RTS}$ and $X \subseteq Act$ then $X\phi \in \mathcal{L}_{RTS}$.
  4. $p \models X\phi$ if $I(p) = X$ and $p \models \phi$.

Note that traces, complete traces, failure pairs, etc. can be obtained as the corresponding kind of simulation formulas without the operator $\wedge$.

By means of the formulas defined above one can define the finitary versions of *completed simulation equivalence, ready simulation equivalence*, etc. It is obvious that failure trace simulation equivalence coincides with failure simulation equivalence and ready trace simulation equivalence with ready simulation equivalence ($p \models X\phi \Leftrightarrow p \models X \wedge \phi$). Also it is not difficult to see that failure simulation equivalence and ready simulation equivalence coincide. For finitely branching processes the finitary versions of these two equivalences coincide with the following infinitary versions.

DEFINITION: A *ready simulation* is a binary relation R on processes, satisfying, for $a \in Act$:

- if pRq and $p \xrightarrow{a} p'$, then $\exists q'$: $q \xrightarrow{a} q'$ and $p'Rq'$;
- if pRq then $I(p) = I(q)$.

Two processes p and q are *ready simulation equivalent* if there exists a ready simulation R with pRq and a ady simulation S with qSp.

ROPOSITION 11: $RT \prec RS$ and $S \prec RS$.
ROOF: For "$RT \preccurlyeq RS$" it suffices to show that $RT(p)$ can be expressed in terms of $RS(p)$:

$$\sigma \in RT(p) \Leftrightarrow \sigma T \in RS(p).$$

$S \preccurlyeq RS$" is even simpler: $\sigma \in S(p) \Leftrightarrow \sigma \in RS(p)$.
$RT \not\succcurlyeq RS$" follows from Counterexample 5, using "$S \preccurlyeq RS$";
$S \not\succcurlyeq RS$" follows from Counterexample 1, using "$CT \preccurlyeq RS$". □

n alternative and maybe more natural testing scenario for finitary ready simulation semantics (or simu- tion semantics) can be obtained by exchanging the duplicator for an *undo*-button on the (ready) trace achine (Figure 5).

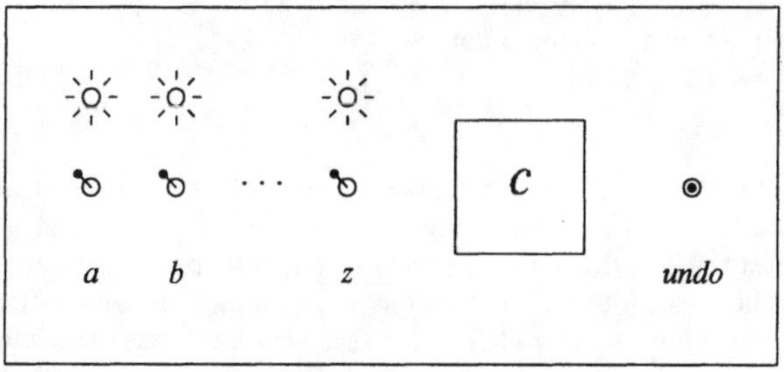

FIGURE 5. *The ready simulation machine*

is assumed that all intermediate states that are past through during a run of a process are stored in a emory inside the black box. Now pressing the *undo*-button causes the machine to shift one state back- ards. In case the button is pressed during the execution of an action, this execution will be interrupted nd the process assumes the state just before this action began. In the initial state pressing the button as no effect. An observation now consists of a (ready) trace, enriched with *undo*-actions. Such observa- ons can easily be translated in (ready) simulation formulas.

10. Refusal (simulation) semantics. In the testing scenarios presented so far, a process is considered to erform actions and make choices autonomously. The investigated behaviours can therefore be classified *generative processes.* The observer merely restricts the spontaneous behaviour of the generative achine by cutting off some possible courses of action. An alternative view of the investigated processes an be obtained by considering them to react on stimuli from the environment and be passive otherwise. *eactive machines* can be obtained out of the generative machines presented so far by replacing the witches by buttons and the display by a green light. Initially the process waits patiently until the bserver tries to press one of the buttons. If the observer tries to press an *a*-button, the machine can eact in two different ways: if the process can not start with an *a*-action the button will not go down and e observer may try another one; if the process can start with an *a*-action it will do so and the button oes down. Furthermore the green light switches on. During the execution of *a* no buttons can be ressed. As soon as the execution of *a* is completed the light switches off, so that the observer knows at the process is ready for a new trial. Reactive machines as described above originate from MILNER 5, 26].

Next I will discuss the equivalences that originate from the various reactive machines. First consider e reactive machine that resembles the failure trace machine, thus without menu-lights and *undo*-button. n observation on such a machine consists of a sequence of accepted and refused actions. Such a equence can be modelled as a failure trace where all refusal sets are singletons. For finitely branching rocesses the resulting equivalence is exactly the equivalence that originates from PHILLIPS notion of *fusal testing* [30]. There it is called *refusal equivalence.* The following proposition shows that for finitely ranching processes refusal equivalence coincides with failure equivalence.

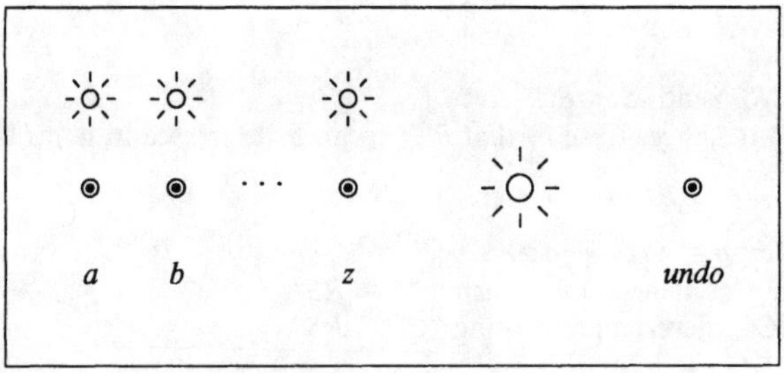

FIGURE 6. *The reactive ready simulation machine*

PROPOSITION 12: Let $p \in \mathbf{A}$ and $\sigma \in FT(p)$. Put $Cont(\sigma) = \{a \in Act \mid \sigma a \in FT(p)\}$.
i. Then, for $X \subseteq Act$, $\sigma X \rho \in FT(p) \Leftrightarrow \sigma(X \cap Cont(\sigma))\rho \in FT(p)$.
ii. If p is finitely branching then $Cont(\sigma)$ is finite.
iii. $\sigma(X \cup Y)\rho \in FT(p) \Leftrightarrow \sigma XY\rho \in FT(p)$.
PROOF: Straightforward. □

If the menu-lights are added to the reactive failure trace machine considered above one can observe ready trace sets, and the green light is redundant. If the green light (as well as the menu-lights) are removed one can only test trace equivalence, since any refusal may be caused by the last action not being ready yet. Reactive machines seem to be unsuited for testing completed trace and failure equivalence. If the menu-lights and the *undo*-button are added to the reactive failure trace machine one gets ready simulation again and if only the *undo*-button is added one obtains an equivalence that may be called *refusal simulation equivalence* and coincides with ready simulation equivalence on the domain of finitely branching processes. The following *refusal simulation formulas* originate from BLOOM, ISTRAIL & MEYER [8].

DEFINITION: The *refusal simulation formulas* and the corresponding satisfaction relation are defined by adding to the definitions of Subsection 2.8 the following extra clauses:
4. If $a \in Act$ then $\neg a \in \mathfrak{L}_{CS}$.
4. $p \models \neg a$ if $a \notin I(p)$.

An alternative family of testing scenarios with reactive machines can be obtained by allowing the observer to try to depress more than one button at a time. In order to influence a particular choice, the observer could already start exercising pressure on buttons during the execution of the preseeding action (when no button can go down). When this preseeding action is finished, at most one of the buttons will go down. These testing scenarios are equipotent with the generative ones: putting pressure on a button is equivalent to setting the corresponding switch on 'free'.

2.11. 2-nested simulation semantics. *2-nested simulation equivalence* popped up naturally in GROOTE & VAANDRAGER [15] as the coarsest congruence with respect to a large and general class of operators that is finer than completed trace equivalence. In order to obtain a testing scenario for this equivalence one has to introduce the rather unnatural notion of a *lookahead* [15]: The *2-nested simulation machine* is a variant of the ready trace machine with duplicator, where in an idle state the machine not only tells which actions are on the menu, but even which simulation formulas are satisfied in the current state.

DEFINITION: A *2-nested simulation* is a simulation contained in simulation equivalence ($\leftrightarrows$). p and q are *2-nested simulation equivalent* if there exists a 2-nested simulation R with pRq and a 2-nested simulation S with qSp.

2.12. Bisimulation semantics. The testing scenario for bisimulation semantics, as presented in MILNER [25] is the oldest and most powerful testing scenario, from which most others have been derived by omitting some of its features. It was based on a reactive failure trace machine with duplicator, but dditionally the observer is equipped with the capacity of *global testing*. Global testing is described in BRAMSKY [1] as: "the ability to enumerate all (of finitely many) possible 'operating environments' at ach stage of the test, so as to guarantee that all nondeterministic branches will be pursued by various opies of the subject process". MILNER [25] implemented global testing by assuming that

) It is the *weather* which determines in each state which a-move will occur in response of pressing the a-button (if the process under investigation is capable of doing an a-move at all);

i) The weather has only finitely many states - at least as far as choice-resolution is concerned;

ii) We can control the weather.

ow it can be ensured that all possible moves a process can perform in reaction on an a-experiment will e investigated by simply performing the experiment in all possible weather conditions. Unfortunately, s remarked in MILNER [26], the second assumption implies that the amount of different a-moves an nvestigated process can perform is bounded by the number of possible weather conditions; so for gen-ral application this condition has to be relaxed.

A different implementation of global testing is given in LARSEN & SKOU [23]. They assumed that every ansition in a transition system has a certain probability of being taken. Therefore an observer can with n arbitrary high degree of confidence assume that all transitions have been examined, simply by repeat-ng an experiment many times.

As argued among others in BLOOM, ISTRAIL & MEYER [8], global testing in the above sense is a rather nrealistic testing ability. Once you assume that the observer is really as powerful as in the described cenarios, in fact more can be tested then only bisimulation equivalence: in the testing scenario of Milner lso the correlation between weather conditions and transitions being taken by the investigated process an be recovered, and in that of Larsen & Skou one can determine the relative probabilities of the vari-us transitions.

An observation in the global testing scenario can be represented as a formula in *Hennessy-Milner logic* 7] (*HML*). An HML formula is a simulation formula in which it is possible to indicate that certain ranches are not present.

EFINITION: The *HML-formulas* and the corresponding satisfaction relation are defined by adding to the efinitions in Subsection 2.8 the following extra clauses:

If $\phi \in \mathcal{L}$ then $\neg\phi \in \mathcal{L}$.

$p \models \neg\phi$ if $p \not\models \phi$.

et $HML(p)$ denote the set of all HML-formula that are satisfied by the process p: $HML(p) = \{\phi \in \mathcal{L} \mid p \models \phi\}$. Two processes p and q are *HML-equivalent* if $HML(p) = HML(q)$.

or finitely branching processes HENNESSY & MILNER [17] provided the following characterization of this quivalence.

EFINITION: Let $p, q \in \mathbb{A}$ be finitely branching processes. Then:

$p \sim_0 q$ is always true.

$p \sim_{n+1} q$ if for all $a \in Act$:

- $p \xrightarrow{a} p'$ implies $\exists q'$: $q \xrightarrow{a} q'$ and $p' \sim_n q'$;
- $q \xrightarrow{a} q'$ implies $\exists p'$: $p \xrightarrow{a} p'$ and $p' \sim_n q'$.

p and q are *observational equivalent*, notation $p \sim q$, if $p \sim_n q$ for every $n \in \mathbb{N}$.

ROPOSITION 13: Let $p, q \in \mathbb{A}$ be finitely branching processes. Then $p \sim q \Leftrightarrow HML(p) = HML(q)$.

ROOF: In HENNESSY & MILNER [18]. □

s observed by PARK [29], for finitely branching processes observation equivalence can be reformulated bisimulation equivalence.

EFINITION: A *bisimulation* is a binary relation R on processes, satisfying, for $a \in Act$:

if pRq and $p \xrightarrow{a} p'$, then $\exists q'$: $q \xrightarrow{a} q'$ and $p'Rq'$;

if pRq and $q \xrightarrow{a} q'$, then $\exists p'$: $p \xrightarrow{a} p'$ and $p'Rq'$.

wo processes p and q are *bisimilar*, notation $p \leftrightarrow q$, if there exists a bisimulation R with pRq.

The relation $\leftrightarrows$ is again a bisimulation. As for similarity, one easily checks that bisimilarity is an equivalence on $\mathbf{A}$. Hence the relation will be called *bisimulation equivalence.* Finally note that the concept of bisimulation does not change if in the definition above the action relations $\xrightarrow{a}$ were replaced by generalized action relations $\xrightarrow{\sigma}$.

PROPOSITION 14: Let $p,q \in \mathbf{A}$ be finitely branching processes. Then $p \leftrightarrows q \Leftrightarrow p \sim q$.
PROOF: "$\Rightarrow$": Straightforward with induction. "$\Leftarrow$" follows from Theorem 5.6 in MILNER [25]. □

For infinitely branching processes $\sim$ is coarser then $\leftrightarrows$ and will be called *finitary bisimulation equivalence.*

Another characterization of bisimulation semantics can be given by means of ACZEL's universe $\mathcal{V}$ of non-well-founded sets [3]. This universe is an extension of the Von Neumann universe of well-founded sets, where the axiom of foundation (every chain $x_0 \ni x_1 \ni \cdots$ terminates) is replaced by an *anti-foundation axiom.*

DEFINITION: Let B denote the unique function $\mathcal{B}:\mathbf{A} \rightarrow \mathcal{V}$ satisfying $\mathcal{B}(p)=\{<a, \mathcal{B}(q)> | p \xrightarrow{a} q\}$ for all $p \in \mathbf{A}$. Two processes p and q are *branching equivalent* if $B(p)=B(q)$.

It follows from Aczel's anti-foundation axiom that such a solution exists. In fact the axiom amounts to saying that systems of equations like the one above have unique solutions. In [3] there is also a section on communicating systems. There two processes are identified iff they are branching equivalent.

A similar idea underlies the semantics of DE BAKKER & ZUCKER [6], but there the domain of processes is a complete metric space and the definition of B above only works for finitely branching processes, and only if $=$ is interpreted as *isometry*, rather then equality, in order to stay in well-founded set theory. For finitely branching processes the semantics of De Bakker and Zucker coincides with the one of Aczel and also with bisimulation semantics. This is observed in VAN GLABBEEK & RUTTEN [13], where also a proof can be found of the next proposition, saying that bisimulation equivalence coincides with branching equivalence.

PROPOSITION 15: Let $p,q \in \mathbf{A}$. Then $p \leftrightarrows q \Leftrightarrow B(p)=B(q)$.

PROPOSITION 16: $RS \prec B$.
PROOF: For "$RS \preccurlyeq B$" it suffices to show that each bisimulation is a ready simulation. This follows since $p \leftrightarrows q \Rightarrow I(p)=I(q)$. "$RS \not\succcurlyeq B$" follows from the following counterexample. □

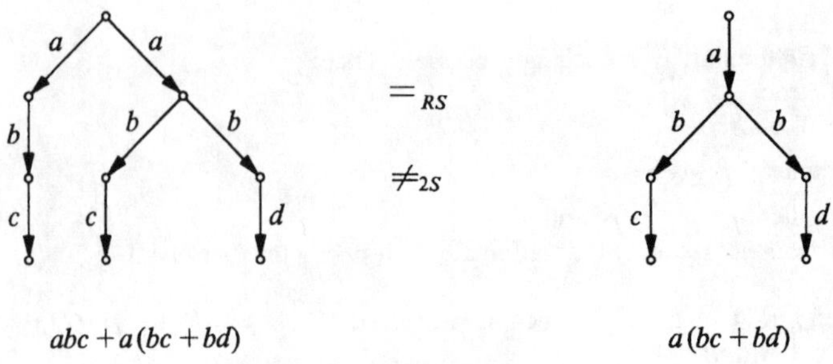

COUNTEREXAMPLE 6

2.13. THEOREM 1: *For all semantics $\mathcal{S}$ and $\mathcal{T}$ on $\mathbf{G}$ defined so far, the formula $\mathcal{S} \preccurlyeq \mathcal{T}$ holds iff there is a path $\mathcal{S} \rightarrow \cdots \rightarrow \mathcal{T}$ in Figure 1.*

PROOF: Most of the implications and counterexamples have been given already. The positions of possible future semantics and 2-nested simulation semantics in the spectrum are treated in the full version of this paper. There it is also shown that the counterexamples are counterexamples indeed. □

COMPLETE AXIOMATIZATIONS

1. A language for finite, concrete, sequential processes. Consider the following basic CCS- and CSP-like nguage BCCSP for finite, concrete, sequential processes over a finite alphabet *Act*:

action: 0 (called *nil* or *stop*) is a constant, representing a process that refuses to do any action.
tion: a is a unary operator for any action $a \in Act$. The expression ap represents a process, starting with an a-action and proceeding with p.
oice: $+$ is a binary operator. $p+q$ represents a process, first being involved in a choice between its summands p and q, and then proceeding as the chosen process.

he set $\mathbb{P}$ of (closed) *process expressions* or terms over this language is defined as usual:
$0 \in \mathbb{P}$,
$ap \in \mathbb{P}$ for any $a \in Act$ and $p \in \mathbb{P}$,
$p+q \in \mathbb{P}$ for any $p,q \in \mathbb{P}$.
ibterms $a0$ may be abbreviated by a.

n $\mathbb{P}$ action relations $\xrightarrow{a}$ for $a \in Act$ are defined as the predicates on $\mathbb{P}$ generated by the *action rules* of able 1. Here a ranges over Act and p and q over $\mathbb{P}$.

$$ap \xrightarrow{a} p \qquad \frac{p \xrightarrow{a} p'}{p+q \xrightarrow{a} p'} \qquad \frac{q \xrightarrow{a} q'}{p+q \xrightarrow{a} q'}$$

TABLE 1

ow all semantic equivalences of Section 2 are well-defined on $\mathbb{P}$, and for each of the semantics it is etermined when two process expressions denote the same process.

2. Axioms. In Table 2 complete axiomatizations for nine from the eleven semantics of this paper that iffer on BSSCP can be found. Axioms for 2-nested simulation and possible-futures semantics are more imbersome, and the corresponding testing notions are less plausible. Therefore they have been omitted. ı order to formulate the axioms, variables have to be added to the language as usual. In the axioms ıey are supposed to be universally quantified. Most of the axioms are axiom schemes, in the sense that ıere is one axiom for each substitution of actions from *Act* for the parameters a,b,c. Some of the xioms are conditional equations, using an auxiliary operator I. Thus provability is defined according to ıe standards of either first-order logic with equality or conditional equational logic. I is a unary operaɔr that calculates the set of initial actions of a process expression, coded as a process expression again.

HEOREM 2: *For each of the semantics $O \in \{T, S, CT, F, R, FT, RT, RS, B\}$ two process expressions $q \in \mathbb{P}$ are O-equivalent iff they can be proved equal from the axioms marked with '+' in the column for O Table 2. The axioms marked with 'v' are valid in O-semantics but not needed for the proof.*

ROOF: For *F*, *R* and *B* the proof is given in BERGSTRA, KLOP & OLDEROG [7] by means of *graph ansformations*. A similar proof for RT can be found in BAETEN, BERGSTRA & KLOP [4]. For the remaing semantics a proof can be given along the same lines. □

ONCLUDING REMARKS

ı this paper various semantic equivalences for concrete sequential processes are defined, motivated, comared and axiomatized. Of course many more equivalences can be given then the ones presented here. he reason for selecting just these, is that they can be motivated rather nicely and/or play a role in the terature on semantic equivalences. In ABRAMSKY & VICKERS [2] the observations which underly many

	B	RS	RT	FT	R	F	CT	S	T
$x+y = y+x$	+	+	+	+	+	+	+	+	+
$(x+y)+z = x+(y+z)$	+	+	+	+	+	+	+	+	+
$x+x = x$	+	+	+	+	+	+	+	+	+
$x+0 = x$	+	+	+	+	+	+	+	+	+
$I(x) = I(y) \Rightarrow a(x+y) = ax+a(x+y)$		+	v	v	v	v	v	v	v
$I(x) = I(y) \Rightarrow ax+ay = a(x+y)$			+	+	v	v	v		v
$ax+ay = ax+ay+a(x+y)$				+		v	v		v
$a(bx+u)+a(by+v) = a(bx+by+u)+a(bx+by+v)$					+	+	v		v
$ax+a(y+z) = ax+a(x+y)+a(y+z)$						+	v		v
$a(bx+u)+a(cy+v) = a(bx+cy+u+v)$							+		v
$a(x+y) = ax+a(x+y)$								+	v
$ax+ay = a(x+y)$									+
$I(0) = 0$	+	+	+	+	+	+	+	+	+
$I(ax) = a0$	+	+	+	+	+	+	+	+	+
$I(x+y) = I(x)+I(y)$	+	+	+	+	+	+	+	+	+

TABLE 2

of the semantics in this paper are placed in a uniform algebraic framework, and some general complet ness criteria are stated and proved.

It is left for a future occasion to give (and apply) criteria for selecting between these equivalences fo particular applications (such as the complexity of deciding if two finite-state processes are equivalent, c the range of useful operators for which they are congruences). The work in this direction reported so fa includes [8] and [15].

Also the generalization of this work to a setting with silent moves and/or with parallelism is left fo the future. In this case the number of equivalences that are worth classifying is much larger. Howeve in many papers parts of a classification can be found already (see for instance [32]).

A generalization to preorders, instead of equivalences, can be obtained by replacing conditions lik $O(p)=O(q)$ by $O(p)\subseteq O(q)$. Since preorders are often useful for verification purposes, it seems to b worthwhile to have to classify them as well.

Furthermore it would be interesting to give explicit representations of the equivalences, by representin processes as sets of observations instead of equivalence classes of process graphs, and defining operator like action prefixing and choice directly on these representations, as has been done for failure semantic in [9] and for readiness semantics in [28].

REFERENCES

[1] S. ABRAMSKY (1987): *Observation equivalence as a testing equivalence.* TCS 53, pp. 225-241.

[2] S. ABRAMSKY & S. VICKERS (1990): *Quantales, observational logic, and process semantics,* unpub lished manuscript.

[3] P. ACZEL (1988): *Non-well-founded sets,* CSLI Lecture Notes No.14, Stanford University.

[4] J.C.M. BAETEN, J.A. BERGSTRA & J.W. KLOP (1987): *Ready-trace semantics for concrete process alge bra with the priority operator.* The Computer Journal 30(6), pp. 498-506.

[5] J.W. DE BAKKER, J.N. KOK, J.-J.CH. MEYER, E.-R. OLDEROG & J.I. ZUCKER (1986): *Contrasting themes in the semantics of imperative concurrency.* In: Current trends in concurrency (J.W. de Bakker, W.-P. de Roever & G. Rozenberg, eds.), LNCS 224, Springer-Verlag, pp. 51-121.

[6] J.W. DE BAKKER & J.I. ZUCKER (1982): *Processes and the denotational semantics of concurrency.* I&C 54(1/2), pp. 70-120.

[7] J.A. BERGSTRA, J.W. KLOP & E.-R. OLDEROG (1988): *Readies and failures in the algebra of communi cating processes.* SIAM Journal on Computing 17(6), pp. 1134-1177.

[8] B. BLOOM, S. ISTRAIL & A.R. MEYER (1988): *Bisimulation can't be traced: preliminary report.* In: Conference Record of the 15th ACM Symposium on Principles of Programming Languages (POPL), San Diego, California, pp. 229-239.

] S.D. BROOKES, C.A.R. HOARE & A.W. ROSCOE (1984): *A theory of communicating sequential processes*. JACM 31(3), pp. 560-599.

0] PH. DARONDEAU (1982): *An enlarged definition and complete axiomatisation of observational congruence of finite processes*. In: Proceedings international symposium on programming: 5th colloquium, Aarhus (M. Dezani-Ciancaglini & U. Montanari, eds.), LNCS 137, Springer-Verlag, pp. 47-62.

1] R. DE NICOLA (1987): *Extensional equivalences for transition systems*. Acta Informatica 24, pp. 211-237.

2] R. DE NICOLA & M. HENNESSY (1984): *Testing equivalences for processes*. TCS 34, pp. 83-133.

3] R.J. VAN GLABBEEK & J.J.M.M. RUTTEN (1989): *The processes of De Bakker and Zucker represent bisimulation equivalence classes*. In: J.W. de Bakker, 25 jaar semantiek, liber amicorum, pp. 243-246.

4] R.J. VAN GLABBEEK, S.A. SMOLKA, B. STEFFEN & C.M.N. TOFTS (1990): *Reactive, generative, and stratified models of probabilistic processes*, to appear in: Proceedings 5th Annual Symposium on Logic in Computer Science (LICS 90), Philadelphia, USA, IEEE Computer Society Press, Washington.

5] J.F. GROOTE & F.W. VAANDRAGER (1988): *Structured operational semantics and bisimulation as a congruence*. Report CS-R8845, Centrum voor Wiskunde en Informatica, Amsterdam, under revision for I&C. An extended abstract appeared in: Proceedings ICALP 89, Stresa (G. Ausiello, M. Dezani-Ciancaglini & S. Ronchi Della Rocca, eds.), LNCS 372, Springer-Verlag, pp. 423-438.

6] M. HENNESSY (1985): *Acceptance trees*. JACM 32(4), pp. 896-928.

7] M. HENNESSY & R. MILNER (1980): *On observing nondeterminism and concurrency*. In: Proceedings ICALP 80 (J. de Bakker & J. van Leeuwen, eds.), LNCS 85, Springer-Verlag, pp. 299-309, a preliminary version of:.

8] M. HENNESSY & R. MILNER (1985): *Algebraic laws for nondeterminism and concurrency*. JACM 32(1), pp. 137-161.

9] C.A.R. HOARE (1978): *Communicating sequential processes*. Communications of the ACM 21(8), pp. 666-677.

0] C.A.R. HOARE (1980): *Communicating sequential processes*. In: On the construction of programs - an advanced course (R.M. McKeag & A.M. Macnaghten, eds.), Cambridge University Press, pp. 229-254.

1] C.A.R. HOARE (1985): *Communicating sequential processes*, Prentice-Hall International.

2] J.K. KENNAWAY (1981): *Formal semantics of nondetermism and parallelism*. Ph.D. Thesis, University of Oxford.

3] K.G. LARSEN & A. SKOU (1988): *Bisimulation through probabilistic testing*. R 88-29, Institut for Elektroniske Systemer, Afdeling for Matematik og Datalogi, Aalborg Universitetscenter, a preliminary report appeared in: Conference Record of the 16th Annual ACM Symposium on Principles of Programming Languages (POPL), Austin, Texas, ACM Press, New York 1989.

4] A.R. MEYER (1985): *Report on the 5th international workshop on the semantics of programming languages in Bad Honnef*. Bulletin of the EATCS 27, pp. 83-84.

5] R. MILNER (1980): *A calculus of communicating systems*, LNCS 92, Springer-Verlag.

6] R. MILNER (1981): *A modal characterisation of observable machine-behaviour*. In: Proceedings CAAP 81 (G. Astesiano & C. Böhm, eds.), LNCS 112, Springer-Verlag, pp. 25-34.

7] R. MILNER (1983): *Calculi for synchrony and asynchrony*. TCS 25, pp. 267-310.

8] E.-R. OLDEROG & C.A.R. HOARE (1986): *Specification-oriented semantics for communicating processes*. Acta Informatica 23, pp. 9-66.

9] D.M.R. PARK (1981): *Concurrency and automata on infinite sequences*. In: Proceedings 5th GI Conference (P. Deussen, ed.), LNCS 104, Springer-Verlag, pp. 167-183.

0] I.C.C. PHILLIPS (1987): *Refusal testing*. TCS 50, pp. 241-284.

1] A. PNUELI (1985): *Linear and branching structures in the semantics and logics of reactive systems*. In: Proceedings ICALP 85, Nafplion (W. Brauer, ed.), LNCS 194, Springer-Verlag, pp. 15-32.

2] L. POMELLO (1986): *Some equivalence notions for concurrent systems. An overview*. In: Advances in Petri Nets 1985 (G. Rozenberg, ed.), LNCS 222, Springer-Verlag, pp. 381-400.

3] W.C. ROUNDS & S.D. BROOKES (1981): *Possible futures, acceptances, refusals and communicating processes*. In: Proceedings 22nd Annual Symposium on Foundations of Computer Science, Nashville, USA 1981, IEEE, New York, pp. 140-149.

A Programming Logic for Formal Concurrent Systems

E. Pascal GRIBOMONT
Philips Research Laboratory
avenue Albert Einstein, 4
1348 Louvain-la-Neuve (Belgium)

email: pascal@prlb.philips.be

Abstract. A simple but general framework for formal design of concurrent systems is presented. The programming notation, called FCS, extends UNITY by introducing notions of process and control flow. This gives rise to a more structured representation of concurrent programs and to a more powerful programming methodology. A relational semantics for FCS is given, that leads to a UNITY-like programming logic and a programming calculus. Special attention is given to the logic itself, but some points of methodology are also briefly discussed.

1 Introduction

Classically, a *transition system* is described by a set Γ of *states* and a set $\mathcal{T}$ of *transitions.* Each transition T_α specifies a binary relation R_α on Γ. A *computation* is a sequence $(\sigma_0, \sigma_1, \ldots, \sigma_n, \ldots)$ of states, satisfying the condition $\forall n \, \exists \alpha \, [\sigma_n R_\alpha \sigma_{n+1}]$. Computations can be finite or infinite, but the last state σ of a finite computation is always a *terminal* state: no state τ exists such that $\sigma R_\alpha \tau$ for some α.

A program can be modelled by a transition system: Γ is the set of memory states, and transitions are (multiple, guarded) assignments. Furthermore, it is also possible to use the formalism of transition systems as a real programming language; this is done, within a rather theoretical framework, in [Kel76,vLS79].

The advantages of modelling concurrent systems by transition systems are, first, the simplicity of the latter concept and, second, the possibility to introduce compositional proof systems in a rather simple way [Lam80,Ger84]. However, a transition system appears as a flat, non-structured set of statements, whereas a concurrent system is (or should be) a highly structured set of statements.

Two main approaches exist for programming within the transition system paradigm. We will summarize them briefly, and give some reasons why a third approach could be more appropriate.

With the first approach, a high level programming language is used, together with a language-independent, transition-based proof system. This has been attempted in [Lam80], [MP83], [LS84], and formalized in [Ger84], [CC89]. This work has led to a new notion of Hoare triple. Classically, the Hoare triple $\{P\}\,X\,\{Q\}$ is interpreted as follows: "Every finite computation of the program X, whose initial state belongs to the set P, ends in a final state belonging to the set Q." The new interpretation of the same triple is: "Every single execution step of X, who starts from a state belonging to the set P, leads to a state belonging to the set Q."
Comment. The symbols P and Q denote sets of system states, although these sets are usually represented by logical assertions.

With this interpretation, compositional Hoare rules can be obtained, such as

$$\frac{\{P\}\, X\, \{Q\}\ ,\ \{P\}\, Y\, \{Q\}}{\{P\}\, X * Y\, \{Q\}} \tag{1}$$

where $*$ denotes some programming construct (sequential composition, for instance). In this framework, program fragments have to be labelled, and special label rules must be introduced. The advantages of this approach are evident: the proof system is compositional and generic (it can be adapted to most of programming languages).

In our opinion, however, the kind of compositionality introduced by rule (1) is not very useful in practice, since assertions P and Q obviously have to be constructed taking into account both X and Y. As a consequence, the designer will not really obtain the program $X * Y$ as the composition of the smaller programs X and Y. This *concurrent Hoare logic* (also called *transition logic* and *generalized Hoare logic*)[1] suggests that a concurrent system is nothing but a flat set of transitions. We do not disagree with this view of concurrent systems but, if it is adopted, the usefulness of composition operators becomes doubtful.

Another critical point of concurrent Hoare logic is the coexistence of labels and composition operators. In fact, when labels exist, operators are no longer needed. For instance, with self-explaining notation, if a system contains the transitions (ℓ, S_1, m) and (m, S_2, n), and if the label m does not appear elsewhere in the system, then S_1 and S_2 will be executed in sequence; no special symbol is needed to denote sequencing.

The second approach consists in programming directly with transition systems, or, more exactly, with sets of assignments [vLS79,CM88]. The latter work clearly demonstrates that programming with sets of assignments (UNITY programs) can be done in a structured way. In UNITY, there are neither labels nor usual composition operators (except set-theoretic union), since the program model has no notions of process and control flow. This point is a definite choice of UNITY's authors, but, in our opinion, these notions can be introduced without destroying the qualities of UNITY and of the programming methodology presented in [CM88]. Besides, from the implementation point of view, processes and control flow have to be specified at last (this is the "mapping problem" evoked in [CM88]). It seems interesting to perform this specification within a formal framework.

We suggest to retain, from the first approach, the new notion of Hoare triple and the use of labels for describing control flow; from the second approach, we retain the absence of the classical operators for sequential or parallel composition of programs. The resulting programming language will be called FCS (for Formal Concurrent Systems).

The paper goes on as follows. In Section 2, the formalism of FCS is introduced, with a relational semantics. Formal programming logic and calculus are developed for FCS in Section 3; it is also shown that the logical system developed for UNITY works for FCS as well. Section 4 briefly discusses some points of methodology, connected with UNITY. Section 5 is devoted to a comparison with related work.

2 Formal concurrent systems

Informally, an FCS is a set of transitions; each transition specifies, first, a transformation of the memory, second, which process(es) perform(s) it, and third, when the execution can take place. This is now defined in a formal way.

---

[1] In fact, these logics are not really equivalent, but the differences between them do not really matter here.

2.1 Syntax

A *formal concurrent system* $\mathcal{S}$ consists of three components. The first one is a set $\mathcal{P}$ of (formal) processes. A *process* is a finite non empty set of *labels*, or *control points*. The processes are pairwise disjoint. Second, the *memory* $\mathcal{M}$ of the system is classically modelled by a set of typed variables. The last component of an FCS is a set $\mathcal{T}$ of *transitions*, which specifies the possible computations of the system. The description and the study of a transition system $\mathcal{S} = (\mathcal{P}, \mathcal{M}, \mathcal{T})$ involve the concepts defined below.

A *system state*, or simply a *state*, is a pair (control state, memory state). A *memory state* associates a value with each variable. A *control state* assigns to each process a label of this process. (This can be viewed as the formal counterpart of the notion of the "program counter", for a physical process.) The set of memory states is denoted Σ and the set of control states is denoted Λ; this last set is (isomorphic to) the cartesian product of the processes. The set of system states is $\Gamma = \Lambda \times \Sigma$.

A *transition* is a list like

$$T = (L,\ C \longrightarrow A,\ M).$$

The *origin* L (entry point) and the *extremity* M (exit point) are *partial control states*, defined on a common domain $type(T) \subset \mathcal{P}$; these partial control states assign to each process *involved* in the transition, i.e., to each $P \in type(T)$, a label $p \in P$.

Comment. As processes are disjoint sets of labels, partial control states can be represented by sets (or sequences) of labels. For instance, the partial control state $\{(P_{i_1}, l_1), \ldots, (P_{i_k}, l_k)\}$ will be written simply $l_1 \ldots l_k$. The *guard* C of the transition T is a boolean expression; the *assignment* A is, as usual, a list like $(x_1, \ldots, x_n) := (t_1, \ldots, t_n)$; the program variables $x_1, \ldots, x_n$ are pairwise distinct, and the type of the expression t_i is the type of the variable x_i.

2.2 Relational semantics

A relational semantics for a programming language is a rule which assigns to each program π written in the language a binary relation $rel(\pi)$ on some fixed set Γ. This relation must be such that the interesting properties of the computations of π should be deducible from $rel(\pi)$. Furthermore, if π is the result of applying some programming construct f to two programs π_1 and π_2, then a function F should exist such that $rel(f(\pi_1, \pi_2)) = F(rel(\pi_1), rel(\pi_2))$.

Throughout this section, $\mathcal{S} = (\mathcal{P}, \mathcal{M}, \mathcal{T})$ denotes some fixed transition system, and $T = (L,\ C \longrightarrow A,\ M)$ is a transition of this system. We will introduce relational semantics for the transition T and the whole system $\mathcal{S}$. These semantics are binary relations on the set of system states $\Gamma = \Lambda \times \Sigma$, and are respectively denoted as $rel(T)$ and $rel(\mathcal{S})$.

Let $r = (L', \rho)$ and $s = (M', \sigma)$ be two states. The couple (r, s) belongs to the relation associated with T if the following requirements are satisfied.

- The partial control states L and M are the restrictions on $type(T)$ of L' and M' respectively.
- The restrictions on $\mathcal{P} \setminus type(T)$ of the control states L' and M' are identical.
- The guard C is true for ρ.
- The assignment A maps memory state ρ onto memory state σ.

This definition results in the formula

$$((L', \rho), (M', \sigma)) \in rel(T) \equiv$$
$$[L = L'_{|type(T)} \wedge M = M'_{|type(T)} \wedge L'_{|\mathcal{P}\setminus type(T)} = M'_{|\mathcal{P}\setminus type(T)} \wedge C(\rho) \wedge \sigma = A(\rho)].$$

The domain $cond(T)$ of the relation $rel(T)$ is the *condition* of the transition T.
In the sequel, we will write X_T for $X_{|type(T)}$ and $X_{\overline{T}}$ for $X_{|\mathcal{P}\setminus type(T)}$; the definition of $rel(T)$

an be rewritten into

$$\begin{aligned}&((L',\rho),(M',\sigma)) \in rel(T) \ \equiv \\ &[L = L'_T \wedge M = M'_T \wedge L'_{\overline{T}} = M'_{\overline{T}} \wedge C(\rho) \wedge \sigma = A(\rho)]\,.\end{aligned} \tag{2}$$

Comments. Formula (2) corresponds to the intuitive idea of executing a transition. For the ontrol state, only the labels corresponding to the processes involved in the transitions are nodified; this is similar to the concept of *joint action* used in [BK88].
The guarded assignment ($C \longrightarrow A$) is viewed as the transition ($\emptyset, C \longrightarrow A, \emptyset$), where $\emptyset$ enotes the partial control state whose domain is empty. The assignment A is viewed as the uarded assignment ($true \longrightarrow A$).
Notation. As a (partial or total) control state is a function, that is, a set of couples, one can vrite $N = N_T \cup N_{\overline{T}}$, where the partial or total control state N and the transition T are rbitrary; the symbol $\cup$ has its usual set-theoretic meaning.

The relational semantics of a set $\mathcal{T}_0$ of transitions, and of the whole system $\mathcal{S}$, are also inary relations on $\Lambda \times \Sigma$. The relations $rel(\mathcal{T}_0)$ and $rel(\mathcal{S})$ are simply the union of the elations associated with the transitions of $\mathcal{T}_0$ and $\mathcal{S}$, respectively. Usually, the notation $r \rightarrow s$ s used instead of $(r,s) \in rel(\mathcal{S})$. The formula

$$r \rightarrow s \quad \equiv \quad \exists T\,[(r,s) \in rel(T)]$$

s a consequence of the definition. The expression $r \rightarrow s$ can be read as "s is an $\mathcal{S}$-successor of ", or "r is an $\mathcal{S}$-predecessor of s". The symbol $\stackrel{*}{\rightarrow}$ denotes the reflexive transitive closure of he relation $rel(\mathcal{S})$. A *computation* of $\mathcal{S}$ is a sequence $(s_n : n = 0,1,\ldots)$ such that $s_n \rightarrow s_{n+1}$ or $n = 0,1,\ldots$; finite computations are allowed, but final states must be terminal states.

Comment. Classically, the relational semantics of a (sequential) program P is the input-output elation of this program; that is, $(r,s) \in rel(P)$ if there exists a finite computation of P vhose initial and final states are respectively r and s. This kind of relational semantics is not onvenient here, for transition systems are not always intended to have finite computations nly. In fact, (r,s) belongs to the input-output relation associated with $\mathcal{S}$ if and only if $r \stackrel{*}{\rightarrow} s$ nd s is a terminal state.

As UNITY programs, FCS programs can be composed by *union* and *superposition*, but this ossibility will not be used in this paper.

3 Programming logic and calculus

n this section, the concurrent Hoare logic for FCS is introduced. We work first in the seman-ical domain, that is, sets of states are considered explicitly, instead of being represented by ssertions. Throughout this section, a formal concurrent system $\mathcal{S} = (\mathcal{P},\mathcal{M},\mathcal{T})$ is fixed; the et of states of $\mathcal{S}$ is noted $\Gamma = \Lambda \times \Sigma$.

3.1 Hoare triples

et P, Q denote sets of states, that is, subsets of $\Gamma = \Lambda \times \Sigma$, and let X denote either an ssignment, a guarded assignment, a transition, a set of transitions or the whole transition ystem $\mathcal{S}$. The formula

$$\forall r \forall s\,[(r \in P \wedge (r,s) \in rel(X)) \Rightarrow s \in Q] \tag{3}$$

is called the *Hoare triple* associated with P, X and Q, and is denoted

$$\{P\}\, X\, \{Q\}\,. \tag{4}$$

Comments. The interpretation of the Hoare triple (4) when X is a whole transition system is "Every single execution step of X, who starts from a state belonging to the set P, leads to a state belonging to the set Q."
In order to simplify the presentation, we have supposed that guarded assignments never fail; more precisely, the guarded assignment $C \longrightarrow A$ may appear in a transition only if A never fails when executed in a state belonging to C.
Hoare logic is adequate to deal with invariants. An *invariant* I is a set of states such that the triple $\{I\}\, S\, \{I\}$ is true. Most programming methodologies are based on invariants (see e.g. [Dij76,Gri81,CM88]); this remains true for FCS. The main point of invariant-based methodologies is as follows. Let $C_0 \subset \Gamma$ be the set of possible initial states. All states of all computations belong to a given set $A \subset \Gamma$ if and only if an invariant I exists such that $C_0 \subset I$ and $I \subset A$.

Hoare triples enjoy useful properties. As a first step, let us mention two properties which connect Hoare triples sharing the same program component X. (The proofs are elementary and omitted.)

THEOREM 1. *If* $(P' \subset P)$, $\{P\}\, X\, \{Q\}$ *and* $(Q \subset Q')$ *hold, then* $\{P'\}\, X\, \{Q'\}$ *holds.*
The triple

$$\{\bigcup_{i\in I} P_i\}\, X\, \{\bigcap_{j\in J} Q_j\}$$

holds if and only if the triples

$$\{P_i\}\, X\, \{Q_j\}$$

hold for all $i \in I$, $j \in J$.
Two interesting special cases are the triples $\{\emptyset\}\, X\, \{Q\}$ and $\{P\}\, X\, \{\Gamma\}$, which are always true.

In a triple $\{P\}\, X\, \{Q\}$, the set P is the *preset* and the set Q is the *postset*. If X and the preset P are fixed, then, for all $Q \subset \Gamma$, the triple is either true or false. Let us define the family

$$POST[P, X] \;=_{def}\; \{Q \subset \Gamma \;:\; \{P\}\, X\, \{Q\}\}\,.$$

THEOREM 2. *The structure* $(POST[P,X], \subset)$ *is a complete lattice. The trivial least upper bound is the whole set of states* Γ*; the greatest lower bound is*

$$slp[P,X] \;=\; \bigcap POST[P,X]\,.$$

Comment. The symbol "*slp*" means "strongest liberal postset".[2]

The dual notion is defined in the same way:

$$PRE[X,Q] \;=_{def}\; \{P \subset \Gamma \;:\; \{P\}\, X\, \{Q\}\}$$

It enjoys similar properties; in particular, the least upper bound of the lattice $(PRE[X,Q], \subset)$ is

$$wlp[X,Q] \;=\; \bigcup PRE[X,Q]\,.$$

(The symbol "*wlp*" means "weakest liberal preset".)
COROLLARY 3. *The following formulas are equivalent:*

$$\begin{array}{l} \{P\}\, X\, \{Q\}\,, \\ Q \in POST[P,X]\,, \\ slp[P,X] \subset Q\,, \\ P \in PRE[X,Q]\,, \\ P \subset wlp[X,Q]\,. \end{array}$$

---

[2]A set S is *stronger* than a set S' if S is *included* in S'. As a consequence, the strongest subset of Γ is the empty set; the weakest subset of Γ is Γ itself.

.2 Structural properties of Hoare triples

Ve are now interested in connections between Hoare triples which differ in their program omponents. A first relation will reduce the case of a guarded assignment to the case of an nguarded one.

HEOREM 4. *The triples* $\{P\}\,(C \rightarrow A)\,\{Q\}$ *and* $\{P \cap C\}\,A\,\{Q\}$ *are equivalent.*

ROOF. Omitted.

Comment. Let us observe that, in the first triple, the symbol C denotes a guard, whereas, in he second triple, the same symbol now denotes the semantics of the guard, that is, the set $\times \{\sigma \in \Sigma\ : C(\sigma)\}$.

OROLLARY 5. *The triple* $\{P \setminus C\}\,(C \rightarrow A)\,\{Q\}$ *is true for all* P, Q.

A second reduction rule connects a transition with the guarded assignment of this transition. wo new pieces of notation are introduced first.

$$t\ L =_{def} \{(L \cup N, \rho) \in \Lambda \times \Sigma\ :\ dom(N) = \mathcal{P} \setminus dom(L)\},$$

$$[at\ M] =_{def} \{(L \cup N, \rho) \in \Lambda \times \Sigma : [dom(L) = dom(M) = \mathcal{P} \setminus dom(N)\ \wedge\ (M \cup N, \rho) \in Q]\}.$$

he formula $Q[at\ M]$ is read "Q restricted at L" or simply "Q at L".

Vhen L is a partial control state, *at* L is the set of system states whose control part is an xtension of L. If Q is a set of system states, then $Q[at\ M]$ is the set of states which differ om a state of $(Q \cap at\ M)$ only in the labels associated with the processes of $dom(M)$. This eads to the identity

$$Q[at\ M] \cap at\ M\ =\ Q \cap at\ M\,.$$

HEOREM 6. *Let* $T\ =\ (L,\ C \rightarrow A,\ M)$.

he triples $\{P\}\,T\,\{Q\}$ *and* $\{P \cap at\ L\}\,(C \rightarrow A)\,\{Q[at\ M]\}$ *are equivalent.*

ROOF. In the following development, all the assertions are equivalent.

$$P\}\,T\,\{Q\},$$

$$L', M', \rho, \sigma\,[((L',\rho) \in P\ \wedge\ ((L',\rho),(M',\sigma)) \in rel(T))\ \Rightarrow\ (M',\sigma) \in Q\,],$$

$$L', M', \rho, \sigma\,[((L',\rho) \in P\ \wedge\ L'_T = L \wedge M'_T = M \wedge L'_{\overline{T}} = M'_{\overline{T}} \wedge C(\rho) \wedge \sigma = A(\rho))\ \Rightarrow\ (M'_T \cup M'_{\overline{T}}, \sigma) \in Q\,],$$

$$L', M', \rho, \sigma\,[((L',\rho) \in P\ \wedge\ (L',\rho) \in at\ L\ \wedge\ L'_{\overline{T}} = M'_{\overline{T}}\ \wedge\ C(\rho)\ \wedge\ \sigma = A(\rho))\ \Rightarrow\ (M \cup M'_{\overline{T}}, \sigma) \in Q\,],$$

$$L', M', \rho, \sigma\,[((L',\rho) \in (P \cap at\ L)\ \wedge\ L'_{\overline{T}} = M'_{\overline{T}}\ \wedge\ C(\rho)\ \wedge\ \sigma = A(\rho))\ \Rightarrow\ (M \cup M'_{\overline{T}}, \sigma) \in Q \cap at\ M\,],$$

$$L', M', \rho, \sigma\,[((L',\rho) \in (P \cap at\ L)\ \wedge\ L' = M'\ \wedge\ C(\rho)\ \wedge\ \sigma = A(\rho))\ \Rightarrow\ (M \cup M'_{\overline{T}}, \sigma) \in Q[at\ M] \cap at\ M\,],$$

$$L', M', \rho, \sigma\,[((L',\rho) \in (P \cap at\ L)\ \wedge\ L' = M'\ \wedge\ C(\rho)\ \wedge\ \sigma = A(\rho))\ \Rightarrow\ (M'_T \cup M'_{\overline{T}}, \sigma) \in Q[at\ M]\,],$$

$$L', M', \rho, \sigma\,[((L',\rho) \in (P \cap at\ L)\ \wedge\ ((L',\rho),(M',\sigma)) \in rel(C \longrightarrow A))\ \Rightarrow\ (M',\sigma) \in Q[at\ M]\,],$$

$$P \cap at\ L\}\,(C \longrightarrow A)\,\{Q[at\ M]\}\,.$$

OROLLARY 7. *The triple* $\{P \setminus at\ L\}\,(L,\ C \rightarrow A,\ M)\,\{Q\}$ *is always true.*

EMMA 8. *If* $dom(L) = dom(M)$ *then the triples*

$P[at\ L] \cap at\ L\}\,(C \longrightarrow A)\,\{Q[at\ M]\}$ *and* $\{P[at\ L]\}\,(C \longrightarrow A)\,\{Q[at\ M]\}$

re equivalent.

ROOF. Omitted.

Theorem 9. *The triples*

$$\{P\}\,(L,\,C\,\to\,A,\,M)\,\{Q\}\,,$$
$$\{P \cap at\ L\}\,(C\,\to\,A)\,\{Q[at\ M]\}\,,$$
$$\{P[at\ L] \cap at\ L\}\,(C\,\to\,A)\,\{Q[at\ M]\}\,,$$
$$\{P[at\ L]\}\,(C\,\to\,A)\,\{Q[at\ M]\}\,,$$
$$\{P[at\ L] \cap C\}\,A\,\{Q[at\ M]\}\,.$$

are equivalent.
Proof. Omitted.

3.3 Programming calculus

The results obtained in the previous paragraphs suggest that, in order to determine the truth-value of a Hoare triple, it is sufficient to know the corresponding weakest preset and/or strongest postset. As a first step, we deduce a more operational definition for the concepts of strongest postset and weakest preset. We have the following theorem.

Theorem 10.

$$\begin{aligned} s \in slp[P;X] \;&\equiv\; \exists\, r\,[(r,s) \in rel(X) \;\wedge\; r \in P]\,,\\ r \in wlp[X;Q] \;&\equiv\; \forall\, s\,[(r,s) \in rel(X) \;\Rightarrow\; s \in Q]\,. \end{aligned}$$

Proof. The triple $\{P\}\,X\,\{slp[X;P]\}$ is always true; this leads to

$$\begin{aligned} &\{P\}\;X\;\{slp[X;P]\}\,,\\ &\forall r \forall s\,[(r \in P \,\wedge\, (r,s) \in rel(X)) \;\Rightarrow\; s \in slp[X;P]]\,,\\ &\forall s\,[\exists r\,(r \in P \,\wedge\, (r,s) \in rel(X)) \;\Rightarrow\; s \in slp[X;P]]\,. \end{aligned}$$

This establishes the inclusion $\{s\,:\,\exists r\,(r \in P \,\wedge\, (r,s) \in rel(X))\} \subset slp[X;P]$.
The converse is proved as follows.

$$\begin{aligned} &\forall r_1 \forall s_1\,[(r_1 \in P \,\wedge\, (r_1,s_1) \in rel(X)) \;\Rightarrow\; (r_1 \in P \,\wedge\, (r_1,s_1) \in rel(X))]\,,\\ &\forall r_1 \forall s_1\,[(r_1 \in P \,\wedge\, (r_1,s_1) \in rel(X)) \;\Rightarrow\; \exists r\,(r \in P \,\wedge\, (r,s_1) \in rel(X))]\,,\\ &\forall r_1 \forall s_1\,[(r_1 \in P \,\wedge\, (r_1,s_1) \in rel(X)) \;\Rightarrow\; s_1 \in \{s\,:\,\exists r\,(r \in P \,\wedge\, (r,s) \in rel(X))\}]\,,\\ &\{P\}\,X\,\{\{s\,:\,\exists r\,(r \in P \wedge (r,s) \in rel(X))\}\}\,,\\ &slp[X;P] \subset \{s\,:\,\exists r\,(r \in P \,\wedge\, (r,s) \in rel(X))\}\,. \end{aligned}$$

The result about *wlp* is proved in a similar way.

Theorem 11. *If* $(C\,\to\,A)$ *is a guarded assignment and if* P *and* Q *are sets of system states, then*

$$\begin{aligned} wlp[(C\,\to\,A);\,Q] \;&=\; (\Gamma \setminus C) \cup wlp[A;Q]\,,\\ slp[P;\,(C\,\to\,A)] \;&=\; slp[P \cap C;\,A]\,. \end{aligned}$$

Proof. Omitted.

Theorem 12. *If* $(L,\,C\,\to\,A,\,M)$ *is a transition and if* P *and* Q *are sets of system states, then*

$$\begin{aligned} wlp[(L,\,C\,\to\,A,\,M);\,Q] \;&=\; (\Gamma \setminus at\ L) \cup wlp[(C\,\to\,A);\,Q[at\ M]]\,,\\ slp[P;\,(L,\,C\,\to\,A,\,M)] \;&=\; (slp[P[at\ L];\,(C\,\to\,A)] \cap at\ M)\,. \end{aligned}$$

PROOF. Only the first identity is proved here; the second one can be proved in a similar way. Let Z be an arbitrary set of system states. In the following development, all formulas are equivalent.

$$\begin{aligned}
&Z \subset wlp[(L,\, C \rightarrow A,\, M);\, Q]\,,\\
&\{Z\}\,(L,\, C \rightarrow A,\, M)\,\{Q\}\,,\\
&\{Z \cap at\ L\}\,(C \rightarrow A)\,\{Q[at\ M]\}\,,\\
&(Z \cap at\ L) \subset wlp[(C \rightarrow A);\, Q]\,,\\
&Z \subset (\Gamma \setminus at\ L) \cup wlp[(C \rightarrow A);\, Q]\,.
\end{aligned}$$

For $Z = wlp[(L,\, C \rightarrow A,\, M);\, Q]$, the development establishes the inclusion

$$wlp[(L,\, C \rightarrow A,\, M);\, Q] \subset (\Gamma \setminus at\ L) \cup wlp[(C \rightarrow A);\, Q]\,;$$

for $Z = (\Gamma \setminus at\ L) \cup wlp[(C \rightarrow A);\, Q]$, the development establishes the inclusion

$$(\Gamma \setminus at\ L) \cup wlp[(C \rightarrow A);\, Q] \subset wlp[(L,\, C \rightarrow A,\, M);\, Q]\,.$$

Comment. These theorems show that the programming calculus for formal concurrent systems reduces to the calculus for assignments; the introduction of processes and labels has not altered the simplicity of transition systems.

Several authors [CC77,vLS79,Cla80,Sif82,Lam87] have introduced notions of *strongest invariant* and *weakest invariant*. Such notions are now adapted for FCS; the proofs are omitted. Let C_0 be a set of states of $\mathcal{S}$. The strongest invariant $sin[C_0, \mathcal{S}]$ is defined as the set of states reachable from a state $s \in C_0$:

$$sin[C_0, \mathcal{S}] \;=_{def}\; \{s : \exists\, s_0\, [s_0 \in C_0 \text{ and } s_0 \xrightarrow{*} s]\}\,.$$

The expression "strongest invariant" comes from the following theorem.

THEOREM 13.

$$sin[C_0, \mathcal{S}] \equiv \bigcap \{I : (C_0 \subset I \text{ and } \{I\}\,\mathcal{S}\,\{I\})\}\,.$$

Fixpoint theory is useful to connect sin with slp. A definition is introduced first.

$$F[C_0, \mathcal{S}](X) \;=_{def}\; (C_0 \cup slp[X;\, \mathcal{S}])\,,$$

for any subset $X \subset \Gamma$. We have the following result.

THEOREM 14.

$$sin[C_0, \mathcal{S}] \equiv \bigcup_{n \in \mathbf{N}} F[C_0, \mathcal{S}]^n(\emptyset)\,.$$

As $F[C_0, \mathcal{S}]$ is easily checked to be $\cup$-additive, $sin[C_0, \mathcal{S}]$ is the least fixpoint (i.e., smallest fixpoint) of the function $F[C_0, \mathcal{S}]$.

A dual notion $win[\mathcal{S}, A_0]$ is useful in the following context. If A_0 is claimed to be an invariance property of the system $\mathcal{S}$, then an invariant is needed to prove this claim. The most interesting one is the weakest, or greatest one (the strongest one is the empty set of states).

$$win[\mathcal{S}, A_0] \;=_{def}\; \{s : \forall\, s_0\, [s \xrightarrow{*} s_0 \;\Rightarrow\; s_0 \in A_0]\}\,.$$

The theorems given above for sin have dual counterparts for win.

Comment. Various symbols and expressions have been introduced to deal with weakest and strongest invariants; the symbols win and sin come from [Lam87].

3.4 Assertional representation

As in classical Hoare logic and programming calculus, it is convenient to represent sets of states by logical formulas, called *assertions*. The correspondence between sets and assertions is easy. Complementation, union and intersection become negation, disjunction and conjunction, respectively. For instance, theorem 12 can be rewritten as

$$\begin{aligned} wlp[(L, C \to A, M); Q] &\equiv (\neg at\ L \vee wlp[(C \to A); Q[at\ M]]), \\ slp[P; (L, C \to A, M)] &\equiv (slp[P[at\ L]; (C \to A)] \wedge at\ M), \end{aligned}$$

where *wlp* and *slp* now have their classical meaning, i.e., weakest liberal precondition and strongest liberal postcondition.
In this framework, *at* L is a *place predicate*, which is true exactly for the elements of the set *at* L. Similarly, if the set Q is represented by an assertion (with the same name Q), then the set $Q[at\ L]$ can also be represented by an assertion (still denoted $Q[at\ L]$); the translation procedure is straightforward.

> Let L be the (partial) control state $\ell_1 \ldots \ell_r$, with $\ell_i \in P_i$ and $P_i \in \mathcal{P}$ for all i; the assertion $Q[at\ L]$ is obtained from the assertion Q as follows. First, each place predicate *at* N occurring in Q is replaced by the conjunction $(at\ n_1 \wedge \cdots \wedge at\ n_q)$, where $N = n_1 \ldots n_q$. Afterwards, for each control point $m \in \pi \in \mathcal{P}$, all occurrences of the place predicate *at* m in Q are replaced
> - by *true* if, for some i, $\pi = P_i$ and $m = \ell_i$;
> - by *false* if, for some i, $\pi = P_i$ but $m \neq \ell_i$;
> - by *at* m (no change) if $\pi \notin \{P_1, \ldots, P_r\}$.

Let us recall that, if the set Σ of memory states is infinite, then there are uncountably many sets of states; as a consequence, some sets cannot be represented by assertions. Especially, *win* and *sin* are usually not representable, except when a formalism stronger than first order logic is used. Such formalisms are μ-calculus (allowing fixpoint computations) and infinitary logic (allowing countable disjunctions); see e.g. [Apt81,Bac81,SRG89] for more details. In the assertional framework, the words "stronger" and "weaker" have their standard meaning; a stronger condition implies a weaker one.

3.5 Deduction system for invariance properties

Invariance properties, or *safety properties*, assert that "something bad cannot happen". An invariance property is modelled by a set of states A_0. The system S with the set of initial states C_0 satisfies the property modelled by the set A_0 if every computation whose initial state belongs to C_0 is included in A_0. (Obviously, C_0 must be a subset of A_0.)

The invariant method to prove invariance properties consists in obtaining an adequate invariant, that is, a set of states I such that

$$\begin{aligned} &C_0 \subset I, \\ &\{I\} S \{I\}, \\ &I \subset A_0 . \end{aligned}$$

Hoare logic is clearly *sound* since the existence of an adequate invariant I guarantees the correctness of the system S with the set of initial states C_0, with respect to the invariance property modelled by A_0. It is also *complete* since, conversely, the correctness guarantees the existence of an adequate invariant. Most of the time, several adequate invariants exists; the strongest one is $sin[C_0, S]$ and the weakest one is $win[S, A_0]$.

The deduction system used to establish properties like $\{P\}\, \mathcal{S}\, \{Q\}$ in FCS is a simplified version of the system presented in [Ger84]. The deduction rules, written in the assertional framework, are given below. The *structural rule* is

$$\frac{\{P\}\, \tau\, \{Q\}\,,\ \text{for all } \tau \in \mathcal{T}}{\{P\}\, \mathcal{S}\, \{Q\}}$$

The *transition rule* is

$$\frac{\{P[at\ L]\}\ (C \longrightarrow A)\ \{Q[at\ M]\}}{\{P\}\ (L,\ C \rightarrow A,\ M)\ \{Q\}}$$

The *guarded assignment* rule is

$$\frac{\{P \wedge C\}\ A\ \{Q\}}{\{P\}\ C \rightarrow A\ \{Q\}}$$

The *assignment rule* asserts that the triple $\{P\}\, A\, \{Q\}$ is true if and only if it is true in classical Hoare logic.

The *label axiom* asserts that, at every time and for each process $P = \{p_1, \ldots, p_n\}$, exactly one of the place predicates $at\ p_1, \ldots, at\ p_n$ is true.

In the assertional framework, Hoare logic remains sound, but additional hypotheses are needed to maintain the completeness. First, as the completeness result relies on the existence of the set of states $sin[C_0, \mathcal{S}]$, this set must be representable. Second, as the Hoare triple $\{P\}\, A\, \{Q\}$ is true if and only if $P \subset wlp[A; Q]$, it is necessary to recognize the validity of formulas like $P \Rightarrow wlp[A; Q]$. The detailed discussion about these hypotheses given in [Apt81] applies without change for FCS. The conclusion is that Hoare logic is "complete in the sense of Cook" : the deductive power of Hoare's rules is sufficient, but Hoare logic might inherit the restrictions due to the weak expressive power of the standard assertional languages, and to the incompleteness of number theory (and other theories about classical data types). This will not be the case if, first, $sin[C_0, \mathcal{S}]$ is representable when C_0 is representable and, second, if every true sentence of number theory expressed in the assertion language is assumed as an axiom.

Let us recall that the restrictions are not harmful in practice. First, useful invariants are representable and, second, relevant facts of number theory should be known by the program designer, even if no formal proof within Peano theory is given.

3.6 Towards a temporal proof system

A *liveness property*, or *progress property*, asserts that "something good eventually happens". Interesting properties of programs can be split in invariance and liveness properties (see [AS87] for a formal presentation). Temporal proof systems can be used to deal not only with invariance properties, but also with liveness properties of programs. Several temporal systems (more or less powerful) has been introduced and can be adapted for FCS. The system introduced in [CM88] for UNITY will be considered here. The main operators of this temporal system are *invariant* and *leads-to*, but they are based on two more elementary operators called *unless* and *ensures*. Let us first recall the definition of *unless* [CM88] :

$$(P\ unless\ Q) =_{def} [\{P \wedge \neg Q\}\, \mathcal{S}\, \{P \vee Q\}]\,.$$

This definition is adopted without change for FCS.

The definition of *ensures* is as follows: $(P\ ensures\ Q)$ is true when $(P\ unless\ Q)$ is true and there exists a transition τ such that $\{P \wedge \neg Q\}\, \tau\, \{Q\}$. This definition depends on the fact that, in UNITY, $cond(\tau)$ is identically true. As this is no longer true in FCS, we replace this definition by the following one, where $cond(\mathcal{T}_0)$ stands for $\bigvee_{\tau\in\mathcal{T}_0} cond(\tau)$:

$$\begin{aligned}&(P \ ensures \ Q) \ =_{def} \\ &(P \ unless \ Q) \ \wedge \ \exists \mathcal{T}_0 \, (\mathcal{T}_0 \subset \mathcal{T} \ \wedge \ [P \ \Rightarrow \ (cond(\mathcal{T}_0) \vee Q)] \ \wedge \ [\{P \wedge \neg Q\} \, \mathcal{T}_0 \, \{Q\}]) \, .\end{aligned} \tag{5}$$

Comment. The operational meaning of $(P \ ensures \ Q)$ is as follows: P remains true until Q is true, and it is possible to reach Q from P in at most one step. Let us mention a particular case: if $P \Rightarrow Q$, then Q is reached immediately, the formula $P \wedge \neg Q$ reduces to *false* and any subset $\mathcal{T}_0$ is acceptable.

The (recursive) definition of $(P \mapsto Q)$ (read "P leads to Q") will be the same in UNITY and in FCS; $(P \mapsto Q)$ is the smallest relation which is closed for the inference rules listed below:

- $\dfrac{(P \ ensures \ Q)}{(P \mapsto Q)}$
- $\dfrac{(P \mapsto R), \ (R \mapsto Q)}{(P \mapsto Q)}$
- $\dfrac{(P_m \mapsto Q) \text{ for all } m \in W \, , \ (P \ \Rightarrow \ \bigvee_{m\in W} P_m)}{(P \mapsto Q)}$

Comment. A stronger programming logic for UNITY has been proposed in [GP89]; as this logic is still based on Hoare triples, it can also be adapted to FCS. In fact, it is already mentioned in [MP83] that a proof system based on linear temporal logic can be obtained for any transition-based programming language. The temporal operator *next* could be defined as follows:

$$(P \ next \ Q) \ =_{def} \ [(P \ \Rightarrow \ cond(\mathcal{T})) \ \wedge \ (\{P\}\mathcal{T}\{Q\})] \, ,$$

meaning that, when P is true at some time, Q is necessarily true the next time.

3.7 Fairness

The intended operational meaning of $(P \mapsto Q)$ is as follows: if P is true, then Q is or will be true. More precisely, for each computation $\mathcal{C} = (s_i : i \in \mathbf{N})$, if $s_n \in P$, then $s_m \in Q$ for some $m \geq n$ (this is sometimes noted $(P \leadsto Q)$).

In order to achieve $(P \mapsto Q) \ \Rightarrow \ (P \leadsto Q)$, it is well-known one has to restrict to *fair* computations. Several notions of fairness can be defined (see [MP83,Fra86]), but the most popular one is *strong fairness.* In FCS, the infinite computation $\mathcal{C}$ is fair if $(s_n, s_{n+1}) \in rel(\tau)$ for infinitely many $n \in \mathbf{N}$, provided that $cond(\tau)$ is true for infinitely many $s_n \in \mathcal{C}$.

Comments. As pointed out in [GP89], the definition does not imply that τ is executed infinitely often, but only that its effect is felt infinitely often.

It is not always appropriate to determine a priori fairness hypotheses; another possibility is to reduce the required liveness properties to a set of elementary fairness requirements.

4 FCS = UNITY + processes + control flow

The language FCS can be viewed as an extension of UNITY. However, an extended language is interesting only if the underlying ideas of the original language and of the corresponding methodology are respected. In this section we investigate whether this is true for FCS.

4.1 UNITY as a subset of FCS

The formal concurrent system $\mathcal{S}$ is *UNITY-like* when $\mathcal{P} = \emptyset$ and $C = true$ for all $(C \longrightarrow A) \in \mathcal{T}$. An immediate consequence is that, in a UNITY-like system, $cond(\tau)$ is identically true for each transition τ. UNITY programs and UNITY-like formal concurrent systems differ only in syntax. Furthermore, the FCS logic nearly reduces to the UNITY logic when only UNITY-like systems are considered. The operator *unless* is defined similarly in UNITY and in FCS. Our definition of *ensures* is slightly less restrictive than the original one: the subset $\mathcal{T}_0$ mentioned in the definition (5) may contain more than one transition. However, this does not induce a difference for the derived operator $\mapsto$. Let us also mention that, for UNITY-like programs, there is no distinction between strong fairness and weak fairness.

4.2 An orderly world for control flow

The language FCS extends the language UNITY by introducing processes and control flow. However, FCS does not really increase the expressive power of UNITY, since control flow can be modelled by special variables ("program counters"). Let us also mention that UNITY programs are cyclic, whereas FCS programs may terminate, but this is not an essential difference. From the theoretical point of view, the extension shows that there exists an orderly world for control flow, just as there exists one for assignments (see [CM88], Chapter 1).

The formal logic and the programming calculus introduced in Section 3 extend UNITY logic without destroying the simplicity of this logic. In our opinion, this result means that explicit control flow can be introduced in Chandy and Misra's methodology, but it does not imply that something is gained by doing so. The sequel of this section presents some cases where explicit control flow might be convenient.

4.3 Control flow introduction

In [CM88], processes and control flow matters are considered at the end of the design only. As FCS provides a formal and clean way to deal with these notions, it is possible to integrate them in the design, and even to deal with processes and control flow early. Chandy and Misra suggest with some insistence that this is not appropriate, but we feel that it can be in some cases. The most adequate way to sustain this claim is to consider substantial examples taken from [CM88], but this would be a bit long in this context. We propose a fragment of the design of a very simple system instead. (A less elementary example is considered in detail in [Grb90b]; a short version is [Grb90].)

Let us suppose we have to realize a system for mutual exclusion, and that we adopt the following elementary idea. The entities competing for access are identified by a "ticket"; tickets are organized in a queue, and access is granted to entities having put their ticket on the waiting queue. We concentrate on the following case: entity p has completed its access, and entity q is the next to be served.

At an abstract level, the next transition could be

$$(q_1 p_2,\ \neg needed_p \wedge q \in E \longrightarrow (INCS, E) := (q, E \setminus \{q\}),\ q_2 p_0)$$

where, for entity p, p_0 is the idle state, p_1 the waiting state and p_2 the critical state. (E is the queue and $INCS$ records the ticket of the entity currently in its critical section.)

At a "CSP-like" level, entities are (real) processes, and so is the controller. The corresponding transitions now are

$$(p_2 C,\ \neg needed_p \longrightarrow skip,\ p_0 C'_p),$$
$$(q_1 C'_p,\ q \in E \longrightarrow (INCS, E) := (q, E \setminus \{q\}),\ q_2 C)$$

where *at* C means that the controller is idle, whereas *at* C'_p means that the controller has received an "end-of-access" message from process p.

At a "network" level, entities are stations and communicate with the controller by asynchronous messages. With self-explanatory notation, the transitions become

$$\begin{array}{rl} (p_2,\ \neg needed_p & \longrightarrow\ END_p := 1\,,\ p_0)\,, \\ (C,\ END_p = 1 & \longrightarrow\ END_p := 0\,,\ C'_p)\,, \\ (C'_p,\ q \in E & \longrightarrow\ (INCS, E, OK_q) := (q, E \setminus \{q\}, 1)\ ,\ C)\,, \\ (q_1,\ OK_q = 1 & \longrightarrow\ OK_q := 0\ ,\ q_2)\,. \end{array}$$

One can see that the only difficulty here comes from the control flow, and to deal with it is the crucial part of the design. Synchronization problems often turn out to reduce to control flow problems. This example also demonstrates that the stepwise refinement paradigm is not restricted to data and variables, but works for control flow as well. In fact, FCS is appropriate for stepwise refinement design involving both data and control, and there is no clear interest to delay control consideration. Even in the case where control flow problems are delayed as long as possible, they have still to be considered; this is the "mapping problem" introduced in [CM88]. FCS can be a good formalism to deal with this problem.

4.4 Compositionality

In order to verify $\{P\}S\{Q\}$, we have to verify $\{P\}\tau\{Q\}$ for each transition τ. In many cases P and Q will be a global invariant I, that is, a rather long formula. As mentioned in [CC89], one could look for a better situation, where P and Q would only be "components" of I. In fact, there is no interest to split I into syntactic components; only two points matter here: it is difficult to *discover* long formulas, and it is difficult to *handle* long formulas.

The solution we propose for the first point is to discover I by stepwise refinement, as in [CM88], but including also "control flow refinement" (see next paragraph). As invariants of more abstract versions are shorter than invariants of less abstract versions, stepwise refinement makes invariant discovery easier.

Concerning the second point, that is, the verification of the triple $\{I\}\tau\{I\}$ for a transition $\tau = (L, C \to A, M)$, where I is a rather long global invariant, we observe first that I is usually a (conjunctive) set of assertions. Most of the time, only few assertions are concerned by transition τ; only these assertions have to be taken into account. Second, in FCS, the transition rule reduces the problem to the verification of the triple $\{I[at\ L]\}C \to A\{I[at\ M]\}$; usually, $I[at\ L]$ and $I[at\ M]$ are simpler than I.

For instance, an assertion J_p of the global invariant of the network version (§4.3) is

$$\begin{array}{rl} & [\,at\ p_1 = (at\ C_p + REQ_p + OK_p + p \in E)] \\ \wedge & [(INCS = p) = (at\ C'_p + OK_p + END_p + at\ p_2)]\,, \end{array}$$

but the formula $J_p[at\ p_0C'_p]$ is simply

$$\begin{array}{rl} & [\,false = (false + REQ_p + OK_p + p \in E)] \\ \wedge & [(INCS = p) = (true + OK_p + END_p + false)]\,, \end{array}$$

which reduces to

$$REQ_p = OK_p = END_p = 0 \wedge p \notin E \wedge INCS = p\,.$$

We have the benefit from "factoring" the invariant into assertions attached with control states, as in older methods, without the drawbacks of these methods.

.5 Atomicity refinement

Vhen a designer refines a program, that is, replaces it by a more efficient one, the validity f the transformation has to be proved. More often than not, this proof consists in selecting . "refined invariant", usually weaker than the original one. As mentioned in [CM88], the election of a refined invariant is a critical design decision. It is therefore tempting to use ,eneric "adaptation procedure" for this purpose. An interesting particular case is the atomicity efinement, that is, the replacement of a coarser-grained version of a system by a finer-grained ne. Such refinements are easy to imagine but, as they are not always valid, they must be erified carefully. Several authors have proposed techniques for adapting the invariant, and our urpose here is to emphasize the role of explicit control flow in this respect.

Let us consider a system S and an invariant I of this system. The transition

$$T = (p,\ C \ \rightarrow\ A\ ,\ q)\,,$$

nvolving a single process $\pi = \{p, q, \ldots\}$, is to be replaced by two new transitions

$$\begin{aligned} T' &= (p,\ C \ \longrightarrow\ A'\ ,\ r)\,, \\ T'' &= (r,\ \mathit{true}\ \longrightarrow\ A''\ ,\ q)\,, \end{aligned}$$

where r is a new label added into process π, and A', A'' are such that $A'; A''$ is sequentially quivalent to A.

Explicit control flow is useful here, since the difference between the old and the new versions s summarized as follows: the place predicate *at* r can be true in the new version, and is dentically false in the old one.

Furthermore, the usefulness of the explicit way to denote control flow is not only for description, but also for design. The designer will look for a new invariant I' which reduces to I when *at* r does not hold. More specifically, a formula J must be obtained, such that

$$I' =_{def} [\neg at\ r \ \Rightarrow\ I] \wedge [at\ r\ \Rightarrow\ J]$$

s an invariant of the refined version.

The formula J is constrained first by

$$\begin{aligned} &\{I[at\ p] \wedge C\}\ A'\ \{J\}\,, \\ &\{J\}\ A''\ \{I[at\ q]\}\,. \end{aligned}$$

(It is supposed that $J = J[at\ r]$; this is not a restriction.) These constraints give rise to a stronger bound $slp[I[at\ p] \wedge C;\ A']$ and to a weaker bound $wlp[A'';\ I[at\ q]]$.

Second, a formula satisfying these constraints has to be selected, which satisfies also the additional constraints

$$\{J\}\,\tau\,\{J\}\,,\quad \text{for all } \tau : \pi \notin type(\tau)\,.$$

(More details are given in [Grb90b]; the relevant point here is that place predicates play a central role in atomicity refinement.)

5 Related work

Starting from [Kel76], much work has been devoted to transition and assignment systems and their use in concurrent programming. The DO-construct introduced in [Dij76] is the ancestor of the transition-based approaches in concurrency, and therefore of languages like UNITY and FCS. The programming calculus also originates from Dijkstra [Dij76,dBa80]; our formalism is based more specifically on [Sif82].

Hoare's logic is a classical tool for sequential programming [Hoa69,Dij76,Apt81,Gri81]. As already mentioned, Hoare's logic has been adapted for concurrent programming in several ways (see e.g. [Lam80,FS81,LS84,Ger84,CC89]). The Guarded Command language and the programming methodology introduced by Dijkstra have influenced many contributions in this area, including ours. Another similar language is the formalism of action system developed in [BK88].

Several authors have suggested that programming with labels is inadequate. Lamport has demonstrated in several papers that this statement is not true (see e.g. [Lam83]). This induced us to give a more prominent role to labels in concurrent programming, and to get rid of any other means for specifying processes and control flow.

The recent book of Chandy and Misra has demonstrated that a programming language with an elementary formal semantics can be useful in practice; this led us to look for an elementary semantics for FCS, and also to think that languages with a complex semantics are not appropriate. Furthermore, the notion of mapping introduced in [CM88] has suggested an application area for FCS.

Acknowledgment. The author is indebted to an anonymous referee for providing many constructive suggestions, including a better proof of theorem 12.

References

[AS87] B. ALPERN and F.B. SCHNEIDER, "Recognizing safety and liveness", *Distributed Computing*, **2**, pp. 117-126, 1987.

[Apt81] K.R. APT, "Ten Years of Hoare's Logic: A Survey – Part I", *ACM Trans. on Programming Languages and Systems*, **3**, pp. 431-483, 1980.

[Bac81] R.J.R. BACK, "Proving Total Correctness of Nondeterministic Programs in Infinitary Logic" *Acta Informatica*, **15**, pp. 233-249, 1981.

[BK88] R.J.R. BACK and R. KURKI-SUONIO, "Distributed Cooperation with Action Systems", *ACM Trans. on Programming Languages and Systems*, **10**, pp. 513-554, 1988.

[CM88] K.M. CHANDY and J. MISRA, "Parallel Program Design: A Foundation", Addison-Wesley, 1988.

[Cla80] E.M. CLARKE, "Synthesis of Resource Invariants for Concurrent Programs", *ACM Trans. on Programming Languages and Systems*, **2**, pp. 338-358, 1980.

[CC77] P. COUSOT and R. COUSOT, "Abstract interpretation: a unified lattice model for static analysis of programs by construction or approximation of fixpoints", *Proc. 4th ACM Symp. on Principles of Programming Languages*, pp. 238-252, 1977.

[CC89] P. COUSOT and R. COUSOT, "A Language Independent Proof of The Soundness and Completeness of Generalized Hoare Logic", *Information and Computation*, **80**, pp. 165-191, 1989.

[dBa80] J.W. de BAKKER, "Mathematical Theory of Program Correctness", Prentice Hall, 1980.

[Dij76] E.W. DIJKSTRA, "A discipline of programming", Prentice Hall, 1976.

[FS81] L. FLON and N. SUZUKI, "The Total Correctness of Parallel Programs", *SIAM J. on Computing*, **10**, pp. 227-246, 1981.

[Fra86] N. FRANCEZ, "Fairness", Springer Verlag, 1986.

[Ger84] R. GERTH, "Transition Logic", *Proc. 16th ACM Symp. on Theory of Computing*, pp. 39-50, 1984.

[GP89] R. GERTH and A. PNUELI, "Rooting UNITY", *Proc. 5th IEEE Workshop on Software Specification and Design*, 1989.

[Grb87] E.P. GRIBOMONT, "Design and proof of communicating sequential processes", *Lecture Notes on Computer Science*, vol. 259, pp. 261-276, Springer-Verlag, 1987.

[Grb90] E.P. GRIBOMONT, "Development of concurrent systems by incremental transformation", *Lecture Notes on Computer Science*, vol. 432, pp. 161-176, Springer-Verlag, 1990.

[Grb90b] E.P. GRIBOMONT, "Development of concurrent systems by incremental transformation", Report M346, 30p., Philips Research Laboratory Belgium, May 1990.

[Gri81] D. GRIES, "The Science of Programming", Springer-Verlag, 1981.

[Hoa69] C.A.R. HOARE, "An axiomatic basis for computer programming", *CACM*, **12**, pp. 576-583, 1969.

[Hoa78] C.A.R. HOARE, "Communicating Sequential Processes", *CACM*, **21**, pp. 666-677, 1978.

oa85] C.A.R. HOARE, "Communicating Sequential Processes", Prentice Hall, 1985.

el76] R.M. KELLER, "Formal Verification of Parallel Programs", *Comm. ACM*, **19**, pp. 371-384, 1976.

am80] L. LAMPORT, "The 'Hoare Logic' of Concurrent Programs", *Acta Informatica*, **14**, pp. 21-37, 1980.

am83] L. LAMPORT, "An Assertional Correctness Proof of a Distributed Algorithm", *Science of Computer Programming*, **2**, pp. 175-206, 1983.

am87] L. LAMPORT, "*win* and *sin*: Predicate Transformers for Concurrency", Technical Report 17, Digital Systems Research Center, Palo Alto, 1987.

S84] L. LAMPORT and F.B. SCHNEIDER, "The 'Hoare Logic' of CSP, and All That", *ACM Trans. on Programming Languages and Systems*, **6**, pp. 281-296, 1984.

IP83] Z. MANNA and A. PNUELI, "How to cook a temporal proof system for your pet language", *Proc. 10th ACM Symp. on Principles of Programming Languages*, pp. 141-154, 1983.

if82] J. SIFAKIS, "A unified approach for studying the properties of transition systems", *Theoretical Computer Science*, **18**, pp. 227-259, 1982.

RG89] F.A. STOMP, W.P. de ROEVER and R.T. GERTH, "The μ-Calculus as an Assertion-Language for Fairness Arguments", *Information and Computation*, **82**, pp. 278-322, 1989.

LS79] A. van LAMSWEERDE and M. SINTZOFF, "Formal derivation of strongly correct concurrent programs", *Acta Informatica*, **12**, pp. 1-31, 1979.

A New Strategy for Proving ω-Completeness applied to Process Algebra

Jan Friso Groote

Centre for Mathematics and Computer Science

P.O. Box 4079, 1009 AB Amsterdam, The Netherlands

Email jfg@cwi.nl

Abstract

A new technique for proving ω-completeness based on proof transformations is presented. This technique is applied to axiom systems for finite, concrete, sequential processes. It turns out that the number of actions is important for these sets to be ω-complete. For the axiom systems for bisimulation and completed trace semantics one action suffices and for traces 2 actions are enough. The ready, failure, ready trace and failure trace axioms are only ω-complete if an infinite number of actions is available. We also consider process algebra with parallelism and show several axiom sets containing the axioms of standard concurrency ω-complete.

1 Introduction

An equational theory E over a signature Σ is called ω-complete iff for all open terms t_1 t_2:

$$\text{for all closed substitutions } \sigma: \; E \vdash \sigma(t_1) = \sigma(t_2) \quad \Leftrightarrow \quad E \vdash t_1 = t_2.$$

Not all equational theories are ω-complete: a well known example is the commutativity of the + in Peano arithmetic. Another example is the three-element groupoid of MURSKII [13], who showed that for an ω-complete specification of the groupoid an infinite number of equations is necessary.

Also in process algebra several theories are not ω-complete, and up till now this was more or less ignored (exceptions are MILNER [11] and MOLLER [12]). But there are several reasons why ω-completeness should not be neglected. In the first place equations between open terms play an important role in process algebra. For instance, processes are often described with sets of (open) equations. A complete set of axioms (not necessarily ω-complete) gives no guarantee that such sets of equations can be dealt with in a satisfactory manner. An example of this situation are the so-called 'axioms of standard concurrency' [2] in ACP, which had to be introduced in addition to the 'complete' set of axioms in order to prove the expansion theorem [3]. The status of these axioms became clear only

The author is supported by the European Communities under RACE project no. 1046, Specification and Programming Environment for Communication Software (SPECS). This article also appears as report CS-R90XX, Centrum voor Wiskunde en Informatica, Amsterdam

fter MOLLER [12] showed that in CCS with interleaving, but without communication, ome of the axioms of standard concurrency are required for ω-completeness.

Furthermore, ω-completeness is also useful for theorem provers [8, 9, 15]. In [14] he so-called method 'proof by consistency' is introduced which can be applied to show nductive theorems equationally provable if ω-completeness of the axioms has been shown. n HEERING [6] it is argued that ω-completeness is desirable for the partial evaluation of rograms. If $P(x,y)$ is a program with parameters x and y, and x has fixed value c, then he program $P_c(y)$ $(=P(c,y))$ should be evaluated as far as possible. In general this can nly be achieved if the evaluation rules are ω-complete.

A more or less standard technique for proving ω-completeness is the following: given a et of axioms E over a signature Σ, find 'normal forms' and show that every open term is rovably equal to a normal form. Then prove that for all pairs of different normal forms, losed instantiations can be found that differ in a model $\mathcal{M}$ for E. E does not necessarily ave to be complete with respect to $\mathcal{M}$. This last step shows that the equivalence of hese instantiations cannot be derived from E. From this ω-completeness of E follows lirectly. We prove the ω-completeness of the trace and completed trace axioms in this vay. This technique has some disadvantages. The proofs are in general quite long and t is often difficult to find a suitable normal form.

In this paper we present an alternative technique that employs transformations of roofs. It is explained in section 3. With this method proofs of ω-completeness turn ut to be shorter and for the major part straightforward. Moreover, no reference to model is necessary. Unfortunately, this new technique cannot always be used. We pply our method to five sets of axioms, which are taken from [4], for finite, concrete, equential processes. Among the proofs we give there is an ω-completeness proof of isimulation semantics of which an earlier and longer version is given in [12]. It turns ut that the number of actions is important for the axiom sets to be ω-complete. We eed an infinite number of actions for the ready trace, failure trace, ready and failure xioms. For the bisimulation and the completed trace axioms at least one action is equired whereas for the trace axioms two actions are necessary. Then we study axiom ets for finite, concrete process algebra with interleaving without communication (also lone in [12]) and interleaving with communication. We give straightforward proofs of he ω-completeness of these sets.

Acknowledgements. I thank Rob van Glabbeek for several fruitful discussions and Alban Ponse for his detailed and constructive comments.

2 Preliminaries

Throughout this text we assume the existence of a countably infinite set V of variables with typical elements x, y, z. A (one sorted) *signature* $\Sigma = (F, rank)$ consists of a set of *function names* F, disjunct with V, and a rank function $rank : F \rightarrow \mathbb{N}$, denoting the rity of each function name in F. $T(\Sigma)$ is the set of *closed terms* over signature Σ and $\mathbb{T}(\Sigma)$ is the set of *open terms* terms over Σ and V. We use the symbol $\equiv$ for syntactic equality between terms. Furthermore, we have *substitutions* $\sigma, \rho : V \rightarrow \mathbb{T}(\Sigma)$ mapping variables to terms. Substitutions are in the standard way extended to functions from erms to terms. An expression of the form $t = u$ $(t, u \in \mathbb{T}(\Sigma))$ is called an *equation* over

$x = x$ (reflexivity)	$\dfrac{x = y}{y = x}$ (symmetry)	$\dfrac{x = y \quad y = z}{x = z}$ (transitivity)
$\dfrac{x_i = y_i \quad 1 \leq i \leq rank(f)}{f(x_1, \ldots, x_{rank(f)}) = f(y_1, \ldots, y_{rank(f)})}$ for all $f \in F$ (congruence)		

Table 1: The inference rules of equational logic

Σ. The letter e is used to range over equations. An expression of the form

$$\frac{e_1, \ldots, e_n}{e}$$

is called an *inference rule.* We call $e_1, \ldots, e_n$ the *premises* and e the *conclusion* of th inference rule. Substitutions are extended to equations and inference rules as expected.

An *equational theory* over a signature Σ is a set E containing equations over Σ. Thes equations are called *axioms.* An equation e can be *proved* from a theory E, notatio $E \vdash e$, if e is an instantiation of an axiom in E or if e is the conclusion of an instantiatio of an inference rule r in table 1 of which all (instantiated) premises can be proved. I it is clear from the context what E is, we sometimes write only e instead of $E \vdash e$. W write $E_1 \vdash E_2$ if $E_1 \vdash e$ for all $e \in E_2$. Note that if $E \vdash t = u$ for $t, u \in T(\Sigma)$, then $t =$ can be proved using closed instantiated axioms and inference rules only.

An equational theory E is *ω-complete* if for all equations e: $E \vdash e$ iff $E \vdash \sigma(e)$ for al substitutions $\sigma : V \rightarrow T(\Sigma)$. Note that the implication from left to right is trivial. So in general we only prove the implication from the right-hand side to the left-hand side.

3 The general proof strategy

Let $\Sigma = (F, rank)$ be a signature and let E be an equational theory over Σ. We present a technique to show that E is ω-complete. Assume $t = t'$ is an equation between open terms that can be proved for all its closed instantiations by the axioms of E. We transform $t = t$ to a closed equation by a substitution $\rho : V \rightarrow T(\Sigma)$ that maps each variable in t and t to a unique closed (sub)term representing this variable. By assumption $E \vdash \rho(t) = \rho(t')$ We transform the proof of this fact to a proof for $E \vdash t = t'$ by a translation R which replaces each subterm representing a variable by the variable itself. This transformation yields the desired proof if requirements (1), (2) and (3) below are satisfied. (1) says that the translation of $\rho(t) = \rho(t')$ must yield $t = t'$ (or something provably equivalent). In general this only works properly if each subterm representing a variable is unique for that variable and cannot be confused with other subterms. Requirements (2) and (3) guarantee that the transformed proof is indeed a proof. This is most clearly stated in equation (5), which is a consequence of (2) and (3).

- For $u \equiv t$ or $u \equiv t'$:

$$E \vdash R(\rho(u)) = u. \tag{1}$$

- For each $f \in F$ with $rank(f) > 0$ and $u_1, ..., u_{rank(f)}, u'_1, ..., u'_{rank(f)} \in T(\Sigma)$:

$$E \cup \{u_i = u'_i,\ R(u_i) = R(u'_i) | 1 \leq i \leq rank(f)\} \vdash \\ R(f(u_1, ..., u_{rank(f)})) = R(f(u'_1, ..., u'_{rank(f)})). \tag{2}$$

- For each axiom $e \in E$ and closed substitution $\sigma : V \to T(\Sigma)$:

$$E \vdash R(\sigma(e)). \tag{3}$$

Theorem 3.1. *Let E be an equational theory over signature Σ. If for each pair of terms $t, t' \in \mathbb{T}(\Sigma)$ that are provably equal for all closed instantiations, there exist a substitution $\rho : V \to T(\Sigma)$ and a mapping $R : T(\Sigma) \to \mathbb{T}(\Sigma)$ satisfying (1),(2) and (3), then E is ω-complete.*

Proof. Let $t, t' \in \mathbb{T}(\Sigma)$ such that for each substitution $\sigma : V \to T(\Sigma)$:

$$E \vdash \sigma(t) = \sigma(t'). \tag{4}$$

We must prove that $E \vdash t = t'$. This is an immediate corollary of the following statement:

$$E \vdash u = u' \text{ for } u, u' \in T(\Sigma) \quad \Rightarrow \quad E \vdash R(u) = R(u'). \tag{5}$$

It follows from (4) that $E \vdash \rho(t) = \rho(t')$. Using (5) this implies $E \vdash R(\rho(t)) = R(\rho(t'))$. By (1) it follows that $E \vdash t = t'$.

Statement (5) is shown by induction on the proof of $E \vdash u = u'$. As u and u' are closed terms, we may assume that the whole proof of $E \vdash u = u'$ consists of closed terms. First we consider the inference rules without premises. There are two possibilities. In the first case $u = u'$ has been shown by the inference rule $x = x$, i.e. $u \equiv \sigma(x) \equiv u'$ for some substitution $\sigma : V \to T(\Sigma)$. Clearly, $E \vdash R(u) = R(u')$ using the same inference rule and a substitution $\sigma' : V \to \mathbb{T}(\Sigma)$ defined by $\sigma'(x) = R(\sigma(x))$. Otherwise, $u = u'$ is an instantiation $\sigma(e)$ of an axiom $e \in E$. Using (3) it follows immediately that $E \vdash R(\sigma(e))$.

We check here the inference rules with premises. First we deal with the rule for transitivity. So assume $E \vdash u = u'$ has been proved using $E \vdash u = u''$ and $E \vdash u'' = u'$. By induction we know that there are proofs for $E \vdash R(u) = R(u'')$ and $E \vdash R(u'') = R(u')$. Applying the inference rule for transitivity again we have that $E \vdash R(u) = R(u')$. The rule for symmetry can be dealt with in the same way. Now suppose that $E \vdash f(u_1, ..., u_{rank(f)}) = f(u'_1, ..., u'_{rank(f)})$ has been proved using $E \vdash u_i = u'_i$ ($1 \leq i \leq rank(f)$). By induction we know that $E \vdash R(u_i) = R(u'_i)$. Using (2), it follows immediately that $E \vdash R(f(u_1, ..., u_{rank(f)})) = R(f(u'_1, ..., u'_{rank(f)}))$. □

This new proof strategy cannot always be applied. This is illustrated by the following example.

Example 3.2. Suppose we have an axiomatization for the natural numbers with a function *max* giving the maximum of any pair of numbers. In the signature we have a 0, a successor function S and *max*. The following set E_{max} of axioms is easily seen to be complete with respect to the standard interpretation.

$$\begin{aligned} &max(x, 0) = x, \\ &max(0, x) = x, \\ &max(S(x), S(y)) = S(max(x, y)). \end{aligned}$$

Clearly, E_{max} is not ω-complete as for instance the general associativity and commutativity of max is not derivable although each closed instance of them is.

It is impossible to use our technique to prove any extension of E_{max} ω-complete. This can be seen by considering the following two terms:

$$t_1 = max(S(0), x) \text{ and}$$
$$t_2 = x.$$

We can see that these terms are not provably equal because with $x = 0$, the first term is equal to $S(0)$ and the second is equal to 0. Note that this is the only way to see the difference. If any term that is not equal to 0 is substituted for x then both terms are equivalent.

Suppose we would like to apply our technique in this case. If we define ρ such that $\rho(x) = 0$ then we must define the translation R such that $R(0) = x$. But then $R(\rho(t_1)) = max(S(x), x)$ which cannot be shown equal to $max(S(0), x)$. If ρ would be chosen such that $\rho(x) \neq 0$ and R could be defined such that $E_{max} \vdash R(t_i) = t_i$ $(i = 1, 2)$ then equation (5), which follows from (2) and (3), cannot hold because it implies that $E_{max} \vdash t_1 = t_2$.

So, this example shows that the new technique is not generally applicable, but as will be shown in the next sections, there are enough cases where the application of this technique leads to attractive proofs.

4 Applications in finite, concrete, sequential process algebra

In the remainder of this paper we apply our technique to prove completeness of several axiom systems. In this section sets given for BCCSP in [4] are studied. BCCSP is a basic CCS and CSP-like language for finite, concrete, sequential processes. It is parameterized by a set *Act* of actions representing the elementary activities that can be performed by processes. We write $|Act|$ for the number of elements in *Act* ($|Act| = \infty$ if *Act* has an infinite number of elements). The language BCCSP contains a constant δ, which is comparable to **0** or NIL in CCS and to STOP in CSP. We call δ *inaction* or sometimes *deadlock*. There is an *alternative composition* operator $+$ with its usual meaning and, furthermore, there is an *action prefix* operator $a :$ for each action a in *Act*.

In the sequel we will often use sums of arbitrary finite size. It is convenient to have a notation for these. Therefore we introduce the abbreviation:

$$\sum_{i \in I} t_i = t_{i_1} + ... + t_{i_n}$$

where $I = \{i_1, ..., i_n\}$ is a finite index set and $t_i \in \mathbb{T}(\text{BCCSP})$ $(i \in I)$. We take $\sum_{i \in \emptyset} t_i = \delta$. Note that this notation is only justified if $+$ is commutative, associative. We only use this notation when this is the case.

The depth $|t|$ of a term $t \in \mathbb{T}(\text{BCCSP})$ is inductively defined as follows:

$$|\delta| = 0, \qquad |x| = 0 \text{ for all } x \in V,$$
$$|a : t| = 1 + |t| \text{ for all } a \in Act, \qquad |t_1 + t_2| = \max(|t_1|, |t_2|).$$

In table 2 we find several axiom systems corresponding to several semantics given in [4]. We will investigate the ω-completeness of these sets. On the top line of this table we

	B	RT	FT	R	F	CT	T
$x+y=y+x$	+	+	+	+	+	+	+
$(x+y)+z=x+(y+z)$	+	+	+	+	+	+	+
$x+x=x$	+	+	+	+	+	+	+
$x+\delta=x$	+	+	+	+	+	+	+
(see (6) in text)		+	+	v	v	v	v
$a:x+a:y=a:x+a:y+a:(x+y)$			+		v	v	v
$a(b:x+u)+a:(b:y+v)=$ $a:(b:x+b:y+u)+a:(b:x+b:y+v)$				+	+	v	v
$a:x+a:(y+z)=a:x+a:(x+y)+a:(y+z)$					+	ω	v
$a:(b:x+u)+a:(c:y+v)=a:(b:x+c:y+u+v)$						+	v
$a:x+a:y=a:(x+y)$							+

Table 2: Axioms for several process algebra semantics

and their abbreviations: B stands for *Bisimulation*, RT for *Ready Trace*, FT for *Failure Trace*, R for *Ready* and F for *Failure semantics*, CT for *Completed Traces* and finally T represents *Trace* semantics. The axioms that are necessary for ready trace semantics (besides the axioms for bisimulation) are given by the following scheme:

$$a:(\sum_{i\in I} a_i:x_i+y)+a:(\sum_{i\in J} a_i:x_i+y)=a:(\sum_{i\in I\cup J} a_i:x_i+y) \tag{6}$$

where $\{a_i|i\in I\}=\{a_i|i\in J\}$, and $x_i,y\in V$ $(i\in I\cup J)$. This scheme differs from the axiomatization given in [4], where an additional function name I and a conditional axiom were used to axiomatize ready trace semantics. We do not want to introduce these concepts here. Both axiomatizations prove exactly the same open equations.

Let X stand for any of the semantics B,RT,... The symbol 'v' in a column of semantics X indicates that an axiom is derivable from the other axioms valid for X. The symbol '+' means that the axiom is required for a complete axiomatization of the models given in [4] and 'ω' means that the axiom is only necessary for an ω-complete axiomatization. It follows immediately that:

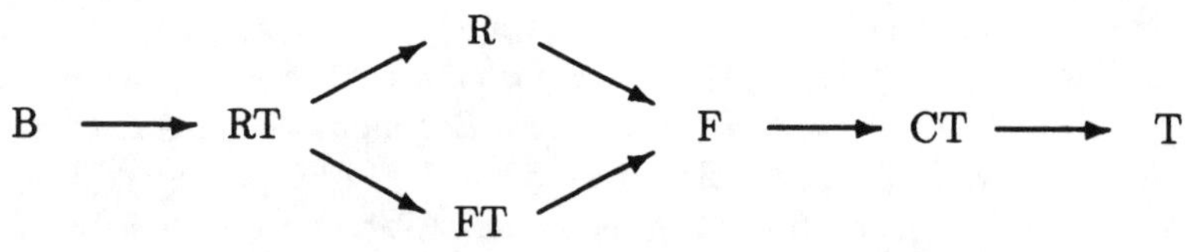

where the semantics to the left are finer than the semantics to the right. The semantics FT and R are incomparable [4]. The abbreviation for a semantics will also be used to denote the set of axioms necessary for its ω-complete axiomatization.

Lemma 4.1. *Let $t,u\in\mathbb{T}(BCCSP)$. If $T\vdash t=u$, then $|t|=|u|$.*

Proof. Direct with induction on the proof of $t=u$. □

As T $\vdash$ B, T $\vdash$ RT etc. it immediately follows from the last lemma that 'X $\vdash t = u \Rightarrow |t| = |u|$', where X is any of the sets B, RT, etc.

4.1 The semantics B

We start considering the axioms for bisimulation semantics. If *Act* contains at least one element, then B is ω-complete. This fact has already been shown in [12] where a traditional technique was used. Note that it makes no sense to investigate the situation where $Act = \emptyset$, because in that case all closed terms will have the form δ, $\delta+\delta$, $\delta+\delta+\ldots$ and therefore they are equal and we only require the axiom $x = y$ for an ω-complete axiomatization.

Theorem 4.1.1. *If $|Act| \geq 1$ then the axiom system B is ω-complete.*
Proof. As $|Act| \geq 1$, *Act* contains at least one action a. This action will play an important role in this proof. We follow the lines set out in theorem 3.1. So, assume we have two terms $t, t' \in \mathbb{T}(\text{BCCSP})$. Select a natural number $m > \max(|t|, |t'|)$ and define $\rho : V \to T(\text{BCCSP})$ by:

$$\rho(x) = a^{n(x)\cdot m} : \delta$$

where $a^k : \delta$ is an abbreviation of k applications of $a :$ to δ and $n : V \to \mathbf{N}\setminus\{0\}$ is a function assigning a unique natural number to each variable in x. Define $R : T(\text{BCCSP}) \to \mathbb{T}(\text{BCCSP})$ as follows:

$$\begin{aligned}
&R(\delta) = \delta,\\
&R(t+u) = R(t) + R(u),\\
&R(b:t) = b : R(t) \text{ if } b \neq a \text{ or } |b:t| \neq m\cdot n(x) \text{ for all } x \in V,\\
&R(a:t) = x \text{ if } |a:t| = m \cdot n(x) \text{ for some } x \in V.
\end{aligned}$$

We will now check conditions (1), (2) and (3) of theorem 3.1. We prove (1) with induction on a term $u \in \mathbb{T}(\text{BCCSP})$ provided $|u| < m$. Note that this is sufficient as $|t| < m$ and $|t'| < m$.

$$\begin{aligned}
&R(\rho(\delta)) = \delta,\\
&R(\rho(x)) = R(a^{n(x)\cdot m} : \delta) = x,\\
&R(\rho(u_1+u_2)) = R(\rho(u_1)) + R(\rho(u_2)) = u_1 + u_2,\\
&R(\rho(b:u)) = b : R(\rho(u)) = b : u \text{ if } b \neq a,\\
&R(\rho(a:u)) = R(a:\rho(u)) =^* a : R(\rho(u)) = a : u.
\end{aligned}$$

$=^*$ follows directly from the observation that $|a : \rho(u)| \neq m \cdot n(x)$ for all $x \in V$. In order to see this, first note that $1 \leq |a : u| < m$. If u does not contain variables, it is clear that $1 \leq |a : \rho(u)| < m$ and hence, $|a : \rho(u)| \neq m \cdot n(x)$. So, suppose u contains variables. By applying ρ to u each variable x is replaced by $a^{n(x)\cdot m} : \delta$. So $|a : \rho(u)| = p + n(x)\cdot m$ where x is a variable in u such that there is no other variable y in u with $n(y) > n(x)$ and p ($1 \leq p < m$) is the 'depth' of the deepest occurrence of x in u. As $1 \leq p < m$, $|a : \rho(u)| \neq n(x) \cdot m$ for each $x \in V$.

Now we check (2). Assume B $\vdash u_i = u_i'$ and B $\vdash R(u_i) = R(u_i')$ for $u_i, u_i' \in T(\text{BCCSP})$ and $i = 1, 2$. We find that:

$$\begin{aligned}
&\text{B} \vdash R(u_1+u_2) = R(u_1) + R(u_2) = R(u_1') + R(u_2') = R(u_1'+u_2').\\
&\text{B} \vdash R(b:u_1) = b : R(u_1) = b : R(u_1') = R(b:u_1') \text{ if } b \neq a.
\end{aligned}$$

$$B \vdash R(a:u_1) =^* a:R(u_1) = a:R(u_1') =^+ R(a:u_1)$$
if $|a:u_1| \neq m \cdot n(x)$ for all $x \in V$.

$=^*$ follows directly from the condition. As $B \vdash u_1 = u_1'$ it follows that $|a:u_1| = |a:u_1'|$ cf. lemma 4.1) and hence, $|a:u_1'| \neq m \cdot n(x)$ for all $x \in V$. This justifies $=^+$.

$$B \vdash R(a:u_1) = x =^* R(a:u_1') \text{ if } |a:u_1| = m \cdot n(x) \text{ for some } x \in V.$$

t follows that $|a:u_1'| = m \cdot n(x)$ explaining $=^*$.

Finally, we must check (3). This is trivial as the axioms do not contain actions. We nly check the axiom $x + y = y + x$. The other axioms can be dealt with in the same vay. Let $\sigma : V \to T(\text{BCCSP})$ be a substitution, then:

$$B \vdash R(\sigma(x+y)) = R(\sigma(x)) + R(\sigma(y)) = R(\sigma(y)) + R(\sigma(x)) = R(\sigma(y+x)).$$

□

.2 The semantics RT,FT,R and F

Ve will show that the sets of axioms RT,FT,R and F are all ω-complete in case *Act* is nfinite. If *Act* is finite, we have the following identity:

$$a : \sum_{i \in J} a_i : \delta + a : (x + \sum_{i \in J} a_i : \delta) = a : (x + \sum_{i \in J} a_i : \delta) \tag{7}$$

vhere $\{a_i | i \in J\} = Act$. Each closed instance of this identity is derivable from the axioms f RT,FT,R or F. However, (7) is not derivable in its general form: if (7) were derivable, hen it would also hold if *Act* would be extended by a 'fresh' action $b \notin \{a_i | i \in J\}$. Define a substitution σ satisfying $\sigma(x) = b : \delta$. Applying σ to (7) yields:

$$a : \sum_{i \in J} a_i : \delta + a : (b : \delta + \sum_{i \in J} a_i : \delta) = a : (b : \delta + \sum_{i \in J} a_i : \delta).$$

but this equation does not hold in the failure model [4]. Hence, it is not derivable from F and therefore it can certainly not be derived from RT,FT or R.

So, in order to prove RT,FT,R and F ω-complete, *Act* must at least be countably nfinite. The following theorem shows that this condition is also sufficient.

Theorem 4.2.1. *If $|Act|$ is infinite, then the axiom sets RT,FT,R and F are ω-complete.*

Proof. Take two terms t, t'. Define a substitution $\rho : V \to T(\text{BCCSP})$ by:

$$\rho(x) = a_x : \delta$$

where a_x is a unique action for each $x \in V$ and a_x must not occur in either t or t'. Note that these actions can always be found as $|Act| = \infty$. Define $R : T(\text{BCCSP}) \to T(\text{BCCSP})$ as follows:

$R(\delta) = \delta,$
$R(a : u) = a : R(u)$ if $a \neq a_x$ for each $x \in V$,
$R(a_x : u) = x,$
$R(u_1 + u_2) = R(u_1) + R(u_2).$

Condition (1) of theorem 3.1 can be checked by induction on the structure of open terms not containing action prefix operators a_x :.

$$\begin{aligned}
&R(\rho(\delta)) = \delta,\\
&R(\rho(x)) = R(a_x : \delta) = x,\\
&R(\rho(a : u)) = R(a : \rho(u)) = a : R(\rho(u)) = a : u \text{ as } a \neq a_x \text{ for each } x \in V,\\
&R(\rho(u_1 + u_2)) = R(\rho(u_1)) + R(\rho(u_2)) = u_1 + u_2.
\end{aligned}$$

Condition (2) can be checked in the same straightforward manner. Suppose X $\vdash R(u_i) = R(u_i')$ for $u_i, u_i' \in T(\text{BCCSP})$ and $i = 1,2$. X may be replaced by either RT,FT,R or F. Then:

$$\begin{aligned}
&\text{X} \vdash R(a : u_1) = a : R(u_1) = a : R(u_1') = R(a : u_1') \text{ if } a \neq a_x \text{ for each } x \in V.\\
&\text{X} \vdash R(a_x : u_1) = x = R(a_x : u_1').\\
&\text{X} \vdash R(u_1 + u_2) = R(u_1) + R(u_2) = R(u_1') + R(u_2') = R(u_1' + u_2').
\end{aligned}$$

Finally, we check (3). We restrict ourselves to the ready trace axiom scheme. All other axioms can be dealt with in the same way. First we assume that $a = a_x$. Let $\sigma : V \to T(\text{BCCSP})$ be a substitution. Then RT $\vdash$:

$$\begin{aligned}
&R(a_x : (\sum_{i\in I} a_i : \sigma(x_i) + \sigma(y)) + a_x : (\sum_{i \in J} a_i : \sigma(x_i) + \sigma(y))) =\\
&x + x = x =\\
&R(a_x : (\sum_{i\in I\cup J} a_i : \sigma(x_i) + \sigma(y))).
\end{aligned}$$

In case $a \neq a_x$ for each $x \in V$, we have that RT proves:

$$\begin{aligned}
&R(a : (\sum_{i\in I} a_i : \sigma(x_i) + \sigma(y)) + a : (\sum_{i\in J} a_i : \sigma(x_i) + \sigma(y))) =\\
&a : (\sum_{i\in I} R(a_i : \sigma(x_i)) + R(\sigma(y))) + a : (\sum_{i\in J} R(a_i : \sigma(x_i)) + R(\sigma(y))) =\\
&a : (\sum_{i\in I\setminus\{i\in I|a_i=a_x\}} a_i : R(\sigma(x_i)) + \sum_{x\in\{x|a_x=a_i\wedge i\in I\}} x + R(\sigma(y)))+\\
&a : (\sum_{i\in J\setminus\{i\in J|a_i=a_x\}} a_i : R(\sigma(x_i)) + \sum_{x\in\{x|a_x=a_i\wedge i\in J\}} x + R(\sigma(y))) =^*\\
&a : (\sum_{i\in (I\cup J)\setminus\{i\in I\cup J|a_i=a_x\}} a_i : R(\sigma(x_i)) + \sum_{x\in\{x|a_x=a_i\wedge i\in J\}} x + R(\sigma(y))) =\\
&R(a : (\sum_{i\in I\cup J} a_i : \sigma(x_i) + \sigma(y))).
\end{aligned}$$

$=^*$ follows from the observations that $\{a_i | i \in I,\ a_i \neq a_x \text{ for some } x \in V\} = \{a_i | i \in J,\ a_i \neq a_x \text{ for some } x \in V\}$ and $\{x | a_x = a_i \wedge i \in I\} = \{x | a_x = a_i \wedge i \in J\}$ which follow directly from the fact that $\{a_i | i \in I\} = \{a_i | i \in J\}$. □

4.3 The completed trace axioms

We now show the ω-completeness for the axiom set CT. However, it is not possible to use the technique presented in the beginning. This will be shown in example 4.3.4. Therefore, we will use a more traditional technique. Hence, it is necessary to explicitly define the ompleted trace semantics for BCCSP. In CT the meaning of a process is its set of traces hat end in inaction.

Definition 4.3.1. The interpretation $[\![.]\!]_{CT} : T(\text{BCCSP}) \rightarrow 2^{Act^*}$ (the set of subsets of trings over *Act*) is defined as follows:

$$[\![\delta]\!]_{CT} = \emptyset,$$
$$[\![a : t]\!]_{CT} = \{a \star s | s \in [\![t]\!]_{CT}\} \cup \{a | [\![t]\!]_{CT} = \emptyset\},$$
$$[\![t_1 + t_2]\!]_{CT} = [\![t_1]\!]_{CT} \cup [\![t_2]\!]_{CT}.$$

We say that $t_1, t_2 \in T(\text{BCCSP})$ are *completed trace equivalent*, notation $t_1 =_{CT} t_2$, iff $[\![t_1]\!]_{CT} = [\![t_2]\!]_{CT}$.

Lemma 4.3.2. *(Soundness) Let $t_1, t_2 \in T(BCCSP)$:*

$$\text{CT} \vdash \text{t}_1 = \text{t}_2 \quad \Rightarrow \quad \text{t}_1 =_{CT} \text{t}_2.$$

Proof. Straightforward using the definitions. □

For completed trace semantics the following theorem states the completeness of the axioms with respect to the given model. Moreover, as t_1 and t_2 may be open terms, ω-completeness is implied also.

Theorem 4.3.3. *If $|Act| \geq 1$ then for all $t_1, t_2 \in \mathbb{T}(BCCSP)$, we have that:*

$$\forall \sigma : V \rightarrow T(BCCSP)\ \sigma(t_1) =_{CT} \sigma(t_2) \quad \Rightarrow \quad \text{CT} \vdash \text{t}_1 = \text{t}_2.$$

Proof. We skip the long and tedious proof of this theorem in which we had to use the standard technique as shown by the next example. □

Example 4.3.4. Consider the following two BCCSP-terms.

$$t_1 = a : x + a : (a : \delta + x),$$
$$t_2 = a : (a : \delta + x).$$

These two terms are clearly different in CT as for a substitution σ with $\sigma(x) = \delta$, $\sigma(t_1)$ has a completed trace a which is not available in $\sigma(t_2)$. For every substitution σ' with $\sigma'(x) \neq \delta$, $\sigma'(t_1) =_{CT} \sigma'(t_2)$. Hence, using the same arguments as in example 3.2, we cannot apply our new technique.

4.4 The trace axioms

Again we do not use the new technique as in this case the 'standard' technique is more convenient to use. We must give the trace semantics explicitly. In trace semantics each process is characterized by its set of prefix closed traces:

Definition 4.4.1. The interpretation $[\![.]\!]_{T} : T(\text{BCCSP}) \rightarrow 2^{Act^*}$ is defined as follows:

$$[\![\delta]\!]_{T} = \emptyset,$$
$$[\![a : t]\!]_{T} = \{a \star \sigma | \sigma \in [\![t]\!]_{T}\} \cup \{a\},$$
$$[\![t_1 + t_2]\!]_{T} = [\![t_1]\!]_{T} \cup [\![t_2]\!]_{T}.$$

We say that $t_1, t_2 \in T(\text{BCCSP})$ are *trace equivalent*, notation $t_1 =_{\mathrm{T}} t_2$, iff $[\![t_1]\!]_{\mathrm{T}} = [\![t_2]\!]_{\mathrm{T}}$

Lemma 4.4.2. *(Soundness) Let $t_1, t_2 \in T(BCCSP)$:*

$$\mathrm{T} \vdash t_1 = t_2 \quad \Rightarrow \quad t_1 =_{\mathrm{T}} t_2.$$

Proof. Straightforward using the definitions. □

For trace semantics we need two actions in order to prove T ω-complete. If $|Act| =$ then the following axiom is valid:

$$x + a : x = a : x.$$

This can easily be seen by proving $\mathrm{T} \vdash t + a : t = a : t$ for all $t \in T(\text{BCCSP})$ wit induction on t if $|Act| = 1$. The axiom $x + a : x = a : x$ is in general not derivable fro T, because instantiating x with $b : \delta$ yields $b : \delta + a : b : \delta \neq_{\mathrm{T}} a : b : \delta$ where $a, b \in Ac$ are two different actions. In the next theorem we show that if $|Act| \geq 2$ then the axio set T is ω-complete. First we define the notion of a syntactic summand. This notion i only used in this section.

Definition 4.4.3. Let $t, u \in \mathbb{T}(\text{BCCSP})$. t is a *syntactic summand* of u, notation $t \sqsubseteq u$ if:

- $t \equiv a : t'$ and $u \equiv a : t'$ for some $t' \in \mathbb{T}(\text{BCCSP})$ or,
- $u \equiv u_1 + u_2$ and $t \sqsubseteq u_1$ or $t \sqsubseteq u_2$.

Lemma 4.4.4. *Let $t_1, t_2 \in \mathbb{T}(BCCSP)$. If for each syntactic summand $u \in \mathbb{T}(BCCSP)$*

$$u \sqsubseteq t_1 \quad \Leftrightarrow \quad u \sqsubseteq t_2$$

then $B \vdash t_1 = t_2$.
Proof. Straightforward. □

Theorem 4.4.5. *If $|Act| \geq 2$ then for each $t_1, t_2 \in \mathbb{T}(BCCSP)$, we have that:*

$$\forall \sigma : V \to T(BCCSP) : \ \sigma(t_1) =_{\mathrm{T}} \sigma(t_2) \quad \Rightarrow \quad \mathrm{T} \vdash t_1 = t_2.$$

Proof. We use the abbreviation $a_1 \star \ldots \star a_n : t$ with $a_1 \star \ldots \star a_n \in Act^\star$ for $a_1 : \ldots : a_n : t$. For $s \in Act^\star$, we define $|s|$ to be $|s : \delta|$, i.e. the length of trace s. For traces $s_1, s_2 \in Act^\star$ we write $s_1 \leq s_2$ if for some $r \in Act^\star$, $s_1 \star r = s_2$ or $s_1 = s_2$. In this case s_1 is a *prefix* of s_2.

First we define a T-*normal form*, which plays a crucial role in this proof. A term $t \in \mathbb{T}(\text{BCCSP})$ is a T-normal form if

$$t \equiv \sum_{i \in I} s_i : \delta + \sum_{i \in J} s_i : x_i$$

with $s_i \in Act^\star$ ($i \in I \cup J$), satisfying:

1) for each s_j ($j \in I \cup J$) with $|s_j| > 1$, there is a $i \in I$ such that $s_i \star a = s_j$ for some $a \in Act$.

2) for each s_j ($j \in J$) with $|s_j| > 0$, there is a $i \in I$ such that $s_j = s_i$.

Fact 1. *Let $t \in \mathbb{T}(BCCSP)$. Then there is a T-normal form t' such that:*

$$\mathrm{T} \vdash t = t'.$$

Proof of fact. Straightforward with induction on t. □

Fact 2. *Let t and t' be two T-normal forms such that for some u, $u \sqsubseteq t$, $u \not\sqsubseteq t'$ or vice versa. Then there is a substitution $\sigma : V \to T(BCCSP)$ such that:*

$$\sigma(t) \neq_{\mathrm{T}} \sigma(t').$$

Proof of fact. By symmetry it is sufficient to consider only the case where $u \sqsubseteq t$ and $u \not\sqsubseteq t'$. We can distinguish between:

(1) $u \equiv s : \delta$ with $s \in Act^\star$. Define $\sigma(x) = \delta$ for all $x \in V$. Note that $s \in [\![\sigma(t)]\!]_{\mathrm{T}}$. Moreover, in this case it holds that $s \in [\![\sigma(t')]\!]_{\mathrm{T}}$ iff $s : \delta \sqsubseteq t'$. Note that the conditions (1) and (2) are required to prove this. Hence, as $s : \delta \not\sqsubseteq t'$, $s \notin [\![\sigma(t')]\!]_{\mathrm{T}}$.

(2) $u \equiv s : x$ for some $x \in V$ and $s \in Act^\star$. Let m be a natural number such that $m > \max(|t|, |t'|)$. Define $\sigma(x) = a^m : b : \delta$ where $a, b \in Act$ are two different actions and $\sigma(y) = \delta$ if $y \not\equiv x$. Clearly, $s \star a^m \star b \in [\![\sigma(t)]\!]_{\mathrm{T}}$. We will show that $s \star a^m \star b \notin [\![\sigma(t')]\!]_{\mathrm{T}}$. Therefore we write $t' \equiv \sum_{i \in I} s_i : \delta + \sum_{i \in J} s_i : y_i$ in the following way:

$$\sum_{i \in I} s_i : \delta + \sum_{i \in K_1} s_i : y_i + \sum_{i \in K_2} s_i : x + \sum_{i \in K_3} s_i : x + \sum_{i \in K_4} s_i : x$$

where

$$\begin{aligned} K_1 &= \{i | i \in J \text{ and } y_i \not\equiv x\}, \\ K_2 &= \{i | i \in J,\ y_i \equiv x \text{ and } |s_i| < |s|\}, \\ K_3 &= \{i | i \in J,\ y_i \equiv x \text{ and } |s_i| = |s|\}, \\ K_4 &= \{i | i \in J,\ y_i \equiv x \text{ and } |s_i| > |s|\}. \end{aligned}$$

Note that $J = K_1 \cup K_2 \cup K_3 \cup K_4$. We will show that $s \star a^m \star b$ cannot originate from any of these components. We deal with all five cases separately:

(a) For any $r \in [\![\sum_{i \in I} s_i : \delta]\!]_{\mathrm{T}}$, $|r| < m$ and therefore $r \neq s \star a^m \star b$.

(b) For any $r \in [\![\sum_{i \in K_1} s_i : \sigma(y_i)]\!]_{\mathrm{T}}$, $|r| < m$ because $\sigma(y_i) = \delta$. Hence, $r \neq s \star a^m \star b$.

(c) For any $r \in [\![\sum_{i \in K_2} s_i : \sigma(x)]\!]_{\mathrm{T}}$, $|r| \leq |s_i| + m + 1 < |s| + m + 1 = |s \star a^m \star b|$. Hence, $r \neq s \star a^m \star b$.

(d) For any $r \in [\![\sum_{i \in K_3} s_i : \sigma(x)]\!]_T$, $r \leq s_i \star a^m \star b$ for some $i \in K_3$. If $|r| <$ $|s| + m + 1$, clearly, $r \neq s \star a^m \star b$. If $|r| = |s| + m + 1$, then $r = s_i \star a^m \star b$. As $s : x \not\sqsubseteq t'$, $s_i \neq s$. Therefore $r \neq s \star a^m \star b$.

(e) Let for some $r \in Act^\star$, $r[i]$ be the ith symbol in r. For any $r \in [\![\sum_{i \in K_4} s_i :$ $\sigma(x)]\!]_T$, $r \leq s_i \star a^m \star b$ for some $i \in K_4$. If $|r| \leq |s| + m$, then clearly $r \neq s \star a^m \star b$. If $|r| > |s| + m$, consider $r[|s| + m + 1]$. As $|s_i \star a^m \star b| > |s \star a^m \star b| > |s_i|$, $r[|s| + m + 1] = a$. But, $s \star a^m \star b[|s| + m + 1] = b$. Hence, if $|r| > |s| + m$, it also holds that $r \neq s \star a^m \star b$.

This finishes the proof of the second fact. □

Using both facts it follows almost immediately that T is ω-complete with respect to $=_T$. Suppose $t, t' \in \mathbb{T}(\text{BCCSP})$ such that for each substitution $\sigma : V \to T(\text{BCCSP})$, it holds that $\sigma(t) =_T \sigma(t')$. Both t and t' are provably equal to T-normal forms u and u' (fact 1). If u and u' have different syntactic summands, then by the second fact $\rho(u) \neq_T \rho(u')$ for some substitution $\rho : V \to T(\text{BCCSP})$. This is a contradiction. Hence, by lemma 4.4.4, $B \vdash u = u'$ and therefore:

$$\mathrm{T} \vdash t = u = u' = t'.$$

□

5 Extensions with the parallel operator

We extend the signature BCCSP with operators for parallelism.

5.1 Interleaving without communication

First, we study BCCSP with the merge and the leftmerge, but without communication. The resulting signature is called $\text{BCCSP}_{\lfloor\!\lfloor}$. We will study $\text{BCCSP}_{\lfloor\!\lfloor}$ in the setting of bisimulation where $|Act| = \infty$. The upper half of table 3 contains a complete set of axioms. The completeness follows immediately from the completeness of the axiom set B for BCCSP because any closed term over the signature $\text{BCCSP}_{\lfloor\!\lfloor}$ can be rewritten to a term over the signature BCCSP.

In order to have an ω-complete set of axioms, we add two new axioms (see the lower squares of table 3). These axioms are derivable for all closed instances. Therefore they are valid in bisimulation semantics. The complete set of axioms in table 3 is called $\mathrm{B}_{\lfloor\!\lfloor}$. The following theorem concerns the ω-completeness of $B_{\lfloor\!\lfloor}$.

$x + y = y + x$	$x \parallel y = x \lfloor\!\lfloor y + y \lfloor\!\lfloor x$
$(x + y) + z = x + (y + z)$	$\delta \lfloor\!\lfloor x = \delta$
$x + x = x$	$a : x \lfloor\!\lfloor y = a : (x \parallel y)$
$x + \delta = x$	$(x + y) \lfloor\!\lfloor z = x \lfloor\!\lfloor z + y \lfloor\!\lfloor z$
$x \lfloor\!\lfloor \delta = x$	$x \lfloor\!\lfloor (y \parallel z) = (x \lfloor\!\lfloor y) \lfloor\!\lfloor z$

Table 3: The axioms for BCCSP with the leftmerge

Theorem 5.1.1. *The set of axioms in table 3 is ω-complete if Act contains an infinite number of actions.*

Proof. Suppose two terms $t, t' \in \mathbb{T}(\mathrm{BCCSP}_{\|\!\!\lfloor})$ are given. Define $\rho : V \to T(\mathrm{BCCSP}_{\|\!\!\lfloor})$ by $\rho(x) = a_x : \delta$ where a_x is a unique action for each $x \in V$ and a_x does neither occur in t nor in t'. Define $R : T(\mathrm{BCCSP}_{\|\!\!\lfloor}) \to \mathbb{T}(\mathrm{BCCSP}_{\|\!\!\lfloor})$ as follows:

$R(\delta) = \delta,$
$R(a : t) = a : R(t)$ where $a \neq a_x$ for all $x \in V$,
$R(a_x : t) = x \mathbin{\|\!\!\lfloor} R(t),$
$R(t + u) = R(t) + R(u),$
$R(t \parallel u) = R(t) \parallel R(u),$
$R(t \mathbin{\|\!\!\lfloor} u) = R(t) \mathbin{\|\!\!\lfloor} R(u).$

In order to show the axioms in table 3 ω-complete we must check properties (1), (2) and (3) of theorem 3.1.

(1) We show that $\mathrm{B}_{\|\!\!\lfloor} \vdash R(\rho(u)) = u$ with induction on $u \in \mathbb{T}(\mathrm{BCCSP}_{\|\!\!\lfloor})$, provided u does not contain actions of the form a_x.
$R(\rho(x)) = x \mathbin{\|\!\!\lfloor} \delta = x,$
$R(\rho(\delta)) = \delta,$
$R(\rho(t + u)) = R(\rho(t)) + R(\rho(u)) = t + u,$
$R(\rho(a : t)) = R(a : \rho(t)) =^* a : R(\rho(t)) = a : t.$
$=^*$ follows from the fact that $a \neq a_x$ for all $x \in V$.

(2) For the $+$-operator the proof is straightforward: $\mathrm{B}_{\|\!\!\lfloor} \cup \{R(t_i) = R(u_i) | i = 1, 2\} \vdash R(t_1 + t_2) = R(t_1) + R(t_2) = R(u_1) + R(u_2) = R(u_1 + u_2)$. The function names $\|\!\!\lfloor$ and $\parallel$ can be dealt with in the same way. The action prefix case is slightly more complicated. $R(t_1) = R(u_1) \vdash R(a : t_1) = a : R(t_1) = a : R(u_1) = R(a : u_1)$ if $a \neq a_x$ for all $x \in V$. In the other case $R(t_1) = R(u_1) \vdash R(a_x : t_1) = x \mathbin{\|\!\!\lfloor} R(t_1) = x \mathbin{\|\!\!\lfloor} R(u_2) = R(a_x : u_1)$.

(3) It is straightforward to check the axioms that do not explicitly refer to actions. Here we only check the axiom $a : x \mathbin{\|\!\!\lfloor} y = a : (x \parallel y)$. Let $\sigma : V \to T(\Sigma)$ be defined such that $\sigma(x) = t$ and $\sigma(y) = u$. $\mathrm{B}_{\|\!\!\lfloor} \vdash R(a : t \mathbin{\|\!\!\lfloor} u) = a : R(t) \mathbin{\|\!\!\lfloor} R(u) = a : (R(t) \parallel R(u)) = R(a : (t \parallel u))$ if $a \neq a_x$ for all $x \in V$. In the other case $\mathrm{B}_{\|\!\!\lfloor} \vdash R(a_x : t \mathbin{\|\!\!\lfloor} u) = (x \mathbin{\|\!\!\lfloor} R(t)) \mathbin{\|\!\!\lfloor} R(u) = x \mathbin{\|\!\!\lfloor} (R(t) \parallel R(u)) = R(a_x : (t \parallel u))$.

□

In many cases it is easy to show the ω-completeness of the axioms of new features introduced in $\mathrm{BCCSP}_{\|\!\!\lfloor}$. As examples we introduce the silent step τ into $\mathrm{BCCSP}_{\|\!\!\lfloor}$ and we will consider $\mathrm{BCCSP}_{\|\!\!\lfloor}$ in trace semantics.

Example 5.1.2. We add a constant τ (*the silent step* or *internal move*) to $\mathrm{BCCSP}_{\|\!\!\lfloor}$. The new signature is called $\mathrm{BCCSP}^{\tau}_{\|\!\!\lfloor}$. The internal step has been axiomatized in different ways. In [10] τ is characterized by three τ-laws. This characterization is often called *weak bisimulation*.

$$\begin{aligned} a : \tau : x &= a : x, \\ \tau : x + x &= \tau : x, \\ a : (\tau : x + y) &= a : (\tau : x + y) + a : x. \end{aligned}$$

If one adds these laws to $B_{\llcorner\!\llcorner}$, obtaining $B^{\tau}_{\llcorner\!\llcorner}$, we have to add the following two axioms in order to make $B^{\tau}_{\llcorner\!\llcorner}$ ω-complete. Axioms of this form already appeared in [7].

$$\begin{array}{rcl} x \llcorner\!\llcorner \tau : y & = & x \llcorner\!\llcorner y, \\ x \llcorner\!\llcorner (\tau : y + z) & = & x \llcorner\!\llcorner (\tau : y + z) + x \llcorner\!\llcorner y. \end{array}$$

Both new axioms are derivable for all closed instances, and therefore valid in any model for $B^{\tau}_{\llcorner\!\llcorner}$.

In [5] τ is axiomatized by the single equation:

$$a : (\tau : (x + y) + x) = a : (x + y).$$

This variant is called *branching bisimulation*. The set $B_{\llcorner\!\llcorner}$, together with this axiom is called $B^{b}_{\llcorner\!\llcorner}$. The single axiom:

$$x \llcorner\!\llcorner (\tau : (y + z) + y) = x \llcorner\!\llcorner (y + z)$$

suffices to make $B^{b}_{\llcorner\!\llcorner}$ ω-complete. This axiom is derivable for all closed instances, and therefore it holds in any model for $B^{b}_{\llcorner\!\llcorner}$.

We do not give the ω-completeness proofs as they can easily be given along the lines of the proof of theorem 5.1.1. In fact it suffices to only check condition (3) for the new axioms, because conditions (1) and (2) are provable in exactly the same way.

Example 5.1.3. Here we study the ω-completeness of $BCCSP_{\llcorner\!\llcorner}$ in trace semantics. As any term over the signature $BCCSP_{\llcorner\!\llcorner}$ can be rewritten to a term over the signature BCCSP by the axioms in $B_{\llcorner\!\llcorner}$, and T is complete for the signature BCCSP in trace semantics, $B_{\llcorner\!\llcorner} \cup T$ is complete for $BCCSP_{\llcorner\!\llcorner}$ in trace semantics. For ω-completeness we must add the equation:

$$x \llcorner\!\llcorner y + x \llcorner\!\llcorner z = x \llcorner\!\llcorner (y + z),$$

which is derivable from $B_{\llcorner\!\llcorner} \cup T$ for all its closed instances. The proof of this fact follows the lines of the proof of theorem 5.1.1.

5.2 Interleaving with communication

In this section the signature BCCSP is extended with the merge, the leftmerge and the *communication merge* ($|$). The signature obtained in this way is called $BCCSP_{|}$. Its properties are described by the axioms in table 4 which are taken from [2]. We have an additional associative and commutative operator $|: Act \times Act \rightarrow Act$ on actions. We assume that *Act* is closed under $|$. In fact $(Act, |)$ is an abelian semigroup. The axioms in the upper two squares of table 4 combined with the condition that $|$ on actions is commutative and associative, are already complete for $BCCSP_{|}$-terms in the bisimulation model. This can again easily be seen by the fact that any term over the signature $BCCSP_{|}$ can be rewritten to a term over BCCSP. For BCCSP the four axioms in the left upper corner of table 4 are complete in the bisimulation model. The axioms in the lower squares are necessary for an ω-complete axiomatization. We call the axiom system in table 4 $B_{|}$.

Example 5.2.1. The following facts are derivable from $B_{|}$. We leave the proofs to the reader.

$x + y = y + x$	
$(x + y) + z = x + (y + z)$	
$x + x = x$	
$x + \delta = x$	
$x \parallel y = x \lfloor\!\lfloor y + y \lfloor\!\lfloor x + x \mid y$	$x \mid y = y \mid x$
$a : x \lfloor\!\lfloor y = a : (x \parallel y)$	$a : x \mid b : y = (a \mid b) : (x \parallel y)$
$\delta \lfloor\!\lfloor x = \delta$	$\delta \mid x = \delta$
$(x + y) \lfloor\!\lfloor z = x \lfloor\!\lfloor z + y \lfloor\!\lfloor z$	$(x + y) \mid z = x \mid z + y \mid z$
$(x \lfloor\!\lfloor y) \lfloor\!\lfloor z = x \lfloor\!\lfloor (y \parallel z)$	$(x \mid y) \mid z = x \mid (y \mid z)$
$x \lfloor\!\lfloor \delta = x$	$x \mid (y \lfloor\!\lfloor z) = (x \mid y) \lfloor\!\lfloor z$

Table 4: The axioms for $\mathrm{BCCSP}_{|}$

$x \parallel y = y \parallel x$,
$(x \parallel y) \parallel z = x \parallel (y \parallel z)$,
$(a_1 \mid ... \mid (a_i \mid a_{i+1}) \mid ... \mid a_n) : x = (a_1 \mid ... \mid (a_{i+1} \mid a_i) \mid ... \mid a_n) : x$,
$(a_1 \mid ... \mid (a_i(a_{i+1} \mid a_{i+2})) \mid ... \mid a_n) : x = (a_1 \mid ... \mid ((a_i \mid a_{i+1}) \mid a_{i+2}) \mid ... \mid a_n) : x$.

The last two identities show that it is not necessary to include axioms for the commutativity and the associativity of $\mid$ on actions in $\mathrm{B}_{|}$.

Theorem 5.2.2. $\mathrm{B}_{|}$ *is ω-complete if Act contains an infinite number of actions.*
Proof. This proof has the same structure as the proof of theorem 5.1.1. We will only give the non-trivial steps of the proof. Suppose two terms $t, t' \in \mathbb{T}(\mathrm{BCCSP}_{|})$ are given. Define $\rho : V \to T(\mathrm{BCCSP}_{|})$ as follows:

$$\rho(x) = a_x : \delta$$

where a_x is unique for each $x \in V$ and actions a_x do not occur in t or t'. We define $R : T(\mathrm{BCCSP}_{|}) \to \mathbb{T}(\mathrm{BCCSP}_{|})$ by:

$R(\delta) = \delta$,
$R((a_1 \mid ... \mid a_n) : t) = (a_1 \mid ... \mid a_n) : R(t)$ if $a_i \neq a_x$ for $1 \leq i \leq n$ and $x \in V$,
$R(a_x : t) = x \lfloor\!\lfloor R(t)$,
$R((a_x \mid a_1 \mid ... \mid a_n) : t) = x \mid R((a_1 \mid ... \mid a_n) : t)$ for $n \geq 1$,
$R((a_1 \mid a_2 \mid ... \mid a_n) : t) = R(a_2 \mid ... \mid a_n \mid a_1) : t)$ for $n \geq 2$ provided $a_1 \neq a_x$ for all $x \in V$,
$R(t + u) = R(t) + R(u)$,
$R(t \parallel u) = R(t) \parallel R(u)$,
$R(t \lfloor\!\lfloor u) = R(t) \lfloor\!\lfloor R(u)$,
$R(t \mid u) = R(t) \mid R(u)$.

For ρ and R we now check properties (1), (2) and (3) of theorem 3.1.

(1) Straightforward. In this step the axiom $x \lfloor\!\lfloor \delta = x$ plays an essential role.

(2) Straightforward for almost all cases, the only exception being the action prefix operator $(a_1 \mid ... \mid a_n) : x$ where for some a_i $(1 \leq i \leq n)$, $a_i = a_x$ with

$x \in V$. Assuming that $B_| \vdash R(t) = R(u)$ for $t, u \in T(BCCSP_|)$, we show that $B_| \vdash R((a_1 \mid ... \mid a_n) : t) = R((a_1 \mid ... \mid a_n) : u)$.

$R((a_1 \mid ... \mid a_n) : t) =$

(a) $x_j \mid (... \mid (x_{j'} \mid ((a_k \mid ... \mid a_{k'}) : R(t)))...) =$
$x_j \mid (... \mid (x_{j'} \mid ((a_k \mid ... \mid a_{k'}) : R(u)))...) =$
$R((a_1 \mid ... \mid a_n) : u)$ if there is a $1 \leq i \leq n$ such that $a_i \neq a_x$ for all $x \in V$.

(b) $x_1 \mid (... \mid (x_{n-1} \mid (x_n \mathbin{\lfloor\!\lfloor} R(t)))...) =$
$x_1 \mid (... \mid (x_{n-1} \mid (x_n \mathbin{\lfloor\!\lfloor} R(u)))...) = R((a_1 \mid ... \mid a_n) : u)$, otherwise.

(3) Only the axioms containing occurrences of the action prefix operator are non trivial to check. So we consider the axioms $a : (x \mathbin{\lfloor\!\lfloor} y) = a : (x \parallel y)$ and $a : x \mid b : y = (a \mid b) : (x \parallel y)$. We start off with the first one. Let $a = (a_1 \mid ... \mid a_n)$ and let σ be a closed substitution such that $\sigma(x) = t$ and $\sigma(y) = u$. Three cases must be considered.

(a) $a_i \neq a_x$ for all $1 \leq i \leq n$ and $x \in V$.
$B_| \vdash R(a : t \mathbin{\lfloor\!\lfloor} u) = a : R(t) \mathbin{\lfloor\!\lfloor} R(u) = a : (R(t) \parallel R(u)) = R(a : (t \parallel u))$.

(b) $a_i = a_{x_i}$ for each $1 \leq i \leq n$ and $x_i \in V$.
$R((a_1 \mid ... \mid a_n) : t \mathbin{\lfloor\!\lfloor} u) =$
$(x_1 \mid (... \mid (x_{n-1} \mid (x_n \mathbin{\lfloor\!\lfloor} R(t)))...)) \mathbin{\lfloor\!\lfloor} R(u) =$
$(((x_1 \mid ... \mid x_{n-1}) \mid x_n) \mathbin{\lfloor\!\lfloor} R(t)) \mathbin{\lfloor\!\lfloor} R(u) =$
$((x_1 \mid ... \mid x_{n-1}) \mid x_n) \mathbin{\lfloor\!\lfloor} (R(t) \parallel R(u)) =$
$(x_1 \mid (... \mid (x_{n-1}) \mid |(x_n \mathbin{\lfloor\!\lfloor} (R(t) \parallel R(u))))...)) =$
$R((a_1 \mid ... \mid a_n) : (t \parallel u))$.

(c) For some $1 \leq i \leq n$, $a_i \neq a_x$ for all $x \in V$ and for some $1 \leq i \leq n$, $a_i = a_x$.
$R((a_1 \mid ... \mid a_n) : t \mathbin{\lfloor\!\lfloor} u) =$
$(x_j \mid (... \mid (x_{j'} \mid ((a_{k_1} \mid ... \mid a_{k'}) : R(t)))...)) \mathbin{\lfloor\!\lfloor} R(u) =$
$(x_j \mid ... \mid x_{j'}) \mid ((a_{k_1} \mid ... \mid a_{k'}) : R(t) \mathbin{\lfloor\!\lfloor} R(u)) =$
$(x_j \mid ... \mid x_{j'}) \mid ((a_k \mid ... \mid a_{k'}) : (R(t) \parallel R(u))) =$
$x_j \mid (... \mid (x_{j'} \mid ((a_k \mid ... \mid a_{k'}) : (R(t) \parallel R(u))))...) =$
$R((a_1 \mid ... \mid a_n) : (t \parallel u))$.

We now check the axiom $a : x \mid b : y = (a \mid b) : (x \parallel y)$. We can distinguish 9 cases (cf. checking the axiom $a : x \mathbin{\lfloor\!\lfloor} y = a : (x \parallel y)$). We will not discuss all of these, but restrict ourselves to the case where some of the actions, but not all, in a and b have the form a_x.

$R(a : t \mid b : u) =$
$(x_{j_1} \mid (... \mid (x_{j_1'} \mid (a_{k_1} \mid ... \mid a_{k_1'}) : R(t))...)) \mid (y_{j_2} \mid (... \mid (y_{j_2'} \mid (b_{k_2} \mid ... \mid b_{k_2'}) : R(u))...)) =$
$(x_{j_1} \mid ... \mid x_{j_1'} \mid y_{j_2} \mid ... \mid y_{j_2'}) \mid ((a_{k_1} \mid ... \mid a_{k_1'}) : R(t) \mid (b_{k_2} \mid ... \mid b_{k_2'}) : R(u)) =$
$(x_{j_1} \mid (... \mid (x_{j_1'} \mid (y_{j_2} \mid (... \mid (y_{j_2'} \mid ((a_{k_1} \mid ... \mid a_{k_1'}) \mid (b_{k_2} \mid ... \mid b_{k_2'})) : (R(t) \parallel R(u)))...)))...)) =$
$R((a_1 \mid ... \mid a_n \mid b_1 \mid ... \mid b_n) : (t \parallel u))$.

In the last step we used example 5.2.1 to rearrange the actions.

□

References

[1] J.A. Bergstra and J. Heering. Which data types have ω-complete initial algebra specifications? Technical Report CS-R8958, Center for Mathematics and Computer Science, Amsterdam, December 1989.

[2] J.A. Bergstra and J.W. Klop. Process algebra for synchronous communication. *Information and Computation*, 60(1/3):109–137, 1984.

[3] J.A. Bergstra and J.V. Tucker. Top down design and the algebra of communicating processes. *Science of Computer Programming*, 5(2):171–199, 1984.

[4] R.J. van Glabbeek. The linear time - branching time spectrum. In J.C.M. Baeten and J.W. Klop, editors, *Proceedings Concur90, LNCS*, Amsterdam, 1990. Springer Verlag.

[5] R.J. van Glabbeek and W.P. Weijland. Branching time and abstraction in bisimulation semantics (extended abstract). In G.X. Ritter, editor, *Information Processing 89*, pages 613–618. Elsevier Science Publishers B.V. (North Holland), 1989.

[6] J. Heering. Partial evaluation and ω-completeness of algebraic specifications. *Theoretical Computer Science*, 43:149–167, 1986.

[7] M. Hennessy. Axiomatising finite concurrent processes. *SIAM Journal of Computing*, 17(5):997–1017, 1988.

[8] D. Kapur and D.R. Musser. Proof by consistency. *Artificial Intelligence*, 31:125–157, 1987.

[9] A. Lazrek, P. Lescanne, and J.-J. Thiel. Tools for proving inductive equalities, relative completeness, and ω-completeness. *Information and Computation*, 84:47–70, 1990.

[10] R. Milner. *A Calculus of Communicating Systems*, volume 92 of *LNCS*. Springer-Verlag, 1980.

[11] R. Milner. A complete axiomatisation for observational congruence of finite-state behaviours. Technical Report ECS-LFCS-86-8, University of Edinburgh, Edinburgh, July 1986.

[12] F. Moller. *Axioms for Concurrency*. PhD thesis, University of Edinburgh, July 1989.

[13] V.L. Murskiĭ. The existence in three-valued logic of a closed class with finite basis, not having a finite complete system of identities. *Doklady Akademii Nauk SSSR*, 163:815–818, 1965. English translation in: *Soviet Mathematics Doklady*, 6:1020-1024, 1965.

[14] D.L. Musser. On proving inductive properties of abstract data types. In *Proceedings, 7th ACM Symp. on Principles of Programming Languages*, pages 154–162, New York, 1980. ACM.

[15] E. Paul. Proof by induction in equational theories with relations between constructors. In B. Courcelle, editor, 9th *Coll. on Trees in Algebra and Programming*, pages 211–225, Bordeaux, France, 1984. Cambridge University Press, London.

Transition System Specifications with Negative Premises

(extended abstract)

Jan Friso Groote

Centre for Mathematics and Computer Science

P.O. Box 4079, 1009 AB Amsterdam, The Netherlands

Email: jfg@cwi.nl

In this article the general approach to Plotkin style operational semantics of [7] is extended to Transition System Specifications (TSS's) with rules that may contain negative premises. Two problems arise: firstly the rules may be inconsistent, and secondly it is not obvious how such rules determine a transition relation. We present a general method, based on the stratification technique of logic programming, to show consistency of a set of rules. Then we show how a specific transition relation can be associated with each consistent TSS in a very natural way. A special format for the rules, the *pure ntyft/ntyxt*-format, is defined. For this format three important theorems hold: firstly, bisimulation is a congruence if all operators are defined using this format, secondly under certain natural restrictions a TSS in *ntyft*-format can be added conservatively to a TSS in *pure ntyft/ntyxt*-format and finally, the trace congruence for image finite processes induced by our format is precisely bisimulation equivalence.

1. Introduction

In recent years, many process calculi, programming languages and specification languages have been provided with an operational semantics in Plotkin style [12]. Often, definitions of semantics in Plotkin style use negative premises in rules. For instance *deadlock detectors* [8, 11], *priority operators* [6, 7], several forms of (semi) synchronous *communication operators* [13] and *sequencing operators* [5, 10] are defined using negative premises. In [5] it is even observed that sequencing can only be defined using negative premises.

Two problems arise when rules have negative premises. Often these problems seem to be ignored.

1. It is possible to give an inconsistent set of rules. This means that one can derive with the rules that a process can perform an action if and only if it cannot do so. In this case the rules do not define an operational semantics.
2. Even if the rules are consistent, it is not immediately obvious how these rules determine an operational semantics. The normal notion of provability of transitions where the rules are used as inference rules is not satisfactory.

A first solution to these problems is to avoid negative premises in rules. Often using additional labels, function names and rules an operational semantics can be given using only positive premises. But then there are many auxiliary transitions that do not correspond to the (positive) behavior of the system that is modeled or specified. Moreover, definitions of an operational semantics become more complex than necessary. This means that an important property of operational semantics in Plotkin style, namely simplicity, is violated. Therefore, we give another solution.

We deal with the first problem by formulating an easy method of checking whether a

The research of the author was supported by ESPRIT project no. 432, An Integrated Formal Approach to Industrial Software Development (METEOR), and by RACE project no. 1046, Specification and Programming Environment for Communication Software (SPECS). A full version of this paper appeared as report CS-R8950, Centrum voor Wiskunde en Informatica, Amsterdam.

transition relation is consistent. This method is based on the *local stratifications* [2, 14] that are used in logic programming. The other problem is solved by formulating an explicit definition of the transition relation. We argue that our choice is a very natural one.

A format of rules that allows negative premises is the GSOS-format of BLOOM, ISTRAIL & MEYER [5]. The GSOS-format is incompatible with the (pure) *tyft*/*tyxt*-format [7] that allows *lookahead* but no negative premises. Our method is applicable to the combination of these two formats. We call this combination the *(pure) ntyft/ntyxt-format*. Here *n* indicates that negative premises are allowed.

2. TRANSITION SYSTEM SPECIFICATIONS AND STRATIFICATIONS

In this paper we assume the presence of a countably infinite set V of *variables* with typical elements x,y,z. A *(single sorted) signature* is a structure $\Sigma=(F,r)$ where F is a set of *function names* disjoint with V and $r:F\rightarrow\mathbb{N}$ is a *rank function* which gives the arity of a function name. $T(\Sigma,W)$ is the set of Σ-terms over $W\subseteq V$. We write $T(\Sigma)$ for $T(\Sigma,\varnothing)$ and $\mathbb{T}(\Sigma)$ for $T(\Sigma,V)$. $Var(t)\subseteq V$ is the set of variables in a term $t\in\mathbb{T}(\Sigma)$. A *substitution* σ is a mapping in $V\rightarrow T(\Sigma)$.

Below we formalize the notion of a Plotkin style definition of an operational semantics.

2.1. DEFINITION. A *TSS* (*Transition System Specification*) is a triple $P=(\Sigma,A,R)$ with $\Sigma=(F,r)$ a signature, A a set of *labels* and R a set of *rules* of the form:

$$\frac{\{t_k \xrightarrow{a_k} t_k' \mid k\in K\}\cup\{t_l \xrightarrow{b_l}\!\!\!\!\!/\;\; \mid l\in L\}}{t\xrightarrow{a} t'}$$

with K and L finite index sets, $t_k,t_k',t_l,t,t'\in\mathbb{T}(\Sigma)$, $a_k,b_l,a\in A$ ($k\in K$, $l\in L$). An expression of the form $t\xrightarrow{a}t'$ is called a (*positive*) *literal*. Here t is called the *source* and t' the *target* of the literal. $t\xrightarrow{a}\!\!\!\!\!/\;\;$ is called a *negative literal* with as intended meaning that t cannot perform an a step. ϕ,ψ,χ are used to range over literals. The literals above the line are called the *premises* and the literal below the line is called the *conclusion*. An *axiom* $\dfrac{\varnothing}{t\xrightarrow{a}t'}$ is often written as $t\xrightarrow{a}t'$. The notions 'substitution' and 'Var' extend to terms, literals and rules as expected.

The index sets K and L are chosen to be finite to avoid some notational complexity. All results in this paper also hold when K and L are infinite.

The purpose of a TSS is to define a *transition relation* $\longrightarrow\subseteq Tr(\Sigma,A)$ where $Tr(\Sigma,A)=T(\Sigma)\times A\times T(\Sigma)$. Elements (t,a,t') of a transition relation $\longrightarrow$ are written as $t\xrightarrow{a}t'$. We say that a positive literal ψ *holds* in $\longrightarrow$, notation $\longrightarrow\models\psi$, if $\psi\in\longrightarrow$. A negative literal $t\xrightarrow{a}\!\!\!\!\!/\;\;$ *holds* in $\longrightarrow$, notation $\longrightarrow\models t\xrightarrow{a}\!\!\!\!\!/\;\;$, if for no $t'\in T(\Sigma)$: $t\xrightarrow{a}t'\in\longrightarrow$.

2.2. For a TSS without negative premises the transition relation that must be associated with it is evident: it simply contains all provable (closed) literals. For a TSS with negative premises it is not so clear which transition relation should be associated with it. BLOOM, ISTRAIL and MEYER [5] require that a transition relation *agrees with* a TSS. We think that this should at least be the case. We repeat their definition here, using our own notation.

2.2.1. DEFINITION. Let $P=(\Sigma,A,R)$ be a TSS. A transition relation $\longrightarrow\subseteq Tr(\Sigma,A)$ *agrees with* P if:

$$\psi\in\longrightarrow \;\Leftrightarrow\; \exists\frac{\{\chi_k\mid k\in K\}}{\chi}\in R \text{ and } \exists\sigma:V\rightarrow T(\Sigma) \text{ such that } \sigma(\chi)=\psi \text{ and } \forall k\in K:\ \longrightarrow\models\sigma(\chi_k).$$

Unfortunately, for a given TSS P it is not guaranteed that a transition relation that agrees with P exists and if it exists it need not be unique. We give two examples illustrating these points.

2.2.2. EXAMPLE. It is possible to give a TSS P that does not define a transition relation. Let P consist of one constant f, one label a and one rule

$$\frac{f \overset{a}{\not\rightarrow}}{f \overset{a}{\rightarrow} f}.$$

For any transition relation $\rightarrow$ that agrees with P, $f \overset{a}{\rightarrow} f \in \rightarrow$ iff $f \overset{a}{\rightarrow} f \notin \rightarrow$. Clearly, such a transition relation does not exist.

2.2.3. EXAMPLE. This example shows that if a transition relation that agrees with a TSS exists, it need not be unique. Take for example a TSS with as only rule:

$$\frac{f \overset{a}{\rightarrow} f}{f \overset{a}{\rightarrow} f}$$

Both the empty transition relation and the transition relation $\{f \overset{a}{\rightarrow} f\}$ agree with this TSS.

2.3. Problems with the existence of a transition relation only occur when the absence of a certain transition is a prerequisite for its presence. Often this is not the case as can be shown by constructing a 'stratification'. This very handy technique stems from logic programming [2, 14].

2.3.1. DEFINITION. Let $P=(\Sigma,A,R)$ be a TSS. A function $S:Tr(\Sigma,A)\rightarrow\alpha$, where α is an ordinal, is called a *stratification* of P if for every rule

$$\frac{\{t_k \overset{a_k}{\rightarrow} t_k' \mid k\in K\} \cup \{t_l \overset{b_l}{\not\rightarrow} \mid l\in L\}}{t \overset{a}{\rightarrow} t'} \in R$$

and every substitution $\sigma:V\rightarrow T(\Sigma)$ it holds that:

for all $k\in K$: $S(\sigma(t_k \overset{a_k}{\rightarrow} t_k')) \leqslant S(\sigma(t \overset{a}{\rightarrow} t'))$

for all $l\in L$ and $t_l'\in T(\Sigma)$: $S(\sigma(t_l \overset{b_l}{\rightarrow} t_l')) < S(\sigma(t \overset{a}{\rightarrow} t'))$

For $\beta<\alpha$, $S_\beta=\{\phi \mid S(\phi)=\beta\}$ is called a *stratum*. P is *stratifiable*, if we can construct a stratification for P. The use of ordinals to index the strata turns out to be useful in certain applications and necessary to prove theorem 6.4. People who don't like ordinals may, at the price of some generality, read natural numbers instead.

2.4. Given a stratification S for a TSS P we construct a relation $\rightarrow_{P,S}$ which we call the transition relation *associated with P* (*based on S*). The idea is as follows: a literal ϕ with $S(\phi)=0$ is in $\rightarrow_{P,S}$ if and only if it can be 'derived' using rules of P, which do not have negative premises. Thus we know which literals ϕ with $S(\phi)=0$ are not in $\rightarrow_{P,S}$. We use this information to find which literals ϕ with $S(\phi)=1$ are in $\rightarrow_{P,S}$. In this way we continue for all strata.

We think that in general this relation is the one people have in mind when specifying a TSS with negative premises. Because the definition of $\rightarrow_{P,S}$ is important in this paper, we provide a formal definition:

2.4.1. DEFINITION. Let $P=(\Sigma,A,R)$ be a TSS with a stratification $S:Tr(\Sigma,A)\rightarrow\alpha$ for some ordinal α. The transition relation $\rightarrow_{P,S}$ *associated with P* (and *based on S*) is defined as:

$$\rightarrow_{P,S} = \bigcup_{0\leqslant i<\alpha} \rightarrow_i^P.$$

where transition relations $\rightarrow_i^P \subseteq Tr(\Sigma,A)$ $(0\leqslant i<\alpha)$, $\rightarrow_{ij}^P \subseteq Tr(\Sigma,A)$ $(0\leqslant i<\alpha,\ 0\leqslant j<\omega)$ are inductively defined by:

$\rightarrow_i^P = \bigcup_{0\leqslant j<\omega} \rightarrow_{ij}^P$ for $1\leqslant i<\alpha$,

$\rightarrow_{ij}^P = \{\phi \mid S(\phi)=i,$

$$\exists \frac{\{\chi_k \mid k \in K\}}{\chi} \in R,\ \exists \sigma: V \rightarrow T(\Sigma):$$

$$\sigma(\chi) = \phi \text{ and}$$

$$\forall k \in K\ [\chi_k \text{ is positive} \Rightarrow \bigcup_{0 \leq j' < j} \longrightarrow^{P}_{ij'} \cup \bigcup_{0 \leq i' < i} \longrightarrow^{P}_{i'} \models \sigma(\chi_k)] \text{ and}$$

$$[\chi_k \text{ is negative} \Rightarrow \bigcup_{0 \leq i' < i} \longrightarrow^{P}_{i'} \models \sigma(\chi_k)]\}$$

for $0 \leq i < \alpha$ and $0 \leq j < \omega$.

The relation $\longrightarrow_{P,S}$ has the following nice properties:

2.4.2. THEOREM. *Let $P=(\Sigma,A,R)$ be a TSS with stratification $S: Tr(\Sigma,A) \rightarrow \alpha$ for some ordinal α. Then there is a transition relation, namely $\longrightarrow_{P,S}$, that agrees with P.*

2.4.3. LEMMA. *Let P be a TSS which is stratified by stratifications S and S'. The transition relation associated with P and based on S is equal to the transition relation associated with P and based on S'.*

This last lemma allows us to drop the stratification as a subscript in the transition relation $\longrightarrow_{P,S}$ associated to a stratifiable TSS P. Furthermore, it provides the following technique to give an operational semantics in Plotkin style when there are negative premises around: define a TSS P and prove with a convenient stratification that P is stratifiable. Then P alone determines the transition relation $\longrightarrow_P$.

3. SOME EXAMPLES SHOWING THE USE OF STRATIFICATIONS

The techniques of the previous section are introduced to show that specifications using negative premises define a transition relation in a neat way. Here three examples illustrate the use of these techniques.

3.1. EXAMPLE. Assume we have a TSS P without negative premises. A stratification can be given by putting all literals in the same stratum.

3.2. EXAMPLE. Here a stratification for TSS's in GSOS-format is defined. This format differs slightly from the GSOS-format as given by BLOOM, ISTRAIL and MEYER [5] because we do not consider a special rule for guarded recursion. Suppose we have a TSS P with signature $\Sigma=(F,r)$, labels A and rules of the form

$$\frac{\{x_k \xrightarrow{a_{kl}} y_{kl} \mid k \in K_1, l \in L_1\} \cup \{x_k \stackrel{a_{kl}}{\not\longrightarrow} \mid k \in K_2, l \in L_2\}}{f(x_1, \ldots, x_{r(f)}) \xrightarrow{a} t}$$

with $f \in F$, $x_1, \ldots, x_{r(f)}, y_{kl}$ variables, $K_1, K_2 \subseteq \{1, \ldots, r(f)\}$, L_1, L_2 finite index sets and $t \in \mathbb{T}(\Sigma)$. For each P in GSOS-format there is a stratification $S: Tr(\Sigma,A) \rightarrow \omega$:

$$S(t \xrightarrow{a} t') = n \quad \text{if } t \text{ contains } n \text{ function names.}$$

S is a stratification as the source in the conclusion of any rule contains more function names than any source in the premises. This means that for any P in GSOS-format the associated transition relation $\longrightarrow_P$ exists.

3.3. EXAMPLE. In [4] a priority operator is defined on process graphs. In [7] an operational definition is given of this operator using rules with negative premises. However, the combination of unguarded recursion, the priority operator and renaming [3] gives rise to inconsistencies. We show that simple conditions on either the renaming operator or recursion can circumvent this problem.

We base this example on the rules for $BPA_{\epsilon\delta}$ as given in [7]. The TSS $P_{prio}=(\Sigma_{prio},A_{prio},R_{prio})$ with $\Sigma_{prio}=(F_{prio},r_{prio})$ contains constant names a for all $a\in Act$ where Act is a finite set of *atomic actions*. We assume that there is a partial ordering $<$ on Act: $a<b$ if a has lower priority than b. Furthermore, the signature contains constant names ϵ for the *empty process*, and δ representing *inaction*. δ resembles *NIL* in CCS

There is a unary function name θ, the *priority operator*. If x can perform several actions, say $x\xrightarrow{a}x'$ and $x\xrightarrow{b}x''$ then $\theta(x)$ allows only the transitions with highest priority. So if $a>b$ then $\theta(x)\xrightarrow{a}\theta(x')$ is an allowed transition while $\theta(x)\xrightarrow{b}\theta(x'')$ is not possible. We have another unary function name ρ_f, the *renaming operator*. f is a renaming function from Act to Act. $\rho_f(x)$ renames the labels of the transitions of x by f. There are two binary operators: *sequential composition* $(\cdot)$ and *alternative composition* $(+)$.

For recursion there is some given set Ξ of *process names*. E is a set of *process declarations* of the form $X\Leftarrow t_X$ for all process names $X\in\Xi$ $(t_X\in T(\Sigma_{prio}\cup\Xi))$ where t_X is the *body* of process name X.

The labels in A_{prio} are given by $Act_{\surd}$ $(=Act\cup\{\surd\})$. $\surd$ is an auxiliary symbol that is introduced to represent termination of a process. The rules are given in table 1. Here a,b range over $Act_{\surd}$. In rule 9 of table 1 we use the abbreviation $\forall b>a\ x\overset{b}{\not\rightarrow}$ in the premises. It means that for all $b>a$, there is a premise $x\overset{b}{\not\rightarrow}$.

1.	$a\xrightarrow{a}\epsilon$	$a\neq\surd$	2.	$\epsilon\xrightarrow{\surd}\delta$
3.	$\dfrac{x\xrightarrow{a}x'}{x+y\xrightarrow{a}x'}$		4.	$\dfrac{y\xrightarrow{a}y'}{x+y\xrightarrow{a}y'}$
5.	$\dfrac{x\xrightarrow{a}x'}{x\cdot y\xrightarrow{a}x'\cdot y}$	$a\neq\surd$	6.	$\dfrac{x\xrightarrow{\surd}x'\quad y\xrightarrow{a}y'}{x\cdot y\xrightarrow{a}y'}$
7.	$\dfrac{x\xrightarrow{a}x'}{\rho_f(x)\xrightarrow{f(a)}\rho_f(x')}$	$a\neq\surd$	8.	$\dfrac{x\xrightarrow{\surd}x'}{\rho_f(x)\xrightarrow{\surd}\rho_f(x')}$
9.	$\dfrac{x\xrightarrow{a}x'\quad\forall b>a\ x\overset{b}{\not\rightarrow}}{\theta(x)\xrightarrow{a}\theta(x')}$	$a,b\neq\surd$	10.	$\dfrac{x\xrightarrow{\surd}x'}{\theta(x)\xrightarrow{\surd}\theta(x')}$
11.	$\dfrac{t_X\xrightarrow{a}x'}{X\xrightarrow{a}x'}$	for $X\Leftarrow t_X\in E$		

TABLE 1

With these rules we have the following inconsistency. Define

$$X\Leftarrow\theta(\rho_f(X)+b)$$

with $f(b)=a$, $f(a)=c$, $f(d)=d$ for all $d\in Act-\{a,b\}$ and $a>b$. Now $X\xrightarrow{b}\epsilon$ iff $X\overset{b}{\not\rightarrow}$.

As a first solution for this problem we consider only renaming functions satisfying the requirement that if $a>b$ then not $f(b)=a$ for all $a,b\in Act$, i.e. we may not rename actions to ones with higher priority. It is now easy to see that a transition relation associated with P_{prio} exists using the following stratification of P_{prio}. Define $rk(a)$ for all $a\in A_{prio}$ by $rk(a)=max(\{rk(b)+1\mid a<b\})$ for $a\in Act$ where $max(\varnothing)=0$ and $rk(\surd)=0$. Define

$S:Tr(\Sigma_{prio},A_{prio})\rightarrow\omega$ by:

$$S(t \xrightarrow{a} t')=rk(a)+1$$

(it is straightforward to check that S is a stratification of P_{prio}).

Another solution is to forbid that priority operators appear in the body of a process name. In this case a stratification can be given by:

$$S(t \xrightarrow{a} t')=n \quad \text{where } n \text{ is the total number of occurrences of } \theta\text{'s in } t.$$

A last possibility is obtained by forbidding unguarded recursion in the bodies of process definitions. A stratification can now be constructed as follows: suppose one has a literal $t \xrightarrow{a} t'$. Let n be the number of θ's in t. Moreover, let m be the number of the θ's in the bodies t'' of all process names X'' $(X''\Leftarrow t_{X''}\in E)$ that occur unguarded in t. Then we define a stratification $S:Tr(\Sigma_{prio},A_{prio})\rightarrow\omega$ by $S(t \xrightarrow{a} t')=n+m$. One can easily check that S is a stratification of P_{prio}.

4. Strong bisimulation equivalence, the ntyft/ntyxt-format and the congruence theorem

Often bisimulation equivalence is considered as the finest extensional equivalence that one wants to impose. If bisimulation equivalence is not a congruence then one can distinguish bisimilar processes by putting them in appropriate contexts. Therefore it is a nice property of a format if it guarantees that all operators definable by this format respect bisimulation.

For a TSS without negative premises, the *tyft/tyxt*-format [7] is the most general format that respects bisimulation as a congruence. Here we introduce the *ntyft/ntyxt*-format as the most general extension of the *tyft/tyxt*-format with negative premises. Bisimulation is again a congruence for operators defined in this format (provided that the rule system is stratifiable).

4.1. Definition. Let $P=(\Sigma,A,R)$ be a stratifiable TSS and let $\longrightarrow_P$ be the transition relation associated with P. A relation $R\subseteq T(\Sigma)\times T(\Sigma)$ is a *(strong) P-bisimulation* if it satisfies:

1. whenever $s\,R\,t$ and $s \xrightarrow{a}_P s'$ then, for some $t'\in T(\Sigma)$, we have $t \xrightarrow{a}_P t'$ and $s'R\,t'$,
2. conversely, whenever $s\,R\,t$ and $t \xrightarrow{a}_P t'$ then, for some $s'\in T(\Sigma)$, we have $s \xrightarrow{a}_P s'$ and $s'R\,t'$.

We say that two terms $t,t'\in T(\Sigma)$ are *(P-)bisimilar*, notation $t \leftrightarrow_P t'$, if there is a P-bisimulation relation R such that $t\,R\,t'$. Note that $\leftrightarrow_P$ is an equivalence relation.

4.2. Definition. Let $\Sigma=(F,r)$ be a signature. Let $P=(\Sigma,A,R)$ be a TSS. A rule $r\in R$ is in *ntyft-format* if it has the form:

$$\frac{\{t_k \xrightarrow{a_k} y_k \mid k\in K\} \cup \{t_l \overset{b_l}{\not\longrightarrow} \mid l\in L\}}{f(x_1,\ldots,x_{r(f)}) \xrightarrow{a} t}$$

where y_k,x_i $(1\leqslant i\leqslant r(f))$ are different variables, $f\in F$ and t_k, t_l, $t\in\mathbb{T}(\Sigma)$. A rule $r\in R$ is in *ntyxt-format* if it fits:

$$\frac{\{t_k \xrightarrow{a_k} y_k \mid k\in K\} \cup \{t_l \overset{b_l}{\not\longrightarrow} \mid l\in L\}}{x \xrightarrow{a} t}$$

where y_k,x are different variables and $t_k,t_l,t\in\mathbb{T}(\Sigma)$. P is in *ntyft-format* if all its rules are in *ntyft*-format and P is in *ntyft/ntyxt-format* if all its rules are in *ntyft-* or in *ntyxt*-format.

4.3. We need a same well foundedness restriction on the premises of the rules as was necessary to prove the congruence theorem for the *tyft/tyxt*-format [7]. Let $t_k \longrightarrow y_k$ be a positive literal of a rule r in *ntyft/ntyxt*-format. If $x \in var(t_k)$ then we say that y_k *depends on* x. If a variable x depends on variable y and y depends on a variable z then we also say that x depends on z. We say that r is *well founded* if no variable y_k $(k \in K)$ depends on itself. A TSS P is *well founded* if all its rules are well founded.

4.3.1. EXAMPLE. The dependencies of the variables in the premises $\{f(x',y_1) \xrightarrow{a} y_2, g(x,y_2) \xrightarrow{a} y_1\}$ of a rule r are given in figure 1. r is not well founded because the graph contains a cycle.

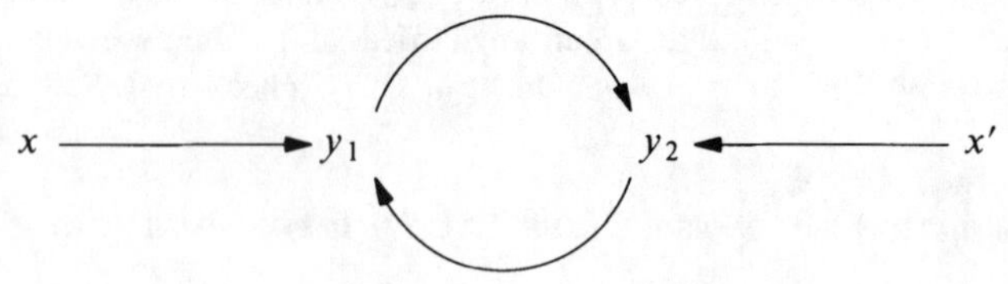

FIGURE 1

Next, we state the congruence theorem.

4.4. THEOREM. *Let P be a well founded, stratifiable TSS in ntyft/ntyxt-format. Then $\leftrightarrows_P$ is a congruence relation for all function symbols in the signature of P.*

5. MODULAR PROPERTIES OF TSS'S

Often one wants to extend a given TSS with new constants and functions. Therefore the *sum* of two TSS's is introduced [7]. The combination of two TSS's P_0 and P_1 is denoted by $P_0 \oplus P_1$. With negative premises care is needed to guarantee that $P_0 \oplus P_1$ still defines a transition relation.

If P_1 is added to P_0 it would be nice if all literals with source $t \in T(\Sigma_0)$ in $\longrightarrow_{P_0 \oplus P_1}$ are exactly the literals in $\longrightarrow_{P_0}$. In this case we say that $P_0 \oplus P_1$ is a *conservative extension* of P_0. A TSS can be conservatively extended under the same conditions as hold for TSS's without negative premises except for a separate check showing that $P_0 \oplus P_1$ is stratifiable [7].

5.1. DEFINITION. Let $\Sigma_i = (F_i, r_i)$ $(i=0,1)$ be two signatures such that $f \in F_0 \cap F_1 \Rightarrow r_0(f) = r_1(f)$. The *sum* of Σ_0 and Σ_1, notation $\Sigma_0 \oplus \Sigma_1$, is the signature:

$$\Sigma_0 \oplus \Sigma_1 = (F_0 \cup F_1, \lambda f. \textit{if } f \in F_0 \textit{ then } r_0(f) \textit{ else } r_1(f)).$$

The *sum* of P_0 and P_1, notation $P_0 \oplus P_1$, is the TSS:

$$P_0 \oplus P_1 = (\Sigma_0 \oplus \Sigma_1, A_0 \cup A_1, R_0 \cup R_1).$$

We say that $P_0 \oplus P_1$ is a *conservative extension* of P_0 if $P_0 \oplus P_1$ is stratifiable and if for all $t \in T(\Sigma_0)$, $a \in A$ and $t' \in T(\Sigma_0 \oplus \Sigma_1)$:

$$t \xrightarrow{a} t' \in \longrightarrow_{P_0 \oplus P_1} \quad \Leftrightarrow \quad t \xrightarrow{a} t' \in \longrightarrow_{P_0}.$$

5.2. If $P_0 \oplus P_1$ is a conservative extension of $P_0 = (\Sigma_0, A_0, R_0)$, then it follows immediately that for all $t, u \in T(\Sigma_0)$: $t \leftrightarrows_{P_0} u \Leftrightarrow t \leftrightarrows_{P_0 \oplus P_1} u$.

Theorem 5.5 gives conditions under which a TSS P_1 can be added conservatively to P_0. In [7] a number of examples have been given to show that the theorem cannot be extended in a straightforward way for TSS's in which only rules with positive premises are allowed. Here we give the same general theorem for negative premises. Only the constraint that $P_0 \oplus P_1$ must be stratifiable is new, but the following example shows that this constraint is necessary. It is not hard to give less trivial examples that also lead to inconsistencies.

5.3. EXAMPLE. Consider a TSS $P_0 = (\Sigma_0, A_0, R_0)$ with an empty signature and a rule

$$\frac{x \xrightarrow{a}\!\!\!\!\!/\;\;}{x \xrightarrow{a} x}.$$

P_0 is stratifiable because there are no Σ_0-terms. If one adds a TSS P_1 to P_0 containing only one constant f and no rules, one gets the situation of example 2.2.2 where we had an inconsistency. Hence, $P_0 \oplus P_1$ is not stratifiable.

5.4. DEFINITION. Let $P = (\Sigma, A, R)$ be a TSS. Let $r \in R$ be a rule. A variable x is called *free* in r if it occurs in r but not in the source of the conclusion or in the target of a positive premise. The rule r is called *pure* if it is well founded and does not contain free variables. P is called *pure* if all rules in R are pure.

5.5. THEOREM. *Let $P_0 = (\Sigma_0, A_0, R_0)$ be a TSS in pure ntyft/ntyxt format and let $P_1 = (\Sigma_1, A_1, R_1)$ be a TSS in ntyft format such that there is no rule in R_1 containing a function name from Σ_0 in the source of its conclusion. Let $P = P_0 \oplus P_1$ be defined and stratifiable. Then P_1 can be added conservatively to P_0.*

6. THE TRACE CONGRUENCE GENERATED BY THE NTYFT/NTYXT FORMAT

In this section we show that when we define operators using the pure *ntyft/ntyxt*-format such that the associated transition relation is image finite, then the trace congruence generated by this format is exactly (strong) bisimulation (theorem 6.4). First we define the notions of a trace congruence generated by a format and of an image finite transition relation.

6.1. DEFINITION. Let $P = (\Sigma, A, R)$ be a TSS and let $\longrightarrow_P$ be the transition relation associated with P. Let $s \in T(\Sigma)$. A sequence $a_1 \star \cdots \star a_n \in A^*$ is a *(P-)trace* of s if there are terms $s_1, \ldots, s_n \in T(\Sigma)$ for some $n \in \mathbb{N}$ such that $s \xrightarrow{a_1}_P s_1 \xrightarrow{a_2}_P, \ldots, \xrightarrow{a_n}_P s_n$. $T(s)$ is the set of all traces from s. Two process terms $s, s' \in T(\Sigma)$ are *trace equivalent with respect to P* if $T(s) = T(s')$. This is also denoted as $s \equiv_P^T s'$.

6.2. DEFINITION. Let $\mathfrak{F}$ be some format of TSS rules. Let $P = (\Sigma, A, R)$ be a stratifiable TSS in $\mathfrak{F}$ format. Two terms $t, t' \in T(\Sigma)$ are *trace congruent with respect to* $\mathfrak{F}$ rules, notation $t \equiv_{\mathfrak{F}}^T t'$, if for every TSS $P' = (\Sigma', A', R')$ in $\mathfrak{F}$ format which can be added conservatively to P and for every $\Sigma \oplus \Sigma'$-context $C[]$: $C[t] \equiv_{P \oplus P'}^T C[t']$.

6.3. DEFINITION. Let $P = (\Sigma, A, R)$ be a stratifiable TSS. Let $\longrightarrow_P$ be the transition relation associated with P. $\longrightarrow_P$ is called *image finite* if for all $s \in T(\Sigma)$ and $a \in A$ the set $\{t \mid s \xrightarrow{a}_P t\}$ is finite.

6.4. THEOREM. *Let $P = (\Sigma, A, R)$ be a stratifiable TSS in pure ntyft/ntyxt-format such that $\longrightarrow_P$ is image finite:*

$$t \equiv_{pure\ ntyft/ntyxt}^T u \;\Leftrightarrow\; t \leftrightarrows_P u \quad (t, u \in T(\Sigma)).$$

PROOF (sketch). The implication from right to left follows immediately using theorem 4.4. The proof from left to right makes use of n-bounded bisimilarity (bisimilarity till depth n) [9]. We use the notation $\overset{n}{\leftrightarrows}_P$ for this. It is a standard fact that for image finite transition relations bisimulation coincides with n-bounded bisimulation for all n. So we must prove that $t \equiv_{\mathfrak{F}}^T u \Rightarrow \forall n\; t \overset{n}{\leftrightarrows}_P u$. By contraposition it is sufficient to show that if for some n, $t \not\leftrightarrows_P^n u$ then

$t \not\equiv_{\mathfrak{F}}^{T} u$. We show this using the bisimulation tester P_T of P. This tester is an alternative to the bisimulation testers of [1,5] and does not use *global testing* operators.

6.4.1. DEFINITION. Let $P=(\Sigma,A,R)$ be a stratifiable TSS. The *bisimulation tester of P*, $P_T=(\Sigma_T,A_T,R_T)$, is a TSS with signature $\Sigma_T=(F_T,r_T)$ containing binary function names B^n and Q_a^n for all $n \in \mathbb{N}$, $a \in A$ and with constant δ. The labels of P_T are $A_T = A \cup \{ok,yes,no\}$. We assume $\Sigma \cap \Sigma_T = \emptyset$ and $ok,yes,no \notin A$. The rules in R_T are given in table 2.

$B^0(x,y) \xrightarrow{yes} \delta$		1
$\dfrac{y \xrightarrow{a} y' \quad B^{n-1}(x',y') \xrightarrow{yes} z}{Q_a^n(x',y) \xrightarrow{ok} \delta}$	for $n>0$, $a \in A$	2
$\dfrac{x \xrightarrow{a} x' \quad Q_a^n(x',y) \not\xrightarrow{ok}}{B^n(x,y) \xrightarrow{no} \delta}$	for $n>0$, $a \in A$	3
$\dfrac{B^n(x,y) \not\xrightarrow{no} \quad B^n(y,x) \not\xrightarrow{no}}{B^n(x,y) \xrightarrow{yes} \delta}$	for $n>0$	4

TABLE 2

The rules are based on the following meaning of the transitions $\xrightarrow{yes}$, $\xrightarrow{no}$ and $\xrightarrow{ok}$:

- $B^n(x,y) \xrightarrow{yes} \delta$ if x and y are n-bounded bisimilar.
- $B^n(x,y) \xrightarrow{no} \delta$ $(n>0)$ if x can perform a step that cannot be simulated by y such that the results are $(n-1)$-bounded bisimilar.
- $Q_a^n(x,y) \xrightarrow{ok} \delta$ $(n>0)$ means that y can perform an a-step such that the result is $(n-1)$-bounded bisimilar with x.

For any stratifiable TSS $P=(\Sigma,A,R)$ in pure *ntyft/ntyxt*-format $P \oplus P_T$ is a conservative and stratifiable extension of P. Moreover, we have the following property for $t,u \in T(\Sigma)$:

$$B^n(t,u) \xrightarrow{yes}_{P \oplus P_0} \delta \Leftrightarrow t \overset{n}{\leftrightarrows}_P u.$$

We had to show that if for some n $t \not\leftrightarrows_P^n u$ then $t \not\equiv_{\mathfrak{F}}^{T} u$. If $t \not\leftrightarrows_P^n u$ then put t and u in the context $B^n(t,[\cdot])$. We have that $B^n(t,t) \xrightarrow{yes}_{P \oplus P_0} \delta$ but $B^n(t,u) \not\xrightarrow{yes}_{P \oplus P_0}$. Hence t and u have different traces in the context $B^n(t,[\cdot])$. So $t \not\equiv_{\text{pure } ntyft/ntyxt}^{T} u$. □

REFERENCES

[1] S. ABRAMSKY (1987): *Observation equivalence as a testing equivalence.* Theoretical Computer Science 53, pp. 225-241.

[2] K.R. APT (1988): *Introduction to logic programming.* Report CS-R8826, Centrum voor Wiskunde en Informatica, Amsterdam, to appear in: Handbook of Theoretical Computer Science, (J. van Leeuwen, Managing Editor), North-Holland.

[3] J.C.M. BAETEN & J.A. BERGSTRA (1988): *Global renaming operators in concrete process algebra.* I&C 78(3), pp. 205-245.

[4] J.C.M. BAETEN, J.A. BERGSTRA & J.W. KLOP (1986): *Syntax and defining equations for an interrupt mechanism in process algebra.* Fund. Inf. IX(2), pp. 127-168.

[5] B. BLOOM, S. ISTRAIL & A.R. MEYER (1988): *Bisimulation can't be traced: preliminary report.* In: Conference Record of the 15th ACM Symposium on Principles of Programming Languages (POPL), San Diego, California, pp. 229-239.

[6] R. CLEAVELAND & M. HENNESSY (1988): *Priorities in process algebra.* In: Proceedings 3th Annual Symposium on Logic in Computer Science (LICS), Edinburgh, pp. 193-202.

[7] J.F. GROOTE & F.W. VAANDRAGER (1988): *Structured operational semantics and bisimulation as a congruence.* Report CS-R8845, Centrum voor Wiskunde en Informatica, Amsterdam, under revision for I&C. An extended abstract appeared in: Proceedings ICALP 89, Stresa (G. Ausiello, M. Dezani-Ciancaglini & S. Ronchi Della Rocca, eds.), LNCS 372, Springer-Verlag, pp. 423-438.

[8] R. LANGERAK (1989): *A testing theory for LOTOS using deadlock detection.* In: Proceedings 9th IFIP WG6.1 International Symposium on Protocol Specification, Testing, and Verification, June 1989, Enschede, The Netherlands (E. Brinksma, G. Scollo & C.A. Vissers, eds.).

[9] R. MILNER (1989): *Communication and concurrency*, Prentice-Hall International.

[10] F. MOLLER (1989): *Axioms for concurrency.* Ph.D. Thesis, Report CST-59-89, Department of Computer Science, University of Edinburgh.

[11] I.C.C. PHILLIPS: *CCS with Broadcast Stability.* Unpublished manuscript.

[12] G.D. PLOTKIN (1981): *A structural approach to operational semantics.* Technical Report DAIMI FN-19, Computer Science Department, Aarhus University.

[13] A. PNUELI (1985): *Linear and branching structures in the semantics and logics of reactive systems.* In: Proceedings ICALP 85, Nafplion (W. Brauer, ed.), LNCS 194, Springer-Verlag, pp. 15-32.

[14] T.C. PRZYMUSINSKI (1987): *On the declarative semantics of deductive databases and logic programs.* In: Foundations of Deductive Databases and Logic Programming (Jack Minker, ed.), Morgan Kaufmann Publishers, Inc., Los Altos, California, pp. 193-216.

Delay-insensitive circuits: an algebraic approach to their design

Mark B. Josephs
Programming Research Group
Oxford University Computing Laboratory
Jan Tijmen Udding
Departments of Computer Science
Groningen University, the Netherlands, and
Washington University, St. Louis, U.S.A.

A novel process algebra is presented; algebraic expressions specify delay-insensitive circuits in terms of voltage-level transitions on wires. The algebraic laws make it possible to specify circuits concisely and facilitate the verification of designs. Individual components can be composed into circuits in which signals along internal wires are hidden from the environment.

This algebraic approach has been applied to the design of non-blocking arbiters. Our designs have subsequently been checked by automatic verifiers, which had to examine approximately 600 states.

0 Introduction

A circuit is connected to its environment by a number of wires. If the circuit functions correctly irrespective of the propagation delays in these wires, the circuit is called *delay-insensitive.* Delay-insensitive circuits are attractive because they can be designed in a modular way; indeed no timing constraints have to be satisfied in connecting such circuits together. As a result of the latest Turing Award Lecture [14], the design of delay-insensitive circuits has drawn renewed interest.

The design of delay-insensitive circuits is made difficult by the need to consider situations in which a signal (voltage-level transition) has been transmitted at one end of a wire but has not yet been received at the other end. The algebraic notation presented in this paper may be helpful in the following ways:

1. The functional behaviour of primitive delay-insensitive components can be captured by algebraic expressions.

2. All possible behaviours of the circuit that results when such components are connected together can be determined by symbolic manipulation.

3. The algebra facilitates the precise specification of the circuit that the designer has to build, including obligations to be met by the environment.

4. The algebra supports verification of the design against the specification.

The algebra is based upon Hoare's CSP notation [7]. It adapts the theory of asynchronous processes [8, 9] to the special case of delay-insensitive circuits. The possibilities of *transmission interference* and *computation interference*, characterized by Udding [15, 16], are faithfully modelled in the algebra; the designer is able to reason about these errors, and so avoid them. Underpinning the algebra is a denotational semantics similar to those given in [2, 9]; the semantics is compatible with the failures-divergences model of CSP [1, 7].

The remainder of this paper provides a step-by-step introduction to the algebra. We demonstrate a hierarchical approach to circuit design by first specifying a non-blocking arbiter and then decomposing the specification into a number of components operating in parallel. Traditional methods based on state-graphs would be wholly inadequate for a problem of this size.

1 Basic Notions and Operators

A process is a mathematical model at a certain level of abstraction of the way in which a delay-insensitive circuit interacts with its environment. Typical names for processes are P and Q. A circuit receives signals from its environment on its input wires and sends signals to its environment on its output wires. Thus with each process are associated an alphabet of input wires and an alphabet of output wires. These alphabets are finite and disjoint. Typical names for input wires are a and b; typical names for output wires are c and d. The time taken by a signal to traverse a wire is indeterminate. Assuming that a wire is low initially, it will carry signals that are alternately high and low-going voltage-level transitions.

In the remainder of this section and section 2, we consider processes with a particular alphabet I of input wires and a particular alphabet O of output wires.

The process P is considered to be "just as good" as the process Q ($P \sqsubseteq Q$) if no environment, which is simply another process, can when interacting with P determine that it is not interacting with Q. (This is the refinement ordering of CSP [1, 7], also known as the *must* ordering [6].) Two processes are considered to be equal when they are just as good as each other.

The refinement ordering is intimately connected with nondeterministic choice. The process $P \sqcap Q$ is allowed to behave either as P or as Q. It reflects the designer's freedom to implement such a process by either P or Q. (Thus $\sqcap$ is obviously commutative, associative and idempotent.) Now $P \sqcap Q = Q$ exactly when $P \sqsubseteq Q$.

A wire cannot accommodate two signals at the same time; they might interfere with one another in an undesirable way. This and any other error are modelled by the process $\perp$ (Bottom or Chaos). The environment must ensure that a process never gets into such a state. The process $\perp$ is considered to be so undesirable that any other process must be an improvement on it:

Law 0. $P \sqsubseteq \perp$

It follows that $\sqcap$ has $\perp$ as a null element.

We shall mostly be concerned with recursively-defined processes. The meaning of the recursion $\mu X.F(X)$ is the least fixpoint of F. Its successive approximations are $\perp$, $F(\perp)$, $F(F(\perp))$, *etc.* All the operators that we shall use to define processes are continuous (and therefore monotonic); all except for recursion are distributive (with respect to nondeterministic choice).

In earlier approaches to an algebra for delay-insensitive circuits, *cf.* [15, 2, 5], a particular input signal is allowed only when this is explicitly indicated, and otherwise is assumed to lead to interference. This is in contrast with the algebra presented here: an input need not result in interference even though the possibility of such an input has not been made explicit in the algebraic expression. This follows the approach taken in [9] and is more convenient in algebraic manipulation, even though at first it may appear less natural.

Thus we write $a?; P$ to denote a process that must wait for a signal to arrive on $a \in I$ before it can behave like P. It is quite permissible for the environment to send a signal along any other input wire b. Such a signal is ignored at least until a signal is sent along a (or a second signal is sent along b causing interference).

A process that waits for input on a and then for input on b before being able to do anything is actually just waiting for inputs on both a and b, their order being immaterial:

Law 1. $a?; b?; P = b?; a?; P$

Complementary to input-prefixing $a?; P$ is output-prefixing $c!; P$, where $c \in O$. This is a process that outputs on c and then behaves like P. The environment may send a signal on any input wire even before it receives the signal on c; whether or not it can do so safely depends on P.

Two outputs by a process on the same wire, one after the other, is unsafe because of the danger of the two signals interfering with one another before they reach the environment. Also, since any output of a process may arrive at the environment an arbitrary time later, the order in which outputs are sent is immaterial. Therefore, we have the following two laws:

Law 2. $c!; c!; P = \bot$

Law 3. $c!; d!; P = d!; c!; P$

Example 0 Law 2. allows us to prove that $c!; \bot = \bot$. This should not be surprising: the process $c!; \bot$ behaves like $\bot$ after it has output on c; a wholly undesirable state results even before the output has reached the environment.

$$
\begin{aligned}
& c!; \bot \\
= \quad & \{ \text{ Law 2. } \} \\
& c!; c!; c!; \bot \\
= \quad & \{ \text{ Law 2. } \} \\
& \bot
\end{aligned}
$$

■

Finally, as in CSP, prefixing is distributive (with respect to nondeterministic choice). For both input and output-prefixing we have the law

Law 4. $x; (P \sqcap Q) = (x; P) \sqcap (x; Q)$

Example 1 We are now in a position to specify a number of elementary delay-insensitive components, *viz.* the Wire, the Fork, the Toggle and the C-element.

Consider a circuit with one input wire a and one output wire c. In response to each signal on a, the circuit should produce a signal on c. The precise behaviour of this circuit is given by the following algebraic expression:

$$\mu X.\ a?; c!; X$$

which we shall refer to as the process W because it can be readily implemented by a wire. Now unfolding the recursion, we have that $W = a?; c!; W$. As in CSP, this equation itself uniquely defines W because its right hand side is guarded.

Next consider the process, with one input wire a and two output wires c and d, defined by the equation $F = a?; c!; d!; F$. This models the behaviour of a fork.

The Toggle [5, 14] also has one input wire and two output wires, but alternates between outputting on c and outputting on d, *i.e.*, $T = a?; c!; a?; d!; T$.

Finally, the Muller C-element repeatedly waits for inputs on wires a and b before outputting on c. It is defined by $C = a?; b?; c!; C$. ■

A more general form of input-prefixing is input-guarded choice. Such a choice allows a process to take different actions depending upon the input received. The choice is made between a number of alternatives of the form $a? \rightarrow P$. For S a finite set of alternatives, the guarded choice $[S]$ selects one of them. An alternative $a? \rightarrow P$ can be selected only if a signal has been received on a. The choice cannot be postponed indefinitely once one or more alternatives become selectable.

Choice with only one alternative is no real choice at all:

Law 5. $[a? \rightarrow P] = a?; P$

The environment cannot safely send a second signal along an input wire until the first signal has been acknowledged. Thus the result of sending two signals on a to the process $a?; a?; P$ is $\perp$ rather than P. The process is as useless as a choice with no alternative:

Law 6. $a?; a?; P = [\,]$

If two alternatives are guarded on a, then either may be chosen after input has been supplied on a. Indeed, the designer has the freedom to implement only one of the two. This is captured in the following law, where the symbol $\square$ separates the various alternatives:

Law 7.

$$\begin{aligned} & [a? \rightarrow P \;\square\; a? \rightarrow Q \;\square\; S] \\ = \; & [a? \rightarrow (P \sqcap Q) \;\square\; S] \\ = \; & [a? \rightarrow P \;\square\; S] \sqcap [a? \rightarrow Q \;\square\; S] \end{aligned}$$

Example 2 With the above law we can prove the following absorption theorem. An alternative guarded on a is absorbed by $a? \rightarrow \perp$:

$$[a? \rightarrow \perp \;\square\; a? \rightarrow P \;\square\; S]$$

$= \quad$ { combining alternatives using Law 7. }

$$[a? \rightarrow (\perp \sqcap P) \;\square\; S]$$

$= \quad$ { $\perp$ is the null element of $\sqcap$ }

$$[a? \rightarrow \perp \;\square\; S]$$

■

Until a process acknowledges receipt of an input signal, a second signal on the same wire can result in interference. So, for S_0 and S_1 sets of alternatives, we have

Law 8. $[a? \rightarrow [S_0] \,\square\, S_1] = [a? \rightarrow [a? \rightarrow \bot \,\square\, S_0] \,\square\, S_1]$

Example 3 As a matter of fact, we can replace $\bot$ in Law 8. by any process P.

$$[a? \rightarrow [a? \rightarrow \bot \,\square\, S_0] \,\square\, S_1]$$

$=$ { absorption law derived in Example 2 }

$$[a? \rightarrow [a? \rightarrow \bot \,\square\, a? \rightarrow P \,\square\, S_0] \,\square\, S_1]$$

$=$ { Law 8. }

$$[a? \rightarrow [a? \rightarrow P \,\square\, S_0] \,\square\, S_1]$$

■

Indeed, if it is unsafe for the environment to send a signal along a particular input wire, it remains unsafe at least until an output has been received. Therefore, we also have the following absorption law.

Law 9. $[a? \rightarrow \bot \,\square\, b? \rightarrow [a? \rightarrow P \,\square\, S_0] \,\square\, S_1] \;=\; [a? \rightarrow \bot \,\square\, b? \rightarrow [S_0] \,\square\, S_1]$

Example 4 With the input-guarded choice we can model somewhat more interesting delay-insensitive components, such as the Merge, the Sequencer and the Decision-Wait element.

The Merge is a circuit with two input wires a and b and one output wire c. In response to a signal on either a or b, it will output on c:

$$M = [a? \rightarrow c!; M \,\square\, b? \rightarrow c!; M]$$

We shall discover, in the next section, that this definition implies that it is unsafe for the environment to supply input on both a and b before receiving an output on c.

The Sequencer [5] responds to each request on r_i with a grant on g_i ($i = 0, 1$). These are sequenced by the control wire c:

$$Seq = c?; [r_0? \rightarrow g_0!; Seq \;\Box\; r_1? \rightarrow g_1!; Seq]$$

Finally, we can define the Decision-Wait element (a 2 × 1 matrix in this case). It expects one input change in its row and one input change in its column. It produces as output the single entry which is indicated by the two changing inputs – there are two entries in this case:

$$\begin{aligned} DW \;=\; & [r_0? \rightarrow [r_1? \rightarrow \bot \;\Box\; c? \rightarrow e_0!; DW] \\ & \Box\; r_1? \rightarrow [r_0? \rightarrow \bot \;\Box\; c? \rightarrow e_1!; DW]\,] \end{aligned}$$

(A C-element can be viewed as a 1 × 1 Decision-Wait element.) ∎

2 More Advanced Constructs

In the last section we provided enough operators to allow us to specify many interesting delay-insensitive components from which larger circuits might be constructed. To better understand these specifications we need to be able to determine how a circuit will behave after some signals have been exchanged with its environment. Before we can do this, it turns out that we need a more general form of guarded choice which allows for *skip* guards as well as input guards.

Recall that a guarded choice $[S]$ consists of a set S of alternatives of the form $a? \rightarrow P$. We shall now allow S to include also alternatives of the form $skip \rightarrow P$. Such an alternative can be selected whether or not any input is supplied. (As in CSP, if no input is supplied, it must eventually be selected.)

The laws of the last section remain valid, but in addition we have several laws involving *skip* guards. (These are also laws in occam [13].) As before, a choice with one alternative is no real choice at all:

Law 10. $[skip \rightarrow P] = P$

Nondeterministic choice can be regarded as a special case of guarded choice:

Law 11. $[skip \rightarrow P \;\Box\; skip \rightarrow Q] = P \sqcap Q$

The selection of a *skip*-guarded alternative is an internal (unobservable) action of a process. This gives rise to the following three laws. In the first, a nondeterministic choice arises after input has been supplied on *a* because the signal may arrive before the selection of a *skip*-guarded alternative:

Law 12.
$$[a? \rightarrow P \;\Box\; skip \rightarrow [a? \rightarrow Q \;\Box\; S_0] \;\Box\; S_1]$$
$$= [skip \rightarrow [a? \rightarrow (P \sqcap Q) \;\Box\; S_0] \;\Box\; S_1]$$

The second law states that a nested *skip*-guarded alternative can be selected in preference to any other alternative:

Law 13. $[skip \rightarrow [skip \rightarrow P \;\Box\; S_0] \;\Box\; S_1] = [skip \rightarrow P \;\Box\; S_0 \;\Box\; S_1]$

The third law is a convexity property of guarded choice:

Law 14. $[skip \rightarrow [S_0] \;\Box\; skip \rightarrow [S_0 \;\Box\; S_1] \;\Box\; S_2] = [skip \rightarrow [S_0] \;\Box\; S_1 \;\Box\; S_2]$

Example 5 Two *skip*-guarded alternatives can be combined together:

$$[skip \rightarrow P \;\Box\; skip \rightarrow Q \;\Box\; S]$$

$=$ { nesting the *skip* guards using Law 13. }

$$[skip \rightarrow [skip \rightarrow P \;\Box\; skip \rightarrow Q] \;\Box\; S]$$

$=$ { $\sqcap$ as guarded choice, Law 11. }

$$[skip \rightarrow (P \sqcap Q) \;\Box\; S]$$

■

Example 6 With impunity we can extend the set of alternatives in a guarded choice with that guarded choice itself as a new *skip*-guarded alternative:

$$[S]$$

$=$ { one choice is no choice, Law 10. }

$$[skip \rightarrow [S]\,]$$

$$= \quad \{\text{ convexity, Law 14. }\}$$

$$[skip \rightarrow [S] \;\Box\; S]$$

■

Example 7 A *skip*-guarded alternative can always be chosen, and so no other alternative need be offered, *i.e.*, $P \subseteq [skip \rightarrow P \;\Box\; S]$.

$$P \sqcap [skip \rightarrow P \;\Box\; S]$$

$$= \quad \{\; \sqcap \text{ as guarded choice, Law 11. }\}$$

$$[skip \rightarrow P \;\Box\; skip \rightarrow [skip \rightarrow P \;\Box\; S]\,]$$

$$= \quad \{\text{ unnesting the } skip \text{ guards, Law 13. }\}$$

$$[skip \rightarrow P \;\Box\; S]$$

In particular, $[skip \rightarrow \bot \;\Box\; S] = \bot$ by Law 0. ■

Example 8 As another example of interference between two consecutive outputs on a wire, we have

$$c!; [skip \rightarrow c!; P \;\Box\; S]$$

$$\supseteq \quad \{\text{ Example 7 and monotonicity of prefixing }\}$$

$$c!; c!; P$$

$$= \quad \{\text{ interference between outputs, Law 2. }\}$$

$$\bot$$

which means that $c!; [skip \rightarrow c!; P \;\Box\; S] = \bot$, by Law 0. ■

We are now able to define how a process behaves after the environment has sent input to it or received output from it. In this paper we shall only consider the (more useful) after-input case. The reader is referred to [10] for the definition of after-output.

The process $P/a?$ behaves like P after its environment has sent it a signal on $a \in I$. Notice that this does not mean that P has received this input yet; the signal may still be on its way. Indeed it is impossible to tell whether P has received the signal until some acknowledging signal has been received from P.

The first two laws for after-input are concerned with undesirable behaviour. A process which has entered an unsafe state remains unsafe. Also, sending two signals in a row on an input wire may cause interference and is therefore unsafe.

Law 15. $\bot/a? = \bot$

Law 16. $P/a?/a? = \bot$

The order in which signals are sent does not determine the order in which they are received, and so

Law 17. $P/a?/b? = P/b?/a?$

An output-prefixed process can do nothing but output, even when sent input:

Law 18. $(c!; P)/a? = c!; (P/a?)$

After-input (on a) distributes through the alternatives in a guarded choice, except for those alternatives guarded on a. Those become *skip*-guarded. Furthermore, an extra alternative is required to indicate that interference can result after a second input on a.

Law 19. $[S]/a? = [a? \rightarrow \bot \;\Box\; S']$,
where S' is formed by substituting for each alternative $A \in S$ the new alternative $A/a?$, defined by

$$(\mathit{skip} \rightarrow P)/a? = \mathit{skip} \rightarrow (P/a?)$$
$$(a? \rightarrow P)/a? = \mathit{skip} \rightarrow P$$
$$(b? \rightarrow P)/a? = b? \rightarrow (P/a?), \text{ for } b \neq a.$$

As a consequence of Laws 5. and 19., we have that

$$(a?; P)/a? = [a? \rightarrow \bot \;\Box\; \mathit{skip} \rightarrow P]$$

and

$$(b?; P)/a? = [a? \rightarrow \bot \ \Box \ b? \rightarrow (P/a?)].$$

Example 9 When $\bot$ is guarded on a, the environment must not supply input on a.

$[a? \rightarrow \bot \ \Box \ S]/a?$

$=$ { after through guarded choice, Law 19.,
S' being some set of alternatives derived from S }

$[a? \rightarrow \bot \ \Box \ skip \rightarrow \bot \ \Box \ S']$

$=$ { unguarded $\bot$, Example 7 }

$\bot$

■

The following law allows us to expand the set of alternatives in a guarded choice to make the behaviour after a particular input explicit:

Law 20. $[S] = [a? \rightarrow [S]/a? \ \Box \ S]$

Example 10 Here is a case in which *skip* can be eliminated:

$[a? \rightarrow \bot \ \Box \ skip \rightarrow [S]]$

$=$ { adding an a-guarded alternative with Law 20. }

$[a? \rightarrow \bot \ \Box \ skip \rightarrow [a? \rightarrow ([S]/a?) \ \Box \ S]]$

$=$ { postponing alternative until after *skip*, Law 12. }

$[skip \rightarrow [a? \rightarrow (([S]/a?) \sqcap \bot) \ \Box \ S]]$

$=$ { one choice is no choice and $\bot$ is null element of $\sqcap$ }

$[a? \rightarrow \bot \ \Box \ S]$

Example 11 Now we can determine how components such as the C-element and the Merge behave after they have interacted with their environment. For the C-element we derive

$$(a?; b?; c!; C)/a?$$

$=$ { after through input-prefixing, corollary to Law 19. }

$$[a? \rightarrow \perp \;\square\; \textit{skip} \rightarrow b?; c!; C]$$

$=$ { one choice is no choice and eliminating *skip* as in Example 10 }

$$[a? \rightarrow \perp \;\square\; b? \rightarrow c!; C]$$

Hence, a C-element that has been sent one input on a must not be sent another (until a signal on c has been received). Before it will output on c, however, it has to input on b. This is exactly how we want a C-element to behave after being sent a signal on a. We can now also compute $C/a?/b?$. A little calculation shows that the result is $[a? \rightarrow \perp \;\square\; b? \rightarrow \perp \;\square\; \textit{skip} \rightarrow c!; C]$, as desired. (It turns out to be the case that $C = C/a?/b?/c!$.)

A more interesting example of the possibility of interference is seen in the specification of Merge. Once a signal has been sent on either input wire, both input wires become unsafe to use. In this case we compute

$$M/a?$$

$=$ { definition of M }

$$[a? \rightarrow c!; M \;\square\; b? \rightarrow c!; M]/a?$$

$=$ { after through choice and prefixing, Laws 18. and 19. }

$$[a? \rightarrow \perp \;\square\; \textit{skip} \rightarrow c!; M \;\square\; b? \rightarrow c!; (M/a?)]$$

$=$ { $M/a?$ is of the form $[\textit{skip} \rightarrow c!; P \;\square\; S]$, Example 8 }

$$[a? \rightarrow \perp \;\square\; b? \rightarrow \perp \;\square\; \textit{skip} \rightarrow c!; M]$$

This shows that signals on both a and b should be withheld until the Merge outputs on c. ∎

3 Composition

In this section we define a parallel composition operator. With it we can determine the overall behaviour of a circuit from the individual behaviour of its components. It is understood that if the output wire of one component has the same name as the input wire of another, then these wires are supposed to be joined together; any signals transmitted along such a connection are hidden from the environment. The parallel composition operator is fundamental to a hierarchical approach to circuit design. It permits an initial specification to be decomposed into a number of components operating in parallel, and each of these components can be designed independently of the rest.

The simplicity of the laws enjoyed by parallel composition is one of the main attractions of our algebra. Indeed, in [15] certain restrictions had to be placed on processes before their composition could even be considered; and in [2] the fixed-point definition of parallel composition was rather unwieldy.

Parallel composition is denoted by the infix binary operator $\|$. All the operators we have met so far do not affect the input and output alphabets of their operands; so, for example, in the nondeterministic choice $P \sqcap Q$, we insist that the input alphabet of P is the same as that of Q, and declare that it is the same as that of $P \sqcap Q$. In the parallel composition $P \| Q$, however, the input alphabet of P should be disjoint from that of Q; likewise, the output alphabet of P should be disjoint from that of Q. (These rules prohibit fan-in and fan-out of wires; the explicit use of Merges and Forks is required.) The input alphabet of $P \| Q$ then consists of those input wires of each process P and Q which are not output wires of the other. Similarly, the output alphabet of $P \| Q$ consists of those output wires of each process which are not input wires of the other.

Parallel composition is commutative. It is also associative, provided we ensure that a wire named in the alphabets of any two processes being composed is not in the alphabets of a third process. If one process in a parallel composition is in an undesirable state, then the overall state is undesirable:

Law 21. $P \| \bot = \bot$

When an output-prefixed process $c!; P$ is composed with another process Q, the output is transmitted along c. Depending on whether or not c is in the input alphabet of Q, the signal on c is sent to Q or to the environment:

Law 22. $(c!; P) \parallel Q = \begin{cases} P \parallel (Q/c?) & \text{if } c \text{ is in the input alphabet of } Q \\ c!; (P \parallel Q) & \text{otherwise} \end{cases}$

It remains only to consider parallel composition of guarded choices. The following law specifies the alternatives in the resulting guarded choice.

Law 23. $[S_0] \parallel [S_1] = [S]$,
where S is formed from the alternatives in S_0 and S_1 in the following way. For each alternative in S_0 of the form $skip \rightarrow P$, we have $skip \rightarrow (P \parallel [S_1])$ in S. For each alternative in S_0 of the form $a? \rightarrow P$ with a not in the output alphabet of $[S_1]$, we have $a? \rightarrow (P \parallel [S_1])$ in S. The alternatives in S_1 contribute to the alternatives in S in a similar way.

Example 12 If one component is able to send a signal that it is unsafe for the other to receive, then their parallel composition is $\perp$.

$$(a!; P) \parallel [a? \rightarrow \perp \,\Box\, S]$$

$=$ { internal communication on a, Law 22. }

$$P \parallel [a? \rightarrow \perp \,\Box\, S]/a?$$

$=$ { Example 9 and $\perp$ null element of parallel composition, Law 21. }

$$\perp$$

■

Example 13 We compute a number of simple compositions in this example. Although the resulting behaviours are well-known, it has never previously been possible to give a straightforward algebraic derivation.

Consider first connecting two wires W_0 and W_1 together. Let $W_0 = a?; b!; W_0$ and $W_1 = b?; c!; W_1$. Then, in their parallel composition, signals on b are hidden from the environment.

$$W_0 \parallel W_1$$

$=$ { definitions of W_0 and W_1 }

$(a?; b!; W_0) \parallel (b?; c!; W_1)$

$=$ { one choice is no choice and parallel composition through guarded choice, Law 23., using that b is internal }

$a?; ((b!; W_0) \parallel (b?; c!; W_1))$

$=$ { internal communication on b, Law 22. }

$a?; (W_0 \parallel (b?; c!; W_1)/b?)$

$=$ { after through prefixing }

$a?; (W_0 \parallel [b? \rightarrow \bot \ \Box \ skip \rightarrow c!; W_1])$

$=$ { substituting for W_0 and applying Law 23., parallel composition through guarded choice, using that b is internal }

$$a?; \ [\ \ a? \rightarrow ((b!; W_0) \parallel [b? \rightarrow \bot \ \Box \ skip \rightarrow c!; W_1]) \\ \Box \ skip \rightarrow (W_0 \parallel (c!; W_1)) \ ]$$

$=$ { one choice is no choice and absorption as in Example 3 }

$a?; [skip \rightarrow (W_0 \parallel (c!; W_1))]$

$=$ { one choice is no choice and external communication with Law 22. }

$a?; c!; (W_0 \parallel W_1)$

Since this recursion is guarded, we conclude that $W_0 \parallel W_1 = W$.

A more interesting example is the composition of a "one-hot" C-element and a Fork in the following way. The C-element is specified by $C = a?; b?; c!; C$ and the Fork by $F = c?; a!; d!; F$. This is a circuit involving feedback.

$C/a? \parallel F$

$=$ { Example 11 and definition of F }

$[a? \rightarrow \bot \ \Box \ b? \rightarrow c!; C] \parallel (c?; a!; d!; F)$

$=$ { one choice is no choice and parallel composition through guarded choice, using that a and c are internal }

$b?; ((c!; C) \parallel (c?; a!; d!; F))$

$=$ { internal communication on c }

$b?; (C \parallel (c?; a!; d!; F)/c?)$

$=$ { definition of C and after through prefixing }

$b?; ((a?; b?; c!; C) \parallel [c? \rightarrow \perp \,\square\, skip \rightarrow a!; d!; F])$

$=$ { one choice is no choice, parallel composition through guarded choice, using that a and c are internal, and definition of C }

$b?; [skip \rightarrow (C \parallel (a!; d!; F))]$

$=$ { one choice is no choice, internal communication on a and external communication on d }

$b?; d!; (C/a? \parallel F)$

By uniqueness of guarded recursion, this combination of C-element and Fork behaves just like a wire. Although the Fork signals on a and d "in parallel", this did not lead to a doubling of the number of states which we had to analyse. We could deal with a entirely before d was pulled out of the parallel composition. This technique can be more generally applied and that is why these algebraic manipulations do not lead to a state explosion. ■

4 A Case Study: Non-Blocking Arbiters

The following challenge was recently posted on an electronic mailing list for the discussion of asynchronous circuits.

> Design a speed-independent "NACKing arbiter", also called (by Carl Ebeling) a non-blocking arbiter.

The component has to arbitrate between two processes competing for use of a resource. Process i $(= 0, 1)$ signals a request on wire r_i. The arbiter either grants the request by acknowledging on a_i or rejects it by signalling on n_i. Once a request has been rejected, the process must repeat the request if it still wants the resource.

Note that in speed-independent design no assumptions can be made about the time it takes for components to produce outputs in response to inputs; communication between components is, however, assumed to be instantaneous. David Dill and Steven Nowick of Stanford, and Steven Burns of Cal Tech, presented solutions. Martin's approach [11, 12] was successfully applied, but even writing the initial specification in that style was found to be a non-trivial exercise.

We took up the challenge and found it very easy to specify the arbiter. Actually the informal decription above is ambiguous and both four-phase ack/two-phase nack and full four-phase protocols are possible. Either way, we were able to follow a straightforward top-down appproach to arrive at strictly delay-insensitive designs. These designs are presented below.

Jerry Burch of Carnegie Mellon University reports that his trace theory based automatic verifier took about 16 seconds to construct the state machines for the components and the specification of our first design, and about 28 seconds to do the actual verification. It examined 608 states. Steven Nowick reports that his automatic verifier examined 438 states and 620 states, respectively, for the two designs. It took about 2 seconds. (One approach to automatic verification of circuits is described in [3, 4].)

Four-phase ack/two-phase nack protocol

We can think of the arbiter as being in one of three quiescent states, *i.e.*, states in which no output will be produced until further input is supplied. The initial state is A and a request on r_i will be acknowledged on a_i. The arbiter enters state G_i which indicates that the resource is granted to process i. Process i gives up the resource by signalling on r_i, which again is acknowleged on a_i, and the arbiter returns to state A. This completes the four-phase ack. While in state G_i, requests by the other process are rejected: this constitutes the two-phase nack. The specification is thus

$$A = [r_0? \rightarrow a_0!; G_0 \,\square\, r_1? \rightarrow a_1!; G_1]$$

where

$$G_0 = [r_0? \rightarrow a_0!; A \,\square\, r_1? \rightarrow n_1!; G_0]$$

$$G_1 = [r_1? \rightarrow a_1!; A \;\Box\; r_0? \rightarrow n_0!; G_1].$$

The simplicity of this specification is deceptive. Implicit in it are such complex behaviours as the following: process 0 sends a request and receives an acknowledgement; both processes then send further requests. It can be shown that this results in a state in which it is nondeterministic whether an a_1 or a n_1 is output; either must apparently be acceptable. (The reason is that wire delays make it possible for the signal on r_1 to arrive either before or after the second signal on r_0.)

The specification does, however, guide us to our first design decision. We introduce three wires c_0, c_1 and c_2 as feedback loops that indicate the next state G_0, G_1 and A, respectively. This enables us to remove recursion and parallelize, in the same sort of way Alain Martin "compiles" his initial CSP specifications. The result (Figure 0) is a kind of synchronizer Sy with three mutually exclusive input wires (the c_i), two other input wires (r_0 and r_1) and six output wires ($q_{i,0}$ and $q_{i,1}$). The rest of the circuit consists of six Forks and five Merges. The parallel composition of these components, after a signal has been sent on c_2 (*e.g.*, by inserting an inverter), exhibits the same behaviour as A.

$$Sy = [c_0? \rightarrow Sy_0 \;\Box\; c_1? \rightarrow Sy_1 \;\Box\; c_2? \rightarrow Sy_2]$$

where

$$Sy_i = [c_0? \rightarrow \bot \;\Box\; c_1? \rightarrow \bot \;\Box\; c_2? \rightarrow \bot \;\Box\; r_0? \rightarrow q_{i,0}!; Sy \;\Box\; r_1? \rightarrow q_{i,1}!; Sy]$$

$q_{0,0}$ is forked to u_0 and v_0; $q_{1,1}$ is forked to u_1 and v_1; $q_{0,1}$ is forked to n_1 and w_0; $q_{1,0}$ is forked to n_0 and w_1; $q_{2,0}$ is forked to s_0 and t_0; $q_{2,1}$ is forked to s_1 and t_1.

s_0 and u_0 are merged to a_0; s_1 and u_1 are merged to a_1; t_0 and w_0 are merged to c_0; t_1 and w_1 are merged to c_1; v_0 and v_1 are merged to c_2.

This completes the first step of our design. The second step is to implement Sy. Our approach is to sequence the r_j so that each is paired with a c_i. The implementation (Figure 1) then consists of a Sequencer, a Decision-Wait element, three Forks and one Merge. Recall the Sequencer from Example 4

$$Seq = c?; [r_0? \rightarrow g_0!; Seq \;\Box\; r_1? \rightarrow g_1!; Seq]$$

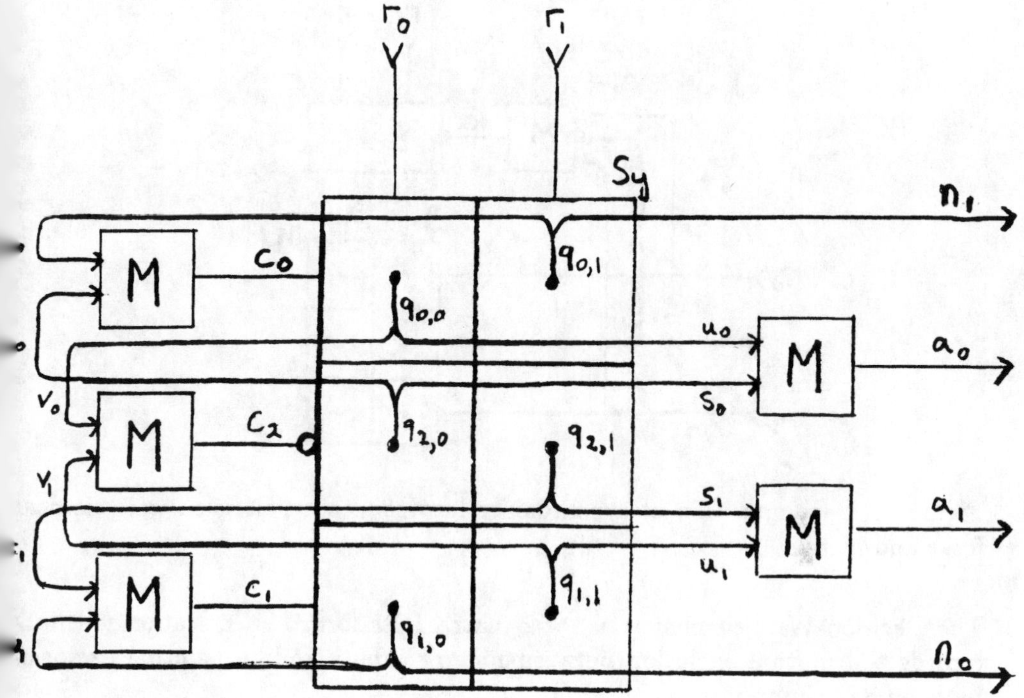

Figure 0: Decomposition of arbiter into synchronizer, forks and merges

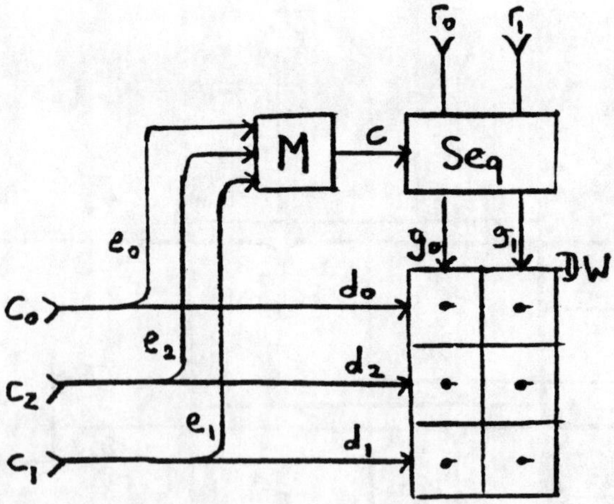

Figure 1: Decomposition of synchronizer into sequencer, decision-wait element forks and merge

The Decision-Wait element is a 3×2 matrix. Its formal specification faithfully records the mutual exclusion between its row inputs and the mutual exclusion between its column inputs.

$$\begin{aligned} DW \;=\; & [g_0? \rightarrow Col_0 \;\Box\; g_1? \rightarrow Col_1 \\ & \Box\; d_0? \rightarrow Row_0 \;\Box\; d_1? \rightarrow Row_1 \;\Box\; d_2? \rightarrow Row_2] \end{aligned}$$

where

$$\begin{aligned} Col_j \;=\; & [g_0? \rightarrow \bot \;\Box\; g_1? \rightarrow \bot \\ & \Box\; d_0? \rightarrow E_{0,j} \;\Box\; d_1? \rightarrow E_{1,j} \;\Box\; d_2? \rightarrow E_{2,j}] \end{aligned}$$

$$\begin{aligned} Row_i \;=\; & [g_0? \rightarrow E_{i,0} \;\Box\; g_1? \rightarrow E_{i,1} \\ & \Box\; d_0? \rightarrow \bot \;\Box\; d_1? \rightarrow \bot \;\Box\; d_2? \rightarrow \bot] \end{aligned}$$

$$\begin{aligned} E_{i,j} \;=\; & [g_0? \rightarrow \bot \;\Box\; g_1? \rightarrow \bot \\ & \Box\; d_0? \rightarrow \bot \;\Box\; d_1? \rightarrow \bot \;\Box\; d_2? \rightarrow \bot \\ & \Box\; skip \rightarrow q_{i,j}!; DW] \end{aligned}$$

Each c_i is forked to d_i and e_i; the three e_i's are merged to c. This completes the design.

Full four-phase protocol

This protocol is slightly more complex than the previous one. This time, after rejecting a request from a particular process, the arbiter will always reject the second request from that process, so completing a four-phase nack. The new specification is

$$A = [r_0? \rightarrow a_0!; G_0 \Box\, r_1? \rightarrow a_1!; G_1]$$

where

$$G_0 = [r_0? \rightarrow a_0!; A \Box\, r_1? \rightarrow n_1!; G_0^1]$$

$$G_1 = [r_1? \rightarrow a_1!; A \Box\, r_0? \rightarrow n_0!; G_1^0]$$

$$G_0^1 = [r_0? \rightarrow a_0!; A^1 \Box\, r_1? \rightarrow n_1!; G_0]$$

$$G_1^0 = [r_1? \rightarrow a_1!; A^0 \Box\, r_0? \rightarrow n_0!; G_1]$$

$$A^0 = [r_0? \rightarrow n_0!; A \Box\, r_1? \rightarrow a_1!; G_1^0]$$

$$A^1 = [r_1? \rightarrow n_1!; A \Box\, r_0? \rightarrow a_0!; G_0^1].$$

We can collapse these seven quiescent states into three by distinguishing between the signal on r_i in phase one and the signal on r_i in phase three in every four-phase communication. This leads us to suggest the following design.

The implementation consists of a Sequencer, two Toggles, a Decision-Wait element, eleven Forks and eight Merges.

The Sequencer is the same as before. After each input on c, an input on either r_0 or r_1 is output on g_0 or g_1, respectively.

Each Toggle alternates between outputting on s_i and t_i in response to input on g_i $(i = 0, 1)$.

The Decision-Wait element is a 4×3 matrix. It expects one input change in its row and one input change in its column. Our version responds to the two changing inputs as follows. The two combinations of inputs on s_0 and d_0 and on s_1 and d_1 are impossible and we leave the result undefined; the outputs produced by the remaining ten combinations are tabulated below.

$$\begin{array}{llll}
s_0 & d_1 & \Longrightarrow & q_0 \\
t_0 & d_1 & \Longrightarrow & q_0 \\
s_1 & d_0 & \Longrightarrow & q_1 \\
t_1 & d_0 & \Longrightarrow & q_1 \\
s_0 & d_2 & \Longrightarrow & q_2 \\
t_0 & d_0 & \Longrightarrow & q_3 \\
t_0 & d_2 & \Longrightarrow & q_4 \\
s_1 & d_2 & \Longrightarrow & q_5 \\
t_1 & d_1 & \Longrightarrow & q_6 \\
t_1 & d_2 & \Longrightarrow & q_7
\end{array}$$

Each q_j ($j = 0, \ldots, 7$) is forked into u_j and v_j; u_0 and u_4 are merged into n_0; u_1 and u_7 are merged into n_1; u_2 and u_3 are merged into a_0; u_5 and u_6 are merged into a_1; v_1 and v_2 are merged into e_0; v_0 and v_5 are merged into e_1; v_3, v_4, v_6 and v_7 are merged into e_2; each e_k ($k = 0, 1, 2$) is forked into c_k and d_k; c_0, c_1 and c_2 are merged into c.

Initially there is a signal on wire e_2.

5 Conclusion

An algebraic approach has been taken to the specification and verification of delay-insensitive circuits. It has not been necessary to express explicitly all the states that such a circuit can enter; instead the possibility of them arising can be deduced using algebraic laws. This has lead to more concise specifications and shorter proofs than would be possible using other methods. Another simplifying factor has been that, following [15], we do not distiguish between high and low-going transitions; this exposes many symmetries that would not otherwise be apparent. The main advantage of our approach is the ease with which we can compute the parallel composition of components. We have worked through many examples in which we used algebraic laws either to prove further laws or to investigate the behaviour of specified circuits.

We concluded with a non-trivial case study in which we used our algebra to assist us in the design of a non-blocking arbiter. Here the conciseness of our specifications was all important. It is very easy to see that our designs are correct, though a step-by-step algebraic proof is rather tedious and has been omitted.

That others have mechanically verified our designs by brute-force state analysis techniques is reassuring.

Acknowledgements

We are most grateful to Tony Hoare and Tom Verhoeff for their interest and encouragement. Thanks are also due to Jerry Burch and Steven Nowick for verifying our designs. The hospitality of the Department of Computer Science at Washington University helped make it possible for us to collaborate over this research. The work was partially funded by the Science and Engineering Research Council of Great Britain and the ESPRIT Basic Research Action CONCUR.

References

[1] S. D. Brookes and A. W. Roscoe. An Improved Failures Model for Communicating Sequential Processes. *Lect. Notes in Comp. Sci. 197*, 281–305, 1984.

[2] W. Chen, J. T. Udding, and T. Verhoeff. Networks of Communicating Processes and their (De)composition. In J. L. A. van de Snepscheut, editor, *The Mathematics of Program Construction*, number 375 in Lecture Notes in Computer Science, 174–196. Springer-Verlag, 1989.

[3] D. L. Dill and E. M. Clarke. Automatic Verification of Asynchronous Circuits Using Temporal Logic. In H. Fuchs, editor, *1985 Chapel Hill Conference on Very Large Scale Integration*, Computer Science Press, 127–143, 1985.

[4] D. L. Dill. *Trace Theory for Automatic Hierarchical Verification of Speed-Independent Circuits.* PhD thesis, CMU-CS-88-119, Dept. of C.S., Carnegie-Mellon Univ., 1988.

[5] J. C. Ebergen. *Translating Programs into Delay-Insensitive Circuits.* PhD thesis, Dept. of Math. and C.S., Eindhoven Univ. of Technology, 1987.

[6] M. Hennessy. *Algebraic Theory of Processes.* Series in Foundations of Computing. The MIT Press, Cambridge, Mass., 1988.

[7] C. A. R. Hoare. *Communicating Sequential Processes.* Prentice-Hall, 1985.

[8] He Jifeng, M. B. Josephs and C. A. R. Hoare. A Theory of Synchrony and Asynchrony. In *Proceedings IFIP Working Conference on Programming Concepts and Methods*, (to appear), 1990.

[9] M. B. Josephs, C. A. R. Hoare, and He Jifeng. A Theory of Asynchronous Processes. *J. ACM*, (submitted), 1989.

[10] M. B. Josephs and J. T. Udding. An Algebra for Delay-Insensitive Circuits. Technical Report WUCS-89-54, Washington University, St. Louis, and in *Proceedings DIMACS/IFIP Workshop on Computer-Aided Verification*, (to appear), 1990.

[11] A. J. Martin. Compiling Communicating Processes into Delay-Insensitive VLSI Circuits. *Distributed Computing*, 1:226–234, 1986.

[12] A. J. Martin. Programming in VLSI: From Communicating Processes to Delay-Insensitive Circuits. Caltech-CS-TR-89-1, Department of Computer Science, California Institute of Technology, 1989.

[13] A. W. Roscoe and C. A. R. Hoare. The laws of occam programming. *Theor. Comp. Sci. 60,* 2:177-229, 1988.

[14] I. E. Sutherland. Micropipelines. 1988 Turing Award Lecture. *Communications of the ACM*, 32(6):720–738, 1989.

[15] J. T. Udding. *Classification and Composition of Delay-Insensitive Circuits.* PhD thesis, Dept. of Math. and C.S., Eindhoven Univ. of Technology, 1984.

[16] J. T. Udding. A formal model for defining and classifying delay-insensitive circuits. *Distributed Computing*, 1(4):197–204, 1986.

Equivalences, Congruences, and Complete Axiomatizations for Probabilistic Processes

Chi-Chang Jou Scott A. Smolka*
Department of Computer Science
SUNY at Stony Brook
Stony Brook, NY 11794-4400
USA

We study several notions of process equivalence—viz. trace, failure, ready, and bisimulation equivalence—in the context of *probabilistic labeled transition systems.* We show that, unlike nondeterministic transition systems, "maximality" of traces and failures does not increase the distinguishing power of trace and failure equivalence, respectively. Thus, in the probabilistic case, trace and maximal trace equivalence coincide, and failure and ready equivalence coincide.

We then propose a language PCCS for communicating probabilistic processes, and present its operational semantics. We show that in PCCS, trace equivalence and failure equivalence are not congruences, whereas Larsen-Skou probabilistic bisimulation is. Furthermore, we prove that trace congruence, the largest congruence contained in trace equivalence, lies between failure equivalence and bisimulation equivalence in terms of its distinguishing strength.

Finally, we stress the similarity between classical process algebra and probabilistic process algebra by exhibiting sound and complete axiomatizations of bisimulation equivalence for finite and finite state probabilistic processes, which are natural extensions of the classical ones (R. Milner, "A complete inference system for a class of regular behaviours," *Journal of Computer and System Science,* Vol. 28, 1984). Of particular interest is the rule for eliminating unguarded recursion, which characterizes the possibility of infinite syntactic substitution as a zero-probability event.

Introduction

this paper, we investigate the impact of *probabilistic behavior* on process algebra theory. This theory dresses the issue of behavioral equivalence of processes and, within the context of some algebraic ecification language, the issues of congruence and complete axiomatization. We begin in Section 2.1 replacing nondeterministic branching in the standard notion of labeled transition system with obabilistic branching, such that the sum of the probabilities of a state's outgoing transitions is 1. The sulting model is one of *probabilistic processes* in contrast to the nondeterministic processes of classical ocess algebra theories such as CCS [Mi89], CSP [Ho85], and ACP [BK86]. In Section 2.2, we lift the finitions of a number of process equivalences from these classical theories—viz. trace [Ho85], maximal ace [BW82], failure [BHR84], maximal failure, ready [OH83], and bisimulation [Mi89] equivalence— the probabilistic case and prove the following result: unlike before, "maximality" of traces and lures does not increase the distinguishing power of trace and failure equivalence, respectively. The

*Professor Smolka's research was supported by the National Science Foundation under Grant CCR-8704309.

intuition here is that "maximality" (or the lack thereof) will be reflected in the probability distribution that a process induces on its traces, for example, and therefore need not be considered explicitly. The situation in the probabilistic case can be summarized as follows (Theorem 2.1):

$$\stackrel{T}{\equiv} = \stackrel{MT}{\equiv} \prec \stackrel{F}{\equiv} = \stackrel{MF}{\equiv} = \stackrel{R}{\equiv} \prec \stackrel{B}{\equiv}$$

where $\preceq$ is a partial order on equivalence relations such that $\equiv_i \preceq \equiv_j$ means $\equiv_j$ is a refinement of $\equiv_i$ ($\stackrel{T}{\equiv}$ denotes trace equivalence, $\stackrel{MT}{\equiv}$ denotes maximal trace equivalence, etc.).

In Section 3, we then consider PCCS [GJS89], a calculus for reasoning about communicating probabilistic processes. PCCS, a dialect of Milner's synchronous calculus SCCS [Mi83], provides us with a setting in which to study the issues of congruence and complete axiomatization. In Section 4 we show that of trace, failure, and bisimulation equivalence, only the last is a congruence in PCCS. Additionally, using a technique similar to one in [BKO88], we prove (Proposition 4.1) that failure equivalence (or equivalently, ready equivalence) is weaker than trace congruence ($\stackrel{T}{\cong}$), the largest congruence contained in trace equivalence. We also prove that trace congruence is weaker than bisimulation equivalence (Proposition 4.2). Thus, with the addition of PCCS trace congruence, we have (Theorem 4.1):

$$\stackrel{T}{\equiv} = \stackrel{MT}{\equiv} \prec \stackrel{F}{\equiv} = \stackrel{MF}{\equiv} = \stackrel{R}{\equiv} \prec \stackrel{T}{\cong} \prec \stackrel{B}{\equiv}$$

Section 5 presents sound and complete axiomatizations of probabilistic bisimulation equivalence for finite and finite-state PCCS processes, which are natural extensions of the classical ones [Mi84]. Of particular interest is the rule for eliminating unguarded recursion, which characterizes the possibility of infinite syntactic substitution as a zero-probability event. The completeness proof for the finite-state case assumes rational probabilities, and uses a new technique for proving PCCS terms equivalent based on the greatest common divisor of the probabilities of equivalent subterms.

Section 6 concludes and also highlights an interesting open problem concerning "weak probabilistic bisimulation."

Related Work

Glabbeek [Gl86, Gl90] and Bergstra et al. [BKO88] have thoroughly investigated the question of "relative distinguishing strength of equivalences" in the context of classical process algebra theory. They have neatly formulated their results as lattices of equivalence relations.

In [LS89], Larsen and Skou consider the testing of probabilistic processes, and exhibit a testing algorithm that, with probability $1 - \epsilon$, where ϵ is arbitrarily small, can distinguish processes that are not bisimilar. Bloom and Meyer [BM89] further show that if nondeterministic bounded-branching processes P and Q are bisimilar, then there is an assignment of probabilities to the edges of P and Q, yielding processes P' and Q' such that (1) P' and Q' are probabilistic bisimilar, and (2) P' and Q' have the same probability of producing a given outcome under every test.

Christoff [Ch90] also considers the testing of probabilistic processes. He proposes three probabilistic testing equivalences, and outlines an algorithm for the verification of these equivalences. Finally, Jones and Plotkin [JP89] investigate a probabilistic powerdomain of evaluations which they use to give the semantics of a language with a probabilistic parallel construct.

Equivalence of Probabilistic Processes

this section, we first define probabilistic labeled transition systems as an extension of the nondeter-inistic labeled transition system from classical process algebra theory. We then lift the definitions of : process equivalences recently proposed in the literature—viz. trace, maximal trace, failure, maxi-al failure, ready, and bisimulation equivalence—to the probabilistic case and compare their relative stinguishing power. Our comparison is in the style of [Gl86] and [BKO88] who showed that, for ndeterministic processes,

$$\overset{T}{\equiv} \prec \overset{MT}{\equiv} \prec \overset{F}{\equiv} \prec \overset{MF}{\equiv} = \overset{R}{\equiv} \prec \overset{B}{\equiv}$$

.1 Basic Definitions

efinition 2.1 A *probabilistic labeled transition system (PLTS)* is a triple $< Pr, \Sigma, \mu >$ where:

- Pr is the set of all *processes*;
- Σ is the set of all *atomic actions*, and 0 is a special symbol not in Σ called the *zero action*;
- $\mu : (Pr \times (\Sigma \cup \{0\}) \times Pr) \longrightarrow [0,1]$ is a total function called the *probabilistic transition function*, satisfying the following restriction: $\forall P \in Pr$,

$$\sum_{\substack{a \in \Sigma \cup \{0\}, \\ Q \in Pr}} \mu(P, a, Q) = 1$$

A PLTS is similar to a *probabilistic automaton* [Ra63] and to a Larsen-Skou process [LS89], in at all exhibit stochastic behavior. However, for a PLTS process, the sum of the probabilities of all itgoing transitions is 1, while for a Larsen-Skou process, the sum of the probabilities of all outgoing ansitions with the same label is 1. This difference can be viewed operationally: a Larsen-Skou process *reactive* in nature, as when confronted with a "button pushing" experiment by the observer, it either icceeds with probability 1 or else fails. A PLTS is *generative* in nature, as the PLTS itself decides hich action to execute next according to a prescribed probability distribution. This generative model more general in that, besides encoding the relative probabilities of transitions with the same label, also encodes the relative probabilities of transitions with different labels. For a more complete scussion of these models, see [GSST90].

The intended meaning of $\mu(P, a, Q) = p$ is that process P with probability p can perform the action a id then behave the same as process Q. Alternatively, we write $\mu_a(P, Q) = p$ whenever $\mu(P, a, Q) = p$, r all actions a in $\Sigma \cup \{0\}$. With $\mu_\epsilon(P, P) = 1$ and $\mu_\epsilon(P, P') = 0$, for all $P' \neq P$, as the basis, we then ductively extend the notion of μ_a to any finite action sequence $s = a_1 a_2 ... a_n \in (\Sigma \cup \{0\})^*$ as follows:

$$\mu_s(P, Q) = \sum_{Q' \in Pr} \mu_{a_1 ... a_{n-1}}(P, Q') \times \mu_{a_n}(Q', Q)$$

ere the value of $\mu_s(P, Q)$ is the sum of the probabilities of all paths labeled by s from P to Q. The :cumulated probabilities are similar to transition probabilities in probabilistic automata and Markov hains.

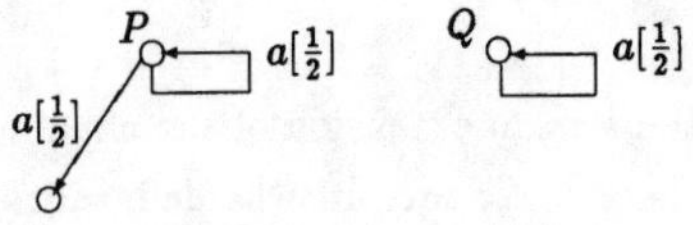

Figure 1: Effects of Probability on Equivalence

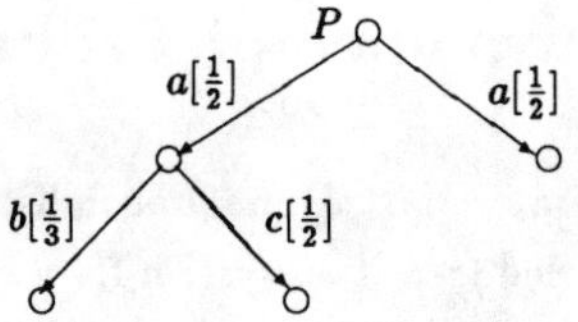

Figure 2: A Probabilistic Labeled Transition Syste

Use of the zero action is restricted to modeling termination. Let $\mathbf{0}$ be the *zero* or *terminat* *process* whose only transition is $\mu(\mathbf{0},0,\mathbf{0}) = 1$. Then, for any process P,

$$\sum_{\substack{a \in \Sigma, \\ Q \in Pr}} \mu(P,a,Q) = p \quad \text{implies} \quad \mu(P,0,\mathbf{0}) = 1-p$$

That is, if P executes non-zero actions from Σ with probability p, then it terminates with probabili $1-p$.

We now lift the classical definitions of process equivalence to probabilistic labeled transition sy tems. The basic idea is that a probabilistic process P may have multiple distinct transition path each labeled s, that lead P to the same refusal set (for failure equivalence), initial set (for ready equi alence), etc. As in computing transition probabilities, we sum the probabilities of occurrence of the transition paths.

Definition 2.2 The *trace function* $T : Pr \longrightarrow (\Sigma^* \longrightarrow [0,1])$ is given by

$$T(P)(s) = \sum_{P' \in Pr} \mu_s(P,P')$$

We say two processes P and Q are *trace equivalent* (written $P \overset{T}{\equiv} Q$) if $T(P) = T(Q)$.

Because of the possibility of zero actions, $T(P)$ is not a probability distribution on action sequenc from Σ^* of the same length. For example, for traces of length 2 in process P of Figure 1, we hav $T(P)(aa) = \frac{1}{2}$, and that is the sum of probabilities of traces of length 2.

A *maximal trace* is a sequence from Σ^* that terminates, i.e., leads to the zero process. We no define an equivalence that takes maximal traces into account. We characterize the maximal traces a probabilistic process P as a function from Σ^* to $[0,1]$, such that $MT(P)(s) = p$ if $s \cdot 0$ is a "trace of the process P with probability p.

Definition 2.3 The *maximal trace function* $MT : Pr \longrightarrow (\Sigma^* \longrightarrow [0,1])$ is given by

$$MT(P)(s) = \mu_{s \cdot 0}(P,\mathbf{0})$$

We say two processes P and Q are *maximal trace equivalent* (written $P \overset{MT}{\equiv} Q$) if $T(P) = T(Q)$ an $MT(P) = MT(Q)$.

With the addition of probability, the nature of process equivalence changes dramatically. F example, ignoring probabilities, P and Q in Figure 1 both have the trace set $\{\epsilon,\ a,\ aa,\ aaa, ...\}$, an are thus trace equivalent. We have to consider maximal trace sets to distinguish them: the maxim trace set of P is $\{a, aa, aaa, ...\}$, while that of Q is $\emptyset$. In the probabilistic case, they can be distinguishe by their trace function alone: $T(P)(aa) = \frac{1}{2}$, $T(P)(aaa) = \frac{1}{4}, ...,$ while $T(Q)(aa) = 1$, $T(Q)(aaa) =$ etc. In the next section we will take a closer look at the effects of probability on process equivalence

efinition 2.4 The *initial set* of a probabilistic process P is defined by

$$Init(P) = \{\, a \mid a \in \Sigma \text{ and } \exists P' \in Pr \text{ such that } \mu(P,a,P') > 0 \,\}$$

In the following, we make use of conditional summations of probabilities, where summands are cluded only if they satisfy some condition. We assign null summation, where no summand satisfies e condition, the value 0.

efinition 2.5 The *ready function* $R : Pr \longrightarrow ((\Sigma^* \times 2^\Sigma) \longrightarrow [0,1])$ is given by

$$R(P)((s,X)) = \sum_{Init(P')=X} \mu_s(P,P')$$

e say two processes P and Q are *ready equivalent* (written $P \overset{R}{\equiv} Q$) if $R(P) = R(Q)$.

efinition 2.6 The *failure function* $F : Pr \longrightarrow ((\Sigma^* \times 2^\Sigma) \longrightarrow [0,1])$ is given by

$$F(P)((s,X)) = \sum_{Init(P') \cap X = \emptyset} \mu_s(P,P')$$

e say two processes P and Q are *failure equivalent* (written $P \overset{F}{\equiv} Q$) if $F(P) = F(Q)$.

efinition 2.7 The *maximal failure function* $MF : Pr \longrightarrow ((\Sigma^* \times 2^\Sigma) \longrightarrow [0,1])$ is given by

$$MF(P)((s,X)) = \sum_{largest(X,P')} \mu_s(P,P'),$$

here $largest(X,P')$ means X is the largest set of actions that satisfies the condition $Init(P') \cap X = \emptyset$. e say two processes P and Q are *maximal failure equivalent* (written $P \overset{MF}{\equiv} Q$) if $MF(P) = MF(Q)$.

To illustrate the above definitions, suppose $\Sigma = \{a,b,c\}$ and consider process P of Figure 2 (0 tions are not drawn). Leaving out zero probabilities, we construct the following tables for P:

s	ϵ	a	ab	ac
$T(P)$	1	1	$\frac{1}{6}$	$\frac{1}{4}$
$MT(P)$	0	$\frac{7}{12}$	$\frac{1}{6}$	$\frac{1}{4}$

(s,X)	$(\epsilon,\{a\})$	$(a,\emptyset)$	$(a,\{b,c\})$	$(ab,\emptyset)$	$(ac,\emptyset)$
$R(P)$	1	$\frac{1}{2}$	$\frac{1}{2}$	$\frac{1}{6}$	$\frac{1}{4}$

(s,X)	$(\epsilon,\{b,c\})$	$(a,\{a,b,c\})$	$(a,\{a\})$	$(ab,\{a,b,c\})$	$(ac,\{a,b,c\})$
$MF(P)$	1	$\frac{1}{2}$	$\frac{1}{2}$	$\frac{1}{6}$	$\frac{1}{4}$

(s,X)	$(\epsilon,\{b,c\})$	$(\epsilon,\{b\})$	$(\epsilon,\{c\})$	$(\epsilon,\emptyset)$				
$F(P)$	1	1	1	1				
(s,X)	$(a,\{a,b,c\})$	$(a,\{a,b\})$	$(a,\{b,c\})$	$(a,\{a,c\})$	$(a,\{a\})$	$(a,\{b\})$	$(a,\{c\})$	$(a,\emptyset)$
$F(P)$	$\frac{1}{2}$	$\frac{1}{2}$	$\frac{1}{2}$	$\frac{1}{2}$	1	$\frac{1}{2}$	$\frac{1}{2}$	1
(s,X)	$(ab,\{a,b,c\})$	$(ab,\{a,b\})$	$(ab,\{b,c\})$	$(ab,\{a,c\})$	$(ab,\{a\})$	$(ab,\{b\})$	$(ab,\{c\})$	$(ab,\emptyset)$
$F(P)$	$\frac{1}{6}$	$\frac{1}{6}$	$\frac{1}{6}$	$\frac{1}{6}$	$\frac{1}{6}$	$\frac{1}{6}$	$\frac{1}{6}$	$\frac{1}{6}$
(s,X)	$(ac,\{a,b,c\})$	$(ac,\{a,b\})$	$(ac,\{b,c\})$	$(ac,\{a,c\})$	$(ac,\{a\})$	$(ac,\{b\})$	$(ac,\{c\})$	$(ac,\emptyset)$
$F(P)$	$\frac{1}{4}$	$\frac{1}{4}$	$\frac{1}{4}$	$\frac{1}{4}$	$\frac{1}{4}$	$\frac{1}{4}$	$\frac{1}{4}$	$\frac{1}{4}$

To define Larsen and Skou's [LS89] probabilistic bisimulation equivalence, we lift the notion of "cumulative probabilistic derivation of a process" to "cumulative probabilistic derivation of a *set* of processes." This is captured by the following probabilistic transition function:

$$\nu : (Pr \times \Sigma \times 2^{Pr}) \longrightarrow [0,1] \text{ such that } \nu(P,a,S) = \sum_{Q \in S} \mu(P,a,Q)$$

Intuitively, $\nu(P,a,S) = q$ means that process P, with total probability q, can do action a and then behave the same as one of the processes in S. We sometimes write $P \xrightarrow{a[q]} S$ to indicate that $\nu(P,a,S) = q$. Denoting the set of equivalence classes of Pr induced by an equivalence relation $\mathcal{R}$ as $Pr/\mathcal{R}$, we have

Definition 2.8 (LS89) An equivalence relation $\mathcal{R} \subseteq Pr \times Pr$ is called a *probabilistic bisimulation* if whenever $(P,Q) \in \mathcal{R}$, then

$$\forall a \in \Sigma, \forall S \in Pr/\mathcal{R}, \;\; \nu(P,a,S) = \nu(Q,a,S) \qquad (*)$$

Two processes P and Q are called *probabilistic bisimulation equivalent* (written $P \overset{B}{\equiv} Q$) if there exists a probabilistic bisimulation $\mathcal{R}$ such that (P,Q) is in $\mathcal{R}$, i.e.,

$$\overset{B}{\equiv} \; = \; \bigcup \{ \mathcal{R} \mid \mathcal{R} \text{ is a probabilistic bisimulation}\}.$$

Similar to the case of classical bisimulation, if $\mathcal{R}_1$ and $\mathcal{R}_2$ are probabilistic bisimulations, then their transitive closure $(\mathcal{R}_1 \cup \mathcal{R}_2)^*$ is again a probabilistic bisimulation. Furthermore, probabilistic bisimulation equivalence, $\overset{B}{\equiv}$, is the largest probabilistic bisimulation. Its existence can be demonstrated by defining a function $\mathcal{F}$ over the set of all equivalence relations $\mathcal{R}$, such that $\mathcal{F}(\mathcal{R}) \subseteq Pr \times Pr$ is the equivalence relation that satisfies $(*)$. $\mathcal{F}$ is monotonic, and $\mathcal{R}$ is a probabilistic bisimulation iff $\mathcal{R} \subseteq \mathcal{F}(\mathcal{R})$. We then have that $\overset{B}{\equiv}$ is the largest fix-point of $\mathcal{F}$.

2.2 Comparison of Equivalences of Probabilistic Processes

The most noteworthy result on equivalences of probabilistic processes is that "maximality" of traces and failures does not increase the distinguishing power of trace and failure equivalence, respectively. That is, trace and maximal trace equivalence coincide, and failure and maximal failure equivalence coincide. Otherwise the situation is the same as in the case of nondeterministic transition systems.

We start with trace and maximal trace equivalence. Let S_T be the set of mappings from Σ^* to $[0,1]$ that are trace functions (see Definition 2.2), and let f map Pr to S_T such that $f(P)$ is the

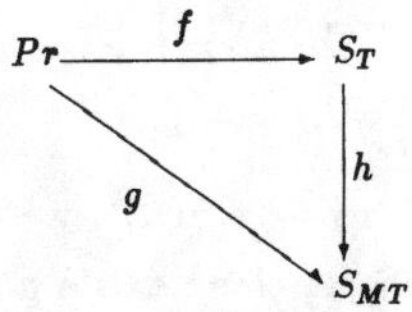

Figure 3: Diagram for Proposition 2.1

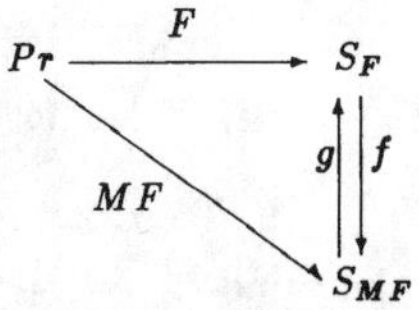

Figure 4: Diagram for Proposition 2.2

ıce function of P. Similarly, S_{MT} is the set of all mappings from Σ^* to $([0,1])^2$ such that the first ɔjection is a process's trace function and the second its maximal trace function (see Definition 2.3). lditionally, let g map Pr to S_{MT} such that $g(P)$ is the pair consisting of P's trace and maximal ıce functions. We show that there exists a function $h : S_T \rightarrow S_{MT}$, such that $g = h \circ f$. This implies at if $P \overset{T}{\equiv} Q$, then $P \overset{MT}{\equiv} Q$, i.e., $\overset{MT}{\equiv} \preceq \overset{T}{\equiv}$.

roposition 2.1 For probabilistic labeled transition systems, the diagram in Figure 3 commutes, i.e., $\circ f = g$.

oof (Sketch): We construct h as follows: For any trace s and any mapping k_T in S_T, suppose $(s) = p$, and that the sum of the probabilities of strings of the form $s \cdot \alpha$, for all α in Σ, is q $\sum k_T(s \cdot \alpha) = q$). Then let $h(k_T) = k_{MT}$ such that k_{MT} maps s to the pair $(p, p-q)$. Since p is the obability of trace s, and q is the sum of the probabilities of all non-zero actions that may occur after $p - q$ is the probability of maximal trace s. Therefore, k_{MT} is an element of S_{MT}. Additionally, for ıy process $P \in Pr$, if $f(P) = k_T$, then $g(P) = h(k_T) = h \circ f(P)$. Thus, $g = h \circ f$. □

emma 2.1 For probabilistic labeled transition systems, trace equivalence coincides with maximal ace equivalence, i.e., $\overset{T}{\equiv} = \overset{MT}{\equiv}$.

roof: $\overset{T}{\equiv} \preceq \overset{MT}{\equiv}$ is from the definition of $\overset{T}{\equiv}$ and $\overset{MT}{\equiv}$, and $\overset{MT}{\equiv} \preceq \overset{T}{\equiv}$ is from Proposition 2.1. □

Consider now failure and maximal failure equivalence. Let S_F be the subset of mappings from $\Sigma^* \times 2^\Sigma$) to $[0,1]$ that are failure functions (see Definition 2.6). Similarly, S_{MF} is the set of all aximal failure functions (see Definition 2.7). We show that there exist functions $f : S_F \rightarrow S_{MF}$ and $: S_{MF} \rightarrow S_F$, such that $MF = f \circ F$ and $F = g \circ MF$.

roposition 2.2 For probabilistic labeled transition systems, the diagram in Figure 4 commutes, i.e., $IF = f \circ F$ and $F = g \circ MF$.

roof (Sketch): To show $MF = f \circ F$, for any failure mapping h_F in S_F and any string s in Σ^*, we ɔnstruct the set $P_s = \{(X, p_X) \mid X \subseteq \Sigma, p_X \in [0,1]$, and $h_F((s,X)) = p_X\}$. The elements in P_s form complete partial order (*cpo*) based on the subset inclusion relation on the set X. Since h_F is an ement of S_F, there exists a process P such that $F(P) = h_F$. From Definition 2.6, p_X is the sum of the robabilities of traces s which reach some process P' satisfying the condition $Init(P') \cap X = \emptyset$. Note ıat if $Init(P') \cap X = \emptyset$, then $\forall X' \subseteq X, Init(P') \cap X' = \emptyset$. For each X, starting from the top of the ɔo, we first calculate the sum q_Y, for all $Y \supset X$, of the probabilities of the traces s which reach some rocesses P' with Y as the largest failure set. Then the difference between p_X and the sum of the above q_Ys is the sum of the probabilities of the traces s which reach some processes P' with X as the largest

failure set. We thus compute a mapping h_{MF} such that for all $A \subseteq \Sigma$, $h_{MF}(s,A) = MF(P)((s,A$
We then let $f(h_F) = h_{MF}$. Therefore, if $F(P) = h_F$, then $MF(P) = h_{MF} = f(h_F) = f \circ F(P)$.

To show $F = g \circ MF$, for any mapping h_{MF} in S_{MF} and any string s in Σ^*, we compute for subsets A of Σ the value $r_{s,A} = \sum_{A \subseteq X} h_{MF}(s,X)$. Since h_{MF} is in S_{MF}, there exists a process P s that $MF(P) = h_{MF}$. From Definition 2.7, $h_{MF}(s,X)$ is the sum of the probabilities of the tra s which reach some process P' with X as the largest failure set. Therefore, $r_{s,A}$ is the sum of probabilities of the traces s which reach some process P' satisfying the condition $Init(P') \cap A =$ i.e., $r_{s,A} = F(P)((s,A))$. We thus compute a mapping h_F such that for each string s and each $A \subseteq$ $h_F(s,A) = r_{s,A} = F(P)((s,A))$. We then let $g(h_{MF}) = h_F$. Therefore, if $MF(P) = h_{MF}$, th $F(P) = h_F = g(h_{MF}) = g \circ MF(P)$.

Lemma 2.2 For probabilistic labeled transition systems, failure equivalence coincides with maxin failure equivalence (or equivalently, ready equivalence), i.e., $\stackrel{F}{\equiv} \; = \; \stackrel{MF}{\equiv}$.

Proof: Directly from Proposition 2.2.

Besides the coincidence of trace and maximal trace equivalence, and failure and maximal failu equivalence, the relationships between the other equivalences are the same as in the case of nondet ministic transition systems. For example, consider ready and bisimulation equivalence.

Proposition 2.3 For probabilistic labeled transition systems, ready equivalence is strictly weak than bisimulation equivalence, i.e., $\stackrel{R}{\equiv} \prec \stackrel{B}{\equiv}$.

Proof: By extending the definition of $P \xrightarrow{a[q]} S$ from actions to action sequences, we can prove t following result by induction on the length of s:

$$P \stackrel{B}{\equiv} Q \text{ implies } \forall s \in \Sigma^* - \{\epsilon\}, \forall S \in Pr/\stackrel{B}{\equiv}, P \xrightarrow{s[p]} S \text{ iff } Q \xrightarrow{s[p]} S. \qquad ($$

To prove $\stackrel{R}{\equiv} \preceq \stackrel{B}{\equiv}$, it suffices to show that $P \stackrel{B}{\equiv} Q$ implies $\forall s \in \Sigma^*, X \subseteq \Sigma, R(P)((s,X))$ $R(Q)((s,X))$.

- $s = \epsilon$: From $P \stackrel{B}{\equiv} Q$, P and Q have the same initial set. Thus $R(P)((\epsilon, Init(P))) = R(Q)((\epsilon, Ini$ 1, and $R(P)((\epsilon,X)) = R(Q)((\epsilon,X)) = 0$, for any other set X.

- $s \in \Sigma^* - \{\epsilon\}$: From $(*)$ above, suppose $\forall S_k \in Pr/\stackrel{B}{\equiv}, P \xrightarrow{s[p_k]} S_k$ and $Q \xrightarrow{s[p_k]} S_k$. Let all process in any equivalence class S_k have initial set $I(S_k) = Init(P')$ for some P' in S_k. By summi up the probabilities p_k of those equivalence classes with the same initial set X, we have th $R(P)((s,X)) = R(Q)((s,X)) = \sum_{I(S_k)=X} p_k$.

The last example in Figure 5 shows that $\stackrel{B}{\equiv}$ is a *strict* refinement of $\stackrel{R}{\equiv}$.

The examples in Figure 5 illustrate the relationships among the probabilistic process equivalence In summary we have the following:

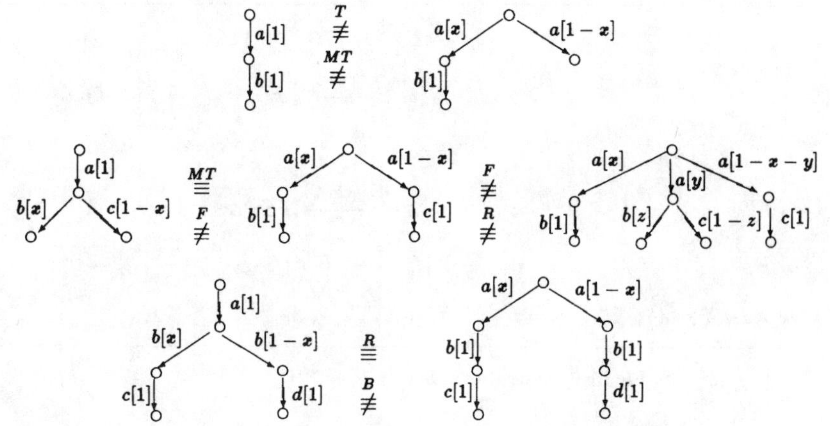

Figure 5: Comparison of Probabilistic Process Equivalences

heorem 2.1 For probabilistic labeled transition systems, maximality does not bring more distinuishing power, i.e.,

$$\overset{T}{\equiv} \; = \; \overset{MT}{\equiv} \; \prec \; \overset{F}{\equiv} \; = \; \overset{MF}{\equiv} \; = \; \overset{R}{\equiv} \; \prec \; \overset{B}{\equiv}$$

□

3 A Calculus of Communicating Probabilistic Processes

this section, we introduce the language PCCS [GJS89] and present its semantics in terms of probbilistic labeled transition systems. In Section 4, we then examine which of the equivalence relations efined in Section 2 are congruences in PCCS. Section 5 contains complete axiomatizations of probailistic bisimulation for finite and finite state PCCS terms.

As mentioned in the introduction, the syntax of PCCS is identical to that of SCCS, except for one mall but significant change: SCCS expressions of the form $\sum_{i \in I} E_i$ are now written $\sum_{i \in I} [p_i]E_i$ where p_i is he probability of behaving like expression E_i.

As in SCCS, let $(Act, \times, 1)$ be the Abelian monoid of *atomic actions.* Intuitively, actions of the rm $\alpha \times \beta$ represent the simultaneous, atomic execution by a process of the actions α and β. We will ften use juxtaposition to denote products of actions, e.g., $\alpha\beta$. The action 1 is called the "idle action" nd represents one unit of delay. It is convenient to assume that *Act* is generated freely from the union f the sets Λ of *names* (e.g., a, b, joe) and $\overline{\Lambda}$ of *co-names* (e.g., $\overline{a}, \overline{b}, \overline{sue}$) such that:

$$\forall \alpha \in \Lambda, \; \exists \, \overline{\alpha} \in \overline{\Lambda} : \alpha \times \overline{\alpha} = \overline{\alpha} \times \alpha = 1, \text{ and vice versa.}$$

Ve then have that $(Act, \times, 1, \overline{})$ is an Abelian group.

Names and co-names play a dual role in the calculus as they denote both the *particulate actions* hat processes can perform, and the *ports* where these actions can occur. The product action $\alpha \times \overline{\alpha} = 1$ epresents a *synchronized communication* between two processes at complementary ports.

Let α, β range over *Act*, A be a subset of *Act* such that $1 \in A$, and $f : Act \rightarrow Act$ be a monoid norphism. We presuppose an infinite set of *process variables* $Var = \{X_1, X_2, ...\}$. The syntax of PCCS

$$\alpha : E \xrightarrow{\alpha[1]}_1 E$$

$$\begin{array}{lcll}
E_j \xrightarrow{\alpha[q]}_k E' & \Longrightarrow & \sum\limits_{i \in I} [p_i] E_i \xrightarrow{\alpha[p_j \cdot q]}_{j.k} E' & (j \in I) \\
E \xrightarrow{\alpha[p]}_i E',\ F \xrightarrow{\beta[q]}_j F' & \Longrightarrow & E \times F \xrightarrow{\alpha\beta[p \cdot q]}_{(i,j)} E' \times F' & \\
E \xrightarrow{\alpha[p]}_i E' & \Longrightarrow & E \lceil A \xrightarrow{\alpha[p/\nu(E,A)]}_i E' \lceil A & (\alpha \in A) \\
E \xrightarrow{\alpha[p]}_i E' & \Longrightarrow & E[f] \xrightarrow{f(\alpha)[p]}_i E'[f] & \\
E_i\{fix\, \widetilde{X}\widetilde{E}/\widetilde{X}\} \xrightarrow{\alpha[p]}_k E' & \Longrightarrow & fix_i\, \widetilde{X}\widetilde{E} \xrightarrow{\alpha[p]}_k E' &
\end{array}$$

Figure 6: Operational Semantics of PCCS

is given by:

$$E ::= \mathbf{0} \mid X_i \mid \alpha : E \mid \sum_{i \in I} [p_i] E_i, \text{ where } p_i \in (0,1], \sum_{i \in I} p_i = 1 \mid E \times F \mid E \lceil A \mid E[f] \mid fix_i \widetilde{X}\widetilde{E}$$

An expression having no free variables is called a *process*, and Pr is the set of all PCCS processe Intuitively, $\mathbf{0}$ has no transitions and represents terminated or deadlocked process, while $\alpha : E$ perform action α with probability 1 and then behaves like E. A summation expression offers a probabilisti choice among its constituent behaviors, while product represents synchronized process compositio The restricted expression $E \lceil A$, may only perform actions from A, while morphism specifies a relabelin of actions. Finally the recursion expression denotes the ith entry in the solution of a system of mutuall recursive process equations.

Operationally, the p_i appearing in a summation expression define a probability distribution on th set of possible transitions of $\sum_{i \in I} [p_i] E_i$. That is, if we know that E_i can perform an action α wit probability q to become the expression E', then we can infer that $\sum_{i \in I} [p_i] E_i$ can perform the action with probability $p_i \cdot q$ to become the expression E'.

The structural operational semantics of PCCS is given in Figure 6 as a set of inference rules, in th style of Plotkin [Pl81] and Milner [Mi83, Mi89]. Of particular interest is the rule for product, whic interprets interprocess synchronization / communication as simultaneous occurrence of independen random events; and the rule for restriction, which expresses the probabilistic transitions of $E \lceil A$ i terms of *conditional probabilities* of E.

The indices appearing on the arrows are used to distinguish different *occurrences* of the *sam* probabilistic transition, and are constructed so that every probabilistic transition of an expression ha a unique index. The following example is illustrative:

$$a : \mathbf{0} \xrightarrow{a[1]}_1 \mathbf{0} \qquad ([\tfrac{1}{2}]a : \mathbf{0} + [\tfrac{1}{2}]a : \mathbf{0}) \xrightarrow{a[\frac{1}{2}]}_{1.1} \mathbf{0} \qquad ([\tfrac{1}{2}]a : \mathbf{0} + [\tfrac{1}{2}]a : \mathbf{0}) \xrightarrow{a[\frac{1}{2}]}_{2.1} \mathbf{0}$$

In the rule for restriction, the function ν computes the *normalization factor*, where $\nu(E, A)$ i the sum of the probabilities of the actions of E labeled by symbols from A. The condition $\alpha \in$ guarantees that $\nu(E, A)$ will not to be equal to 0 when the normalization is actually performed. Th formal definition of ν, using $\{\!|, |\!\}$ as multi-set brackets, is given by

$$\nu(E, A) = \sum \{\!| p_i \mid E \xrightarrow{\alpha[p_i]}_i E_i \ , \ \alpha \in A |\!\}$$

The probabilistic transitions of a restricted expression are computed in a stratified fashion: the ansitions of the (largest) innermost restriction-free subexpression are computed first. These are used compute the normalization factor and hence the transitions of the innermost restricted subexpres- on. By repeating this procedure, the transitions of an expression having arbitrary restriction depth n be computed.

A PCCS process is said to be *stochastic* if the sum of the probabilities of the transitions that n be inferred from the rules of Figure 6 is 1. Otherwise, when this sum is strictly less than 1, the ocess is said to be *substochastic*, and therefore possesses a non-zero probability of termination. For ample, $[p]a:\mathbf{0}+[1-p]\mathbf{0}$ is substochastic. Consider a substochastic process E with probability q of rmination. Then, E is also assumed to possess a zero-transition that, with probability q, leads E the zero process. Thus, the semantics of every PCCS process is a PLTS of the form given in ection 2. [1]

A more complete description of PCCS, as well as an application of the language, can be found in JS89].

Congruences in PCCS

this section, we determine whether the probabilistic equivalences defined in Section 2 are congruences ith respect to the language PCCS. As discussed above, the operational semantics of PCCS defines a obabilistic labeled transition system for each PCCS expression; therefore, the definitions of Section 2 n be used with no change. After defining trace congruence as the largest congruence contained trace equivalence, we then compare the relative distinguishing power of trace equivalence, ready uivalence (or equivalently, failure equivalence), trace congruence and bisimulation equivalence.

act 4.1 Trace equivalence and ready equivalence are not congruences in PCCS.

We exhibit trace-equivalent processes that are no longer trace-equivalent when placed in a context restriction. In particular, processes P and Q in Figure 7 are trace-equivalent since $T(P)$ and $T(Q)$ ap: $\epsilon \rightarrow 1$, $a \rightarrow 1$, $aa \rightarrow \frac{1}{3}$, $ab \rightarrow \frac{1}{3}$, $ac \rightarrow \frac{1}{3}$, and all other action sequences to 0. However, $(P\lceil\{a,c\})$ maps: $aa \rightarrow \frac{1}{2}$, $ac \rightarrow \frac{1}{2}$, while $T(Q\lceil\{a,c\})$ maps: $aa \rightarrow \frac{8}{15}$, $ac \rightarrow \frac{7}{15}$.

We exhibit ready-equivalent processes that are no longer ready-equivalent when placed in a context restriction. In particular, processes P and Q in Figure 7 are ready-equivalent, since $R(P)$ and (Q) map: $(\epsilon,\{a\}) \rightarrow 1$, $(a,\{a,b,c\}) \rightarrow 1$, $(aa,\emptyset) \rightarrow \frac{1}{3}$, $(ab,\emptyset) \rightarrow \frac{1}{3}$, $(ac,\emptyset) \rightarrow \frac{1}{3}$, and the rest 0. However, $R(P\lceil\{a,c\})$ maps: $(aa,\emptyset) \rightarrow \frac{1}{2}$, $(ac,\emptyset) \rightarrow \frac{1}{2}$, while $R(Q\lceil\{a,c\})$ maps: $(aa,\emptyset) \rightarrow$ $\frac{8}{15}$, $(ac,\emptyset) \rightarrow \frac{7}{15}$. Therefore, $P\lceil\{a,c\}$ and $Q\lceil\{a,c\}$ are not ready equivalent.

emma 4.1 Larsen and Skou's probabilistic bisimulation equivalence is a congruence in PCCS.

roof: By structural induction on the context. We present here only the cases for product and restric- on; the other cases are similar.

[1]This is not quite true as the PCCS operational semantics define indexed transitions. The indices are for "bookkeep- g" purposes only (see Section 3) and can be abstracted away by summing the probabilities of indexed transitions that therwise look the same, e.g. $[\frac{1}{3}]a:\mathbf{0}+[\frac{2}{3}]a:\mathbf{0} \xrightarrow{a[1]} \mathbf{0}$.

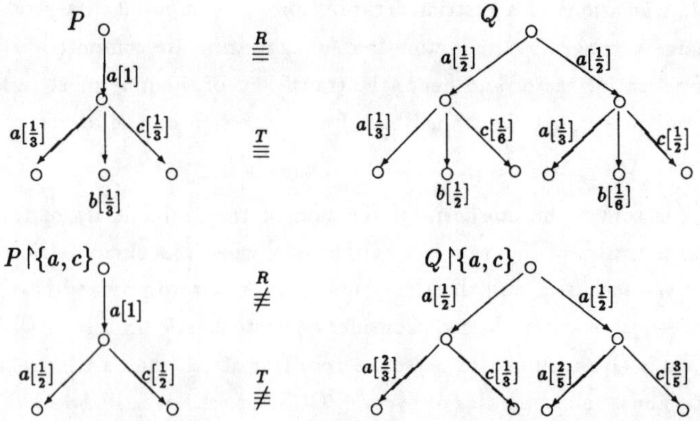

Figure 7: Trace Equivalence and Ready Equivalence are not congruences

- Product:

  We show that $P \overset{B}{\equiv} Q$ implies $P \times R \overset{B}{\equiv} Q \times R$, for any process $R \in Pr$. Let (P,Q) in a probabilis bisimulation $\mathcal{R}_0$. We show that the relation $\mathcal{R} = \{(P \times R,\ Q \times R) \mid (P,Q) \in \mathcal{R}_0, R \in Pr\} \cup Id$ is a probabilistic bisimulation. By assumption, for any action $a \in Act$ and equivalence cla $\mathcal{S} \in Pr/\mathcal{R}_0$, $P \xrightarrow{a[p]} \mathcal{S}$ iff $Q \xrightarrow{a[p]} \mathcal{S}$. Now suppose $R \xrightarrow{b[r]} R'$. Then $P \times R \xrightarrow{ab[pr]} \mathcal{S} \times R'$ $Q \times R \xrightarrow{ab[pr]} \mathcal{S} \times R'$, where $\mathcal{S} \times R'$ is defined to be $\{(P', R') \mid P' \in \mathcal{S}\}$. By definition, $\mathcal{S} \times R'$ an equivalence class of $\mathcal{R}$. Therefore, $\mathcal{R}$ is a probabilistic bisimulation.

- Restriction:

  We show that $P \overset{B}{\equiv} Q$ implies $P{\restriction}A \overset{B}{\equiv} Q{\restriction}A$, for any $A \subseteq \Sigma$. Let (P,Q) in a probabilist bisimulation $\mathcal{R}_0$. We show that the relation $\mathcal{R} = \{(P{\restriction}A,\ Q{\restriction}A) \mid (P,Q) \in \mathcal{R}_0, A \subseteq \Sigma\} \cup Id_{Pr}$ a probabilistic bisimulation. This follows from the fact that $P \overset{B}{\equiv} Q$ implies $\forall A \subseteq \Sigma : \nu(P,A)$ $\nu(Q,A)$. $\mathcal{R}$ is therefore a probabilistic bisimulation.

Trace congruence ($\overset{T}{\cong}$) is defined as the largest congruence contained in trace equivalence ($\overset{T}{\equiv}$ We now compare the distinguishing power of trace congruence, failure equivalence and bisimulati equivalence. Unlike the non-probabilistic case, where $\overset{F}{\equiv} = \overset{T}{\cong}$ [BKO88], we have the following:

Proposition 4.1 In PCCS, failure equivalence is strictly weaker than trace congruence, i.e., $\overset{F}{\equiv} \prec$

Proof: Using a technique similar to one in [BKO88], we prove $\overset{F}{\equiv} \preceq \overset{T}{\cong}$ by showing that if $P \overset{F}{\not\equiv} Q$, th there exists a context $\mathcal{C}[\,]$ such that $\mathcal{C}[P] \overset{T}{\not\equiv} \mathcal{C}[Q]$.

Let $P \overset{F}{\not\equiv} Q$. Assume without loss of generality that for some trace s and some subset X of $F(P)((s,X)) \neq F(Q)((s,X))$. Suppose s has length n, and S is the set of actions appearing in s. L N be the process consisting of n sequential 1 actions, $z \neq 1$ be a new action that does not appear P or Q, and Xz be the set $\{xz \mid x \in X\}$. Consider the context $\mathcal{C}[\,] = ([\,] \times (N : z)){\restriction}(S \cup Xz)$.

Since $F(P)((s,X))$ is the sum of probabilities of all strings s with initial set $Y \subseteq \Sigma - X$, $P \times (N : z)$, it is also the sum of probabilities of all strings s with initial set $Yz \subseteq \Sigma - (S \cup Xz)$. Thu

$P)((s,X))$ is also the sum of probabilities of all terminated strings s in $\mathcal{C}[P]$, the context that allows y actions in S or Xz. Thus, $MT(\mathcal{C}[P])(s) = F(P)((s,X))$. Similarly, $MT(\mathcal{C}[Q])(s) = F(Q)((s,X))$. erefore, $MT(\mathcal{C}[P])(s) \neq MT(\mathcal{C}[Q])(s)$, and $\mathcal{C}[P] \overset{MT}{\not\equiv} \mathcal{C}[Q]$. Since $\overset{T}{\equiv} = \overset{MT}{\equiv}$, $\mathcal{C}[P] \overset{T}{\not\equiv} \mathcal{C}[Q]$. Thus, $\overset{T}{\not\equiv} Q$.

That failure equivalence is *strictly* weaker than trace congruence is demonstrated by processes P d Q of Figure 7. They are failure equivalent, however, as shown in Fact 4.1, $P\lceil\{a,c\} \overset{T}{\not\equiv} Q\lceil\{a,c\}$. at is, P and Q are not trace congruent. □

oposition 4.2 In PCCS, trace congruence is strictly weaker than bisimulation equivalence, i.e., $\prec \overset{B}{\equiv}$.

oof: Suppose $P \overset{B}{\equiv} Q$, then for all contexts $\mathcal{C}[\,]$, $\mathcal{C}[P] \overset{B}{\equiv} \mathcal{C}[Q]$. Since $\overset{T}{\equiv} \prec \overset{B}{\equiv}$, we have that $\mathcal{C}[P] \overset{T}{\equiv} \mathcal{C}[Q]$. us, if $P \overset{B}{\equiv} Q$, then $P \overset{T}{\cong} Q$. That is, $\overset{T}{\cong} \preceq \overset{B}{\equiv}$.

That trace congruence is *strictly* weaker than bisimulation equivalence is demonstrated by the fact at the last two processes in Figure 5 are trace congruent but not bisimulation equivalent. □

eorem 4.1 In PCCS,

$$\overset{T}{\equiv} = \overset{MT}{\equiv} \prec \overset{F}{\equiv} = \overset{MF}{\equiv} = \overset{R}{\equiv} \prec \overset{T}{\cong} \prec \overset{B}{\equiv}$$

□

Complete Axiomatizations of Probabilistic Bisimulation

1 A Complete Axiom System for Finite Processes

nite probabilistic processes are defined to be the set of processes constructed without the recursion erator [Mi89]. *Finite serial* probabilistic processes are defined to be the set of finite probabilistic ocesses with only two operators: summation and prefix. We use $\sum_i [p_i]a_i : E_i$ to be the *standard form* finite serial probabilistic processes.

mma 5.1 Using axiom system $\mathcal{A}_0$ of Appendix A, any finite probabilistic process can be transmed into a probabilistically bisimilar finite serial probabilistic process.
oof: By structural induction. □

We show that the axiom system $\mathcal{A}_1$, formed by supplementing $\mathcal{A}_0$ with the following equational les, yields a sound and complete axiomatization of probabilistic bisimulation equivalence for finite ocesses. We consider binary summation only. Derived forms such as $[p]P + [q]Q + [r]R$ are to be ken as an abbreviation for one side or the other of axiom $S2$ below.

- *equivalence*:

  $E1:\ E = E$

  $E2:\ \text{if } E = F \text{ then } F = E$

  $E3:\ \text{if } E = F \text{ and } F = G \text{ then } E = G$

- *summation:*

  $S1: [p]P + [q]Q = [q]Q + [p]P$

  $S2: [p]P + [q{+}r]([\frac{q}{q+r}]Q + [\frac{r}{q+r}]R) = [p{+}q]([\frac{p}{p+q}]P + [\frac{q}{p+q}]Q) + [r]R$

  $S3: [p]P + [1-p]P = P$

Unlike SCCS, we do not have a rule of the form $[p]P + [1-p]\mathbf{0} = P$ as the left-hand side proce is more substochastic than the right-hand side process.

First, the rules are sound, i.e., if two processes can be derived from $\mathcal{A}_1$, they are probabilistical bisimilar. Since all finite probabilistic processes can be transformed into finite serial probabilist processes, to show the axiomatization complete, we have only to prove the following:

Lemma 5.2 (Completeness) For finite serial probabilistic processes P and Q, if $P \overset{B}{\equiv} Q$, then $\mathcal{A}_1 \vdash P = Q$.

Proof: By induction on the depth of P and Q.

5.2 A Complete Axiom System for Finite State Processes

A *finite state* probabilistic process is constructed using only three operators: prefix, summation an recursion [Mi89]. We now show that the axiom system $\mathcal{A}_2$, formed by supplementing $\mathcal{A}_1$ with th following equational rules, is a sound and complete axiomatization of probabilistic bisimulation fo finite state probabilistic processes.

- *congruence:*

  $C1:$ if $\widetilde{E} = \widetilde{E}'$ then $F\{\widetilde{E}/\widetilde{X}\} = F\{\widetilde{E}'/\widetilde{X}\}$

- *recursion:*

  $R1: fixXE = fixY(E\{Y/X\})$, where Y is not free in $fixXE$ or in E

  $R2: fixXE = E\{fixXE/X\}$

  $R3: fixX([p]E + [1-p]X) = fixXE$

  $R4:$ if $E = F\{E/X\}$ and X is guarded in F then $E = fixXF$

It is easy to show that the above rules respect probabilistic bisimulation equivalence, i.e. they ar sound. The rest of this section shows that they are complete for the case of finite state probabilisti processes with rational probabilities. Basically, the proof follows the line of Milner's in [Mi84]; w will note the instances where probabilities play a prominent role. In the following, we use "$\vdash$" as a abbreviation for "$\mathcal{A}_2\vdash$".

Theorem 5.1 (Completeness) Let E and F be finite state PCCS expressions having rational prob abilities only. If E and F have no free variables and $E \overset{B}{\equiv} F$ then $\vdash E = F$.

Proof (Sketch): Following [Mi84], we can show that: (1) for any expression Exp, there exist mutuall recursive expressions $Exp_1, ..., Exp_n$ such that each Exp_i can be proven equal to an expression in stan dard summation form, possibly with some unguarded terms; and Exp can be proven to be equivalen

Exp_1. (2) if two sets of expressions satisfy the same set of recursive equations, then they can be oven equal.

From (1), we can assume that there are two sets of mutually recursive expressions without free riables $\{E_1, ..., E_n\}$ and $\{F_1, ..., F_m\}$ such that $\vdash E = E_1$ and $\vdash F = F_1$. Our goal is to find a single t of recursive equations that has both $\{E_1, ..., E_n\}$ and $\{F_1, ..., F_m\}$ as solutions. Then from (2), $E_1 = F_1$.

From $E \overset{B}{\equiv} F$, $\vdash E = E_1$, $\vdash F = F_1$ and the soundness of the axiom system $\mathcal{A}_2$, we have $E_1 \overset{B}{\equiv} F_1$. rom $E_1 \overset{B}{\equiv} F_1$, we can build a relation $\mathcal{R} \subseteq \{E_1, ..., E_n\} \times \{F_1, ..., F_m\}$ in which each E_i will be paired ith some F_j such that $E_i \overset{B}{\equiv} F_j$, and vice versa. For each pair (E_i, F_j) in $\mathcal{R}$, we can build a relation $_{ij}$ on their subterms such that $([p_{ik}]a_{ik} : E_{ik}, [q_{jl}]b_{jl} : F_{jl}) \in \mathcal{R}_{ij}$ iff $a_{ik} = b_{jl}$ and $(E_{ik}, F_{jl}) \in \mathcal{R}$. From $_{ij}$, we can split matched subterms of E_i and F_j into multiple subterms with equal probabilities based 1 the greatest common divisor of their probabilities. Thus, for each pair $(E_i, F_j) \in \mathcal{R}$ we can find 1 equation that both E_i and F_j satisfy. $\{E_1, ..., E_n\}$ and $\{F_1, ..., F_m\}$ satisfy the system of recursive quations built this way.

Therefore, from $\vdash E = E_1$, $\vdash F = F_1$ and $\vdash E_1 = F_1$, we can infer $\vdash E = F$. □

Conclusions

Ve have systematically investigated the issues of equivalence, congruence, and complete axiomatization or probabilistic processes. In particular, we augmented labeled transition systems with probabilities nd lifted the definitions of six well-known process equivalences appropriately. We showed that, unlike eterministic transition systems, maximality of traces and failures does not increase the distinguishing ower of trace and failure equivalence, respectively. We have also shown that in the calculus PCCS, race equivalence and ready equivalence are not congruences, whereas Larsen-Skou probabilistic bisimlation is. Finally, we presented sound and complete axiomatizations of probabilistic bisimulation for nite and finite state PCCS processes.

Currently, we are developing the notion of a metric space for PCCS processes. This will permit a hift in attention from equivalent processes to probabilistically similar processes, which may be a more easonable expectation of processes in practice.

We close with an interesting open problem. The notion of probabilistic bisimulation considered in his paper is "strong" in the sense that the idle action 1, which is observable only if one can measure lelay, is not distinguished from the other actions in *Act*. The situation is analogous to the one in CCS. There a notion of weak bisimulation equivalence has been defined [Mi89] which largely abstracts away rom the effect of the unobservable action τ. Thus, it is natural to define a notion of *weak probabilistic isimulation* in PCCS. The issues of equivalence, congruence, and complete axiomatization remain pen in this context.

Acknowledgements: The authors are grateful to Rob van Glabbeek, Kim Larsen, Robin Milner, Bernhard Steffen, and Chris Tofts for valuable discussions on probabilistic processes.

Appendix A Axiom System $\mathcal{A}_0$

All PCCS processes can be transformed into the standard form $\sum_i [p_i]\alpha_i : P_i$ through the equation system $\mathcal{A}_0$ given by:

Axiom system $\mathcal{A}_0$:

1. $\sum_i [p_i](\sum_j [q_j]\alpha_{ij} : P_{ij}) = \sum_{ij} [p_i \cdot q_j]\alpha_{ij} : P_{ij}$
2. $\sum_i [p_i]\alpha_i : P_i \times \sum_j [q_j]\beta_j : Q_j = \sum_{ij} [p_i \cdot q_j]\alpha_i\beta_j : (P_i \times Q_j)$
3. $(\sum_i [p_i]\alpha_i : P_i) \upharpoonright A = \sum_j [\frac{p_j}{r}]\alpha_j : (P_j \upharpoonright A), \quad \alpha_j \in A$ and $r = \sum_{\alpha_j \in A} p_j$.
4. $(\sum_i [p_i]\alpha_i : P_i)[f] = \sum_i [p_i]f(\alpha_i) : (P_i[f])$
5. $fix\ X\ (\sum_i [p_i]\alpha_i : E_i) = \sum_i [p_i]\alpha_i : (fix\ X\ E_i)$

References

[BHR84] S. Brookes, C.A.R. Hoare, A. Roscoe, "A theory of communicating sequential processes" *Journal of ACM*, 31:3, 560-599 (1984).

[BK86] J.A. Bergstra, J.W. Klop, "Algebra of communicating processes," in CWI Monographs, *Proceedings of the CWI Symposium on Mathematics and Computer Science* (eds. J.W. de Bakker, M Hazewinkel and J.K. Lenstra), 89-138, North Holland, Amsterdam (1986).

[BKO88] J.A. Bergstra, J.W. Klop, E.R. Olderog, "Readies and failures in the algebra of communicating processes," *SIAM Journal of Computing,* Vol. 17, 1134-1177 (1988).

[BM89] B. Bloom, A.R. Meyer, "A remark on bisimulation between probabilistic processes," *Logik a Botik,* LNCS 363, eds. Meyer & Tsailin, Springer-Verlag (1989).

[BW82] M. Broy, M. Wirsing, "On the algebraic specification of finitary infinite communicating sequential processes," *Working Conference on Formal Description of Programming Concept II,* (ed D. Björner), Germisch, June 1982, 171-196, North-Holland, Amsterdam (1982).

[Ch90] Ivan Christoff, "Testing equivalences and fully abstract models for probabilistic," this volume.

[JP89] C. Jones, G. D. Plotkin, "A Probabilistic powerdomain of evaluations", *Proceedings of 4th Annual Symposium on Logic in Computer Science*, 186-195 (1989).

[Gl86] R.J. van Glabbeek, "Notes on the methodology of CCS and CSP," *Technical Report CS-R8624*, Centre for Mathematics and Computer Science, The Netherlands.

[Gl90] R.J. van Glabbeek, "The linear time - branching time spectrum," this volume.

[GJS89] A. Giacalone, C. Jou, S. A. Smolka, "Algebraic reasoning for probabilistic concurrent systems", *Proceedings of Working Conference on Programming Concepts and Methods*, IFIP TC 2, Sea of Gallilee, Israel, April 1990.

SST90] R.J. van Glabbeek, S. A. Smolka, B. Steffen, C. Tofts, "Reactive, generative, and stratified models of probabilistic processes," *IEEE Symp. on Logic in Computer Science,* Philadelphia, PA., USA, June 1990.

to85] C.A.R. Hoare, *Communicating Sequential Processes,* Prentice-Hall International (1985).

S89] K.G. Larsen, A. Skou, "Bisimulation through probabilistic testing," *Proceedings of 16th ACM Symp. on Principles of Programming Languages,* Austin, TX (1989).

Ii83] R. Milner, "Calculi for synchrony and asynchrony," *Theoret. Comput. Science,* Vol. 25, 267-310 (1983).

Ii84] R. Milner, "A complete inference system for a class of regular behaviours," *Journal of Computer and System Science,* Vol. 28, 439-466 (1984).

Ii89] R. Milner, *Communication and Concurrency,* Prentice Hall International Series in Computer Science, United Kingdom (1989).

)H83] E. R. Olderog and C.A.R. Hoare, "Specification-oriented semantics for communicating processes," *Proceedings of 10th ICALP*, Barcelona, ed. J. Diaz, LNCS 154, 561-572, Springer-Verlag (1983).

'a81] D. Park, "Concurrency and automata on infinite sequences", *Proc. 5th GI Conference*, LNCS 104, Springer-Verlag (1981).

'l81] G. D. Plotkin, "A structural approach to operational semantics," *Technical Report DAIMI FN-19,* Computer Science Department, Aarhus University (1981).

ta63] M.O. Rabin, "Probabilistic automata," *Information and Control,* Vol. 6, 230-245 (1963).

Rewriting as a Unified Model of Concurrency*

José Meseguer
SRI International, Menlo Park, CA 94025, and
Center for the Study of Language and Information
Stanford University, Stanford, CA 94305

1 Introduction

The main goal of this paper is to propose a general and precise answer to the question:

What is a concurrent system?

It seems fair to say that this question has not yet received a satisfactory answer, and that th resulting situation is one of *conceptual fragmentation* within the field of concurrency. A relate problem is the *integration of concurrent programming with other programming paradigms*, suc as functional and object-oriented programming. Integration attempts typically graft an existin concurrency model on top of an existing language, but such *ad hoc* combinations often lead t monstrous deformities which are extremely difficult to understand. Instead, this paper propose a *semantic integration* of those paradigms based on a common *logic and model theory*.

The logic, called *rewriting logic*, is implicit in term rewriting systems but has passed fo the most part unnoticed due to our overwhelming tendency to associate term rewriting wit equational logic. Its proof theory exactly corresponds to (truly) concurrent computation, an the model theory proposed for it in this paper provides the general concept of concurrent systen that we are seeking.

This paper also proposes rewrite rules as a very high level language to program concurren systems. Specifically, a language design based on rewriting logic is presented containing a *func tional sublanguage* entirely similar to OBJ3 [10] as well as more general *system modules*, an also *object-oriented modules* that provide notational convenience for object-oriented application but are reducible to system modules [24]. The language's semantics is directly based on th model theory of rewriting logic and yields the desired semantic integration of concurrency wit functional and object-oriented programming.

The resulting notion of concurrent system is indeed very general and specializes to a wid variety of existing notions in a very natural way. Section 5 discusses the specializations to labelle transition systems, Petri nets and concurrent object-oriented programming in some detail, an summarizes several others; however, space limitations preclude a more comprehensive discussion for which we refer the reader to [25].

Acknowledgements. I specially thank Prof. Joseph Goguen for our long term collaboratio on the OBJ and FOOPS languages [10, 11], concurrent rewriting [15] and its implementation o the RRM architecture [9, 13], all of which have directly influenced this work; he has also pro vided many positive suggestions for improving a previous version of this paper. I specially thank Prof. Ugo Montanari for our collaboration on the semantics of Petri nets [26, 27]; the algebrai ideas that we developed in that work have been a source of inspiration for the more genera ideas presented here. Mr. Narciso Martí-Oliet deserves special thanks for our collaboration o the semantics of linear logic and its relationship to Petri nets [22, 21], which is another source o

*Supported by Office of Naval Research Contracts N00014-90-C-0086, N00014-88-C-0618 and N00014-86-C 0450, and NSF Grant CCR-8707155.

spiration for this work; he also provided very many helpful comments and suggestions for im-roving the exposition. I also thank all my fellow members of the OBJ and RRM teams, past and resent, and in particular Dr. Claude Kirchner, Dr. Sany Leinwand, and Mr. Timothy Winkler, ho deserves special thanks for his many very good comments about the technical content as ell as for his kind assistance with the pictures. I also wish to thank Prof. Pierre-Louis Curien, rof. Pierpaolo Degano, Prof. Brian Mayoh, Prof. Robin Milner, Dr. Mark Moriconi, Prof. Peter losses and Dr. Axel Poigné, all of whom provided helpful comments and encouragement.

Concurrent Rewriting

he idea of concurrent rewriting is very simple. It is the idea of *equational simplification* that e are all familiar with from our secondary school days, *plus* the obvious remark that we can do any of those simplifications independently, i.e., in *parallel.* Consider for example the following AT module written in a notation similar to that of OBJ:

```
fmod NAT is
  sort Nat .
  op 0 : -> Nat .
  op s_ : Nat -> Nat .
  op _+_ : Nat Nat -> Nat [comm] .
  vars N M : Nat .
  eq N + 0 = N .
  eq (s N) + (s M) = s s (N + M) .
endfm
```

```
mod NAT-CHOICE is
  extending NAT .
  op _?_ : Nat Nat -> Nat .
  vars N M : Nat .
  rl N ? M => N .
  rl N ? M => M .
endm
```

AT defines the Peano natural numbers. It begins with the keyword `fmod` followed by the mod-le's name, and ends with the keyword `endfm`. The sort `Nat` is declared using the keyword ort. Each of the functions provided by the module, as well as the sorts of their arguments and he sort of their result, is introduced using the keyword op. The syntax is user-definable, and ermits specifying function symbols in "prefix," (for example, `s_`), "infix" (`_+_`) or any "mixfix" ombination as well as standard parenthesized notation. Variables to be used for defining equa-ions are declared with their corresponding sorts, and then equations are given (in this example, quations for addition); such equations provide the actual "code" of the module.

To compute with this module, one performs equational simplification by using the equations rom left to right until no more simplifications are possible. Note that this can be done *con-urrently*, i.e., applying several equations at once, as in the example of Figure 1, in which the laces where the equations have been matched at each step are marked. Notice that the function ymbol `_+_` was declared to be commutative by the attribute[1] `[comm]`. This not only asserts that he equation `N + M = M + N` is satisfied in the intended semantics, but it also means that when oing simplification we are allowed to apply the rules for addition not just to *terms* —in a purely yntactic way— but to *equivalence classes* of terms *modulo* the commutativity equation. In the xample of Figure 1, the equation `N + 0 = N` is applied (modulo commutativity) with `0` both n the right *and* on the left.

The above module has equations that are Church-Rosser and terminating and is therefore *unctional*; its mathematical semantics is given by the *initial algebra*[2] associated to the syntax nd equations in the module [14], i.e., associated to the *equational theory* that the module epresents. Up to now, most work on term rewriting has dealt with that case. However, the true ossibilities of the concurrent rewriting model are by no means restricted to this case. Indeed,

[1]In OBJ it is possible to declare several attributes of this kind for an operator, including also associativity and dentity, and then do rewriting modulo such attributes.

[2]In the above example, the initial algebra is of course the natural numbers with successor and addition.

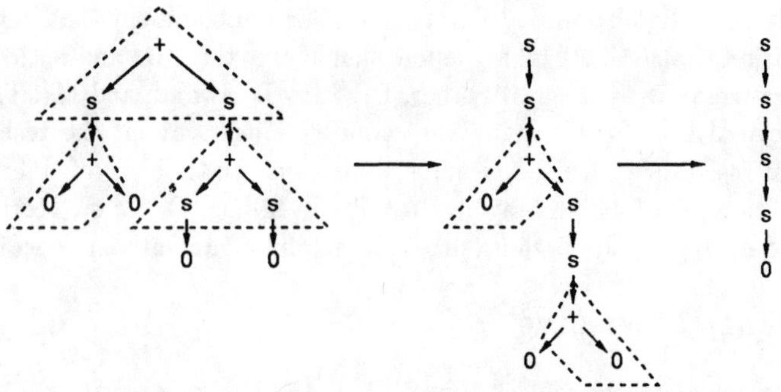

Figure 1: Concurrent rewriting of an arithmetic expression

most of the applications discussed in this paper, including many well known concurrency models are not functional. Consider the NAT-CHOICE module above, which adds a nondeterministic choice operator to the natural numbers. The intuitive *operational behavior* of this module is quite clear. Natural number addition remains unchanged and is computed using the two rules in the NAT module. Notice that any occurrence of the choice operator in an expression can be eliminated by choosing either of the arguments. In the end, we can reduce any ground expression to a natural number in Peano notation. The *mathematical semantics* of the module is much less clear. If we adopt an initial algebra semantics, it follows by the rules of equational deduction with the two equations in NAT-CHOICE that `N = M`, i.e., everything collapses to one point. Therefore the declaration `extending NAT`, whose meaning is that two distinct natural numbers are not identified by the new equations introduced in NAT-CHOICE, is violated in the worse possible way by this semantics; yet, the operational behavior in fact respects such a declaration. To indicate that this is not the semantics intended, I have used the keyword `mod` (instead of the previous `fmod`) to indicate that this module *is not functional*. Similarly, I have written a new keyword `rl` —instead of the usual `eq` before each equation— and replaced the equal sign by the new sign "`=>`" to suggest that `rl` declarations must be understood as "rules" and not as equations in the usual sense. At the operational level the equations introduced by the keyword `eq` in a functional module are also implemented as rewrite rules; the difference however lies in the *mathematical semantics* given to the module, which for modules like the one above should *not* be the initial algebra semantics. My proposal is to seek a logic and a model theory that are the perfect match for this problem. For this solution to be in harmony with the old one, the new logic and the new model theory should *generalize* the old ones.

2.1 Basic Universal Algebra

For the sake of making the exposition simpler, I treat the *unsorted* case; the many-sorted and order-sorted cases can be given a similar treatment. Therefore, a set Σ of function symbols is a ranked alphabet $\Sigma = \{\Sigma_n \mid n \in \mathbb{N}\}$. A Σ-algebra is then a set A together with an assignment of a function $f_A : A^n \longrightarrow A$ for each $f \in \Sigma_n$ with $n \in \mathbb{N}$. I denote by T_Σ the Σ-algebra of ground Σ-terms, and by $T_\Sigma(X)$ the Σ-algebra of Σ-terms with variables in a set X. Similarly, given a set E of Σ-equations, $T_{\Sigma,E}$ denotes the Σ-algebra of equivalence classes of ground Σ-terms modulo the equations E (i.e., modulo provable equality using the equations E); in the same way, $T_{\Sigma,E}(X)$ denotes the Σ-algebra of equivalence classes of Σ-terms with variables in X modulo the equations E. We let $t =_E t'$ denote the congruence modulo E of two terms t, t', and $[t]_E$ or just $[t]$ denote the E-equivalence class of t.

Given a term $t \in T_\Sigma(\{x_1, \ldots, x_n\})$, and a sequence of terms $u_1, \ldots, u_n$, $t(u_1/x_1, \ldots, u_n/x_n)$

notes the term obtained from t by *simultaneously substituting* u_i for x_i, $i = 1, \ldots, n$. To nplify notation, I denote a sequence of objects $a_1, \ldots, a_n$ by $\overline{a}$, or, to emphasize the length of e sequence, by $\overline{a}^n$. With this notation, $t(u_1/x_1, \ldots, u_n/x_n)$ is abbreviated to $t(\overline{u}/\overline{x})$.

2 Rewriting Logic

'e are now ready to introduce the new logic that we are seeking, which I call *rewriting logic*. *signature* in this logic is a pair (Σ, E) with Σ a ranked alphabet of function symbols and E a t of Σ-equations. Rewriting will operate on equivalence classes of terms modulo a given set of uations E. In this way, we free rewriting from the syntactic constraints of a term representation ıd gain a much greater flexibility in deciding what counts as a *data structure*; for example, string writing is obtained by imposing an associativity axiom, and multiset rewriting by imposing sociativity and commutativity. Of course, standard term rewriting is obtained as the particular se in which the set E of equations is empty. The idea of rewriting in equivalence classes is well ıown (see, e.g., [17, 5].)

Given a signature (Σ, E), the *sentences* are sequents of the form $[t]_E \longrightarrow [t']_E$ with t, t' -terms, where t and t' may possibly involve some variables from the countably infinite set $= \{x_1, \ldots, x_n, \ldots\}$. A *theory* in this logic, called a rewrite theory, is a slight generalization of e usual notion of theory —which is typically defined as a pair consisting of a signature and a t of sentences for it— in that, in addition, we allow rules to be labelled. This is very natural r many applications, and customary for automata —viewed as labelled transition systems— ıd for Petri nets, which are both particular instances of our definition (see Section 5.)

efinition 1 A (*labelled*) *rewrite theory*[3] $\mathcal{R}$ is a 4-tuple $\mathcal{R} = (\Sigma, E, L, R)$ where Σ is a ranked phabet of function symbols, E is a set of Σ-equations, L is a set called the set of *labels*, and R is set of pairs $R \subseteq L \times (T_{\Sigma,E}(X)^2)$ whose first component is a label and whose second component a pair of E-equivalence classes of terms, with $X = \{x_1, \ldots, x_n, \ldots\}$ a countably infinite set variables. Elements of R are called *rewrite rules*[4]. We understand a rule $(r, ([t], [t']))$ as a belled sequent and use for it the notation $r : [t] \longrightarrow [t']$. To indicate that $\{x_1, \ldots, x_n\}$ is the t of variables occurring in either t or t', we write[5] $r : [t(x_1, \ldots, x_n)] \longrightarrow [t'(x_1, \ldots, x_n)]$, or in ›breviated notation $r : [t(\overline{x}^n)] \longrightarrow [t'(\overline{x}^n)]$. □

Given a rewrite theory $\mathcal{R}$, we say that $\mathcal{R}$ *entails* a sequent $[t] \longrightarrow [t']$ and write $\vdash [t] \longrightarrow [t']$ if and only if $[t] \longrightarrow [t']$ can be obtained by finite application of the following *rules of deduction*:

1. **Reflexivity**. For each $[t] \in T_{\Sigma,E}(X)$,

$$\overline{[t] \longrightarrow [t]}$$

[3] I consciously depart from the standard terminology, that would call $\mathcal{R}$ a *rewrite system*. The reason for this parture is very specific. I want to keep the term "rewrite system" for the *models* of such a theory, which will be fined in Section 3 and which really are systems with a dynamic behavior. Strictly speaking, $\mathcal{R}$ is not a system; is only a static, linguistic, *presentation* of a class of systems —including the initial and free systems that most rectly formalize our dynamic intuitions about rewriting.

[4] To simplify the exposition, in this paper I consider only *unconditional* rewrite rules. However, all the results esented here have been extended to conditional rules in [25] with very general rules of the form

$$r : [t] \longrightarrow [t'] \text{ if } [u_1] \longrightarrow [v_1] \wedge \ldots \wedge [u_k] \longrightarrow [v_k].$$

his of course increases considerably the expressive power of rewrite theories.

[5] Note that, in general, the set $\{x_1, \ldots, x_n\}$ will depend on the representatives t and t' chosen; therefore, we low any possible such qualification with explicit variables.

2. **Congruence.** For each $f \in \Sigma_n$, $n \in \mathbb{N}$,

$$\frac{[t_1] \longrightarrow [t'_1] \quad \dots \quad [t_n] \longrightarrow [t'_n]}{[f(t_1, \dots, t_n)] \longrightarrow [f(t'_1, \dots, t'_n)]}$$

3. **Replacement.** For each rewrite rule $r : [t(x_1, \dots, x_n)] \longrightarrow [t'(x_1, \dots, x_n)]$ in R,

$$\frac{[w_1] \longrightarrow [w'_1] \quad \dots \quad [w_n] \longrightarrow [w'_n]}{[t(\overline{w}/\overline{x})] \longrightarrow [t'(\overline{w'}/\overline{x})]}$$

4. **Transitivity.**

$$\frac{[t_1] \longrightarrow [t_2] \qquad [t_2] \longrightarrow [t_3]}{[t_1] \longrightarrow [t_3]}$$

Equational logic (modulo a set of axioms E) is obtained from rewriting logic by adding the following rule:

5. **Symmetry.**

$$\frac{[t_1] \longrightarrow [t_2]}{[t_2] \longrightarrow [t_1]}$$

With this new rule, sequents derivable in equational logic are *bidirectional*; therefore, in this case we can adopt the notation $[t] \leftrightarrow [t']$ throughout and call such bidirectional sequents *equations*.

In rewriting logic a sequent $[t] \longrightarrow [t']$ should not be read as "$[t]$ *equals* $[t']$," but as "$[t]$ *becomes* $[t']$." Therefore, rewriting logic is a logic of *becoming* or *change*, not a logic of equality in a static Platonic sense. Adding the symmetry rule is a *very strong* restriction, namely assuming that *all change is reversible*, thus bringing us into a timeless Platonic realm in which "before" and "after" have been identified. A related observation is that $[t]$ should not be understood as a *term* in the usual first-order logic sense, but as a *proposition* —built up using the *logical connectives* in Σ— that asserts being in a certain *state* having a certain *structure*. The rules of rewriting logic are therefore rules to reason about *change in a concurrent system*. They allow us to draw valid conclusions about the evolution of the system from certain basic types of change known to be possible thanks to the rules R.

2.3 Concurrent Rewriting

A nice consequence of having defined rewriting logic is that concurrent rewriting, rather than emerging as an operational notion, actually *coincides* with deduction in such a logic.

Definition 2 Given a rewrite theory $\mathcal{R} = (\Sigma, E, L, R)$, a (Σ, E)-sequent $[t] \longrightarrow [t']$ is called:

- a *0-step concurrent $\mathcal{R}$-rewrite* iff it can be derived from $\mathcal{R}$ by finite application of the rules 1 and 2 of rewriting deduction (in which case $[t]$ and $[t]'$ necessarily coincide);

- a *one-step concurrent $\mathcal{R}$-rewrite* iff it can be derived from $\mathcal{R}$ by finite application of the rules 1-3, with at least one application of rule 3; if rule 3 was applied exactly once, we then say that the sequent is a one-step *sequential* $\mathcal{R}$-rewrite;

- a *concurrent $\mathcal{R}$-rewrite* (or just a *rewrite*) iff it can be derived from $\mathcal{R}$ by finite application of the rules 1-4.

We call the rewrite theory $\mathcal{R}$ *sequential* if all one-step $\mathcal{R}$-rewrites are necessarily sequential. A sequential rewrite theory $\mathcal{R}$ is in addition called *deterministic* if for each $[t]$ there is at most one one-step (necessarily sequential) rewrite $[t] \longrightarrow [t']$. The notions of sequential and deterministic rewrite theory can be made relative to a given subset $S \subseteq T_{\Sigma,E}(X)$ by requiring that the corresponding property holds for each $[t']$ "reachable from S," i.e., for each $[t']$ such that for some $[t] \in S$ there is a concurrent $\mathcal{R}$-rewrite $[t] \longrightarrow [t']$. □

The usual notions of confluence, termination, normal form, etc., as well as the well known urch-Rosser property of confluent rules remain unchanged when considered from the perective of concurrent rewriting [25]. Indeed, concurrent rewriting is a more convenient way of nsidering such notions than the traditional way using sequential rewriting.

Semantics

such, a rewrite theory $\mathcal{R} = (\Sigma, E, L, R)$ is a *static description* of what a system can do. The eaning should be given by a model of its actual *behavior.* I construct below a model in which havior exactly corresponds to deduction.

Given a rewrite theory $\mathcal{R} = (\Sigma, E, L, R)$, the model that we are seeking is a category $\mathcal{T}_{\mathcal{R}}(X)$ ose objects are the equivalence classes of terms $[t] \in T_{\Sigma,E}(X)$ and whose morphisms are equivance classes of terms representing proofs in rewriting deduction, i.e., concurrent $\mathcal{R}$-rewrites. The les for generating such terms, with the specification of their respective domain and codomain, e given below. Note that in the rest of this paper I always use "diagrammatic" notation for orphism composition, i.e., $\alpha;\beta$ always means the composition of α *followed by* β.

1. **Identities.** For each $[t] \in T_{\Sigma,E}(X)$,

$$\frac{}{[t] : [t] \longrightarrow [t]}$$

2. **Σ-structure.** For each $f \in \Sigma_n$, $n \in \mathbb{N}$,

$$\frac{\alpha_1 : [t_1] \longrightarrow [t'_1] \quad \ldots \quad \alpha_n : [t_n] \longrightarrow [t'_n]}{f(\alpha_1, \ldots, \alpha_n) : [f(t_1, \ldots, t_n)] \longrightarrow [f(t'_1, \ldots, t'_n)]}$$

3. **Replacement.** For each rewrite rule $r : [t(\overline{x}^n)] \longrightarrow [t'(\overline{x}^n)]$ in R,

$$\frac{\alpha_1 : [w_1] \longrightarrow [w'_1] \quad \ldots \quad \alpha_n : [w_n] \longrightarrow [w'_n]}{r(\alpha_1, \ldots, \alpha_n) : [t(\overline{w}/\overline{x})] \longrightarrow [t'(\overline{w'}/\overline{x})]}$$

4. **Composition.**

$$\frac{\alpha : [t_1] \longrightarrow [t_2] \qquad \beta : [t_2] \longrightarrow [t_3]}{\alpha;\beta : [t_1] \longrightarrow [t_3]}$$

For the case of equational logic we can also define a similar model as a category $\mathcal{T}_{\mathcal{R}}^{\leftrightarrow}(X)$ ctually a *groupoid*[6]) by using the rule of symmetry to generate additional terms:

5. **Inversion.**

$$\frac{\alpha : [t_1] \longrightarrow [t_2]}{\alpha^{-1} : [t_2] \longrightarrow [t_1]}$$

onvention and Warning. In the case when the same label r appears in two different rules R, the "proof terms" $r(\overline{\alpha})$ can sometimes be *ambiguous*. I will always assume that such nbiguity problems *have been resolved* by disambiguating the label r in the proof terms $r(\overline{\alpha})$ if ecessary. With this understanding, I adopt the simpler notation $r(\overline{\alpha})$ to ease the exposition.

Each of the above rules of generation defines a different operation taking certain proof terms arguments and returning a resulting proof term. In other words, proof terms form an algebraic ructure $\mathcal{P}_{\mathcal{R}}(X)$ consisting of a graph with identity arrows and with operations f (for each $\in \Sigma$), r (for each rewrite rule), and $\_;\_$ (for composing arrows). Our desired model $\mathcal{T}_{\mathcal{R}}(X)$ is e quotient of $\mathcal{P}_{\mathcal{R}}(X)$ modulo the following equations[7]:

[6] A category $\mathcal{C}$ is called a *groupoid* iff any morphism $f : A \longrightarrow B$ in $\mathcal{C}$ has an inverse morphism $f^{-1} : B \longrightarrow A$ ch that $f;f^{-1} = 1_A$, and $f^{-1};f = 1_B$.

[7] In the expressions appearing in the equations, when compositions of morphisms are involved, we always plicitly assume that the corresponding domains and codomains match.

1. **Category.**
   (a) *Associativity.* For all α, β, γ $\quad (\alpha;\beta);\gamma = \alpha;(\beta;\gamma)$
   (b) *Identities.* For each $\alpha : [t] \longrightarrow [t']$ $\quad \alpha;[t'] = \alpha$ *and* $[t];\alpha = \alpha$

2. **Functoriality of the Σ-algebraic structure.** For each $f \in \Sigma_n$, $n \in \mathbb{N}$,
   (a) *Preservation of composition.* For all $\alpha_1, \ldots, \alpha_n, \beta_1, \ldots, \beta_n$,

$$f(\alpha_1;\beta_1, \ldots, \alpha_n;\beta_n) = f(\alpha_1, \ldots, \alpha_n); f(\beta_1, \ldots, \beta_n)$$

   (b) *Preservation of identities.* $\quad f([t_1], \ldots, [t_n]) = [f(t_1, \ldots, t_n)]$

3. **Axioms in E.** For $t(x_1, \ldots, x_n) = t'(x_1, \ldots, x_n)$ an axiom in E, for all $\alpha_1, \ldots, \alpha_n$,

$$t(\alpha_1, \ldots, \alpha_n) = t'(\alpha_1, \ldots, \alpha_n)$$

4. **Exchange.** For each $r : [t(x_1, \ldots, x_n)] \longrightarrow [t'(x_1, \ldots, x_n)]$ in R,

$$\frac{\alpha_1 : [w_1] \longrightarrow [w'_1] \quad \ldots \quad \alpha_n : [w_n] \longrightarrow [w'_n]}{r(\overline{\alpha}) = r(\overline{[w]}); t'(\overline{\alpha}) = t(\overline{\alpha}); r(\overline{[w']})}$$

Similarly, the groupoid $\mathcal{T}_{\mathcal{R}}^{\leftrightarrow}(X)$ is obtained by identifying the terms generated by rules 1-5 modulo the above equations plus the additional:

5. **Inverse.** For any $\alpha : [t] \longrightarrow [t']$ in $\mathcal{T}_{\mathcal{R}}^{\leftrightarrow}(X)$, $\quad \alpha;\alpha^{-1} = [t]$ *and* $\alpha^{-1};\alpha = [t']$

Note that the set X of variables is actually a parameter of these constructions, and we need not assume X to be fixed and countable. In particular, for $X = \emptyset$, I adopt the notations $\mathcal{T}_{\mathcal{R}}$ and $\mathcal{T}_{\mathcal{R}}^{\leftrightarrow}$, respectively. The equations in 1 make $\mathcal{T}_{\mathcal{R}}(X)$ a category, the equations in 2 make each $f \in \Sigma$ a functor, and 3 forces the axioms E. The exchange law states that any rewriting of the form $r(\overline{\alpha})$ —which represents the *simultaneous* rewriting of the term at the top using rule r *and* "below," i.e., in the subterms matched by the rule— is equivalent to the sequential composition $r(\overline{[w]}); t'(\overline{\alpha})$ corresponding to first rewriting on top with r and then below on the matched subterms. The exchange law also states that rewriting at the top by means of rule r and rewriting "below" are processes that are independent of each other and therefore can be done in any order. Therefore, $r(\overline{\alpha})$ is also equivalent to the sequential composition $t(\overline{\alpha}); r(\overline{[w']})$. Since $[t(x_1, \ldots, x_n)]$ and $[t'(x_1, \ldots, x_n)]$ can be regarded as functors $\mathcal{T}_{\mathcal{R}}(X)^n \longrightarrow \mathcal{T}_{\mathcal{R}}(X)$, the exchange law just asserts that r is a *natural transformation* [20], i.e.,

Lemma 3 For each $r : [t(x_1, \ldots, x_n)] \longrightarrow [t'(x_1, \ldots, x_n)]$ in R, the family of morphisms

$$\{r(\overline{[w]}) : [t(\overline{w}/\overline{x})] \longrightarrow [t'(\overline{w}/\overline{x})] \mid \overline{[w]} \in T_{\Sigma,E}(X)^n\}$$

is a natural transformation $r : [t(x_1, \ldots, x_n)] \Longrightarrow [t'(x_1, \ldots, x_n)]$ between the functors $[t(x_1, \ldots, x_n)], [t'(x_1, \ldots, x_n)] : \mathcal{T}_{\mathcal{R}}(X)^n \longrightarrow \mathcal{T}_{\mathcal{R}}(X)$. □

What the exchange law provides in general is a way of *abstracting* a rewriting computation by considering immaterial the order in which rewrites are performed "above" and "below" in the term; further abstraction among proof terms is obtained from the functoriality equations. The equations 1-4 provide in a sense the *most abstract* view of the computations of the rewrite theory $\mathcal{R}$ that can reasonably be given. In particular, we can prove that all terms have an equivalent expression as step sequences or as interleaving sequences:

emma 4 For each $[\alpha] : [t] \longrightarrow [t']$ in $\mathcal{T}_{\mathcal{R}}(X)$, either $[t] = [t']$ and $[\alpha] = [[t]]$, or there is an $\in \mathbb{N}$ and a chain of morphisms $[\alpha_i]$, $0 \leq i \leq n$ whose terms α_i describe one-step (concurrent) writes

$$[t] \xrightarrow{\alpha_0} [t_1] \xrightarrow{\alpha_1} \ldots \xrightarrow{\alpha_{n-1}} [t_n] \xrightarrow{\alpha_n} [t']$$

ch that $[\alpha] = [\alpha_0; \ldots; \alpha_n]$. In addition, we can always choose all the α_i corresponding to quential rewrites. □

The category $\mathcal{T}_{\mathcal{R}}(X)$ is just one among many *models* that can be assigned to the rewriting eory $\mathcal{R}$. The general notion of model, called an $\mathcal{R}$-system, is defined as follows:

efinition 5 Given a rewrite theory $\mathcal{R} = (\Sigma, E, L, R)$, an *$\mathcal{R}$-system $\mathcal{S}$* is a category $\mathcal{S}$ together ith:

- a (Σ, E)-algebra structure, i.e., for each $f \in \Sigma_n$, $n \in \mathbb{N}$, a functor $f_{\mathcal{S}} : \mathcal{S}^n \longrightarrow \mathcal{S}$, in such a way that the equations E are satisfied, i.e., for any $t(x_1, \ldots, x_n) = t'(x_1, \ldots, x_n)$ in E we have an identity of functors $t_{\mathcal{S}} = t'_{\mathcal{S}}$, where the functor $t_{\mathcal{S}}$ is defined inductively from the functors $f_{\mathcal{S}}$ in the obvious way.

- for each rewrite rule $r : [t(\overline{x})] \longrightarrow [t'(\overline{x})]$ in R a natural transformation $r_{\mathcal{S}} : t_{\mathcal{S}} \Longrightarrow t'_{\mathcal{S}}$.

An *$\mathcal{R}$-homomorphism* $F : \mathcal{S} \longrightarrow \mathcal{S}'$ between two $\mathcal{R}$-systems is then a functor $F : \mathcal{S} \longrightarrow \mathcal{S}'$ ich that it is a Σ-algebra homomorphism —i.e., $f_{\mathcal{S}} * F = F^n * f_{\mathcal{S}'}$, for each f in Σ_n, $n \in \mathbb{N}$— nd such that "F preserves R," i.e., for each rewrite rule $r : [t(\overline{x})] \longrightarrow [t'(\overline{x})]$ in R we have ie identity of natural transformations $r_{\mathcal{S}} * F = F^n * r_{\mathcal{S}'}$, where n is the number of variables ppearing in the rule. This defines a category $\underline{\mathcal{R}\text{-}Sys}$ in the obvious way.

An *$\mathcal{R}$-groupoid* is an $\mathcal{R}$-system $\mathcal{S}$ whose category structure is actually a groupoid. This efines a full subcategory $\underline{\mathcal{R}\text{-}Grpd} \subseteq \underline{\mathcal{R}\text{-}Sys}$. □

What the above definition captures formally is the idea that the models of a rewrite theory re *systems*. By a "system" I of course mean a machine-like entity that can be in a variety of tates, and that can change its state by performing certain *transitions*. Such transitions are of ourse transitive, and it is natural and convenient to view states as "idle" transitions that do ot change the state. In other words, a system can be naturally regarded as a *category*, whose bjects are the states of the system and whose morphisms are the system's transitions.

For *sequential* systems, this is in a sense the end of the story (see Section 5.1.) As I will rgue and justify more fully with examples in Section 5, what makes a system *concurrent* is recisely the existence of an additional *algebraic structure*. Ugo Montanari and I first observed his fact for the particular case of Petri nets for which the algebraic structure is precisely that f a commutative monoid [26, 27]. However, this observation holds in full generality for *any lgebraic structure whatever*. What the algebraic structure captures is twofold. Firstly, *the tates themselves are distributed according to such a structure*; for Petri nets the distribution akes the form of a *multiset* that we can visualize with tokens and places; for a functional rogram involving just syntactic rewriting, the distribution takes the form of a *labelled tree tructure* which can be spatially distributed in such a way that many transitions (i.e., rewrites) an happen concurrently in a way analogous to the concurrent firing of transitions in a Petri et. Secondly, *concurrent transitions are themselves distributed according to the same algebraic tructure*; this is what the notion of $\mathcal{R}$-system captures, and is for example manifested in the oncurrent firing of Petri nets and, more generally, in any type of concurrent rewriting.

The expressive power of rewrite theories to specify concurrent transition systems[8] is greatly ncreased by the possibility of having not only transitions, but also *parameterized transitions*,

[8]Such expressive power is further increased by allowing *conditional* rewrite rules, a more general case to which ll that is said in this paper has been extended in [25].

System	$\longleftrightarrow$	*Category*
State	$\longleftrightarrow$	*Object*
Transition	$\longleftrightarrow$	*Morphism*
Procedure	$\longleftrightarrow$	*Natural Transformation*
Distributed Structure	$\longleftrightarrow$	*Algebraic Structure*

Figure 2: The mathematical structure of concurrent systems

i.e., *procedures*. This is what rewrite rules —with variables— provide. The family of states t which the procedure applies is given by those states where a component of the (distributed state is a substitution instance of the lefthand side of the rule in question. The rewrite rule i then a *procedure*[9] which transforms the state *locally*, by replacing such a substitution instanc by the corresponding substitution instance of the righthand side. The fact that this can tak place concurrently with other transitions "below" is precisely what the concept of a *natura transformation* formalizes. The table of Figure 2 summarizes our present discussion.

A detailed proof of the following theorem on the existence of initial and free $\mathcal{R}$-systems fo the more general case of conditional rewrite theories is given in [25], where the soundness an completeness of rewriting logic for $\mathcal{R}$-system models is also proved.

Theorem 6 $\mathcal{T}_{\mathcal{R}}$ is an initial object in the category $\underline{\mathcal{R}\text{-}Sys}$, and $\mathcal{T}_{\mathcal{R}}^{\leftrightarrow}$ is an initial object in the cat egory $\underline{\mathcal{R}\text{-}Grpd}$. More generally, $\mathcal{T}_{\mathcal{R}}(X)$ has the following universal property: Given an $\mathcal{R}$-systen $\mathcal{S}$, each function $F : X \longrightarrow Obj(\mathcal{S})$ extends uniquely to an $\mathcal{R}$-homomorphism $F^{\natural} : \mathcal{T}_{\mathcal{R}}(X) \longrightarrow \mathcal{S}$ $\mathcal{T}_{\mathcal{R}}^{\leftrightarrow}(X)$ has the same universal property with respect to $\mathcal{R}$-groupoids. □

3.1 Equationally Defined Classes of Models

Since $\mathcal{R}$-systems are an "essentially algebraic" concept[10], we can consider classes Θ of $\mathcal{R}$-system defined by the satisfaction of additional equations. Such classes give rise to full subcategor inclusions $\Theta \hookrightarrow \underline{\mathcal{R}\text{-}Sys}$, and by general universal algebra results about essentially algebrai theories (see, e.g., [2]) such inclusions are *reflective* [20], i.e., for each $\mathcal{R}$-system $\mathcal{S}$ there is a $\mathcal{R}$-system $R_{\Theta}(\mathcal{S}) \in \Theta$ and an $\mathcal{R}$-homomorphism $\rho_{\Theta}(\mathcal{S}) : \mathcal{S} \longrightarrow R_{\Theta}(\mathcal{S})$ such that for any $\mathcal{R}$ homomorphism $F : \mathcal{S} \longrightarrow \mathcal{D}$ with $\mathcal{D} \in \Theta$ there is a unique $\mathcal{R}$-homomorphism $F^{\diamond} : R_{\Theta}(\mathcal{S}) \longrightarrow \mathcal{D}$ such that $F = \rho_{\Theta}(\mathcal{S}); F^{\diamond}$. The full subcategory $\underline{\mathcal{R}\text{-}Grpd} \subseteq \underline{\mathcal{R}\text{-}Sys}$ is also reflective, but it is no equationally definable. The situation generalizes that of the inclusion of the category of group into the category of monoids. What we have in this case is an inclusion that is a *forgetful functo* from a category of algebras with additional operations (in this case the inversion operation. However, for any equationally definable (full) subcategory $\Theta \subseteq \underline{\mathcal{R}\text{-}Sys}$, defined by a collection o equations H, the intersection $\Theta \cap \underline{\mathcal{R}\text{-}Grpd}$ has a very simple description, since it is just the ful subcategory of $\underline{\mathcal{R}\text{-}Grpd}$ definable by the equations H.

Therefore, we can consider subcategories of $\underline{\mathcal{R}\text{-}Sys}$ or of $\underline{\mathcal{R}\text{-}Grpd}$ that are defined by certai equations and be guaranteed that they have initial and free objects, that they are closed b subobjects and products, etc. Consider for example the following conditional equations:

$$\forall f, g \in Arrows,\ f = g \ \ if \ \ \partial_0(f) = \partial_0(g) \wedge \partial_1(f) = \partial_1(g)$$

$$\forall f, g \in Arrows,\ f = g \ \ if \ \ \partial_0(f) = \partial_1(g) \wedge \partial_1(f) = \partial_0(g).$$

[9] Its *actual parameters* are precisely given by a substitution.

[10] In the precise sense of being specifiable by an "essentially algebraic theory" or a "sketch" [2]; see [25].

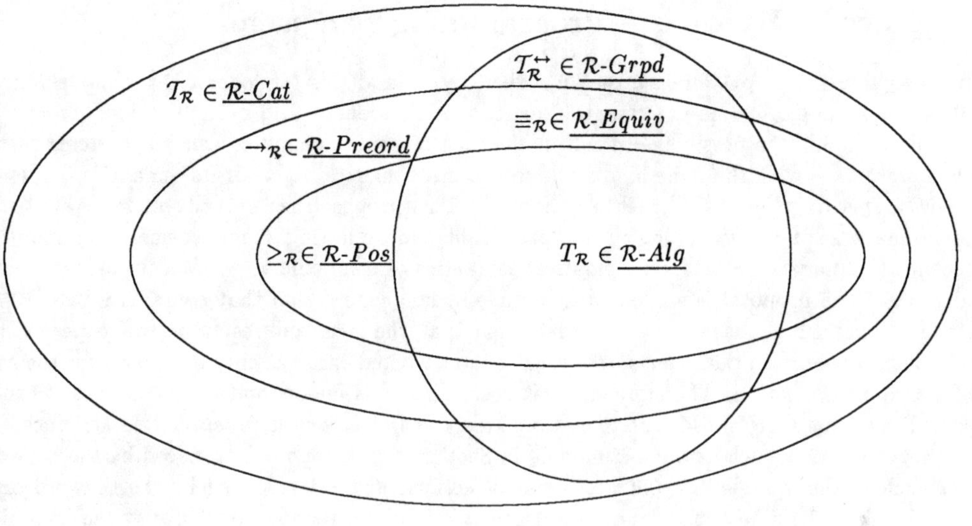

Figure 3: Subcategories of $\underline{\mathcal{R}\text{-}Sys}$ and their initial objects

he first equation forces a category to be a preorder, and the addition of the second requires this reorder to be a poset. By imposing the first one, or by imposing both, we get full subcategories

$$\underline{\mathcal{R}\text{-}Pos} \subseteq \underline{\mathcal{R}\text{-}Preord} \subseteq \underline{\mathcal{R}\text{-}Sys}.$$

routine inspection of $\underline{\mathcal{R}\text{-}Preord}$ for $\mathcal{R} = (\Sigma, E, L, R)$ reveals that its objects are preordered -algebras $(A, \leq)$ (i.e., preordered sets with a Σ-algebra structure such that all the operaons in Σ are monotonic) that satisfy the equations E and such that for each rewrite rule : $[t(\overline{x})] \longrightarrow [t'(\overline{x})]$ in R and for each $\overline{a} \in A^n$ we have, $t_A(\overline{a}) \geq t'_A(\overline{a})$. The poset case is entirely nalogous, except that the relation $\leq$ is a partial order instead of being a preorder. The reflection inctor associated to the inclusion $\underline{\mathcal{R}\text{-}Preord} \subseteq \underline{\mathcal{R}\text{-}Sys}$, sends $\mathcal{T}_{\mathcal{R}}(X)$ to the familiar $\mathcal{R}$*-rewriting elation*[11] $\rightarrow_{\mathcal{R}(X)}$ on E-equivalence classes of terms with variables in X. It is easy to show that ewriting logic remains complete when we restrict the models to be preorders [25]. Similarly, the eflection associated to the inclusion $\underline{\mathcal{R}\text{-}Pos} \subseteq \underline{\mathcal{R}\text{-}Sys}$ maps $\mathcal{T}_{\mathcal{R}}(X)$ to the partial order $\geq_{\mathcal{R}(X)}$ btained from the preorder $\rightarrow_{\mathcal{R}(X)}$ by identifying any two $[t], [t']$ such that $[t] \rightarrow_{\mathcal{R}(X)} [t']$ and $'] \rightarrow_{\mathcal{R}(X)} [t]$. Again, rewriting logic remains complete for poset models [25].

Intersecting $\underline{\mathcal{R}\text{-}Pos}$ and $\underline{\mathcal{R}\text{-}Preord}$ with the category $\underline{\mathcal{R}\text{-}Grpd}$ we get two subcategories defin-ble by the first equation or by both, but now in the context of $\underline{\mathcal{R}\text{-}Grpd}$. Combining the notions f a groupoid and a preorder we get exactly the notion of an *equivalence relation* and therefore subcategory $\underline{\mathcal{R}\text{-}Equiv}$ whose initial object is the usual congruence $\equiv_{\mathcal{R}}$ on ground terms modlo provable equality generated by the rules in $\mathcal{R}$ when regarded as equations. A poset that also a groupoid yields a *discrete category* whose only arrows are identities, i.e., a set. It is herefore easy to see that the subcategory obtained by intersecting $\underline{\mathcal{R}\text{-}Pos}$ with $\underline{\mathcal{R}\text{-}Grpd}$ is just he familiar category $\underline{\mathcal{R}\text{-}Alg}$ of ordinary Σ-algebras that satisfy the equations $E \cup$ *unlabel*(R), here the *unlabel* function removes the labels from the rules and turns the sequent signs " $\longrightarrow$ " nto equality signs. Similarly, the reflection functor into $\underline{\mathcal{R}\text{-}Alg}$ maps $\mathcal{T}_{\mathcal{R}}(X)$ to $T_{\mathcal{R}}(X)$, the free -algebra on X. Figure 3 summarizes the relationships among all these categories.

[11] It is perhaps more suggestive to call $\rightarrow_{\mathcal{R}(X)}$ the *reachability relation* of the system $\mathcal{T}_{\mathcal{R}}(X)$.

4 Rewrite Rules as a Programming Language

In this paper I have put forward the view that, by generalizing the logic and the model theor of equational logic to those of rewriting logic, a much broader field of applications for rewrit rule programming is possible —based on the idea of programming *concurrent systems* rathe than *algebras*— with the same high standards of mathematical rigor for its semantics. I presen below a specific proposal for such a semantics. This proposal has two advantages. First, th functional case of equational logic is kept as a sublanguage having a more specialized semantics second, the operational and mathematical semantics of a module are related in a particularl nice way. The proposal is embodied in Maude, a language design that contains OBJ3 [10] a its functional sublanguage. As already mentioned, all the ideas and results in this paper exten without problem[12] to the *order-sorted* case[13]; the unsorted case has only been used for the sak of a simpler exposition. Therefore, all that is said below is understood in the context of order sorted rewriting logic. In Maude there are three kinds of *modules*: *functional* —introduced b the keyword `fmod`, such as the `NAT` module in Section 2—, *system* —introduced by the keywor `mod` such as the module `NAT-CHOICE`— and *object-oriented* —introduced by the keyword `omo` (See Section 5.3.) The semantics of object-oriented modules reduces to that of system modules therefore, in this section we focus on the functional and system cases. Functional and system modules are respectively of the form `fmod` $\mathcal{R}$ `endfm`, and `mod` $\mathcal{R}'$ `endm`, for $\mathcal{R}$ and $\mathcal{R}'$ rewritin theories[14]. Their semantics is given in terms of a *machine* linking the module's operationa semantics with its denotational semantics. The general notion of a machine is as follows.

Definition 7 For $\mathcal{R}$ a rewrite theory and $\Theta \hookrightarrow \underline{\mathcal{R}\text{-}Sys}$ a reflective full subcategory, an *$\mathcal{R}$-machin over* Θ is an $\mathcal{R}$-homomorphism $[\![\_]\!] : \mathcal{S} \longrightarrow \mathcal{M}$ —called the machine's *abstraction map*— wit $\mathcal{S}$ an $\mathcal{R}$-system and $\mathcal{M} \in \Theta$. Given $\mathcal{R}$-machines over Θ, $[\![\_]\!] : \mathcal{S} \longrightarrow \mathcal{M}$ and $[\![\_]\!]' : \mathcal{S}' \longrightarrow \mathcal{M}'$ ar $\mathcal{R}$-machine *homomorphism* is a pair of $\mathcal{R}$-homomorphisms (F, G), $F : \mathcal{S} \longrightarrow \mathcal{S}'$, $G : \mathcal{M} \longrightarrow \mathcal{M}'$ such that $[\![\_]\!]; G = F; [\![\_]\!]'$. This defines a category $\underline{\mathcal{R}\text{-}Mach/\Theta}$; it is easy to check that the initia object in this category is the unique $\mathcal{R}$-homomorphism $\mathcal{T}_{\mathcal{R}} \longrightarrow R_{\Theta}(\mathcal{T}_{\mathcal{R}})$ □

The intuitive idea behind a machine $[\![\_]\!] : \mathcal{S} \longrightarrow \mathcal{M}$ is that we can use a *system* $\mathcal{S}$ to *comput* a result relevant for a *model* $\mathcal{M}$ of interest in a class Θ of models. What we do is to perform a certain computation in $\mathcal{S}$, and then output the result by means of the abstraction map $[\![\_]\!]$ A very good example is an *arithmetic machine* with $\mathcal{S} = \mathcal{T}_{\mathtt{NAT}}$, for `NAT` the rewriting theory of the Peano natural numbers corresponding to the module `NAT`[15] in Section 2, with $\mathcal{M} = \mathbb{N}$, and with $[\![\_]\!]$ the unique homomorphism from the initial `NAT`-system $\mathcal{T}_{\mathtt{NAT}}$; i.e., this is the initia machine in $\underline{\mathtt{NAT}\text{-}Mach/\mathtt{NAT}\text{-}Alg}$. To compute the result of an arithmetic expression t, we perform a terminating rewriting and output the corresponding number, which is an element of $\mathbb{N}$.

Each choice of a reflective full subcategory Θ as a category of models yields a different semantics. As already implicit in the arithmetic machine example, the *semantics of a functiona module*[16] `fmod` $\mathcal{R}$ `endfm` is the initial machine in $\underline{\mathcal{R}\text{-}Mach/\mathcal{R}\text{-}Alg}$. For the *semantics of a system module* `mod` $\mathcal{R}$ `endm` not having any functional submodules[17] I propose the initial machine in $\underline{\mathcal{R}\text{-}Mach/\mathcal{R}\text{-}Preord}$, but other choices are also possible. On the one hand, we could choose to be as concrete as possible and take $\Theta = \underline{\mathcal{R}\text{-}Sys}$ in which case the abstraction map is the identity homomorphism for $\mathcal{T}_{\mathcal{R}}$. On the other hand, we could instead be even more abstract, and choose

[12]Exercising of course the well known precaution of making explicit the universal quantification of rules.

[13]I.e., there is not just one sort, but a partially ordered set of sorts —with the ordering understood as type inclusion— and the function symbols can be overloaded [12].

[14]This is somewhat inaccurate in the case of system modules having functional submodules, which is discussed below, because we have to "remember" that the submodule is functional.

[15]In this case E is the commutativity attribute, and R consists of the two rules for addition.

[16]For this semantics to behave well, the rules R in the functional module $\mathcal{R}$ should be *confluent* modulo E.

[17]See below for a discussion of submodule issues.

= $\underline{\mathcal{R}\text{-}Pos}$; however, this would have the unfortunate effect of collapsing all the states of a cyclic writing, which seems undesirable for many "reactive" systems. If the machine $\mathcal{T}_{\mathcal{R}} \longrightarrow \mathcal{M}$ is the mantics of a functional or system module with rewrite theory $\mathcal{R}$, then we call $\mathcal{T}_{\mathcal{R}}$ the module's *erational semantics*, and $\mathcal{M}$ its *denotational semantics*.

In Maude a module can have *submodules*. Functional modules can only have functional bmodules, but system modules can have both functional and system submodules. For example, T was declared a submodule of NAT-CHOICE. The meaning of submodule relations in which e submodule and the supermodule are both of the same kind is the obvious one, i.e., we gment the signature, equations, labels, and rules of the submodule by adding to them the rresponding ones in the supermodule; we then give semantics to the module so obtained cording to its kind, i.e., functional or system. The semantics of a system module having functional submodule is somewhat more delicate. Suppose that the rewrite theory of the nctional submodule[18] is $\mathcal{R} = (\Sigma, E, L, R)$ and that of the system supermodule plus its system bmodules is $\mathcal{R}' = (\Sigma', E', L', R')$; as before we can form $\mathcal{R}\cup\mathcal{R}' = (\Sigma\cup\Sigma', E\cup E', L\cup L', R\cup R')$, t the semantics of the module is now given by the initial machine in the category

$$\underline{(\mathcal{R}\cup\mathcal{R}')\text{-}Mach/(\Sigma\cup\Sigma', E\cup E'\cup unlabel(R), L', R')\text{-}Preord}.$$

otice that $\underline{(\Sigma\cup\Sigma', E\cup E'\cup unlabel(R), L', R')\text{-}Preord}$ is an equationally definable full subcatory of $\underline{(\mathcal{R}\cup\mathcal{R}')\text{-}Preord}$, namely the one defined by the equations $t(\overline{x}) = t'(\overline{x})$ for each rewrite le $r : [t(\overline{x})] \longrightarrow [t'(\overline{x})]$, and therefore is also reflective.

Given a preorder $\mathcal{M}$ in $\underline{(\Sigma\cup\Sigma', E\cup E'\cup unlabel(R), L', R')\text{-}Preord}$ we can forget about R' d the labels and view it as an $\mathcal{R}$-algebra $\mathcal{M}|_{\mathcal{R}}$. Given a system module mod $\mathcal{R}'$ endm having od $\mathcal{R}$ endfm as its functional submodule and $\mathcal{T}_{\mathcal{R}\cup\mathcal{R}'} \longrightarrow \mathcal{M}$ as its semantics, we say that this bmodule relation is *extending* if the unique $\mathcal{R}$-homomorphism $h : T_{\mathcal{R}} \longrightarrow \mathcal{M}|_{\mathcal{R}}$ is injective; milarly, we say that it is *protecting* if h is an isomorphism. We leave for the reader to check at the extending relation asserted for the importation of NAT in NAT-CHOICE does in fact hold.

As OBJ, Maude has also *theories* to specify semantic requirements for interfaces and to make gh level assertions about modules; they can be functional, system, or object-oriented. Also as BJ, Maude has *parameterized modules* —again of the three kinds— and *views* that are theory terpretations relating theories to modules or to other theories. Details for all these aspects of e language will appear elsewhere[19]. Finally, note that Maude is a *logic programming language* the general axiomatic sense made precise in [23].

Unifying Models of Concurrency

abelled transition systems, Petri nets and concurrent object-oriented programming are dissed as specializations of concurrent rewriting; other specializations are also discussed briefly.

.1 Labelled Transition Systems

This is the particularly simple case of rewrite theories $\mathcal{R} = (\Sigma, E, L, R)$ such that $E = \emptyset$, $= \Sigma_0$, i.e., Σ only involves constants, and all the rules in $\mathcal{R}$ are of the form $r : a \longrightarrow b$ for $, b$ constants. For example, the transition system of Figure 4 corresponds to the rewrite theory the sytem module LTS in the same figure. Since Σ contains only constants and the rules ave no variables, the rules 1-5 of rewriting logic specialize to very simple rules. The rule of ongruence becomes a trivial subcase of reflexivity, and the rule of replacement just yields each le $r : a \longrightarrow b$ in $\mathcal{R}$ as its own consequence. Thus, we just have reflexivity and transitivity with

[18] We assume that, if several functional submodules have been declared, we have already taken their union.
[19] Some basic results about views and parameterization for system modules have already been given in [25].

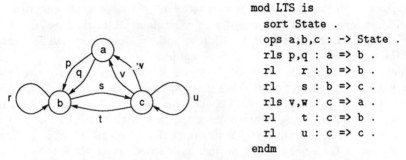

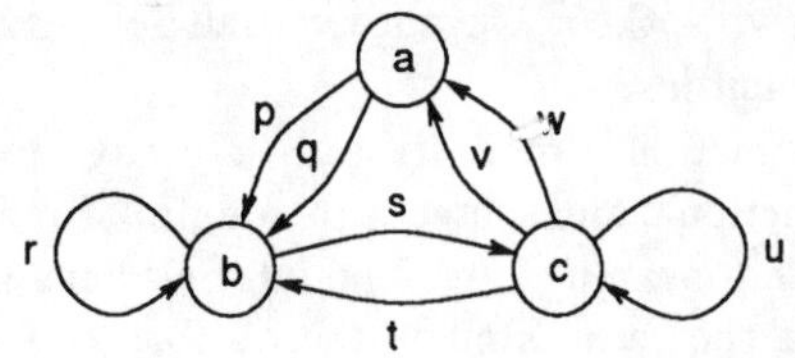


```
mod LTS is
  sort State .
  ops a,b,c : -> State .
  rls p,q : a => b .
  rl    r : b => b .
  rl    s : b => c .
  rls v,w : c => a .
  rl    t : c => b .
  rl    u : c => c .
endm
```

Figure 4: A labelled transition system and its code in Maude

the rules $r : a \longrightarrow b$ in $\mathcal{R}$ as basic axioms. Therefore, $\mathcal{T}_{\mathcal{R}}$ is just the *free category* —also calle the *path category*— generated by the labelled transition system when regarded as a graph. Mor generally, *any* $\mathcal{R}$-system with $\mathcal{R}$ a labelled transition system is just a *category* $\mathcal{C}$ together wit the assignment of an object of $\mathcal{C}$ to each constant in Σ and a morphism in $\mathcal{C}$ for each rule in R i a way consistent with the assignment of objects. In other words, such systems are just *sequenti systems*, and their sequentiality is precisely due to the absence of any operations other tha constants. In fact, labelled transition systems are intrinsically *sequential* as rewrite theories, i the precise sense of Definition 2. However, since several transitions are in general possible fron a given state, they exhibit a form of *nondeterminism.*

Interleaving approaches to concurrency restrict themselves to labelled transition systems o similar sequential structures. We can always sequentialize a concurrent computation (see Lemm 4) and therefore much valuable work can be and has been done in this context. However, th context as such is intrinsically sequential and forces a form of *indirect reasoning* when considerin concurrency aspects; therefore, it seems quite limited. Plato's analogy of the cave[20] may provid an apt metaphor for this situation, with labelled transition systems being the wall of the cav on which the shadows of true concurrency are reflected. The metaphor seems apt because i agrees with the mathematical facts; for $\mathcal{R}$ an arbitrary rewrite theory, the descent into the cav is precisely the forgetful functor $\underline{\mathcal{R}\text{-}Sys} \longrightarrow \underline{Cat}$.

5.2 Petri Nets

This is one of the most basic models of concurrency. It has the great advantage of exhibitin concurrency *directly*, not through the indirect mediation of interleavings. Its relationship t concurrent rewriting can be expressed very simply. It is just the particular case of rewrit theories $\mathcal{N} = (\Sigma, E, L, R)$ with $\Sigma_0 = \Delta \uplus \{\lambda\}$, $\Sigma_2 = \{\otimes\}$, with all the other Σ_n empty, wit $E = ACI$ —with ACI the axioms of *associativity* and *commutativity* for $\otimes$ and *identity* λ fo $\otimes$— and with all terms in the rules R ground terms. Consider for example the Petri net i Figure 5, which represents a machine to buy subway tickets. With a dollar we can buy a ticke $t1$ by pushing the button $b-t1$ and get two quarters back; if we push $b-t2$ instead, we get longer distance ticket $t2$ and one quarter back. Similar buttons allow purchasing the tickets wit quarters. Finally, with one dollar we can get four quarters by pushing *change*. The correspondin rewrite theory is that of the TICKET module in the same figure.

The rules of deduction specialize as follows. The congruence rule applies just to $\otimes$ an yields instances of reflexivity for the constants. Since the rewrite rules have no variables, th replacement rule yields each of the rewrite rules as axioms. Interpreting $\otimes$ as conjunction i linear logic [8], this specialization yields sound and complete rules of deduction for the linea

[20] Republic, Bk. VII, 514-517.

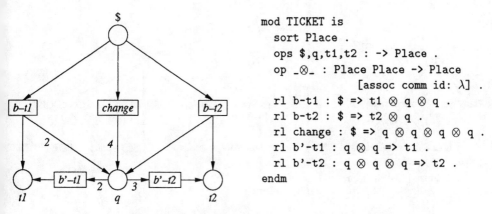

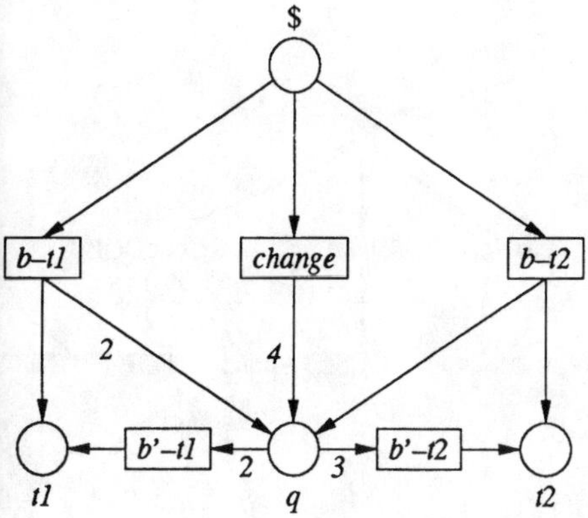


```
mod TICKET is
  sort Place .
  ops $,q,t1,t2 : -> Place .
  op _⊗_ : Place Place -> Place
                [assoc comm id: λ] .
  rl b-t1 : $ => t1 ⊗ q ⊗ q .
  rl b-t2 : $ => t2 ⊗ q .
  rl change : $ => q ⊗ q ⊗ q ⊗ q .
  rl b'-t1 : q ⊗ q => t1 .
  rl b'-t2 : q ⊗ q ⊗ q => t2 .
endm
```

Figure 5: A Petri net and its code in Maude

…eory having each of the rewrite rule sequents as axioms; i.e., rewriting logic specializes in this …se to *conjunctive linear logic*. The models of rewrite theories $\mathcal{N}$ of this kind are categories with commutative monoid structure in which we have chosen certain objects —the "places"— and …rtain morphisms —the "transitions." The initial system $\mathcal{T}_{\mathcal{N}}$ is exactly the category $\mathcal{T}[\mathcal{N}]$ that …go Montanari and I associated to a Petri net as its semantics in [26, 27]. Narciso Martí-Oliet …nd I later studied the connection of this model with models for linear logic in [22, 21] and …btained in this way a systematic triangular correspondence between Petri nets, linear logic and …tegories which is a particular instance of the more general triangular correspondence between …ncurrent systems, rewriting logic and categories developed in this paper.

.3 Concurrent Object-Oriented Programming

…he basic syntax for objects and messages is given by the following order-sorted rewrite signature:

sorts Object, Attribute, Attributes, Message, Configuration, Data, Value .
sorts OId, CId, AId . `*** object, class and attribute identifiers`
subsorts Object, Message < *Configuration* .
subsorts Attribute < *Attributes* .
subsorts OId, Data < *Value* .
op $\langle \_ : \_ \mid \_ \rangle$: *OId CId Attributes* ⟶ *Object* .
op $(\_ : \_)$: *AId Value* ⟶ *Attribute* .
op $\_ , \_$: *Attributes Attributes* ⟶ *Attributes* [*assoc comm id: nil*] .
op $\_\,\_$: *Configuration Configuration* ⟶ *Configuration* [*assoc comm id:* λ] .

…here the operators $\_\,\_$ and $\_ , \_$ are both associative and commutative with respective identities … and *nil*. With this syntax, an *object* is represented as a term $\langle O : C \mid a_1 : v_1, \ldots, a_n : v_n \rangle$ where … is the object's name, C is its class, the a_i's are the names of the object's *attributes*, and the …'s are their corresponding *values*. The *configuration* is the distributed state of the concurrent …bject-oriented system and is represented as a multiset of objects and messages. The system …volves by concurrent rewriting (modulo ACI) of the configuration by means of rewrite rules …pecific to each particular system, whose lefthand and righthand sides may in general involve …atterns for several objects and messages. For example, objets in a class *Accnt* of bank accounts, …ach having a *bal*(ance) attribute, may receive messages for crediting or debiting the account …nd evolve according to the rules:

$$credit(B, M)\ \langle B : Accnt \mid bal : N \rangle \longrightarrow \langle B : Accnt \mid bal : N + M \rangle$$
$$debit(B, M)\ \langle B : Accnt \mid bal : N \rangle \longrightarrow \langle B : Accnt \mid bal : N - M \rangle.$$

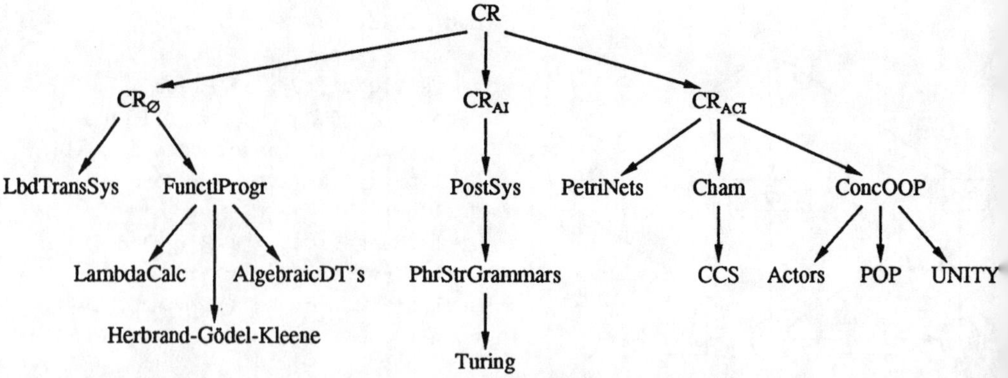

Figure 6: The Big Picture

Concurrent object-oriented systems can be defined in Maude by means of *object-oriented module definitions* of the form omod $\mathcal{O}$ endom which provide special syntax taking advantage of the structural properties common to all such systems. However, *the semantics of object-oriented modules is entirely reducible to that of system modules*, i.e., we can systematically translate an object-oriented module omod $\mathcal{O}$ endom into a corresponding system module mod $\mathcal{O}^\flat$ endm whose $\mathcal{O}^\flat$-machine semantics is the object-oriented module's intended semantics. Maude's object-oriented modules are discussed in detail in [24]; such modules share some similarities with those of FOOPS [11], and the idea of transforming objects by rewrite rules goes back to [9]. However, in comparison with FOOPS, both the treatment of concurrency and the semantics are new.

5.4 The Big Picture

Space limitations preclude a detailed discussion of other models of concurrency to which concurrent rewriting specializes (see [25].) However, we can summarize such specializations using Figure 6, where CR stands for concurrent rewriting, the arrows indicate specializations, and the subscripts $\emptyset$, AI, and ACI stand for syntactic rewriting, rewriting modulo associativity and identity, and ACI rewriting respectively. *Functional programming* (in particular Maude's functional modules) corresponds to the case of *confluent*[21] rules, and includes the λ-calculus (in combinator form) and the Herbrand-Gödel-Kleene theory of recursive functions. Rewriting modulo AI yields Post systems and related grammar formalisms, including Turing machines. Rewriting modulo ACI includes Berry and Boudol's *chemical abstract machine* [3] (which itself specializes to CCS [28]), as well as actors [1] and Unity's model of computation [4] which can both be regarded as special cases of concurrent object-oriented programming with rewrite rules; a third special case is Engelfriet et al.'s POPs and POTs higher level Petri nets [6, 7].

6 Concluding Remarks

Within the space constraints of this paper it is impossible to do justice to the wealth of related literature on term rewriting, abstract data types, concurrency, Petri nets, linear and equational logic, ordered, continuous and nondeterministic algebras, etc. The paper [25] contains 85 such references. I would however like to mention Huet's lecture notes [16], which contains a brief discussion of rules for rewriting logic, and also work on applications of 2-categories to rewriting and to domain-theoretic and categorical approximations, including work by Rydeheard and Stell [30] and Pitts [29], whose relationship to this work is studied in [25].

[21] Although not reflected in the picture, rules confluent *modulo* equations E are also functional.

I conclude pointing out that the model theory of rewriting logic presented here —besides elding the general notion of concurrent system that we were seeking and providing the se-antic basis for the integration of the concurrent, functional and object-oriented computational radigms— does also establish a general *triangular correspondence* between logic, categories d concurrent systems that can be summarized as follows:

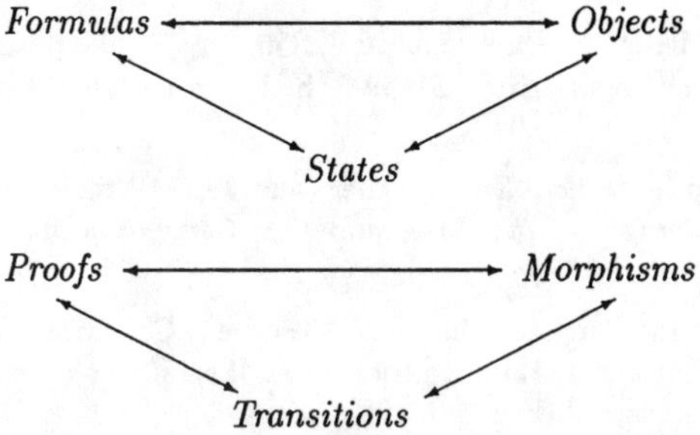

his generalizes to arbitrary rewrite systems the triangular correspondence between linear logic, etri nets and linear categories previously developed in joint work with Narciso Martí-Oliet 2]. In particular, the correspondence between logic and categories is a Lambek-Lawvere corre-pondence [18, 19], a type of correspondence more abstract and general than the Curry-Howard omorphism.

References

[1] G. Agha. *Actors.* MIT Press, 1986.

[2] M. Barr and C. Wells. *Toposes, Triples and Theories.* Springer-Verlag, 1985.

[3] Gérard Berry and Gérard Boudol. The Chemical Abstract Machine. In *Proc. POPL'90*, pages 81–94. ACM, 1990.

[4] K. Many Chandy and Jayadev Misra. *Parallel Program Design: A Foundation.* Addison-Wesley, 1988.

[5] N. Dershowitz and J.-P. Jouannaud. Rewrite systems. In *Handbook of Theoretical Computer Science, Vol. B.* North-Holland, 1990.

[6] J. Engelfriet. Net-based description of parallel object-based systems, or POTs and POPs. Technical report, Noordwijkerhout FOOL Workshop, May 1990.

[7] J. Engelfriet, G. Leih, and G. Rozenberg. Parallel object-based systems and Petri nets, I and II. Technical Report 90-04-5, Dept. of Computer Science, University of Leiden, February 1990.

[8] Jean-Yves Girard. Towards a geometry of interaction. In J.W. Gray and A. Scedrov, editors, *Proc. AMS Summer Research Conference on Categories in Computer Science and Logic, Boulder, Colorado, June 1987*, pages 69–108. American Mathematical Society, 1989.

[9] J.A. Goguen and J. Meseguer. Software for the rewrite rule machine. In *Proceedings of the International Conference on Fifth Generation Computer Systems, Tokyo, Japan*, pages 628–637. ICOT, 1988.

[10] Joseph Goguen, Claude Kirchner, Hélène Kirchner, Aristide Mégrelis, José Meseguer, and Timothy Winkler. An introduction to OBJ3. In Jean-Pierre Jouannaud and Stephane Kaplan, editors, *Proceedings, Conference on Conditional Term Rewriting, Orsay, France, July 8-10, 1987*, pages 258–263. Springer-Verlag, Lecture Notes in Computer Science No. 308, 1988.

[11] Joseph Goguen and José Meseguer. Unifying functional, object-oriented and relational programming with logical semantics. In Bruce Shriver and Peter Wegner, editors, *Research Directions in Object-Oriented Programming*, pages 417–477. MIT Press, 1987. Preliminary version in *SIGPLAN Notices*,

Volume 21, Number 10, pages 153-162, October 1986; also, Technical Report CSLI-87-93, Center fc the Study of Language and Information, Stanford University, March 1987.

[12] Joseph Goguen and José Meseguer. Order-sorted algebra I: Partial and overloaded operations, error and inheritance. Technical Report SRI-CSL-89-10, SRI International, Computer Science Lab, Jul 1989. Given as lecture at Seminar on Types, Carnegie-Mellon University, June 1983. Submitted fc publication.

[13] Joseph Goguen, José Meseguer, Sany Leinwand, Timothy Winkler, and Hitoshi Aida. The rewrit rule machine. Technical Report SRI-CSL-89-6, SRI International, Computer Science Lab, Marc 1989.

[14] Joseph Goguen, James Thatcher, Eric Wagner, and Jesse Wright. Initial algebra semantics an continuous algebras. *Journal of the Association for Computing Machinery*, 24(1):68–95, Januar 1977.

[15] Joseph A. Goguen, Claude Kirchner, and José Meseguer. Concurrent term rewriting as a model c computation. In R. Keller and J. Fasel, editors, *Proc. Workshop on Graph Reduction, Santa Fe, Ne Mexico*, pages 53–93. Springer LNCS 279, 1987.

[16] G. Huet. *Formal Structures for Computation and Deduction.* INRIA, 1986.

[17] Gerard Huet. Confluent reductions: Abstract properties and applications to term rewriting systems *Journal of the Association for Computing Machinery*, 27:797–821, 1980. Preliminary version in 18t Symposium on Mathematical Foundations of Computer Science, 1977.

[18] Joachim Lambek. Deductive systems and categories II. In *Category Theory, Homology Theory an their Applications I.* Springer Lecture Notes in Mathematics No. 86, 1969.

[19] F.W. Lawvere. Adjointness in foundations. *Dialectica*, 23(3/4):281–296, 1969.

[20] Saunders MacLane. *Categories for the working mathematician.* Springer, 1971.

[21] Narciso Martí-Oliet and José Meseguer. An algebraic axiomatization of linear logic models. Technica Report SRI-CSL-89-11, SRI International, Computer Science Lab, December 1989. To appear in G.M. Reed, A.W. Roscoe and R. Wachter (eds.), *Proceedings of the Oxford Symposium on Topolog in Computer Science*, Oxford University Press, 1990.

[22] Narciso Martí-Oliet and José Meseguer. From Petri nets to linear logic. In D.H. Pitt et al., editor *Category Theory and Computer Science*, pages 313–340. Springer Lecture Notes in Computer Science Vol. 389, 1989. Full version to appear in *Mathematical Structures in Computer Science.*

[23] José Meseguer. General logics. In H.-D. Ebbinghaus et al., editor, *Logic Colloquium'87*, pages 275–329. North-Holland, 1989.

[24] José Meseguer. A logical theory of concurrent objects. In *ECOOP-OOPSLA'90 Conference on Object-Oriented Programming, Ottawa, Canada, October 1990.* ACM, 1990.

[25] José Meseguer. Rewriting as a unified model of concurrency. Technical Report SRI-CSL-90-02, SRI International, Computer Science Laboratory, February 1990. Revised June 1990.

[26] José Meseguer and Ugo Montanari. Petri nets are monoids. Technical report, SRI International, Computer Science Laboratory, January 1988. Revised June 1989; to appear in *Information and Computation.*

[27] José Meseguer and Ugo Montanari. Petri nets are monoids: A new algebraic foundation for net theory. In *Proc. LICS'88*, pages 155–164. IEEE, 1988.

[28] Robin Milner. *Communication and Concurrency.* Prentice Hall, 1989.

[29] A. Pitts. An elementary calculus of approximations. Unpublished manuscript, University of Sussex, December 1987.

[30] D.E. Rydeheard and J.G. Stell. Foundations of equational deduction: A categorical treatment of equational proofs and unification algorithms. In *Proceedings of the Summer Conference on Category Theory and Computer Science, Edinburgh, Sept. 1987.* Springer LNCS 283, 1987.

A Temporal Calculus of Communicating Systems

Faron Moller* Chris Tofts
Department of Computer Science
University of Edinburgh

Abstract

In this paper, we introduce a calculus of communicating systems which allows for the expression and analysis of timing constraints, for example as is important for real-time processes. We present the language, along with its formal semantics, and derive algebraic laws for reasoning about processes in the language. Though the core language is simple, we show that the language has several powerful derived operators which we demonstrate to be useful in several examples.

Introduction

ıe analysis of the temporal properties of concurrent processes gives insight into some interesting pects of concurrent programming, such as timeout in fault-tolerant systems (eg, protocols), and ıration control in critical real-time systems (eg, radiation exposure). There have been several ıdies made to provide a formalism within which these concepts can be expressed (*e.g.,* [Bae89], ro90], [Jef89], [Koy83], [Ree86]). Many of these assume synchrony, which results in some the more interesting temporal properties of processes being inexpressible, or they assume a obal clock and require actions to happen at precise moments measured by that global clock, approach which also has its drawbacks.

The earlier report [Tof88] provided an extension to CCS, Milner's *Calculus of Communicating ocesses*, which admitted a notion of timing. That approach was unsatisfactory in that the count it gave of time was somewhat eccentric. Processes could only evolve simultaneously a communication, whilst time and action were otherwise interleaved. In order to give a fuller count of time for asynchronous processes, we rewrite the theory to derive our language TCCS, *Temporal Calculus of Communicating Systems*, in which time is allowed to pass independent the functional aspects of a processes.

In order to give this fuller account, the process state transition system upon which the tuition of the calculus is built has semantically been split into two orthogonal parts, one scribing the functional aspect of the process, and the other describing its temporal aspect. This lows for the separation of functional and temporal concerns in analysing process behaviour, but so there are sound physical motivations for imposing this separation. Computation involves ergy changes, and by a result of quantum mechanics (*cf.*, [Dir58], [Sch82]) energy changes d time cannot be measured simultaneously. Thus it seems reasonable when producing an servation-based model of time and computation not to permit the simultaneous observation these two activities. We therefore assume that actions have no duration, although we could

*Research supported by ESPRIT BRA Grant No. 3006 — CONCUR

model within our language actions with duration in the manner of [Cas87], or more simply b attaching a duration directly onto actions, requiring a process to take some amount of time i stabilising into a new state.

In this report we present our calculus, which is an extension of that discussed in [Tof89]. W present both its syntax and formal semantics, and give several interesting examples to demon strate its utility. We proceed as well to axiomatise our defined semantic equivalence, and prese sound and complete axiomatisations for different subcalculi of finite process terms.

The present paper is an abridged version of [Mol89], and this latter report should be referre to for all proofs not included in the present paper.

2 The Language TCCS

Our language TCCS will be a timed extension of CCS, Milner's *Calculus of Communicatin Systems* of [Mil89]. To define our language, we first presuppose a set Λ of atomic action symbo not containing τ or ε, and we define $\mathrm{Act} = \Lambda \cup \{\tau\}$. We assume that Λ can be partitioned into tw equinumerous sets with a "complementation" bijection $\bar{\cdot}$ between them extended by $\overline{\overline{a}} = a$. Thes complementary actions form the basis of the communication method in our calculus, analogou to that in CCS. We also presuppose some set Var of process variables, and take $\mathcal{T} = \{1, 2, 3, \ldots$ to represent divisions in time. Hence we are restricting our representation of time to the discret integral model rather than a more generous treatment such as for example the full set of positiv reals. For most of the presentation, we can consider the time domain to be continuous, but w require time to be discrete for the completeness proof of our axiomatisation. We comment o this choice of time domain and the extension to a continuous domain more fully in the conclusio of the paper.

The collection of TCCS expressions, ranged over by P, is then defined by the following BN expression where we take $a \in \mathrm{Act}$, $X \in \mathrm{Var}$, $t \in \mathcal{T}$, and S ranging over relabelling function those $S : \mathrm{Act} \to \mathrm{Act}$ such that $\overline{S(a)} = S(\overline{a})$ and $S(\tau) = \tau$.

$$P \ ::= \ \mathbf{0} \ \big| \ X \ \big| \ a.P \ \big| \ (t).P \ \big| \ \delta.P \ \big| \ P \oplus P \ \big| \ P + P \ \big| \ P \,|\, P \ \big| \ P \backslash a \ \big| \ P[S] \ \big| \ \mu_i \tilde{x}.\tilde{P}$$

The informal interpretation of these terms is as follows.

- $\mathbf{0}$ represents the nil process, which can neither proceed with any action, nor proceed throug time.
- X represents the process bound to the variable X in a given environment.
- $a.P$ represents the process which can perform the action a and evolve into the process P upon so doing; it cannot progress through time before performing the action a.
- $(t).P$ represents the process which will evolve into the process P after exactly t units c time.
- $\delta.P$ represents the process which behaves as the process P, but is willing to wait an amount of time before actually proceeding. The understanding of this process is that it i in fact wanting to synchronise or communicate with its environment, but is willing to wai until such time as the environment is ready to participate in such a communication.

- $P + Q$ represents a choice between the two processes P and Q. The process will behave as the process P or the process Q, with the choice being made at the time of the first action. Thus for instance any initial passage of time must be allowed by both P and Q. We shall refer to this operator as *strong* choice.

- $P \oplus Q$ represents a slightly different choice between the two processes P and Q. The process will behave as the process P or the process Q, with the choice being made at the time of the first action, or else at the occurrence of a passage of time when only one of the operands may allow the time passage to occur. In this case, the second *"stopped"* process will be dropped from the computation. We shall refer to this operator as *weak* choice.

- $P \mid Q$ represents the parallel composition of the two processes P and Q. Each of the processes may do any actions independently, or they may synchronise on complementary actions, resulting in a τ action. Any passage of time must be allowed and recorded by each of P and Q.

- $P\backslash a$ represents the process P with the action $a \in \Lambda$ restricted away, that is, not allowed to occur.

- $P[S]$ represents the process P with its actions relabelled by the relabelling function S.

- $\mu_i \widetilde{x}.\widetilde{P}$ represents the solution x_i taken from the solutions to the mutually recursive definitions of the processes $\widetilde{x} = \{x_1, x_2, \ldots, x_n\}$ defined as particular solutions to the equations $\widetilde{x} = \widetilde{P}$.

me points worth noting which arise from the above informal description are as follows.

- The process $\mathbf{0}$ acts as a deadlock process in that it cannot perform any actions, nor witness any passage of time. Hence, by the definitions of the strong and weak choice operators $+$ and $\oplus$, and the parallel composition operator $|$, the constant $\mathbf{0}$ acts as an annihilator with respect to strongly adding to or composing with time-guarded processes, and as a unit with respect to the weak choice operator $\oplus$. Thus in particular, local temporal deadlock will imply global deadlock — if only time derivations are possible from each component of a parallel composition involving $\mathbf{0}$, then the whole composite process is deadlocked. Hence, of interest is the derived *nontemporal deadlock* process $\delta.\mathbf{0}$, which will allow any time to pass, but can never perform any actions. This process thus stands as a unit with respect to the strong choice operator $+$ and the parallel composition operator $|$.

- The definition of the delay prefix δ is such that it is only meaningful to follow it with action terms. $\delta.P$ represents a process which is delaying the *actions* of P until the environment in which the process is executing will allow the actions to proceed. Thus for instance, the process $\delta.(1).a.\mathbf{0}$ can never perform its action a, as it can never get past the delaying δ. Hence, this process will be identified with $\delta.\mathbf{0}$ the nontemporal deadlock.

As a point of notation, we shall occasionally omit the dot when applying the prefix operators, d also drop trailing $\mathbf{0}$'s, thus for instance rendering $\delta.a.\mathbf{0} + (t).b.\mathbf{0}$ as $\delta a + (t)b$. Also, we shall low the prefix operators to take equal precedence over the concurrency operator, which will ke precedence over the recursion operator, which finally in turn will take precedence over the oice operators; outside of this, the binary operators will be left associative, and we shall freely se parentheses to override the precedence rules. Finally, we shall allow ourselves to specify

processes definitionally, by providing recursive definitions of processes. For example, we sha
write $A \stackrel{\text{def}}{=} a.A$ rather than $A \stackrel{\text{def}}{=} \mu x.a.x$.

We can define some useful derived operators as follows. Firstly, we can define a *loose tin prefix* as follows.

$$[t].P \stackrel{\text{def}}{=} (t).\delta.P$$

This allows a process to wait for a certain fixed time to pass before being willing to participa in a communication, and then to await any length of time afterwards for another process in th environment to participate in the communication.

Next we can define a *process prefix* operator as follows.

$$\text{INIT}_t(P) \stackrel{\text{def}}{=} P \mid (t).\mathbf{0}$$

This allows a process to execute normally for a certain fixed time t, and then die when it cann perform any actions without possibly first allowing some further amount of time to pass.

Finally we can define a *timeout* operator as follows.

$$\text{TIMEOUT}_t(P,Q) \stackrel{\text{def}}{=} \delta.P + (t).Q$$

This allows a process P to commence execution through a communication with the environmer at any instant over a certain fixed time, but then allows another process Q to execute afte that time, thus preempting the first possibility. This operator is especially useful for modellin timeouts in fault-tolerant systems such as protocols.

For the remainder of this section we shall present several examples which will demonstrat the utility of the calculus. (Further examples can be found in [Tof89]).

Example 2.1 *Suppose we have some collection of (distinct) times $t_i \in \mathcal{T}$ $(1 \leq i \leq n)$, and w define the simple processes H_i by*

$$H_i \stackrel{\text{def}}{=} (t_i).finish_i.\mathbf{0}.$$

These processes can be viewed as horses competing in a race; the time which horse i takes t cross the finish line is given by t_i.

Following this interpretation, if we denote then a race to be the parallel composition of th horse processes, that is,

$$Race \stackrel{\text{def}}{=} \prod_{1 \leq i \leq n} H_i,$$

then we discover that

$$Race = (t_m).finish_m.\mathbf{0},$$

ere we interpret equality by "having the same transition graph", and where m is such that $< t_i$ for each $i \neq m$. Hence the effect of a race is to report the winning horse.

If we now denote another process *Race'* by

$$Race' \stackrel{\text{def}}{=} \sum_{1 \leq i \leq n} H_i,$$

en we can discover in fact that

$$Race = Race'.$$

his is a natural consequence of the fact that elements of a choice pass time together until the oice is actually made. This anomoly, where parallel computation is confused with nondeter- inistic choice, has an analogy in computation theory, where the process Race is viewed as the uring Machine concept of nondeterministic acceptance — try each path in parallel, and accept one of the paths succeeds.

Note that if we altered the definition of our horses slightly by defining the processes as

$$\widetilde{H}_i \stackrel{\text{def}}{=} (t_i).finish_i.\delta.\mathbf{0},$$

en this analogy breaks down. The corresponding race processes $\widetilde{Race}$ and $\widetilde{Race}'$ no longer have e same transition sequences, as $\widetilde{Race}'$ still only reports the winning horse, whereas $\widetilde{Race}$ reports e times of all subsequent horses.

xample 2.2 *Consider the following two recursively-defined processes.*

$$A \stackrel{\text{def}}{=} a.A' \oplus (1).B$$
$$B \stackrel{\text{def}}{=} b.B' \oplus (1).A$$

ere we are representing a system which is allowing one of two possible computation paths to be llowed, and furthermore allowing the choice to be determined by time, in that the first path can e followed only at even time units, whilst the second path can be followed only at odd time units. his system can be considered for example as the description of a multiplexing system based on olling or sampling, or as the basis of a mutual exclusion algorithm for two independent processes haring a common critical section. We can see this by interpreting the actions a and b as the rst and second processes entering the critical section, respectively. Then the processes A' and B' can be reasonably defined by $A' \stackrel{\text{def}}{=} c.B$ and $B' \stackrel{\text{def}}{=} d.A$, where the actions c and d represent the rst and second processes exiting the critical section, respectively.

We can trivially generalise this system to give a simple mutual exclusion algorithm for any umber of processes sharing a common resource.

xample 2.3 *In this example, we describe a timed implementation of the Alternating Bit Pro- ocol.*

There are three components to our protocol: the sender, the receiver, and the medium. These hree components are described as follows. First the sender S has the following definition.

$$S \stackrel{\text{def}}{=} S_0 \qquad S_b \stackrel{\text{def}}{=} \delta.send.S'_b$$
$$S'_b \stackrel{\text{def}}{=} \overline{s_b}.\text{TIMEOUT}_{t_r}(rack_b.S_{1-b} + rack_{1-b}.S'_b,\ S'_b)$$

Here, t_r is the amount of time which the sender will wait after sending a message before assumin the message had been lost and retransmitting. So the sender at some point in time receives request to send a message (the "send" action), immediately sends the message to the mediu tagged with an appropriate binary number (the "$\overline{s_b}$" action), and then waits for the appropriat acknowledgement to arrive from the medium (the "$rack_b$" action). If it receives the wrong ac knowledgement, or is left waiting for the full duration of the retry time t_r, then it shall retransm the message.

Next the receiver R has the following definition.

$$R \stackrel{\text{def}}{=} \delta.r_0.R_0 \qquad R_b \stackrel{\text{def}}{=} receive.R'_b$$
$$R'_b \stackrel{\text{def}}{=} \overline{sack_b}.\text{TIMEOUT}_{t_r}(r_{1-b}.R_{1-b} + r_b.R'_b,\ R'_b)$$

This definition is completely symmetric to the definition of the sender, with the exception tha it starts in a state where it is awaiting the first message to appear from the medium (the actio "r_0").

Finally, the medium has the following definition.

$$M \stackrel{\text{def}}{=} \delta.s_0.(t_t).\overline{r_0}.M + \delta.s_1.(t_t).\overline{r_1}.M + \delta.sack_0.(t_t).\overline{rack_0}.M + \delta.sack_1.(t_t).\overline{rack_1}.M$$

Here, t_t represents the transmission time. Hence the medium at some point in time receives eithe a message from the sender or an acknowledgement from the receiver, suitably tagged, and passe the message or acknowledgement straight through after a delay of t_t, the transmission time.

We can then define our protocol as follows.

$$\text{ABP} \stackrel{\text{def}}{=} (S \mid M \mid R)\backslash\{s_0, s_1, r_0, r_1, sack_0, sack_1, rack_0, rack_1\}$$

With these definitions, under the assumption that $t_r > 2t_t$ (that is, that the sender and receive will wait at least twice the transmission time, the amount of time required for a message to ge across the medium and have an acknowledgement return, before timing out and retransmittin its message or acknowledgement), we can show that

$$\text{ABP} = \text{SPEC},$$

where

$$\text{SPEC} \stackrel{\text{def}}{=} \delta.send.(t_t).receive.(t_t).\text{SPEC}.$$

and equality again interpreted as "having the same transition graph", but now modulo the occur rence of any number of τ transitions.

The Semantics of TCCS

this section, we define the operational semantics of our language. The semantics is *transition* ;*sed*, structurally presented in the style of [Plo81], outlining what actions and time delays a ocess can witness. In order to define our semantics however, we must first define a syntactic edicate which will allow us to describe when a process must stop delaying its computation thin a particular amount of time. This is done using the following function on TCCS terms.

efinition 3.1 *The function* $|\cdot|_{\mathcal{T}} : \text{TCCS} \rightarrow \{0, 1, 2, \ldots, \omega\}$ *defines the maximum delay which process may allow before forcing a computation (or deadlock) to occur. Formally, this function defined as follows:*

$$
\begin{array}{lll}
|\mathbf{0}|_{\mathcal{T}} = 0 & |P \oplus Q|_{\mathcal{T}} = \max(|P|_{\mathcal{T}}, |Q|_{\mathcal{T}}) & |P \backslash a|_{\mathcal{T}} = |P|_{\mathcal{T}} \\
|X|_{\mathcal{T}} = 0 & |P + Q|_{\mathcal{T}} = \min(|P|_{\mathcal{T}}, |Q|_{\mathcal{T}}) & |P[S]|_{\mathcal{T}} = |P|_{\mathcal{T}} \\
|a.P|_{\mathcal{T}} = 0 & |P \,|\, Q|_{\mathcal{T}} = \min(|P|_{\mathcal{T}}, |Q|_{\mathcal{T}}) & |\mu_i \tilde{x}.\tilde{P}|_{\mathcal{T}} = |P_i\{{}^{\mu\tilde{x}.\tilde{P}}/_{\tilde{x}}\}|_{\mathcal{T}} \\
|(s).P|_{\mathcal{T}} = s + |P|_{\mathcal{T}} & & \\
|\delta.P|_{\mathcal{T}} = \omega & &
\end{array}
$$

his definition is well defined as long as all recursive variables are guarded by action or time :efixes.

In Figure 1, we present the operational rules for our language. They are presented in a natural ɜduction style, and are to be read as follows: if the transition(s) above the inference line can be ferred, then we can infer the transition below the line. Our transitional semantics over TCCS ıen is given by the least relations $\longrightarrow \subseteq \text{TCCS} \times \text{Act} \times \text{TCCS}$ and $\rightsquigarrow \subseteq \text{TCCS} \times \mathcal{T} \times \text{TCCS}$ vritten $P \xrightarrow{a} Q$ and $P \overset{t}{\rightsquigarrow} Q$ respectively) satisfying the rules laid out in Figure 1. Notice that ıese rules respect the informal description of the constructs given in the previous section.

We can now define an equivalence relation $\sim$ on closed terms of TCCS based on Park's notion ˜ a *bisimulation* ([Par81]) as follows.

efinition 3.2 $P \sim Q$ *if and only if for all* $a \in \text{Act}$ *and for all* $t \in \mathcal{T}$,

> *(i) if* $P \xrightarrow{a} P'$ *then* $Q \xrightarrow{a} Q'$ *for some* Q' *with* $P' \sim Q'$;
> *(ii) if* $Q \xrightarrow{a} Q'$ *then* $P \xrightarrow{a} P'$ *for some* P' *with* $P' \sim Q'$;
> *(iii) if* $P \overset{t}{\rightsquigarrow} P'$ *then* $Q \overset{t}{\rightsquigarrow} Q'$ *for some* Q' *with* $P' \sim Q'$;
> *(iv) if* $Q \overset{t}{\rightsquigarrow} Q'$ *then* $P \overset{t}{\rightsquigarrow} P'$ *for some* P' *with* $P' \sim Q'$.

We can easily confirm that this definition gives us an equivalence relation over TCCS terms. Ioreover, this semantic interpretation respects the informal description of the maximum delay ınction $|\cdot|_{\mathcal{T}}$ defined above. That is, we can easily confirm the following proposition, which tells s that a process can idle for as long as the maximum delay function allows.

roposition 3.3 $P \overset{t}{\rightsquigarrow} P'$ *iff* $|P|_{\mathcal{T}} \geq t$.

Using this proposition, we can also demonstrate that this equivalence is in fact substitutive, ıus giving the following result.

roposition 3.4 $\sim$ *is a congruence with respect to the operators of* TCCS.

$$\frac{}{a.P \xrightarrow{a} P}$$

$$\frac{P \xrightarrow{a} P'}{\delta.P \xrightarrow{a} P'}$$

$$\frac{P \xrightarrow{a} P'}{P + Q \xrightarrow{a} P'}$$

$$\frac{Q \xrightarrow{a} Q'}{P + Q \xrightarrow{a} Q'}$$

$$\frac{P \xrightarrow{a} P'}{P \oplus Q \xrightarrow{a} P'}$$

$$\frac{Q \xrightarrow{a} Q'}{P \oplus Q \xrightarrow{a} Q'}$$

$$\frac{P \xrightarrow{a} P'}{P \mid Q \xrightarrow{a} P' \mid Q}$$

$$\frac{Q \xrightarrow{a} Q'}{P \mid Q \xrightarrow{a} P \mid Q'}$$

$$\frac{P \xrightarrow{a} P', \; Q \xrightarrow{\overline{a}} Q'}{P \mid Q \xrightarrow{\tau} P' \mid Q'}$$

$$\frac{P \xrightarrow{a} P'}{P \backslash L \xrightarrow{a} P' \backslash L} (a, \overline{a} \notin L)$$

$$\frac{P \xrightarrow{a} P'}{P[S] \xrightarrow{S(a)} P'[S]}$$

$$\frac{P_i\{{}^{\mu\widetilde{x}.\widetilde{P}}/_{\widetilde{x}}\} \xrightarrow{a} P'}{\mu_i\widetilde{x}.\widetilde{P} \xrightarrow{a} P'}$$

$$\frac{}{\delta.P \stackrel{t}{\rightsquigarrow} \delta.P}$$

$$\frac{}{(s+t).P \stackrel{s}{\rightsquigarrow} (t).P}$$

$$\frac{}{(t).P \stackrel{t}{\rightsquigarrow} P}$$

$$\frac{P \stackrel{s}{\rightsquigarrow} P'}{(t).P \stackrel{s+t}{\rightsquigarrow} P'}$$

$$\frac{P \stackrel{t}{\rightsquigarrow} P', \; Q \stackrel{t}{\rightsquigarrow} Q'}{P + Q \stackrel{t}{\rightsquigarrow} P' + Q'}$$

$$\frac{P \stackrel{t}{\rightsquigarrow} P'}{P \oplus Q \stackrel{t}{\rightsquigarrow} P'} (|Q|_{\mathcal{T}} < t)$$

$$\frac{Q \stackrel{t}{\rightsquigarrow} Q'}{P \oplus Q \stackrel{t}{\rightsquigarrow} Q'} (|P|_{\mathcal{T}} < t)$$

$$\frac{P \stackrel{t}{\rightsquigarrow} P', \; Q \stackrel{t}{\rightsquigarrow} Q'}{P \oplus Q \stackrel{t}{\rightsquigarrow} P' \oplus Q'}$$

$$\frac{P \stackrel{t}{\rightsquigarrow} P', \; Q \stackrel{t}{\rightsquigarrow} Q'}{P \mid Q \stackrel{t}{\rightsquigarrow} P' \mid Q'}$$

$$\frac{P \stackrel{t}{\rightsquigarrow} P'}{P \backslash L \stackrel{t}{\rightsquigarrow} P' \backslash L}$$

$$\frac{P \stackrel{t}{\rightsquigarrow} P'}{P[S] \stackrel{t}{\rightsquigarrow} P'[S]}$$

$$\frac{P_i\{{}^{\mu\widetilde{x}.\widetilde{P}}/_{\widetilde{x}}\} \stackrel{t}{\rightsquigarrow} P'}{\mu_i\widetilde{x}.\widetilde{P} \stackrel{t}{\rightsquigarrow} P'}$$

Figure 1: Operational Rules

$$(A_1)\ \delta\delta x = \delta x \qquad (A_2)\ \delta(t)x = \delta\mathbf{0}$$

$$(A_3)\ (s)(t)x = (s+t)x$$

$$(\oplus_1)\ (x \oplus y) \oplus z = x \oplus (y \oplus z) \qquad (\oplus_2)\ x \oplus y = y \oplus x$$

$$(\oplus_3)\ x \oplus x = x \qquad (\oplus_4)\ x \oplus \mathbf{0} = x$$

$$(\oplus_5)\ \delta x \oplus \delta y = \delta(x \oplus y) \qquad (\oplus_6)\ (t)x \oplus (t)y = (t)(x \oplus y)$$

$$(\oplus_7)\ \delta\mathbf{0} = (t)\delta\mathbf{0} \qquad (\oplus_8)\ \delta a x = a x \oplus (1)\delta a x$$

Figure 2: Equational Theory for $\oplus$

The Equational Theory of TCCS

n this section, we present equational laws for the calculus TCCS which are sound with respect o our semantic congruence $\sim$ defined above. We shall use $=$ to represent derivability in our quational theory, and $\equiv$ to represent syntactic identity modulo associativity and commutativity f the $\oplus$ operator.

As a start, we restrict our attention to the finite sequential subset of TCCS with only the eak choice operator, that is, that subset given by the following syntax.

$$P \ ::= \ \mathbf{0} \ \big| \ a.P \ \big| \ (t).P \ \big| \ \delta.P \ \big| \ P \oplus P$$

Ve shall refer to this subcalculus as TCCS_0.

Our equational theory for TCCS_0 is presented in Figure 2. We can straightforwardly show hat these laws are sound with respect to our observational congruence $\sim$ over the full calculus CCS. Hence we get the following result.

roposition 4.1 (Soundness) $p = q \implies p \sim q$.

Next, we want to show that these laws are complete for reasoning over TCCS_0. We can ccomplish this using the following normal forms.

efinition 4.2

- *A term is a* <u>*NF term*</u> *(or is* <u>*in NF*</u>*) if it is in* $\bigoplus NF$*, tNF, or* δNF*.*
- *A term is a* <u>$\bigoplus$ *NF term*</u> *(or is* <u>*in* $\bigoplus NF$</u>*) if it is of the form* $\bigoplus_{1 \le i \le n} a_i.p_i$ *where each of the* p_i*'s are in NF. In particular, taking* $n = 0$ *gives* $\mathbf{0}$ *as a* $\bigoplus NF$*.*

- *A term is a* <u>*tNF term*</u> *(or is* <u>*in tNF*</u>*) if it is of the form* $p \oplus (1).q$ *where* p *is in* $\bigoplus NF$ *an* q *is in NF. We shall also call* $(1)p$ *where* p *is in NF a tNF term, viewing it as shorthan notation for* $\mathbf{0} \oplus (1)p$.
- *A term is a* <u>*δNF term*</u> *(or is* <u>*in δNF*</u>*) if it is of the form* $\delta.p$ *where* p *is in* $\bigoplus NF$.

Our first task in proving completeness is to show that we can derive a normal form term fo every term $p \in \mathrm{TCCS}_0$.

Proposition 4.3 *Every term* $p \in \mathrm{TCCS}_0$ *can be equated to a term in NF using the laws o Figure 2.*

So for our completeness result, we only need to show that for *NF* terms P and Q, if $P \sim Q$ then $P = Q$. We prove this result by dividing it up depending on the types of normal form tha P and Q are in.

Proposition 4.4

i) If $P \sim Q$ *where* $P \equiv \bigoplus_{1 \leq i \leq m} a_i.p_i$ *is in* $\bigoplus NF$ *and* Q *is in NF, then* $Q \equiv \bigoplus_{1 \leq j \leq n} b_j.q_j$ *i in* $\bigoplus NF$*, and* $P = Q$.

ii) If $P \sim Q$ *where* $P \equiv P_0 \oplus (1).P_1$ *and* $Q \equiv Q_0 \oplus (1).Q_1$ *are in tNF, then* $P = Q$.

iii) If $P \sim Q$ *where* $P \equiv P_0 \oplus (1).P_1$ *is in tNF and* $Q \equiv \delta.Q_0$ *is in δNF, then* $P = Q$.

iv) If $P \sim Q$ *where* $P \equiv \delta.P_0$ *and* $Q \equiv \delta.Q_0$ *are in δNF, then* $P = Q$.

Finally we get our completeness result as a corollary of the preceeding lemmata.

Corollary 4.5 (Completeness) $p \sim q$ *implies* $p = q$.

We now consider adding the strong choice operator + to the language. What we find is tha the extended language is completely axiomatised by the laws of Figure 2 along with those o Figure 3. All of these laws are easily confirmed to be sound over the whole calculus. Notice in particular the soundness of the distributive law $(+\oplus_2)$. The opposite corresponding distributive law, namely

$$x \oplus (y + z) \;=\; (x \oplus y) + (x \oplus z)$$

is in fact not sound, as for instance,

$$(1).a \oplus \big((1).b + c\big) \;\not\sim\; \big((1).a \oplus (1).b\big) \;+\; \big((1).a \oplus c\big)$$

since the first term evolves in time 1 into the term a, whereas the second term evolves in time 1 into the term $a \oplus b$.

Not all of these laws are necessary for our completeness result; many are in fact derivable from the earlier $\oplus$ laws. However they are all included for two reasons: firstly, to note that the

$(+_1)$	$(x+y)+z = x+(y+z)$	$(+_2)$	$x+y = y+x$
$(+_3)$	$x+x = x$	$(+_4)$	$x+\delta\mathbf{0} = x$
$(+_5)$	$\mathbf{0}+(t)x = \mathbf{0}$	$(+_6)$	$\mathbf{0}+ax = ax$
$(+_7)$	$\mathbf{0}+\delta x = \mathbf{0}+x$	$(+_8)$	$ax+(t)y = ax$
$(+_9)$	$ax+\delta y = ax+y$	$(+_{10})$	$\delta x+\delta y = \delta(x+y)$
$(+_{11})$	$(t)x+(t)y = (t)(x+y)$	$(+_{12})$	$\delta x+(t)\delta x = \delta x$
$(+\oplus_1)$	$a.x+b.y = a.x\oplus b.y$	$(+\oplus_2)$	$x+(y\oplus z) = (x+y)\oplus(x+z)$

Figure 3: Equational Theory for +

sual expected laws are valid; and secondly to state another completeness theorem, namely that e laws $(A_1)\rightarrow(A_3)$ with $(+_1)\rightarrow(+_{12})$ completely characterise the sublanguage of sequential rms involving only the strong choice operator $+$ without the weak choice operator $\oplus$.

To accomplish the task of proving the present completeness result, we need simply justify e following proposition.

roposition 4.6 *For any two NF terms p and q, there is a term r not involving + such that* $+q=r$.

ence, strong choice is in fact a nonprimitive operator, definable in the language with only the xpressive weak choice operator.

When we add the parallel operator to the finite language, we get a complete theory by adding the above laws those of Figure 4. To do this we need only justify the following proposition.

roposition 4.7 *For any two NF terms p and q, there is a term r not involving | such that* $\mid q=r$.

ence, concurrency is also a nonprimitive operator, definable in the language with only the xpressive weak choice operator. However, this result was expected, as we are working within an terleaving semantic model.

We have not attempted to find a complete set of laws for the theory with strong choice and arallelism, without the weak choice operator. This is due to the weak expressive power of the rong choice operator. For instance, the term

$$(1)a\mid\delta b$$

Let $X = \bigoplus_{1\le i\le m} a_i x_i$ and $Y = \bigoplus_{1\le j\le n} b_j y_j$

$$(E_1)\quad X \mid Y = \bigoplus_{1\le i\le m} a_i(x_i \mid Y) \oplus \bigoplus_{1\le j\le n} b_j(X \mid y_j) \oplus \bigoplus_{a_i=\bar{b}_j} \tau(x_i \mid y_j)$$

$$(E_2)\quad X \mid (Y \oplus (1)y) = \bigoplus_{1\le i\le m} a_i(x_i \mid (Y \oplus (1)y)) \oplus \bigoplus_{1\le j\le n} b_j(X \mid y_j) \oplus \bigoplus_{a_i=\bar{b}_j} \tau(x_i \mid y_j)$$

$$(E_3)\quad (X \oplus (1)x) \mid Y = \bigoplus_{1\le i\le m} a_i(x_i \mid Y) \oplus \bigoplus_{1\le j\le n} b_j((X \oplus (1)x) \mid y_j)) \oplus \bigoplus_{a_i=\bar{b}_j} \tau(x_i \mid y_j)$$

$$\begin{aligned}(E_4)\quad (X \oplus (1)x) \mid (Y \oplus (1)y) = & \bigoplus_{1\le i\le m} a_i(x_i \mid (Y \oplus (1)y)) \\ & \oplus \bigoplus_{1\le j\le n} b_j((X \oplus (1)x) \mid y_j)) \\ & \oplus \bigoplus_{a_i=\bar{b}_j} \tau(x_i \mid y_j) \oplus (1)(x \mid y)\end{aligned}$$

$$(E_5)\quad \delta X \mid Y = \bigoplus_{1\le i\le m} a_i(x_i \mid Y) \oplus \bigoplus_{1\le j\le n} b_j(\delta X \mid y_j) \oplus \bigoplus_{a_i=\bar{b}_j} \tau(x_i \mid y_j)$$

$$(E_6)\quad X \mid \delta Y = \bigoplus_{1\le i\le m} a_i(x_i \mid \delta Y) \oplus \bigoplus_{1\le j\le n} b_j(X \mid y_j) \oplus \bigoplus_{a_i=\bar{b}_j} \tau(x_i \mid y_j)$$

$$\begin{aligned}(E_7)\quad \delta X \mid (Y \oplus (1)y) = & \bigoplus_{1\le i\le m} a_i(x_i \mid (Y \oplus (1)y)) \oplus \bigoplus_{1\le j\le n} b_j(\delta X \mid y_j) \\ & \oplus \bigoplus_{a_i=\bar{b}_j} \tau(x_i \mid y_j) \oplus (1)(\delta X \mid y)\end{aligned}$$

$$\begin{aligned}(E_8)\quad (X \oplus (1)x) \mid \delta Y = & \bigoplus_{1\le i\le m} a_i(x_i \mid \delta Y) \oplus \bigoplus_{1\le j\le n} b_j((X \oplus (1)x) \mid y_j)) \\ & \oplus \bigoplus_{a_i=\bar{b}_j} \tau(x_i \mid y_j) \oplus (1)(x \mid \delta Y)\end{aligned}$$

$$(E_9)\quad \delta X \mid \delta Y = \delta\Big(\bigoplus_{1\le i\le m} a_i(x_i \mid \delta Y) \oplus \bigoplus_{1\le j\le n} b_j(\delta X \mid y_j) \oplus \bigoplus_{a_i=\bar{b}_j} \tau(x_i \mid y_j)\Big)$$

Figure 4: Equational Theory for Parallel Terms

s no equivalent term in this sublanguage which does not involve the parallel operator. This is not the case with the weak choice operator, which can be used to eliminate all occurrences the strong choice and parallel operators from any (finite) expression.

Conclusions and Future Extensions

this paper, we have introduced a model for describing and reasoning about concurrent and mmunicating processes which incorporates the analysis of the temporal properties of such ocesses. We have defined our calculus, an extension of the calculus CCS, and have provided veral examples to demonstrate its utility. We furthermore have presented for it a formal mantics, based on the strong bisimulation equivalence of CCS, and have derived a sound and mplete equational axiomatisation for reasoning about finite processes.

We have ignored the restriction and relabelling operators in our study of the equational eory as they add no complications. Both operators distribute through the sequential operators expected (*cf* the analogous laws in CCS), so the theory for finite processes with these two tra operators can be trivially completely axiomatised as well. We could as well present more neral Expansion Theorem laws, as is the case in CCS.

In developing our calculus within as general a framework as we could, we eventually decided to strict ourselves to using the positive integers as our time frame. We could equally have defined ır calculus over some dense set of times, for instance the positive rationals, or the positive reals. owever, there were two reasons for which we elected to work within a discrete time model. The rst point is a pragmatic one: a discrete time model often intuitively more closely matches the stem model. For example, digital hardware can only execute in discrete time steps, the period the time step being dictated by the speed at which the hardware components stabilise into ew states. The second point is theoretical: the theory is much simpler. Had we used a dense me, we would have run into problems with our completeness proofs. For the present theory, we ave a simple graph model based on our transitional semantics with which to found our proofs; is model would not be anything as simple if we had used a dense time (more specifically, a me domain which did not contain a least time).

Working over some dense time, we would maintain the soundness of most, but not all of our xioms. For instance, for the sequential weak choice language, the law

$$\oplus_8 \quad \delta ax \;=\; ax \oplus (1)\delta ax$$

ould be invalid, as the term on the left hand side would be able to perform an a action after /2 unit of time, whereas this is not true for the term on the right hand side. We would in fact onjecture that replacing this single law by the two laws

$$\begin{aligned} \oplus_{8_1} \quad \delta ax &= \delta ax \oplus ax \\ \oplus_{8_2} \quad \delta ax &= \delta ax \oplus (t)\delta ax \end{aligned}$$

the theory presented in Figure 2 would give us a sound and complete theory for reasoning ver dense times. Note that these two new laws are sound over the discrete time model, as they re derivable from the equational theory using the law $\oplus_8$, but that they cannot replace the law $_8$ in the discrete model.

Furthermore, we believe the soundness and completeness result for the strong choice languag to remain valid for the dense time model as well, and that adding this to our refined theory fc the weak choice operator gives a sound and complete axiomatisation for the full finite sequenti language with dense times. Finally, we conjecture that a sound and complete axiomatisatio for this language with the parallel operator is provided by including the expansion theorem law (E_1), (E_2), (E_3), (E_5), (E_6) and (E_9) of Figure 4 refined by replacing each "(1)" time prefix b the more general "(t)" time prefix, along with somewhat more complicated refinements of th laws (E_4), (E_7) and (E_8).

This work represents the first steps in the formal development of a framework for reasonin about the temporal properties of processes defined within a mathematical algebra, and as such there are many future extensions which we are presently pursuing. Firstly, we have not tackle here the problem of abstracting away from silent τ actions, as hinted at in the example the alternating bit protocol. We are presently attempting to find the right approach to th abstraction of the internalised functional behaviour of processes, ideally in a fashion closel related to the successful approach taken in CCS. Furthermore we mention the possible extensio of the calculus to actual timing analyses. Ideally, we would like to define certain preorders whic would be defined between functionally equivalent terms but would distinguish between thei relative speeds. This presents another aspect of the calculus which we are currently pursuing.

Bibliography

[Bae89] Baeten, J.C.M., J.A. Bergstra, *"Real Time Process Algebra"*, Preliminary Draft, 10/20/89 1989.

[Cas87] Castellani, I., M. Hennessy, *"Distributed Bisimulation"*, University of Sussex, Departmen of Computer Science Report No. 5/87, July 1987.

[Dir58] Dirac, P.A.M., *Principles of Quantum Mechanics, 4th Edition*, Oxford, 1958.

[Gro90] Groote, J.F., *"Specification and Verification of Real Time Systems in ACP"*, preprint o the Centre for Mathematics and Computer Science, 1990.

[Jef89] Jeffrey, A., *"Synchronous CSP"*, Oxford University, to appear.

[Koy83] Koymans, R., J. Vytopil, W.P. de Roever, *"Real-Time and Asynchronous Message Passing* Technical Report RUU-CS-83-9, University of Eindhoven, 1983.

[Mil80] Milner, R., *A Calculus of Communicating Systems*, Lecture Notes in Computer Science 92 Springer-Verlag, 1980.

[Mil89] Milner, R., *Communication and Concurrency*, Prentice–Hall International, 1989.

[Mol89] Moller, F., C. Tofts, *"A Temporal Calculus of Communicating Systems"*, University o Edinburgh Report No. LFCS-89-104, 1989.

[Par81] Park, D.M.R., *"Concurrency and Automata on Infinite Sequences"*, Proceedings of the 5th G.I. Conference, Lecture Notes in Computer Science 104, Springer–Verlag, 1981.

[Plo81] Plotkin, G.D., *"A Structured Approach to Operational Semantics"*, DAIMI FN-19, Computer Science Department, Aarhus University, 1981.

e86] Reed, G.M., A.W. Roscoe, *"A Timed Model for CSP*, Proceedings ICALP '86, Lecture Notes in Computer Science 226, 1986.

h82] Schiff, L.I., *Quantum Mechanics, 3rd Edition*, McGraw–Hill, 1982.

f88] Tofts, C., *"Temporal Ordering for Concurrency*, University of Edinburgh Report No. LFCS-88-49, 1988.

f89] Tofts, C., *"Timing Concurrent Processes"*, University of Edinburgh Report No. LFCS-89-103, 1989.

Proving termination of communicating programs

Paweł Pączkowski *

LFCS, Department of Computer Science
University of Edinburgh
King's Buildings, Edinburgh EH9 3JZ, UK

Abstract

We demonstrate an assertional proof technique for showing termination of CSP-like programs. We introduce the notion of *annotation* which generalizes the idea of annotated program or proof outline. Our annotation is defined as a branching structure of predicates of the assertion language. The framework of annotations allows us to adopt the method of well founded counters. The resulting methodology for doing termination proofs is sound and complete for arithmetical interpretations. Moreover, we do not need to introduce auxiliary variables into the verified programs. Instead of using auxiliary variables to encode references to control flow, what is normally done, we factor out the reasoning on the flow of control from the assertional correctness proof. The separated reasoning on the flow of control can be easily mechanized. An example is worked out in which termination of a program for set partitioning is proved. The presented approach is, in fact, only a part of a more comprehensive methodology which is developed in the full version of this paper.

1 Introduction

In this paper we demonstrate an assertional proof technique for showing termination of CSP-like programs. In fact, we present only a part of a more comprehensive methodology for proving partial correctness, deadlock freedom and termination of concurrent programs.

We leave aside approaches to program verification that are based on history variables [Misra Chandy 81, Soundararajan 83, ZRE 85]. Such proof techniques are natural for many message-passing programs but there are examples, like set partitioning [Dijkstra 82], where introducing the whole history of communication into reasoning would seem excessive. Proof systems more suitable for this case have been proposed in [AFR 80, Levin Gries 81]. Termination is dealt with in [Levin Gries 81] but completeness of the proof system is not investigated there. The proof system of [AFR 80] can be also modified so that it can be used to prove total correctness [Apt 83]. Again, no completeness result for so extended proof system was shown. Moreover, both proof systems contain the auxiliary variable rule which permits adding auxiliary variables and statements to programs for the purpose of verification and which is necessary even for doing partial correctness proofs. The role of auxiliary variables is to encode some information on the flow of control or history of the computation. So encoded information when used in assertions becomes entangled in state predicates and can obscure the clarity of argument.

*on leave from Institute of Mathematics, University of Gdańsk, 80-952 Gdańsk, Poland.

Our proof methodology does not require using auxiliary variables. We introduce the notion of *nnotation* which generalizes the idea of annotated program or proof outline. Annotation is de- ned as a transition system with predicates of the assertion language attached to configurations. he framework of annotations allows us to adopt the method of well founded counters. The result- g methodology for doing termination proofs is sound and complete for arithmetical interpretations. nnotations can be also used for verification of program properties other than termination as demon- rated in the full version of this paper [Pączkowski 89].

Additionally, our proof technique can be viewed as a modest step towards mechanization of the rogram verification process. Rather than encoding references to control flow in auxiliary variables we ctor out the reasoning on the flow of control from the assertional correctness proof. The separated asoning on control flow can be easily mechanized.

A concept similar to annotations was exploited by in [Brookes 86] to build proof systems for partial rrectness and deadlock freedom for CSP. Brookes' concern, however, is compositionality resulting more complex annotations than presented here. He also does not tackle total correctness. Our pproach is not compositional in this sense that proving a property of parallel composition $S_1 \parallel S_2$ is ot decomposed into proofs of properties of S_1 and S_2. We believe that hierarchical development of roofs can be recovered by introducing action refinement. We do not study this issue in this paper ut we indicate some potential for action refinement in the example we work out in the last section.

The paper is organized as follows. After some preliminary definitions we specify the CSP-like nguage we are going to use. Its operational semantics is presented next, where structural description f control flow receives a special attention. Then annotations are introduced and, in Section 7, a roof technique for termination is developed. Soundness and completeness proofs are presented. We nd with an example applying our proof technique to the set partitioning program.

2 Preliminaries

labelled transition system (lts in short) is a triple $(C, Act, \longrightarrow)$, where C is a set of configurations, *ct* is a set of actions and $\longrightarrow \subset C \times Act \times C$ is a transition relation. We write $c \xrightarrow{\alpha} c'$ if $(c, \alpha, c') \in \longrightarrow$. A *path* in an lts is a finite or infinite sequence of its transitions $c_i \xrightarrow{\alpha_i} c_i'$, $i = 0, 1, \ldots$ ich that $c_i' = c_{i+1}$.

We will be also considering *extended* lts's $(C, Act, \longrightarrow, I)$, where $(C, Act, \longrightarrow)$ is an lts and I a set of distinguished initial configurations, $I \subset C$. An (extended) lts is called finite if its sets of onfigurations and actions (and, hence, transition relation) are finite.

We introduce the following important notion

efinition Let $A_i = (C_i, Act_i, \longrightarrow_i, I_i)$, for $i = 1, 2$. We say that ρ is a *simulation* from A_1 to A_2 ρ is a function from configurations of A_1 into configurations of A_2 such that

1. ρ preserves initial configurations, i.e. if $c \in I_1$ then $\rho(c) \in I_2$
2. if $c \xrightarrow{\alpha}_1 c'$ then $\rho(c) \xrightarrow{\alpha}_2 \rho(c')$

Ve also say that A_2 simulates A_1 if there is a simulation from A_1 to A_2.

In the following we assume some first order assertion language $\mathcal{P}$ with equality. The terms and uantifier free formulas of our assertion language will be used as, respectively, assignable expressions nd boolean expressions of the programming language we will be dealing with. We also assume ome fixed interpretation of the assertion language and whenever we talk about semantical validity r satisfaction of formulas we mean it relative to our fixed interpretation.

Valuations of variables, i.e. functions from the set of variables of $\mathcal{P}$ to the domain of the fixe interpretation, will be called states. St will denote the set of all states and σ will range over it. F a predicate p belonging to $\mathcal{P}$ $[\![p]\!]$ will denote the set $\{\sigma \in St \mid \sigma \models p\}$. Σ will range over subsets St. We write $\Sigma \models p$ when $\forall \sigma \in \Sigma \ \sigma \models p$.

3 Programming language

The proposed program verification method will be applied to a CSP-like programming language which provides message passing through channels. Below follows the abstract syntax of its statements S and guarded statements G, where a belongs to a class of atomic statements which include assignments, b ranges over boolean expressions and c ranges over communication statements whic as usually in CSP, can be of two kinds: $ch!t$ or $ch?x$, where ch is a channel name, t a term and x variable. For simplicity, no scoping rules are assumed for channels.

$$G \;=\; b \rightarrow S \mid b;c \rightarrow S \mid G \,[\!]\, G$$

$$S \;=\; a \mid c \mid \mathbf{do}\ G\ \mathbf{od} \mid \mathbf{if}\ G\ \mathbf{fi} \mid S;S \mid S \parallel S$$

We introduce the notion of *atomic action* (not to be confused with atomic statement). The set atomic actions consists of atomic statements, boolean expressions and the constructs $b_1;c_1 \parallel b_2;c$ representing performed communications, where one of c_i is an input action and the other is an outpu action on the same channel (such communication actions are called *matching*).

The structural operational semantics for the above programming language can be defined in standard way along the lines of [Plotkin 81]. We will depart slightly from this standard procedur by defining first a transition system which describes only possible flows of control in programs. Th actual operational semantics will be obtained by interpreting semantically transitions of control flow We proceed in such a way because anyhow we will need the definition of control flow later on.

4 Control flow

We define an lts called *CF* representing possible control flows of programs. The set of configuration of *CF* consists of the statements of S plus the additional symbol ε for the finished computation. transition $S \xrightarrow{\alpha} T$ is to be understood as a possibility of statement S to perform an action α afte which statement T will remain to be executed. The transitions of *CF* are defined in two steps. Firs an auxiliary transition relation $\longrightarrow_p$ of potential transitions is defined, then $\longrightarrow_p$ is restricted t give $\longrightarrow$, the actual transition relation of *CF*. Transitions $\longrightarrow$ are labelled with atomic action of S, transitions $\longrightarrow_p$ can be additionally labelled with "unmatched" communication guards $b;c$ Table 1 presents the axioms and rules for deriving the potential transitions of *CF*. The rule belov restricts $\longrightarrow_p$ to $\longrightarrow$.

$$\frac{S \xrightarrow{\alpha}_p S' \mid \varepsilon}{S \xrightarrow{\alpha} S' \mid \varepsilon} \quad \text{for } \alpha \text{ an atomic action of } S$$

As is usually done, $\dfrac{S \xrightarrow{\alpha} T_1 \mid T_2}{S' \xrightarrow{\beta} T_1' \mid T_2'}$ abbreviates the two rules $\dfrac{S \xrightarrow{\alpha} T_1}{S' \xrightarrow{\beta} T_1'}, \dfrac{S \xrightarrow{\alpha} T_2}{S' \xrightarrow{\beta} T_2'}$.

Note, that guarded statements G and the additional symbol *fail* are not configurations of *CF* although they can appear in derivations of transitions.

$$(b \rightarrow S) \xrightarrow{b}_p S \qquad (b \rightarrow S) \xrightarrow{\neg b}_p \mathit{fail}$$

$$(b\,;c \rightarrow S) \xrightarrow{b;c}_p S \qquad (b\,;c \rightarrow S) \xrightarrow{\neg b}_p \mathit{fail}$$

$$\frac{G_1 \xrightarrow{\alpha}_p S}{G_1 \mathbin{[\!]} G_2 \xrightarrow{\alpha}_p S} \qquad \frac{G_2 \xrightarrow{\alpha}_p S}{G_1 \mathbin{[\!]} G_2 \xrightarrow{\alpha}_p S}$$

$$\frac{G_1 \xrightarrow{b_1}_p \mathit{fail} \qquad G_2 \xrightarrow{b_2}_p \mathit{fail}}{G_1 \mathbin{[\!]} G_2 \xrightarrow{b_1 \wedge b_2}_p \mathit{fail}}$$

$$a \xrightarrow{a}_p \varepsilon \qquad c \xrightarrow{true;c}_p \varepsilon$$

$$\frac{S \xrightarrow{\alpha}_p S' \mid \varepsilon}{S\,;T \xrightarrow{\alpha}_p S';T \mid T}$$

$$\frac{G \xrightarrow{\alpha}_p S}{\textbf{if } G \textbf{ fi} \xrightarrow{\alpha}_p S}$$

$$\frac{G \xrightarrow{\alpha}_p S \mid \mathit{fail}}{\textbf{do } G \textbf{ od} \xrightarrow{\alpha}_p S\,;\textbf{do } G \textbf{ od} \mid \varepsilon}$$

$$\frac{S \xrightarrow{\alpha}_p S' \mid \varepsilon}{S \parallel T \xrightarrow{\alpha}_p S' \parallel T \mid T} \qquad \frac{T \xrightarrow{\alpha}_p T' \mid \varepsilon}{S \parallel T \xrightarrow{\alpha}_p S \parallel T' \mid S}$$

$$\frac{S_1 \xrightarrow{b_1;c_1}_p S_1' \mid \varepsilon \qquad S_2 \xrightarrow{b_2;c_2}_p S_2'}{S_1 \parallel S_2 \xrightarrow{b_1;c_1 \parallel b_2;c_2}_p S_1' \parallel S_2' \mid S_2'} \quad \text{for matching } c_1, c_2$$

$$\frac{S_1 \xrightarrow{b_1;c_1}_p S_1' \mid \varepsilon \qquad S_2 \xrightarrow{b_2;c_2}_p \varepsilon}{S_1 \parallel S_2 \xrightarrow{b_1;c_1 \parallel b_2;c_2}_p S_1' \mid \varepsilon} \quad \text{for matching } c_1, c_2$$

Table 1. Inference system for deriving potential transitions. S, T, possibly indexed and primed, range over statements of $\mathcal{S}$ (not over ε); G, G_1, G_2 range over guarded statements (not *fail*).

5 Semantics

By operational semantics we mean an lts whose configurations are pairs $\langle\sigma\,,\,S\rangle$, where σ is a stat and S a configuration of *CF*. Transitions of operational semantics are obtained by interpreting se mantically the transitions of control flow:

$$\frac{S \xrightarrow{\alpha} S' \mid \varepsilon \qquad\qquad (\sigma,\sigma') \in [\![\alpha]\!]}{\langle\sigma\,,\,S\rangle \xrightarrow{\alpha} \langle\sigma'\,,\,S'\rangle \mid \langle\sigma'\,,\,\varepsilon\rangle}$$

where $[\![\alpha]\!] \subset St \times St$ is a relational semantics of atomic action α, defined as follows for the specifi cases of assignment, boolean test and communication action:

$$\begin{array}{lll} (\sigma,\sigma') \in [\![x := t]\!] & \text{iff} & \sigma' = \sigma[\sigma(t)/x] \\ (\sigma,\sigma') \in [\![b]\!] & \text{iff} & \sigma \models b \text{ and } \sigma' = \sigma \\ (\sigma,\sigma') \in [\![b_1; c_1 \parallel b_2\,; c_2]\!] & \text{iff} & \sigma \models b_1 \wedge b_2 \text{ and } \sigma' = \sigma[\sigma(t)/x], \\ & & \text{where one of } c_i \text{ is } ch!t \text{ and the other } ch?x \end{array}$$

For each atomic action a other than assignment we assume some predefined *total* relation $[\![a]\!]$.

In the following definitions the two transition systems *CF* and operational semantics are restricte to describe just single programs.

Definition Control flow *of a program* S, denoted $CF(S)$, is an extended lts obtained by adding to C the single initial configuration S and restricting *CF* to the reachable part, i.e. to the configuratior and transitions that can be reached from the initial configuration.

Definition For a program S and a set of states Σ the *behaviour of S from initial states* Σ, denote $Beh(S,\Sigma)$, is defined as an extended lts obtained by adding to the operational semantics the set c initial configurations $\{\langle\sigma\,,\,S\rangle \mid \sigma \in \Sigma\}$ and restricting this lts to the reachable part.

Proposition 1 *$CF(S)$ is a finite extended lts for any program S.*

Proof. It can be shown inductively that only a finite number of configurations and transitions i reachable from S in *CF*. □

Proposition 2 *$CF(S)$ simulates $Beh(S,\Sigma)$ for any S,Σ.*

Proof. First we check that if $\langle\sigma\,,\,T\rangle$ is a configuration of $Beh(S,\Sigma)$ then T is a configuratio of $CF(S)$, i.e. T is reachable in *CF* from the configuration S. This can be done by induction o the length of a minimal path from any initial configuration of $Beh(S,\Sigma)$ to $\langle\sigma\,,\,T\rangle$. Consequently $\rho(\langle\sigma\,,\,T\rangle) = T$ is a well defined function and it is easy to check that ρ is a simulation. □

6 Annotations

As the name suggests, annotations are a generalization of annotated programs, or proof outlines which frequently appear in Hoare style program correctness proofs. The central role of proof out lines as a basis for Hoare style logics for concurrency was exposed in [de Roever 85] and [Schneide Andrews 86].

Definition Let *Ind* be a set of indices. An annotation is a finite extended lts $A = (P_A, Act_A, \longrightarrow, i$ whose configurations are indexed predicates of the assertion language, $P_A \subset \mathcal{P} \times Ind$, the action

Act_A are contained in the set of the atomic actions of the considered programming language and i_A is a distinguished initial configuration.

The indices are used to allow the same predicate to appear in different configurations of an annotation. Otherwise the indices are not important; in particular we say that a configuration is satisfied on a state σ when the predicate-part of the configuration is satisfied on σ.

Annotations can be thought of as finitary characterizations of infinite, in general, behaviours. Such characterizations are fine enough to enable termination proofs.

7 Termination

Definition $Beh(S, \Sigma)$ is *terminating* if there are no infinite paths in it.

We adopt the familiar idea of well founded loop counters. We assume that the assertion language and its interpretation are such that counters ranging over a well founded set can be defined, i.e. there are

a well founded set WF in the domain of the interpretation

a unary predicate wf such that $\sigma \models wf(l)$ iff $\sigma(l) \in WF$

a binary predicate $\prec$ interpreted as the ordering relation on WF

We assume that all predicates appearing in configurations of annotations used for termination proofs have a designated free variable l which will be interpreted over WF (to assure this, each predicate p has a conjunct $wf(l)$).

Definition An annotation A is *decreasing in* l if for each transition $p \xrightarrow{\alpha} q$ of A the partial correctness triple $\{p\}\,\alpha\,\{\exists l'\ l' \prec l \wedge q[l'/l]\}$ is valid. The later condition means that $\forall w \in WF\ \exists v \in WF\ v \prec w \wedge \{p[w/l]\}\,\alpha\,\{q[v/l]\}$.

For simplicity the Hoare triple is understood here semantically. Instead we could assume some proof system for partial correctness of atomic actions. For example, such a proof system would contain the familiar axiom for the assignment $\{p[t/x]\}\ x := t\ \{p\}$ and the consequence rule.

Proposition 3 *If A simulates $Beh(S, [\![p]\!])$, A is decreasing in l and $p \supset \exists l\ i_A$ is satisfied then $Beh(S, [\![p]\!])$ is terminating.*

Proof. Consider $\langle\sigma_0, S\rangle \xrightarrow{\alpha_1} \langle\sigma_1, S_1\rangle \xrightarrow{\alpha_2} \cdots$, a sequence of transitions of $Beh(S, [\![p]\!])$. Since A simulates $Beh(S, [\![p]\!])$ and A is a decreasing in l there exist elements $w_i \in WF$ and predicates p_i in A such that $i_A \xrightarrow{\alpha_1} p_1 \xrightarrow{\alpha_2} \cdots$ and $\sigma_0 \models i_A[w_0/l]$, $\sigma_i \models p_i[w_i/l]$ and $w_{i+1} \prec w_i$. If $\langle\sigma_0, S\rangle \xrightarrow{\alpha_1} \langle\sigma_1, S_1\rangle \xrightarrow{\alpha_2} \cdots$ were an infinite path then the sequence $w_0, w_1, \ldots$ would be infinite, which is excluded. Thus there are no infinite paths in $Beh(S, \Sigma)$ □

Proposition 3 sets a pattern for doing termination proofs. The flow of control in computations of a program S is represented by the branching structure of an annotation A. This is formalized by postulating that A simulates $CF(S)$. Then, the rest of the proof is reduced to checking that A is decreasing in l, that is, to characterizing transitions of A, where no references to the flow of control are needed.

By Proposition 2, a simulation from $CF(S)$ to A induces a simulation from $Beh(S, \Sigma)$ to A which justifies the following

Corollary 1 *If A simulates $CF(S)$, A is decreasing in l and $p \supset \exists l\, i_A$ is satisfied then $Beh(S, [\![p$ is terminating.* □

The corollary above is an important observation because it allows to mechanize a part of th verification process.

Proposition 4 *There is an algorithm for checking whether an annotation simulates control flow a program.*

Proof. Follows from the fact that only finite transition systems are involved. □

Corollary 1 summarizes the proof technique we propose and ensures its soundness. Comparin this proof technique to approaches based on proof outlines we observe that in order to define a pro outline assertions are attached to points in the text of a program. However, there is a correspo dence between points in program's text and configurations of control flow and in some proofs th correspondence has to be taken into consideration and encoded in assertions with the help of auxi iary variables. We chose to annotate directly structures that correspond to control flows of verifie programs and establish this correspondence (i.e. simulation) in a separate step of the terminatio proof.

Next we show completeness of our proof technique for arithmetical interpretations. Our pro follows the idea used in [Apt 84] to deal with nondeterministic programs. For the definition arithmetical interpretation see [Harel 79]. Here we briefly summarize that the assertion language assumed to include the symbols of arithmetic which get their standard meaning in the arithmetic interpretation. Additionally, there exists a formula in $\mathcal{P}$ which provides the ability to encode finit sequences of elements of the interpretation domain in single elements.

Theorem 1 (Completeness for arithmetical interpretations) *Assume that the interpretation is arit metical. If $Beh(S, [\![p]\!])$ is terminating then there exists an annotation A such that A is decreasin in l, A simulates $CF(S)$ and $p \supset \exists l\, i_A$ is satisfied.*

Proof. Obviously, the numerals contained in the domain of the arithmetical interpretation are take as the well founded set needed for termination proofs. Let us replace *wf* and $\prec$ by more suggesti symbols *nat*, $<$.

Denote by x the vector of all free variables that appear in S or p. Let n, l be fresh variable ranging over naturals. For each configuration T of $CF(S)$ we will define a predicate $comp_T(x, n$ which is satisfied of a state σ if and only if a path of length $\sigma(n)$ starts from $\langle \sigma , T\rangle$ in $Beh(S, [\![p]\!])$

The required annotation A can be then obtained by attaching to each configuration T of $CF(S$ a predicate

$$p_T = nat(l) \wedge (\forall n > l \quad \neg comp_T(x, n)).$$

In other words, the above formula is satisfied of a state σ if there are no paths of length greater tha $\sigma(l)$ from $\langle \sigma , T\rangle$. The configurations of $CF(S)$ play the role of indices in A, the transition relatio and the initial configuration of A are induced from $CF(S)$.

Let us check that $p \supset \exists l\, p_S$ (p_S is the initial predicate of A). Let $\sigma \models p$. $Beh(S, [\![p]\!])$ is terminatin so $Beh(S, \{\sigma\})$ also does not have infinite paths from $\langle \sigma , S\rangle$. Thus $Beh(S, \{\sigma\})$ is a tree (has n loops). Since $Beh(S, \{\sigma\})$ is finitely branching, by Kõnig's lemma, $Beh(S, \{\sigma\})$ is finite and as result there exists a bound on the length of a path in $Beh(S, \{\sigma\})$, so $\sigma \models \exists l\, p_S$.

It is easy to see that A is decreasing in l. It remains to define $comp_T$. We give the mai idea (details can be found in [Pączkowski 89]). First we can define an auxiliary predicate $trans(u, z$

xpressing 'there is a transition in $Beh(S, [p])$ from the configuration encoded in z to the configuration ncoded in u', where some simple encoding of the behaviour's configurations is assumed. Now, in an rithmetical interpretation finite sequences can be encoded in single variables so computation paths onsisting of transitions described by $trans(u, z)$ can be represented by single variables. This enables s to define $comp_T$. □

Note, that in the proof above we did not need to extend the program S with auxiliary statements. 'his shows that auxiliary variables are indeed not needed in our proof technique.

8 Example

Ve work out one of standard examples, partitioning of a set [Dijkstra 82, AFR 80, Barringer 85]. iven two disjoint sets of integers S_0 and T_0, $S_0 \cup T_0$ has to be partitioned into two subsets S and T uch that $|S| = |S_0|$, $|T| = |T_0|$ and $max(S) < min(T)$.

The program *Set_Part* presented in Table 2 is a solution of the above problem. In contrast o [Dijkstra 82, AFR 80] we do not adopt the distributed termination convention but we achieve n equivalent effect by using the same boolean condition for termination of loops in both parallel omponents, i.e. we claim that our solution behaves in the same way as its counterpart which uses he distributed termination convention.

```
Set_Part  ≡  Small || Large

Small  ≡  mx := max(S) ;                          Large  ≡
          ch ! mx ;                                          ch ? y ;
          S := S − {mx} ;                                    T := T ∪ {y} ; mn := min(T) ;
          ch ? x ;                                           ch ! mn ;
          S := S ∪ {x} ; mx := max(S) ;                      T := T − {mn} ;
          do                                                 do
          mx > x ; ch ! mx →                                 mx > x ; ch ? y →
                S := S − {mx} ;                                    T := T ∪ {y} ; mn:= min(T) ;
                ch ? x ;                                           ch ! mn ;
                S := S ∪ {x} ; mx := max(S)                        T := T − {mn}
          od                                                 od
```

Table 2. Program for set partitioning.

In order to show that $Beh(Set\_Part, [true])$ is terminating we follow the pattern set in Corollary 1. The transitions of the required annotation A are given in Figure 1. It is a mechanizable part of the roof to check that A simulates $CF(Set\_Part)$.

A counter l ranging over the well founded set of integers that are greater than -9 will appear n predicates of A . The idea behind A is that during execution of *Set_Part* either the action $S := S - \{mx\}$ decreases the number of elements of S which are greater than $min(T)$, or the next xecution of the loop guards terminates both loops. The predicates appearing in A handle those two ases differently.

In order to shorten the notation define $d(S,T) = 8 \cdot |\{s \in S \mid s > min(T)\}|$, where $|X|$ denotes he cardinality of a set X. The predicates of A are listed below.

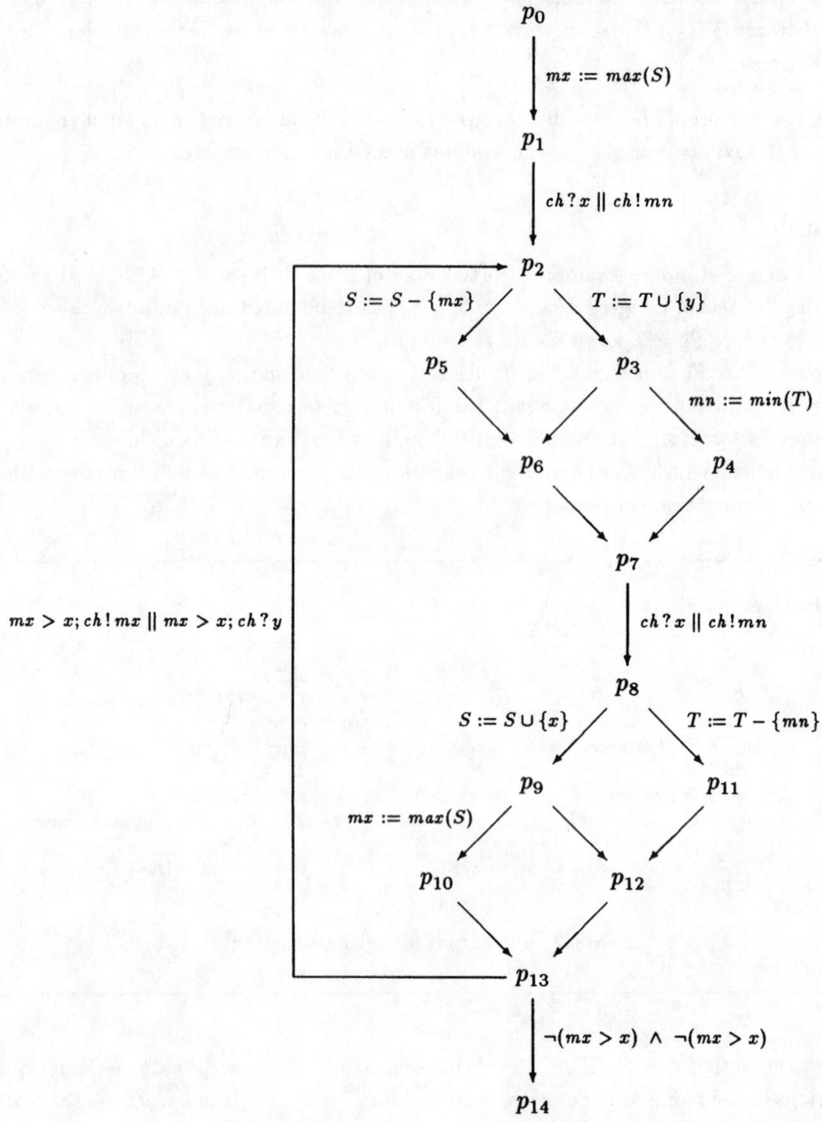

Figure 1. Transitions of annotation A. The arrows not marked with labels are assumed to have the same labels as the arrows parallel to them. p_0 is the initial predicate.

$$
\begin{array}{lll}
p_0 & \equiv & l = d(S,T) + 2 \\
p_1 & \equiv & l = d(S,T) + 1 \ \wedge \ mx = max(S) \\
p_2 & \equiv & l = d(S,T) \qquad \wedge \ mx = y = max(S) \\
p_3 & \equiv & l = d(S,T) - 1 \ \wedge \ mx = max(S) \\
p_4 & \equiv & l = d(S,T) - 2 \ \wedge \ mx = max(S) \ \wedge \ mn = min(T) \\
p_5 & \equiv & (l = d(S,T) + 7 \ \vee \ (max(S) \leq min(T) \ \wedge \ l = -1)) \ \wedge \ y > max(S) \\
p_6 & \equiv & (l = d(S,T) + 6 \ \vee \ (max(S) \leq min(T) \ \wedge \ l = -2)) \\
p_7 & \equiv & (l = d(S,T) + 5 \ \vee \ (max(S) \leq min(T) \ \wedge \ l = -3)) \ \wedge \ mn = min(T) \\
p_8 & \equiv & (l = d(S,T) + 4 \ \vee \ (max(S) \leq x \ \wedge \ l = -4)) \ \wedge \ x = mn = min(T) \\
p_9 & \equiv & (l = d(S,T) + 3 \ \vee \ (max(S) = x \ \wedge \ l = -5)) \\
p_{10} & \equiv & (l = d(S,T) + 2 \ \vee \ (mx = x \ \wedge \ l = -6)) \ \wedge \ mx = max(S) \\
p_{11} & \equiv & (l = d(S,T) + 3 \ \vee \ (max(S) \leq x \ \wedge \ l = -5)) \ \wedge \ x < min(T) \\
p_{12} & \equiv & (l = d(S,T) + 2 \ \vee \ (max(S) = x \ \wedge \ l = -6)) \\
p_{13} & \equiv & (l = d(S,T) + 1 \ \vee \ (mx = x \ \wedge \ l = -7)) \ \wedge \ mx = max(S) \\
p_{14} & \equiv & l = -8
\end{array}
$$

Annotation A defined above is decreasing in l. For example, let us check that for the transition $p_2 \stackrel{S:=S-\{mx\}}{\longrightarrow} p_5$ the triple $\{p_2\}\ S := S - \{mx\}\ \{\exists l'\ l' < l \ \wedge \ p_5[l'/l]\}$ is satisfied, i.e.

$$\{l = d(S,T) \ \wedge \ mx = y = max(S)\}$$
$$S := S - \{mx\}$$
$$\{\exists l'\ l < l' \ \wedge \ (l' = d(S,T) + 7 \ \vee \ (max(S) < min(T) \ \wedge \ l' = -1)) \ \wedge \ y > max(S)\}.$$

This can be readily verified by considering two cases: either $d(S,T) > 0$ and then the statement $S := S - \{mx\}$ decreases $d(S,T)$ by 8 or $d(S,T) = 0$ which implies that $max(S) < min(T)$ after the execution of $S := S - \{mx\}$ ensuring that l' can be taken equal to -1.

Finally, for each $i = 1, \ldots, 14$ $\ p_i \supset (l > -9)$ is satisfied which ensures that l ranges over a well founded set.

We end by indicating briefly the potential for action refinement in the example above. Suppose we would like to replace one of the atomic actions, say $T := T \cup \{y\}$, with a composite statement R. In order to establish that the termination property of *Set_Part* is preserved by such an action refinement it is enough to ensure that variables used in R are disjoint with variables of the process *Small* and that R actually implements $T := T \cup \{y\}$, in particular, that R is terminating.

Acknowledgements

I would like to thank Colin Stirling for bringing total correctness to my attention and also for many useful discussions and detailed comments on the drafts of this paper. Also, the comments of anonymous referees helped to improve the paper. This work was supported by University of Edinburgh Studentship and ORS Award. University of Gdańsk granted a leave from didactic duties.

References

[Apt 83] K.R. Apt, Formal justification of a proof system for communicating sequential processes, *Journal of the ACM* 30(1), pp. 197-216 (1983).

[Apt 84] K.R. Apt, Ten years of Hoare's Logic: a survey — Part II: nondeterminism, *Theoretical Computer Science* 28, pp. 83-109 (1984).

[AFR 80] K.R. Apt, N. Francez, W.P. de Roever, A proof system for communicating sequentia processes, *TOPLAS* 2(3), pp. 359–384 (1980).

[Barringer 85] H. Barringer, *A Survey of Verification Techniques for Parallel Programs*, LNCS 191 Springer-Verlag, 1985.

[Brookes 86] S.D. Brookes, A semantically based proof system for partial correctness and deadloc in CSP, in: *Proceedings 1986 LICS*, pp. 58-65.

[Dijkstra 82] E.W. Dijkstra, A correctness proof for communicating processes — A small exercise in: *Selected writtings on Computing: A Personal Perspective.* Springer-Verlag, 1982.

[Harel 79] D. Harel, *First-Order Dynamic Logic*, LNCS 68, Springer-Verlag, 1979.

[Levin Gries 81] G.M. Levin, D. Gries, A proof technique for communicating sequential processes *Acta Informatica* 15, pp. 159–172 (1981).

[Misra Chandy 81] J. Misra, K.M. Chandy, Proofs of Networks of Processes, *IEEE Transactions on Software Engineering*, vol. SE-7 (4), pp. 417–426, 1981.

[Pączkowski 89] P. Pączkowski, Proving correctness of concurrent programs without using auxiliary variables, ECS-LFCS-89-100, University of Edinburgh, 1989.

[Plotkin 81] G.D. Plotkin, A Structural Approach to Operational Semantics, DAIMI Report FN–19 Aarhus University, 1981.

[de Roever 85] W.P. de Roever, The quest for compositionality: a survey of assertion based proo systems for concurrent programs. Part 1, Technical Report RUU-CS-85-2, University of Utrecht 1985.

[Schneider Andrews 86] F.B. Schneider, G.R. Andrews, Concepts for concurrent programming, in *Current Trends in Concurrency. Overviews and Tutorials.* (J.W. de Bakker, W.P. de Roever, G Rozenberg, Eds.), LNCS 224, Springer-Verlag, 1986.

[Soundararajan 83] N. Soundararajan, Correctness proofs of CSP programs, *Theoretical Computer Science* 24(2), pp. 131–141 (1983).

[ZRE 85] J. Zwiers, W.P. de Roever, P. van Emde Boas, Compositionality and concurrent networks soundness and completeness of a proof system, in: *Proc. of ICALP 85*, pp. 509–519, LNCS 194, Springer-Verlag 1985.

Factorization of Finite State Machines under Observational Equivalence*

Huajun Qin Philip Lewis

Department of Computer Science
SUNY at Stony Brook
Stony Brook, NY 11794, USA

Abstract

A usual approach to designing a complex concurrent system is to follow the top-down design methodology: the abstract specification of the system is decomposed into a network of communicating modules such that the behavior of the modules in composition is equivalent to the behavior of the system specification.

The factorization problem is to construct the specification of a submodule X when the specifications of the system and all submodules but X are given. It is usually described by the equation $A|X \stackrel{e}{=} B$ where A and X are submodules of system B, | is a composition operator, and $\stackrel{e}{=}$ is the equivalence criterion.

In this paper we use a finite state machine (FSM) model consistent with CCS and study the factorization problem $A \| X \approx B$ where $\|$ is a derived CCS composition operator and $\approx$ represents observational equivalence. An algorithm is presented and proved correct to find the most general specification of submodule X for $A \| X \approx B$ with B deterministic. This paper extends and is based on the work of M.W. Shields.

Introduction

oncurrent systems are hard to design, verify and debug. A usual approach to designing complex system is to follow the top-down design methodology: the abstract specifica-on of the system is decomposed into a network of communicating modules such that ie *behavior* of the modules in *composition* is *equivalent* to the behavior of the system pecification. This process is repeated for each module until the specification of the ıbmodules can be easily implemented.

A problem that arises naturally in this methodology is the factorization problem (or ıbmodule construction problem), which is to construct the specification of a submodule when the specifications of the system and all submodules but X are given. This roblem can be described by the equation

$$A|X \stackrel{e}{=} B$$

here A and X are modules of B, | is a composition operator, and $\stackrel{e}{=}$ is the equiva-nce criterion. Different factoring methods might be necessary depending on the form

*Partially supported by the National Science Foundation under grant number CCR8822839

of the composition operators, parallelism models (asynchronous or synchronous), an equivalence criteria.

In this paper we use a finite state machine (FSM) model consistent with CCS [Mi80 and consider the factorization problem $A \| X \approx B$ where $\|$ is a derived CCS compositio operator and $\approx$ represents observational equivalence. An algorithm is presented an proved correct to find the most general solution R to $A \| X \approx B$ if solutions exist, when B and R are under certain constraints (specifically B is deterministic and R is rigid). Thi paper extends and is based on the work of M.W. Shields [Sh89]. The key observation i to use a set of pairs $[p, q]$, where p is a state of A and q is a state of B, to represent a state of X.

The paper is organized as follows: In section 2 we describe our finite state machin (FSM) model for specification. An algorithm to solve $A \| X \approx B$ is presented and prove correct in section 3. We discuss the related work in section 4 and conclude the pape with a proposal for future work in section 5.

2 Preliminaries

We view the specifications of a concurrent system and its submodules as finite stat processes and express them as FSMs. In this section we give the definitions for ou machine model and discuss the notation to be used in later sections.

Definition 1 *A finite state machine (FSM) is a quadruple* $M = (S, s_0, \Sigma, \Delta)$ *where*

(i) *S is a finite nonempty set of states,*

(ii) *s_0 is the initial state,*

(iii) *Σ is a finite nonempty set of action symbols with $\tau \notin \Sigma$,*

(iv) *$\Delta : S \times (\Sigma \cup \{\tau\}) \rightarrow 2^S$ is called the transition relation.*

We write $p \xrightarrow{a} q$ if $q \in \Delta(p, a)$. Also we write $p \xrightarrow{a} \star$ if $p \xrightarrow{a} q$ for some q; $p \not\xrightarrow{a} \star$ if there is no $q \in S$ such that $p \xrightarrow{a} q$. We define $\overset{\tau}{\Rightarrow}$ as $(\xrightarrow{\tau})^*$ and $\overset{a}{\Rightarrow}$ as $\overset{\tau}{\Rightarrow}\overset{a}{\Rightarrow}\overset{\tau}{\Rightarrow}$ for $a \neq \tau$.

When we use an FSM to model a finite state process, Σ is the set of the actions that the process can perform, and τ represents an (invisible) internal action.

For a set of action symbols Σ we define $\hat{\Sigma} = \Sigma \cup \{\overline{u} | u \in \Sigma\}$. We postulate $u = \overline{\overline{u}}$ for any action symbol. Next we formally define the combined behavior of a network of communicating FSMs:

Definition 2 *The* composition *of two FSMs* $M_1 = (S, s_0, \Sigma_1, \Delta_1)$ *and* $M_2 = (T, t_0, \Sigma_2, \Delta$ *is the machine* $M = M_1 \| M_2 = (S \times T, (s_0, t_0), \Sigma, \Delta)$ *where* $\Sigma = (\Sigma_1 \cup \Sigma_2) - (\hat{\Sigma_1} \cap \hat{\Sigma_2})$ *and* Δ *is defined as follows:*

1. *if* $s_i \xrightarrow{\lambda} s_j$ *in* M_1, *and* $\lambda \in (\Sigma \cup \{\tau\})$, *then* $(s_j, t) \in \Delta((s_i, t), \lambda)$ *for all* $t \in T$;
2. *if* $t_k \xrightarrow{\mu} t_l$ *in* M_2, *and* $\mu \in (\Sigma \cup \{\tau\})$, *then* $(s, t_l) \in \Delta((s, t_k), \mu)$ *for all* $s \in S$;
3. *if* $s_i \xrightarrow{\lambda} s_j$ *in* M_1, $t_k \xrightarrow{\mu} t_l$ *in* M_2, *and* $\lambda = \overline{\mu}$, *then* $(s_j, t_l) \in \Delta((s_i, t_k), \tau)$.

For states $p \in S$ and $q \in T$, we say *p and q have a λ communication* (in $M_1 \| M_2$) if $\xrightarrow{\lambda}\star$ and $q \xrightarrow{\overline{\lambda}}\star$. With respect to the composition machine $M_1 \| M_2$, we use the notation $Ex(M_1)$ to denote the set of symbols of M_1 appearing in $M_1 \| M_2$ and $Com(M_1)$ to denote the set of symbols of M_1 not appearing in $M_1 \| M_2$. Formally, $Ex(M_1) = \Sigma_1 - (\hat{\Sigma_1} \cap \hat{\Sigma_2})$, $Ex(M_2) = \Sigma_2 - (\hat{\Sigma_1} \cap \hat{\Sigma_2})$, $Com(M_1) = \Sigma_1 - Ex(M_1)$, and $Com(M_2) = \Sigma_2 - Ex(M_2)$. We identify a machine with its states that are reachable from its initial state. Readers familiar with CCS [Mi80] will notice that $M_1 \| M_2$ is the same as $(M_1 \mid M_2)\backslash L$ in CCS with $L = \hat{\Sigma_1} \cap \hat{\Sigma_2}$).

Observational equivalence was introduced in [Mi80] and is defined below as in [Mi88]:

Definition 3 (Milner) *A binary relation $\mathcal{R}$ on states is a weak bisimulation relation if for each $(p,q) \in \mathcal{R}$ and each $a \in \Sigma \cup \{\tau\}$, (1) and (2) hold:*
(1). Whenever $p \xrightarrow{a} p'$ then for some q', $q \overset{a}{\Rightarrow} q'$ and $(p',q') \in \mathcal{R}$;
(2). Whenever $q \xrightarrow{a} q'$ then for some p', $p \overset{a}{\Rightarrow} p'$ and $(p',q') \in \mathcal{R}$.
States p and q are observationally equivalent (we write $p \approx q$) if (p,q) is in a weak bisimulation relation.

Proposition 1 (Milner) $\approx$ *equals $\bigcup\{\mathcal{R} \mid \mathcal{R}$ is a weak bisimulation relation on state space $S\}$ and is an equivalence relation on S.*

Definition 4 *Two FSMs M and M′ are observationally equivalent if there exists a weak bisimulation $\mathcal{R}$ that contains the pair (s_0, s_0') of their initial states.*

Given below are some constraints on FSMs and their properties which are useful in section 3:

Definition 5 *An FSM M is* rigid *if it contains no τ transitions.*
An FSM is M is deterministic *if (1) M is rigid and (2) for every state p and every $u \in \Sigma_M$, $p \xrightarrow{u} p'$ and $p \xrightarrow{u} p''$ imply $p' = p''$.*

Proposition 2 *Let q be deterministic and suppose $p \approx q$:*

1. *if $p \xrightarrow{u} p'$ for some $u \neq \tau$, then (a) there must exist q' such that $q \xrightarrow{u} q'$ and $p' \approx q'$; (b) for every q' such that $q \xrightarrow{u} q'$, $p' \approx q'$.*
2. *for every p' such that $p \xrightarrow{\tau} p'$, $p' \approx q$.*

3 Factorization under Observational Equivalence

In this section we first give the definition of a *weak factorization equation* $A \| X \approx B$ and then present an algorithm that produces a solution R to the equation $A \| X \approx B$ if one exists, or reports failure otherwise. The correctness of the algorithm is proved and the solution produced by the algorithm is proved to be the most general one among all the possible solutions.

Definition 6 *A weak factorization equation is an expression of the form $A \| X \approx B$ where B is deterministic.*
R is a solution to a weak factorization equation $A \| X \approx B$ iff R is rigid , $Ex(R) = \Sigma_B - \Sigma_A$, $Com(R) = \{\overline{\lambda} \mid \lambda \in \Sigma_A - \Sigma_B\}$ and $A \| R \approx B$.

We observe that if FSMs B and B' are observationally equivalent and B is deterministic then any FSM R is a solution to weak factorization equation $A\|X \approx B$ if and only if R is a solution to $A\|X\approx B'$. Therefore the algorithm presented below for solving weak factorization equation can be applied to find solutions to the larger class of equations $A\|X\approx B$ with B observationally equivalent to a deterministic FSM in two steps: (1) find one determinstic FSM B' that is observationally equivalent to B (this can be achieved in polynomial time) and (2) apply the algorithm to solve $A\|X\approx B'$.

We consider only rigid FSMs as solutions to weak factorization equations, because for any R such that $A\|R\approx B$, a (rigid) solution R' can derived from R by collapsing states that are reachable through τ transitions into a single state and deleting all τ transitions. Furthermore if there exists a nondeterministic solution R (i.e. there exists state p in R such that $p \xrightarrow{u} p'$ and $p \xrightarrow{u} p''$ but $p' \neq p''$), a deterministic solution R' can be obtained from R by deleting all but one u transitions from R for each state and for each action symbol u.

Suppose R is a solution to $A\|X\approx B$. We represent the states in the composed machine $A\|R$ by pairs (s_A, s_R) where $s_A \in A$ and $s_R \in R$. If $(s_A, s_R)\approx s_B$ for some state $s_B \in B$ A at s_A can transition to a next state under an external or a τ action without R at s_R knowing it. Therefore R must anticipate and handle the consequence of asynchrony of A. This motivates the following definition and proposition:

Definition 7 *Define $(s_A, s_B) \rightarrow_I (s'_A, s'_B)$ iff one of the following holds:*

1. $u \in Ex(A), u \in \Sigma_B$, $s_A \xrightarrow{u} s'_A$ *and* $s_B \xrightarrow{u} s'_B$*;*

2. $u = \tau$, $s_A \xrightarrow{\tau} s'_A$ *and* $s_B = s'_B$.

Let $\Rightarrow_I$ to be the reflexive and transitive closure of relation $\rightarrow_I$.
Define $I(s_A, s_B) = \{[s'_A, s'_B] | (s_A, s_B) \Rightarrow_I (s'_A, s'_B)\}$.

Intuitively if $[s'_A, s'_B] \in I(s_A, s_B)$ then there must exist a sequence $\sigma \in (Ex(A) \cup \{\tau\})^*$ such that $s_A \xrightarrow{\sigma} s'_A$ and $s_B \overset{\sigma}{\Rightarrow} s'_B$.

Proposition 3 *If R is a solution to $A\|X \approx B$, then for any s_A, s_R, s_B satisfying $(s_A, s_R)\approx s_B$, it is true that $(s'_A, s_R)\approx s'_B$ for each $[s'_A, s'_B] \in I(s_A, s_B)$.*

Proof: Since $[s'_A, s'_B] \in I(s_A, s_B)$, there must exist a finite sequence of state pairs such that $(s^1_A, s^1_B) \rightarrow_I (s^2_A, s^2_B) \rightarrow_I \cdots \rightarrow_I (s^n_A, s^n_B)$ with $n \geq 0, (s_A, s_B) = (s^1_A, s^1_B)$ and $(s'_A, s'_B) = (s^n_A, s^n_B)$. Since $(s_A, s_B) \rightarrow_I (s^2_A, s^2_B)$ we have by definition: (1) $s_A \xrightarrow{u} s^2_A$ and $s_B \xrightarrow{u} s^2_B$ or (2) $s_A \xrightarrow{\tau} s^2_A$ and $s_B = s^2_B$. In case 1, $(s_A, s_R)\approx s_B$, $(s_A, s_R) \xrightarrow{u} (s^2_A, s_R)$ and $s_B \xrightarrow{u} s^2_B$ for some u. By proposition 2.1.b, we have $(s^2_A, s_R)\approx s^2_B$. In case 2, $(s_A, s_R)\approx s_B$ and $(s_A, s_R) \xrightarrow{\tau} (s^2_A, s_R)$. By proposition 2.2, we have $(s^2_A, s_R)\approx s^2_B$. Applying the above reasoning $n-1$ times, we can prove that $(s^n_A, s_R)\approx s^n_B$ which is $(s'_A, s_R)\approx s'_B$. □

The above proposition indicates that a state s_R must participate in the bisimulation with all pairs in $I(s_A, s_B)$ if we are to have $(s_A, s_R)\approx s_B$. This leads to the idea of using state pairs in $I(s_A, s_B)$ to identify a state s_R in R.

Next we present our algorithm to find a solution R to $A\|X \approx B$. In the algorithm description, we use a box (or a set) of state pairs $[s_A, s_B]$ to represent a state of the FSM R. The algorithm starts from the initial box containing state pairs in $I(s_{A_0}, s_{B_0})$, and creates all the boxes reachable through the transitions created along the way. It tries to aintain the invariant for each box X_i: if $[s_A, s_B] \in X_i$, then $s_A \| X_i \approx s_B$. Iteratively it ecks that if the invariant is maintained for all the boxes reachable from the initial box, nd deletes the transitions to those 'bad' boxes that can not maintain the invariant.

If the initial box is 'bad', the algorithm reports failure. Otherwise it outputs a solution achine that takes all the remaining 'good' boxes as its states.

The solution R may include some DON'T-CARE conditions in which in some state r some action symbol, the machine can go to any state. A DON'T-CARE transition R will never be enabled in $A \| R$. Since DON'T-CARE transitions can be instantiated go to any state, they are are useful in the minimization of solution machines. A state at is the target of a DON'T-CARE transition is called a DON'T-CARE state.

.1 An Algorithm to generate a solution to $A \| X \approx B$

tep 1 Generate all state pairs $[s_A, s_B]$ for all $s_A \in A$ and $s_B \in B$. Mark $[s_A, s_B]$ 'bad' if $s_A \xrightarrow{u} \star$ for some $u \in Ex(A)$ but $s_B \not\xrightarrow{u} \star$.

Let $Ex(X) = \Sigma_B - \Sigma_A$ and $Com(X) = \{\overline{\lambda} | \lambda \in \Sigma_A - \Sigma_B\}$.

tep 2 For all $s_A \in A, s_B \in B$, generate $I(s_A, s_B) = \{[s'_A, s'_B] | (s_A, s_B) \Rightarrow_I (s'_A, s'_B)\}$. For every $[s'_A, s'_B] \in I(s_A, s_B)$, create an edge labeled u (or a u link) from state $[s'_A, s'_B]$ to state $[s''_A, s''_B] \in I(s_A, s_B)$ if one of the following holds:

(1) $u \in Ex(A), u \in \Sigma_B, s'_A \xrightarrow{u} s''_A$ and $s'_B \xrightarrow{u} s''_B$;

(2) $u = \tau, s'_A \xrightarrow{\tau} s''_A$ and $s'_B = s''_B$.

tep 3 Create initial box X_0 containing state pairs in $I(s_{A_0}, s_{B_0})$ where s_{A_0} and s_{B_0} are the initial states of A and B. Mark box X_0 'unprocessed'. Associate with each box X a set of ports labeled with actions in $Ex(X)$ and $Com(X)$.
Do the following while there is an 'unprocessed' box X_i:

(1) Check all state pairs in box X_i: if there is a state pair $[s_A, s_B]$ in X_i marked 'bad' then mark box X_i 'bad' and 'processed'; else do (2) and (3) below, and mark X_i 'processed'.

(2) For each symbol $u \in Ex(X)$, do the following:
Let $X_i(u) = \bigcup\{I(s_A, s'_B) | [s_A, s_B] \in X_i, s_B \xrightarrow{u} s'_B \text{ in } B\}$.
If there exists $[s_A, s_B]$ in X_i such that $s_B \not\xrightarrow{u} \star$, then mark port u on box X_i 'bad' else do (a) and (b) below:

(a) Check if there is a box X_j containing exactly all state pairs in $X_i(u)$. If not, create such a box X_j and mark it 'unprocessed'.

(b) Create a transition (X_i, u, X_j). Create a link from $[s_A, s_B]$ in X_i to $[s_A, s'_B]$ in X_j labeled with u.

(3) For each symbol $\overline{\lambda}$ in $Com(X)$ do the following:
Let $X_i(\lambda) = \bigcup\{I(s'_A, s_B) | [s_A, s_B] \in X_i, s_A \xrightarrow{\lambda} s'_A \text{ in } A\}$.
if $X_i(\lambda) = \emptyset$ then create a new state called DON'T-CARE and a new transition $(X_i, \overline{\lambda}$, DON'T-CARE) else do (a) and (b) below:

(a) Check if there is a box X_j containing exactly all state pairs in $X_i(\lambda)$. If not, create such a box and mark it 'unprocessed'.

(b) Create a transition $(X_i, \overline{\lambda}, X_j)$. If $[s_A, s_B] \in X_i$ and $s_A \xrightarrow{\lambda} s'_A$, create a link from $[s_A, s_B]$ in X_i to $[s'_A, s_B]$ in X_j labeled $\overline{\lambda}$.

Step 4 repeat

(1) If there is a u transition ($u \in Ex(X) \cup Com(X)$) from a box X_j to a 'bad' box X_i, mark port u on box X_j 'bad'. Delete all u transitions from X_j. Delete all u links from state pairs in X_j to state pairs in X_i.

(2) Mark box X_i 'bad' if there exists a state pair $[s_A, s_B] \in X_i$ such that $s_B \xrightarrow{u} s'_B$ for some $u \in \Sigma_B$, but there does not exist a sequence of action links such that $[s_A, s_B] \xrightarrow{\lambda_1} [s^1_A, s_B] \xrightarrow{\lambda_2} \cdots \xrightarrow{\lambda_n} [s^n_A, s_B] \xrightarrow{u} [s'_A, s'_B]$ for some $n \geq 0$ and $\lambda_k \in Com(X) \cup \{\tau\}$ for $1 \leq k \leq n$

until there are no new bad boxes found in (1) and (2) above.

Step 5 If initial box X_0 is 'bad', report '$A \| X \approx B$ is not solvable'. Otherwise construct an FSM R as following:

1. The states of R are DON'T-CARE and the remaining 'good' (or not 'bad') boxes;
2. The action symbol alphabet Σ_R is $Ex(X) \cup Com(X)$;
3. The initial state of R is X_0;
4. The transitions of R are those remaining transitions between 'good' boxes and the transitions $(X_i, \overline{\lambda}, \text{DON'T-CARE})$ if they exist.

Output R and report 'R is a solution to $A \| X \approx B$'.

3.2 Solving Equation $A \| X \approx B$ — An Example

We now show by an example how to use our algorithm to solve the equation $A \| X \approx B$ where FSMs A and B are shown in figure 1. The solution generated by our algorithm is also shown in figure 1. We use $s \xrightarrow{\overline{\beta}} \star$ to denote the transition $s \xrightarrow{\overline{\beta}}$DON'T-CARE.

Below we give the traces of our algorithm in the course of obtaining solution R.

At step 1, the following state pairs are generated: $[A_0, B_0]$, $[A_0, B_1]$, $[A_0, B_2]$, $[A_1, B_0]$, $[A_1, B_1]$, $[A_1, B_2]$, $[A_2, B_0]$, $[A_2, B_1]$, $[A_2, B_2]$. $[A_0, B_1]$ is marked 'bad' because A_0 can make a transition under a but B_1 cannot.

At step 2, I sets are generated. $I(A_0, B_0) = \{[A_0, B_0], [A_1, B_1]\}$ and $I(A_0, B_2) = \{[A_0, B_2], [A_1, B_0]\}$. There is an '$a$' link from $[A_0, B_0]$ to $[A_1, B_1]$ for the states in $I(A_0, B_0)$, and an 'a' link from $[A_0, B_2]$ to $[A_1, B_0]$ for the states in $I(A_0, B_2)$. $I(p, q) = \{[p, q]\}$ when (p, q) is one of the following: (A_0, B_1), (A_1, B_0), (A_1, B_1), (A_1, B_2), (A_2, B_0), (A_2, B_1), (A_2, B_2).

At step 3, the initial box X_0 and all other boxes X_1–X_9 are created iteratively.

Boxes X_1, X_5, and X_8 are marked 'bad' because there are 'bad' state pairs in these boxes. Transitions between boxes and links between state pairs in different boxes are created according to the algorithm. They are shown in figure 1.

The small cells on the sides or tops of boxes contain the port labels b, λ and β. A 'bullet' sign • on a label or a state pair in a box means that the port or the state pair is marked 'bad'.

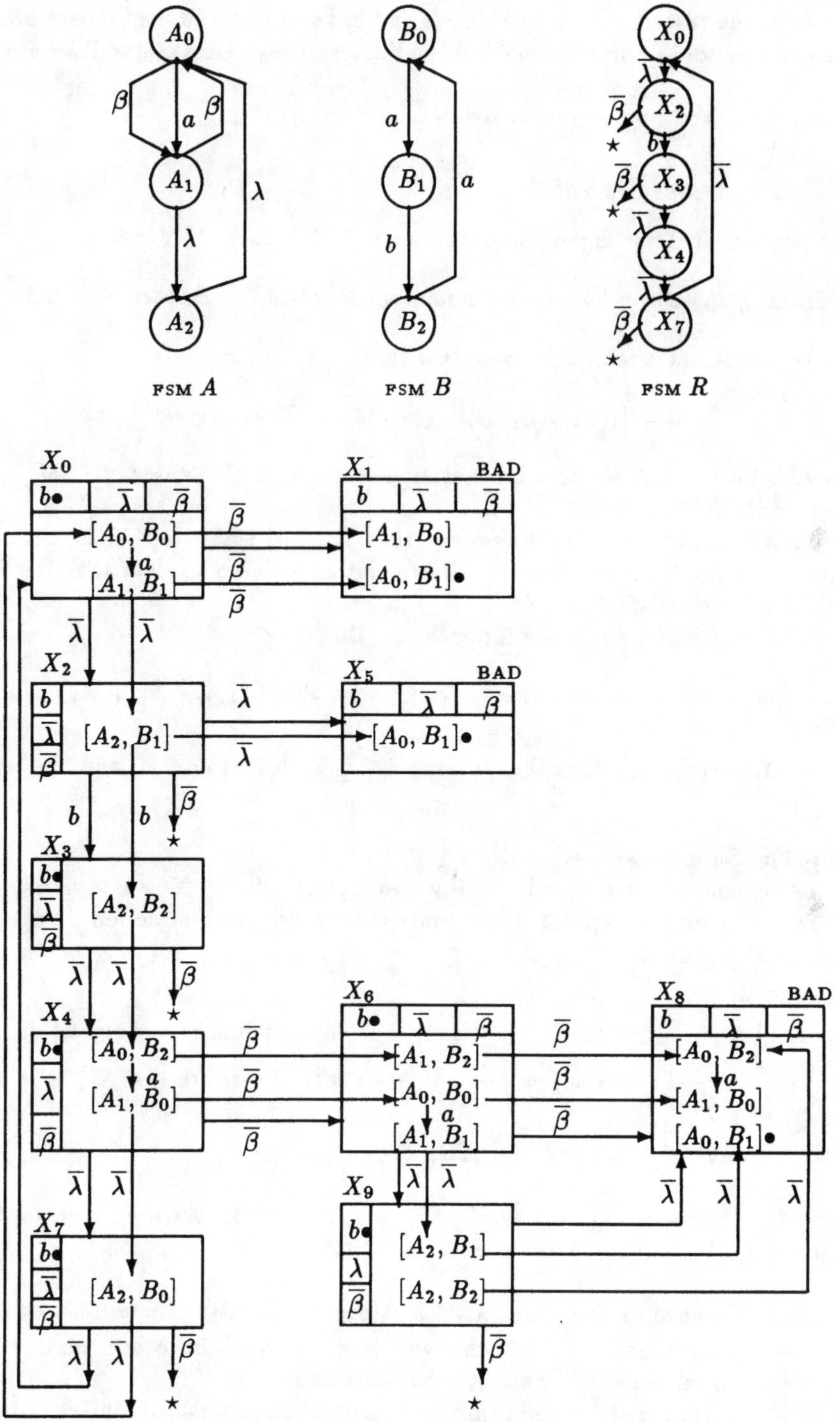

Figure 1: FSMs A, B and a solution R to $A \| X \approx B$

At step 4, the transitions to 'bad' boxes and links to states in 'bad' boxes are deleted and the corresponding ports are marked 'bad'. Iteratively boxes X_9 and X_6 are marked 'bad'.

At step 5 the solution R is reported as shown in figure 1.

3.3 Correctness Proof

We first prove that if our algorithm outputs an FSM R then $A \| R \approx B$.

Theorem 1 *If the algorithm reports a solution R to $A \| X \approx B$, then $A \| R \approx B$.*

Proof: We construct a relation *Bisim* such that:

$$Bisim = \{((s_A, X_i), s_B) | [s_A, s_B] \in X_i, X_i \text{ is a 'good' box}\}$$

and prove below that *Bisim* is a bisimulation relation on the states of FSM B and the composed FSM $A \| R$.

We observe that if (s_A, X_i) *Bisim* s_B, then $[s_A, s_B] \in X_i$ and X_i is 'good' by the definition of *Bisim*. To prove *Bisim* is a bisimulation we need to show that the following two conditions are satisfied:

Suppose (s_A, X_i) *Bisim* s_B. Then for each $u \in \Sigma_B \cup \{\tau\}$,

(1) if $s_B \xrightarrow{u} s'_B$, then there must exist (s'_A, X'_i) such that $(s_A, X_i) \stackrel{u}{\Rightarrow} (s'_A, X'_i)$ and (s'_A, X'_i) *Bisim* s'_B;

(2) if $(s_A, X_i) \xrightarrow{u} (s'_A, X'_i)$, then there must exist s'_B such that $s_B \stackrel{u}{\Rightarrow} s'_B$ and (s'_A, X'_i) *Bisim* s'_B.

Proof of (1): Suppose $s_B \xrightarrow{u} s'_B$ with $u \in \Sigma_B$.
$[s_A, s_B] \in X_i$ and X_i is a 'good' box by assumption. Since X_i was not marked as a 'bad' box at algorithm step 4.2, there must exist a sequence of action links such that $[s_A, s_B] \xrightarrow{\lambda_1} [s_A^1, s_B] \xrightarrow{\lambda_2} \cdots \xrightarrow{\lambda_n} [s_A^n, s_B] \xrightarrow{u} [s'_A, s'_B]$ for some $n \geq 0$, with $[s'_A, s'_B]$ in some X'_i and X'_i a good box.

Suppose $[s_A^j, s_B] \in X^j$ for $1 \leq j \leq n$. By the construction of our algorithm, $X_i \xrightarrow{\lambda_1} X^1 \xrightarrow{\lambda_2} X^2 \cdots \xrightarrow{\lambda_n} X^n \xrightarrow{u} X'_i$ and $s_A \xrightarrow{\overline{\lambda_1}} s_A^1 \xrightarrow{\overline{\lambda_2}} s_A^2 \cdots \xrightarrow{\overline{\lambda_n}} s_A^n$. Therefore $(s_A, X_i) \xrightarrow{\tau} (s_A^1, X^1) \xrightarrow{\tau} \cdots \xrightarrow{\tau} (s_A^n, X^n) \xrightarrow{u} (s'_A, X'_i)$.

Hence $(s_A, X_i) \stackrel{u}{\Rightarrow} (s'_A, X'_i)$ and (s'_A, X'_i) *Bisim* s'_B.

Proof of (2): Suppose $(s_A, X_i) \xrightarrow{u} (s'_A, X'_i)$, $u \in \Sigma_B \cup \{\tau\}$. We need to consider cases where $u \in Ex(A), u \in Ex(X)$ and $u = \tau$.

$u \in Ex(A)$: We have $s_A \xrightarrow{u} s'_A$ and $X'_i = X_i$ since $u \notin Ex(X)$. There must exist a state $s'_B \in B$ such that $s_B \xrightarrow{u} s'_B$, otherwise $[s_A, s_B]$ would have been marked a 'bad' state at algorithm step 1 and X_i a bad box at step 3.1.
$[s'_A, s'_B] \in I(s_A, s_B)$ by definition of I, and $[s'_A, s'_B] \in X_i$ by our construction of boxes. Therefore we have $s_B \stackrel{u}{\Rightarrow} s'_B$ and (s'_A, X_i) *Bisim* s'_B.

$\in Ex(X)$: We have $X_i \xrightarrow{u} X_i'$, $s_A' = s_A$ since $u \notin Ex(A)$. There must exist s_B' such that $s_B \xrightarrow{u} s_B'$ and $[s_A, s_B'] \in X_i'$, otherwise no (X_i, u, X_i') transition is created at algorithm step 3.2. Also X_i' must be a 'good' box, otherwise (X_i, u, X_i') transition would have been deleted at algorithm step 4.1. Thus $s_B \stackrel{u}{\Rightarrow} s_B'$ and (s_A, X_i') *Bisim* s_B'.

$= \tau$: A τ transition from (s_A, X_i) in $A \| R$ can be the result of a τ transition in A or a communication between A and R.

(a) the τ transition is a τ transition in A. Then $s_A \xrightarrow{\tau} s_A'$, $X_i = X_i'$.
We have $[s_A', s_B] \in I(s_A, s_B)$. $I(s_A, s_B) \subseteq X_i$ implies that $[s_A', s_B] \in X_i$. Since X_i is a 'good' box, we conclude that $s_B \stackrel{\tau}{\Rightarrow} s_B$ and (s_A', X_i') *Bisim* s_B.

(b) τ transition is caused by some λ communication between A and R. Suppose $s_A \xrightarrow{\lambda} s_A'$, $X_i \xrightarrow{\overline{\lambda}} X_i'$ for some $\lambda \in Com(A)$.
$X_i \xrightarrow{\overline{\lambda}} X_i'$ means that X_i' is a good box. Otherwise the transition $(X_i, \overline{\lambda}, X_i')$ would have been deleted at algorithm step 4.1. $[s_A, s_B] \in X_i$ and $s_A \xrightarrow{\lambda} s_A'$ implies that $[s_A', s_B] \in X_i'$ at algorithm step 3.3. Thus $s_B \stackrel{\tau}{\Rightarrow} s_B$ and (s_A', X_i') *Bisim* s_B.

Summing up all the cases, we have proved that *Bisim* is a bisimulation relation. Be-ause (s_{X_0}, X_0) *Bisim* s_{B_0}, we have $(s_{A_0}, X_0) \approx s_{B_0}$, implying that $A \| R \approx B$.

In addition R is rigid with $Ex(R) = \Sigma_B - \Sigma_A$ and $Com(R) = \{\overline{\lambda} | \lambda \in \Sigma_A - \Sigma_B\}$ by the onstruction in our algorithm. We therefore conclude that R is a solution to $A \| X \approx B$. □

We need to prove the following:

heorem 2 *If there exists a solution R' to $A \| X \approx B$, then our algorithm reports a olution R to $A \| X \approx B$.*

Our algorithm reports a solution whenever X_0 is not marked as 'bad'; hence we must iow that if there exists a solution, X_0 will not be marked 'bad'. Since a box can be iarked 'bad' only at algorithm steps 3.1 or 4.2, we show that X_0 is not marked 'bad' at ther of these steps, whenever there is a solution.

Let predicate $P(s_R, X_i)$ be "for every $[s_A, s_B] \in X_i, (s_A, s_R) \approx s_B$ " where s_R is a state some solution FSM R' to $A \| X \approx B$. We say *a box X_i satisfies predicate P with respect* R' if $P(s_R, X_i)$ is true for some state s_R in R'.

Suppose FSM R' is a solution to $A \| X \approx B$ and let $s_{R_0'}$ be the initial state of R'. efore we prove theorem 2 we show below that $P(s_{R_0'}, X_0)$ is true and all boxes satisfying redicate P with respect to R' are not marked 'bad' in our algorithm.

roposition 4 *If there exists a solution FSM R' to $A \| X \approx B$ then $P(s_{R_0'}, X_0)$ is true.*

roof: If $A \| R' \approx B$ then $(s_{A_0}, s_{R_0'}) \approx s_{B_0}$ where s_{A_0}, $s_{R_0'}$ and s_{B_0} are the initial states of SMs A, R' and B respectively. $X_0 = I(s_{A_0}, s_{B_0})$ by our algorithm and $P(s_{R_0'}, X_0)$ is ue by $(s_{A_0}, s_{R_0'}) \approx s_{B_0}$ and proposition 3. □

roposition 5 *If there exists a solution FSM R' to $A \| X \approx B$ then all boxes satisfying redicate P with respect to R' will not be marked 'bad' at algorithm step 3.1.*

Proof: Consider any box X_i such that predicate $P(s_R, X_i)$ holds for some s_R in R'. We have $(s_A, s_R) \approx s_B$ for each $[s_A, s_B] \in X_i$, where (s_A, s_R) is a state in the composed FSM $A \| R'$.

For any u transition $s_A \xrightarrow{u} s'_A$ with $u \in Ex(A)$, we have $s_B \xrightarrow{u} s'_B$ for some $s'_B \in B$ by proposition 2 because $(s_A, s_R) \approx s_B$ and $(s_A, s_R) \xrightarrow{u} (s'_A, s_R)$. Thus $[s_A, s_B]$ will not be marked 'bad' at algorithm step 1.

Box X_i is not marked 'bad' at algorithm step 3.1 because each $[s_A, s_B] \in X_i$ is not marked 'bad'. □

Proposition 6 *If there exists a solution FSM R' to $A \| X \approx B$ such that there is a state $s_R \in R'$ and $P(s_R, X_i)$ is true, then $s_R \xrightarrow{u} s'_R$ with $u \in Ex(R') \cup Com(R')$ implies that one of the following must hold:*

(a) *$u = \overline{\lambda} \in Com(R')$ and the transition $X_i \xrightarrow{\overline{\lambda}}$DON'T-CARE is created at algorithm step 3;*

(b) *there exists some box X_j such that the transition $X_i \xrightarrow{u} X_j$ is created at algorithm step 3 and $P(s'_R, X_j)$ is true.*

Proof: By proposition 5 X_i is not marked 'bad' at algorithm step 3.1 and thus algorithm steps 3.2 and 3.3 will be performed on box X_i.

Consider the transitions from state s_R:

(1) $s_R \xrightarrow{\overline{\lambda}} s'_R$, $\overline{\lambda} \in Com(R')$.

(a) For every $[s_A, s_B] \in X_i$, there is no s'_A such that $s_A \xrightarrow{\lambda} s'_A$: Then s_A and s_R do not have λ communication. At algorithm step 3.3, the transition $X_i \xrightarrow{\overline{\lambda}}$DON'T-CARE is created.

(b) $s_A \xrightarrow{\lambda} s'_A$ for some $[s_A, s_B] \in X_i$: Then there is a box X_j such that $X_j = X_i(\lambda)$ $= \bigcup\{I(s'_A, s_B) \mid [s_A, s_B] \in X_i,\ s_A \xrightarrow{\lambda} s'_A\}$ and $X_i \xrightarrow{\overline{\lambda}} X_j$ is created at algorithm step 3.3.

s_A and s_R have a λ communication and $(s_A, s_R) \xrightarrow{\tau} (s'_A, s'_R)$ in FSM $A \| R'$. Since $(s_A, s_R) \approx s_B$ we have $(s'_A, s'_R) \approx s_B$ by proposition 2.2. By $(s'_A, s'_R) \approx s_B$ and proposition 3, it holds that $(s''_A, s'_R) \approx s''_B$ for each $[s''_A, s''_B] \in I(s'_A, s_B)$. Thus $(s''_A, s'_R) \approx s''_B$ for each $[s''_A, s''_B] \in X_j$, which is $P(s'_R, X_j)$.

(2) $s_R \xrightarrow{u} s'_R$, $u \in Ex(X)$.

We have $(s_A, s_R) \xrightarrow{u} (s_A, s'_R)$ for each $[s_A, s_B] \in X_i$. Since $(s_A, s_R) \approx s_B$ and $(s_A, s_R) \xrightarrow{u} (s_A, s'_R)$, there must exist s'_B such that $s_B \xrightarrow{u} s'_B$ and $(s_A, s'_R) \approx s'_B$ by proposition 2.1. At algorithm step 3.2 there is a box X_j constructed such that $X_j = X_i(u) = \bigcup\{I(s_A, s'_B) \mid [s_A, s_B] \in X_i, s_B \xrightarrow{u} s'_B\}$ and $X_i \xrightarrow{u} X_j$ is created.

By proposition 3, $(s''_A, s'_R) \approx s''_B$ for each $[s''_A, s''_B] \in I(s_A, s'_B)$. Therefore $(s''_A, s'_R) \approx s''_B$ for each $[s''_A, s''_B] \in X_j$, which is $P(s'_R, X_j)$. □

Proposition 7 *If there exists a solution FSM R' to $A \| X \approx B$ then all boxes satisfying predicate P with respect to R' will not be marked 'bad' at algorithm step 4.2.*

Proof: All boxes satisfying predicate P will not be marked 'bad' at algorithm step 3.1 by proposition 5. We now prove by contradiction that all these boxes will not be marked 'bad' at algorithm step 4.2.

Suppose box X_i is the first among all the boxes satisfying predicate P to be marked 'bad' at algorithm step 4.2. Assume $P(s_R, X_i)$ is true for some state s_R in FSM R'.

X_i is marked 'bad' at step 4.2 only if there exists a state pair $[s_A, s_B] \in X_i$ such that $s_B \xrightarrow{u} s'_B$ for some $u \in \Sigma_B$ but there does not exist a sequence of links $[s_A, s_B] \xrightarrow{\lambda_1} [s^1_A, s_B] \xrightarrow{\lambda_2} \cdots \xrightarrow{\lambda_l} [s^n_A, s_B] \xrightarrow{u} [s'_A, s'_B]$ for some $l \geq 0$. $\lambda_k \in Com(X) \cup \{\tau\}$ for $1 \leq k \leq l$.

Because $(s_A, s_R) \approx s_B$, there must exist $(s_A, s_R) \stackrel{u}{\Rightarrow} (s'_A, s'_R)$ and $(s'_A, s'_R) \approx s'_B$. Let $(s_A, s_R) \xrightarrow{\tau} (s^1_A, s^1_R) \xrightarrow{\tau} \cdots \xrightarrow{\tau} (s^n_A, s^n_R) \xrightarrow{u} (s^{n+1}_A, s^{n+1}_R) \stackrel{\tau}{\Rightarrow} (s'_A, s'_R)$ for some $n \geq 0$. $(s'_A, s'_R) \approx s'_B$ and $(s^{n+1}_A, s^{n+1}_R) \stackrel{\tau}{\Rightarrow} (s'_A, s'_R)$ imply that $(s^{n+1}_A, s^{n+1}_R) \approx s'_B$ by proposition 2. Also $(s^k_A, s^k_R) \approx s_B$ for $1 \leq k \leq n$ by $(s_A, s_R) \approx s_B$ and proposition 2.

Consider τ transition $(s_A, s_R) \xrightarrow{\tau} (s^1_A, s^1_R)$:

a) The τ transition is an internal τ transition in A: Then $s_A \xrightarrow{\tau} s^1_A$ and $s_R = s^1_R$. We have $[s^1_A, s_B] \in X^1 = X_i$. Link $[s_A, s_B] \xrightarrow{\tau} [s'_A, s_B]$ exists in box X_i.

b) The τ transition is a β^1 communication ($\overline{\beta^1} \in Com(R)$) between s_A and s_R: Then $s_A \xrightarrow{\beta^1} s^1_A$ and $s_R \xrightarrow{\overline{\beta^1}} s^1_R$ for some $s^1_A \in A$, $s^1_R \in R$. By proposition 6 and $P(s_R, X_i)$ we know that there is some box X^1 such that $X_i \xrightarrow{\overline{\beta^1}} X^1$ is created and $P(s^1_R, X^1)$ is true.

Link $[s_A, s_B] \xrightarrow{\overline{\beta^1}} [s^1_A, s_B]$ is created with $[s^1_A, s_B]$ in box X^1 at algorithm step 3.3 because $s_A \xrightarrow{\beta^1} s^1_A$.

Proceeding similarly for each τ transition $(s^k_A, s^k_R) \xrightarrow{\tau} (s^{k+1}_A, s^{k+1}_R)$ with $1 \leq k \leq n$, we have $[s_A, s_B] \xrightarrow{\overline{\beta^1}} [s^1_A, s_B] \xrightarrow{\overline{\beta^2}} \cdots \xrightarrow{\overline{\beta^n}} [s^n_A, s_B]$, with $[s^k_A, s_B] \in X^k$ and $P(s^k_R, X^k)$ for some box X^k $(1 \leq k \leq n)$.

Consider u in $(s^n_A, s^n_R) \xrightarrow{u} (s^{n+1}_A, s^{n+1}_R)$. $P(s^n_R, X^n)$ holds by above argument. Also $(s^n_A, s^n_R) \approx s_B$ and $s_B \xrightarrow{u} s'_B$. Consider possible cases of u:

$u \in Ex(A)$: Then $s^n_A \xrightarrow{u} s^{n+1}_A$, $s^n_R = s^{n+1}_R$ and $[s^{n+1}_A, s'_B] \in X^{n+1} = X^n$. The link $[s^n_A, s_B] \xrightarrow{u} [s^{n+1}_A, s'_B]$ is created at algorithm step 2 and exists in X^n. Also $P(s^{n+1}_R, X^{n+1})$ is true.

$u \in Ex(R')$: Then $s^n_R \xrightarrow{u} s^{n+1}_R$, $s^n_A = s^{n+1}_A$. By proposition 6 we know that there exists some box X^{n+1} such that $X^n \xrightarrow{u} X^{n+1}$ is created at algorithm step 3.2 and $P(s^{n+1}_A, X^{n+1})$ is true.

The link $[s^n_A, s_B] \xrightarrow{u} [s^{n+1}_A, s'_B]$ is created at step 3.2 with $[s^{n+1}, s'_B] \in X^{n+1}$ because $s_B \xrightarrow{u} s'_B$.

In summary each X^k $(1 \leq k \leq n+1)$ is not marked 'bad' at algorithm step 3.1 because $P(s^k_R, X^k)$ is true.

Therefore the sequence of links $[s_A, s_B] \xrightarrow{\overline{\beta^1}} [s^1_A, s_B] \xrightarrow{\overline{\beta^2}} \cdots \xrightarrow{\overline{\beta^n}} [s^n_A, s_B] \xrightarrow{u} [s^{n+1}_A, s'_B]$ exists among 'good' boxes. This implies that box X_i will not be marked 'bad' at algorithm step 4.2, a contradiction.

Hence all boxes satisfying predicate P will not be marked 'bad' at algorithm step 4.2. □

Proof of theorem 2: If there is a solution R' to $A \| X \approx B$, then $P(s_{R'_0}, X_0)$ is true by proposition 4. By propositions 5 and 7, X_0 is not a bad box, and hence our algorithm reports a solution. By theorem 1, the FSM R constructed by our algorithm is a solution to $A \| X \approx B$. □

Definition 8 *A binary relation T on states is a strong inclusion relation if the following holds:*
if $p \; T \; q$ and $p \xrightarrow{u} p'$, then either (a) $q \xrightarrow{u}$DON'T-CARE or (b) $p' \neq$ DON'T-CARE, $q \xrightarrow{u} q'$ and $p' \; T \; q'$.

We say q strongly includes p (write $p \sqsubseteq q$) if there exists a strong inclusion relation T such that $(p, q) \in T$.

Intuitively q strongly includes p if q can 'behave like' p. Especially a DON'T-CARE state can behave like any state because it can be instantiated to any state when necessary. Note that a strong inclusion relation becomes a strong simulation relation in [La87] if no DON'T-CARE states are involved.

Definition 9 *A solution R to $A \| X \approx B$ is more general than a solution R' to $A \| X \approx B$ if $s_{R'_0} \sqsubseteq s_{R_0}$.*
A solution R to $A \| X \approx B$ is the most general one if for every solution R' to $A \| X \approx B$, $s_{R'_0} \sqsubseteq s_{R_0}$.

Theorem 3 *The solution R constructed in our algorithm, if it exists, is the most general solution to $A \| X \approx B$.*

Proof: Consider any solution R' to $A \| X \approx B$. $P(s_{R'_0}, X_0)$ is true by proposition 4.

To show $s_{R'_0} \sqsubseteq X_0$ we prove that the relation

$$Incl = \{(s_{R'}, X_i) \mid s_{R'} \text{ is a state in } R' \text{ and } P(s_{R'}, X_i)\}$$

is a strong inclusion relation.

Suppose $s_{R'} \; Incl \; X_i$ and $s_{R'} \xrightarrow{u} s'_{R'}$ for some $u \in Ex(R') \cup Com(R')$. By proposition 6, either $u = \overline{\lambda} \in Com(R')$ and $X_i \xrightarrow{\overline{\lambda}}$DON'T-CARE or $X_i \xrightarrow{u} X_j$ is created at algorithm step 3 and $P(s'_{R'}, X_j)$ is true. $P(s'_{R'}, X_j)$ being true implies X_j being a 'good' box by propositions 5 and 7 and thus the transition $X_i \xrightarrow{u} X_j$ will not be deleted at algorithm step 4.1. Hence $X_i \xrightarrow{u} X_j$ and $s'_{R'} \; Incl \; X_j$.

By definition 8 *Incl* is a strong inclusion relation and thus $s_{R'_0} \sqsubseteq X_0$.

Therefore R is more general than R' by definition 9. We conclude that if there exist solutions to $A \| X \approx B$, the solution constructed in our algorithm is the most general one. □

The time complexity of this algorithm is exponential in the size of A and B because in the worst case an exponential number of boxes can be generated.

Related Work

'his paper extends the work of [Sh89] in which the CCS equations of the form $(A|X)\backslash L{\approx}B$ ith B *weakly determinate* were studied.

In [Sh89] Shields showed that a solution to $(A|X)\backslash L{\approx}B$ exists if and only if there xists an *uncompromised* set S in which each element is a union of some B_I's (similarly efined as set I in our paper). A solution R to $(A|X)\backslash L{\approx}B$ can be constructed from in which each element of S is a state of R and the transitions between states are erived from the transitions of (p,q) pairs in the elements of S. An algorithm presented y Shields is similar in concepts to ours except that, instead of starting with one initial ox corresponding to $I(s_{A_0}, s_{B_0})$ and generating other boxes that can be reached from he initial box, Shields' algorithm starts with a set of boxes each of which contains the tates in $I(s_{A_0}, s_{B_0})$, and then generate other boxes that can be reached by transitions om these boxes. Hence Shields's algorithm deals with many more boxes and requires ore intermediate storage.

In [QL90] the authors of this paper studied the strong factorization equation $A|\!|\!|X{\sim}B$ n addition to the weak factorization equation $A|\!|\!|X \approx B$. An algorithm is presented and roved correct to find the most general solution, if solutions exist, to $A|\!|\!|X \sim B$ with B trongly determinate.

In [Ra89] Ranatunga studied the solutions to CCS equations of the form $(A|X)\backslash L{\sim}B$, here L is arbitrary and a solution R can have transitions under any action symbol in 3. An algorithm is presented and proved correct to obtain a solution if solutions exist. ecause the solutions allowed in the equations in [Ra89] are more general than those llowed in our strong factorization equations, the algorithm and its correctness proof in Ra89] are more complex than those in [QL90] for strong factorization equations.

In [LL89] Larsen *et al* provided a theoretical foundation for solving equation systems f the form

$$C_1(X){\sim}P_1,\ \cdots,\ C_n(X){\sim}P_n \tag{1}$$

here C_i are arbitrary contexts (i.e. derived operators) of some process algebra (say CS in [Mi80]). They have shown that there exist solutions to equation (1) if and only if here exists a *consistency relation* C over a subset of the state set in a *Disjunctive Modal Transition System* T, and a solution can be readily extracted from C and T. Although hey did not present an algorithm that will find a consistent subset from among the xponential number of subsets of the state set in T, such an algorithm can be obtained n a simple way by modifying the partitioning algorithm for bisimulation.

In [Pa87] Parrow presented a tableau transformation method for solving CCS equaions of the form $(A|X)\backslash L{\approx}B$ where B is deterministic and implemented the method as semi-automatic program. The program can run in automatic or semi-automatic mode. n automatic mode, if the program terminates and produces a solution, it will find the nost general solution. However the most general solution may be quite large. In semiutomatic mode, the user can make certain choices which may lead to a less general but maller solution.

In [MB83] and [SA88] the factorization problem was studied in a different model here specifications are expressed in terms of execution sequences and trace equivalence s used as the equivalence criterion. The formula underlying their method defines the nost general specification (allowing for all possible execution sequences) for the missng submodule that together with the other given submodule will satisfy the system specification under trace equivalence. One limitation noted by the authors is that an a tomatically generated specification for a submodule may cause deadlocks because trac equivalence is insensitive to deadlock potentials.

5 Conclusion and Future Research

In this paper we have studied weak factorization equation $A\|X \approx B$. An algorith was presented and proved correct to find the most general solution, if solutions exist, t $A\|X \approx B$ with B deterministic.

As shown in theorem 3, all the solutions to $A\|X \approx B$ are related by strong inclusio to the solution produced in our algorithm. Research is under way to investigate furthe the relationship among all the solutions and to find optimization techniques to obtai solutions that have fewer states and transitions.

Acknowledgments

The authors are grateful to the referees for their helpful suggestions.

References

[La87] K. G. Larsen, "A Context Dependent Equivalence Between Processes", *Theo retical Computer Science*, Vol. 49, pp. 185–215, 1987.

[LL89] K. G. Larsen, X. Liu, "Equation Solving Using Modal Transition Systems" In *Proceedings of the Fifth Annual IEEE Symposium on Logic in Compute Science*, pp. 108–117, June 1990.

[MB83] P. Merlin, G. Bochmann, "On the Construction of Submodule Specifications an Communication Protocols", *ACM Transactions on Programming Language and Systems*, Vol. 5, No. 1, pp. 1–25, 1983.

[Mi80] R. Milner, *Calculus for Communicating Systems*, LNCS 92, Springer Verlag 1980.

[Mi88] R. Milner, "Operational and Algebraic Semantics of Concurrent Processes" *Laboratory for Foundations of Computer Science, Univ. of Edinburgh*, Technica Report, ECS-LFCS-88-46, Feb. 1988.

[Pa87] J. Parrow, "Submodule Construction as Equation Solving in CCS", *Theoretica Computer Science*, Vol. 68, pp. 175–202, 1989.

[QL90] H. Qin, P. Lewis, "Factorization of Finite State Machines under Strong an Observational Equivalences", (submitted to *Journal of Formal Aspects of Com puting)*

[Ra89] L. P. Ranatunga, "Process Synthesis by Solving an Equation in CCS", (sub mitted for publication), 1989.

[SA88] D.P. Sidhu, J. Aristizabal, "Constructing Submodule Specifications and Network Protocols", *IEEE Transactions on Software Engineering*, Vol. 14, No. 11, pp. 1565–1577, November 1988.

[Sh89] M.W. Shields, "Implicit System Specification and the Interface Equation", *The Computer Journal*, Vol. 32, No. 5, pp. 399-412, 1989.

Partial Order Logics for Elementary Net Systems: State- and Event-approaches

Sinachopoulos A.
Laboratoire d'Informatique Théorique
Université Libre de Bruxelles
CP, Boul. du Triomphe
B-1050 Bruxelles
Belgium

Work supported by th
Basic Research Espri
Project no 3148 (DEMON

Abstract: We give two kinds of specific axiomatics, one describing cases and the other describing actions of en-systems. These two axiomatics do not have the same expressive power, since contacts are not expressible in the action-based approach. The problem of globality and locality, as well as extensions of the given axiomatics to axiomatics describing processes are discussed.

Keywords: Partial Order Logics, Next Operator, Petri Nets, Elementary Net Systems, Specification, Case Graphs, Action Graphs, Specific Axiomatics, State Approach, Action Approach.

1 Introduction

Partial Orders, PO's, are the most suitable tool for the description of concurrent systems. By means of PO's we can indeed express concurrency as well as non-determinism [Rei5].

The standard axiomatics for PO's is an extension of the Lewis's axiomatics S_4 enriched by past operators. Causal dependencies between two adjacent components of a PO are expressed by a next operator which is not self-dual as it is the case for linear orders [Krö]. Since there is no axiom in logics denoting antisymmetry [Bur], axioms for PO's describe pre-orders as well. Taking advantage of this fact we will use the same axiomatics for both pre- and partial orders.

General axiomatics are not able to describe exactly a concrete concurrent system. Even a local specification of a general axiomatics cannot express in full detail all properties of such a system. Thus intending to describe a concrete system we have to use specific axiomatics, ie. a set of postulates [Ch] reflecting the properties of the system considered [OL,MP]. These postulates are added to the general underlying logic which expresses characteristic features of PO's. Hence general axiomatics and postulates form a specific theory corresponding to a given concurrent system. Then we can no more speak about completeness of our specific theory w.r.t. the corresponding models, since completeness can only be required from general axiomatics. What we can do with such specific theories is to examine their models. In the best case these models are defined up to isomorphism [Ch].

General axiomatics for PO's describe the ordering of states, or that of actions or both of them. As states behave quite differently than actions, specific axiomatics describing both actions and states would be complicated and unwieldy. An axiomatic description both states and actions can be achievd by Propositional Dynamic Logic and μ-Calculi :o] but, as far as we know, only for linear orders. Therefore we prefer to give two parate sets of postulates, the one for states and the other for actions. In both cases e use the same general axiomatics for PO's where the semantic through the valuation nction determines whether the model consists of states or actions. Hence postulates of ır formalism describe either actions or states, not both of them.

The objective of this work is to define axiomatically a consequence operator $\vdash$, par- lel to but stronger than $\models$, the standard semantical one, allowing syntactic reasonig and examine the expressive power of both actions- and state- oriented logics characterizing ementary net systems (en-systems). Our specific theories for en-systems are expressed the monadic second order logic [Cou,Pel] enriched by modal operators. By means of ır formalism we can choose the degree of globality and locality we want to express, since ır theories are able to describe both global and local situations, Rem. 16, Def. 19, Rem.), Def. 9, [SD]. We consider only finite systems.

The second section provides the general PO axiomatics, its semantics and the relation etween semantics and axiomatics.

In the third section we examine specific axiomatics describing cases of a given en-system . The new elements in our approach are specific axioms describing exactly the cases of Σ well as connectedness and lack of causality [SD]. Our model is the case graph of Σ. By eans of our formalism we are able to characterize phenomena like contacts and conflicts ıd relations like the Independency relation between events. Moreover, for contact-free stems the formalism we use becomes simpler, Th.7.

The fourth section contains axiomatics corresponding to actions, ie. events and steps hich can be activated in a concrete en-system Σ. Such axiomatics are modelled by action aphs, Def. 9. (We do not use event structures like [Pen], because the conflict heredity aracterizing event structures is of a rather restrictive nature. More about this point ın be found in the sixth section.) Since action graphs are not generally connected, Th.), the models of our axiomatics cannot be defined up to isomorphism. It turns out that ıe Independency relation as well as conflicts are expressible in our formalism, Th. 13 , ut contacts are not, since the action graph of an en-system and that of its completion oincide, Th. 12.

In the fifth section we consider axiomatics for a process π of Σ. The corresponding odel is the structure of cuts of events of π. This structure is connected, hence our models re defined up to isomorphism.

In the last section we examine the relationship of our approaches to other approaches, pecially those of [Pen,BC].

The reader is supposed to be familiar with Petri nets, elementary net systems and basic otions of temporal logic as well. For more details see [Rei1], [Ro,Thi] and [Pen,Rei2,Rei4] spectively.

General Axiomatics

PO logics are modelled by Kripke models of the kind $\mathcal{M} = (G, \leq, *, V)$ where G is the t of the possible worlds of $\mathcal{M}$, $\leq$ is a PO in G, $*$ is the next state relation in G, and $: G \mapsto 2^P$ is the valuation function from the elements of G to subsets of the set P of all tomic propositions. The axiomatics for PO's we use is given by the following definition:

efinition 1:
et $P = \{p, q, \ldots\}$ be the set of atomic propositions. The axiomatic system Γ consists of ıe following axioms and rules:
ll axioms of the propositional calculus and Modus Ponens, $p, p \rightarrow q \vdash q$, the Rule f Temporal Generalization, $p \vdash \Box p, \overline{\Box}\, p$ and the axioms

$A_1:$	$\Box(p \to q) \to \Box p \to \Box q$	$\overleftarrow{\Box}(p \to q) \to \overleftarrow{\Box} p \to \overleftarrow{\Box} q$
$A_2:$	$p \to \Box \overleftarrow{\Diamond} p$	$p \to \overleftarrow{\Box} \Diamond p$
$A_3:$	$\Box p \to p$	$\overleftarrow{\Box} p \to p$
$A_4:$	$\Box p \to \Box\Box p$	$\overleftarrow{\Box} p \to \overleftarrow{\Box}\overleftarrow{\Box} p$
$A_5:$	$\Diamond p \leftrightarrow \neg\Box\neg p$	$\overleftarrow{\Diamond} p \leftrightarrow \neg \overleftarrow{\Box} \neg p$
$A_6:$	$\otimes(p \to q) \to (\otimes p \to \otimes q)$	$\overleftarrow{\otimes}(p \to q) \to (\overleftarrow{\otimes} p \to \overleftarrow{\otimes} q)$
$A_7:$	$\bigcirc p \leftrightarrow \neg \otimes \neg p$	$\overleftarrow{\bigcirc} p \leftrightarrow \neg \overleftarrow{\otimes} \neg p$
$A_8:$	$\Box p \to \otimes p$	$\overleftarrow{\Box} p \to \overleftarrow{\otimes} p$
$A_9:$	$p \to \otimes \overleftarrow{\bigcirc} p$	$p \to \overleftarrow{\otimes} \bigcirc p$

■

The axiomatic Γ is an extension of the S_4 axiomatics with past and next operator: strong ones, $\otimes$ and $\overleftarrow{\otimes}$, and weak ones, $\bigcirc$ and $\overleftarrow{\bigcirc}$. $\otimes$ is no longer self dual as in linea orders [Krö]: its dual is the weak next operator $\bigcirc$ (axiom A_7). The semantic for the nex operators at a state $S \in G$ of a Kripke model $\mathcal{M} = (G, \leq, *, V)$ is given by:

Definition 2:
$\models_S \otimes a$ iff for all $S' \in G$ s.t. $S * S'$ we have $\models_{S'} a$
$\models_S \overleftarrow{\otimes} a$ iff for all $S' \in G$ s.t. $S' * S$ we have $\models_{S'} a$
$\models_S \bigcirc a$ iff for some $S' \in G$ s.t. $S * S'$ we have $\models_{S'} a$
$\models_S \overleftarrow{\bigcirc} a$ iff for some $S' \in G$ s.t. $S' * S$ we have $\models_{S'} a$

■

The semantic interpretation of $\Box, \overleftarrow{\Box}, \Diamond, \overleftarrow{\Diamond}$ is the standard one, eg. [Pen,Rei2,Sin1] Validity of a proposition a in a model $\mathcal{M} = (G, \leq, *, V)$, $\mathcal{M} \models a$, is provided iff for al elements S of G we have $\models_S a$.

The intuition behind our next operators is similar to that of POTL, [PW], but th next operators in POTL must cooperate with unseparable branching time operators, eg $\forall\Diamond$, which characterise the linear order of paths, see also [BPM,ES].

Γ is suitable for the axiomatization of PO's with a next state relation, as it is sound an complete w.r.t. PO's provided with a next state relation [Sin1]. Since there is no axion characterizing antisymmetry, Γ characterize pre-orders as well. A completeness proof fo Γ w.r.t. pre-orders with a next state relation follows from the completeness proof w.r.t PO's [Sin1].

3 A State-based Approach

We recall some concepts of the net theory. It is supposed that the notions of a net pre-set $\bullet x$, post-set $x\bullet$, occurrence net, slices, *co* and *li* relations and co-sets are known For more details see [Rei1].

Definition 3:
$\Sigma = (B, E, F, c_{in})$ is an elementary net system, (en-system), iff (B, E, F) is a net so that for all $x \in B \cup E$ we have $\bullet x \cup x\bullet \neq \emptyset$ and $\bullet x \cap x\bullet = \emptyset$ and $c_{in} \subseteq B$. B contains the conditions and E the events of Σ.
Let be $e, e' \in E$. $Ind(e, e')$, e and e' are independent, iff $(\bullet e \cup e\bullet) \cap (\bullet e' \cup e'\bullet) = \emptyset$.
u is a step of Σ iff $u \subseteq E$ and for all $e_1, e_2 \in u$, $e_1 \neq e_2$, we have $Ind(e_1, e_2)$.
Then single events of E are steps as well.

:t be $c \subseteq B$ and u be a step. u is enabled at c iff $\bullet u \subseteq c$ and $u \bullet \cap c = \emptyset$.
: is the set of cases of Σ iff c_Σ is the least set of subsets of B satisfying: $c_{in} \in c_\Sigma$ and $c \in c_\Sigma$ and $c[u\rangle c'$ then $c' \in c_\Sigma$.
he case graph of Σ is the (edge-labelled) graph $\mathcal{G}_\Sigma := (c_\Sigma, Y, \phi)$ where $Y := \{(c,c') \mid c' \in c_\Sigma$ *and there exists* $u \in \mathcal{U}$ *s.t.* $c[u\rangle c'\}$ and ϕ is the labelling function $\phi : Y \longmapsto \mathcal{U}$ t. if $(c,c') \in Y$ and $c[u\rangle c'$ then $\phi(c,c') = u$.

■

Intending to axiomatize sets of conditions of Σ we consider the conditions of Σ as our omic propositions, ie. for the set of atomic propositions P we have $P = B$ (a kind of osed world assumption). Then we need the following notation:

iotation 4:
A is a set of formulas then:
$A = \vee_{a\in A}\, a\,,\quad \bigvee\neg A = \vee_{a\in A}\, \neg a\,,\quad \bigwedge A = \wedge_{a\in A}\, a\,,$
$\neg A = \wedge_{a\in A}\, \neg a\,,\quad \bigvee\emptyset = \bot\,,\quad \bigwedge\emptyset = \top,$
here $\bot$ is the formula $a \wedge \neg a$ (false) and $\top$ is the formula $a \vee \neg a$ (true).

■

Although this notation is sometimes regarded as complicated and unwieldy, it is the andard way of expressing net properties as formulas of temporal logics, see also [BCG].

efinition 5:
et $\Sigma = (B, E, F, c_{in})$ be an en-system, c_Σ its cases and $\mathcal{U}$ its set of steps. Γ_Σ is the tension of Γ by the following axioms:

A_1: For all $c \in c_\Sigma$ we have $\Gamma_\Sigma \vdash \overleftarrow{\Diamond}\,\Diamond(\bigwedge c \wedge \bigwedge\neg(B-c)) \vee \Diamond\,\overleftarrow{\Diamond}\,(\bigwedge c \wedge \bigwedge\neg(B-c))$

A_2: For all $c \in c_\Sigma$ and for all $u \in \mathcal{U}$ if $c[u\rangle c'$ then

$$\Gamma_\Sigma \vdash \bigwedge c \wedge \bigwedge\neg(B-c) \rightarrow \bigcirc(\bigwedge c' \wedge \bigwedge\neg(B-c'))$$

$$\Gamma_\Sigma \vdash \bigwedge c' \wedge \bigwedge\neg(B-c') \rightarrow \overleftarrow{\bigcirc}\,(\bigwedge c \wedge \bigwedge\neg(B-c))$$

A_3: For all $c, c' \in c_\Sigma$ if there is no $u \in \mathcal{U}$ s.t. $c[u\rangle c'$ then

$$\Gamma_\Sigma \vdash \bigwedge c \wedge \bigwedge\neg(B-c) \rightarrow \otimes\neg(\bigwedge c' \wedge \bigwedge\neg(B-c'))$$

$$\Gamma_\Sigma \vdash \bigwedge c' \wedge \bigwedge\neg(B-c') \rightarrow \overleftarrow{\otimes}\,\neg(\bigwedge c \wedge \bigwedge\neg(B-c))$$

A_4 : $\Gamma_\Sigma \vdash \vee_{c\in c_\Sigma}(\bigwedge c \wedge \bigwedge\neg(B-c))$

A_5 : For all $d \subseteq B$, if there is no $c \in c_\Sigma$ such that $d = c$ we have: $\Gamma_\Sigma \vdash \neg \bigwedge d \vee \bigvee(B-d)$

■

The prefix $\Gamma_\Sigma \vdash$ will be omitted.

$NA_1 - NA_5$ is the specification characterizing cases of Σ. NA_1, NA_2 and NA_3 de-cribe the pre-order structure and NA_4, NA_5 describe the logical structure of the cases of :. NA_1 expresses not only the fact that cases are connected but it describes the kind of onnectedness as well, ie. in every moment all cases are reachable forwards or backwards, Rei2]. NA_2 expresses the change from case to case. NA_3 points out that if two cases re not related by the enabling of a step, then they cannot be related by the strong next tate operator. NA_4 denotes that in every moment at least one case holds. NA_5 forbids he derivation of a conjunction of elements of B which do not form a case. The three ormer axioms express the pre-ordering of cases and the two latter ones describe exactly

what cases consist of and which subsets of the conditions of B form the cases.
Γ_Σ is modelled by the case graph of Σ: Γ_Σ is sound w.r.t. the case graph of Σ Since this graph is connected by NA_1, the models of Γ_Σ are defined up to isomorphism [Cou,Pel,Sin1].

Γ_Σ is intended to describe the case graph of Σ, ie. it has a model. That means tha there is no formula a for which $\Gamma_\Sigma \vdash a$ and $\Gamma_\Sigma \vdash \neg a$. Hence Γ_Σ is consistent.
Moreover we require that our atomic propositions are the conditions of Σ, namely Γ_Σ : able to describe nothing but Σ. In practice we are working with Kripke models for whic the valuation function V is the function id_B. Then, even if Σ_1 and Σ_2 differ only in th label of their conditions, eg.

Γ_{Σ_1} and Γ_{Σ_2} are different axiomatics because they refer to different proposition and the corresponding languages are different.
Obviously if B_{Σ_1} is isomorphic to B_{Σ_2}, where $\phi : B_{\Sigma_1} \mapsto B_{\Sigma_2}$ is an isomorphism, we ca identify the corresponding atomic propositions modifying the valuation functions, eg. i Σ_1 and Σ_2 are state space similar, [Ro], then the stuctures $(c_{\Sigma_2}, \leq, \phi)$ and $(c_{\Sigma_1}, \leq, \phi^{-1}$ are models of Γ_{Σ_1} and Γ_{Σ_2} respectively.
If $c[e_1\rangle c'$ and $c[e_2\rangle c'$, eg. if e_1, e_2 have exactly the same pre- and post conditions, Γ cannot distinguish between e_1 and e_2, which is expected, since the formulas derivable i Γ_Σ consist of conditions of Σ.

By means of our axiomatics we can express contacts, conflicts and the Independenc relation.

Definition 6:
1. c contains a contact at $e \in E$ iff $\bullet e \subseteq c$ and $c \cap e\bullet \neq \emptyset$.
2. Σ is contact-free iff for all $c \in c_\Sigma$ there is no $e \in E$ s.t. c contains a contact at e
3. e_1, e_2 are in conflict at c iff $c[e_1\rangle$, $c[e_2\rangle$ but $\{e_1, e_2\}$ is not enabled at c.

■

By means of Def. 5 we have the following theorem showing the expressive power of Γ_Σ

Theorem 7:
Let $\Sigma = (B, E, F, c_{in})$ be an en-system, c_Σ its cases, $c \in c_\Sigma$ and e, e_1, $e_2 \in E$ s.t. $e_1 \neq e_2$
Then:
1. $Ind(e_1, e_2)$ iff $\vdash \bigwedge((\bullet e_1 \cup e_1\bullet) \cap (\bullet e_2 \cup e_2\bullet))$.
2. c is in contact at e iff there exists an $a \in e\bullet \cap c$ such that
$\vdash \Diamond \overleftarrow{\Diamond} (\bigwedge \bullet e \wedge a \wedge \bigwedge(c - \bullet e - a) \wedge \bigwedge \neg(B - c)) \vee \overleftarrow{\Diamond} \Diamond(\bigwedge \bullet e \wedge a \wedge \bigwedge(c - \bullet e - a) \wedge \bigwedge \neg(B - c))$
3. Let $c[e_1\rangle c_1$ and $c[e_2\rangle$. e_1, e_2 are in conflict at c iff
there exists an $a \in \bullet e_2$ such that
$\vdash a \wedge \bigwedge(\bullet e_1 - a) \wedge \bigwedge(c - \bullet e_1) \wedge \bigwedge \neg(B - c) \rightarrow \bigcirc(c_1 \wedge \neg a \wedge \bigwedge \neg(B - c_1 - a))$
or
there exists an $a \in e\bullet$ such that
$\vdash \bigwedge \bullet e_1 \wedge \bigwedge(c - \bullet e_1) \wedge \neg a \wedge \bigwedge \neg(B - c - a) \rightarrow \bigcirc(\bigwedge(c_1 - a) \wedge a \wedge \bigwedge \neg(B - c_1))$.

Proof outline:
1. By Def.6, Not.5 and soundness of Γ_Σ w.r.t. the case graph of Σ.
2. By NA_1, Def.6 and soundness of Γ_Σ w.r.t. the case graph of Σ.

3. By NA_2, Def.6 and soundness of Γ_Σ w.r.t. the case graph of Σ. ■

Let $c_1[e_1\rangle \cdots [e_k\rangle c_{k+1}$ be a firing sequence of Σ. By NA_2 and the derivable in Γ formula)$a \rightarrow \Diamond a$ we can derive the following general progress formula for the above mentioned ring sequence:
$(\bigwedge c_1 \wedge \bigwedge \neg(B - c)) \rightarrow \Diamond(\bigwedge c_{k+1} \wedge \bigwedge \neg(B - c_{k+1}))$ (1)
loreove, by the tautologie $a \wedge b \rightarrow a$ and the derivable in Γ formula $\bigcirc(a \wedge b) \rightarrow \bigcirc a \wedge \bigcirc b$ e can simplify the right-hand side of (1) and get:
$\vdash (\bigwedge c_1 \wedge \bigwedge \neg(B - c)) \rightarrow \Diamond(\bigwedge e\bullet_k)$
If we want to simplify the left-hand side of implications like (1), we have no formal ieans for doing this. This can be done only for contact- free en-systems:

heorem 8:
Σ is a contact-free en-system then for all cases c of Σ we have:

$$\vdash \bigwedge c \wedge \bigwedge \neg(B - c) \leftrightarrow \bigwedge c$$

roof-outline:
he $\rightarrow$ direction is clear.
he $\leftarrow$ direction follows in two steps:
1. We prove indirectly that if Σ is contact-free, $c \in c_\Sigma$ and $a \in B - c$ then $c \cup \{a\} \notin c_\Sigma$.
2. We prove by NA_5 and the previous step that if Σ is contact-free then for all cases c of Σ we have $\vdash \neg(\bigwedge c \wedge \bigvee(B - c)), \bigwedge c \rightarrow \bigwedge \neg(B - c)$ ■

4 An Action-based Approach

When we intend to give postulates characterizing actions of a concrete en-system Σ, he question arises why we do not use event structures [Pen,NPW,W] as models. This is ue to the following reasons:

- The axiom of conflict heredity in [NPW,W,Pen] is restrictive since it rules out events which are in conflict and have a common post-condition, see also [BC]. Moreover, by means of event structures we cannot directly describe events of Σ, but only express properties of the unfolding of Σ [NPW,W].
- In the Petri net theory, conflicts may occur in a pre-order structure. In partial orders, ie. processes of Σ, all conflicts are solved and we have no conflict occurence.
- We want to take advantage of the fact that pre- and partial orders are characterized by the same axiomatics.
- We wish to express steps as well.

Hence we are working with pre-orders and steps. In our models the conflict relation and the occurrence ordering $\leq$ between events are in general no longer disjoint as in Pen], since our possible worlds, the actions, are pre-ordered. We consider actions to be ll steps of Σ which can be enabled by some case of Σ:

Definition 9:
et Σ be an en-system, c_Σ its cases and $\mathcal{U}$ its steps. Let $\bar{\mathcal{U}} = \{u \in \mathcal{U} \mid$ there exists $c \in c_\Sigma$ .t. $c[u\rangle \}$ be the set of actions of Σ. The action graph of Σ is the graph $G_E = (\bar{\mathcal{U}}, f)$

where f is a relation in $\bar{\mathcal{U}} \times \bar{\mathcal{U}}$ and $(u_1, u_2) \in f$ iff there exist $c_1, c_2 \in c_\Sigma$ such that $c_1[u_1\rangle c_2$ and $c_2[u_2\rangle$.

■

According to Def. 9 we are not only working with the reduction of Σ, [Ro], but also with the reduction of $\mathcal{U}$ to $\bar{\mathcal{U}}$, ie. we consider only the active steps of Σ. In the following we will always consider reduced en-systems.

Then we provide G_E with a pre-order relation $\leq$:

For u_1, $u_2 \in \bar{\mathcal{U}}$ we define $u_1 \leq u_2$ *iff* $u_1 = u_1$ *or* there exist for Σ a step sequence $c_1[u_1\rangle \cdots c_k[u_2\rangle$. Obviously $\leq$ is a pre-order. Let $*$ be the next action relation in G_E. Then G_E can be considered as a model for the axiomatics Γ, where the set P of atomic propositions consists of the events of Σ which can be enabled. More precisely, a model for Γ is the Kripke model $(\bar{\mathcal{U}}, \leq, *, V)$ where $V : \bar{\mathcal{U}} \longmapsto 2^P$ is the valuation function from the elements of $\bar{\mathcal{U}}$ to the powerset of P.

In our model the conflict relation and the $\leq$ relation are not disjoint, since the occurrence ordering $\leq$ is a pre-order. Moreover, the action graph of Σ is not in general connected:

Theorem 10:
Let $\Sigma = (B, E, F; c_{in})$ be an en-system and c_Σ its cases. If there exist no $e_1, e_2 \in c_\Sigma$ such that $c_{in}[e_1\rangle$, $c_{in}[e_2\rangle$ and e_1, e_2 in conflict at c_{in} then Σ is connected iff G_E is connected.

Proof outline:
($\Rightarrow$:) If u_1, $u_2 \in \bar{\mathcal{U}}$ and e_1 e_2 are not connected by a step sequence, then there exist u_1', $u_2' \in \bar{\mathcal{U}}$ such that $c_{in}[u_1'\rangle$ and $c_{in}[u_2'\rangle$ and u_1, u_1' and u_2, u_2' are connected. The hypothesis that $u_1' \cup u_2' \notin \bar{\mathcal{U}}$ leads to a contradiction. Then G_E is connected.
($\Leftarrow$:) By its definition Σ has no isolated elements.

■

For the axiomatization of the actions of Σ we will follow the method described in [SD] and also followed by the axiomatization of the cases of Σ. Our atomic propositions are the active events of Σ:

Definition 11:
Let $\Sigma = (B, E, F, c_{in})$ be an en-system, c_Σ its cases and $\bar{\mathcal{U}}$ its set of active steps. Γ^Σ is the extension of Γ by the following specific axioms:

NB_2: For all u, $u' \in \bar{\mathcal{U}}$ if there exist c, $c' \in c_\Sigma$ such that $c[u\rangle c'[u'\rangle$ then

$$\Gamma^\Sigma \vdash \bigwedge u \wedge \bigwedge \neg(E - u) \rightarrow \bigcirc(\bigwedge u' \wedge \bigwedge \neg(E - u'))$$

$$\Gamma^\Sigma \vdash \bigwedge u' \wedge \bigwedge \neg(E - u') \rightarrow \overleftarrow{\bigcirc} (\bigwedge u \wedge \bigwedge \neg(E - u))$$

NB_3: For all $u, u' \in \bar{\mathcal{U}}$ if for all c, $c' \in c_\Sigma$ from $c'[u\rangle c$ follows that *non* $c[u'\rangle$ then

$$\Gamma^\Sigma \vdash \bigwedge u \wedge \bigwedge \neg(E - u) \rightarrow \otimes\neg(\bigwedge u' \wedge \bigwedge \neg(E - u'))$$

$$\Gamma^\Sigma \vdash \bigwedge u' \wedge \bigwedge \neg(E - u) \rightarrow \overleftarrow{\otimes} \neg(\bigwedge u \wedge \bigwedge \neg(E - u))$$

$B_4: \quad \Gamma^\Sigma \vdash \bigvee_{u \in \bar{\mathcal{U}}} (\bigwedge u \wedge \bigwedge \neg (E - u))$

$B_5:$ For all $d \subseteq E$ if there is no $u \in \bar{\mathcal{U}}$ such that $d = u$ then $\quad \Gamma^\Sigma \vdash \neg \bigwedge d \vee \bigvee (E - d)$

Moreover, if Σ is connected and there is no conflict at its initial case then we can extend $^\Sigma$ by the following specific axiom denoting connectedness:

B_1: For all $u \in \bar{\mathcal{U}}$ we have $\Gamma^\Sigma \vdash \overleftarrow{\Diamond} \Diamond (\bigwedge u \wedge \bigwedge \neg (E - u)) \vee \Diamond \overleftarrow{\Diamond} (\bigwedge u \wedge \bigwedge \neg (E - u))$

■

If Σ is not reduced we replace E by E', where E' is the set of active elements of Σ. Obviously Γ^Σ is sound w.r.t. the action graph of Σ.

Connected graphs are definable up to isomorphism [Cou,SD]. Hence if Γ^Σ contains B_1 then its models are defined up to isomorphism. If the action graph G_E is not con- ected then Γ^Σ axiomatizes all graphs consisting of one or more components of G_E [SD].

As in corresponding event approaches, [Pen,NPW], Γ^Σ provides no information about onditions and about the number of pre- and post conditions of events and steps. Hence $^\Sigma$ cannot always distinguish between common pre- and post conditions in the case of ck of independence between two events:
Γ^Σ is given by $\quad e_1 \wedge \neg e_2 \rightarrow \otimes \neg (e_2 \wedge \neg e_1), \quad e_2 \wedge \neg e_1 \rightarrow \otimes \neg (e_1 \wedge \neg e_2), (NB_3),$
$e_1 \wedge \neg e_2) \vee (e_2 \wedge \neg e_1), (NB_4), \quad \neg (e_1 \wedge e_2), (NB_5),$
e have three possibilities:

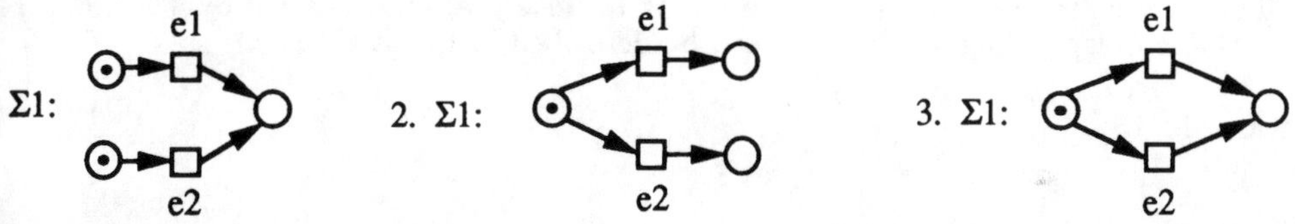

nd Γ^Σ cannot distinguish between them.
By means of Γ^Σ we cannot express contacts. Γ^Σ describes the action graph G_E of Σ nd G_E and G'_E, the action graph of the S-complementation Σ' of Σ, [Thi], coincide:

heorem 12:
et $\Sigma = (B, E, F; c_{in})$ be an en-system and $\Sigma' = (B', E, F'; c'_{in})$ its S-complementation. hen G_E and G'_E, the corresponding action graphs, coincide.
roof outline:
$G_E = (\bar{\mathcal{U}}, f)$, $G'_E = (\bar{\mathcal{U}}', f')$ are the corresponding action graphs and $\mathcal{U}$, $\mathcal{U}'$ the sets of teps respectively, by [Thi] we have $\mathcal{U} = \mathcal{U}'$. Moreover, $u \in \bar{\mathcal{U}}$ iff $u \in \bar{\mathcal{U}}'$ and $(u, u') \in f$ iff $u, u') \in f'$.

■

Then we cannot expect to describe contacts by means of Γ^Σ. However independency f two different events is expressible in Γ^Σ:

heorem 13:

1. If $\vdash \bigwedge u \wedge \bigwedge \neg (E - u) \rightarrow \bigcirc (e_1 \wedge \bigwedge \neg (E - e_1))$, $\bigwedge u \wedge \bigwedge \neg (E - u) \rightarrow \bigcirc (e_2 \wedge \bigwedge \neg (E - e_2))$ then
   $Ind(e_1, e_2) \quad$ iff $\quad \vdash \bigwedge u \wedge \bigwedge \neg (E - u) \rightarrow \bigcirc (e_1 \wedge e_2 \wedge \bigwedge \neg (E - e_1 - e_2))$

2. If $\vdash e_1 \wedge \bigwedge \neg(E - e_1) \rightarrow \bigcirc(\bigwedge u \wedge \bigwedge \neg(E-u))$, $e_2 \wedge \bigwedge \neg(E-e_2) \rightarrow \bigcirc(\bigwedge u \wedge \bigwedge \neg(E-u$
then

$$Ind(e_1, e_2) \quad \text{iff} \quad \vdash e_1 \wedge e_2 \wedge \bigwedge \neg(E - e_1 - e_2) \rightarrow \bigcirc(\bigwedge u \wedge \bigwedge \neg(E - u))$$

3. *non* $Ind(e_1, e_2)$ iff $\vdash \neg e_1 \vee \neg e_2 \vee \bigvee(E - e_1 - e_2)$

Proof outline:
1. $(\Rightarrow)$: By the hypothesis $\{e_1, e_2\} \in \mathcal{U}$ and there exist $c, c' \in c_\Sigma$ such that $c[u\rangle c'[e_1\rangle$ an
$c[u\rangle c'[e_2\rangle$. Then $c'[\{e_1, e_2\}\rangle$ and $\{e_1, e_2\} \in \bar{\mathcal{U}}$.
2.$(\Rightarrow)$: $\{e_1, e_2\} \in \mathcal{U}$ and if $c[u\rangle$ then $((c - \{e_1\bullet, e_2\bullet\}) \cup \{\bullet e_1, \bullet e_2\})[\{e_1, e_2\}\rangle$.
1. and 2. $(\Leftarrow:)$ By the semantics and the fact that if a set of propositions has a mod
then it is contradiction-free.
3. By NB_5.

■

Since conflicts are characterized by lack of independency we have the following theoren

Theorem 14:
1. If e_1, e_2 are in conflict at c then $\vdash \neg e_1 \vee \neg e_2 \vee \bigvee(E - e_1 - e_2)$.
2. If $u \in \bar{\mathcal{U}}$, $c[u\rangle$, $\vdash \bigwedge u \wedge \bigwedge \neg(E - u) \rightarrow \bigcirc(e_1 \wedge \bigwedge \neg(E - e_1))$, $\vdash \bigwedge u \wedge \bigwedge \neg(E - u) -$
$\bigcirc(e_2 \wedge \bigwedge \neg(E - e_2))$ and $\vdash \neg e_1 \vee \neg e_2 \vee \bigvee(E - e_1 - e_2)$ then e_1, e_2 are in conflict at c

Proof outline:
By Th. 13.

■

Th. 14.2 expresses conflicts corresponding to the situation described by Th. 13.1. Th
situation corresponding to Th. 13.2 can be described in the same way.

Example 15:

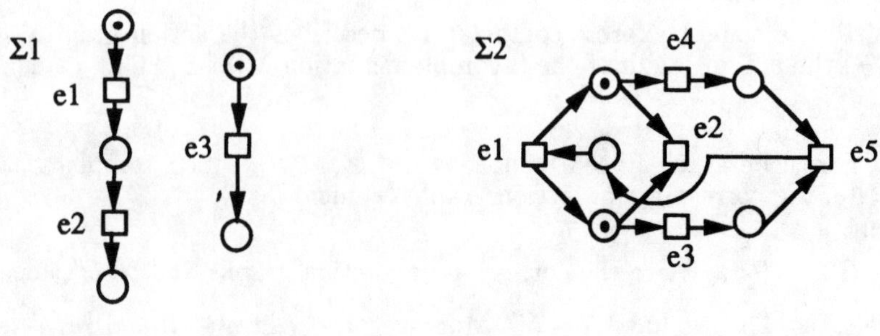

ne corresponding action graphs:

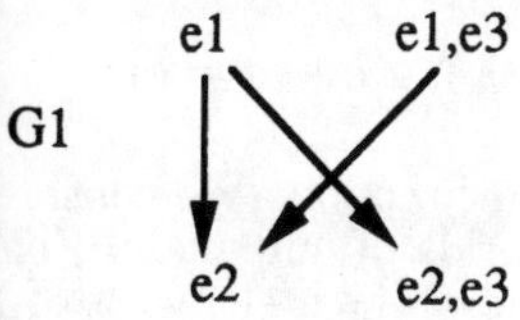

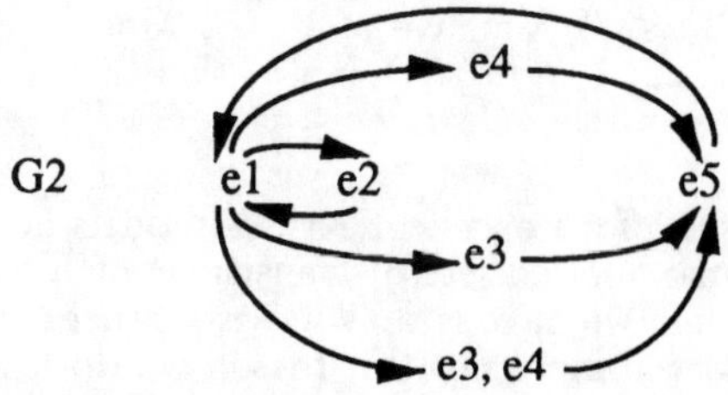

he connectedness axiom NB_1 is not valid in G_1: It is not the case that
$_{e_1,e_3\}} \Diamond \overleftarrow{\Diamond} (e_2 \wedge e_3 \wedge \neg e_1) \vee \overleftarrow{\Diamond} \Diamond (e_2 \wedge e_3 \wedge \neg e_1)$.

■

emark 16:
et $\Sigma = (B, E, F; c_{in})$ be an en-system and $u, e \in \mathcal{U}$. If $c[u\rangle c'[e\rangle$ we have by NB_2 that

$$\vdash \bigwedge u \wedge \bigwedge \neg(E - u) \rightarrow \bigcirc(e \wedge \bigwedge \neg(E - e)) \quad (1)$$

ıd as in Γ_Σ we get $\vdash \bigwedge u \wedge \bigwedge \neg(E - u) \rightarrow \bigcirc e$. In Γ^Σ we cannot simplify the left-hand de of implications like (1). To simplify would mean (as in Th. 8) that the formula $(\bigwedge u \wedge \bigvee(E - u))$ (2) were derivable. Let Σ and G its action graph be as follows:

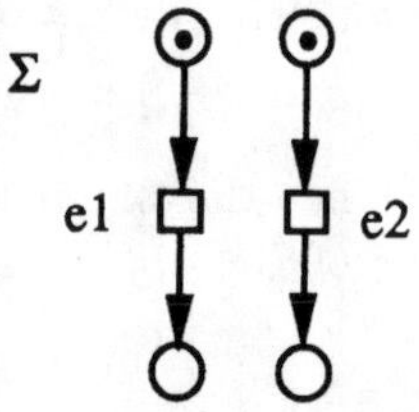

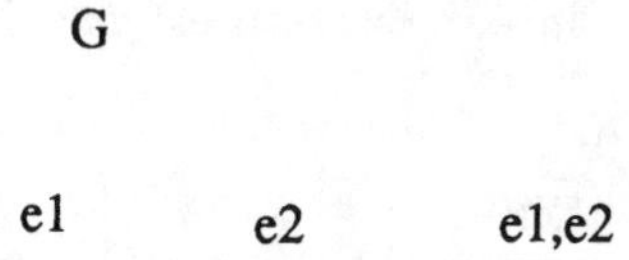

he active steps are $\{e_1\}, \{e_2\}, \{e_1, e_2\}$. Let $u = e_1$. If the formula (2) were derivable ıen $\neg(e_1 \wedge e_2)$ or equivalently $\neg e_1 \vee \neg e_2$ would be derivable. But $\{e_1, e_2\}$ is an active ep and already denoted in the action graph. Γ^Σ has a model, it is contradiction-free. ence (2) cannot be derivable.
he intuition behind the above mentioned problematic is that in Γ^Σ and especially by B_4 and NB_5 we can get rid of subformulas in the left-hand side of implications like (1) ıly if the described active steps are the maximal (w.r.t. set-theoretical conclusion) sets ıaracterized by the specific axioms of Γ^Σ. We were able to have the corresponding Th. for Γ_Σ only for a contact-free en-system Σ because cases of contact-free en-systems are ctually the maximal subsets of B having the properties expressed by Γ_Σ. ■

Axiomatizing Events of Processes

he most useful model for the behaviour of en-systems are processes [Ro,Rei2]. Formally:

efinition 17:
et $\Sigma = (B, E, F, c_{in})$ be an en-system and c_Σ its cases. π is a process of Σ iff $\pi = (K, p)$,

where $K = (B_K, T_K, F_K)$ is an occurrence net, SL its slices and $p : K \longmapsto \Sigma$ satisfies:
1. There is a slice S of K s.t. $p(S) \in c_\Sigma$.
2. For all $t \in T_K$ we have $p(t) \in E$. Moreover $\bullet p(t) = p(\bullet t)$ and $p(t)\bullet = p(t\bullet)$.
3. For all $x, y \in K$ if $p(x) = p(y)$ then $x\ li\ y$.
4. There is no $e \in E$ s.t. $\bullet e \subseteq p(\{s \in SL \mid s\bullet = \emptyset\})$ or $e\bullet \subseteq p(\{s \in SL \mid \bullet s = \emptyset\})$.

■

Processes afford excellent Kripke models having p as valuation function. We conside cuts (ie. maximal co-sets of transitions of K) as the possible worlds of our model. Th main problem we have is that if π is a process of Σ, we do not know which elements of Σ are represented in π. For this reason we work with the π-restriction of Σ:

Definition 18:
Let $\Sigma = (B, E, F; c_{in})$ be an en-system, c_Σ its cases and $\pi = (K, p)$, $K = (S_K, T_K, F_K)$ $T_K \neq \emptyset$, a process of Σ. We define:

$$c_{\Sigma|} := \{c \in c_\Sigma \mid \textit{there is } S \in SL_\pi \textit{ s.t. } p(S) = c\}$$

$$B_| := \{a \in B \mid \textit{there is } s \in S_K \textit{ s.t. } p(s) = a\}$$

$$E_| := \{e \in E \mid \textit{there are } c_1, c_2 \in c_{\Sigma_|} \textit{ s.t. } c_1[e\rangle c_2\}$$

$$F_| := F\ |_{(B_| \times E_|) \cup (E_| \times B_|)}$$

$\Sigma_| := (B_|, E_|, F_|, c_{in})$ is the π-restriction of Σ. ■

$\Sigma_|$ is an en-system. Moreover, $c_{\Sigma|}$ are the cases of $\Sigma_|$ and all elements of $\Sigma_|$ are represented in π. Since the structure of cuts is connected, we can give the following definition

Definition 19:
Let Σ be an en-system, π a process of Σ and $\Sigma_|$ the π-restriction of Σ. Then the axiomati system Γ_π^Σ is the axiomatic system Γ extended by the specific axioms of Def. 11, wher $\bar{\mathcal{U}}$ is replaced by $\mathcal{U}' = \{u \in \bar{\mathcal{U}} \mid$

$\mid$ *there exists* $c \in c_{\Sigma_|}$ *s.t.* $c[u\rangle$ *and for all* $c \in c_{\Sigma_|}$ *and* $u' \in \bar{\mathcal{U}}$ *if* $c[u\rangle$ *and* $c[u'\rangle$ *then* $u \not\subset u'\}$,
restricted on $\Sigma_|$:

$$\Gamma_\pi^\Sigma = \Gamma \cup (\{NB_1, NB_2, NB_3, NB_4, NB_5\}\ |_{\Sigma_|})$$

■

Remark 20:
We had to modify the set of active steps of the Def. 11, because we wanted to axiomatiz only the cuts of π. If we want to axiomatize the events of Σ which are represented in π we can modify Def. 19 replacing $\mathcal{U}'$ by the set $E_|$. Then we obtain the specific axiomatics $\Gamma^{\Sigma_|}$ corresponding to event structures like [Pen], where the conflict relation $\sharp$ is empty.
■

Since cuts of π are maximal, we have the following theorem:

Theorem 21:
If Σ is an en-system, π a process of Σ and $u \in \mathcal{U}'$, then $\Gamma_\pi^\Sigma \vdash \bigwedge u \wedge \bigwedge \neg(E_| - u) \leftrightarrow \bigwedge u$.
Proof outline:
As in Th. 8.

■

Relationship to Other Approaches

Γ_Σ and Γ^Σ can be considered as extensions of standard approaches like [MP,MM]. Our ecific axiomatics are not only enriched by axioms like NA_1 or NB_1, NA_3 or NB_3 and A_5 or NB_5, but the occurrence of negations of conditions or events in our formalism lows our models to be defined up to isomorphism. However, local specifications of gen- al axiomatics like that of [Pen] cannot carry the same amount of information carried by $_\Sigma$ and Γ^Σ.

The intuition behind Γ_Σ and Γ^Σ is the same as in [Pel]. Both $\phi_N(\bar{Y})$ of [Pel] and $^\Sigma_\tau$ attempt to characterize the concurrent behaviour of en-systems. However [Pel] has a ther language theoretical point of view and expresses properties of contracted processes s in [Ro]).

Like [BC] we do not use event structures as in [Pen], because the axiom of conflict eredity is restrictive and because we want to describe steps as well. It is not so clear [Pen] how steps can be expressed and how we can distinguish between globality and cality, while by means of Γ^Σ we are able to describe both globality and locality and distinguish between them by modifying Γ^Σ to Γ^{Σ_l}, thus expressing locality, Rem. 20. Ioreover, we regard unfoldings of en-systems rather as models of a branching time logic xpressing runs [Rei3]. This will be the aim of a future work.

[BC] give an algebraic calculus describing events as in Γ^Σ. The lack of steps is counter- alanced by the relation $\sharp$ expressing conflicts. However the relationship between algebraic O approaches and logical ones has not been yet explored.

Discussion

In this work we extended the PO axiomatic system Γ to the specifc theories Γ_Σ and $^\Sigma$ describing cases and actions respectively of an en-system Γ. Γ^Σ was modified to Γ^Σ_π escribing cuts of a process π of Σ. For a comparison between state- and event-approaches e also [Br1,Br2].

By means of our formalisms we could express the independency relation, conflicts nd contacts. By means of Th. 7.2 and Th. 12 we proved that an en-system Σ cannot lways be considered as a contact-free one, since Γ^Σ, describing actions, cannot distinguish etween Σ and the S-complementation of Σ, in contrast to Γ_Σ, describing cases in a very recise way, which does distinguish between Σ and its S-complementation.

There are still a lot of problems to be solved. We do not know whether there exists an xiomatic system containing binary operators like *until* which is complete w.r.t. partial rders. We also do not know what kind of Induction Principle we can use in PO's.

The contribution of PO logic to compositionality will be the aim of a next work.

References

BPM] BenAri,Pnueli,Manna: The Temporal Logic of Branching Time. Acta Informatica 20, 1983.

BF] Best,Fernandez: Notations and Terminology on Petri Net Theory. Arbeitspapiere der GMD 195, 1986.

BC] Boudol,Castellani: Concurrency and Atomicity. TCS 59 (1988).

[BCG] Browe,Clarke,Grümberg: Characterizing Finite Kripke Structures by Tempora Logic. TCS 59 (1988).

[Br1] Broy M. : Spezifikation und Entwurf komplexer, kausal vernetzter System Informatik-Fachberichte 187, Springer-Verlag 1988.

[Br2] Broy M. : Requirement and Design Specification for Distributed Systems. LNC 335, Springer-Verlag 1988.

[Bur] Burgess J. : Basic Tense Logic. In: Handbook of Philosophical Logic, Vol. 2, Gabba and Guenther (eds.), D. Reidel Publishing Company, 1984.

[Cou] Courcelle B.: The Monadic Second Order Logic of Graphs: Definable Sets of Finit Graphs. LNCS 344, 1989.

[Ch] Church A.: Introduction to Mathematical Logic, Vol. 1. Princeton University Press 1956.

[ES] Emerson,Srinivasan: Branching Time Temporal Logic. Linear Time, Branching Tim and Partial Order in Logics and Models for Concurrency, Goos,Hartmanis (eds) LNCS 354, 1989.

[Ko] Kozen D. : Results on the Propositional μ-calculus. TCS 27, 1983.

[Krö] Kröger F. : Temporal Logic of Programs. EATCS Monographs on Theoretical Com puter Science, Vol. 8, Springer-Verlag, 1987.

[MM] Masini,Maggiolo-Schettini: Local and Global Time Logic. Universita degli Studi d Pisa, Dipartimento di Informatica, TR-5/88.

[MP] Manna,Pnueli : Verification of Concurrent Programs: The Temporal Framework. In The Correctness Problem in Computer Science. Boyer, Moore (eds.), Internationa Lecture Series in Computer Science, Academic Press, New York, 1981.

[NPW] Nielsen,Plotkin,Winskel : Petri Nets, Event Structures and Domains. TCS 13 (1981).

[OL] Owicki,Lamport : Proving Liveness Properties of Concurrent Programs. ACM, Vol 4, No. 3, July 1982.

[Pel] Pelz E. : About the concurrent behaviour of EN systems: definability and closure results. 9th European Workshop on Application and Theory of Petri Nets. Venice, Italy, June 1988.

[Pen] Penczek W. : A Temporal Logic for Event Structures. Fund. Inf. 11, 1988.

[PW] Pinter,Wolper : A Temporal Logic for Reasoning about Partially Ordered Computations. Proceedings of the third ACM Symposium on Principles of Distributed Computing, Vancouver, Canada, 1984.

[Rei1] Reisig W.: Petri Nets. An Introduction. EATCS Monographs in Computer Sciences, Vol. 4, Springer-Verlag, 1985.

[Rei2] Reisig W.: Towards a Temporal Logic for True Concurrency. Arbeitspapiere der GMD 277, 1987.

[Rei3] Reisig W. : Towards a temporal logic of causality and choice. Linear Time, Branching Time and Partial Order in Logics and Models for Concurrency, Goos,Hartmanis (Eds), LNCS 354, 1989.

[Rei4] Reisig W. : Temporal Logic and Causality in Concurrent Systems. LNCS 335, Concurrency 88, Springer-Verlag, 1988.

[Rei5] Reisig W. : A Report on the REX School/Workshop on Linear Time, Branching Time and Partial Order in Logics and Models for Concurrency. Bulletin of EATCS, No 36, October 1988.

[Ro] Rozenberg G. : Behaviour of Elementary Net Systems. LNCS 254, Springer-Verlag, 1987.

[Sin1] Sinachopoulos A. : Temporal Logics for Elementary Net Systems. Arbeitspapiere der GMD 353, 1988.

[SD] Sinachopoulos,Devillers : Partial Order Logics for Axiomatizing Concurrent Systems. Submitted for publication, 1990.

[Thi] Thiagarajan P. : Elementary Net Systems. LNCS 254, Springer-Verlag, 1987.

[W] Winskel G.: Event structures. Proc. Advances in Petr Nets '86, LNCS 255, 1987.

Priority as Extremal Probability

Scott A. Smolka*
Department of Computer Science
SUNY at Stony Brook
Stony Brook, NY 11794-4400
USA
sas@sbcs.sunysb.edu

Bernhard Steffen
Aarhus University
Datalogisk Afdeling
Ny Munkegade 116 - 8000 Aarhus C
Denmark
bus@daimi.dk

1 Introduction

In [vGSST90], three models of probabilistic processes are presented. The *reactive model*, due to Larsen and Skou [LS89], is an adaptation of the reactive viewpoint embodied in classical process algebra [Pnu85]: for each action symbol α, a process either possesses no α-transitions, or else the sum of the probabilities of all α-transitions is 1. The *generative model* is an extension of the reactive model in which the sum of the probabilities of all transitions of a process is at most 1. Besides encoding the relative frequency of transitions with the same label, a generative process also encodes the relative frequency of transitions with different labels.

The *stratified model* is an extension of the generative model that captures the branching structure of the *purely probabilistic choices* made by a process. In contrast to the generative model where relative frequency is preserved in the context of restriction, in the stratified model restriction may alter relative frequency, albeit in a structured, level-wise way. For example, consider the process P defined by the following recursive expression:

$$P \overset{def}{=} fix_X(\tfrac{1}{3}a\,.\,X + \tfrac{2}{3}(\tfrac{1}{2}b\,.\,X + \tfrac{1}{2}c\,.\,X))$$

P may be interpreted as a scheduler that services task a exactly one-third of the time, and equally services the remaining tasks (a and b) the other two-thirds of the time. Placing P in a restriction context that precludes the execution of c yields process $fix_X(\frac{1}{3}a\,.\,X + \frac{2}{3}b\,.\,X)$ in the stratified model and process $fix_X(\frac{1}{2}a\,.\,X + \frac{1}{2}b\,.\,X)$ in the generative model. Thus the above interpretation is retained only in the stratified model. In practice, task a might correspond to a special system process, such as a garbage collector, while b and c would correspond to the user processes.

In this paper, we extend the stratified model to obtain a very general notion of *process priority*. This can be done simply by additionally considering the extremal case, where a probability guard is 0. For example, the expression

$$1P + 0(1Q + 0R)$$

gives priority to process P over Q and R, and priority to Q over R. Thus process R can only be executed in a restriction context that excludes P and Q.

*Research supported by NSF Grant CCR-8704309.

Our extension to the stratified model results in what we term the *model of extremal probability.* This model retains all the properties of the stratified model, and it covers all the essential features of the process algebra models for priority that we know of. Thus it provides a uniform framework in which to reason about probabilities and priorities. Also, as we shall show, a simple abstraction of the extremal probability model yields a customized framework for reasoning exclusively about priority.

Three features seem important when modeling priority. They concern the priority relation that may be defined between different actions:

1. The relation should be *globally dynamic*, i.e. it should allow an action a to have priority over an action b in one state of the system, and the converse in some other state.

2. It should be possible to define an arbitrary *partial order* at each system state. In particular, the relation should not be constrained to a "leveled" priority structure in which two incomparable actions are required to have the same priority relation with respect to all other actions.

3. The relation should be modeled *transparently* and *intuitively*, i.e. without using any complicated encodings.

Whereas the importance of the third feature is evident, the first two features can be illustrated and motivated by means of a "real life" example. Recall the Roman empire and their system of distributing governing power. Mainly, there were two kinds of authorities, the senate and the consuls. In times of peace, the power of the consuls was minimal. Every major decision was taken by the senate. In our interpretation this would mean that the activities of the senate are of higher priority than those of the consuls. However, in times of crisis this order was reversed: all power was transferred to the consuls because, during a crisis, decisions had to be made quickly, as opposed to democratically. This illustrates the necessity of having a globally dynamic priority structure—in a more technical setting, the handling of exceptions may cause a similar dynamic change in the priority structure, in order to handle run-time errors.

To illustrate the second feature, consider two Roman senators. They are independent in that neither one has priority over the other. However, if we assume that one of them is corrupted by a third party, then one of these senators is dependent on a person of which the other senator is independent. This results in a nonlinear, or non-leveled, priority structure—in a more technical setting, the handling of an exception and an interrupt may be mutually independent actions, however they serve as control entities for different types of external events.

The central approaches to priority in process algebra are the ones by Baeten, Bergstra and Klop [BBK86], Cleaveland and Hennessy [CH88], and Camilleri [Cam89]. Whereas the approach in [BBK86] is not globally dynamic, Cleaveland and Hennessy do not address the issue of a general partial order structure for priorities. Camilleri satisfies both of these features. However transitions in his model must be augmented with rather elaborate contextual information in order to encode the priority structure. Thus his approach lacks the third feature.

Our approach satisfies all three features. It is globally dynamic, it allows for an arbitrary partial order of priorities, and it transparently models priority. The third property is achieved by explicitly representing the priority structure in terms of probability transitions. These are transitions labeled with real numbers from the interval $[0, 1]$, indicating the relative frequency of choice at a certain state. The principle is the same as presented in [vGSST90]; only the handling of 0-probabilities is new. In fact, it is just this addition that provides our framework with the power to express priority.

Summary of Technical Results

We will be working within the framework of *PCCS*, a specification language for probabilistic processes introduced in [GJS90]. PCCS is derived from Milner's SCCS [Mil83] by replacing the operator

of nondeterministic process summation with a probabilistic counterpart. Several PCCS expressio
have appeared above, which should give the flavor of the language.

Three semantic models for PCCS are presented in [vGSST90]. This paper extends the stratifi
model to deal with probability branches whose "a priori" probability is 0. Such a branch can be tak
only if all other alternatives are precluded by context constraints. We will refer to this extension
$PCCS_\zeta$, where the ζ signifies the possibility of *zero*-probability transitions. In particular, we prese
the following results for $PCCS_\zeta$:

- A *structural operational semantics* is given as a set of inference rules in the style of Plotk [Plo81] and Milner [Mil89]. These inference rules constitute a *semantic functional* from the s of $PCCS_\zeta$ expressions, Pr_ζ, to a domain of probabilistic labeled transition systems.

- A notion of *bisimulation equivalence* is developed along the lines of [LS89]. There, Larsen a Skou introduced *probabilistic bisimulation*, a natural and elegant extension of strong bisimulati [Par81, Mil83]. We prove that our version of bisimulation equivalence is a congruence with respe to all $PCCS_\zeta$ operators. Thus it provides a compositional notion of semantics for $PCCS_\zeta$ th is consistent with its operational semantics.

- The expressive power of $PCCS_\zeta$ is demonstrated by investigating two sublanguages of PCC and one abstraction.

  $PCCS_\sigma$: the sublanguage of $PCCS_\zeta$ where sums either do not have any 0-probability guarc or are of the form $1\,P + 0\,Q$.

  $PCCS_S$: the sublanguage of $PCCS_\sigma$ without any 0-probability guards.

  $PCCS_\pi$: the abstraction of $PCCS_\sigma$, which results from stripping away all probability guar that are different from 1 or 0.

  This hierarchy of languages demonstrates the effect of extending the stratified model of [vGSST9 here represented as $PCCS_S$, in order to deal with the extremal case of 0 probability guards. T extension maintains all the properties of $PCCS_S$, and it covers all the essential features of t process algebra models for priority that we know of. The discussion will concentrate on PCC and its relationship to the approaches of Baeten, Bergstra and Klop [BBK86], Cleaveland a Hennessy [CH88], and Camilleri [Cam89].

2 Syntax of $\mathbf{PCCS}_\zeta$

As in SCCS, the *atomic actions* of $PCCS_\zeta$ form an Abelian monoid $(Act, \cdot, 1)$ that is generated free
from the set Λ of *particulate actions*. Intuitively, actions of the form $\alpha \cdot \beta$ represent the simultaneou
atomic execution by a process of the actions α and β. We will often use juxtaposition to denote acti
product, e.g. $\alpha\beta$.

Let X be a variable, $\alpha \in Act$, A a subset of Act, and $f : Act \to Act$. Then the syntax of PCC
is given by:

$$E ::= \mathbf{0} \mid X \mid \alpha . E \mid \sum [p_i]\, E_i, \text{ where } p_i \in [0,1],\ \sum p_i = 1$$
$$\mid E \times F \mid E \lceil A \mid E[f] \mid fix_X E$$

An expression having no free variables is called a *process*, and Pr_ζ is the set of all $PCCS_\zeta$ process
Intuitively, $\mathbf{0}$ is the *zero process* having no transitions, while $\alpha . E$ performs action α with probabili
1 and then behaves like E. The expression $\sum[p_i]\, E_i$ offers a probabilistic choice among its constitue
behaviors E_i, while $E \times F$ represents synchronized product. The restricted expression $E\lceil A$ can on

$$\alpha . E \xrightarrow{\alpha} E$$

$$E \xrightarrow{\alpha} E',\ F \xrightarrow{\beta} F' \quad \Longrightarrow \quad E \times F \xrightarrow{\alpha\cdot\beta} E' \times F'$$

$$E \xrightarrow{\alpha} E',\ \alpha \in A \quad \Longrightarrow \quad E\lceil A \xrightarrow{\alpha} E'\lceil A$$

$$E \xrightarrow{\alpha} E' \quad \Longrightarrow \quad E\,[f] \xrightarrow{f(\alpha)} E'\,[f]$$

$$E\{fix_X E/X\} \xrightarrow{\alpha} E' \quad \Longrightarrow \quad fix_X E \xrightarrow{\alpha} E'$$

Figure 1: Inference Rules for Action Transitions of $PCCS_\zeta$

rform actions from the set A, and $E[f]$ specifies a relabeling of actions. Finally, $fix_X E$ defines a cursive process.

In this paper, we require summation expressions to be finite, write the binary version of summation $[p]\,E + [1-p]\,F$, and often omit the square brackets around the probabilities.

The only difference between the syntax of $PCCS_\zeta$ and the syntax of PCCS presented in [vGSST90] the consideration of 0-probability guards in summation expressions. The effect of restriction is a ormalization that proportionally distributes the probabilities of the restricted transitions over the maining transitions. This requires a somewhat arbitrary decision when normalizing a process like $P + 0\,Q$, which we resolve by distributing the probabilities evenly.

The Extremal Probability Model

this section, we first lift the operational semantics of the stratified model of probabilistic processes obtain the *extremal probability model.* This is accomplished by augmenting the range of normaled probabilities with the bottom element $\perp$. The presence of $\perp$ permits the distinction between a stricted process that has no permitted transitions, such as $(\frac{1}{3}\,a.\mathbf{0} + \frac{2}{3}\,b.\mathbf{0})\lceil\{c\}$; versus a restricted rocess that does have permitted action transitions but with accompanying probability transitions 0, e.g. $(1\,a.\mathbf{0} + 0\,b.\mathbf{0})\lceil\{b\}$. In our semantics, the former process will be equivalent to $\mathbf{0}$ and the tter process will be equivalent to $1\,b.\mathbf{0}$. We then equip the extremal probability model with a notion f probabilistic bisimulation, and show that the resulting equivalence relation is a congruence with espect to $PCCS_\zeta$.

.1 Operational Semantics of $\mathbf{PCCS}_\zeta$

s for the stratified model, the structural operational semantics of $PCCS_\zeta$ in the extremal probability odel is given in terms of inference rules that define two types of transition relations: *action transitions,* s in SCCS, and *probability transitions.* Action transitions are of the form $P \xrightarrow{\alpha} Q$. Probability ransitions are of the form $P \overset{p}{\longmapsto} Q$, meaning that P, with probability p, can behave as process Q. This separation of action and probability permits the branching structure of probabilistic choices to e captured explicitly. The rules for action transitions are given in Figure 1. They are the same as SCCS [Mil83] with the exception that there is no rule for process summation—in the stratified nd extremal probability models, the only choice mechanism is probabilistic. The inference rules for robability transitions appear in Figure 2.

Except for the second and the fourth rule, all of the inference rules of Figure 2 are straightforward daptations of the standard SCCS rules. (Another difference from SCCS is that probability transitions re indexed, a point we discuss below.) The second rule is needed to avoid deadlock in a synchronous roduct that is caused by a difference in depth of the purely probabilistic branching structures of the rgument processes. For example, we do not want $(\frac{1}{2}\,a.\mathbf{0} + \frac{1}{2}\,b.\mathbf{0}) \times c.\mathbf{0}$ to deadlock simply because

$$
\begin{array}{lcrcl}
 & & \sum_i [p_i]E_i & \overset{p_i}{\longmapsto}_i & E_i \\
E \overset{\alpha}{\longrightarrow} E' & \Longrightarrow & E & \overset{1}{\longmapsto}_1 & E \\
E \overset{p}{\longmapsto}_i E',\ F \overset{q}{\longmapsto}_j F' & \Longrightarrow & E \times F & \overset{p \cdot q}{\longmapsto}_{(i,j)} & E' \times F' \\
E \overset{p}{\longmapsto}_i E',\ \nu_S(E', A) \neq \bot & \Longrightarrow & E \lceil A & \xrightarrow{\ \rho_\zeta(E,A,p)\ }_i & E' \lceil A \\
E \overset{p}{\longmapsto}_i E' & \Longrightarrow & E\,[f] & \overset{p}{\longmapsto}_i & E'\,[f] \\
E\{fix_X E / X\} \overset{p}{\longmapsto}_i E' & \Longrightarrow & fix_X E & \overset{p}{\longmapsto}_i & E'
\end{array}
$$

Figure 2: Inference Rules for Probability Transitions of $PCCS_\zeta$

there does not exist a probability transition in the right argument. Deadlock is avoided by the seco rule, which provides the missing probability-1 transition.

The fourth rule deals with the restriction operator, and expresses the probability transitions $E \lceil A$ in terms of conditioned probability transitions of E. Intuitively, $E \lceil A$ behaves like E, where a probability transitions to subexpressions that necessarily require the execution of a restricted action a eliminated. The probabilities associated with these transition are proportionately distributed amo the remaining probability transitions. As discussed in greater detail below, the second hypothesis the restriction rule is what realizes this pruning away of probability transitions.

The term $\rho_\zeta(E, A, p)$ gives the conditional probability of the derived transition for expression E a restriction context that permits only actions from the set A. It is defined by

$$
\rho_\zeta(E,A,p) = \begin{cases} \bot & \text{if } \nu_\zeta(E,A) = \bot \\ \frac{1}{n} & \text{if } \nu_\zeta(E,A) = 0 \text{ and } n = |\{E \overset{0}{\longmapsto}_i E_i \mid \nu_\zeta(E_i, A) \neq \bot\}| \\ p/\nu_\zeta(E,A) & \text{otherwise} \end{cases}
$$

Here $\nu_\zeta(E, A)$ is the "normalization factor" of expression E in restriction context A, and is define formally below. A normalization factor of $\bot$ (the first clause of ρ_ζ) indicates that the restricte expression possesses no transitions, and ρ_ζ is also defined to be $\bot$ in this case. By the second hypothes of the restriction rule, ρ_ζ is only applied in situations where $\nu_\zeta(E', A) \neq \bot$, which implies $\nu_\zeta(E, A)$ $\bot$ also. Therefore, the purpose of the first clause is merely to provide a complete and consiste definition of ρ_ζ.

A normalization factor of 0 (the second clause of ρ_ζ) means that only probability-0 transitio remain unrestricted and that there is at least one of them. Let there be $n \geq 1$ of these transition Then they are transformed into probability-$\frac{1}{n}$ transitions, thus enabling the zero-probability portic of the process for execution. The second clause therefore yields equations like

$$
(1\, a.P \,+\, 0\, b.Q \,+\, 0\, c.R) \lceil \{b, c\} \;=\; \tfrac{1}{2}\, b.Q \,+\, \tfrac{1}{2}\, c.R
$$

meaning that processes of lower priority may only be executed when all processes of higher priorit are excluded by the context.

The final clause of ρ_ζ corresponds to a value for $\nu_\zeta(E, A)$ in the interval $(0, 1]$, which is used t compute the conditional probabilities of the transitions of $E \lceil A$ under the assumption of A. For ex- ample,

$$
(\tfrac{1}{3}\, a.P \,+\, \tfrac{1}{3}\, b.Q \,+\, \tfrac{1}{3}\, c.R) \lceil \{a, b\} \;=\; \tfrac{1}{2}\, a.P + \tfrac{1}{2}\, b.Q
$$

Using $\{\!|, |\!\}$ as multi-set brackets, and adopting the convention that empty summation is $\bot$ (and not

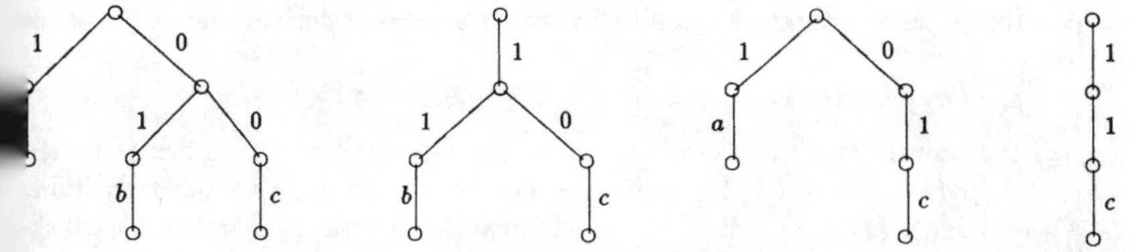

1) P 2) $P\lceil\{b,c\}$ 3) $P\lceil\{a,c\}$ 4) $P\lceil\{c\}$

Figure 3: Extremal Probability Transition Systems

ısual), the formal definition of ν_ζ is given by:

$$\nu_\zeta(E,A) = \begin{cases} 1 & \text{if } E \xrightarrow{\alpha}, \alpha \in A \\ \bot & \text{if } E \xrightarrow{\beta}, \beta \notin A \\ \sum_i \{\!| p_i \mid E \overset{p_i}{\longmapsto}_i E_i \ , \ \nu_\zeta(E_i, A) \neq \bot |\!\}, & \text{otherwise} \end{cases}$$

The condition $\nu_\zeta(E', A) \neq \bot$ in the premise of the restriction inference rule means that derivative of E is capable of performing an action transition from the set A, perhaps with probability 0. ısequently, the probability transition of the premise should remain after restriction, conditioned ler the assumption of A.

To illustrate both the inference rule for restricted expressions and the way in which priority is tured consider the process

$$P \stackrel{def}{=} 1\,a.\mathbf{0} + 0\,(1\,b.\mathbf{0} + 0\,c.\mathbf{0})$$

the following, P is "partially evaluated" with respect to a representative sampling of the relevant triction contexts. The results of these evaluations are the restriction-free expressions on the right-ıd side.

$$\begin{array}{lll} 1) & P\lceil\{a,b,c\} & \rightsquigarrow & 1\,a.\mathbf{0} + 0\,(1\,b.\mathbf{0} + 0\,c.\mathbf{0}) \\ 2) & P\lceil\{b,c\} & \rightsquigarrow & 1\,(1\,b.\mathbf{0} + 0\,c.\mathbf{0}) \\ 3) & P\lceil\{a,c\} & \rightsquigarrow & 1\,a.\mathbf{0} + 0\,1\,c.\mathbf{0} \\ 4) & P\lceil\{c\} & \rightsquigarrow & 1\,(1\,c.\mathbf{0}) \\ 5) & P\lceil\{\,\} & \rightsquigarrow & \mathbf{0} \end{array}$$

ther evaluation of these expressions would yield the stratified transition systems of Figure 3.

As mentioned above, all probability transitions are indexed. The indices are used to distinguish erent *occurrences* of the *same* probability transition, and are constructed so that every probability nsition of an expression has a unique index. (The indices were used above in computing the malization factor of a restricted expression, and will be used in the next section to define *cumulative* bability distributions.) The following examples are illustrative:

$$a.\mathbf{0} \overset{1}{\longmapsto}_1 a.\mathbf{0} \qquad (\tfrac{1}{2}a.\mathbf{0} + \tfrac{1}{2}a.\mathbf{0}) \overset{\frac{1}{2}}{\longmapsto}_1 a.\mathbf{0} \qquad (\tfrac{1}{2}a.\mathbf{0} + \tfrac{1}{2}a.\mathbf{0}) \overset{\frac{1}{2}}{\longmapsto}_2 a.\mathbf{0}$$

The inference rules of Figures 1 and 2 define the semantics of $PCCS_\zeta$ expressions in terms of nsition systems from the extremal probability model. Such transition systems are *stochastic* in sense that for each non-deadlocked state, the sum of the probabilities of its outgoing probability nsitions is 1.

2 Bisimulation Semantics

simulation in the extremal probability model can now be defined in a fashion analogous to stratified imulation [vGSST90]. As prerequisites, we need to define μ_ζ, the *cumulative probability distribution*

function (PDF), which computes the total probability by which a process derives a set of process
μ_ζ is of the form

$$\mu_\zeta : (Pr_\zeta \times (Act \cup \{*\}) \times \mathcal{P}(Pr_\zeta)) \longrightarrow [0,1] \cup \{\bot\}$$

where $\mathcal{P}(Pr_\zeta)$ denotes the power set of Pr_ζ, and $*$ is a dummy symbol used to mark probabili transitions. For $\alpha \in Act$, $\mu_\zeta(P, \alpha, S) \in \{1, \bot\}$ indicates whether or not P has an α–transition some process in S. Otherwise, $\mu_\zeta(P, *, S) \in [0,1] \cup \{\bot\}$ indicates the total probability by which can behave as some process in S. The case of $\bot$ indicates that P possesses no probability transiti to any process in S. Recalling that the empty sum of probabilities is $\bot$, we have:

Definition 1 (Cumulative PDF) $\mu_\zeta : (Pr_\zeta \times (Act \cup \{*\}) \times \mathcal{P}(Pr_\zeta)) \longrightarrow [0,1] \cup \{\bot\}$ *is the to function given by :* $\forall \alpha \in Act \cup \{*\}, \forall P \in Pr_\zeta, \forall S \subseteq Pr_\zeta$.

$$\mu_\zeta(P, \alpha, S) = \begin{cases} 1 & \text{if } \exists Q \in S.\ P \xrightarrow{\alpha} Q \\ \sum \{\!| p_i \mid Q \in S\ ,\ P \overset{p_i}{\longmapsto}_i Q |\!\} & \text{if } \alpha = * \\ \bot & \text{otherwise} \end{cases}$$

Given an equivalence relation $\mathcal{R}$ on Pr_ζ, we denote its corresponding set of equivalence classes $Pr_\zeta/\mathcal{R}$, and define:

Definition 2 *An equivalence relation* $\mathcal{R} \subseteq Pr_\zeta \times Pr_\zeta$ *is a* ζ-bisimulation *if* $(P, Q) \in \mathcal{R}$ *impl* $\forall S \in Pr_\zeta/\mathcal{R}$, $\forall \alpha \in Act \cup \{*\}$,

$$\mu_\zeta(P, \alpha, S) = \mu_\zeta(Q, \alpha, S)$$

Two processes P, Q *are* ζ-bisimulation equivalent *(written* $P \overset{\zeta}{\sim} Q$*) if there exists a* ζ*-bisimulation such that* $(P, Q) \in \mathcal{R}$.

If $\mathcal{R}_1$ and $\mathcal{R}_2$ are ζ-bisimulations, then the transitive closure of their union $(\mathcal{R}_1 \cup \mathcal{R}_2)^*$ is agair ζ-bisimulation. Thus, as in the case of classical bisimulation, ζ-bisimulation equivalence is the larg ζ-bisimulation, *i.e.*

$$\overset{\zeta}{\sim} = \bigcup\{\, \mathcal{R} \mid \mathcal{R} \text{ is a } \zeta\text{-bisimulation} \,\}$$

and can be found by a straightforward adaptation of the fixed-point iteration technique of [Mil89].

Like strong bisimulation does for SCCS or CCS, ζ-bisimulation equivalence provides a compositior notion of semantics for $PCCS_\zeta$ that is consistent with the operational semantics defined in the la section. Specifically:

Theorem 1 (Congruence) *For* $P, Q \in Pr_\zeta$:

$$P \overset{\zeta}{\sim} Q \quad \textit{implies} \quad \textit{for all } PCCS_\zeta \textit{ contexts } C[\ ]: \quad C[P] \overset{\zeta}{\sim} C[Q]$$

Proof: The proof proceeds by induction on the structure of contexts. The case for the empty enviro ment is trivial. Thus it only remains to investigate the cases corresponding to the top-most operat of a context. We present the cases of product and restriction here; the other cases are similar (or ev easier). For an equivalence relation $\mathcal{R}$ and a process P, we write $[P]_\mathcal{R}$ to denote the equivalence cla of which P is a member.

oduct:

Let us first show that $P \stackrel{\zeta}{\sim} Q$ implies $P \times R \stackrel{\zeta}{\sim} Q \times R$. Thus assume that (P, Q) belongs to the ζ-bisimulation $\mathcal{R}_0$. It is sufficient to show that the relation

$$\mathcal{R} = \{(P \times R, Q \times R) \mid (P, Q) \in \mathcal{R}_0, R \in Pr_\zeta\} \cup Id_{Pr_\zeta}$$

is a ζ-bisimulation. First note that it is an equivalence relation. Now suppose $(P \times R, Q \times R) \in \mathcal{R}$, and let $r = \mu_\zeta(P \times R, \gamma, [P' \times R']_{\mathcal{R}})$ $(= \mu_\zeta(P \times R, \gamma, [P']_{\mathcal{R}_0} \times R')\,)$. Then there are two cases to be distinguished.

$\gamma \in Act$:

In this case we have

$$r = \mu_\zeta(P, \alpha, [P']_{\mathcal{R}_0}) = \mu_\zeta(R, \beta, R') = 1, \text{ and } \gamma = \alpha \cdot \beta$$

for unique actions α and β. Thus, $P \stackrel{\zeta}{\sim} Q$ yields $\mu_\zeta(Q, \alpha, [P']_{\mathcal{R}_0}) = 1$, and therefore $\mu_\zeta(Q \times R, \gamma, [P' \times R']_{\mathcal{R}}) = 1$ as desired.

$\gamma = *$:

In this case we have

$$\mu_\zeta(P, *, [P']_{\mathcal{R}_0}) = p, \;\; \mu_\zeta(R, *, R') = q, \text{ and } r = pq$$

for unique $p, q \in [0, 1]$. Thus $P \stackrel{\zeta}{\sim} Q$ yields $\mu_\zeta(Q, *, [P']_{\mathcal{R}_0}) = p$, and therefore $\mu_\zeta(Q \times R, *, [P' \times R']_{\mathcal{R}}) = r$, which completes the proof in this case.

Thus $\mathcal{R}$ is a ζ-bisimulation. The symmetric result that $P \stackrel{\zeta}{\sim} Q$ implies $R \times P \stackrel{\zeta}{\sim} R \times Q$, which can be proven exactly in the same way, completes the argument.

estriction:

Here we must show that $P \stackrel{\zeta}{\sim} Q$ implies $P \lceil A \stackrel{\zeta}{\sim} Q \lceil A$, for $A \subseteq Act$. Thus let (P, Q) belong to the ζ-bisimulation $\mathcal{R}_0$. It is enough to show that the relation

$$\mathcal{R} = \{(P \lceil A, Q \lceil A), \mid (P, Q) \in \mathcal{R}_0\} \cup Id_{Pr_\zeta}$$

is a ζ-bisimulation. This follows almost immediately from the following fact, whose proof is straightforward:

$$P \stackrel{\zeta}{\sim} Q \text{ implies } \forall A \subseteq Act : \nu_\zeta(P, A) = \nu_\zeta(Q, A)$$

□

4 Sublanguages and Abstractions

o illustrate the expressive power of $PCCS_\zeta$, we consider two sublanguages of $PCCS_\zeta$ and one abstraction.

- the sublanguage $PCCS_\sigma$ of $PCCS_\zeta$, where sums either do not have any 0-probability guards, or are of the form $1\,P + 0\,Q_1 + \cdots + 0\,Q_n$. We call this the language of *separated priorities* because of the separation of priorities from standard probabilities.
- the sublanguage $PCCS_S$ of $PCCS_\sigma$ without 0-probability guards. This language is just the version of PCCS that is defined by the stratified model in [vGSST90].

- the abstraction $PCCS_{\pi}$ of $PCCS_{\sigma}$, which results from stripping away all probability guards tha are different from 1 or 0. This language is essentially SCCS enriched by a mechanism to expre a very general notion of priority.

This hierarchy of languages demonstrates the expressive power of our approach. $PCCS_{\zeta}$ retains a the properties of $PCCS_S$ discussed in the introduction, and even the abstraction $PCCS_{\pi}$ covers a the essential features of the process algebra models of priority that we know of. It should be noted tha the definition of ζ-bisimulation in Section 3 can be easily adapted to the three simplified language and that the congruence result is maintained as well.

In the following, we will concentrate on $PCCS_{\pi}$ and its relationship to the approaches of Baeter Bergstra and Klop [BBK86], Cleaveland and Hennessy [CH88], and Camilleri [Cam89]. Whereas thei approaches are based on asynchronous languages, our approach is based on the synchronous languag SCCS. This difference does not play a prominent role in our comparison of the approaches. Insteac we will concentrate on the way priority is modeled, in particular, with respect to the three feature introduced in Section 1.

The Baeten-Bergstra-Klop Approach

The approach of Baeten, Bergstra and Klop [BBK86], which we will henceforth refer to as th BBK-approach, is based on a partial ordering $<$ on atomic processes (*i.e.* actions). This partial orde globally expresses priority *potential*. That is, at every state there are two possibilities: either, th priority relation is the one expressed by the partial order on the action set, which is the case if th state occurs within the scope of the "priority evaluator" θ, or no priority structure is imposed. Thu the BBK-approach is static in that the same priority relation on actions holds at all possible state For example, if $a < b$ at some state of the system, then $a > b$ cannot be valid at any other state.

The BBK-approach therefore satisfies the second of the three features we have proposed, but no the first. Finally, there is a significant difference between expressing priority potential, as is done i the BBK-approach, versus explicitly specifying a process that consists of two subprocesses with on having priority over the other. Thus the third feature, which concerns the transparent modeling c priority, rather than priority potential, does not apply to the BBK-approach.

The Cleaveland-Hennessy Approach

The approach of Cleaveland and Hennessy [CH88], which we will refer to as the CH-approach, i based on the idea of providing a prioritized version $\underline{a}$ and a standard version a for each actio $a \in Act$. This allows the priority relation between actions $a, b \in Act$ to vary from one state t another by writing $\underline{a} + b$ at one state, and $a + \underline{b}$ at another. In this respect, the CH-approach i more general than the BBK-approach. However, it is less general in terms of the priority relation that can be expressed. In particular, the CH-approach distinguishes between two categories of action prioritized and unprioritized. As pointed out in [CH88], this approach can be easily extended t a multi-level priority structure. However an extension to the generality of the BBK-approach, wher arbitrary partial orders are allowed, is less obvious.

An example of such a general priority relation is given by $a > b$ and $a > c$ and $d > c$. Here a ha priority over both b and c, and d has priority only over c. To express partial orders of this kind, th CH-approach would require a rather complicated structure on its action set involving a different set o indices for each potential partial order. Our approach naturally covers such general priority relation For example the priority relation defined above could be expressed by: $(1\,a + 0\,b) + (1\,(a+d) + 0\,c)$ In summary, the CH-approach possesses the first and the third feature, but not the second.

The Camilleri Approach

In his process algebra, Camilleri [Cam89] expresses priority in essentially the same way as do we *i.e.* ignoring details, $P \mathbin{+\!\!)} Q$ and $1\,P + 0\,Q$ have the same meaning. Thus, the Camilleri-approach ha e same degree of generality as ours: it is globally dynamic and at each state the priority structures ay constitute an arbitrary partial order.

However, there is a significant difference between Camilleri's approach and ours at the semantic odel (transition system) level. Whereas we simply maintain the priority structure in terms of probility (priority) transitions, Camilleri indirectly expresses this structure essentially by recording all ntext-transitions pairs. More technically, transitions are augmented with *refusal sets* in his model, hich characterize the nature of the contexts in which a transition is possible. In order to obtain a ngruence result similar to Theorem 1 transitions must also be decorated with acceptance sets of the ocess under consideration. Thus, Camilleri's approach lacks the third feature.

Conclusion

Ve have presented a language that uniformly captures both probability and priority in a very general ay. This has been achieved simply by extending the syntax of the language $PCCS_S$ to allow probality guards of 0, in conjunction with a careful treatment of 0-probability transitions at the semantic vel. The expressive power of the resulting language $PCCS_\zeta$ has been demonstrated: already a strong mplification of $PCCS_\zeta$ covers the essential features of the known process algebra models for priority.

As future work, we intend to investigate further the sublanguages $PCCS_\sigma$, $PCCS_\pi$ and $PCCS_S$ of CCS_ζ, for example, to obtain fully abstract models for them with respect to some notion of testing. dditionally, we are developing an equational theory for $PCCS_\pi$ based on its accompanying notion bisimulation. This theory will be constructed along the lines of [BBK86], but will also reflect the resence of globally dynamic priority relations.

cknowledgements: The authors would like to thank Chris Tofts for his early contributions to this ork. They are also grateful to Ivan Christoff, Rob van Glabbeek, Kim Larsen, and Robin Milner for aluable discussions on probability and priority in process algebra.

eferences

BK86] J. C. M. Baeten, J. A. Bergstra, and J. W. Klop. Syntax and defining equations for an interrupt mechanism in process algebra. *Fundamenta Informaticae IX*, pages 127–168, 1986.

am89] J. Camilleri. Introducing a priority operator to ccs. Technical report, Computer Laboratory, University of Cambridge, Cambridge, England, May 1989.

CH88] R. Cleaveland and M. Hennessy. Priorities in process algebras. In *Proceedings of 3rd Annual IEEE Symposium on Logic in Computer Science*, pages 193–202, Edinburgh, Scotland, July 1988.

GJS90] A. Giacalone, C.-C. Jou, and S. A. Smolka. Algebraic reasoning for probabilistic concurrent systems. In *Proceedings of Working Conference on Programming Concepts and Methods*, Sea of Gallilee, Israel, April 1990. IFIP TC 2.

LS89] K. G. Larsen and A. Skou. Bisimulation through probabilistic testing. In *Proceedings of 16th Annual ACM Symposium on Principles of Programming Languages*, 1989.

Mil83] R. Milner. Calculi for synchrony and asynchrony. *Theoretical Computer Science*, 25:267–310, 1983.

[Mil89] R. Milner. *Communication and Concurrency.* International Series in Computer Scienc
Prentice Hall, 1989.

[Par81] D. M. R. Park. Concurrency and automata on infinite sequences. In *Proceedings of 5th G. Conference on Theoretical Computer Science*, volume 104 of *Lecture Notes in Comput Science*, pages 167–183. Springer-Verlag, 1981.

[Plo81] G. D. Plotkin. A structural approach to operational semantics. Technical Report DAIM FN-19, Computer Science Department, Aarhus University, 1981.

[Pnu85] A. Pnueli. Linear and branching structures in the semantics and logics of reactive system In *Proceedings of 12th ICALP*, Lecture Notes in Computer Science, pages 15–32. Spring Verlag, 1985.

[vGSST90] R. van Glabbeek, S. A. Smolka, B. Steffen, and C. M. N. Tofts. Reactive, generative, an stratified models of probabilistic processes. In *Proceedings of 5th Annual IEEE Symposiu on Logic in Computer Science*, pages 130–141, Philadelphia, PA, June 1990.

A Synchronous Calculus of Relative Frequency

Chris Tofts
Laboratory for the Foundations of Computer Science
Department of Computer Science
Edinburgh University

Abstract

We present a weighted synchronous calculus that can be interpreted as reasoning over probabilistic processes [LS89,SST89,GSST90]. The abstraction from absolute weight to relative frequency is obtained semantically. We also add a notion of dominance which can be interpreted as priority. This notion is shown to be dual to that of "zero probability" [SST89,GSST90] and can be used to construct arbitrary priority structures. Finally, an equational system for reasoning about the weighted processes is presented.

Introduction

-e properties of probabilistic computation, both sequential [Wel83,Rab76] and concurrent JS90,LS89,GSST90,SST89], are interesting for several reasons. The notion of probability was veloped to give an account of non-deterministic systems and is a natural quantification for sys-ns with choice. Computational methodologies which are probabilistic in nature are frequently re efficient and natural than conventional programming methods. We wish to produce a culus for reasoning about such systems.

We hope that by giving an account of *relative frequency* of choice we can address some of the ues of fairness in concurrent systems. In particular we will be looking at the choice operator our calculus. Conventionally, we write a choice as

$$A \ + \ B$$

d interpret it as a process that has the capabilities of both the processes A and B. This choice underspecified. In an unbiased environment — one which lets the process decide which choice take — we could see behaviours ranging from infinite selection of one process over the other some fixed ratio of choice between the processes.

There is a large body of theory about Markov chains [Kei]. In our study of probabilistic ncurrent systems, we will be able to describe our processes as Markov systems and then use e extant techniques.

Our calculus is synchronous, since we shall be quantifying the relative frequency of free nultaneous choice. In an asynchronous system choices may be resolved at arbitrary times, thus giving a choice which may not be between equally free objects. This calculus simplifie the probabilistic system of [SST89,GSST90] and is based upon Milner's *Synchronous Calculus* *Communicating Systems* [Mil83]. We choose not to quantify our choices directly with probabilitie but to use *weights* which we can interpret as probabilistic specification via the concept of relativ frequency; our notion of equality will allow us to make that interpretation. Our intention is tha processes such as the following will be identified:

$$1P \;+\; 2Q \text{ and } 3P \;+\; 6Q.$$

That is to say that we shall see 1 occurrence of P for every 2 of Q in the left hand process, an shall see 3 occurrences of P to every 6 of Q in the right hand process. Since the relative frequenc of P and Q is the same in both systems we would like to equate these process descriptions.

Weights are operationally more convenient than probabilities since we do not have to ensur that they normalize to any particular value. This normalization problem is the cause of th complexity of the derivation law for permission in probabilistic systems. Using weights th permission rule is literally the pruning of sub-processes that will never execute as none of thei actions are permitted.

As in [SST89,GSST90,Tof89,Tof90a] we shall work with a multi-coloured transition system That is to say we have more than one notion of evolution. We intend to use two differen mechanisms of state change, one being a weighted choice, the other computation. This metho is used because it makes the underlying mechanism of state change clear and to obtain a ric system which is capable of expressing most of the processes we wish to study.

In this paper we define the calculus WSCCS, *a weighted synchronous calculus of communicat ing systems.* We will define the natural relative equality via the notion of a direct equality, an show that both of these equalities are substitutive. We introduce a special weight which denote dominance and show that it is consistent with our calculus. Expressions using this weight can b interpreted as giving processes a priority structure and we give an interpretation operator in th style of [BBK86]. We present an axiom system and demonstrate that it is sound and complet for finite process terms.

This paper is a reduced version of [Tof90b] which contains a value passing version of th calculus.

2 The Language WSCCS

Our language WSCCS is an extension of Milner's SCCS [Mil83]. To define our language w presuppose an abelian group *Act* of atomic action symbols with identity **1** and the inverse of being $\overline{a}$. As in SCCS, the complements a and $\overline{a}$ form the basis of communication. We also take set of weights $\mathcal{W}$, denoted by w_i, which are the positive natural numbers $\mathcal{P}$ and a set of proces variables Var.

The collection of WSCCS expressions ranged over by E is defined by the following BN expressions, where $a \in Act$, $X \in Var$, $w_i \in \mathcal{W}$, S ranging over renaming functions, thos $S : Act \longrightarrow Act$ such that $S(\mathbf{1}) = \mathbf{1}$ and $\overline{S(a)} = S(\overline{a})$, action sets $A \subseteq Act$, with $\mathbf{1} \in A$, an arbitrary *finite* indexing sets I:

$$E ::= X \mid a.E \mid \Sigma_{i\in I} w_i E_i \mid E \times E \mid E\lceil A \mid E[S] \mid \mu_i \tilde{X}\tilde{E}.$$

e let Pr denote the set of closed expressions, and add **0** to our syntax, which is defined by $\stackrel{def}{=} \Sigma_{i\in\emptyset} w_i E_i$.

The informal interpretation of our operators is as follows:

- **0** represents a process which cannot proceed;
- X represents the process bound to the variable X;
- $a.E$ represents a process which can perform the action a thereby becoming the process represented by E;
- $\Sigma_{i\in I} w_i E_i$ repressents the *weighted* choice between the processes E_i, the weight of the outcome E_i being determined by w_i. We think in terms of repeated experiments on this process and we expect to see over a large number of experments the process E_i being chosen with a relative frequency of $\frac{w_i}{\Sigma_{i\in I} w_i}$;
- $E \times F$ repressents the synchronous parallel composition of the two processes E and F. At each step each process must perform an action, the composition performing the product (in *Act*) of the individual actions;
- $E\lceil A$ represents a process where we only permit actions in the set A. This operator is used to enforce communication and bound the scope of actions;
- $E[S]$ represents the process E relabelled by the function S;
- $\mu_i \tilde{X}\tilde{E}$ represents the solution X_i taken from solutions to the mutually recursive equations $\tilde{X} = \tilde{E}$.

Often we shall omit the dot when applying prefix operators; also we drop trailing **0**, and will e a binary plus instead of the two (or more) element indexed sum, thus writing $\Sigma_{1,2}\{1_1 a.\mathbf{0},\ 2_2 b.\mathbf{0}\}$ $1a + 2b$. Finally we allow ourselves to specify processes definitionally, by providing recursive finitions of processes. For example, we write $A \stackrel{def}{=} a.A$ rather than $A \stackrel{def}{=} \mu x.ax$.

We now give some example processes.

1 Gambler's Ruin

e take the actions *1*, *2*, *3*, *4*, *5*, *6* and describe a fair die as the following process:

$$Die \stackrel{def}{=} 1\overline{1}.Die + 1\overline{2}.Die + 1\overline{3}.Die + 1\overline{4}.Die\ 1\overline{5}.Die + 1\overline{6}.Die,$$

d a gambler with an amount of money n, who wins 6 units of money when a six is thrown and ses his stake (of size one) on any other outcome as the following process (we add the actions *in* and *lose*):

$$\begin{aligned} Gamb(0) &\stackrel{def}{=} \mathbf{0} \\ Gamb(n) &= 1\mathit{1}lose.Gamb(n-1) + 1\mathit{2}lose.Gamb(n-1) \\ &\quad 1\mathit{3}lose.Gamb(n-1) + 1\mathit{4}lose.Gamb(n-1) \\ &\quad 1\mathit{5}lose.Gamb(n-1) + 1\mathit{6}win.Gamb(n+6) \end{aligned}$$

The process

$$Game(n) \stackrel{def}{=} (Gamb(n) \times Die)\lceil\{\mathbf{1},\ win,\ lose\}$$

describes a betting game in which the gambler starts with n units of money. We can use thi description to find the Markov system that the above process describes and we could predicat such parameters as the expected time until our gambler is bankrupt.

We can make our die unfair by changing the weighting of the various outcomes; such a unfair die can be described as follows.

$$Die \stackrel{def}{=} 21.Die + 12.Die + 13.Die + 14.Die + 25.Die + 16.Die;$$

This die only produces a six one time in eight and thus our gambler would run out of mone sooner. If in the group *Act* the operation on number actions is addition, then the process

$$Die \times Die$$

describes rolling two dice and adding their outcomes.

2.2 Spontaneous Failure

Most systems are subject to spontaneous failure due to hardware faults or disruption of electricit supply, but most descriptions of systems assume that they do not have behaviour of this nature The process

$$Fail \stackrel{def}{=} m\mathbf{1}.Fail + 1 fail.\mathbf{0}$$

describes a system that emits the action *fail* with weight 1 at each turn and then becomes the nil process. If we take an idealized process M describing a system without spontaneous failure then the behaviour of the process

$$M \times Fail$$

will be that of M except on a *fail* action where the process will subsequently stop. Note that there is no communication structure allowing M to prevent the occurrence of a fail or even influence its likelihood.

Whilst a calculus built from natural weights may at first seem unnatural as a way of expressing probabilities, we can consider the calculus as describing rational probabilities, which we have forced into naturals by multiplication by a suitable factor. We shall see later that multiplication of weights by a constant factor is conserved by our interpretation of WSCCS processes.

3 The Semantics of WSCCS

In this section we define the operational semantics of WSCCS. The semantics is transition based, structurally presented in the style of [Plo81], defining the actions that a process can perform and

$$\frac{}{a.E \xrightarrow{a} E} \qquad \frac{}{a.E \xrightarrow{1} a.E} \qquad \frac{}{\sum_{i\in I} w_i E_i \overset{w_i}{\longmapsto} E_i}$$

$$\frac{E \xrightarrow{a} E' \quad F \xrightarrow{b} F'}{E \times F \xrightarrow{ab} E' \times F'} \qquad \frac{E \overset{w}{\longmapsto} E' \quad F \overset{v}{\longmapsto} F'}{E \times F \overset{wv}{\longmapsto} E' \times F'}$$

$$\frac{E \xrightarrow{a} E' \quad a \in A}{does_A(E)} \qquad \frac{E \overset{w}{\longmapsto} E' \quad does_A(E')}{does_A(E)}$$

$$\frac{E \xrightarrow{a} E' \quad a \in A}{E\lceil A \xrightarrow{a} E'\lceil A} \qquad \frac{E \overset{w}{\longmapsto} E' \quad does_A(E')}{E\lceil A \overset{w}{\longmapsto} E'\lceil A}$$

$$\frac{E \xrightarrow{a} E'}{E[S] \xrightarrow{S(a)} E'[S]} \qquad \frac{E \overset{w}{\longmapsto} E'}{E[S] \overset{w}{\longmapsto} E'[S]}$$

$$\frac{E_i\{\mu_i \tilde{x} \tilde{E}/\tilde{x}\} \xrightarrow{a} E'}{\mu_i \tilde{x}.\tilde{E} \xrightarrow{a} E'} \qquad \frac{E_i\{\mu_i \tilde{x} \tilde{E}/\tilde{x}\} \overset{w}{\longmapsto} E'}{\mu_i \tilde{x}.\tilde{E} \overset{w}{\longmapsto} E'}$$

Figure 1: Operational Rules for WSCCS

ne weight with which a state can be reached. In Figure 1 we present the operational rules of VSCCS. They are presented in a natural deduction style. The transitional semantics of WSCCS given by the least relation $\longrightarrow \subseteq WSCCS \times Act \times WSCCS$ and the least multi-relation $\longmapsto \subseteq bag(WSCCS \times \mathcal{W} \times WSCCS)$ [1], which are written $E \xrightarrow{a} F$ and $E \overset{w}{\longmapsto} F$ respectively, atisfying the rules laid out in Figure 1. These rules respect the informal description of the perators given earlier. The reason that processes are multi-related by weight is that we may pecify more than one way to choose the same process with the same weight and we have to etain all the copies. For example, the process

$$1P + 1P + 1Q$$

an evolve to the process P with commulative weight 2, so that we have to retain both evolutions.

The predicate $does_A(E)$ is well defined since we have only permitted finitely branching choice xpressions. The action of the permission operator is to prune from the choice tree those processes hat can no longer perform any action.

.1 Direct Bisimulation

)ur bisimulations will be based on the accumulation technique of Larsen and Skou [LS89]. We tart by defining accumulations of evolutions for both types of transition.

)efinition 3.1 *Let S be a set of processes then:*

- $P \overset{w}{\longmapsto} S$ *iff* $w = \sum\{w_i \,:\, P \overset{w_i}{\longmapsto} Q, Q \in S\}$;

[1]Where $\longmapsto \subseteq Bag(WSCCS \times \mathcal{W} \times WSCCS)$ is the bag whose elements are those of the set $WSCCS \times \mathcal{W} \times$ $VSCCS$, with the usual notion of bag.

- $P \xrightarrow{a} S$ *iff* $P \xrightarrow{a} Q$ *for some* $Q \in S$.

We define a form of bisimulation that identifies two processes if the total weight of evolvi[ng] into any equivalent states is the same. This is not quite the indentification we wish to make, b[ut] we will make such an identification later.

Definition 3.2 *An equivalence relation* $R \subseteq Pr \times Pr$[2] *is a* direct bisimulation *if* $(P, Q) \in$ *implies for all* $S \in Pr/R$ *that:*

- *for all* $w \in \mathcal{W}$, $P \xmapsto{w} S$ *iff* $Q \xmapsto{w} S$;
- *for all* $a \in Act$, $P \xrightarrow{a} S$ *iff* $Q \xrightarrow{a} S$.

Two processes are direct bisimulation equivalent, *written* $P \overset{d}{\sim} Q$, *if there exists a direct bisim[u]lation* R *between them.*

Definition 3.3

$$\overset{d}{\sim} \equiv \bigcup\{R \ : \ R \text{ is a direct bisimulation }\}.$$

That $\overset{d}{\sim}$ is an equivalence follows immediately from it being a union of equivalences.

Lemma 3.4 *Let* P *and* Q *be processes such that* $P \overset{d}{\sim} Q$. *Then for all action sets* A, $does_A(P)$ *iff* $does_A(Q)$.

Proposition 3.5 *Direct equivalence is substitutive for finite processes. Thus, given* $P \overset{d}{\sim} Q$ *an[d]* $P_i \overset{d}{\sim} Q_i$ *for all* $i \in I$ *then:*

1. $a.P \overset{d}{\sim} a.Q$; 2. $\Sigma_{i\in I} w_i P_i \overset{d}{\sim} \Sigma_{i \in I} w_i Q_i$;
3. $P \times E \overset{d}{\sim} Q \times E$; 4. $P\lceil A \overset{d}{\sim} Q\lceil A$;
5. $P[S] \overset{d}{\sim} Q[S]$.

We proceed by the usual technique of pointwise extension to define our equivalence for infini[te] processes.

Definition 3.6 *Let* $\tilde{E}$ *and* $\tilde{F}$ *be expressions containing variables at most* $\tilde{X}$. *Then we will s[ay]* $\tilde{E} \overset{d}{\sim} \tilde{F}$ *if for all process sets* $\tilde{P}$, $\tilde{E}\{\tilde{P}/\tilde{X}\} \overset{d}{\sim} \tilde{F}\{\tilde{P}/\tilde{X}\}$.

Proposition 3.7 *If* $\tilde{E} \overset{d}{\sim} \tilde{F}$ *then* $\mu_i\tilde{X}.\tilde{E} \overset{d}{\sim} \mu_i\tilde{X}.\tilde{F}$.

[2] We denote the equivalence class of a processes P with respect to R by $[P]_R$. When it is clear from the conte[xt] to which equivalence we are refering, we will omit the subscript.

2 Relative Bisimulation

nfortunately, the congruence given by direct bisimulation is too strong; it does not capture ır notion of relative frequency, but captures total frequency. Since we would like to be able to uate processes such as,

$$2P \;+\; 3Q \text{ and } 4P \;+\; 6Q,$$

e need to weaken our notion of equality. The basic idea is that in order to show two processes uivalent, for each pair of equivalent states we can choose a *constant* factor such that the total eight of equivalent immediate derivatives is related by multiplication by that factor. If we can this for all potentially equivalent states then we will say that the processes are the same in rms of relative frequency. Since the constant factor may well need to be a rational (and we ish to keep our numbers as simple as possible) we will actually use two constants in comparing lative frequency. This allows us to use a symmetrical defintion.

efinition 3.8 *We say an equivalence relation $R \subseteq Pr \times Pr$ is a* relative bisimulation *(its quivalence classes being Pr/R) if $(P,Q) \in R$ implies that:*

1. *There are $c_1, c_2 \in \mathcal{P}$ such that for all $S \in Pr/R$ and for all $w \in \mathcal{W}$, $P \stackrel{w}{\longmapsto} S$ iff $Q \stackrel{v}{\longmapsto} S$ and $c_1 w = c_2 v$,*

2. *For all $S \in Pr/R$ and for all $a \in Act$, $P \stackrel{a}{\longrightarrow} S$ iff $Q \stackrel{a}{\longrightarrow} S$;*

vo processes are relative bisimular equivalent, *written $P \stackrel{r}{\sim} Q$ if there exists a relative bisimula- ion R between them.*

We have chosen to use multiplication by a constant rather than division as this permits us to tay within the natural numbers. We could have normalized so that the total weight actions of ny state is 1, and then we would have had an equivalence that is identical to that of stratified isimulation [SST89,GSST90].

efinition 3.9

$$\stackrel{r}{\sim} \equiv \bigcup\{R \;:\; R \textit{ is a relative bisimulation}\}.$$

roposition 3.10 *Let P and Q be processes such that $P \stackrel{d}{\sim} Q$, then $P \stackrel{r}{\sim} Q$.*

efinition 3.11 *Let $\tilde{E}$ and $\tilde{F}$ be expressions containing variables at most $\tilde{X}$. Then we will say $\tilde{E} \stackrel{r}{\sim} \tilde{F}$ if for all process sets $\tilde{P}$, $\tilde{E}\{\tilde{P}/\tilde{X}\} \stackrel{r}{\sim} \tilde{F}\{\tilde{P}/\tilde{X}\}$.*

roposition 3.12 $\stackrel{r}{\sim}$ *is a congruence for finite and infinite processes.*

Whilst we could axiomatize this system and these notions of equality we prefer to wait until fter our consideration of the addition of dominance in the next section.

4 Adding Infinite Frequency

The notion of priority, the arbitrary choice of one possibility over another, is a natural contr mechanism for computational machinery. The paper [CH88] provides an operational semantic for a single level of priority, and in [BBK86] an equational formulation is given for arbitrar priority over actions.

In this system we will represent priority as the dominance of one choice of future state ove another, in other words, in terms of processes. This is interpreted as having a state whos relative frequency is infinite with respect to other states into which it can evolve. If we wer to experiment on such a process, in a free context, we would only ever see occurrences of th dominant state.

There are two approaches to introducing this concept to our system of weights: a zero weigh or an infinite weight. There are

$$0P \; + \; 3Q \text{ and } 1P \; + \; \omega Q,$$

both being natural candidates for a process where the state Q has priority over the state P. W choose to use the latter form as it matches more closely our notion of dominance, or infinit frequency. Thus we extend our weight set $\mathcal{W}$ to be $\mathcal{P} \cup \{\omega\}$ with the following multiplicatio and addition rules for ω:

$$\omega + n = n + \omega = \omega + \omega = \omega,$$
$$\omega * n = n * \omega = \omega * \omega = \omega.$$

Note that addition and multiplication are still associative and commutative in the extende weight set.

Proposition 4.1 $\stackrel{d}{\sim}$ *is a congruence for processes defined over the augmented weight set.*

> **Proof:** We simply take account of the possibility of ω weighted states, and since they are distinct and stay distinct under composition then the equivalence will be substitutive.

For the relative case we have to be a little more careful. In the definition of $\stackrel{r}{\sim}$ we are onl allowed to use normalizing factors that are in $\mathcal{P}$. If we weaken this to $\mathcal{W}$, in the case of ou augmented weight set, then we have that for $n, m, r, s \in \mathcal{P}$:

$$nP \; + \; mQ \sim \omega P \; + \omega Q \sim rP \; + sQ,$$

but not necessarily that

$$nP \; + \; mQ \sim rP \; + \; sQ,$$

contradicting the requirement that our relation be an equivalence.

Proposition 4.2 $\stackrel{r}{\sim}$ *is a congruence for processes over the augmented weight set.*

$$\frac{E \stackrel{\omega}{\longmapsto} E'}{\Theta(E) \stackrel{1}{\longmapsto} \Theta(E')} \qquad \frac{E \stackrel{a}{\longrightarrow} E'}{\Theta(E) \stackrel{a}{\longrightarrow} \Theta(E')} \qquad \frac{E \stackrel{w}{\longmapsto} E' \quad E \stackrel{\omega}{\not\longmapsto}}{\Theta(E) \stackrel{w}{\longmapsto} \Theta(E')}$$

Figure 2: The priority interpretation operator Θ

The result of not allowing factors from the whole of $\mathcal{W}$ is that the following two processes annot be identified:

$$nP \text{ and } \omega P.$$

ore generally, we should like to say, in a free context, that the following processes are identical:

$$\omega P + nQ + mR \text{ and } 1P.$$

order to enable us to make the comparison and to interpret our priority information, we define priority interpretation operator Θ in the style of [BBK86]. Its operational rules are given in igure 2. It should be noted that since we have restricted ourselves to finitely branching processes ne negated evolution in the definition of Θ is decidable.

roposition 4.3 *Let P and Q be processes such that $P \stackrel{d}{\sim} Q$, then $\Theta(P) \stackrel{d}{\sim} \Theta(Q)$.*

roposition 4.4 *Let P and Q be processes such that $P \stackrel{r}{\sim} Q$, then $\Theta(P) \stackrel{r}{\sim} \Theta(Q)$.*

The following demonstrates the expressive power of our notion of priority.

efinition 4.5 *Let $\mathcal{T}$ be an* n-ary tree *of processes, with $P^r_{\mathcal{T}}$ is root, and sub-trees $\mathcal{T}_1, \ldots, \mathcal{T}_n$, en the process $P_{\mathcal{T}}$ respects a process with the priority structure given by the tree. We define $_{\mathcal{T}}$ recursively over the structure of the tree:*

$$P_{\mathcal{T}} \stackrel{def}{=} \omega P^r_{\mathcal{T}} + 1P_{\mathcal{T}_1} + \ldots + 1P_{\mathcal{T}_n}.$$

roposition 4.6 *Let $\mathcal{T}$ be a tree of processes with $P^r_{\mathcal{T}}$ its root, then $\Theta(P_{\mathcal{T}}) \stackrel{r}{\sim} 1P^r_{\mathcal{T}}$*

We now present some example uses of the priority structure.

.1 Broadcasting

n some systems, such as distributed system of machines each of which can fail, it is necessary o be able to send a signal to as many processes as will listen. This we can represent as a bias owards sending as many copies as possible of an action. Let us say we wish to broadcast $\overline{a}$. hen we can define a process which transmits to up to n other processes:

$$Brd(n) = \omega\overline{a}^n.Brd(n) + 1(\omega\overline{a}^{n-1}.Brd(n) + 1(\ldots + 11.Brd(n))).$$

4.2 A logon controller

We wish to define a system with the following properties:

- There are three types of user p, s and l;
- Logoffs are always dealt with first;
- When they are fewer than five users no one has priority;
- When there are five to ten users then those logging on as p or l have priority to those logging on as s;
- When there are more than ten users then p will have priority over s and l, and l will have priority over s.

We call our process *Log* and paramaterize it by the number of users.

$$\begin{aligned}
Log(0) &= \omega(1p.Log(n+1) + 1s.Log(n+1) + 1l.Log(n+1)) + 11.Log(n)\\
Log(1 \leq n \leq 5) &= \omega off.Log(n-1) + 1(\omega(1p.Log(n+1) + 1s.Log(n+1) +\\
&\qquad 1l.Log(n+1)) + 11.Log(n))\\
Log(6 \leq n \leq 10) &= \omega off.Log(n-1) + 1(\omega(1p.Log(n+1) + 1l.Log(n+1)) +\\
&\qquad 1(\omega s.Log(n+1) + 11.Log(n)))\\
Log(11 \leq n) &= \omega off.Log(n-1) + 1(\omega p.Log(n+1) + 1(\omega l.Log(n+1) +\\
&\qquad 1(\omega s.Log(n+1) + 11.Log(n))))
\end{aligned}$$

5 Equational Theory

We develop a set of equational laws for WSCCS, in a similar fashion to [CJS90,MT89], which are sound with respect to $\stackrel{d}{\sim}$ defined above. We shall use $=$ to represent derivability in out equational theory and $\equiv$ to represent syntactic identity. We postpone the development of an equational system for $\stackrel{r}{\sim}$ until the end of this section.

As a start we restrict our attention to the finite subset given by the following syntax:

$$E ::= a.E \mid \Sigma_{i \in I} w_i E_i,$$

and we refer to this language as WSCCS_0.

Our equational theory is presented in Figure 3. We can straightforwardly demonstrate that these laws are sound with respect to $\stackrel{d}{\sim}$. Hence:

Proposition 5.1 $p = q \implies p \stackrel{d}{\sim} q$

Definition 5.2 *We define when an expression is in normal form.*

- *A term is in* **NF** *if it is in* **pNF** *or* **ΣNF** .
- *A term is in* **pNF** *if it is of the form* a.**NF** .

$(\Sigma_1)\ \Sigma_{i\in I}w_iE_i = \Sigma_{j\in J}v_jE_j$	$\begin{cases}\text{there is a surjection } f: I \longmapsto J \text{ with} \\ v_j = \Sigma\{w_i \mid i \in I \wedge f(i) = j\}, \text{ and for all } i \text{ with } f(i) = j \text{ then} \\ E_i = E_j.\end{cases}$

Figure 3: Equational rules for WSCCS_0

$(Exp_1)\ a.E \times b.F = ab.(E \times F)$	$(Exp_2)\ a.E \times \Sigma_{j\in J}v_jF_j = \Sigma_{j\in J}v_j(a.E \times F_j)$
$(Exp_3)\ \Sigma_{i\in I}w_iE_i \times \Sigma_{j\in J}w_jF_j = \Sigma_{(i,j)\in(I\times J)}w_iv_j(E_i \times F_j)$	

Figure 4: Expansion theorems

- *A term is in* **ΣNF** *if it is of the form* $\Sigma_{i\in I}w_iE_i$ *with* $E_i \neq E_{i'}$ *for* $i \neq i'$ *and all* E_i *in* **NF** .

Definition 5.3 *The* stratification depth *of a process, written* $s(E)$ *is defined recursively as follows:*

$$s(a.E) = 0; \quad s(\Sigma_{i\in I}w_iE_i) = 1 + max\{s(E_i)\}.$$

Proposition 5.4 *Every term in* $WSSCS_0$ *can be equated to a term in* **NF** *using the laws of Figure 3.*

We now demonstrate that our normal forms are distinct.

Proposition 5.5

- *If* $P \overset{d}{\sim} Q$ *and* P *is in* **pNF** *then* Q *is in* **pNF** .
- *If* $P \overset{d}{\sim} Q$ *and* P *is in* **ΣNF** *then* Q *is in* **ΣNF** .

Proposition 5.6

1. *If* $P \overset{d}{\sim} Q$ *and both* P *and* Q *are in* **pNF** *then* $P = Q$.
2. *If* $P \overset{d}{\sim} Q$ *and both* P *and* Q *are in* **ΣNF** *then* $P = Q$.

We extend our language from WSCCS_0 to WSCCS_1 by adding the operator $\times$. The expansion laws needed to axiomatize $\times$ are presented in Figure 4; as all closed instances of the laws for associativity and commutativity can be derived, these laws have been omitted.

Proposition 5.7 *For any two NF terms* p *and* q *there is a term* r *not involving* $\times$ *such that* $p \times q = r$.

Further we extend from WSCCS_1 to WSCCS_2 by adding the permission operator. We need to axiomatize the predicate $does_A(E)$ in order to give a complete axiomatic presentation. The laws of Figure 5 axiomatize permission.

$$(Res_1)\ (a.E)\lceil A = \begin{cases} a.(E\lceil A) \text{ if } a \in A \\ \mathbf{0} \text{ otherwise.} \end{cases}$$

$$(Res_2)\ (\Sigma_{i\in I} w_i E_i)\lceil A = \Sigma_{j\in J} w_j (E_j \lceil A) \text{ where } J = \{i \in I | d_A(E_i)\}$$

Figure 5: Axiomatization of permission

$$(\Theta_1)\ \Theta(a.E) = a.\Theta(E)$$

$$(\Theta_2)\ \Theta(\Sigma_{i\in I} w_i E_i) = \begin{cases} \Sigma_{j\in J} 1.\Theta(E_j) \text{ where } J = \{i \in I | w_i = \omega\} \text{ and } J \neq \emptyset, \\ \Sigma_{i\in I} w_i \Theta(E_i) \text{ if } J = \emptyset \end{cases}$$

Figure 6: Priority Axiomatization

Definition 5.8 *Let A be an action set then the predicate, $d_A(E)$, expressing the fact that E ca[n] perform an action in A, is defined recursively as follows:*

- *If $a \in A$ then $d_A(a.E)$;*
- *If there exists $i \in I$ with $d_A(E_i)$ then $d_A(\Sigma_{i\in I} w_i E_i)$.*

Proposition 5.9 *Let E be in* **NF** *then $d_A(E)$ iff $does_A(E)$*

Proposition 5.10 *For any* **NF** *term p there is an* **NF** *term r not using the permission operato[r] such that $p\lceil A = r$.*

Finally we extend from $WSCCS_2$ to $WSCCS_3$ by adding the priority interpretation operato[r] Θ. The laws required are presented in Figure 6.

Proposition 5.11 *Let p be a term in* **NF** *then there is a term r not involving Θ such tha[t] $r = \Theta(p)$.*

Proposition 5.12 *(Completeness) Our axiom system Σ_1, Exp_1, Exp_2, Exp_3 , Res_1, Res_2 Θ_1, Θ_2 is complete for the language $WSCCS_3$.*

We have axiomatized $\overset{d}{\sim}$ and not $\overset{r}{\sim}$; we can add the axiom of Figure 7. Along with the earlie[r] axioms this gives a complete and sound axiomatization of WSCCS with respect to $\overset{r}{\sim}$.

Proposition 5.13 *Let $p \overset{r}{\sim} q$ with p and q in* Σ**NF** *then $p = q$.*

$$(Ren)\ \Sigma_{i\in I} w_i E_i = \Sigma_{i\in I} n w_i E_i \text{ where } n \in \mathcal{P}$$

Figure 7: Renormalization axiom

Conclusions and further work

e have presented a calculus of relative frequency that is both elegant and simple. We can press any degree of relative frequency between processes; for example:

$$nA + mB \qquad \omega A + nB \qquad \omega A + \omega B$$

press respectively that A and B occur, in a fixed ratio, with A dominant, and in any ratio. e notion of priority developed is very flexible and gives us a high degree of expressive power. nlike the work of [BBK86] as we have encoded priority into our processes it is dynamic with spect to particular actions.

Currently the author is using this calculus to represent the behaviour of communal insects, study in concurrency in the large. We would like to extend our axiom system to finite state pressions as in [GJS90]. Although WSCCS is elegant, in order to express some of the approximate equalities of [CJS90] we would need some complex normalizations encoded into our uality.

Acknowledgements

would like to thank Joachim Parrow and David Pym for some very helpful discussions on the ntent of this work. The concurrency club especially David Walker, Colin Stirling and Faron oller for their interest. Scott Smolka and Berhard Steffen for interesting me, once again, in the oblem of probabilistic concurrent computation.

Bibliography

86] J. Baeten, J. Bergstra and J. Klop, Syntax and defining equations for an interrupt mechanism in process algebra, Fundamenta Informatica IX, pp 127-168, 1986

88] R. Cleveland and M. Hennessey, Priorites in Process Algebras, proceedings LICS 1988.

90] A. Giaclone, C.-C. Jou, and S. A. Smolka, Algebraic reasoning for probabilistic concurrent systems, proceedings of working conference on programming concepts and methods IFIP TC 2, 1990.

90] R. van Glabbek, S. A. Smolka, B. Steffen and C.Tofts, Reactive, Generative and Stratified Models of Probabilistic Processes, proceedings LICS 1990.

89] C. Jones and G. D. Plotkin, A probabilistic powerdomain of evaluations, proceedings LICS 1989.

ei] J. Keilson, Markov Chain Models - Rarity and exponentiality, Applied Mathematical Sciences 28, Springer Verlag.

89] K. G. Larsen and A. Skou. Bisimulation through probabilistic testing. proceedings POPL 1989.

[Mil83] R. Milner, Calculi for Synchrony and Asynchrony, Theoretical Computer Science 25(3), 267-310, 1983.

[Mil89] R. Milner, Communication and Concurrency, Prentice Hall, 1989.

[MT89] F. Moller and C.Tofts, a Temporal Calculus of Communicating Systems, LFCS-89-1 University of Edinburgh.

[Par81] D. Park, Concurrency and Automata on infinite sequences, Springer LNCS 104.

[Plo81] G. D. Plotkin, A structured approach to operational semantics. Technical report Da Fn-19, Computer Science Department, Aarhus University. 1981

[Rab76] M. Rabin, Probabilistic algorithmns, Algorithmns and Complexity, recent results and n directions, J. Traub ed Academic press, New York, 1976 pp 21-39.

[SST89] S. Smolka, B. Steffen and C. Tofts, unpublished notes. Working title, Probability + F striction ⇒ priority.

[Tof89] C. Tofts, Timing Concurrent Processes, LFCS-89-103, University of Edinburgh.

[Tof90a] C. Tofts, Proof Methods and Pragmatics for Parallel Programming, Thesis in examinati

[Tof90b] C. Tofts, Relative Frequency in a Synchronous Calculus, LFCS-90-108, University of E inburgh.

[Wel83] D. Welsh, Randomised algorithmns, Discrete applied mathematics 5, 1983 pp 133-145.

ON THE COMPOSITIONAL CHECKING OF VALIDITY

(Extended Abstract)

Glynn Winskel
Computer Science Department, Aarhus University, Denmark

ntroduction

his paper is concerned with deciding whether or not assertions are valid of a parallel process ing methods which are directed by the way in which the process has been composed. The sertions are drawn from a modal logic with recursion, capable of expressing a great many roperties of interest [EL]. The processes are described by a language inspired by Milner's CCS d Hoare's CSP, though with some modifications. The choice of constructors allows us to andle a range of synchronisation disciplines and ensures that the processes denoted are finite ate. The operations are prefixing, a non-deterministic sum, product, restriction, relabelling d a looping construct. Arbitrary parallel compositions are obtained by using a combination product, restriction and relabelling.

/e are interested in deciding whether or not an assertion A is valid of a process t. If it is valid, the sense that every reachable state of t satisfies A, we write $\models A : t$. Rather than perform e check $\models A : t$ monolithically, on the whole transition system denoted by the term t, we ould often rather break the verification down into parts, guided by the composition of t. For stance if t were a sum $t_0 + t_1$ we can ask what assertions A_0 and A_1 should be valid of t_0 and respectively to ensure that A is valid of $t_0 + t_1$. This amounts to requiring assertions A_0, A_1 ch that

$$\models A : t_0 + t_1 \text{ iff } \models A_0 : t_0 \text{ and } \models A_1 : t_1.$$

nce the assertions A_0 and A_1 are found, a validity problem for $t_0 + t_1$ is reduced to a problem to with t_0 and another with t_1. Further, if the assertions can be found routinely only knowing e top-level operation, that e.g. the process is a sum, we are also told how to construct a rocess as a sum for which the assertion A is valid: first find components t_0 and t_1 making A_0 d A_1 valid respectively.

his paper investigates the extent to which the composition of t can guide methods for deciding $A : t$. It formulates new compositional methods for deciding validity, and exposes some ndamental difficulties. Algorithms are provided to reduce validity problems for prefixing, m, relabelling, restriction and looping to validity problems for their immediate components— l these reductions depend only on the top-level structure of terms. The existence of these ductions rests on being able to 'embed' the properties of a term in the properties, or products properties, of its immediate subterms. Because there is not such a simple embedding for e product construction of terms, as might be expected, similar reductions become much more omplicated for products; although there are general results, and the reductions can be simple in ecial cases, the general treatment for products meets with fundamental difficulties. Whereas ductions for products always exist for this finite state language, they demonstrably no longer ist depend on the top-level (product structure) of the term; in particular, a simple assertion exhibited for which the size of the reduction must be quadratic in the number of states of he process. An attempt is thus made to explain what makes product different from the other perations with respect to compositional reasoning, and to delimit the obstacles to automated ompositional checking of validity on parallel processes.

1 Transition systems and properties

The syntax, presented formally in the next section, will consist of process terms and assertions.

Process terms will denote labelled transition systems with distinguished initial states. A *labelled transition system* is a structure $(S, i, L, tran)$ where S is a set of *states* containing a distinguished state i, L is a set of *labels*, and $tran \subseteq S \times L \times S$ is a set of *transitions*; as normal, we often write $s \xrightarrow{\alpha} s'$ if $(s, \alpha, s') \in tran$. A state of a labelled transition system is *reachable* iff it can be obtained as the end state of a sequence of transition beginning at the initial state.

A closed assertion is to denote a *property* of a labelled transition system, *i.e.*a subset of its reachable states. We write $P(T)$ for the set of properties of a labelled transition system T.

We construct labelled transition systems using the constructions of *prefixing, sum, product restriction, relabelling* and *looping* starting from the *nil* process. These operations form the basis of our syntax for processes. We now describe these constructions. As has been stated properties of a labelled transition system are identified with subsets of reachable states. The constructions in our language of transition systems are associated with maps. These prove useful in importing properties of immediate components of a term into a property of the term itself Such mappings between properties are a key to compositional reasoning about processes. We introduce them alongside the constructions with which they are associated.

nil: The *nil* transition system is $(\{i\}, i, \emptyset, \emptyset)$.

Prefixing: For a label α and a labelled transition system $T = (S, i, L, tran)$ the *prefix* αT is obtained by adjoining a new initial state and introducing an α-transition from it to the old initial state. More concretely:

$$\alpha\, T = (S', \emptyset, L \cup \{\alpha\}, tran')$$

where $S' = \{\{s\} \mid s \in S\} \cup \{\emptyset\}$, and

$$\begin{aligned}(s_1, \beta, s_1') \in tran' \text{ iff } & (s_1 = \emptyset \ \&\ \beta = \alpha \ \&\ s_1' = \{i\}) \text{ or} \\ & (s_1 = \{s\} \ \&\ s_1' = \{s'\} \ \&\ (s, \beta, s') \in tran, \text{ for some } s, s'\end{aligned}$$

There is map $S \rightarrow S'$ taking $s \in S$ to the corresponding state $\{s\} \in S'$. It extends to a map on properties $P(T) \rightarrow P(\alpha T)$. It is convenient to name this map on properties after the prefixing operation and we define

$$\alpha(-) : P(T) \rightarrow P(\alpha T)$$

by taking $\alpha U = \{\{s\} \mid s \in U\}$ for $U \in P(T)$.

Sum: Let $T_0 = (S_0, i_0, L_0, tran_0)$ and $T_1 = S_1, i_1, L_1, tran_1)$ be labelled transition systems. Our nondeterministic sum operation $T_0 + T_1$ is a little different from Milner's. It identifies disjoint copies of the transition systems at their initial states. We define

$$T_0 + T_1 = ((S_0 \times \{i_1\}) \cup (\{i_0\} \times S_1), (i_0, i_1), L_0 \cup L_1, tran')$$

where

$$((s, i_1), \alpha, (s', i_1)) \in tran' \text{ iff } (s, \alpha, s') \in tran_0$$

and

$$((i_0, s), \alpha, (i_0, s')) \in tran' \text{ iff } (s, \alpha, s') \in tran_1.$$

, the sum construction is obtained by juxtaposing disjoint copies of the transition systems , T_1 but identified at their initial states. The difference with Milner's sum are illustrated by is example:

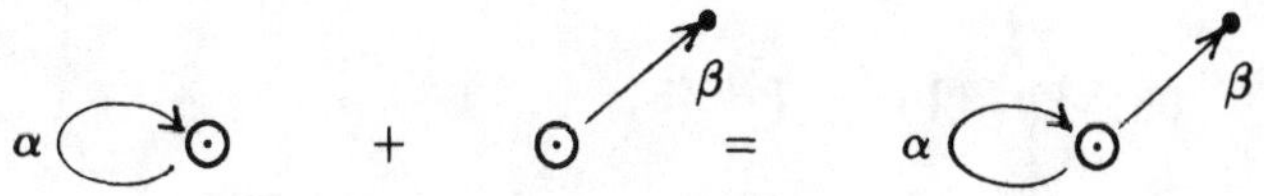

the sum it is possible to arbitrarily many α transitions from one component and then do a β ansition; this is impossible for Milner's sum where once a transition occurs in one component a sum then all future transitions must be from the same component. (The introduction of ew states demanded by Milner's sum would complicate the reduction.)

sum $T_0 + T_1$ is associated with two injection functions on states:

$$inj_i : S_i \rightarrow (S_0 \times \{i_1\}) \cup (\{i_0\} \times S_1), i = 0, 1$$

ith $inj_0(s) = (s, i_1)$ and $inj_1(s) = (i_0, s)$. They induce a map between properties:

$$(- + -) : P(T_0) \times P(T_1) \longrightarrow P(T_0 + T_1)$$

iven by $V_0 + V_1 = \{inj_0(s) \mid s \in V_0\} \cup \{inj_1(s) \mid s \in V_1\}$ on $V_0 \in P(T_0), V_1 \in P(T_1)$.

roduct: Let $T_0 = (S_0, i_0, L_0, tran_0)$ and $T_1 = (S_1, i_1, L_1, tran_1)$ be labelled transition systems. heir *product* $T_0 \times T_1$ consists of states $S_0 \times S_1$ with initial state (i_0, i_1), labels $L_0 \times_* L_1$ defined be

$$\{(\alpha_0, *) \mid \alpha \in L_0\} \cup \{(\alpha_0, \alpha_1) \mid \alpha_0 \in L_0, \alpha_1 \in L_1\} \cup \{(*, \alpha_1) \mid \alpha_1 | \in L_1\},$$

nd transitions $((s_0, s_1), (a_0, a_1), (s_0', s_1'))$ provided these satisfy:

$$a_0 \neq * \Rightarrow (s_0, a_0, s_0') \in tran_0 \text{ and } a_0 = * \Rightarrow s_0 = s_0', \text{ and}$$
$$a_1 \neq * \Rightarrow (s_1, a_1, s_1') \in tran_1 \text{ and } a_1 = * \Rightarrow s_1 = s_1'.$$

ntuitively, the product allows arbitrary synchronisations between pairs of transitions in two omponents, allowing too for the possibility of a transition in one component proceeding indeendently of the other.

product $T_0 \times T_1$ is associated this map on properties:

$$(- \times -) : P(T_0) \times P(T_1) \rightarrow P(T_0 \times T_1)$$

here $V_0 \times V_1$ is the cartesian product $\{(s_0, s_1) \mid s_0 \in V_0, s_1 \in V_1\}$.

estriction: Let $T = (S, i, L, tran)$ be a labelled transition system. Given a subset of labels we can restrict the transitions of T to those with labels in Λ. Define the *restriction* $T \restriction \Lambda =$ $S, i, L \cap \Lambda, tran')$ where $tran' = \{(s, \alpha, s') \in tran \mid \alpha \in \Lambda\}$.

he reachable states of $T \restriction \Lambda$ are cut-down from those of T. There is an associated map on roperties:

$$(- \restriction \Lambda) : P(T) \longrightarrow P(T \restriction \Lambda)$$

here $V \restriction \Lambda = \{s \in V \mid s \text{ reachable in } T \restriction \Lambda\}$.

Relabelling: It is often useful to relabel the transitions of a labelled transition system. Let T = $(S, i, L, tran)$. Let Ξ be a relabelling function from L to labels. Define the relabelled transition system $T\{\Xi\}$ to be $(S, i, \Xi L, tran')$ where where $tran' = \{(s, \Xi(\alpha), s) \mid (s, \alpha, s') \in tran\}$.

Relabelling leaves the states unaffected. Consequently any property of T can be regarded as property of $T\{\Xi\}$. Define

$$(-\{\Xi\}); P(T) \longrightarrow P(T\{\Xi\})$$

by taking $V\{\Xi\} = V$.

Looping: Let $T = (S, i, L, tran)$ be a labelled transition system. Assume U denotes a property of T. Then by $T/\alpha, U$ we mean the transition system obtained from T by introducing a transition (s, α, i) for each s satisfying the property U. More concretely $T/\alpha, U$ is the $(S, i, L \cup \{\alpha\}, tran$ where $tran' = tran \cup \{(s, \alpha, i) \mid s \in U\}$.

Like relabelling, the looping construct also leaves the states unaffected. Define

$$(-/\alpha, J) : P(T) \rightarrow P(T/\alpha, J)$$

by taking $(V/\alpha, J) = V$.

Parallel compositions: We can represent a variety of different parallel composition through a combined use of product, restriction and relabelling. For example, assuming a distinguished atomic label τ and a bijection $\alpha \mapsto \bar{\alpha}$ between non-τ atomic labels such that $\bar{\bar{\alpha}} = \alpha$, we can represent the parallel composition of CCS: take it to be

$$T_0 \mid T_1 =_{def} (T_0 \times T_1 \restriction \Lambda)\{\Xi\}$$

where the restricting set Λ consists of labels $(\alpha, *), (*, \alpha), (\alpha, \bar{\alpha}), (\tau, *), (*, \tau)$ where α ranges over all labels but for the distinguished label τ, and the relabelling Ξ acts so $\Xi(\alpha, *) = \Xi(*, \alpha) =$ and $\Xi(\tau, *) = \Xi(*, \tau) = \Xi(\alpha, \bar{\alpha}) = \tau$.

2 Languages

2.1 Syntax

Terms t denote labelled transition systems with distinguished initial states. Assertions A denote their properties. In fact an assertion A will only sensibly denote a property of t when a well-formedness judgement $A : t$ holds. The "raw" syntax of terms and assertions, ignoring for the moment their well-formedness, is mutually dependent and given as follows:

Terms:

$$t ::= nil \mid \alpha t \mid t_0 + t_1 \mid t_0 \times t_1 \mid t \restriction \Lambda \mid t\{\Xi\} \mid (t/\alpha, A)$$

Assertions:

$$\begin{aligned} A :=\, & I \mid T \mid F \mid A_0 \wedge A_1 \mid \neg A \mid \\ & \langle a \rangle A \mid \overline{\langle a \rangle} A \mid \\ & X \mid \nu X.A \mid \\ & \alpha A \mid A_0 + A_1 \mid A_0 \times A_1 \mid A \restriction \Lambda \mid A\{\Xi\} \mid (A_0/\alpha, A_1) \mid \\ & (\vdash A : t) \end{aligned}$$

ere t is a term, α is a label, Λ is a subset of labels, Ξ is a relabelling function, a is a label ossibly the 'idling' label $*$), and X is an assertion variable. It is convenient to assume assertion riables belong to unique terms and we write $\mathrm{Var}(t)$ for the countably infinite set of assertion riables associated with the term t; so $\mathrm{Var}(t)$ and $\mathrm{Var}(t')$ are disjoint if t and t' are distinct. ocess terms t are associated with a set of labels $Labels(t)$ defined by structural induction:

$$Labels(nil) = \emptyset, \qquad Labels(\alpha t) = Labels(t) \cup \{\alpha\},$$
$$Labels(t_0 + t_1) = Labels(t_0) \cup Labels(t_1), \qquad Labels(t_0 \times t_1) = Labels(t_0) \times_* Labels(t_1),$$
$$Labels(t \upharpoonright \Lambda) = Labels(t) \cap \Lambda, \qquad Labels(t\{\Xi\}) = \Xi Labels(t),$$
$$Labels(t/\alpha, A) = Labels(t) \cup \{\alpha\}.$$

e use $Labels_*(t)$ to mean $Labels(t) \cup \{*\}$.

he assertion language is essentially a modal ν-calculus with recursion. There are 'forwards' d 'backwards' modalities—the latter are useful in obtaining reductions for the product. The sertion I will be used to refer to initial states; $I : t$ will denote the property holding just at e initial state of the transition system denoted by the process term t. In addition, assertions clude constructions on properties with meanings described in the last section; these are used to uild properties of a term from properties of its immediate components. It has another unusual nstruction: a *validity assertion* $(\vdash A : t)$ which will, in effect, be true or false according to hether or not A is valid in t, with respect to a particular interpretation of its free variables as roperties.

e shall employ some standard abbreviations, and write $A_0 \wedge A_1$ for $\neg(\neg A_0 \vee \neg A_1)$, $A_0 \rightarrow A_1$ r $\neg A_0 \vee A_1$, $A_0 \leftrightarrow A_1$ for $(A_0 \rightarrow A_1) \wedge (A_1 \rightarrow A_0)$. The minimum fixed point $\mu X.A$ stands for $\nu X.\neg A[\neg X/X]$. As regards substitution, we assume the usual renaming of bound variables to void the capture of free variables. In some reductions we use a nonstandard *directed* conditional

$$B \rightarrow A_0 | A_1$$

abbreviate $(B \wedge A_0) \vee A_1$. This is unusual; one would expect $(B \wedge A_0) \vee (\neg B \wedge A_1)$. The onstandard choice is taken to avoid problems with monotonicity in the bodies of recursive efinitions. Besides the directed conditional is always used in a context where the right arm A_1 is logically stronger than the left A_0; then the directed conditional $B \rightarrow A_0|A_1$ is logically quivalent to $(B \wedge A_0) \vee (\neg B \wedge A_1)$.

he raw syntax allows assertions which are not sensible. For example, in the construct $\nu X.A$ are must be taken that the body A determines a monotonic operator on sets of states. A ufficient condition for this is that all occurrences of the variable X are positive, i.e. under an ven number of negations; otherwise the recursive assertion is not well-formed. The judgement $A : t$ says when an assertion A is well-formed as well as when it expresses a sensible property of term t, once given properties for its free assertion variables. The well-formedness judgement given by rules which are reminiscient of typing rules. This is consistent with the view that process term is regarded as a type of properties. Well-formedness of assertions affects well-ormedness of terms because the looping construct on terms $(t/\alpha, A)$ involves an assertion which e insist is closed and such that $A : t$.

Well-formedness rules:

$$I:t \qquad T:t \qquad F:t \qquad \frac{A_0:t \quad A_1:t}{A_0 \vee A_1:t} \qquad \frac{A:t}{\neg A:t}$$

$$\frac{A:t \quad a \in Labels_*(t)}{\langle a \rangle A:t} \qquad \frac{A:t \quad a \in Labels_*(t)}{\langle \overline{a} \rangle A:t}$$

$$X:t \quad \text{if } X \in \mathrm{Var}(t) \qquad \frac{X:t \quad A:t \quad X \text{ +ve in } A}{\nu X.A:t}$$

$$\frac{A:t}{\alpha A:\alpha t} \qquad \frac{A_0:t_0 \quad A_1:t_1}{A_0+A_1:t_0+t_1} \qquad \frac{A_0:t_0 \quad A_1:t_1}{A_0 \times A_1:t_0 \times t_1}$$

$$\frac{A:t}{A\lceil\Lambda:t\lceil\Lambda} \qquad \frac{A:t}{A\{\Xi\}:t\{\Xi\}} \qquad \frac{A_0:t \quad A_1:t \quad A_1 \text{ is closed}}{(A_0/\alpha, A_1):t/\alpha, A_1}$$

$$\frac{A:t_0}{(\vdash A:t_0):t}$$

Validity assertions, of the form $\vdash A : t$ will play a transient, though important, role in the reductions. Although the reductions will often introduce validity assertions, they can be removed so that subsequent reductions work on assertions free of them.

Definition: An assertion which does not contain any validity assertions will be called *pure*.

2.2 Semantics

From the previous section, we understand each of the constructions in our language of process terms and so the denotation of a process term by a labelled transition system; in the case of the looping we will need to rely on the semantics of closed assertions as properties, made precise shortly.

Notation: In our subsequent work we will adopt the convention that a term t denotes a labelled transition system

$$(S_t, i_t, L_t, tran_t),$$

and write, for instance, $s \xrightarrow{\alpha}_t s'$ to signify a transition in the transition system denoted by t. We shall write $P(t)$ for the set properties of (the transition system denoted by) t.

We give semantics to assertions A accompanied by a judgement $A : t$. To cope with the possibility of free assertion variables in A, we use environments. Together assertion variables form the set

$$\mathrm{Var} = \bigcup\{\mathrm{Var}(t) \mid t \in \mathrm{Term}\}.$$

An *environment* ρ is a function

$$\rho : \mathrm{Var} \to \bigcup\{P(t) \mid t \in \mathrm{Term}\}$$

such that $\rho(X) \in P(t)$ for $X \in \mathrm{Var}(t)$. Define Env to be the set of environments. The denotation of $A : t$ will be $[\![A : t]\!]$ of type $\mathrm{Env} \to P(t)$. Define:

$$
\begin{array}{lll}
I : t \rrbracket & = & \lambda\rho.\{i_t\} \\
T : t \rrbracket & = & \lambda\rho.S_t \\
F : t \rrbracket & = & \lambda\rho.\emptyset \\
A_0 \vee A_1 : t \rrbracket & = & \lambda\rho.\llbracket A_0 : t \rrbracket\rho \cup \llbracket A_1 : t \rrbracket\rho \\
\neg A : t \rrbracket & = & \lambda\rho.(S_t \setminus \llbracket A : t \rrbracket\rho) \\
\langle\alpha\rangle A : t \rrbracket & = & \lambda\rho.\{s \in S_t \mid \exists s'.\, s \xrightarrow{\alpha}_t s' \,\&\, s' \in \llbracket A : t \rrbracket\rho\} \\
X : t \rrbracket & = & \lambda\rho.\rho(X) \\
\nu X.A : t \rrbracket & = & \lambda\rho.(\nu U.\llbracket A : t \rrbracket\rho[U/X]) \\
 & & \text{the greatest fixed point of the function } U \mapsto \llbracket A : t \rrbracket\rho[U/X] \\
\alpha A : \alpha t \rrbracket & = & \lambda\rho.\alpha(\llbracket A : t \rrbracket\rho) \\
A_0 + A_1 : t_0 + t_1 \rrbracket & = & \lambda\rho.\llbracket A_0 : t_0 \rrbracket\rho + \llbracket A_1 : t_1 \rrbracket\rho \\
A_0 \times A_1 : t_0 \times t_1 \rrbracket & = & \lambda\rho.\llbracket A_0 : t_0 \rrbracket\rho \times \llbracket A_1 : t_1 \rrbracket\rho \\
A \restriction \Lambda : t \restriction \Lambda \rrbracket & = & \lambda\rho.\llbracket A : t \rrbracket\rho\lceil\Lambda \\
A\{\Xi\} : t\{\Xi\} \rrbracket & = & \lambda\rho.(\llbracket A : t \rrbracket\rho)\{\Xi\} \\
\llbracket A_0/\alpha, A_1 : (t/\alpha, A_1) \rrbracket & = & \lambda\rho.(\llbracket A_0 : t \rrbracket\rho/\alpha, \llbracket A_1 \rrbracket\rho) \\
(\vdash A : t_0) : t \rrbracket & = & \lambda\rho.(\llbracket A : t_0 \rrbracket\rho = S_{t_0} \rightarrow S_t \mid \emptyset)
\end{array}
$$

efinition: *(Validity)* Let $A : t$ be an assertion. Define

$$\models A : t \text{ iff } \llbracket A : t \rrbracket\rho = S_t \text{ for all environments } \rho.$$

How to do reductions

Ve first motivate the technique by considering the reduction for the looping construct. Let t be term, $J : t$ a closed assertion. Then $(t/\alpha, J)$ denotes a transition system like that of t but with xtra α transitions from all the states satisfying J to the initial state. Suppose A is a closed ssertion of the pure ν-calculus (with no mention of validity assertions) so that $A : (t/\alpha, J)$. We escribe how to produce an assertion $B : t$ such that

$$\models A : (t/\alpha, J) \text{ iff } \models B : t$$

nd in this way reduce the validity problem for a term $(t/\alpha, J)$ to one for t. The assertion B ill be defined by structural induction on A.

the course of the structural induction we will generally encounter assertions which have free ariables. To cope with this the reduction is done with respect to a substitution σ transforming ariables $X : (t/\alpha, J)$ to assertions $(Y/\alpha, J) : (t/\alpha, J)$. In order not to introduce spurious ependencies it will be assumed that the free variables Y do not appear free in any assertion eing reduced and that σ yields distinct Y for distinct X. With respect to such a change of ariables σ, we will consider a few clauses of the reduction, and indicate how it can be proved hat if A is pure with $A : t/\alpha, J$ and $red(A : t/\alpha, J; \sigma) = B$ with $B : t$ and

$$\models A[\sigma] \leftrightarrow B/\alpha, J : (t/\alpha, J).$$

his means that for all environments assigning properties to the free assertion variables, the ssertions A and $B/\alpha, J$ denote the same property of $t/\alpha, J$. It follows that, whenever $A : t/\alpha, J$ closed,

$$(\models A : t/\alpha, J) \text{ iff } (\models B : t).$$

this sense, a validity problem for a term $t/\alpha, J$ is reduced to one for t. We present a few lauses of the reduction:

$$\begin{array}{lcl} red(\langle\alpha\rangle A : t/\alpha, J; \sigma) & = & (\vdash I \rightarrow B : t) \rightarrow (((\langle\alpha\rangle B) \vee J) \mid \langle\alpha\rangle B \\ & & \text{where } red(A : t/\alpha, J; \sigma) = B \\ red(X : t/\alpha, J; \sigma) & = & Y \text{ when } \sigma(X) = Y/\alpha, J \\ red(\nu X.A : t/\alpha, J; \sigma) & = & \nu Y.red(A : t/\alpha, J; \sigma) \text{ where } \sigma(X) = Y/\alpha, J. \end{array}$$

The second and third clauses express little more than a renaming of free variables. To understand the first reduction, assume inductively that

$$\models A[\sigma] \leftrightarrow B/\alpha, J : t/\alpha, J$$

and argue, for a state s of $t/\alpha, J$ and arbitrary environment ρ, that

$$\begin{array}{l} s \in [\![\langle\alpha\rangle A[\sigma] : (t/\alpha, J)]\!]\rho \iff \exists s'.s \xrightarrow{\alpha}_{t/\alpha,J} s' \;\&\; s' \in [\![A[\sigma] : (t/\alpha, J)]\!]\rho \\ \qquad \iff \exists s'.s \xrightarrow{\alpha}_{t/\alpha,J} s' \;\&\; s' \in [\![B : t]\!]\rho \quad \text{by induction} \\ \iff \begin{cases} (\exists s'.s \xrightarrow{\alpha}_t s' \;\&\; s' \in [\![B : t]\!]\rho) \text{ or } s \in [\![J : t]\!]\rho & \text{if } i_t \in [\![B : t]\!]\rho, \\ (\exists s'.s \xrightarrow{\alpha}_t s' \;\&\; s' \in [\![B : t]\!]\rho) & \text{if } i_t \notin [\![B : t]\!]\rho, \end{cases} \\ \qquad \text{directly from the looping construction,} \\ \qquad \iff \begin{cases} s \in [\![((\langle\alpha\rangle B) \vee J : t]\!]\rho & \text{if } [\![\vdash I \rightarrow B : t]\!]\rho = S_t \\ s \in [\![\langle\alpha\rangle B : t]\!]\rho & \text{if } [\![\vdash I \rightarrow B : t]\!]\rho = \emptyset \end{cases} \\ \qquad \iff s \in [\![(\vdash I \rightarrow B : t) \rightarrow (((\langle\alpha\rangle B) \vee J) \mid \langle\alpha\rangle B : t]\!]\rho \end{array}$$

There remains however one hitch. The reduction, like that for the other term constructors, works on pure assertions—those which do not contain validity assertions. As is clear from some of the clauses above, reductions can sometimes yield assertions with validity assertions. If we are now to continue the reduction (using the structure of t) we must show how to prepare such validity assertions so they can be handled by these further reductions.

Look at one clause where validity assertions are introduced:

$$red(\langle\alpha\rangle A : t/\alpha, J; \sigma) = (\vdash I \rightarrow B : t) \rightarrow (((\langle\alpha\rangle B) \vee J) \mid \langle\alpha\rangle B$$

where $red(A : t/\alpha, J; \sigma) = B$. If B is closed there are no difficulties: we check for the smaller term t whether or not $I \rightarrow B$ is valid and if it is return the left, and otherwise the right branch of the conditional as the appropriate reduction. Validity assertions $\vdash B : t$ cause no difficulties when B is closed. But in general B will contain free assertion variables. However, ultimately we are concerned with reducing a closed assertion, which will mean that all free variables in validity assertions are bound by an enclosing recursive definition. The following fact means that, for a closed assertion denoting a property of t, the internal validity assertions introduced by its reduction can be made closed, and so benign because they refer to proper subterms of t:

Lemma 1 *(The closure lemma)*
Let $C[\ ]$ be a context such that $C[Y] : t$ and Y occurs positively in $C[Y] : t$, for any variable $Y : t_0$. Suppose $B : t_0$ and $[\![B : t_0]\!]\rho = \emptyset$ or $[\![B : t_0]\!]\rho = S_{t_0}$, for any environment ρ. Let X be a variable such that $\nu X. C[B] : t$. Then

$$\models \nu X. C[B] \leftrightarrow \nu X. C[B[\nu X. C[T]/X]] : t.$$

As an illustration, consider the reduction of $\nu X. \langle\alpha\rangle X : t/\alpha, J$. This should yield an assertion of t true at those reachable states of t which become able to do arbitrarily many α-transitions nce the loops of the construction $t/\alpha, J$ are introduced. Assume the change of variables takes X to $Y/\alpha, J$. Then, following the reductions above, we get

$$red(\nu X.\ \langle\alpha\rangle X : t/\alpha, J; \sigma) = \nu Y.\ ((\vdash I \rightarrow Y : t) \rightarrow (((\langle\alpha\rangle Y) \vee J) | \langle\alpha\rangle Y).$$

By the closure lemma we can close the validity assertion, to obtain the equivalent

$$\nu Y.(\vdash I \rightarrow \nu Y.\ (((\langle\alpha\rangle Y) \vee J)) : t) \rightarrow ((\langle\alpha\rangle Y) \vee J | \langle\alpha\rangle Y.$$

Thus $red(\nu X.\ \langle\alpha\rangle X : t/\alpha, J; \sigma)$ is equivalent to

1. $\nu Y.\ \langle\alpha\rangle Y$ if $\not\vdash I \rightarrow \nu Y.(((\langle\alpha\rangle Y) \vee J)$, and to
2. $\nu Y.\ (((\langle\alpha\rangle Y) \vee J)$ if $\vdash I \rightarrow \nu Y.\ (((\langle\alpha\rangle Y) \vee J)$.

In other words, a state in $t/\alpha, J$ can do arbitrarily many α transitions

1. if the corresponding state in t can, or
2. it can reach a state in J through α-transitions and the initial state of t can either do unboundedly many α-transitions or itself reach a state in J via α-transitions.

This is the kind of result one could write down informally, except one might forget a case in 2. The informal argument is helped enormously through there being a simple reading of the recursive assertion. The reductions work for all manner of recursive assertions. I hope this indicates how the reductions perform rather complicated inference steps.

To illustrate more fully the issues involved in performing reductions we will consider the case of sums. Provided $A_0 : t_0$ and $A_1 : t_1$ then $A_0 + A_1 : t_0 + t_1$. This sum constructor on assertions reflects the operation we have seen for obtaining a property of a sum from properties of its components. For a closed assertion $A : t_0 + t_1$ we are interested in how to produce assertions $A_0 : t_0, A_1 : t_1$ so that

$$\models A \leftrightarrow A_0 + A_1 : t_0 + t_1.$$

Provided we can ensure in addition that the pair $A_0 : t_0$, $A_1 : t_1$ is *balanced* in the sense that

$$i_{t_0} \in [\![A_0 : t_0]\!]\rho \quad \text{iff} \quad i_{t_1} \in [\![A_1 : t_1]\!]\rho,$$

for all environments ρ, then

$$\models A : t_0 + t_1 \quad \text{iff} \quad (\models A_0 : t_0 \text{ and } \models A_1 : t_1).$$

(Without the additional requirement of the assertions being balanced the "only if" direction of the statement can fail because only one of A_0 and A_1 can be true at the initial state.) The method for producing A_0, A_1 will work by induction on the structure of A, in the course of which we cannot hope to always deal with closed assertions. In particular, how are we to reduce a variable $X : t_0 + t_1$? The answer rests on the fact that the reduction will take place relative to a change of variables, in which variables like X are replaced by $Y_0 + Y_1$ for distinct variables $Y_0 : t_0$, $Y_1 : t_1$.

To illustrate the mechanism of the reduction it is shown how, for a closed assertion $\nu X.A : t_0 + t_1$, a balanced pair of assertions $B_0 : t_0$ and $B_1 : t_1$ can be found such that

$$\models \nu\, X.A \ \leftrightarrow\ B_0 + B_1 : t_0 + t_1.$$

Of key importance are the maps between properties $P(t_0+t_1)$ and $P(t_0) \times P(t_1)$. The change of variables is associated with the map

$$in : P(t_0) \times P(t_1) \to P(t_0+t_1) \text{ where } in(V_0, V_1) = U_0 + U_1.$$

On the other hand, the reduction is associated with a map *out* in the converse direction

$$out : P(t_0+t_1) \to P(t_0) \times P(t_1) \text{ where } out(U) = (out_0(U), out_1(U))$$

which projects a property U of $t_0 + t_1$ to a pair of properties

$$out_0(U) = \{s \in S_{t_0} \mid inj_0(s) \in U\},$$
$$out_1(U) = \{s \in S_{t_1} \mid inj_1(s) \in U\}.$$

It is easy to see that the maps are monotonic with respect to inclusion and that *out* is an*embedding* of the properties $P(t_0+t_1)$ in $P(t_0) \times P(t_1)$ in the sense that $in \circ out = 1_{P(t_0+t_1)}$. These facts are important because they fit into a general pattern for transforming fixed points:

Lemma 2 *(The embedding lemma)*
Suppose D and E are complete lattices for which $in : D \to E$ and $out : E \to D$ are monotonic, with $in \circ out = 1_E$. Suppose $\phi : E \to E$ is monotonic. Defining

$$\psi = out \circ \phi \circ in$$

we obtain a monotonic function $\psi : D \to D$ for which $\nu\phi = in(\nu\psi)$.

An assertion $A : t_0+t_1$ with a single free variable $X : t_0+t_1$ determines a function $\phi : P(t_0+t_1) \to P(t_0+t_1)$ from a property U of t_0+t_1 to a property $[\![A : t_0+t_1]\!]\rho[U/X]$ of t_0+t_1. Suppose we have already obtained a reduction of A to a balanced pair of assertions A_0, A_1 with respect to a change of variables taking X to $Y_0 + Y_1$, i.e.

$$\models A[Y_0+Y_1/X] \leftrightarrow A_0 + A_1$$

Then, equivalently, we can see the reduction as giving a syntactic expression of the embedding:

$$[\![A_0 : t_0]\!]\rho = out_0([\![A[Y_0+Y_1/X] : t_0+t_1]\!]\rho)$$
$$[\![A_1 : t_1]\!]\rho = out_1([\![A[Y_0+Y_1/X] : t_0+t_1]\!]\rho).$$

Writing $\psi_0(V_0, V_1) = [\![A_0 : t_0]\!]\rho[V_0/Y_0, V_1/Y_1]$ and $\psi_1(V_0, V_1) = [\![A_1 : t_1]\!]\rho[V_0/Y_0, V_1/Y_1]$, yields

$$\psi_0(V_0, V_1) = out_0(\phi(V_0+V_1)) \quad \text{and} \quad \psi_1(V_0, V_1) = out_1(\phi(V_0+V_1))$$

for all $V_0 \in P(t_0), V_1 \in P(t_1)$. Now, defining ψ by

$$\psi(V_0, V_1) = (\psi_0(V_0, V_1), \psi_1(V_0, V_1)),$$

for $V_0 \in P(t_0), V_1 \in P(t_1)$, we can see that $\psi = out \circ \phi \circ in$. The well-formedness of assertions will ensure monotonicity of ϕ and ψ so we can apply the embedding lemma to obtain:

$$\nu U.\phi(U) = in(\nu(V_0, V_1).(\psi_0(V_0, V_1), \psi_1(V_0, V_1))).$$

By Bekić's theorem

$$\nu U.\phi(U) = in(\nu V_0.\psi_0(V_0, \nu V_1.\psi_1(V_0, V_1)), \nu V_1.\psi_1(\nu V_0.\psi_0(V_0, V_1), V_1)).$$

With an eye back to syntax, this means:

$$\models \nu X.A \leftrightarrow (\nu Y_0.A_0[\nu Y_1.A_1/Y_1] + \nu Y_0.A_1[\nu Y_0.A_0/Y_0]) : t_0 + t_1.$$

We have thus succeeded in producing a pair of assertions $B_0 : t_0$, $B_1 : t_1$ such that

$$\models \nu X.A \leftrightarrow (B_0 + B_1) : t_0 + t_1.$$

A small additional argument, based on the assumption that $A_0 : t_0, A_1 : t_1$ are balanced, shows that $B_0 : t_0$, $B_1 : t_1$ form a balanced pair. Because $B_0 : t_0$, $B_1 : t_1$ denote fixed points of ψ we see, for an arbitrary environment ρ, that

$$[\![B_0 : t_0]\!]\rho = [\![A_0 : t_0]\!]\rho' \quad \text{and} \quad [\![B_1 : t_1]\!]\rho = [\![A_1 : t_1]\!]\rho'$$

where $\rho' = \rho[[\![B_0 : t_0]\!]\rho/Y_0, [\![B_1 : t_1]\!]\rho/Y_1]$. Now we observe that because the pair $A_0 : t_0, A_1 : t_1$ is balanced, so is $B_0 : t_0$, $B_1 : t_1$.

The reductions for the other constructions follow similar lines. Reductions will express embeddings of properties of a term in the properties, or products of properties, of its immediate components. They will be defined with respect to a change of variables associated with the left inverse to the embedding.

4 Summary of results

We now describe how to perform reductions for all the operations. As with the reduction for the looping construct, we shall need to change variables, so as to transform properties of a term to corresponding properties of its immediate subcomponents. We shall call such transformations *changes of variables*. All such transformations will be achieved through substitutions which introduce only fresh variables over properties.

Definition: A substitution σ is said to be *fresh* for an assertion A if it has the properties:

(i) for all variables X at which σ is defined the free variables in $\sigma(X)$ are disjoint from those in A, and

(ii) for distinct variables X and X', at which σ is defined, the free variables in $\sigma(X)$ and $\sigma(X')$ are disjoint.

Many of the reductions will introduce validity assertions. These are harmless however. They will always be validity assertions with respect to a smaller term than that of immediate interest, and, through the use of the closure lemma, lemma 1, they can be made closed whenever they arise as the reduction of a closed assertion; as such they can be checked, replaced by T or F as appropriate, and hence eliminated.

4.1 The reduction for *nil*

Given a closed, pure assertion $A : nil$, it is a simple matter to see whether or not it is valid at *nil*. The following function yields true in case it is valid, and false otherwise:

$$\begin{array}{lcl}
red(I : nil) & = & \text{true} \\
red(T : nil) & = & \text{true} \\
red(F : nil) & = & \text{false} \\
red(A_0 \vee A_1 : nil) & = & red(A_0 : nil) \text{ or } (red(A_1 : nil) \\
red(\neg A : nil) & = & \text{not } red(A : nil) \\
red(\langle a \rangle A : nil) & = & red(A : nil) \text{ if } a = *, \text{ false otherwise.} \\
red(\nu X.A : nil) & = & red(A[T/X] : nil).
\end{array}$$

Theorem 3 *For $A : nil$ a closed, pure assertion, $red(A : nil)$ iff $\models A : nil$.*

4.2 Reduction for prefixing

The reduction for prefixing is based on an embedding of $P(\alpha t)$ into $P(t) \times P(t)$, for a term t meeting the requirements of the embedding lemma 2. Define $down : P(\alpha t) \to P(t)$ by taking

$$down(U) = \{s \in S_t \mid \{s\} \in U\}.$$

Define $contI : P(\alpha t) \to P(t)$ by taking

$$contI(U) = \{i_t \mid i_{\alpha t} \in U\}.$$

Now we take the embedding to be

$$out = (down, contI) : P(\alpha t) \to P(t)^2$$

so that $out(U) = (down(U), contI(U))$. The converse map arises by taking

$$in : P(t)^2 \to P(\alpha t)$$

where

$$in(V_0, V_1) = \alpha(V_0) \cup \{i_{\alpha t} \mid i_t \in V_1\}.$$

It is easy to check that both *in* and *out* are monotonic and $in \circ out = 1_{P(\alpha t)}$.

The map *in* corresponds to a change of variables, with respect to which we'll define a reduction whose two components correspond to the two components of the embedding *out*.

Definition: Assume $A : \alpha t$. A *change of variables* of $A : \alpha t$ is a substitution σ, with domain $Var(\alpha t)$, which is fresh for A, and such that for all variables $X : \alpha t$ there are distinct variables $Y_0 : t$ and $Y_1 : t$ with

$$\sigma(X) = \alpha(Y_0) \vee ((\vdash I \to Y_1 : t) \wedge I).$$

This change of variables expresses the map *in*:

Proposition 4 *Let $X : \alpha t$, $Y_0 : t$ and $Y_1 : t$ be distinct variables. Suppose*

$$\sigma(X) = \alpha(Y_0) \vee ((\vdash I \to Y_1 : t) \wedge I).$$

Let ρ be an environment. Then, for $V_0, V_1 \in P(t)$

$$in(V_0, V_1) = [\![\sigma(X) : \alpha t]\!]\rho[V_0/Y_0, V_1/Y_1].$$

iven a pure assertion $A:\alpha t$ and σ a change of variables for it we define two functions $ed^0(A:\alpha t;\sigma)$ and $red^1(A:\alpha t;\sigma)$, such that

$$\llbracket red^0(A:\alpha t;\sigma):t\rrbracket\rho = down(\llbracket A[\sigma]:\alpha t\rrbracket\rho)$$
$$\llbracket red^1(A:\alpha t;\sigma):t\rrbracket\rho \cap \{i_t\} = contI(\llbracket A[\sigma]:\alpha t\rrbracket\rho)$$

r any environment ρ.

y structural induction on $A:\alpha t$, with respect to a change of variables for it, define

$$\begin{aligned}
red^0(I:\alpha t;\sigma) &= F\\
red^0(T:\alpha t;\sigma) &= T\\
red^0(F:\alpha t;\sigma) &= F\\
red^0(A_0\vee A_1:\alpha t;\sigma) &= red^0(A_0:\alpha t;\sigma)\vee red^0(A_1:\alpha t;\sigma)\\
red^0(\neg A:\alpha t;\sigma) &= \neg red^0(A:\alpha t;\sigma)\\
red^0(\langle a\rangle A:\alpha t;\sigma) &= \langle a\rangle red^0(A:\alpha t,\sigma)\\
red^0(\overline{\langle a\rangle} A:\alpha t;\sigma) &= \overline{\langle a\rangle} red^0(A:\alpha t;\sigma) \text{ if } a\neq\alpha\\
red^0(\overline{\langle \alpha\rangle} A:\alpha t;\sigma) &= \overline{\langle \alpha\rangle} red^0(A:\alpha t;\sigma)\vee(red^1(A:\alpha t;\sigma)\wedge I)\\
red^0(X:\alpha t;\sigma) &= Y_0 \text{ where}\\
& \sigma(X)=\alpha(Y_0)\vee((\vdash I\rightarrow Y_1:t)\wedge I.)\\
red^0(\nu X.A:\alpha t;\sigma) &= \nu Y_0.\ red^0(A:\alpha t;\sigma)[\nu Y_1.red^1(A:\alpha t;\sigma)/Y_1]) \text{ where}\\
& \sigma(X)=\alpha(Y_0)\vee((\vdash I\rightarrow Y_1:t)\wedge I)\\
red^0(\alpha A:\alpha t;\sigma) &= A
\end{aligned}$$

$$\begin{aligned}
red^1(I:\alpha t;\sigma) &= T\\
red^1(T:\alpha t;\sigma) &= T\\
red^1(F:\alpha t;\sigma) &= F\\
red^1(A_0\vee A_1:\alpha t;\sigma) &= red^1(A_0:\alpha t;\sigma)\vee red^1(A_1:\alpha t;\sigma)\\
red^1(\neg A:\alpha t;\sigma) &= \neg red^1(A:\alpha t;\sigma)\\
red^1(\langle *\rangle A:\alpha t;\sigma) &= red^1(A:\alpha t;\sigma)\\
red^1(\langle a\rangle A:\alpha t;\sigma) &= F \text{ if } a\neq * \,\&\, a\neq\alpha\\
red^1(\langle \alpha\rangle A:\alpha t;\sigma) &= red^0(A:\alpha t;\sigma)\\
red^1(\overline{\langle *\rangle} A:\alpha t;\sigma) &= red^1(A:\alpha t,\sigma)\\
red^1(\overline{\langle a\rangle} A:\alpha t;\sigma) &= F \text{ if } a\neq *\\
red^1(X:\alpha t;\sigma) &= Y_1 \text{ where } \sigma(X)=\alpha(Y_0)\vee((\vdash I\rightarrow Y_1:t)\wedge I)\\
red^1(\nu X.A:\alpha t;\sigma) &= \nu Y_1.(red^1(A:\alpha t;\sigma)[\nu Y_0.red^0(A:\alpha t;\sigma)/Y_0])\\
red^1(\alpha A:\alpha t;\sigma) &= F
\end{aligned}$$

heorem 5 *Assume $A:\alpha t$ with A pure. Suppose σ is a change of variables of $A:\alpha t$. Let $ed^0(A:\alpha t;\sigma)=A_0$ and $red^1(A:\alpha t;\sigma)=A_1$. Then $A_0:t$ and $A_1:t$. Moreover*

$$\llbracket A_0:t\rrbracket\rho = down(\llbracket A[\sigma]:\alpha t\rrbracket\rho) \textit{ and}$$
$$i_t\in\llbracket A_1:t\rrbracket\rho \quad \textit{iff} \quad i_{\alpha t}\in\llbracket A[\sigma]:\alpha t\rrbracket\rho$$

or any environment ρ.
f $A:\alpha t$ is closed then $(\models A:\alpha t)$ iff $(\models A_0\wedge(I\rightarrow A_1):t)$.

4.3 Reduction for sum

The reduction will be based on an embedding of $P(t_0 + t_1)$ in $P(t_0) \times P(t_1)$, for terms t_0, t_1. Define

$$out_0 : P(t_0 + t_1) \to P(t_0),$$
$$out_1 : P(t_0 + t_1) \to P(t_1)$$

by taking $out_0(U) = \{s \in S_{t_0} \mid inj_0(s) \in U\}$ and $out_1(U) = \{s \in S_{t_1} \mid inj_1(s) \in U\}$ for all $U \in P(t_0 + t_1)$. Define the embedding

$$out : P(t_0 + t_1) \to P(t_0) \times P(t_1)$$

by taking $out(U) = (out_0(U), out_1(U))$ for all $U \in P(t_0 + t_1)$. Define its left-inverse

$$in : P(t_0) \times P(t_1) \to P(t_0 + t_1)$$

by taking $in(V_0, V_1) = V_0 + V_1$. Both *in* and *out* are monotonic, and it is easily seen that $in \circ out = 1_{P(t_0+t_1)}$.

The map *in* accompanies a change of variables:

Definition: Let $A : t_0 + t_1$. A *change of variables of* $A : t_0 + t_1$ is a substitution σ with domain $Var(t_0 + t_1)$, which is fresh for A, and such that for any variable $X : t_0 + t_1$ we have $\sigma(X) = Y_0 + Y_1$ for distinct variables $Y_0 : t_0$ and $Y_1 : t_1$.

Proposition 6 *Let $X : t_0 + t_1$, $Y_0 : t_0$ and $Y_1 : t_1$ be distinct variables. Suppose $\sigma(X) = Y_0 + Y_1$. Let ρ be an environment. Then, for $V_0 \in P(t_0)$, $V_1 \in P(t_1)$*

$$in(V_0, V_1) = [\![\sigma(X) : \alpha t]\!]\rho[V_0/Y_0, V_1/Y_1].$$

With respect to a change of variables σ, we can transform an assertion $A : t_0 + t_1$ to the sum of a pair of assertions $A_0 : t_0$ and $A_1 : t_1$ which realise the components of the embedding *out*, i.e.

$$[\![A_0 : t_0]\!]\rho = out_0([\![A[\sigma] : t_0 + t_1]\!]\rho) \quad \text{and} \quad [\![A_1 : t_1]\!]\rho = out_1([\![A[\sigma] : t_0 + t_1]\!]\rho)$$

for any environment ρ.

The reduction is carried out by the pair of functions $red^0(A : t_0 + t_1; \sigma)$, $red^1(A : t_0 + t_1; \sigma)$, acting on an assertion A for which $A : t_0 + t_1$ and a change of variables for it. They are defined by the following structural induction (we omit the clauses for red^1 as they reflect those for red^0):

$$\begin{array}{lcl}
red^0(I : t_0 + t_1; \sigma) & = & I \\
red^0(T : t_0 + t_1; \sigma) & = & T \\
red^0(F : t_0 + t_1; \sigma) & = & F \\
red^0(A \vee B : t_0 + t_1; \sigma) & = & A_0 \vee B_0 \\
 & & \text{where } A_0 = red^0(A : t_0 + t_1; \sigma) \text{ and} \\
 & & \qquad B_0 = red^0(B : t_0 + t_1; \sigma) \\
red^0(\neg A : t_0 + t_1; \sigma) & = & \neg red^0(A : t_0 + t_1; \sigma) \\
red^0(\langle * \rangle A : t_0 + t_1; \sigma) & = & red^0(A : t_0 + t_1; \sigma) \\
red^0(\langle \alpha \rangle A : t_0 + t_1; \sigma) & = & (\vdash I \to \langle \alpha \rangle A_1 : t_1) \to (\langle \alpha \rangle A_0) \vee I \mid \langle \alpha \rangle A_0 \ \text{ if } \alpha \neq *, \\
 & & \text{where } red^0(A : t_0 + t_1; \sigma) = A_0, red^1(A : t_0 + t_1; \sigma) = A_1, \\
red^0(\overline{\langle * \rangle} A : t_0 + t_1; \sigma) & = & red^0(A : t_0 + t_1; \sigma) \\
red^0(\overline{\langle \alpha \rangle} A : t_0 + t_1; \sigma) & = & ((\vdash I \to \overline{\langle \alpha \rangle} A_1 : t_1) \to (\overline{\langle \alpha \rangle} A_0) \vee I \mid \overline{\langle \alpha \rangle} A_0 \ \text{ if } \alpha \neq *, \\
 & & \text{where } red^0(A : t_0 + t_1; \sigma) = A_0, red^1(A : t_0 + t_1; \sigma) = A_1 \\
red^0(X : t_0 + t_1; \sigma) & = & (\vdash I \to Y_1 : t_1) \to Y_0 \vee I | Y_0 \ \text{ where } \sigma(X) = Y_0 + Y_1 \\
red^0(\nu X.A : t_0 + t_1; \sigma) & = & \nu Y_0.A_0[\nu Y_1.A_1/Y_1] \\
 & & \text{where } A_0 = red^0(A : t_0 + t_1; \sigma), A_1 = red^1(A : t_0 + t_1; \sigma) \\
 & & \text{and } \sigma(X) = Y_0 + Y_1 \\
red^0(A_0 + A_1 : t_0 + t_1; \sigma) & = & (\vdash I \to A_1 : t_1) \to (A_0 \vee I) | A_0
\end{array}$$

heorem 7 *Let $A : t_0 + t_1$ be pure. Let $A_0 = red^0(A : t_0 + t_1; \sigma)$ and $A_1 = red^1(A : t_0 + t_1; \sigma)$, r σ a change of variables for A. Then*

$$[\![A_0 : t_0]\!]\rho = out_0([\![A[\sigma] : t_0 + t_1]\!]\rho) \quad \textit{and} \quad [\![A_1 : t_1]\!]\rho = out_1([\![A[\sigma] : t_0 + t_1]\!]\rho)$$

r any environment ρ.
$A : t_0 + t_1$ is closed then $(\models A : t_0 + t_1)$ iff $[(\models A_0 : t_0)$ and $(\models A_1 : t_1)]$.

.4 Reduction for looping.

he properties $P(t/\alpha, J)$ and $P(t)$ are the same, and this time the embedding and its inverse ith respect to which the reduction is performed are both the identity map. The inverse is ealised by a change of variables, which essentially just renames them:

efinition: Assume $A : t/\alpha, J$ with A pure. A *change of variables* of $A : t/\alpha, J$ is a substitution , with domain $Var(t/\alpha, J)$, which is fresh for A, and of the form $\sigma(X) = Y/\alpha, J$ for variables $: t/\alpha, J$ and $Y : t$.

he effort goes into finding an assertion to an assertion $red(A : t/\alpha, J; \sigma)$, such that for a pure $: (t/\alpha, J)$ and a change of variables σ for it,

$$\models A[\sigma] \leftrightarrow red(A : t/\alpha, J; \sigma) : t/\alpha, J.$$

he reduction of a pure assertion $A : t/\alpha, J$, with respect to a change of variables for it, is efined by structural induction:

$$
\begin{array}{lcl}
red(I : t/\alpha, J; \sigma) & = & I \\
red(T : t/\alpha, J; \sigma) & = & T \\
red(F : t/\alpha, J; \sigma) & = & F \\
red(A_0 \vee A_1 : t/\alpha, J; \sigma) & = & red(A_0 : t/\alpha, J; \sigma) \vee red(A_1 : t/\alpha, J; \sigma) \\
red(\neg A : t/\alpha, J; \sigma) & = & \neg red(A : t/\alpha, J; \sigma) \\
red(\langle b \rangle A : t/\alpha, J; \sigma) & = & \langle b \rangle red(A : t/\alpha, J; \sigma) \text{ if } b \neq \alpha \\
red(\langle \alpha \rangle A : t/\alpha, J; \sigma) & = & (\vdash I \rightarrow B : t) \rightarrow ((\langle \alpha \rangle B) \vee J) \mid \langle \alpha \rangle B \\
 & & \text{where } red(A : t/\alpha, J; \sigma) = B \\
red(\overline{\langle b \rangle} A : t/\alpha, J; \sigma) & = & \overline{\langle b \rangle} red(A : t/\alpha, J; \sigma) \text{ if } b \neq \alpha \\
red(\overline{\langle \alpha \rangle} A : t/\alpha, J; \sigma) & = & \neg(\vdash J \rightarrow \neg B : t) \rightarrow ((\overline{\langle \alpha \rangle} B) \vee I) \mid \overline{\langle \alpha \rangle} B \\
 & & \text{where } red(A : t/\alpha, J; \sigma) = B \\
red(X : t/\alpha, J; \sigma) & = & Y \text{ when } \sigma(X) = Y/\alpha, J \\
red(\nu X.A : t/\alpha, J; \sigma) & = & \nu Y.red(A : t/\alpha, J; \sigma) \text{ where } \sigma(X) = Y/\alpha, J \\
red(A/\alpha, J : t/\alpha, J; \sigma) & = & A.
\end{array}
$$

Theorem 8 *Let A be pure with $A : t/\alpha, J$. Let σ be a change of variables of $A : t/\alpha, J$. Let $red(A : t/\alpha, J; \sigma) = B$. Then $B : t$ and moreover*

$$\models A[\sigma] \leftrightarrow B/\alpha, J : t/\alpha, J.$$

If $A : t/\alpha, J$ is closed then $(\models A : t/\alpha, J)$ iff $(\models B : t)$.

4.5 Reduction for restriction

Any property of a restriction $t \restriction \Lambda$ can be regarded as a property of the component t; define

$$out : P(t \restriction \Lambda) \rightarrow P(t)$$

by taking $out(U) = U$. The inverse map takes account of the fact that properties of $t \restriction \Lambda$ consist of states which are reachable via transitions within Λ. It is

$$in : P(t) \rightarrow P(t \restriction \Lambda)$$

defined by $in(V) = V \cap S_{t \restriction \Lambda}$—note this means $in(V) = V \restriction \Lambda$. Both maps are monotonic and together satisfy $in \circ out = 1_{P(t \restriction \Lambda)}$.

As usual the inverse map is associated with a change of variables:

Definition: Assume $A : t \restriction \Lambda$. A *change of variables* of $A : t \restriction \Lambda$ is a substitution σ, with domain $Var(t \restriction \Lambda)$, which is fresh for A, and of the form $\sigma(X) = Y \restriction \Lambda$ for variables $X : t \restriction \Lambda$ with $Y : t$ a variable.

With respect to a change of variables σ we define a reduction of pure assertions $A : t \restriction \Lambda$ to assertions $red(A : t \restriction \Lambda; \sigma) : t$ such that

$$\models A[\sigma] \leftrightarrow red(A : t \restriction \Lambda; \sigma) \restriction \Lambda : t \restriction \Lambda$$

The reduction is related to the embedding *out* in the sense that

$$[\![red(A : t \restriction \Lambda; \sigma) : t]\!] \rho \cap S_{t \restriction \Lambda} = out([\![A[\sigma] : t \restriction \Lambda]\!] \rho).$$

is defined on pure assertion A for which $A : t \upharpoonright \Lambda$ and reductions σ, on variables of them by the following structural induction:

$$
\begin{array}{lcl}
red(I : t \upharpoonright \Lambda; \sigma) & = & I \\
red(T : t \upharpoonright \Lambda; \sigma) & = & T \\
red(F : t \upharpoonright \Lambda; \sigma) & = & F \\
red(A_0 \vee A_1 : t \upharpoonright \Lambda; \sigma) & = & red(A_0 : t \upharpoonright \Lambda, \sigma) \vee red(A_1 : t \upharpoonright \Lambda; \sigma) \\
red(\neg A : t \upharpoonright \Lambda; \sigma) & = & \neg red(A : t \upharpoonright \Lambda; \sigma) \\
red(\langle\alpha\rangle A : t \upharpoonright \Lambda; \sigma) & = & \begin{cases} F & \text{if } \alpha \notin \Lambda \\ \langle\alpha\rangle red(A : t \upharpoonright \Lambda; \sigma) & \text{if } \alpha \in \Lambda \end{cases} \\
red(\overline{\langle\alpha\rangle} A : t \upharpoonright \Lambda; \sigma) & = & \begin{cases} F & \text{if } \alpha \notin \Lambda \\ \overline{\langle\alpha\rangle}(R_\Lambda \wedge red(A : t \upharpoonright \Lambda; \sigma)) & \text{if } \alpha \in \Lambda \end{cases} \\
 & & \text{where } R_\Lambda \equiv \mu X.I \vee \bigvee_{\beta \in \Lambda} \overline{\langle\beta\rangle} X \\
red(X : t \upharpoonright \Lambda; \sigma) & = & Y \text{ where } \sigma(X) = Y \upharpoonright \Lambda \\
red(\nu X.A : t \upharpoonright \Lambda; \sigma) & = & \nu Y.red(A : t \upharpoonright \Lambda; \sigma) \text{ where } \sigma(X) = Y \upharpoonright \Lambda \\
red(A \upharpoonright \Lambda : t \upharpoonright \Lambda; \sigma) & = & A.
\end{array}
$$

Theorem 9 *Let A be a pure assertion such that $A : t \upharpoonright \Lambda$, for which σ is a change of variables. Let $red(A : t{\upharpoonright}\Lambda; \sigma) = B$. Then $B : t$ and*

$$\models A[\sigma] \leftrightarrow red(A : t \upharpoonright \Lambda; \sigma) \upharpoonright \Lambda : t \upharpoonright \Lambda$$

If $A : t{\upharpoonright}\Lambda$ is closed then $(\models A : t{\upharpoonright}\Lambda)$ iff $(\models R_\Lambda \rightarrow B : t)$.

4.6 Reduction for relabelling

Because the properties $P(t)$ and $P(t\{\Xi\})$ are the same, the reductions and change of variables correspond to the identity map and are relatively straightforward.

Definition: Assume $A : t\{\Xi\}$. A *change of variables* of $A : t\{\Xi\}$ is a substitution σ, with domain $Var(t\{\Xi\})$, which is fresh for A, and of the form $\sigma(X) = Y\{\Xi\}$ for variables $X : t\{\Xi\}$ and $Y : t$.

With respect to a change of variables σ, we define a reduction of a pure assertion $A : t\{\Xi\}$ to an assertion $red(A : t\{\Xi\}; \sigma)$ by structural induction:

$$
\begin{array}{lcl}
red(I : t\{\Xi\}; \sigma) & = & I \\
red(T : t\{\Xi\}; \sigma) & = & T \\
red(F : t\{\Xi\}; \sigma) & = & F \\
red(A_0 \vee A_1 : t\{\Xi\}; \sigma) & = & red(A_0 : t\{\Xi\}; \sigma) \vee red(A_1 : t\{\Xi\}; \sigma) \\
red(\neg A : t\{\Xi\}; \sigma) & = & \neg red(A : t\{\Xi\}; \sigma) \\
red(\langle * \rangle A : t\{\Xi\}; \sigma) & = & red(A : t\{\Xi\}; \sigma) \\
red(\langle\alpha\rangle A : t\{\Xi\}; \sigma) & = & \bigvee_{\beta \in \Xi^{-1}\{\alpha\}} \langle\beta\rangle red(A : t\{\Xi\}; \sigma) \\
red(\overline{\langle * \rangle} A : t\{\Xi\}; \sigma) & = & red(A : t\{\Xi\}; \sigma) \\
red(\overline{\langle\alpha\rangle} A : t\{\Xi\}; \sigma) & = & \bigvee_{\beta \in \Xi^{-1}\{\alpha\}} \overline{\langle\beta\rangle} red(A : t\{\Xi\}; \sigma) \\
red(X : t\{\Xi\}; \sigma) & = & Y \text{ where } \sigma(X) = Y\{\Xi\} \\
red(\nu X.A : t\{\Xi\}; \sigma) & = & \nu Y.red(A : t\{\Xi\}; \sigma) \text{ where } \sigma(X) = Y\{\Xi\} \\
red(A\{\Xi\} : t\{\Xi\}; \sigma) & = & A.
\end{array}
$$

Theorem 10 *Let A be a pure assertion for such that $A : t\{\Xi\}$ for which σ is a change of variables. Let $red(A : t\{\Xi\}; \sigma) = B$. Then $B : t$ and*

$$\models A[\sigma] \leftrightarrow B\{\Xi\} : t\{\Xi\}.$$

If $A : t\{\Xi\}$ is closed then $(\models A : t\{\Xi\})$ iff $(\models B : t)$.

4.7 Reduction for product

Looking back at the constructions so far, we see they share a common property, the presence of an embedding from properties of a constructed term to properties of its immediate components which are realised by the reductions on assertions of that term. Indeed, this reduction can be performed without looking at the composition of the immediate components; for instance, the reduction for $t_0 + t_1$ proceeds independently of the composition of t_0 and t_1.

The difficulty in obtaining analogous reductions for parallel compositions stems from there not being such an embedding from properties of products to properties of their components. While there is the map

$$(- \times -) : P(t_0) \times P(t_1) \longrightarrow P(t_0 \times t_1)$$

There is no 1-1 map in the converse direction if one of t_0, t_1 has more than one and the other more than two reachable states—a little arithmetic shows that then the set $P(t_0 \times t_1)$ has more states than $P(t_0) \times P(t_1)$. Reduction for assertions of a product, in general, cannot follow the same scheme as that of the other constructions. We are obliged to look for a different method of embedding and reduction or at special kinds of properties in $P(t_0 \times t_1)$ such as those which can embed in $P(t_0) \times P(t_1)$.

Properties having the shape $V_0 \times V_1$, a cartesian product of $V_0 \in P(t_0), V_1 \in P(t_1)$, are in correspondence with, and so embed in, properties $P(t_0) \times P(t_1)$. By cutting down the properties of a product to those denoted by the following assertions, we can obtain a reduction:

$$A ::= I \mid T \mid B \times C \mid \langle(a,b)\rangle A \mid \overline{\langle(a,b)\rangle} A \mid ((a,b))A \mid \overline{((a,b))}A \mid A \wedge A' \mid X \mid \nu X.\, A$$

where we use $(a)A$ to abbreviate $[a]A \wedge \langle a \rangle T$ and $\overline{(a)}A$ for $\overline{[a]}A \wedge \overline{\langle a \rangle}T$. Any closed assertion A in this class, for which $A : t_0 \times t_1$, has the property that

$$\models A \leftrightarrow A_0 \times A_1 : t_0 \times t_1$$

for simply found $A_0 : t_0$, $A_1 : t_1$. These assertions for the components are obtained with respect to a change of variables for $t_0 \times t_1$ which is a substitution, fresh for A, sending variables $X : t_0 \times t_1$ to $Y_0 \times Y_1$, for distinct variables $Y_0 : t_0, Y_1 : t_1$. Define:

$$\begin{aligned}
red(I : t_0 \times t_1; \sigma) &= (I,\ I) \\
red(T : t_0 \times t_1; \sigma) &= (T,\ T) \\
red(\langle(a,b)\rangle A : t_0 \times t_1; \sigma) &= (\langle a \rangle A_0,\ \langle b \rangle A_1) \\
red(\overline{\langle(a,b)\rangle} A : t_0 \times t_1; \sigma) &= (\overline{\langle a \rangle} A_0,\ \overline{\langle b \rangle} A_1) \\
red(((a,b))A : t_0 \times t_1; \sigma) &= ((a)A_0,\ (b)A_1) \\
red(\overline{((a,b))}A : t_0 \times t_1; \sigma) &= (\overline{(a)}A_0,\ \overline{(b)}A_1) \text{ where } red(A : t_0 \times t_1; \sigma) = (A_0,\ A_1) \\
red(A \wedge A' : t_0 \times t_1; \sigma) &= ((A_0 \wedge A_0'),\ (A_1 \wedge A_1')) \text{ where} \\
red(A : t_0 \times t_1; \sigma) &= (A_0,\ A_1),\ red(A' : t_0 \times t_1; \sigma) = (A_0',\ A_1')
\end{aligned}$$

$$\begin{aligned} red(X : t_0 \times t_1; \sigma) &= (Y_0,\ Y_1) \text{ where } \sigma(X) = Y_0 \times Y_1 \\ red(\nu X.\ A : t_0 \times t_1; \sigma) &= ((\nu Y_0.\ A_0),\ (\nu Y_1.\ A_1)) \\ \text{where } \sigma(X) &= Y_0 \times Y_1 \text{ and } red(A : t_0 \times t_1; \sigma) = (A_0,\ A_1) \end{aligned}$$

While this reduction works for a nontrivial class of assertions the class is limited. In particular, it does not include the 'reachability' assertions $R_\Lambda \equiv \mu X.I \vee \bigvee_{\beta \in \Lambda} \langle \overline{\beta} \rangle X$, true of those states which are reachable from the initial state purely by transitions with labels in Λ. However, in the special case where

$$(\lambda_0, \lambda_1) \in \Lambda \ \&\ \lambda_0 \neq * \Rightarrow (\lambda_0, *) \in \Lambda, \text{ and } (\lambda_0, \lambda_1) \in \Lambda \ \&\ \lambda_1 \neq * \Rightarrow (*, \lambda_1) \in \Lambda \qquad (+)$$

we have that

$$\models R_\Lambda \leftrightarrow R_{\Lambda_0} \times R_{\Lambda_1} : t_0 \times t_1$$

where $\Lambda_0 = \{\lambda_0 \mid \exists \lambda_1.\ (\lambda_0, \lambda_1) \in \Lambda\}$, $\Lambda_1 = \{\lambda_1 \mid \exists \lambda_0.\ (\lambda_0, \lambda_1) \in \Lambda\}$. As we shall see, this has implications for reductions with respect to the parallel composition of Milner's CCS.

Any recursion-free assertion of a product can be routinely transformed into a finite disjunction $\bigvee_{i \in I} B_i \times C_i$ (though conjunctions cause a quadratic 'blow-up' in size and negations an exponential 'blow-up'). As we have seen, there are special cases of recursive assertions where this can be achieved too. Once we have a property of a product $t_0 \times t_1$ expressed in such a form, the following result provides a method for reducing its validity to validities in the components t_0, t_1. Note the result is independent of the composition of t_0 and t_1.

Proposition 11 *Suppose $\bigvee_{i \in I} B_i \times C_i : t_0 \times t_1$ is a finite disjunction.*

$$\models \bigvee_{i \in I} B_i \times C_i : t_0 \times t_1$$

iff

$$(\models \bigvee_{j \in J} B_j : t_0) \text{ or} (\models \bigvee_{k \in K} C_k : t_1)$$

for all partitions $J \dot{\cup} K = I$.

It is useful to generalise the above proposition a little, so we can establish a property of a product relative to assumptions on the states of each component. This paper has concentrated on 'backwards proof'; given a goal for a compound term it has addressed how to reduce this to subgoals for its immediate components. One use of the following proposition is when asking the converse: if $\models B : t_0$ and $\models C : t_1$ does it follow that $\models A : t_0 \times t_1$? The proposition provides a partial answer—it depends on A having been expressed as a finite disjunction $\models \bigvee_{i \in I} B_i \times C_i : t_0 \times t_1$. Then:

Proposition 12 *Suppose $\bigvee_{i \in I} B_i \times C_i : t_0 \times t_1$ is a finite disjunction. Let $B : t_0$, $C : t_1$. Then*

$$\models B \times C \rightarrow \bigvee_{i \in I} B_i \times C_i : t_0 \times t_1$$

iff

$$(\models B \rightarrow \bigvee_{j \in J} B_j : t_0) \text{ or} (\models C \rightarrow \bigvee_{k \in K} C_k : t_1)$$

for all partitions $J \dot{\cup} K = I$.

As a corollary of this proposition, we obtain reduction results for certain parallel compositions including that of CCS. A parallel composition of t_0, t_1 has the form

$$t_0 \parallel t_1 \equiv ((t_0 \times t_1)\lceil\Lambda)\{\Xi\}$$

A problem $\models A : t_0 \parallel t_1$ reduces to

$$\models R_\Lambda \rightarrow A' : t_0 \times t_1$$

where A' is obtained by carrying out the reductions for relabelling, then restriction. (Note th reduction for restriction, as expressed by theorem 9, introduces the reachability assertion R_Λ. Now, in the case of $t_0|t_1$, for CCS, the appropriate restriction is with respect to a subset satisfying (+). It follows that

$$\models R_\Lambda \leftrightarrow R_{\Lambda_0} \times R_{\Lambda_1} : t_0 \times t_1$$

for reachability assertions R_{Λ_0} and R_{Λ_1}. Hence if A' can be expressed as a manageable disjunc tion $\bigvee\!\!\!\bigvee_{i \in I} B_i \times C_i$ we have a reductive way of checking $\models A : t_0|t_1$:

$$\models A : t_0|t_1$$

iff

$$(\models R_{\Lambda_0} \rightarrow \bigvee\!\!\!\bigvee_{j \in J} B_j : t_0) \text{ or} (\models R_{\Lambda_1} \rightarrow \bigvee\!\!\!\bigvee_{k \in K} C_k : t_1)$$

for all partitions $J \dot{\cup} K = I$. Certainly, A' can be so decomposed when the original assertion A contains no variables. Note that even for assertions A without recursion, the statement $\models A : t_0 \times t_1$, being one of validity, can express a nontrivial invariant of $t_0 \times t_1$. It is emphasised that again this reduction does not depend on the structure of t_0 and t_1.

Unfortunately, the important use of restriction in CCS to internalise communication along a channel does not use a restricting set satisfying (+). The problem with checking validity for terms which force internal communication along a channel centres on the difficulty of expressing

$$R_{\{(a,b)\}} \equiv \mu X. \, (I \vee \overline{\langle (a,b) \rangle} X),$$

true of those states in a product which are reachable from the initial state via a sequence of (a,b)-transitions, as a finite, and manageable, disjunction $\bigvee\!\!\!\bigvee_{i \in I} B_i \times C_i : t_0 \times t_1$. Of course, once we know the size of the transition system $t_0 \times t_1$ to be k, we have

$$\models R_{\{(a,b)\}} \leftrightarrow \bigvee\!\!\!\bigvee_{n<k} \overline{\langle a \rangle}^n I \times \overline{\langle b \rangle}^n I : t_0 \times t_1.$$

With luck, the recursion might become stationary at an earlier point, but to be a valid equivalence for *all* transition systems of size k all the k disjuncts have to be included. This reduction is thus quadratic in the size of the transition system. Certainly in this case the reduction can no longer be independent of t_0, t_1, with the assertion language as it stands presently.

Conclusion

General methods have been provided for reasoning compositionally with a modal ν-calculus. These methods are presently being implemented by Henrik Andersen at Aarhus. Henrik has also extended the reductions to cope with a more traditional recursive definition of processes which could be used in place of looping. The introduction of process variables which this entails ould be useful for other reasons. Because all the reductions are directed only by the top-level operation on terms they might well be helpful in synthesising a process satisfying a specification, using process variables to leave parts of term unspecified.

There remain important properties of products which do not seem directly amenable to the techniques outlined here. It is notable though that some nontrivial assertions have reductions which are independent of the structure of the components of a product. It is hoped that the techniques and limitations exposed here will help guide the search for methods of reasoning about parallel processes. Promising leads may be found in [CLM] and [LX].

The approach here can be understood as running the compositional proof system of [W] backwards, and relates to the more modest compositional proof systems of [St] and [W1], and more superficially to [GS]. The reductions however have a fuller treatment of assertion variables than the proof system of [W]; the latter should be redone so that it supports the reasoning given by the reductions in a forwards direction and makes plain the sense in which the reductions correspond to running the proof system backwards.

Acknowledgements

I would like to acknowledge the support of the Esprit Basic Research Actions CEDISYS and CLICS. I've had valuable conversations with Henrik Andersen. Thanks to the referees for their comments.

References

[CLM] Clarke, E., Long, D., and McMillan, K., Compositional model checking. LICS 1989.

[EL] Emerson, A. and Lei, C., Efficient model checking in fragments of the propositional mu-calculus. Proc.LICS, 1986.

[GS] Graf, I., and Sifakis, J., A logic for the description of nondeterministic programs and their properties. Report IMAG RR 551, Grenoble, France, 1985.

[LX] Larsen,K.G. and Xiuxin, L., Compositionality through an operational semantics of contexts. Proc. ICALP 90, 1990.

[St] Stirling, C, A complete modal proof system for a subset of SCCS. LNCS 185, 1985.

[W1] Winskel, G., A complete proof system for SCCS with modal assertions. In the proceedings of Foundations of Software Technology (1985).

[W] Winskel, G., A compositional proof system on a category of labelled transition systems. To appear in Information and Computation, 1990.

Real-Time Behaviour of Asynchronous Agents

Wang Yi
Department of Computer Sciences
Chalmers University of Technology and the University of Göteborg
S-41296 Göteborg, Sweden

Abstract

In this paper, we present a calculus for real-time communicating systems. The calculus is an extension of Milner's CCS with explicit time. In SCCS, $P \xrightarrow{1} Q$ means that if P exists at time r, it will proceed to Q at time $r+1$. The time delay is exactly one unit. We extend this idea to asynchronous agents by allowing arbitrary delays. We write $P \xrightarrow{\epsilon(t)} Q$ to mean that *after t units of time, P will become Q*, where ϵ stands for *idling*. Based on the notion of bisimulation, two equivalence relations over agents are defined. It has been shown that the strong equivalence is a congruence and the weak one is preserved by all operators except summation and recursion [W90]. Various examples are given to illustrate the approach.

1 Motivation

Consider a vending machine with the following informal specification. *You can insert a coin at any time you like. Thereafter, you can choose coffee or tea. After inserting a coin, if you do not take your drink within* 30 *seconds, the machine will turn back automatically to its initial state to collect money.* Using CCS, the machine may be formally described as follows:

$$S_0 \stackrel{def}{=} money.S_1$$
$$S_1 \stackrel{def}{=} \tau.S_0 + \overline{coffee}.S_0 + \overline{tea}.S_0$$

Here timeout is modelled by a τ-action. In fact, we have no way to state that the timeout period after receiving a coin is 30 seconds in CCS. Clearly, this machine may accept money without delivering a drink, $S_0(\stackrel{money}{\Longrightarrow})^* S_0$.

It is well known that process algebras such as CCS and CSP cannot express time delays between events. They deal with the *quantitative* aspects of time in a *qualitative* way. This greatly simplifies the procedures of specification and verification of systems and allows the traditional approaches to be applied to a large family of complex problems. However, as the above example shows, it is not always appropriate to treat time in a purely qualitative way. In many cases, we need to express exact time limits. For example, *"p should be ready to do μ in* 3 *micro-seconds"*, *"every* 10 *seconds, a message should be sent out"*, etc.

Over the years, various attempts have been made [M83, C82, RR86, S90 ,MT89, T89]. In [M83], Milner develops a time-dependent calculus for synchronous agents based on the idea that each atomic action takes one unit of time. The real-time calculus of [C82] is an extension along this line by allowing continuous time instead of discrete time. A denotational semantics for a real-time extension of CSP called timed CSP is given in [RR86]. More recently, Sifakis et al. have proposed a timed extension of ACP [S90]. In [MT89, T89], a temporal calculus for communicating systems is developed. The calculus is along the line of CCS.

The aim of this work is also to investigate the possibility of finding a calculus for real-time communicating systems in the style of CCS. The calculus should allow us to express explicit time delay between events. Our starting point is CCS. We shall follow the original ideas of CCS as far as possible. Time delays will be modelled by a set of events indexed by a given time domain. This set of time events will be treated as a subset of the alphabets of the calculus. Thus the timed calculus will have exactly the same syntax as the time-independent calculus, CCS. Further, in developing a transitional semantics for the calculus, agents shall be forced to proceed whenever they can do so by performing τ-actions. Thus the behaviour of agents will be more deterministic. This important assumption is called *maximal progress* assumption [RH89].

More precisely, we shall develop a timed calculus with the following properties:

- The calculus is parameterized with a time domain. The time domain can be an arbitrary well-ordered domain such as the natural numbers and the possitive real numbers including 0. A dense time domain shall yield a continuous calculus; whereas a time domain like the natural numbers will give rise to a discrete calculus.

- Actions are atomic in the sense that they take no time. But whenever it is necessary, we can put an explicit time delay before an atomic action. This is the minimum time that has to elapse before the action is enabled. This idea appears also in ESTELLE and other work [S90,T89].

- Agents cooperate in the same way as in CCS. They must wait for each other to communicate. In other words, communication between agents is synchronous. But whenever both partners of a communication are ready, they will perform the communication immediately. Agents may also perform actions independently of each other. In this sense, the calculus is asynchronous.

- Concurrency is modelled by "interleaving" in the sense that a composite agent can always be expanded to a non-deterministic one as in CCS.

We shall distinguish two types of atomic actions: *controllable* and *non-controllable*. By a controllable action α, we mean that α needs the external stimulus $\bar{\alpha}$ to occur. So an agent can never complete such a controllable action by itself. But an agent can always influence another agent capable of doing α by supplying $\bar{\alpha}$ to it. As mentioned above, atomic actions take no time. We shall write $P \xrightarrow{\alpha} Q$ to mean that *if the environment is ready to offer* $\bar{\alpha}$, P *may perform* α *immediately and become* Q *in doing so.* The transition takes no time.

The non-controllable actions are τ-actions or complete events as called by Milner [M86, M89] that do not require external stimuli to occur. Thus agents can never influence each other by performing a non-controllable action. A typical non-controllable action is a successful communication of two agents by performing α and $\bar{\alpha}$ simultaneously. Therefore, we give them a unique name τ as in CCS. We shall write $P \xrightarrow{\tau} Q$ to mean that P *may perform* τ *immediately and then behave like* Q. The action, τ is also atomic. So this transition takes no time either.

Intuitively, we may think of an agent as a black box with buttons and each button has a lamp to indicate the readiness of the button. Then the buttons correspond to the controllable actions. When the lamp of a button turns on, it means that the action is *enabled*. You can always push down the button as soon as the lamp is on. The lamps may also turn off automatically when the agent performs non-controllable actions internally. So here the non-controllable action τ is visible in the sense that we can always observe the lamps' switching off. But you can not control when or how these activities occur. This is the so-called non-controllability.

The switching of the lamps makes it possible for one to record the time delay before an action is enabled. For instance, when we start a terminal, it will take a while before the login-prompt appears. The time delay is usually called *response time* in real-time systems. The central idea of real-time systems is to study whether a system is able to perform a certain action within a given time limit. It should not be too early or too late. Traditional formalisms often ignore the time information and only consider the communication capability of systems. This is obviously inadequate to time-critical systems. The idling behaviour of such systems is equally important. So we need an extra action to represent idling. Recall that in SCCS, $P \xrightarrow{1} Q$ means that if P exists at time r, it will idle for one unit of time and then behave like Q at time $r + 1$. The time delay is exactly one unit. We may extend this idea to an arbitrary time delay. Let ϵ denote idling. We shall write

$$P \xrightarrow{\epsilon(t)} Q$$

to mean that P *will idle for* t *units of time and then behave like* Q. Note that this transition takes t units of time. Within t units of time, no non-controllable actions are performed. In the case that P has performed some non-controllable action, it cannot be considered as idle. During the time delay, it may be preparing for an atomic action, for example, creating some messages to send or performing a series of local activities or just be waiting for computing resources.

2 Regular Agents

The so-called regular agents are those that have no explicit parallelism appearing in their structures (no parallel composition). They shall be considered as the fundamental part of real-time behaviour. So we choose to study them first.

2.1 Syntax and Semantics

Given time domain $\mathcal{T}$, let $\delta_{\mathcal{T}}$ be the set $\{\epsilon(t) | t \in \mathcal{T} - \{0\}\}$ of time events assuming that 0 is the least element of $\mathcal{T}$. Each $\epsilon(t)$ of $\delta_{\mathcal{T}}$ is an empty action where t stands for how long exactly an agent will be idle. Note that $\delta_{\mathcal{T}}$ does not include the least element $\epsilon(0)$. In this paper, we shall considser the case when $\mathcal{T}$ consists of the positive real numbers $\mathcal{R}^+$ and 0, i.e. $[0, \infty)$. t, u, v will range over $\mathcal{T}$. We adopt the convention that $u - t$ is always 0 when $t \geq u$.

Let $\mathcal{L}$ be a set of labels denoting controllable actions and τ be a unique symbol representing non-controllable actions. Following Milner, let $\mathcal{L} = \Lambda \cup \bar{\Lambda}$, where Λ is known as the set of names and $\bar{\Lambda} = \{\bar{\alpha} | \alpha \in \Lambda\}$ is the set of co-names. The action $\bar{\alpha}$ is the stimulus or complement of α. α shall also be viewed as the stimulus of $\bar{\alpha}$, i.e. $\bar{\bar{\alpha}} = \alpha$. τ and $\epsilon(t)$ have no complements. However for convenience, we define $\bar{\tau} = \tau$ and $\bar{\epsilon}(t) = \epsilon(t)$. We shall often use the more intuitive notations, $\alpha!$ and $\alpha?$ instead of $\bar{\alpha}$ and α. Let $\mathcal{A}ct = \mathcal{L} \cup \{\tau\} \cup \delta_{\mathcal{T}}$. We use σ, ϕ to range over $\mathcal{A}ct$, α, β to range over $\mathcal{L}$ and μ, ν to range over $\mathcal{L} \cup \{\tau\}$.

Further, assume a set of agent names ranged over by X, Y. The names may be indexed or parameterized like X_i or $X(x_1 ... x_n)$ as in CCS. Let P, Q, R range over agent expressions. Then the set of regular agent expressions is the least set generated by the following combinators together with agent variables and mutually recursive definition, $X_i \stackrel{def}{=} P_i$ where $i \in I$ for some well-understood index set I.

1. Constant: NIL
2. Prefix: $\sigma.P \qquad (\sigma \in \mathcal{A}ct)$
3. Summation: $P + Q$

In the following, we shall use the notation $X \stackrel{def}{=} P$ informally. X shall be understood as a constant agent defined by P as in CCS [M89]. So it is natural that we can infer $X \xrightarrow{\sigma} P'$ from $P \xrightarrow{\sigma} P'$, if X is defined by P, i.e. $X \stackrel{def}{=} P$.

2.1.1 NIL

The simplest agent is the one that can do nothing but idling.

$$\frac{}{\text{NIL} \xrightarrow{\epsilon(t)} \text{NIL}}$$

Remark: NIL idles forever. It models a deadlocked state of systems or a terminated process that does not show any behaviour to its environment in its whole life. □

2.1.2 Prefix

The prefix form is classified into three sub-cases according to different kinds of actions.

1. *Controllable Action*

$$(1) \quad \frac{}{\alpha.P \xrightarrow{\alpha} P}$$

$$(2) \quad \frac{}{\alpha.P \xrightarrow{\epsilon(t)} \alpha.P}$$

Remark: $\alpha.P$ is able to perform a controllable action α immediately. The transition takes no time, since α is atomic. It sounds unrealistic that a state transition takes no time. However, whenever it is necessary, a delay can be put before α, see the following rules for idling. The point is that we want to separate time delay of preparing for the atomic action α from the moment when α is *committed*. This transition corresponds to the *commit*-operation for α. By the second rule, $\alpha.P$ will wait (idle) when its environment is not ready. □

Example 2.1 : *a simple vending machine*

$$S_0 \stackrel{def}{=} money?.S_1$$
$$S_1 \stackrel{def}{=} coffee!.S_0$$

The machine can always accept money (a coin) and then deliver coffee directly. This machine is an ideal one, which is infinitely fast. But the machine will wait when the user is not ready.

(a) $S_0 \xrightarrow{money?} S_1$ and $S_0 \xrightarrow{\epsilon(t)} S_0$ $(t > 0)$

(b) $S_1 \xrightarrow{coffee!} S_0$ and $S_1 \xrightarrow{\epsilon(t)} S_1$ $(t > 0)$

2. *Non-controllable Action*

$$\frac{}{\tau.P \xrightarrow{\tau} P}$$

Remark: τ is autonomous and will occur immediately whenever it is enabled. This embodies the *maximal progress* assumption which requires that an agent will never wait unnecessarily. *So whenever it can proceed by performing a τ-action, it will wait for no time.* Note that $\tau.P$ has only one τ-transition, unlike $\alpha.P$ has a chain of $\epsilon(t)$-transitions. □

3. *Idling*

$$(1)\quad \frac{}{\epsilon(u).P \xrightarrow{\epsilon(u)} P}$$

$$(2)\quad \frac{}{\epsilon(t+u).P \xrightarrow{\epsilon(t)} \epsilon(u).P}$$

$$(3)\quad \frac{P \xrightarrow{\epsilon(t)} P'}{\epsilon(u).P \xrightarrow{\epsilon(t+u)} P'}$$

Remark: Note first that $t, u > 0$. An agent P can be delayed for u units of time. So $\epsilon(u).P$ will be idle within u units of time and then behave like P. Naturally, if P becomes P' after t seconds, $\epsilon(u).P$ will do the same after $t + u$ seconds. □

So far, two types of waiting have been introduced. First, $\alpha.P$ is willing to wait for any units of time, if the environment is not ready to offer $\bar{\alpha}$. By contrast, $\epsilon(u).P$ is forced to wait for u units of time. The later may be called *forced* waiting. However we shall not distinguish these two kinds of waiting, because they are different only in an intensional sense. No environment will be able to discover the difference.

It is important to note that $\epsilon(0)$ has been excluded from both the syntax and semantics of agents. However for convenience, we shall use $\epsilon(0).P$ as a notation for P, i.e. $\epsilon(0).P \equiv P$.

Example 2.2 *a "realistic" vending machine*

In practice, a coffee machine can always accept money, but it will always take a while to deliver coffee. A more realistic machine may behave like the following agent.

$$\begin{aligned} S_0 &\stackrel{def}{=} money?.S_1 \\ S_1 &\stackrel{def}{=} \epsilon(2).S_2 \\ S_2 &\stackrel{def}{=} coffee!.S_0 \end{aligned}$$

First, you can put in a coin whenever you want. Thereafter, you have to wait 2 seconds to get your coffee. For 2 seconds, the coffee-button is locked. In other words, it takes 2 seconds to enable channel *coffee*. Then you can wait for years (if you like) as long as nobody takes your coffee. This may be expressed symbolically as follows.

(a) $S_0 \xrightarrow{money?} S_1$ and $S_0 \xrightarrow{\epsilon(t)} S_0$ $(t > 0)$

(b) $S_1 \xrightarrow{\epsilon(t)} \epsilon(2-t).S_2$ $(t < 2)$ and $S_1 \xrightarrow{\epsilon(2)} S_2$

(c) $S_2 \xrightarrow{coffee!} S_0$ and $S_2 \xrightarrow{\epsilon(t)} S_2$ $(t > 0)$

Example 2.3 : *a simple timer*

$$\begin{aligned} T_0 &\stackrel{def}{=} time?u.T_1(u) \\ T_1(u) &\stackrel{def}{=} \epsilon(u).T_2 \\ T_2 &\stackrel{def}{=} timeout!.T_0 \end{aligned}$$

The timer is first ready to accept a time period u at any time from channel *time*. Then it starts its clock. When the given time period elapses, it times out. However if the environment is not ready to respond to the time-out signal by supplying the complement *timeout*?, it will wait. Note that this timer cannot be reset.

(a) $T_0 \xrightarrow{time?u} T_1(u)$ and $T_0 \xrightarrow{\epsilon(t)} T_0$ $(t > 0)$

(b) $T_1(u) \xrightarrow{\epsilon(t)} \epsilon(u-t).T_2$ $(t < u)$ and $T_1(u) \xrightarrow{\epsilon(u)} T_2$

(c) $T_2 \xrightarrow{timeout!} T_0$ and $T_2 \xrightarrow{\epsilon(t)} T_2$ $(t > 0)$

2.1.3 Summation

The summation combinator is used to describe non-deterministic choices.

$$(1)\quad \frac{P \xrightarrow{\mu} P'}{P+Q \xrightarrow{\mu} P'}$$

$$(2)\quad \frac{Q \xrightarrow{\mu} Q'}{P+Q \xrightarrow{\mu} Q'}$$

$$(3)\quad \frac{P \xrightarrow{\epsilon(t)} P' \qquad Q \xrightarrow{\epsilon(t)} Q'}{P+Q \xrightarrow{\epsilon(t)} P'+Q'}$$

Remark: The first two rules are easy to understand. They are from the standard CCS. Note that $\mu \neq \epsilon(t)$. The third rule says that if both P and Q are idling, then no alternatives can be lost. The sum $P+Q$ is also idling. Perhaps both P and Q are waiting to be chosen by the environment. This shows how the so-called external non-determinism works. On the other hand, if one of the alternatives cannot idle, the sum cannot do it either. For example, $\epsilon(1).P + \tau.Q$ has no $\epsilon(t)$-derivative because $\tau.Q$ must proceed immediately, even though $\epsilon(1).P$ can wait at least for one unit of time. This also embodies the maximal progress assumption. □

Example 2.4 : *a vending machine with timeout*

$$M_0 \stackrel{def}{=} money?.M_1$$
$$M_1 \stackrel{def}{=} \epsilon(2).coffee!.M_0 + \epsilon(3).tea!.M_0 + \epsilon(30).\tau.M_0$$

Here time-out is modelled by a τ-action as we did in describing such a machine using CCS previously. The significant step made here is that the τ-action will not appear within 30 seconds, that has been explicitely stated in the delay-operation $\epsilon(30)$. before τ. So within 30 seconds, a drink is ensured.

This machine can also sell tea. We may imagine that the machine has two buttons for coffee and tea respectively and a slot for money. Further each of the buttons and the slot has a lamp. At the very first moment when the machine is started, the lamp for money turns on. You can insert a coin as long as the money-lamp is on. Further if you are not ready, the machine will wait for you as usual. Formally, $M_0 \xrightarrow{money?} M_1$ and $M_0 \xrightarrow{\epsilon(t)} M_0$ $(t > 0)$.

Thereafter you have to wait 2 seconds before coffee is ready and one more second if you wish to have tea instead. So after 3 seconds, both of the coffee-lamp and the tea-lamp are on. This means that you can choose coffee or tea —you can decide what you want but not both. If you cannot make up your mind within 30 seconds, the lamps for coffee and tea will turn off automatically and the money-lamp turns on again. This tells you that you have lost a coin. These can also be stated formally.

1. $M_1 \stackrel{\epsilon(t)}{\longrightarrow}\stackrel{coffee!}{\longrightarrow} M_0 \ (2 \leq t \leq 30)$

2. $M_1 \stackrel{\epsilon(t)}{\longrightarrow}\stackrel{tea!}{\longrightarrow} M_0 \ (3 \leq t \leq 30)$

3. $M_1 \stackrel{\epsilon(30)}{\longrightarrow}\stackrel{\tau}{\longrightarrow} M_0$

2.2 Properties of Agents

Let $\mathcal{P}^r$ stand for the set of regular agents and $\stackrel{\sigma}{\longrightarrow}$ for each $\sigma \in \mathcal{A}ct$ be the least relation over $\mathcal{P}^r$, satisfying the inference rules developed. This yields a standard labelled transition system: $\langle \mathcal{P}^r, \mathcal{A}ct, \longrightarrow \rangle$. In the following, we study the properties of this system. They shall be considered as fundamental properties of timed agents. So when introducing new combinators, we shall make sure that the composite agents satisfy these properties.

Lemma 2.1 *(maximal progress)* $\exists P_1 : P \stackrel{\tau}{\longrightarrow} P_1 \supset \forall t > 0 : \neg \exists P_2 : P \stackrel{\epsilon(t)}{\longrightarrow} P_2$

Remark: This is the most important property of agents. One shall discover that it determines the whole design of the calculus. Intuitively it says that agents never wait unnecessarily or whenever it can perform a τ-action, it will do so immediately. Therefore there will be no $\epsilon(t)$-derivative in parallel with a τ-derivative for each agent. □

Lemma 2.2 *(determinacy)* $P \stackrel{\epsilon(t)}{\longrightarrow} P_1 \ \& \ P \stackrel{\epsilon(t)}{\longrightarrow} P_2 \supset P_1 \equiv P_2$

Remark: When time goes, if an agent does nothing but idling then it cannot reach different states. "$\equiv$" is the syntactical identity. □

Lemma 2.3 *(continuity)* $P \stackrel{\epsilon(t+u)}{\longrightarrow} P_2 \ \supset\subset \ \exists P_1 : P \stackrel{\epsilon(t)}{\longrightarrow} P_1 \ \& \ P_1 \stackrel{\epsilon(u)}{\longrightarrow} P_2$
where $t, u > 0$ *and* $\supset\subset$ *is the usual logical equivalence.*

Remark: An agent can idle for 10 seconds implies that it can idle for 5 seconds first and then another 5 seconds more and vice versa. This lemma says that if an agent proceeds from one instant to the other, it must reach all the intermediate instants between them. □

Lemma 2.4 *(persistency)* $P \stackrel{\epsilon(t)}{\longrightarrow} P_1 \ \& \ P \stackrel{\alpha}{\longrightarrow} P_2 \supset \exists P_1' : \ P_1 \stackrel{\alpha}{\longrightarrow} P_1'$

Remark: By idling, an agent will not lose any capabilities of performing the actions that it is able to perform originally. Note that when only considering regular agents, it will be the case that $P_1' \equiv P_2$. But this is not true in general when introducing parallel composition. To generalize this result, we distinguish P_1' from P_2. □

So far, we have presented the four main properties of timed agents. The reader may ask why just these four have been chosen. Do we forget some one? We shall see that for each of these four lemmas, there is a corresponding algebraic law. More surprisingly, these four basic laws are the central part in a complete axiomatization for bisimulation equivalence [W90]. This convinces us that the four lemmas are the right candidates.

3 Composite Agents

This section will be mainly concerned with how to construct complex agents based on regular ones by means of new combinators. We start with the most important one, parallel composition.

3.1 Parallel Composition

Assume that P and Q are regular agents. Now we want them to work in parallel. Let $P|Q$ be the parallel composition of P and Q. Then what transitions are possible for $P|Q$?

One may have the first intuition that if P can perform μ alone, it should be able to do it in the context of $P|Q$. Furthermore if P and Q can perform α and $\bar{\alpha}$ simultaneously, they produce a complete event τ at the same time. This suggests the following rules.

$$(1)\quad \frac{P \xrightarrow{\mu} P'}{P|Q \xrightarrow{\mu} P'|Q}$$

$$(2)\quad \frac{Q \xrightarrow{\mu} Q'}{P|Q \xrightarrow{\mu} P|Q'}$$

$$(3)\quad \frac{P \xrightarrow{\alpha} P' \qquad Q \xrightarrow{\bar{\alpha}} Q'}{P|Q \xrightarrow{\tau} P'|Q'}$$

These three rules are from the standard CCS. Now we consider the behaviour of $P|Q$ when P and Q are idling. Suppose that after t units of time, P may evolve to P' and Q may evolve to Q' internally. Naturally $P|Q$ may proceed to $P'|Q'$ since time is fair to everyone.

$$\frac{P \xrightarrow{\epsilon(t)} P' \qquad Q \xrightarrow{\epsilon(t)} Q'}{P|Q \xrightarrow{\epsilon(t)} P'|Q'}$$

Unfortunately this rule destroyes the *maximal progress* property. For instance, let $X \stackrel{def}{=} \alpha.A$ and $Y \stackrel{def}{=} \bar{\alpha}.B$. By the rule given above, one can infer $X|Y \xrightarrow{\epsilon(t)} X|Y$ for arbitrary t, since $X \xrightarrow{\epsilon(t)} X$ and $Y \xrightarrow{\epsilon(t)} Y$. One can also infer $X|Y \xrightarrow{\tau} A|B$ since $X \xrightarrow{\alpha} A$ and $Y \xrightarrow{\bar{\alpha}} B$.

Two consequences arise: $X|Y \xrightarrow{\tau} A|B$ and $X|Y \xrightarrow{\epsilon(t)} X|Y$. Obviously they contradict the maximal progress assumption. In this case, if $X|Y$ is allowed to idle, then unnecessary non-determinism is introduced. One may ask how long time $X|Y$ is allolwed to idle. We shall force $X|Y$ to perform the internal communication involving α and $\bar{\alpha}$ and become $A|B$ immediately. The τ-transition of $X|Y$ precludes idling. So all $\epsilon(t)$-transitions of $X|Y$ should be avoided.

The question is how to forbid a composite agent from idling when it is able to perform a τ-action resulted from an internal communication. The solution is by introducing a notion of *timed sort* which is similar to the notion of *sort* in CCS. Roughly speaking, a timed sort of P is a set of controllable actions that includes all actions of P that are enabled within a given time interval.

Definition 3.1

1. $P \xrightarrow{\alpha}_0 Q$ *iff* $P \xrightarrow{\alpha} Q$

2. $P \xrightarrow{\alpha}_t Q$ *iff* $\exists P' : P \xrightarrow{\epsilon(t)} P'$ *and* $P' \xrightarrow{\alpha} Q$

Definition 3.2 *(timed sort for regular agents) Let P be regular agent and $L \subseteq \mathcal{L}$. P is said to have sort L within time interval t iff $\forall t' < t, \alpha \in \mathcal{L} : \exists P' : P \xrightarrow{\alpha}_{t'} P' \supset \alpha \in L$.*

For example, the whole set $\mathcal{L}$ of labels is a timed sort for each agent within arbitrary time durations. On the other hand, there may exist a least sort of P for each given time interval. This may not be easy to determine. But there is a natural way to assign sorts to agents in terms of their syntactic structures as in CCS [M89].

Definition 3.3 *Let $t, u > 0$ and P be a regular agent. We define $Sort_t(P)$ as the least set satisfying the following equations:*

$$\begin{aligned}
Sort_t(NIL) &= \emptyset \\
Sort_t(\alpha.Q) &= \{\alpha\} \\
Sort_t(\tau.Q) &= \emptyset \\
Sort_t(\epsilon(t).Q) &= \emptyset \\
Sort_t(\epsilon(t+u).Q) &= \emptyset \\
Sort_{t+u}(\epsilon(t).Q) &= Sort_u(Q) \\
Sort_t(Q+R) &= Sort_t(Q) \cup Sort_t(R) \\
Sort_t(X) &= Sort_t(P) \qquad \text{when } X \stackrel{def}{=} P
\end{aligned}$$

Intuitively, $Sort_t(P)$ only includes what controllable actions P is able to offer in the first step within a given time limit. For example, $Sort_t(\tau.Q)$ is empty, because $\tau.Q$ must perform τ first. Consider the vending machine which has been described as follows.

1. $M_0 \stackrel{def}{=} money?.M_1$

2. $M_1 \stackrel{def}{=} \epsilon(2).coffee!.M_0 + \epsilon(3).tea!.M_0 + \epsilon(30).\tau.M_0$

The following can be easily computed.

$$\begin{aligned}
Sort_{10000}(M_0) &= \{money?\} \\
Sort_{2.999}(M_1) &= \{coffee!\} \\
Sort_3(M_1) &= \{coffee!\} \\
Sort_{29.999}(M_1) &= \{coffee!, tea!\} \\
Sort_{31}(M_1) &= \{coffee!, tea!\}
\end{aligned}$$

It is trivial to check that they are really sorts for M_i. We have a general result.

Proposition 3.1 *For each regular agent P and $t > 0$, $Sort_t(P)$ is a timed sort of P for time interval t.*

Remark: The result tells us that for all $t' < t$ if there exists P' such that $P \xrightarrow{\alpha}_{t'} P'$, then $\alpha \in Sort_t(P)$. But the inverse does not hold because of the maximal progress assumption. For example, $Sort_{31}(M_1) = \{coffee!, tea!\}$; but M_1 can never deliver coffee or tea after 30 seconds. □

We define $\bar{L} = \{\bar{\alpha} | \alpha \in \mathcal{L}\}$ for arbitrary $L \subseteq \mathcal{L}$. A more useful result can be established.

Corollary 3.1 *For all regular agents P, Q,*
$$Sort_t(P) \cap \overline{Sort_t(Q)} = \emptyset \supset \forall t' < t : \forall \alpha : \neg \exists P', Q' : P \xrightarrow{\alpha}_{t'} P' \;\&\; Q \xrightarrow{\bar{\alpha}}_{t'} Q'$$

Remark: If $Sort_t(P) \cap \overline{Sort_t(Q)} = \emptyset$, then P and Q cannot perform a controllable action α and its complement $\bar{\alpha}$ simultaneously within t units of time. This implies that P and Q cannot communicate with each other within t units of time. □

We place $Sort_t(P) \cap \overline{Sort_t(Q)} = \emptyset$ as a side-condition in the timing rule for parallel composition. It will guarantee that the composite agent satisfies the maximal progress property. Let P, Q be regular agents.

$$(4)\quad \frac{P \xrightarrow{\epsilon(t)} P' \qquad Q \xrightarrow{\epsilon(t)} Q'}{P|Q \xrightarrow{\epsilon(t)} P'|Q'} \qquad (Sort_t(P) \cap \overline{Sort_t(Q)} = \emptyset)$$

Remark: To infer $P|Q \xrightarrow{\epsilon(t)} P'|Q'$, it is required not only $P \xrightarrow{\epsilon(t)} P'$ and $Q \xrightarrow{\epsilon(t)} Q'$ but also $Sort_t(P) \cap \overline{Sort_t(Q)} = \emptyset$. The side-condition makes sure that $P|Q$ cannot perform a τ-action by an internal communication between them within t units of time. □

As desired, the following holds.

Proposition 3.2 *For all regular agents P, Q, the composite agent $P|Q$ satisfies lemma* 1,2,3 *and* 4.

To see how the maximal progress works for parallel composition, Consider the example $X|Y$ again where $X \stackrel{def}{=} \alpha.A$ and $Y \stackrel{def}{=} \bar{\alpha}.B$. As usual, $X|Y \xrightarrow{\tau} A|B$ since $X \xrightarrow{\alpha} A$ and $Y \xrightarrow{\bar{\alpha}} B$. But $X|Y \xrightarrow{\epsilon(t)} X'|Y'$ for no t, as required because $Sort_t(X) \cap \overline{Sort_t(Y)} = \{\alpha\}$.

3.2 Examples and Further Constructions

This section will present various examples to illustrate how to use the combinators introduced to model real-time problems. Moreover, new combinators shall be motivated.

3.2.1 Concurrency and interleaving

For clarity, NIL shall be omitted from agent expressions when it is understood from the context. A classical example in CCS is that $a|b$ is equivalent to $a.b + b.a$. This says that CCS models concurrency by interleaving. Later we shall show these two agents are also equivalent in the timed semantics. Let us consider $\epsilon(1).\alpha|\epsilon(1).\beta$ and $\epsilon(1).\alpha.\epsilon(1).\beta + \epsilon(1).\beta.\epsilon(1).\alpha$.

$\epsilon(1).\alpha|\epsilon(1).\beta$ may be understood as follows. If the environment choses α first, the α-experiment takes 1 hour and then the β-experiment can be successful directly because the composite agent has started preparing for both α and β one hour ago. On the other hand, if the environment choses β first, the β-experiment takes 1 hour and then the α-experiment will take no time. These may be expressed formally.

1. $\epsilon(1).\alpha|\epsilon(1).\beta \xrightarrow{\epsilon(1)} \alpha|\beta \xrightarrow{\alpha} \text{NIL}|\beta \xrightarrow{\beta} \text{NIL}$

2. $\epsilon(1).\alpha|\epsilon(1).\beta \xrightarrow{\epsilon(1)} \alpha|\beta \xrightarrow{\beta} \alpha|\text{NIL} \xrightarrow{\alpha} \text{NIL}$

Clearly, the above transitions are impossible for $\epsilon(1).\alpha.\epsilon(1).\beta + \epsilon(1).\beta.\epsilon(1).\alpha$. After the first experiment which takes one unit of time, the second one also takes one unit of time. We have the following transitions.

1. $\epsilon(1).\alpha.\epsilon(1).\beta + \epsilon(1).\beta.\epsilon(1).\alpha \xrightarrow{\epsilon(1)} \alpha.\epsilon(1).\beta + \beta.\epsilon(1).\alpha \xrightarrow{\alpha} \epsilon(1).\beta \xrightarrow{\epsilon(1)} \beta \xrightarrow{\beta} \text{NIL}$

2. $\epsilon(1).\alpha.\epsilon(1).\beta + \epsilon(1).\beta.\epsilon(1).\alpha \xrightarrow{\epsilon(1)} \alpha.\epsilon(1).\beta + \beta.\epsilon(1).\alpha \xrightarrow{\beta} \epsilon(1).\alpha \xrightarrow{\epsilon(1)} \alpha \xrightarrow{\alpha} \text{NIL}$

Note that the following transitions are also possible for the composite agent.

1. $\epsilon(1).\alpha|\epsilon(1).\beta \xrightarrow{\epsilon(1)} \alpha|\beta \xrightarrow{\alpha} \text{NIL}|\beta \xrightarrow{\epsilon(t)} \text{NIL}|\beta \xrightarrow{\beta} \text{NIL}$

2. $\epsilon(1).\alpha|\epsilon(1).\beta \xrightarrow{\epsilon(1)} \alpha|\beta \xrightarrow{\beta} \alpha|\mathrm{NIL} \xrightarrow{\epsilon(t)} \alpha|\mathrm{NIL} \xrightarrow{\alpha} \mathrm{NIL}$

After the first experiment, the composite agent can wait for arbitrary t units of time to synchronize with the environment or observer.

The two agents have different real-time behaviour though they have the same communication capability. Hence they cannot be identified in a real-time environment. In fact, $\epsilon(1).\alpha|\epsilon(1).\beta$ is equivalent to $\epsilon(1).\alpha.\beta + \epsilon(1).\beta.\alpha$, that shall be shown later.

We shall develop an expansion theorem for parallel composition by that, each parallel composed agent can be proved to be equivalent to a regular agent. So concurrency in the timed model is modelled by interleaving as in CCS.

3.2.2 Timer and Timeout

A simple timer has been given previously. The problem with that one is that after setting a time period, one can never reset it. An alarm timer is often required to be changeable.

Example 3.1 *an alarm timer*

$$\begin{aligned} T_0 &\stackrel{def}{=} time?t.T_1(t) \\ T_1(t) &\stackrel{def}{=} \epsilon(t).T_2 + time?u.T_1(u) \\ T_2 &\stackrel{def}{=} timeout!.T_0 + time?u.T_1(u) \end{aligned}$$

Now the user can always start a new timeout period.

Example 3.2 *a distributed implementation of the vending machine*

Timeout has been modelled by τ-action when describing the vending machine. Actually, the implementation of such a machine is by using a timer. The timer is a component of it. The rest of the machine is a drink-cooker as follows.

$$\begin{aligned} C_0 &\stackrel{def}{=} money?.time!30.C_1 \\ C_1 &\stackrel{def}{=} \epsilon(2).coffee!.C_0 + \epsilon(3).tea!.C_0 + timeout?.C_0 \end{aligned}$$

This component is responsible for *collecting money, starting the timer, drink-cooking and dealing with timeout.* A vending machine may be constructed by a drink-cooker and a timer through parallel composition. We have a distributed implementation, $C_0|T_0$ for the machine M_0 given before.

As usuall, the machine is capable of accepting a coin at any time and then it will start the timer. So after receiving a coin, the whole machine will perform τ immediately. It is the internal event of starting the timer. Formally,

1. $C_0|T_0 \xrightarrow{\epsilon(t)} C_0|T_0 \ (t > 0)$

2. $C_1|T_0 \xrightarrow{money?} time!30.C_1|T_0 \xrightarrow{\tau} C_1|T_1(30)$

Thereafter, the machine starts to prepare coffee and tea; the timer is counting in parallel with the drink-cooking. It will take 2 seconds for coffee and 3 seconds for tea. When 30 seconds passes away, the timer times out. Then the machine turns back to its initial state to collect money.

Let $C_1(t) \equiv \epsilon(2-t).coffee!.C_0 + \epsilon(3-t).tea!.C_0 + timeout?.C_0$. Then

1. $C_1|T_1(30) \xrightarrow{\epsilon(t)} C_1(t)|T_1(30-t)\ (0 < t \leq 30)$

2. $C_1(t)|T_1(30-t) \xrightarrow{coffee!} C_0|T_1(30-t)\ (2 \leq t \leq 30)$

3. $C_1(t)|T_1(30-t) \xrightarrow{tea} C_0|T_1(30-t)\ (3 \leq t \leq 30)$

4. $C_1(30)|T_2 \xrightarrow{\tau} C_0|T_0$

Observe that anyone can reset the timer such that it can never time out. One can also respond to the time-out signal externally. This is because that channels *time* and *timeout* are controllable by the environment. We may like to hide them so that nobody can access them externally except the machine itself. Let L be a subset of names i.e. $L \subseteq \Lambda$.

$$\frac{P \xrightarrow{\sigma} P'}{P\backslash L \xrightarrow{\sigma} P'\backslash L} \quad (\sigma, \bar{\sigma} \notin L)$$

Now we get a rather realistic implementation of the vending machine.

$(C_0|T_0)\backslash L$ where $L = \{time?t \mid t \geq 0\} \cup \{timeout!\}$

It is easy to check that this more complicated machine has the desired behaviour. In fact, this composite agent has the same behaviour as M_0, (see example 2.4 for M_0) when ignoring τ. Compared with $(C_0|T_0)\backslash L$, M_0 is a more abstract description that shall be viewed as a specification; whereas the complex agent here is much closer to a real implementation of such a vending machine. Later we shall show that they are indeed equivalent in a rigorous sense (weakly bisimulation equivalent).

So far, we have only considered the case when two regular agents working in parallel. This can be easily extended to the more general case that allows several such agents to cooperate together. Actually all one needs to do for this is to add one more equation, $Sort_t(P|Q) = Sort_t(P) \cup Sort_t(Q)$ to definition 3.3.

3.2.3 The WatchDog

This example has been considered by many authors in the literature [KK88, RH89] on real-time systems. The following shows how to model the behaviour of such a timer using timed agents.

A WatchDog is a process whose task is to watch a number of individual processes $P_1 \ldots P_n$ working properly. Each P_i is supposed to send an ok_i signal every t units of time to inform the WatchDog that it is functioning normally. The WatchDog is ready to accept the signal at any time. But if it does not receive the signal from any P_i within d $(d > t)$, it will send out a warning signal *alarm* immediately. Assume that each P_i behaves as follows:

$$P_i \stackrel{def}{=} \epsilon(t).\tau.ok_i!.P_i + \epsilon(t).\tau.fail_i!.\text{NIL}$$

During the time delay t, P_i is doing its job, but may be developing in different ways. In every t units of time if it proceeds on the right way, it will send a signal ok_i, where i indicates the signal is from P_i. But if something goes wrong, it will turn on the lamp $fail_i$ instead.

The WatchDog is implemented by parallel composition of $W_1 \ldots W_n$ where

$$W_i \stackrel{def}{=} ok_i?.W_i + \epsilon(d).\tau.alarm!.W_i$$

W_i is monitoring P_i. If ok_i arrives within d units of time, it keeps watching for the next time period. Otherwise it starts the alarm when d elapses. Then the whole system is $W_1|...|W_n|P_1|...|P_n$.

It is obvious that all W_i have similar structures but different channel names. One may like to make copies of such agents by changing the channel names. This is the task of the relabelling operator. Let S be a bijective function from $\mathcal{A}ct$ to $\mathcal{A}ct$, that satisfies the following conditions, $S(\epsilon(t)) = \epsilon(t)$ for all t, $S(\tau) = \tau$ and $S(\bar{\alpha}) = \overline{S(\alpha)}$.

$$\frac{P \xrightarrow{\sigma} P'}{P[S] \xrightarrow{S(\sigma)} P'[S]}$$

The semantics for relabelling is the same as in CCS.

4 The Calculus: Timed CCS

The preceding sections have presented the main part of the calculus. This section may serve as a summary.

4.1 Syntax

We have introduced a time domain $\mathcal{T}$ that is the possitive real numbers including 0 and a set of actions $\mathcal{A}ct = \mathcal{L} \cup \{\tau\} \cup \delta_{\mathcal{T}}$, where $\mathcal{L} = \Lambda \cup \overline{\Lambda}$ and $\delta_{\mathcal{T}} = \{\epsilon(t) | t > 0\}$. Recall that σ, θ range over $\mathcal{A}ct$, α, β over $\mathcal{L}$ and μ, ν over $\mathcal{L} \cup \{\tau\}$. Furthermore, let L be a subset of Λ and S be a bijection from $\mathcal{A}ct$ to $\mathcal{A}ct$ with $S(\epsilon(t)) = \epsilon(t)$, $S(\tau) = \tau$ and $S(\bar{\alpha}) = \overline{S(\alpha)}$. Such a S is also called a relabelling.

Let E, F be meta-variables ranging over agent expressions that may contain free variables. $\tilde{E}$ will denote a vector $\{E_i\}_{i \in I}$ of expressions. Whenever it is understood from the context, the index $i \in I$ shall be omitted and E_i will denote the ith element of $\tilde{E}$. Then the language has the following syntax.

$$E ::= \mathrm{NIL} \mid X \mid \sigma.E \mid E + F \mid E|F \mid E \backslash L \mid E[S] \mid \mathbf{rec}_i \tilde{X} : \tilde{E}$$

Note that this is precisely the syntax of CCS. But the set of prefix operators is enlarged by the set $\delta_{\mathcal{T}}$ of time events. As usual, $\mathbf{rec}_i \tilde{X} : \tilde{E}$ stands for the ith component of the mutually recursive definition $\{X_i \stackrel{def}{=} E_i\}_{i \in I}$. When the index set I is a singleton, $\mathbf{rec}_i \tilde{X} : \tilde{E}$ is abbreviated by $\mathbf{rec} X.E$. However we shall often use the more readable notation $X \stackrel{def}{=} E$ instead. Moreover $\sum_{i \in I} P_i$ will stand for $P_1 + ... + P_n$ when $I = \{1...n\}$ and $\epsilon(0).P$ for P. As usual, $E(\tilde{P}/\tilde{X})$ will denote the expression where every free occurrence of X_i in E has been replaced by P_i. Closed expressions are defined with $\mathbf{rec}_i$ binding variables and called agents. $\mathcal{P}$ will stand for the set of agents. P, Q, R shall range over $\mathcal{P}$.

4.2 Semantics

First, we assign a timed sort for each agent based on its syntactic structure.

Definition 4.1 *Let $t, u > 0$ and $P \in \mathcal{P}$. $Sort_t(P)$ is defined as the least set satisfying the following equations:*

$$\begin{array}{rcl}
Sort_t(NIL) & = & \emptyset \\
Sort_t(\alpha.Q) & = & \{\alpha\} \\
Sort_t(\tau.Q) & = & \emptyset \\
Sort_t(\epsilon(t).Q) & = & \emptyset \\
Sort_t(\epsilon(t+u).Q) & = & \emptyset \\
Sort_{t+u}(\epsilon(t).Q) & = & Sort_u(Q) \\
Sort_t(Q+R) & = & Sort_t(Q) \cup Sort_t(R) \\
Sort_t(Q|R) & = & Sort_t(Q) \cup Sort_t(R) \\
Sort_t(Q \backslash L) & = & Sort_t(Q) - L \\
Sort_t(Q[S]) & = & \{S(\alpha) | \alpha \in Sort_t(Q)\} \\
Sort_t(\mathbf{rec}_i \tilde{X} : \tilde{E}) & = & Sort_t(E_i(\mathbf{rec}\tilde{X} : \tilde{E}/\tilde{X}))
\end{array}$$

The inference rules for agents are summarized in table 1. Note that $t, u > 0$, $\alpha \in \mathcal{L}, \mu \in \mathcal{L} \cup \{\tau\}$ and $\sigma \in \mathcal{A}ct = \mathcal{L} \cup \{\tau\} \cup \delta_T$.

Let $\xrightarrow{\sigma}$ for each $\sigma \in \mathcal{A}ct$ be the least relation over $\mathcal{P}$, defined by the inference rules. This yields a labelled transition system, $\langle \mathcal{P}, \mathcal{A}ct, \longrightarrow \rangle$. As desired, we have the following.

Proposition 4.1 *All agents in $\mathcal{P}$ satisfy lemma* 1, 2, 3 *and* 4.

5 Bisimulation and Equivalences

To complete the calculus, this section will establish equivalence relations between agents.

5.1 Bisimulation and Strong Equivalence

One may like to identify two agents that cannot be distinguished by their observer (environment) interacting with them. Obviously, this equality depends on the *capability* of an observer. For instance, an observer who cannot measure time will never tell the difference between two programs that terminate in 1 second and 1 year with the same result. This is a typical observer in CCS. We shall require that an observer is able to observe actions including controllable and non-controllable ones. Moreover, the observer is equipped with a clock and it can also record the time delay between events or observe the events $\epsilon(t)$. So the set of observations is $\mathcal{A}ct = \mathcal{L} \cup \{\tau\} \cup \delta_T$. The following are all standard.

Definition 5.1 *(Strong Bisimulation) A binary relation $\mathcal{S} \subseteq \mathcal{P} \times \mathcal{P}$ is a strong bisimulation if $(P, Q) \in \mathcal{S}$ implies for all $\sigma \in \mathcal{A}ct$,*

1. $\forall P' : P \xrightarrow{\sigma} P' \supset \exists Q' : Q \xrightarrow{\sigma} Q' \;\&\; (P', Q') \in \mathcal{S}$
2. $\forall Q' : Q \xrightarrow{\sigma} Q' \supset \exists P' : P \xrightarrow{\sigma} P' \;\&\; (P', Q') \in \mathcal{S}$

Definition 5.2 *(Strong Equivalence) We say P and Q are strongly equivalent denoted $P \sim Q$ iff there exists a strong bisimulation $\mathcal{R}$ such that $(P, Q) \in \mathcal{R}$.*

As usual $\sim$ has the following properties.

Proposition 5.1

1. *$\sim$ is an equivalence relation.*

Constant	$\dfrac{}{\mathrm{NIL} \xrightarrow{\epsilon(t)} \mathrm{NIL}}$	
Prefix	$\dfrac{}{\alpha.P \xrightarrow{\alpha} P}$	$\dfrac{}{\alpha.P \xrightarrow{\epsilon(t)} \alpha.P}$
	$\dfrac{}{\tau.P \xrightarrow{\tau} P}$	
	$\dfrac{}{\epsilon(t+u).P \xrightarrow{\epsilon(t)} \epsilon(u).P}$	$\dfrac{}{\epsilon(u).P \xrightarrow{\epsilon(u)} P}$
	$\dfrac{P \xrightarrow{\epsilon(t)} P'}{\epsilon(u).P \xrightarrow{\epsilon(t+u)} P'}$	
Summation	$\dfrac{P \xrightarrow{\mu} P'}{P + Q \xrightarrow{\mu} P'}$	$\dfrac{P \xrightarrow{\mu} P'}{P + Q \xrightarrow{\mu} Q'}$
	$\dfrac{P \xrightarrow{\epsilon(t)} P' \qquad Q \xrightarrow{\epsilon(t)} Q'}{P + Q \xrightarrow{\epsilon(t)} P' + Q'}$	
Composition	$\dfrac{P \xrightarrow{\mu} P'}{P\|Q \xrightarrow{\mu} P'\|Q}$	$\dfrac{Q \xrightarrow{\mu} Q'}{P\|Q \xrightarrow{\mu} P\|Q'}$
	$\dfrac{P \xrightarrow{\alpha} P' \qquad Q \xrightarrow{\bar{\alpha}} Q'}{P\|Q \xrightarrow{\tau} P'\|Q'}$	
	$\dfrac{P \xrightarrow{\epsilon(t)} P' \qquad Q \xrightarrow{\epsilon(t)} Q'}{P\|Q \xrightarrow{\epsilon(t)} P'\|Q'}$	$(Sort_t(P) \cap \overline{Sort_t(Q)} = \emptyset)$
Restriction	$\dfrac{P \xrightarrow{\sigma} P'}{P \backslash L \xrightarrow{\sigma} P' \backslash L} \quad (\sigma, \bar{\sigma} \notin L)$	
Relabelling	$\dfrac{P \xrightarrow{\sigma} P'}{P[S] \xrightarrow{S(\sigma)} P'[S]}$	
Recursion	$\dfrac{E_i(\mathbf{rec}\tilde{X} : \tilde{E}/\tilde{X}) \xrightarrow{\sigma} P}{\mathbf{rec}_i\tilde{X} : \tilde{E} \xrightarrow{\sigma} P}$	

Table 1: Inference rules for TCCS

2. $\sim = \bigcup\{\mathcal{R} | \mathcal{R}$ *is a strong bisimulation*$\}$.

The later says that $\sim$ is the largest bisimulation. So to prove the equivalence of two agents, we just need to construct a strong bisimulation containing the pair of agents. This gives rise to a simple proof technique. We show this by example.

Example 5.1 : *modelling concurrency by interleaving*

$$\alpha|\beta \sim \alpha.\beta + \beta.\alpha$$

Let $\mathcal{R} = \{(\alpha|\beta, \alpha.\beta + \beta.\alpha)\} \cup \{(\mathrm{NIL}|P, P) | P \in \mathcal{P}\} \cup \{(P|\mathrm{NIL}, P) | P \in \mathcal{P}\}$. It is trivial to check $\mathcal{R}$ is a strong bisimulation. A simple variant of this example is that $\epsilon(1).\alpha|\epsilon(1).\beta \sim \epsilon(1).\alpha.\beta + \epsilon(1).\beta.\alpha$. We leave the proof to the readers.

5.2 Equational Laws

The proofs of the following laws are mostly by exhibiting bisimulations. For instance, $\{(\mathrm{NIL}, \epsilon(t).\mathrm{NIL}) | t \geq 0\}$ is a strong bisimulation. This gives us a simple law.

Proposition 5.2 $\epsilon(t).NIL \sim NIL$

The following are standard CCS-laws. They are also valid in the timed semantics.

Proposition 5.3

1. $P + Q \sim Q + P$
2. $(P + Q) + R \sim P + (Q + R)$
3. $P + P \sim P$
4. $P + NIL \sim P$

5. $P|Q \sim Q|P$
6. $(P|Q)|R \sim P|(Q|R)$
7. $P|NIL \sim P$

8. $(P + Q)\backslash L \sim Q\backslash L + P\backslash L$
9. $(\sigma.P)\backslash L \sim NIL$ $(\sigma \in L \cup \overline{L})$
10. $(\sigma.P)\backslash L \sim \sigma.(P\backslash L)$ $(\sigma \notin L \cup \overline{L})$
11. $NIL\backslash L \sim NIL$

12. $(P + Q)[S] \sim Q[S] + P[S]$
13. $(\sigma.P)[S] \sim S(\sigma).(P[S])$
14. $NIL[S] \sim NIL$

15. $A \sim P$ *when* $A \stackrel{def}{=} P$
16. $\mathbf{rec}_i\tilde{X} : \tilde{E} \sim E_i(\mathbf{rec}\tilde{X} : \tilde{E}/\tilde{X})$

Actually the above laws do not concern very much with timing. The following shall deal with the real-time properties of agents. Each of the four basic lemmas has its corresponding laws.

Proposition 5.4

1. (maximal progress) $\tau.P + \epsilon(t).Q \sim \tau.P$ $(t > 0)$

2. *(determinacy)*

   (a) $\epsilon(t).(P+Q) \sim \epsilon(t).P + \epsilon(t).Q$

   (b) $\epsilon(t).(P|Q) \sim \epsilon(t).P|\epsilon(t).Q$

3. *(continuity)* $\qquad \epsilon(t+u).P \sim \epsilon(t).\epsilon(u).P$

4. *(persistency)* $\qquad \alpha.P + \epsilon(t).\alpha.P \sim \alpha.P$

Proposition 5.5 *(expansion theorem) Let* $P \equiv \sum_{i\in I} \epsilon(t_i).\mu_i.P_i$ *and* $Q \equiv \sum_{j\in J} \epsilon(u_j).\nu_j.Q_j$. *Then*

$$\begin{aligned} P|Q \quad \sim \quad & \textstyle\sum_{\mu_i,\nu_j \in \mathcal{L} \& \mu_i = \bar{\nu}_j} \epsilon(max(t_i,u_j)).\tau.(P_i|Q_j) \\ & + \textstyle\sum_{i\in I} \epsilon(t_i).\mu_i.(P_i|(\sum_{j\in J} \epsilon(u_j - t_i).\nu_j.Q_j)) \\ & + \textstyle\sum_{j\in J} \epsilon(u_j).\nu_j.((\sum_{i\in I} \epsilon(t_i - u_j).\mu_i.P_i)|Q_j) \end{aligned}$$

Remark: Note first that $\epsilon(0).P \equiv P$ and $t - t' = 0$ when $t' \geq t$. The expansion is static in the sense that it ignores the possibility that some of the summands in the right-hand side may be eliminated directly. However the sum may be further reduced by the maximal progress law. □

We have established a number of laws. There are many more. For instance, from $P \sim Q$, we can infer $P + R \sim Q + R$ and $P|R \sim Q|R$. In fact we have a general result that $\sim$ is a congruence.

Now we extend $\sim$ to open expressions in the standard way. Assume E and F include variables $\tilde{X}$ at most.

Definition 5.3 $E \sim F$ *iff for all agents* $\tilde{P}$, $E(\tilde{P}/\tilde{X}) \sim F(\tilde{P}/\tilde{X})$

This leads to the following.

Theorem 5.1 $\sim$ *is a congruence.*

5.3 Bisimulation and Weak Equivalence

In defining strong bisimulation equivalence, it is supposed that the observer is able to observe all actions include controllable and non-controllable ones and also capable of measuring time. Now let us deny the observer the ability of observing non-controllable actions. Then question arises: to what extent, can τ be ignored? We shall define an equivalence of agents in terms of controllable actions and time delays between them; whereas τ will be treated as *idling*.

We introduce a set $\mathcal{E} = \mathcal{L} \cup \Delta_{\mathcal{T}}$ of experiments where $\mathcal{L}$ is the set of controllable actions and $\Delta_{\mathcal{T}} = \{\varepsilon(t) | t \geq 0\}$. Each $\varepsilon(t)$ of $\Delta_{\mathcal{T}}$ is an empty experiment, that lasts for t units of time. ξ shall range over $\mathcal{E}$.

Definition 5.4 *(Experiment Relation)*

1. $P \stackrel{\varepsilon(t)}{\Longrightarrow} Q$ *iff there exist* $n \geq 0, P_0 ... P_n$ *such that*

   $P \stackrel{\varsigma_0}{\longrightarrow} P_0 ... P_n \stackrel{\varsigma_n}{\longrightarrow} Q$ *where* $\varsigma_i \in \{\tau\} \cup \delta_{\mathcal{T}}$ *and* $t = \sum_{\varsigma_i = \epsilon(t_i)} t_i$.

2. $P \stackrel{\alpha}{\Longrightarrow} Q$ *iff there exist* $n \geq 1, P_0 ... P_n$ *such that*

   $P \stackrel{\tau}{\longrightarrow} P_0 ... P_m \stackrel{\tau}{\longrightarrow} P_m \stackrel{\alpha}{\longrightarrow} P_{m+1} \stackrel{\tau}{\longrightarrow} P_{m+2} ... P_n \stackrel{\tau}{\longrightarrow} Q$

This gives rise to a labelled transition system $\langle \mathcal{P}, \mathcal{E}, \Longrightarrow \rangle$. Naturally we can formalize bisimulation in terms of this system.

Definition 5.5 *(Weak Bisimulation) A binary relation $\mathcal{W} \subseteq \mathcal{P} \times \mathcal{P}$ is a weak bisimulation if $(P,Q) \in \mathcal{W}$ implies for all $\xi \in \mathcal{E}$,*

1. $\forall P' : P \stackrel{\xi}{\Longrightarrow} P' \supset \exists Q' : Q \stackrel{\xi}{\Longrightarrow} Q' \;\&\; (P',Q') \in \mathcal{W}$
2. $\forall Q' : Q \stackrel{\xi}{\Longrightarrow} Q' \supset \exists P' : P \stackrel{\xi}{\Longrightarrow} P' \;\&\; (P',Q') \in \mathcal{W}$

Following [M89, M83] of Milner, we have another characterization of weak bisimulation, that is more convenient to work with.

Definition 5.6 *Let $\sigma \in \mathcal{A}ct$, we define $P \stackrel{\hat{\sigma}}{\Longrightarrow} P'$ as follows.*

1. $P \stackrel{\hat{\epsilon}(t)}{\Longrightarrow} P'$ *iff* $P \stackrel{\epsilon(t)}{\Longrightarrow} P'$
2. $P \stackrel{\hat{\tau}}{\Longrightarrow} P'$ *iff* $P \stackrel{\epsilon(0)}{\Longrightarrow} P'$
3. $P \stackrel{\hat{\alpha}}{\Longrightarrow} P'$ *iff* $P \stackrel{\alpha}{\Longrightarrow} P'$

Proposition 5.6 *$\mathcal{R}$ is a bisimulation iff $(P,Q) \in \mathcal{R}$ implies that for all $\sigma \in \mathcal{A}ct$,*

1. $\forall P' : P \stackrel{\sigma}{\longrightarrow} P' \supset \exists Q' : Q \stackrel{\hat{\sigma}}{\Longrightarrow} Q' \;\&\; (P',Q') \in \mathcal{R}$
2. $\forall Q' : Q \stackrel{\sigma}{\longrightarrow} Q' \supset \exists P' : P \stackrel{\hat{\sigma}}{\Longrightarrow} P' \;\&\; (P',Q') \in \mathcal{R}$

Definition 5.7 *(Weak Equivalence) We say that P and Q are* weakly equivalent, *written $P \approx Q$ iff there exists a weak bisimulation $\mathcal{W}$ such that $(P,Q) \in \mathcal{W}$.*

$\approx$ also possesses the following properties.

Proposition 5.7

1. *$\approx$ is an equivalence relation.*
2. $\approx = \bigcup\{\mathcal{W} | \mathcal{W}$ *is a weak bisimulation*$\}$.

As usuall, the equivalence between agents can be proved by constructing a weak bisimulation. We illustrate this by example.

Example 5.2 *proving the vending machine correct*

Recall that two vending-machines have been defined previously M_0 and $(C_0|T_0)\backslash L$ where $L = \{time?d \mid d \in \mathcal{T}\} \cup \{timeout\}$. See example 2.4 for M_0 and example 3.1, 3.2 for T_0, C_0. Now we want to establish that the distributed version of the machine implements the centralized one. This is to show that the two machines are weakly equivalent,

In the following, $A||B$ shall stand for for $(A \mid B)\backslash \{time?d \mid d \in \mathcal{T}\} \cup \{ timeout \}$. We need to prove $C_0 \mid\mid T_0 \approx M_0$. Let

$$\begin{aligned} M_1(t) &\equiv \epsilon(2-t).coffee!.M_0 + \epsilon(3-t).tea!.M_0 + \epsilon(30-t).\tau.M_0 \\ C_1(t) &\equiv \epsilon(2-t).coffee!.M_0 + \epsilon(3-t).tea!.M_0 + timeout?.M_0 \end{aligned}$$

and

$$
\begin{aligned}
S_1 &\equiv \{(C_0 \,||\, T_0, M_0), \\
&\quad\ (time!30.C_1 \,||\, T_0, M_1)\} \\
S_2 &\equiv \{(C_1(t) \,||\, T_1(30-t), M_1(t)) \mid 0 \leq t \leq 30\}
\end{aligned}
$$

It is easy to check that $S_1 \cup S_2$ is a weak bisimulation. Therefore $C_0||T \approx M_0$.

In terms of $\sim$, we have established a sequence of laws. They are also valid for $\approx$ since $\approx$ is larger than $\sim$.

Proposition 5.8 *$\sim\subseteq\approx$ or $P \sim Q$ implies $P \approx Q$.*

In addition, we have the following τ-laws. They are also standard CCS-laws.

Proposition 5.9

1. $\tau.P \approx P$
2. $\tau.P + P \approx \tau.P$
3. $\sigma.(P + \tau.Q) \approx \sigma.(P + \tau.Q) + \sigma.Q$

Unfortunately $\approx$ is not a congruence. This can be shown by a counter example. It is easy to establish $\epsilon(t).\tau.\text{NIL} \approx \epsilon(t).\text{NIL}$. But $\epsilon(t).\tau.\text{NIL} + \alpha.\text{NIL} \not\approx \epsilon(t).\text{NIL} + \alpha.\text{NIL}$. Obviously, the right hand-side $\epsilon(t).\text{NIL} + \alpha.\text{NIL}$ is equivalent to $\alpha.\text{NIL}$. It can always do α after t. But the left hand side $\epsilon(t).\tau.\text{NIL} + \alpha.\text{NIL}$ will perform τ immediately after t units of time and become NIL. This example shows that $\approx$ is not preserved by summation. However we have the following.

Theorem 5.2 *$\approx$ is preserved by all operators except summation and recursion.*

6 Acknowledgements

I would like to thank Uno Holmer, Sören Holmström, K.V.S. Prasad and Bjön von Sydow for the many helpful discussions on this work.

References

[C82] Cardelli L. *Real Time Agents*, LNCS, 140, 1982.

[KK88] Koymans R. and Kuiper R. *Paradigms for Real-time Systems*, LNCS, 331, 1988.

[M80] Milner R. *A calculus of communicating systems*, LNCS 92, 1980.

[M83] Milner R. *Calculi for Synchrony and Asynchrony*, TCS, Vol 25, 1983.

[M86] Milner R. *Process Constructors and Interpretations*, Information Processing 86, H,J. Kugler (ed), Elsevier Science Publishers (Noth-Holland), 1986.

[M89] Milner R. *Communication and Concurrency*, Prentice Hall International Series in Computer Science, 1988.

[MT89] Moller F. and Tofte C. *A Temperal Calculus of Communicating Systems*, LFCS report, University of Edingburgh, December 1989.

[RH89] de Roever W.P. and Hooman J.J.M. *Design and Verification of Real-time Distributed Computing: an Introduction to Compositional Methods*, Proceeding of the 9th IFIP WG 6.1 International Symposium on Protocol Specification, Testing and Verification, 1989.

[RR86] Reed G.M. and Roscoe A.W. *A Timed Model for Communicating Sequential Processes*, LNCS 226, 1986.

[S90] Sifakis J. et al. *ATP: an Algebra for Timed Processes*, Laboratoire de Genie Informatique, IMAG-Campus, B.P.53X, 38041 Grenoble Cedex, France, 1990.

[T89] Tofte C. *Timing Concurrent Processes*, LFCS report, University of Edingburgh, December 1989.

[W90] Wang Y. *A Calculus of Real-Time Communicating Systems* (in preparation), Dept. of Computer Sciences, Chalmers University, Sweden, 1990.

EFFECTIVE SOLUTIONS TO DOMAIN EQUATIONS
AN APPROACH TO EFFECTIVE DENOTATIONAL SEMANTICS

(S.YOCCOZ , CNRS, Université de Bordeaux-1)

Abstract : This paper gives an effective interpretation to the construction of solutions for domain equations in a category of complete metric spaces . Our starting point is the work by P.America and J.Rutten , which we summarizes briefly . We show that their construction can be given an effective content (in the sense that the solutions are recursively presented spaces). We indicate, in the last section , how this approach can be used to estimate the complexity of the operators induced by contexts in some parallel languages .

Address : S.Yoccoz , LABRI, UER Maths-Info, 351 Cours de la Liberation
33400 TALENCE (FRANCE)

E-mail : yoccoz@geocub.greco-prog.fr

0 Introduction

The aim of this paper is to provide mathematical grounds for the development of effective denotational semantics of sequential and parallel languages in the framework of metric spaces . Our starting point was the work done by P.America, J.de Bakker and J.Rutten [AR, A et al.] on the solutions of domain equations in a category of complete metric spaces . These spaces can be a good ground for the development of notions of effectiveness , provided they are "well-behaved" (i.e recursively presented) . We prove indeed that , within some modifications, the solutions to domain equations are, in most cases, recursively presented, giving then an effective content to the denotational semantics they carry . A study of the operators induced by contexts on the domains can shed some light on the complexity of the operators of the language, and suggest some modifications in their operational semantics . Due to the uniformity of the framework we work in , one can furthermore hope to use such an approach to build a good description of the relative complexity of sequential and parallel languages [Y1,Y2]
The remaining of the paper is organized as follows : Section 1 is a brief summary of [AR] ; Section 2 and 3 contain a presentation of the mathematical framework we are working in . Section 4 presents two "case analyses" (CCS with guarded recursion and POOL) and gives a description of the operators induced by contexts in the case of CCS.

Most of the proofs do not appear here (for lack of space) ; they appear in [Y1] , together with further developments.

1 Solutions of domain equations over a category of complete metric spaces

The idea of [AR] is to solve domain equations not (as it is done usually) in a category of ordered structures, but in a category of complete metric spaces . This aim is reached in two steps :

(i) Let C be the category of complete metric spaces, equipped with the following morphisms :

$$(A, d) \xrightarrow{\iota} (B, d') \text{ is a morphism if } \iota = \langle i,j \rangle \text{, where}$$

* $i : A \to B$ is an isometric embedding , $j : B \to A$ is a non distance increasing map
* $j \circ i = id_A$

(intuitively, A is a space of approximants of elements of B)
Let us denote by $\delta(\iota)$ the distance $d'_{B\to B}(i \circ j , id_B) = \max_{y\in B} d(i(j(y)) , y)$
(δ is a measure of the precision of the approximation).

Definition : An ω-chain

$$(D_0 ,d_0) \xrightarrow{\iota_0} (D_1, d_1) \xrightarrow{\iota_1} (D_2 ,d_2) \ldots$$

is a converging tower if

$(\forall \varepsilon > 0)(\exists N)(\forall N \leq n < m) \quad \delta(\iota_{nm}) \leq \varepsilon$,

where ι_{nm} is $\iota_{m-1} \circ \cdots \circ \iota_n$ (for later use , $\iota_{nm} = < i_{nm}, j_{mn} >$).

The following theorem is proved in [AR] :

Theorem 1: Let $(D_0, d_0) \xrightarrow{\iota_0} (D_1, d_1) \xrightarrow{\iota_1} (D_2, d_2) \ldots$ be a converging tower .

Then it has a colimit (direct limit) ((D,d) , γ_n) which is given by :

$D = \{ (x_n)_n \ / \ \forall n \geq 0 \ (x_n \in D_n \ \ \& \ x_n = j_n(x_{n+1}) \}$

$d((x_n)_n , (y_k)_k) = \sup_{n \in N} \{ d_n(x_n, y_n) \}$

and $\gamma_n = < \alpha_n, \beta_n >$ where $\alpha_n : D_n \rightarrow D$ is given by

$$
\begin{aligned}
(\alpha_n(x))_k &= j_{nk}(x) \text{ if } k < n \\
&= x \qquad \text{ if } k = n \\
&= i_{nk}(x) \text{ if } k > n
\end{aligned}
$$

and $\beta_n : D \rightarrow D_n$ is simply the projection $(x_n)_n \rightarrow x_n$.

(This construction is exactly similar to the contruction of Scott domains)

Following the work done in categories of ordered sets, the authors define a contracting functor to be a functor $F : \mathbf{C} \longrightarrow \mathbf{C}$ such that , for some $0 \leq \varepsilon < 1$, and for all $D \xrightarrow{\iota} D'$,

$$\delta(F\iota) \leq \varepsilon \,.\, \delta(\iota) \,.$$

They prove that any contracting functor has a fixed point , using the facts that it preserves the colimit of converging towers , and that the one-point space {p} is a "weak" initial object in **C** (for any complete space X, there is always a map from {p} to X in **C**).

Unicity of fixpoints is then obtained if one replaces contracting functors by strictly contracting functors : a strictly contracting functor is a contracting functor F verifying :

$\forall$ (P,d) $\in$ **C** , $\forall$ (Q,d') $\in$ **C** , $\exists \varepsilon < 1$, $\forall \iota, \iota' \in \mathbf{C}(P,Q)$,

$d_{FP \rightarrow FQ}(F\iota, F\iota') \leq \varepsilon \,.\, d_{P \rightarrow Q}(\iota, \iota')$, where , if $\iota = <i,j>$ and $\iota' = <i',j'>$

$d_{P \rightarrow Q}(\iota, \iota') = \max \{ \sup_{x \in P} (d'(i(x), i'(x))) \ , \ \sup_{y \in Q} (d(j(y) , j'(y))) \}$

(ii) The authors consider then a family Func of functors generated by the grammar :

$$F ::= F_M \ / \ id^{\varepsilon} \ / \ F_1 \rightarrow F_2 \ / \ F_1 \rightarrow^{\iota} F_2 \ / F_1 \cup F_2 \ / \ F_1 x \, F_2 / \ P_{cl}(F_1) / \ F_2 \circ F_1 \,.$$

where : * F_M is the constant functor of value (M,d) on the objects, and $<id_M, id_M>$ on the morphisms .

* id^{ε} ((M ,d)) = (M , min (1 , ε.d)) on the objects, and the identity on the morphims .

* $(F_1 \rightarrow F_2)(P)$ is the space of continuous functions $F_1P \rightarrow F_2P$ equipped with the usual max metric ($d(f,g) = \max_{x \in F_1P} (d_{F_2P}(f(x) ,g(x)))$)

If $P \rightarrow^{\iota} Q$ is a C-morphism, $F_k \iota = <i_k, j_k>$ then

$(F_1 \rightarrow F_2)(\iota) = < \lambda f. (i_2 \circ f \circ j_1), \lambda g . (j_2 \circ g \circ i_1) >$

* $(F_1 \rightarrow^1 F_2)$ is defined in the same way, except that $(F_1 \rightarrow^1 F_2)(P)$ is the space of Lipschitz functions of constant ≤ 1 .
* $(F_1 \cup F_2)(P)$ [resp. $(F_1 x\, F_2)(P)$] is the disjoint union [resp. product] of F_1P and F_2P ; its behaviour on distances and morphisms is the natural one .
* $(P_{cl}(F)(P)$ is the space of closed subsets of FP , equipped with the Hausdorff distance $d'(X,Y) = \max \{ \sup_{x\in X} (\inf_{y\in Y} d(x,y)) , \sup_{y\in Y} (\inf_{x\in X} d(x,y)) \}$

and given on morphisms by : if $P \rightarrow^1 Q$, $F\iota = <i,j>$,
$(P_{cl}(F)(\iota) = < \lambda X. \{i(x) / x \in X\} , \lambda Y. \text{Closure} (\{ j(y) / y \in Y \}) >$.

The authors associate to each functor F in Func a constant c(F) defined inductively by :

$c(F_M) = 0$ $\qquad$ $c(id^{\varepsilon}) = \varepsilon$

$c(F_1 \rightarrow F_2) = \max (\infty . c(F_1) ,c(F_2))$ (where $\infty . x = \infty$ except $\infty . 0 = 0$)

$c(F_1 \rightarrow^1 F_2) = c(F_1) + c(F_2)$ $\qquad$ $c(F_2 \circ F_1) = c(F_2).c(F_1)$

$c(F_1 \cup F_2)$, $c(F_1 x\, F_2) = \max (c(F_1),c(F_2))$

$c(P_{cl}(F)) = c(F)$.

and prove that every functor F in Func such that $c(F) < 1$ is strictly contracting . As a corollary, they can state that every domain equation of the form D = FD , with $F \in$ Func, and $c(F) < 1$ has unique solution in **C** , up to isomorphism .

2 Fixed points of effective functors of a category of complete metric spaces

We recall first two basic definitions from classical and effective descriptive set theory :

Definition : (1) A metric space (X,d) is separable (or of denumerable type) if there exist a denumerable dense subset of X (we will call such a set a dense basis) . It is a Polish space if , moreover, it is perfect, i.e there is no isolated points .

(2) A separable space (X,d) has a recursive presentation if there exists a denumerable set $B(X) = \{ r_0 , r_1 ,...... \}$ of points of X such that

* B(X) is dense in X
* the relations $L(i,j,k,m) \Longleftrightarrow d(r_i , r_j) \leq k /_{m+1}$ and $L'(i,j,k,m) \Longleftrightarrow d(r_i , r_j) < k /_{m+1}$ are recursive .

For later use, we call the structure (B(X) , L, L') a presentation of (X,d) .
Recursive presentations are the foundation of effective descriptive set theory, as they allow to extend the notion of computable functions and sets to almost any Polish space . Our main criterion for the effectivity of the construction of solutions to domain equations will then be the fact that these solutions do have , in many cases, a recursive presentation . On that basis, one can study computation problems within these solutions (see section 4) .

We prove first that converging towers have colimits in the full subcategory of separable complete spaces :

Proposition 1: Let T : $(D_0, d_0) \longrightarrow^{\iota_0} (D_1, d_1) \longrightarrow^{\iota_1} (D_2, d_2) \ldots$
be a converging tower, each (D_i, d_i) being separable . Then its colimit built in theorem 1 is itself separable .

Proof : Let B_i be a dense basis for D_i ; then the set
$\{ \mathbf{x} \in D \,/\, \exists i\, \exists x_i \in D_i\ \mathbf{x} = \alpha_i(x_i) \}$ is a dense basis for D .

We must define now an effective version of the contracting functors presented in section 1 :

Definition: A functor F : **C** → **C** is effective if :
(i) for any separable space X , FX is separable, and if X has a recursive presentation ,then FX has a recursive presentation (uniformly) computable in the presentation of X .
(ii) If $\iota = \langle i,j \rangle$ is a morphism between two separable spaces X and Y such that , for some dense bases B(X) , B(Y) : $B(Y) \supseteq i(B(X))$ and $B(X) \supseteq j(B(Y))$, then $[F\iota]$ is (uniformly) computable in $[\iota]$, where $[\iota]$ is the disjoint sum of
$\{ \langle n,m \rangle \,/\, i(a_n) = b_m,\ a_n \in B(X),\ b_m \in B(Y) \}$ and $\{ \langle p,q \rangle \,/\, j(b_p) = a_q \}$

Remarks : (1) We can always restrict our attention to the behaviour of F on separable spaces , since effectivity implies the existence of a denumerable "base" on which we can compute
(2) As far as the behaviour of functors on morphisms is concerned, we can restrict ourselves to "basis-preserving" morphisms : they are dense in the space of morphisms, and they are the only one that can be considered "computable" when they are defined between two recursively presented spaces .

We will call F effectively contracting if F is effective, and $\delta(F\iota) < \varepsilon \,.\, F(\iota)$ for all ι and for some recursive real $\varepsilon < 1$.

Proposition 2 : Let F be an effectively contracting functor from **C** to **C** . Then F has a fixed point (D,d) with a recursive presentation .

Corollary : Let F be an effective strictly contracting functor . Then the unique fixed point of F has a recursive presentation (uniformly computable in the presentation of F{p})

3 Reflexive domain equations : effectiveness of their solutions

The functors in the class Func built in section 1 are not necessarily effective (in particular, strong conditions have to be imposed on the arrow constructor $F_1 \rightarrow F_2$ to obtain an effective functor, see below) . We have then two solutions :

- either restrict the class Func to a class EFunc of effective functors on the category **C** ; this is sufficient for some domain equations (such as the equations for CCS) but does not work in other cases (POOL-like equations).
- or consider the restriction of Func to a full subcategory **CO** of **C** , and prove then that convergent towers have colimits in **CO**, and that every functor in Func is effective on **CO** .

We present these two approaches below :

a) Construction of the class Efunc :

Let us recall that Func is given by the grammar :

$$F ::= F_M \;/\; id^{\varepsilon} \;/\; F_1 \rightarrow F_2 \;/\; F_1 \rightarrow^1 F_2 \;/ F_1 \cup F_2 \;/\; F_1 \times F_2 / \; P_{cl}(F_1) \,/\; F_2 \circ F_1 \,.$$

We modify step by step this generating system in order to get Efunc ; we prove in each case that the effectivity of the functors is preserved . We will then be able to use the function c(F) built in Section 1 to get a family of domain equations with "effective" solutions. One will notice that this construction is somehow maximal (with respect to effectivity) within the class Func .

(i) F_M : We must restrict these functors to the case where M has a recursive presentation (this includes all "interesting" polish spaces, as well as all the denumerable sets equipped with a discrete metric) .

(ii) id^{ε} : no modification has to be done here , except that ε must be a recursive real: it must be clear that id^{ε} preserves dense sets, and that $id^{\varepsilon}(d)$ is recursive in d for any distance d .

(iii) $F_1 \longrightarrow F_2$: this is the most difficult case , as the space of continuous functions between two separable spaces X and Y is not separable if X is not compact .
The most natural restriction is the following :
* F_1 must be effectively compact , i.e :
- the image by F_1 of any separable space is a compact space .
- for any space P with a recursive presentation , for all n , one can (uniformly) compute a set $\{x_1 , ..., x_{p(n)}\}$ of elements of the dense basis of F_1P (this dense basis is given by F_1) such that F_1P is covered by the open balls $B(x_i , 1/n)$ .This is quite natural : an effective functional depends only on some finite , possibly arbitrary large information, hence the compacity ; the effectivity of the finite coverings is a consequence of the effectivity of F_1 .

* F_2 must verify : for any recursively presented space P , (F_2P,d') verifies the following equivalent conditions (which we will call the "successor property")
- all open balls are clopen balls
- the set V of values of d' (as a subset of **R**) verifies :

for any $x \in V, d \neq 0$, $\{y \in V / y \geq x\}$ is well-ordered by the usual order on **R** .

This is necessary if one wants to characterize "completely" continuous functions from F_1P to F_2P by their behaviour on some dense basis of F_1P . If this condition is not verified, one can check (see below) that $F_1P \to F_2P$ will have only a "semirecursive" presentation : the relation $d(f_i,f_j) \leq m/r+1$ will be recursive , but not the relation $d(f_i,f_j) < m/r+1$.

Proposition 3 : Let F_1 , F_2 be some effective functors, F_1 and F_2 as above . Then $F_1 \to F_2$ is effective .

Sketch of proof : (I) $F_1P \to F_2P$ has a recursive presentation :

We suppose that $F_1P = (X,d)$ with dense basis $\{a_n\}$ and $F_2P = (Y,d')$ with dense basis $\{b_n\}$. We can suppose that for any x, y in X , we have $d(x,y) \leq 1$. One has to remember that, since (X,d) is compact , any continuous function verifies the following : let ε be fixed , and let , for any x in X , $\eta(x)$ be such that

$$\forall y \ \ d(x,y) < 2\eta(x) \Rightarrow d(f(x),f(y)) < \varepsilon/2 .$$

$\{B(x,\eta(x)) / x \in X\}$ is obviously an open cover of X, from which we can extract a finite cover $(B(x_i , \eta(x_i))$; then :

for all x, y in X , $d(x,y) < \min \eta(x_i) \Rightarrow d(f(x), f(y)) < \varepsilon$.

A dense basis for $F_1 P \to F_2 P$ is built as follows (see e.g [B]) :

let G_{mn} be $\{ f / d(x,x') \leq 1/m \Rightarrow d'(f(x), f(x')) \leq 1/n \}$

Notice that G_{mn} is monotone in m, antimonotone in n and that the G_{mn} cover $F_1 P \to F_2 P$.

Let $\{a_{i_1}, ..., a_{i_{p(m)}}\}$ be such that the balls $B(a_{i_j} , 1/m)$ cover X , and let $\phi \in \mathbf{N}^{[1,p(m)]}$.

We define

$$H_\phi = \{ f / f \in G_{mn} \ \& \ d'(f(a_{j_k}) , b_{\phi(k)}) \leq 1/n \} \ \ (\text{the } H_\phi \text{ cover } G_{mn})$$

For any ϕ such that H_ϕ is non-empty, choose a distinguished element g_ϕ of H_ϕ . Then the collection of these g_ϕ's is dense in $F_1 P \to F_2 P$ (basically , for any f in $F_1 P \to F_2 P$, for any n, one can find a g_ϕ such that $d(f,g_\phi) \leq 4/n$). The problem is to find the good g_ϕ's .

(a) Let $\{g_\phi\}$ be an arbitrary collection of functions built as above . Then one can find a collection $\{h_\phi\}$ of functions verifying :

* for all n , for all g_ϕ , there exist an h_ϕ such that $d(g_\phi, h_\phi) \leq 1/n$
* the functions h_ϕ are sending the dense basis $\{a_n\}$ in the dense basis $\{b_n\}$

(this is immediate : we know then that the h_ϕ's form a dense basis of $F_1 P \rightarrow F_2 P$)

b) If (X,d) is compact, for all continuous f : X → Y sending dense basis to dense basis, and for all n, there exist a continuous function α such that :

* $d(f,\alpha) \leq 1/n$
* α is sending dense basis to dense basis and $\{<p,q> / \alpha(a_p) = b_q\}$ is recursive .

(c) Let $\alpha_{\kappa\lambda}$ be the functions built in (b) , with $f = h_\phi$. They form a dense denumerable basis for X →Y ; we can verify that the relations $d(\alpha_{\kappa\lambda}, \alpha_{\mu\nu}) \leq m/r{+}1$ and $d(\alpha_k, \alpha_l) < m/r{+}1$ are recursive using the fact that Y verifies the successor property .

(II) $F_1 \rightarrow F_2$ is effective on morphisms :
This is far more easier , and reduces to the problem of finding a code of the composition of recursive functions, knowing their respective codes .

(iv) $F_1 \rightarrow^1 F_2$: this case receives obviously the same treatment as $F_1 \rightarrow F_2$.
(v) $F_1 \cup F_2$, $F_1 \times F_2$, $F_1 \circ F_2$ are trivial cases, and suffer no restrictions .

(vi) $P_{cl}(F)$: the restrictions that we must put on F are similar to the ones in the functional case :

Proposition 4 : Let F be an effective and effectively compact functor . Then $P_{cl}(F)$ is effective.

Remark : When F is not compact, one can always use $P_{co}(F)(X)$, the space of compact subsets of F(X) , instead of $P_{cl}(F)(X)$: $P_{co}(F)$ is effective whenever F is .

We can summarize the construction above :

Theorem 3 : Let EFunc be the family of functors defined by :
(i) any constant functor F_M , with M recursively presented,and all the functors id^ε for a recursive real ε are in Efunc
(ii) Efunc is closed under union, product, composition of functors, and the compact subsets space formation (P_{co}) .
(iii) if F effectively compact and G with the successor property are in Efunc, then F→G, $F \rightarrow^1 G$ and $P_{cl}(F)$ are in Efunc .

Then any functor F in Efunc is effective . Moreover, if $c(F) < 1$, then the unique solution to the equation X = FX is recursively presented .

b) Restriction of Func to the full subcategory **CO** :

The full subcategory **CO** of **C** is not too difficult to define :
- all (recursively presented) objects in **CO** must be compact (otherwise the functors $F_1 \to F_2$ cannot be effective)
- all (recursively presented) objects in **CO** must verify the successor property (for the same reasons) .

It appears (see the remark below) that this is not totally sufficient , for the colimit in **C** of a convergent tower of spaces verifying the successor property does not necessarily verify the successor property ; we are then led to the following definition and theorem :

Theorem 4 : Let **CO** be the full subcategory of **C**, of which objects are the compact complete metric spaces (X,d) verifying $\{1/n \,/\, n \in \mathbf{N}\} \cup \{0\} = 1/\mathbf{N} \supseteq \mathrm{Range}(d)$. Then the following hold in **CO** :
(1) Any convergent tower T has a colimit
(2) Let Func' be the class of functors from **CO** to **CO** generated by the grammar
$F ::= F_M \,/\, id^{\varepsilon} \,/\, F_1 \to F_2 \,/\, F_1 \to^1 F_2 \,/\, F_1 \cup F_2 \,/\, F_1 x\, F_2 \,/\, P_{cl}(F)$,
with M recursively presented, and ε a recursive real . Then any functor F in Func' is effective, and any domain equation X = FX , with F in Func' and $c(F) < 1$ has a unique , recursively presented, solution in **CO** .

Proof : (1) We just need to prove that the colimit of a converging tower, all of which objects are in **CO** , is still in **CO** . Let T be a converging tower

$$(X_0, d_0) \to^{\iota_0} (X_1, d_1) \to \ldots$$

with $(\forall i)$ (X_i, d_i) is compact and $1/\mathbf{N} \supseteq \mathrm{Range}(d_i)$. Let (D,d) be its colimit in **C** .Then :
- (D,d) is compact : one can show that D is homeomorphic to a closed subset of $\Pi_{i \in N}(X_i, d_i)$ and use then Tychonoff's lemma .
- $1/\mathbf{N} \supseteq \mathrm{Range}(d)$: remember that for any two elements x and y of D ,
$d(x, y) = \max_n d_n(x_n, y_n)$; since $1/\mathbf{N}$ is discrete, this maximum is reached , so $1/\mathbf{N} \supseteq \mathrm{Range}(d)$.

(2) Any functor in Func' is effective : this is simply due to the fact that Func' is included in the restriction of the class EFunc to the category **CO** .

Remark : The restriction $\{1/n \,/\, n \in \mathbf{N}\} \cup \{0\} = 1/\mathbf{N} \supseteq \mathrm{Range}(d)$ can seem rather strange ; one can check however that the colimit of a converging tower of spaces verifying the successor property does not necessarily verify the successor property : let
$(X_n, d_n) = (\underline{\mathbf{N}} \cup \{a_0, \ldots, a_n\}, d_n)$, with d_n equal to $\underline{\mathbf{d}}$ on $\underline{\mathbf{N}}$ and $d_n(x, a_n) = 1 + 1/n$;

let $\iota_n = (i_n, j_n)$ be such that i_n is just the inclusion of X_n in X_{n+1} and $j_n(a_{n+1}) = a_n$. Then The (X_n,d_n) form a converging sequence of compact spaces verifying the successor property, but the colimit (D,d) of this tower verifies Range(d) = $1_{/\mathbf{N}} \cup (1 + 1_{/\mathbf{N}})$, so D does not verify the successor property . On the other hand , it is obviously sufficient , in order to define **CO**, to ask the range of any distance d to be included in a fixed (not necessarily $1_{/\mathbf{N}}$) set verifying the successor property ; $1_{/\mathbf{N}}$ is just the simplest such set that makes sense .

4 Computations in semantic domains

The basic idea of this section is the following : given a programming language (sequential or parallel) , and some denotational semantics for it ,what is the relation between :
(a) the operators on the semantic domain induced by the contexts , and
(b) the computable functions on this domain, provided it is recursively presented .

Let us first give a few basic definitions from effective descriptive set theory (we could suppose that all the semantic domains we will consider are in fact polish spaces, i.e separable, complete metric spaces without isolated points ; unless it is necessary, we will just suppose that they are separable and recursively presented) :

Definitions: (i) Let (X,d) be a recursively presented space, of presentation ($\{r_i\}$, L, L') . A basic system of neighbourhoods for (X,d) is then

$$N(X,i) = B(r_j\, ,\, k_{/\,l+1}) \quad \text{with} \quad i = \langle j\,,k\,,l \rangle$$

(ii) A subset A of X is semirecursive iff there exists a recursive function ϕ from **N** to **N** such that

$$A = \cup_{n \in \mathbf{N}} N(X,\phi(n)) .$$

(iii) A subset A of X is recursive iff A and X - A are both semirecursive

(iv) Let (X,d) and (Y,d') be two recursively presented spaces, and f a map from X to Y ; let G^f be the subset of X x **N** defined by :

$$(x,s) \in G^f \text{ iff } f(x) \in N(Y,s) .$$

Then f is recursive if G^f is recursive (one can uniformly compute arbitrary good approximations of f(x) for any x) . More generally, a function will be Γ-recursive (or in Γ) if its neighborhood graph G^f is in Γ for some complexity class Γ .

A) Models for well-guarded CCS (under SOS semantics) :

Basically, CCS is a system of labelled transitions defined on an algebra of terms . Labels (μ) are taken into the disjoint union (L) of countable sets Δ , $\overline{\Delta}$, $\{\tau\}$, where τ means the internal or silent action , resulting from the synchronized product of complementary actions $\lambda \in \Delta$ and $\overline{\lambda} \in \overline{\Delta}$ (with $\overline{\overline{\lambda}} = \lambda$). Partial injective homomorphisms h on L satisfying $h\tau = \tau$ and $h(\overline{\lambda}) = \overline{(h\lambda)}$ are called renaming functions. A renaming function acts as a restriction outside its domain . For the purpose of the present study, we set as a special requirement that renaming functions act as the identity almost everywhere (whence they may be given an effective syntax) .

The signature of the algebra is $\Sigma = \Sigma_0 \cup \Sigma_1 \cup \Sigma_2$, where $\Sigma_0 = \{ \text{nil} \}$, $\Sigma_1 = \{ \mu / \mu \in \Lambda \} \cup \{ \{h\} / \ h$ a renaming function $\}$ and $\Sigma_2 = \{ +, \mid \}$ - supplying nondeterministic sum and parallel composition . Terms t are either variables $x \in X$ (a set of variables) , or well formed expressions $op_k (t_1 ... t_k)$ where $op_k \in \Sigma_k$, or guardedly well formed recursive terms x_i where $(x_1 = t_1 \ldots x_n = t_n)$, where each x_i is guarded in t_i .

Transitions $t \to^{\mu} t'$ are those provable from the axioms and rules of the structural operational semantics [Pl] . Due to the well-guardedness of recursive definitions, the set of all transitions is recursive, and for each t , the set of transitions $t \longrightarrow^{\mu} t'$ originating from t is finite .

The intended semantics for CCS terms is given by :

$$[t] = \{ < a , [s] > / \ t \to^{a} s \text{ is a provable transition } \} .$$

Putting L = N, the metric space X in which we interpret terms is the solution of the domain equation

$$X = P_{co} (\omega \times X) .$$

Using the general theorem above, we know that this equation has a recursively presented solution , namely the colimit of the converging tower

$$\{p_0\} \to P_{co}(\omega \times \{p_0\}) \to P_{co}(\omega \times P_{co}(\omega \times \{p_0 \})) \to \ldots$$

It must be obvious that the colimit built in the first section would be in this case quite impractical, so it is natural to look for some more natural space isomorphic to it . Indeed we have :

<u>Proposition</u> : $(P_{co} (\omega^{\omega}) , \delta)$ is the unique (up to isomorphism) solution of the domain equation $X = P_{co} (\omega \times X)$, where δ is the Hausdorff metric on $P_{co}(\omega^{\omega})$, ω^{ω} being equipped with the usual Baire metric .

<u>Proof</u> : We just have to build an isometric bijection between $P_{co} (\omega^{\omega})$ and $P_{co} (\omega \times P_{co} (\omega^{\omega}))$. It is easier in fact to build it from $P_{co} (\omega \times P_{co} (\omega^{\omega}))$ to $P_{co} (\omega^{\omega})$: let X be a subset of ω^{ω} , and X_n be its n - section , i.e

$$X_n = \{ m \in \omega / \exists \alpha \in X \ \alpha(n)) = m \} .$$

Then X is compact iff X is closed and X_n is finite for every n . We can then encode the compact subsets of ω^{ω} as follows : let $(C,U) : P_{fin}(\omega) \to \omega$ be the usual bijective, primitive recursive encoding of finite subsets of ω ; let α be defined by

$$\forall n \ \alpha(n) = C(X_n) .$$

Then , for any β in ω^{ω} , $\beta \in X$ iff $(\forall n) \ \beta(n) \in U (\alpha(n))$.

Let now $X = \{ (n , X_{\alpha}) / n \in I , \alpha \in J \}$ be a member of $P_{co} (\omega \times P_{co} (\omega^{\omega}))$;

we define $f(X) = \{ \beta / \exists (n, X_\alpha) \in X\ \beta(0) = \langle n , \alpha(0) \rangle \ \&\ \forall m>0\ \beta(m) = \alpha(m) \}$.
One can then prove that f is an isometry (so f(X) is necessarily compact) .

We can now give our first characterization of contextual operators :

Definition : A CCS context C[ ] is normal iff it is of the form
x_i **where** ($x_1 = C_1$ .... $x_n = C_n$) , where the subcontexts C_i do not contain occurrences of the operator **where** .

Remark : This restriction on the recursion operator is, in our eyes, strenghtened by the theorem below , which does not hold for arbitrary contexts (see the remark following Lemma 4) . Furthermore, given some extra-axiomatization of fixpoints (see e.g [Ba]), one can prove that any CCS term is equivalent to a normal one .

Theorem 4 : Let C be a normal context in CCS , and [C] be its interpretation as a map from
$P_{co}(\omega^\omega)$ to $P_{co}(\omega^\omega)$. Then [C] is a Π^0_1 - recursive map .

Proof : We just have to prove that the interpretations of CCS operators are Π^0_1 - maps , and that the composition of a Π^0_1 - map and a recursive map is still Π^0_1 .

The interpretation of CCS terms is given by :
$$[\,t\,] = f(\, \{ ([a] , [s]) / t \to^a s \text{ is a provable transition } \} \,),$$
f being the isometry built in the previous proof .
The nil operator is interpreted as the empty set , so it has obviously a recursive interpretation .
One must remark that the isometry f is recursive, so we can uniformly compute [t] from the set of ([a],[s]) such that $t \to^a s$ is provable , or compute [s] from [t] if we know that $t \to^a s$ is a provable transition .

Lemma 1 : The unary operators a. and {h} have recursive interpretations .

Proof : We have , for any t , [a.t] = f({ ([a], [t]) }) ; the context a._ is then interpreted as the map :

[a._] : $X \longrightarrow f(\{([a], X)\})$, which is recursive due to the recursive character of f .

Let h be a renaming function with finite support I ; then
$$[t \{h\}] = f(\{ ([h(a)] , [s \{h\}]) / t \to^a s \text{ is a provable transition and } a \in I \}).$$
This is a primitive recursion scheme, so the interpretation [_ {h}] is recursive (recursive functions are closed under primitive recursion , and f is recursive) .

Lemma 2 : The binary operators + and $|$ have recursive interpetations .

We just have to prove that [+] is recursive , since [$|$] can be defined by a primitive recursion scheme from f and [+] . The recursive character of [+] , which is just the set theoretic union , is quite easy to check .

Lemma 3 : The context x_i **where** ($x_1 = \_1$ .... x_n $= \_n$) has a Π^0_1 interpretation .

Proof : We limit ourselves to the case n = 1 , the extension to the general case being straightforward. We can remark first that the substitution has a trivial , recursive interpretation, since :

$$[\, t\,[x/u]\,] = f(\{\ (\,[a]\,,[\,t'\,[x/u]]\,)\,/\,t \to^a t' \text{ is provable}\}) \ \text{ if } \ t \neq x\,,\,[t] \neq \varnothing$$
$$[\,u\,] \text{ if } t = x\,.$$

Let us call $\psi^n{}_x(t)$ the nth unfoldment of x **where** x=t ; then

$$[\ x \textbf{ where } x= t\,] = f(\cup_n\ \{\ ([a],[s])\,/\,\psi^n{}_x(t) \to^a s \text{ is provable }\}\,)$$
$$= \cup_n\ f(\{\ ([a],[s])\,/\,\psi^n{}_x(t) \to^a s \text{ is provable }\}\,)\,;$$

It is quite easy to see that the map Unfold : $([t]\,,n) \to [\psi^n{}_x(t)]$ is given by a primitive recursive scheme , so it is recursive . We have now :

$[\ x$ **where** $x = t\,] \in N_i$ iff $(\forall n)\ [\psi^n{}_x(t)] \in N_i$,

so the neighbourhood graph G of the interpretation of $t \to x$ **where** x =t is defined by

$(X,s) \in G$ iff $(\forall n)\ (\,(X, n)\,,s\,) \in G^{Unfold}$, so G is Π^0_1 .

Lemma 4 : Let f be a Π^0_1 map and g a recursive map between powers of $P_{co}(\omega^\omega)$; then $f_o g$ is a Π^0_1 map.

Remark : It is not true in general that the composition of two Π^0_1 maps is Π^0_1 (the best one can prove is that it is Π^0_2) . If one allows then non-normal contexts, the complexity of their interpretations will be probably much higher than Π^0_1 ; the author does not know of any class of Π^0_n -recursive functions closed under composition .

B) Solving POOL-like domain equations :

We refer to [A et al.] for a presentation of the language POOL ; we just point here at the difficulties raised by the denotational semantics built for it , and indicate a natural way to solve them .

Denotational semantics for POOL :

The core of the denotational semantics for POOL as it is built in [A et al.] is the construction of a domain P of processes ; this construction presents a number of difficulties from our point of view, which we will try to solve below . We give first the construction presented in [A et al.] , with some small (coding) modifications:

Objects : AObj = $\mathbf{N} \times \mathbf{N}$ (these are the active objects ; the first component is just the class name of the object, the second one its "rank" in the class)

$\text{Obj} = \text{AObj} \cup \mathbf{Z} \cup \{ tt, ff \} \cup \{ nil \} \approx \mathbf{N}$ (if one allows some coding)

States : $\Sigma = (\text{AObj} \to \text{IVar} \to \text{Obj}) \times (\text{AObj} \to \text{TVar} \to \text{Obj}) \times P_{fin} (\text{AObj})$

(the first and second components store the values of the variables for each active object , the third one contains the object names currently in use)

Processes : $\text{Send}_P = \text{Obj} \times \text{MName} \times \text{Obj}^* \times (\text{Obj} \to P) \times P$

$\text{Answer}_P = \text{Obj} \times \text{MName} \times (\text{Obj}^* \to (\text{Obj} \to P) \to^1 P)$

$\text{Step}_P = (\Sigma \times P) \cup \text{Send}_P \cup \text{Answer}_P .$

$P = \{p_0\} \cup id^{1/2} (\Sigma \to P_{cl} (\text{Step}_P)) .$

Three main problems arise when one wants to find an effective solution to the above domain equation :

- the presence of functors of the type $X \to Y$, where X is not compact
- the presence of constant functors F_M where M is not separable (e.g F_Σ)
- the presence of the subexpression ($\text{Obj} \to P) \to^1 P$ which should force P to be compact and to verify the successor property .

If one chooses the usual topology for the discrete spaces involved in this equation, P cannot be recursively presented , nor even separable .Two solutions are then available :

- restrict the meaning of the arrow constructor in a constructive sense : this is difficult, as not even a restriction to recursive functions gives rise to a separable space, and quite artificial in a classical setting .
- change the topology of some of the basic spaces under consideration (Obj , Σ) : this seems to be the best and most natural solution so far . It is certainly justified from a mathematical point of view, and we will try to justify it from a more practical point of view .

States : We need Σ to be compact and separable , which it is not .

Even if we restrict Σ to

$(\text{AObj} \to_{fin} (\text{IVar} \to_{fin} \text{Obj})) \times (\text{AObj} \to_{fin} (\text{TVar} \to_{fin} \text{Obj})) \times P_{fin} (\text{AObj})$

it is still not compact (since $P_{fin}(\text{AObj})$ is not compact).The only solution is to change the

topology on Obj and AObj :
let ($\underline{\mathbf{N}}$, $\underline{\mathbf{d}}$) be a one-point compactification of **N** equipped with the discrete metric, i.e :

$$\underline{\mathbf{N}} = \mathbf{N} \cup \{\omega\} \quad \text{and } \underline{\mathbf{d}}(n,m) = 1/_{\min(n,m)+1} \text{ if } n \neq m .$$

We fix then (provided some standard coding) Obj = ($\underline{\mathbf{N}}$, $\underline{\mathbf{d}}$) and AObj = ($2\underline{\mathbf{N}}$, $\underline{\mathbf{d}}$),

$$\text{IVar} = \text{TVar} = (\underline{\mathbf{N}} , \underline{\mathbf{d}}) .$$

Claim : $(\Sigma,d) = (\text{AObj} \rightarrow (\text{IVar} \rightarrow \text{Obj})) \times (\text{AObj} \rightarrow (\text{TVar} \rightarrow \text{Obj})) \times P_{fin} (\text{AObj})$
is compact , recursively presented and $1_{/\mathbf{N}} \supseteq \text{Range } (d)$.

Corollary : The unique solution of the domain equation defining P is compact and recursively presented (using theorem 4).

In order to get a good evaluation of the complexity of the operators induced by POOL contexts, one needs then to build (as it was done for CCS) a manageable presentation of the space P , which is rather uneasy . Alternatively, one could work solely on the fact that the isomorphism defining P ($P \rightarrow (\Sigma \rightarrow P_{cl} (\text{Stepp}))$) is indeed recursive (as we did for CCS contexts) . Using either approach, we have been able to prove (see [Y1]) that the operators induced by POOL contexts are recursive .

5 Conclusion

This paper contains only a very small and preliminary part of the development of effective semantics within the framework of metric spaces . Among problems which we did not address in the present paper , and which we are presently studying , one can quote the following :
1) The fact that the isomorphisms in the category **C** coincide with isometries is probably too strong in most cases when one wants to obtain natural models for some languages (a good example of it is the pure lambda-calculus : we have indeed a good model for it - namely the Cantor space , which is homeomorphic to the space of non-expanding maps on itself - but this model is not a solution of $X \approx X \rightarrow^1 X$ in the present setting). The best thing to do is then to change the notion of morphism so as to obtain homeomorphisms as isomorphisms, without losing the fact that convergent towers have a colimit (see [Y2]) .
2) A case-by-case study can be useful in order to detect contexts (or operators) which increase abnormally the overall complexity of the language ; from a more global point of view, it is certainly necessary to be able to define classes of operators (and languages) which behave (as far as complexity is concerned) in an uniform way (independently of the syntax and the particular space we are working in) .
3) No comparison has been yet made with the theory of effective domains ; the facts that the effectivity of the structures is given internally (by the metric) rather than by some external coding, and that we work in a widely developped field (descriptive set theory and constructive analysis) support us in our belief that the present approach could turn to be richer .

Acknowledgements

I want to thank Philippe Darondeau for drawing my attention to this problem, and Jan Rutten for the conversations we had over a first version of this work .

Bibliography

[A et al] : P.America , J. De Bakker , J.N.Kok , J.Rutten :
Denotational Semantics of a Parallel Object-Oriented Language
(Information and Computation , 83 , 1989)

[AR] : P.America and J.Rutten : Solving reflexive domain equations in a category of complete metric spaces (JCSS., 39 , 1989)

[Ba] E.Badouel : Algebraically closed theories (Proceedings MFCS' 89)

[B] Bourbaki : Topologie generale v.1 et 2 (Hermann)

[Pl] G.Plotkin : A structural approach to operational semantics (Rep Daimi FN19, Aarhus,1981)

[Y1] S.Yoccoz : Effective denotational semantics for sequential and parallel languages, part 1
(Rapport Interne, Labri - Universite de Bordeaux-I , June 1990)

[Y2] S.Yoccoz : Effective denotational semantics for sequential and parallel languages, part 2
(to appear as technical report , Labri-Université de Bordeaux 1, Aug.1990)

Author Index

408: M. Leeser, G. Brown (Eds.),Hardware Specification, Verification Synthesis: Mathematical Aspects. Proceedings, 1989. VI, 402 pages.).

409: A. Buchmann, O. Günther, T. R. Smith, Y.-F. Wang (Eds.), Design Implementation of Large Spatial Databases. Proceedings, 1989. IX, pages. 1990.

410: F. Pichler, R. Moreno-Diaz (Eds.), Computer Aided Systems ory – EUROCAST '89. Proceedings, 1989. VII, 427 pages. 1990.

411: M. Nagl (Ed.), Graph-Theoretic Concepts in Computer Science. ceedings, 1989. VII, 374 pages. 1990.

412: L. B. Almeida, C. J. Wellekens (Eds.), Neural Networks. Proceed-, 1990. IX, 276 pages. 1990.

413: R. Lenz, Group Theoretical Methods in Image Processing. VIII, pages. 1990.

414: A.Kreczmar, A. Salwicki, M. Warpechowski, LOGLAN '88 – ort on the Programming Language. X, 133 pages. 1990.

415: C. Choffrut, T. Lengauer (Eds.), STACS 90. Proceedings, 1990. 312 pages. 1990.

416: F. Bancilhon, C. Thanos, D. Tsichritzis (Eds.), Advances in Data-e Technology – EDBT '90. Proceedings, 1990. IX, 452 pages. 1990.

417: P. Martin-Löf, G. Mints (Eds.), COLOG-88. International Confer-e on Computer Logic. Proceedings, 1988. VI, 338 pages. 1990.

418: K. H. Bläsius, U. Hedtstück, C.-R. Rollinger (Eds.), Sorts and Types in icial Intelligence. Proceedings, 1989. VIII, 307 pages. 1990. (Subseries I).

419: K. Weichselberger, S. Pöhlmann, A Methodology for Uncertainty nowledge-Based Systems. VIII, 136 pages. 1990 (Subseries LNAI).

420: Z. Michalewicz (Ed.), Statistical and Scientific Database Manage-nt, V SSDBM. Proceedings, 1990. V, 256 pages. 1990.

421: T. Onodera, S. Kawai, A Formal Model of Visualization in Comput-Graphics Systems. X, 100 pages. 1990.

422: B. Nebel, Reasoning and Revision in Hybrid Representation tems. XII, 270 pages. 1990 (Subseries LNAI).

423: L. E. Deimel (Ed.), Software Engineering Education. Proceed-s, 1990. VI, 164 pages. 1990.

424: G. Rozenberg (Ed.), Advances in Petri Nets 1989. VI, 524 pages. 0.

425: C. H. Bergman, R. D. Maddux, D. L. Pigozzi (Eds.), Algebraic ic and Universal Algebra in Computer Science. Proceedings, 1988. XI, pages. 1990.

426: N. Houbak, SIL – a Simulation Language. VII, 192 pages. 1990.

427: O. Faugeras (Ed.), Computer Vision – ECCV 90. Proceedings, 0. XII, 619 pages. 1990.

428: D. Bjørner, C. A. R. Hoare, H. Langmaack (Eds.), VDM '90. VDM Z – Formal Methods in Software Development. Proceedings, 1990. I, 580 pages. 1990.

429: A. Miola (Ed.), Design and Implementation of Symbolic Computa-Systems. Proceedings, 1990. XII, 284 pages. 1990.

430: J. W. de Bakker, W.-P. de Roever, G. Rozenberg (Eds.), Stepwise inement of Distributed Systems. Models, Formalisms, Correctness. ceedings, 1989. X, 808 pages. 1990.

431: A. Arnold (Ed.), CAAP '90. Proceedings, 1990. VI, 285 pages. 0.

432: N. Jones (Ed.), ESOP '90. Proceedings, 1990. IX, 436 pages. 0.

433: W. Schröder-Preikschat, W. Zimmer (Eds.), Progress in Distri-ted Operating Systems and Distributed Systems Management. oceedings, 1989. V, 206 pages. 1990.

435: G. Brassard (Ed.), Advances in Cryptology – CRYPTO '89. Pro-edings, 1989. XIII, 634 pages. 1990.

436: B. Steinholtz, A. Sølvberg, L. Bergman (Eds.), Advanced Informa-n Systems Engineering. Proceedings, 1990. X, 392 pages. 1990.

437: D. Kumar (Ed.), Current Trends in SNePS – Semantic Network ocessing System. Proceedings, 1989. VII, 162 pages. 1990. (Sub-ries LNAI).

438: D. H. Norrie, H.-W. Six (Eds.), Computer Assisted Learning – CAL '90. Proceedings, 1990. VII, 467 pages. 1990.

Vol. 439: P. Gorny, M. Tauber (Eds.), Visualization in Human-Computer Interaction. Proceedings, 1988. VI, 274 pages. 1990.

Vol. 440: E.Börger, H. Kleine Büning, M. M. Richter (Eds.), CSL '89. Proceedings, 1989. VI, 437 pages. 1990.

Vol. 441: T. Ito, R. H. Halstead, Jr. (Eds.), Parallel Lisp: Languages and Systems. Proceedings, 1989. XII, 364 pages. 1990.

Vol. 442: M. Main, A. Melton, M. Mislove, D. Schmidt (Eds.), Mathematical Foundations of Programming Semantics. Proceedings, 1989. VI, 439 pages. 1990.

Vol. 443: M. S. Paterson (Ed.), Automata, Languages and Programming. Proceedings, 1990. IX, 781 pages. 1990.

Vol. 444: S. Ramani, R. Chandrasekar, K. S. R. Anjaneyulu (Eds.), Knowledge Based Computer Systems. Proceedings, 1989. X, 546 pages. 1990. (Subseries LNAI).

Vol. 445: A. J. M. van Gasteren, On the Shape of Mathematical Arguments. VIII, 181 pages. 1990.

Vol. 446: L. Plümer, Termination Proofs for Logic Programs. VIII, 142 pages. 1990. (Subseries LNAI).

Vol. 447: J. R. Gilbert, R. Karlsson (Eds.), SWAT 90. 2nd Scandinavian Workshop on Algorithm Theory. Proceedings, 1990. VI, 417 pages. 1990.

Vol. 449: M. E. Stickel (Ed.), 10th International Conference on Automated Deduction. Proceedings, 1990. XVI, 688 pages. 1990. (Subseries LNAI).

Vol. 450: T. Asano, T. Ibaraki, H. Imai, T. Nishizeki (Eds.), Algorithms. Proceedings, 1990. VIII, 479 pages. 1990.

Vol. 451: V. Mařík, O. Štěpánková, Z. Zdráhal (Eds.), Artificial Intelligence in Higher Education. Proceedings, 1989. IX, 247 pages. 1990. (Subseries LNAI).

Vol. 452: B. Rovan (Ed.), Mathematical Foundations of Computer Science 1990. Proceedings, 1990. VIII, 544 pages. 1990.

Vol. 453: J. Seberry, J. Pieprzyk (Eds.), Advances in Cryptology – AUSCRYPT '90. Proceedings, 1990. IX, 462 pages. 1990.

Vol. 454: V. Diekert (Ed.), Combinatorics on Traces. XII, 165 pages. 1990.

Vol. 455: C. A. Floudas, P. M. Pardalos (Eds.), A Collection of Test Problems for Constrained Global Optimization Algorithms. XIV, 180 pages. 1990.

Vol. 456: P. Deransart, J. Maluszyński (Eds.), Programming Language Implementation and Logic Programming. Proceedings, 1990. VIII, 401 pages. 1990.

Vol. 457: H. Burkhart (Ed.), CONPAR '90 – VAPP IV. Proceedings, 1990. XIV, 900 pages. 1990.

Vol. 458: J. C. M. Baeten, J. W. Klop (Eds.), CONCUR '90. Proceedings, 1990. VII, 537 pages. 1990.

This series reports new developments in computer science research an teaching – quickly, informally and at a high level. The type of materi considered for publication includes preliminary drafts of original paper and monographs, technical reports of high quality and broad interes advanced level lectures, reports of meetings, provided they are of ex ceptional interest and focused on a single topic. The timeliness of a manu script is more important than its form which may be unfinished or tentative If possible, a subject index should be included. Publication of Lectur Notes is intended as a service to the international computer science com munity, in that a commercial publisher, Springer-Verlag, can offer a wid distribution of documents which would otherwise have a restricted reac ership. Once published and copyrighted, they can be documented in th scientific literature.

Manuscripts

Manuscripts should be no less than 100 and preferably no more than 500 pages in length.
They are reproduced by a photographic process and therefore must be typed with extreme care. Symbol not on the typewriter should be inserted by hand in indelible black ink. Corrections to the typescrip should be made by pasting in the new text or painting out errors with white correction fluid. Authors receiv 75 free copies and are free to use the material in other publications. The typescript is reduced slightly i size during reproduction; best results will not be obtained unless the text on any one page is kept withi the overall limit of 18 x 26.5 cm (7 x 10½ inches). On request, the publisher will supply special paper wit the typing area outlined.

Manuscripts should be sent to Prof. G. Goos, GMD Forschungsstelle an der Universität Karlsruhe, Haid- un Neu-Str. 7, 7500 Karlsruhe 1, Germany, Prof. J. Hartmanis, Cornell University, Dept. of Computer Science, Ithaca NY/USA 14853, or directly to Springer-Verlag Heidelberg.

Springer-Verlag, Heidelberger Platz 3, D-1000 Berlin 33
Springer-Verlag, Tiergartenstraße 17, D-6900 Heidelberg 1
Springer-Verlag, 175 Fifth Avenue, New York, NY 10010/USA
Springer-Verlag, 37-3, Hongo 3-chome, Bunkyo-ku, Tokyo 113, Japan

ISBN 3-540-53048-7
ISBN 0-387-53048-7